Innovation, Science, a
Institutional Change

Innovation, Science, and Institutional Change

Edited by

Jerald Hage

and

Marius Meeus

with the editorial committee of: Charles Edquist,
J. Rogers Hollingsworth, Gretchen Jordan,
Harro van Lente, and Sue Mohrman

OXFORD

UNIVERSITY PRESS

OXFORD

UNIVERSITY PRESS

Great Clarendon Street, Oxford OX2 6DP
Oxford University Press is a department of the University of Oxford.
It furthers the University's objective of excellence in research, scholarship,
and education by publishing worldwide in

Oxford New York

Auckland Cape Town Dar es Salaam Hong Kong Karachi
Kuala Lumpur Madrid Melbourne Mexico City Nairobi
New Delhi Shanghai Taipei Toronto

With offices in

Argentina Austria Brazil Chile Czech Republic France Greece
Guatemala Hungary Italy Japan Poland Portugal Singapore
South Korea Switzerland Thailand Turkey Ukraine Vietnam

Oxford is a registered trade mark of Oxford University Press
in the UK and in certain other countries

Published in the United States
by Oxford University Press Inc., New York

British Library Cataloguing in Publication Data
Data available

Library of Congress Cataloging in Publication Data
Innovation, science, and institutional change / edited by Jerald Hage and Marius Meeus.
  p. cm.
Based on research papers presented at a conference held in October 2003 at the
Department of Innovation Studies, the University of Utrecht, Netherlands, and a
second conference held in May 2004 at the Keck Graduate Institute, Claremont, California.
  1. New products—Management—Congresses. 2. Product management—Congresses.
  3. Knowledge management—Congresses. 4. Technological innovations—Management—Congresses.
  5. Information technology—Management—Congresses. 6. Organizational change—Management—Congresses.
  I. Hage, Jerald, 1932– II. Meeus, Marius T. H., 1957–
  HF5415.153.I56 2006
  338'.064–dc22
  2006009452

Typeset by SPI Publisher Services, Pondicherry, India
Printed in Great Britain

ISBN 978-0-19-929919-5   978-0-19-957345-5 (pbk.)
1 3 5 7 9 10 8 6 4 2

ACKNOWLEDGMENTS

Many people need to be thanked for their contributions to this book.

The insights for the rethinking of research agendas in the study of innovation originated almost simultaneously, despite their quite disparate disciplines and perspectives, in the minds of two people: Marius Meeus and Terry Shinn.

Marius Meeus was, with Jerry Hage, part of an interdisciplinary team of researchers at the Netherlands Institute for the Advancement of Science (NIAS), which had been formed to work on the problem of innovation during 1998–9. They had adjoining offices. One day, Marius Meeus suggested to Jerry Hage that there was a need for a new handbook on innovation, because the subject matter was fragmented across disciplines. The two of them, sociologists who have studied industrial innovation, but on different continents (the US and Europe), had been struck by the quite dissimilar topics and ways of thinking of the members of the NIAS team, which included an economist, a political scientist, a management scholar, and a historian of science, as well as them. Two of the members of the team, J. Rogers Hollingsworth (historian of science) and Frans van Waarden (political scientist), were asked to write contributions for this book.

Shortly after Marius Meeus's suggestion, Jerry Hage traveled to Paris to visit Terry Shinn, a director of research in the history and sociology of science at La Maison des Sciences de l'Homme. Terry suggested that his discipline had ignored the study of industrial innovation and that he would like to have a conference that would write position papers on this topic. Unfortunately, he was unable to secure funding for this conference.

Several years later, while consulting with Gretchen Jordan on the development of instruments to evaluate research laboratories, Jerry Hage realized that her work, plus other research projects funded by the Strategic Planning Office of Basic Energy Sciences in the US Department of Energy, had been referenced in neither the history and sociology of science nor the study of industrial innovation. He suggested to her that an international conference to review research agendas in the different disciplines would be a useful way for advancing the field, and wrote a brief two-page memorandum that summarized some of the main points provided in the introductory chapter of this book. Gretchen Jordan discussed this memorandum with Bill Valdez, head of the Strategic Planning Office of Basic Energy Sciences; he concurred that such a handbook would be in line with his strategy of rethinking how to evaluate the scientific and technological research. Valdez's Office of Strategic Planning provided initial funding for a planning conference and the first year.

To plan an international and interdisciplinary conference, a committee was formed: its members were Marius Meeus, J. Rogers Hollingsworth, Gretchen Jordan, Sue Mohrman, Charles Edquist, and Jerry Hage. The selection of the committee members was based on ensuring the representation of different knowledge communities—industrial innovation, history and sociology of science, evaluation research, management science, and economics respectively—as well as quite large networks that span multiple regions of the world. They met for several days in December 2002 in Washington, DC. At this conference, some of the integrative themes discussed in the introductory chapter were developed as an aid in selecting individuals who would be asked to write papers for this first conference. In addition, the committee spent some time thinking about how to solve the many problems that plague handbooks that develop from conferences: the lack of intellectual integration consequent on having many contributors, the absence of a concluding chapter, the need for cross-references within chapters, etc. Each of the committee members was responsible for asking individuals from their respective disciplines and perspectives to write papers. In this way, a personal liaison was ensured and, as a consequence, a much higher participation rate.

Over the next six months, the contributors were selected and worked on their papers for the first conference. Most of these were distributed in advance as well as at the conference so that everyone would have their own copy. This initial conference was held October 2003 in Utrecht, the Netherlands, where Marius Meeus and the Department of Innovation Studies at the University of Utrecht were official hosts. The objective of this conference was mainly the presentation of ideas and interchanges that could form the basis for the rewriting of the papers. In particular, John Campbell made a number of useful integrative comments. Given the success of this conference, the Strategic Planning Office provided another grant for the second conference. During the next six months, all the contributors reworked their chapters to incorporate the criticisms that they had received.

The second conference was held in May 2004 at the Keck Graduate Institute, Claremont, CA, where our official host was David Finegold. As well as a critique of the basic matrix provided in the introductory chapter, the objective of this conference was intellectual integration. It was decided that one of the sections should be renamed. Furthermore, it became clear that, because a number of the authors did not have English as their native language, all of the contributions had to be edited before it could be submitted to a publisher. Susan Finegold started this process in the fall of 2004.

But editing was not the only form of review. Members of the editorial committee took responsibility for intellectually reviewing various sections and others joined them in this process: Marius Meeus worked on both the first section and the concluding chapter by Parry Norling; Phil Shapira and Gretchen Jordan helped review the contributions in the second section; the third section was reviewed by Sue Mohrman and Harro van Lente; for the last section, Steve Casper aided Jerry Hage. Where appropriate, everyone was asked to cross-reference chapters by others.

This brief overview of the creation process of this handbook of research agendas makes apparent that many people have contributed; they include:

- NIAS for funding several of the contributors;
- Marius Meeus and Terry Shinn for the basic ideas;
- Bill Valdez, Gretchen Jordan, and the Strategic Planning Office of Basic Energy Science of DoE for the funding of several conferences and the copy-editing work;
- Marius Meeus and David Finegold, and the staff that helped them hosting the conferences;
- the editorial committee of Sue Mohrman, J. Rogers Hollingsworth, Gretchen Jordan, Charles Edquist, Marius Meeus, and Jerry Hage for their planning;
- Sue Mohrman, Harro van Lente, Gretchen Jordan, Phil Shapira, Steve Casper, Marius Meeus, and Jerry Hage for their editorial reviews;
- the many individuals whose comments at each of the conferences have advanced the quality of this handbook.

Many thanks to all!

Jerald Hage and Marius Meeus

TABLE OF CONTENTS

PART IV.   INSTITUTIONS AND INSTITUTIONAL CHANGE

CONTRIBUTORS

NICLAS ADLER, Director, FENIX Center for Research on Knowledge and Business Creation, Stockholm School of Economics

DEEPA ARAVIND, Assistant Professor, College of Staten Island, City University of New York

JOHN L. CAMPBELL, Professor of Sociology, Dartmouth College

STEVEN CASPER, Assistant Professor, Keck Graduate Institute

CRISTINA CHAMINADE, CIRCLE (Centre for Innovation, Research and Competences in the Learning Economy), Lund University

FARIBORZ DAMANPOUR, Professor of Management, Department of Management and Global Business, Rutgers University

CHARLES EDQUIST, Professor of Innovation, Division of Innovation, Lund Institute of Technology, Lund University

JAN FABER, Associate Professor of Innovation, Department of Innovation Studies

DAVID FINEGOLD, Professor, Keck Graduate Institute, Claremont Collegus

JAMES FOSTER, Professor of Environmental Regulation, Center for European Studies, Massachusetts Institute of Technology

JAY R. GALBRAITH, Professor of Management, Marshall School of Business, University of Southern California

LUKE GEORGHIOU, Director, PREST, University of Manchester

JERALD HAGE, Director, Center for Innovation, Department of Sociology, University of Maryland

ARMAND HATCHUEL, Professor of Industrial Design, École des Mines de Paris

MIKAEL HILDÉN, Director, Finnish Environment Institute, Helsinki

J. ROGERS HOLLINGSWORTH, Professor of History and of Sociology, University of Wisconsin–Madison

ERIC JOLIVET, Associate Professor, University of Toulouse

GRETCHEN B. JORDAN, Principal Member of Technical Staff, Sandia National Laboratories

STEFAN KUHLMANN, Professor and Chair, Department of Science, Technology, and Policy Studies, University of Twente, The Netherlands

PASCAL LEMASSON, Researcher, École des Mines de Paris

MARC MAURICE, Professor of Industrial Relations, L.E.S.T., Aix-en-Provence

MARIUS T. H. MEEUS, Professor of Innovation and Organization, Department of Organization Studies, Tilburg University, and is director at the center for Innovation Research

J. STANLEY METCALFE, Co-Director, CRIC, University of Manchester

SUSAN A. MOHRMAN, Senior Research Scientist, Marshall School of Business, University of Southern California

PETER MONGE, Professor of Management, Marshall School of Business, University of Southern California

IKUJIRO NONAKA, Professor of Strategy, Graduate School of International Corporate Strategy, Hitotsubashi University, Xerox Distinguished Faculty Scholar, IMIO, UC Berkeley, and Visiting Dean and Professor, Center for Knowledge and Innovation Research, Helsinki School of Economics

PARRY M. NORLING, Former Manager of Chemical Research, Dupont Corporation

HERMAN OOSTERWIJK, Assistant Professor, Twente University,

VESA PELTOKORPI, COE Project Director, Graduate School of International Corporate Strategy, Hitotsubashi University

WERNER RAMMERT, Professor of Sociology and Technology Studies, Technical University of Berlin

PHILIP SHAPIRA, Professor of Innovation, Management, and Policy, Manchester Institute of Innovation Research, Manchester Business School, University of Manchester, and Professor, School of Public Policy, Georgia Institute of Technology, Atlanta

TERRY SHINN, Director of Research, Centre National dela Recherche Scientifique, Paris

HARRO VAN LENTE, Assistant Professor of Innovation Studies, Department of Innovation Studies, Utrecht University

FRANS VAN WAARDEN, Professor of Public Policy, Utrecht University

BENOIT WEIL, Researcher, École des Mines de Paris

ABBREVIATIONS

A*STAR	Agency for Science, Technology, and Research
AEA	Atomic Energy Authority
AEC	Atomic Energy Commission
ATM	auto-teller machine
BPO	business process outsourcing
BRE	Building Resources Establishment
Bt	*Bacillus thuringiensis*
CAD	computer aided design
CEO	chief executive officer
CERN	Centre Européen de Recherche Nucléaire
CME	coordinated market economy
CPU	central processing unit
DOD	Department of Defense (USA)
DOE	Department of Energy (USA)
DoE	Department of the Environment (UK)
DTI	Department of Trade and Industry
EDB	Economic Development Board
EIS	European Innovation Scoreboard
EPA	Environmental Protection Agency
EPO	European Patent Office
ET	education and training
FAO	Food and Agriculture Organization
FDA	Food and Drug Administration
GDP	gross domestic product
GMO	genetically modified organism
GPT	general-purpose technology
GSM	Global Systems for Mobile
HE	higher education
HPO	high-performance organization
HR	human resources
HSE	high-skill ecosystem
HT	herbicide tolerance
ICE	internal combustion engine
ICT	information and communications technologies
IMCB	Institute of Molecular and Cellular Biology
IMF	International Monetary Fund
IOP	Innovation-Oriented Program
IOR	interorganizational relations
IP	intellectual property
IPO	initial public offering

IPR	intellectual property right
IR	infrared
IT	information technology
LME	liberal market economy
MIT	Massachusetts Institute of Technology
MITI	Ministry of International Trade and Industry (Japan)
MNC	multinational corporation
MNE	multinational enterprises
NAS	National Academy of Sciences
NASDAQ	National Association of Securities Dealers Automated Quotations
NCE	new chemical entity
NGO	non-governmental organization
NIH	National Institutes of Health
NOAA	National Oceanic and Atmospheric Administration
NPD	new product development
NSF	National Science Foundation
NSI	National Systems of Innovation
OECD	Organization for Economic Cooperation and Development
PTT	Post, Telephone, and Telegraph
R&D	research and development
RAFI	Rural Advancement Foundation International
ROA	return on assets
RTD	research technology development
S&E	science and engineering
S&T	science and technology
SECI	Socialization, Externalization, Combination, and Internalization
SI	Systems of Innovation
SME	small and medium-sized enterprises
TQM	total quality management
UMTS	Universal Mobile Telecommunications System
USDA	United States Department of Agriculture
WTO	World Trade Organization
ZEV	zero emission vehicle

Product and Process Innovation, Scientific Research, Knowledge Dynamics, and Institutional Change: An Introduction

Marius T. H. Meeus and Jerald Hage

Introduction

Given the importance of high-tech products in international trade, the centrality of new scientific technology in national security, the discussions about job retraining in the context of globalization, and the continual debate about the best government policies to handle these issues, the study of innovation in products and processes hardly needs justification. What does need justification is the need for a new approach to such study, one that opens a new set of theoretical questions and answers, and that provides the basis of a much more comprehensive, interdisciplinary theory about these issues.

Our rationale for a new approach is based on four contentions:

- the narrow focus of previous research in economics, sociology and management;
- the need for multiple disciplines;
- the advantages of having both Europeans and Americans make contributions in a comparative perspective;
- the major transformations that are occurring in knowledge creation.

In particular, business organizations are becoming more risk averse, thus making universities the prime source of basic scientific research. These new phenomena associated with innovation require the development of new theories in many of these relevant disciplines and, perhaps most fundamentally, a much more complex theory of innovation and of knowledge production.

Our first contention is the narrowness of the focus of most other efforts that primarily concentrate on product and process innovation: we argue that innovation studies in economics, sociology, and management research have been primarily dominated by approaches explaining the determinants of innovation at the organizational level of analysis, ignoring the many other levels such as research laboratories, industrial sectors, and, of course, the national or macro-level system of innovation (Damanpour 1991; Zammuto and O'Connor 1992; Hage 1999). It has become increasingly apparent that there are considerable differences between nations in their rates of industrial innovation, including the sectors in which such innovation is taking place (Nelson 1993; Hall and Soskice 2001). What has made this fact highly visible is the much greater attention that governments have given to measuring, by various methods, their innovation performance (see NSF, Science and Research Indicators, various years: OECD 2001; European Scoreboard 2003). Great disparities exist in some countries between investments in business R&D and innovation performance in the economy. As a way of integrating, at the macro-level, diverse sets of factors such as types of market coordination, patterns of institutional norms, relatedness of actors in the transfer of knowledge, cultural factors, etc., Nelson (1993) suggests the concept of a national system of innovation as one explanation of these disparities. But even a combination of a macro- and meso-analysis of the determinants of innovation does not handle all the potential reasons for the lack of a one-to-one correspondence between investment in R&D and innovation payoff.

Another new development that can offer insight is the increasing importance of scientific research. As already observed, businesses are spending less on basic research and more on applied research, creating issues of integration between these two spheres in what some have called the 'idea-innovation chain' (Kline and Rosenberg 1986) or 'idea-innovation network' (Hage and Hollingsworth 2000). Therefore, this book has a section on issues in scientific research (such as the way it is organized and coordinated) which then influence national innovation performances.

Besides the development of the concept of national systems of innovation, another important new idea relevant to innovation is the new research topic of organizational learning. In particular, the concept of absorptive capacity (Cohen and Levinthal 1990; Zahra and George 2002) has been advanced as a way of explaining why some organizations learn more than others. But this only begins to touch on the critical topic of knowledge dynamics, which reflects upon the various ways in which knowledge moves and transforms as it does. The advantage of this section on knowledge dynamics is that it focuses attention on the horizontal, rather than just the vertical, dimensions implied in the macro- and meso-discussions. As can be appreciated in Part III of this book, a new set of research agendas with this perspective has developed.

Finally, just as the research on product and process innovation has lacked a dynamic qual-

ity, so has the research on the institutional context. Again, we have corrected for this imbalance by having a section on institutional *change*. Not only does this provide a corrective to the institutional literature, but it allows us to examine the various attempts by governments to change their institutional contexts so as to have more innovation, whether in products and services or in scientific advances. Beyond this, we have also emphasized the dynamics of change by arguing that the fundamental shifts are occurring because of three major forces: globalization, post-industrialization, and knowledge specialization.

Therefore, this book expands the focus to new research agendas in the four major, interrelated topics:

- product and process innovation;
- scientific research;
- knowledge dynamics;
- institutional change.

The book devotes a part to each.

Various handbooks touch on these areas but, because they do not provide an equal amount of space for each area, the disparate ideas in these different topics are never highlighted. Across our four parts, the parallel themes—models, networks, coordination modes, state policies, and systemic changes—provide considerable integration, and help the reader home in on a specific theme that is of interest. In addition, there are three cross-cutting themes—knowledge, evolution, and levels of analysis—that appear in many of the contributions, regardless of the parallel theme.

Central to the book is our assumption that a satisfactory theory about any one of the four issues requires a theory that integrates all of them. The proof of this assertion is discussed below under the rubric of cross-cutting themes, specifically the role of knowledge production as one theme that unites the disparate parts of the book. An essential equation connecting the knowledge base, collective learning, and innovation is suggested.

The second contention is that many of the major topic areas within these four parts need to be handled in a multidisciplinary way. Most books on innovation have been dominated by one or two disciplines, and yet one finds major insights in at least five distinct areas of expertise. This book has contributions from a number of economists, sociologists, management specialists, political scientists, and even historians of science, making it intellectually much livelier than the others that are currently on the market or forthcoming (see Table 1.1). One reason to offer a mix of disciplines is to broaden the sources for an explanation as to why there is not a one-to-one relationship between expenditures on R&D and innovation performance measures.

One advantage of a multidisciplinary approach is that different disciplines tend to focus on quite disparate levels of analysis, for example:

- the business organizations and the way in which they are structured, which is a concern of organizational sociologists and management specialists;
- the quality of the basic science, a concern of S&T evaluators and historians of science;
- the sector or industrial level, a concern for both economists and sociologists;
- the policies of the government, a concern of political scientists;

Table 1.1. The diversity of backgrounds of the contributors
The need for multiple disciplines[a]

	Senior authors[b]	All authors
1. Management science	5	13
2. Economics	4	6
3. Political scientists	4	6
4. S&T experts	5	5
5. Sociologists	4	5
Total	22	35

The need for representation from multiple countries[c]

	Senior authors[d]	All authors
1. The Netherlands	3	5
2. France	3	6
3. UK	2	2
4. Germany	2	2
5. Sweden	1	3
6. Japan	1	2
7. Finland	—	1
Total non-US	12	21
Total US	10	14
Total	22	35

[a] Most have multiple disciplines themselves.
[b] Senior author's discipline.
[c] Many have lived in more than one country and speak more than one language; some have worked in international comparative research projects.
[d] Senior author's nationality.

- the macro-institutional environment which is a concern of economists, political scientists, and sociologists.

Another reason to have a mix of disciplines is that we not only establish a number of new research agendas for the study of industrial innovation but, in the various contributions, outline advances in the areas of the sociology of science and institutional analysis, as well as a new area that might be called the study of knowledge production.[1]

A third contention is that Europeans and Americans from within the same disciplines have tended, because of the context of their own research, to reach different theoretical and practical conclusions about product and process innovation, scientific discoveries and technological innovation, knowledge dynamics, and, above all, institutional change. A good example of this is that the Europeans have created interdisciplinary departments of innovation studies—four of our European contributors (Edquist, Meeus, Faber, and van Lente) are located in them—whereas the Americans have been slower to do so. Furthermore, within these continents there are also broad differences. By having eleven European contributions (see Table 1.1) from a total of twenty-one non-American authors and ten American contributions from a total of fourteen American authors, we have a rich diversity of perspectives that forms the best opportunity for developing new theories about innovation and knowledge. As an added bonus, a number of the authors of the contributions are themselves involved either in international teams and/or comparative research. Other books on innovation tend to be dominated by one or another continent.

A fourth contention is that the rapid increase in the rate of product and process innovation, not only in the developed countries but in the developing ones, has generated the need for new theories, not just about innovation but about many other topics. One begins to perceive the greater influence of knowledge as both a resource and a powerful agent of change in many levels and sectors of society. All the ramifications of this growth in knowledge production necessitate exploring an important new research agenda. Indeed, the contribution of this book is that it is presenting new theories, new frameworks, new kinds of comparative data, as well as new research agendas, not just for innovation but also for the larger agenda of knowledge production.

If we abstract one of the central findings from the organizational literature on innovation, it is that you need a diversity of ideas to have an innovative product. Applying these ideas to the development of a book about innovation means that we need the diversity of disciplines and cultures to develop a new perspective. Below we discuss how this diversity is integrated. But first we examine what are the new research agendas that need to be addressed.

New research agendas

In the management and organizational sociology literatures, innovations have traditionally been defined as a new idea or practice. But in the stream of research that has issued over four decades, these disciplines have operationalized innovation as either the creation of new products or services or the adoption of new tools, treatments, or technologies. In the introduction to Part I, on industrial innovation, there is a more extended treatment of the definition of innovation. Similarly extended definitions of the other major concepts, such as science, knowledge dynamics, and institutional change, are provided in each of the introductory sections.

In sociology, the research on organizational innovation began with a theory about the determinants of innovation published in 1965 by Hage, one of the contributors to this volume, and then tested in a panel study of sixteen health and welfare organizations (Hage and Aiken 1967). A large body of research then ensued, and was examined in a meta-analysis by another contributor, Damanpour, in 1991. Finally, in a review article summarizing this research, Hage (1999) observed three consistent findings about the determinants of organizational innovation:

- the complexity of the division of labor or diversity;
- the organic structure (decentralization of organizational decision-making and the elimination of bureaucratic rules) or integration;
- the adoption of high-risk strategies.

This review indicated the need for movement in a number of new directions, including research on science, the inclusion of interorganizational networks, and the importance of the institutional context. But at a broader level, one could argue that this management/organizational sociology literature on innovation (Damanpour 1991; Zammuto and O'Connor 1992; Hage 1999) had the following problems:

- it ignored economic variables;
- it did not study research laboratories;
- it did not have a theory of sector/knowledge differences;
- it did not have a theory of national context differences;
- it did not have a theory of evolution or change.

Beyond this, the multidisciplinary nature of innovation studies became increasingly recognized, as a number of literatures began to merge together in the study of innovation including:

- evolutionary economics, network studies;
- S&T evaluation studies and organizational learning;
- technological regime, typologies of sectors;
- national systems of innovation, national policy studies;
- evolutionary models in the sociology of science, economics, and organizational sociology.

The list order of the lacunae and the merging literatures has been arranged to highlight the obvious parallels. Perhaps the most critical aspect of these new literatures is the recognition of the need for a number of diverse disciplines,

and also that there are multiple levels and sectors of society, including industrial sectors. The development of these new agendas can be simply arranged around the observation of how each part's contributions move beyond the existing literature on organizational innovation.

New agendas in the topic of industrial innovation or new processes and products

Reviews such as those by Damanpour (1991), Hage (1999), and Zammuto and O'Connor (1992) are primarily concerned with the organizational and management literatures and ignore the economics literature, which has also developed a number of theories about the economic determinants of industrial innovation. In Chapter 2, Damanpour and Aravind review this literature, find few consistent findings, and suggest what would be a new agenda for research that might be more promising. Their review combines a number of studies in both Europe and the United States.

A particularly important line of research on the topic of organizational learning, and how tacit knowledge can be made explicit, was instigated by Nonaka and his colleagues in Japan (Nonaka and Takeuchi 1995). This volume has a contribution, by Nonaka and Peltokorpi demonstrating how Japan can produce radical innovations in relatively short time periods. The particular radical innovation, the hybrid car, has enormous implications for international competition and environmental protection.

One of the major new phenomena has been the explosion in the number of new and different kinds of interorganizational relationships (IORs). As a consequence, network studies has become high on the agenda of innovation research. Meeus and Faber's chapter reviews this new literature and suggests some promising new ways to think about the problem of two or more organizations working together to create new products. Given the importance of this topic of interorganizational relations, each of the other parts has additional ways of thinking about this problem, all of which supplement the other. They are discussed in the last chapter (see Table 1.1).

A dramatic intellectual development in economics has been the creation of a new model quite different from the classic neo-equilibrium model. It is called 'evolutionary economics.' This movement has been much stronger in Europe than in America; the fourth contribution in Part I is by Metcalfe, who has contributed the most to this new perspective. The topic of evolution—not just in the economy but also in other aspects of society—is another major leitmotiv of this book and is used as a cross-cutting theme in the last chapter.

A parallel theme ignored in the previous work on organizational innovation is the role of the state and its various industrial policies for creating innovation. In the chapter by Foster, Hildén, and Adler, we have the exploration of the differential success of this policy in stimulating Green approaches in the manufacturing of paper products in Sweden, Finland, Germany, and the US. This comparative study could only have been accomplished with a combined American-European team. Finally, in the last contribution to this part, Chaminade and Edquist expand upon the literature of national systems of innovation with an original contribution about the functions of innovation in the economy.

New agendas as suggested by the topic of science

Much less studied, yet of increasing importance, is the role of scientific research in the development of new products and services. Advances in science—that is, new discoveries, concepts, theories, and empirical findings—are becoming central to the success of the economy and of national security. Governments certainly think so, which is one reason why the study of the state and its policies has to be added to the research agenda on innovation. Government concerns are particularly reflected in the steady growth in investments in basic science, and perhaps even more convincingly

in their attempts to measure performance. Yet, despite its importance, research on both industrial and scientific laboratories has been almost totally absent, hence the need for a part on this, one that also advances research in the sociology of science.

One of the limitations of industrial innovation, as indicated in the brief review provided above, is its focus on a single model, the organic structure, for explaining why some organizations have more innovation. In the first contribution in Part II, Jordan provides a theory of four different styles of management of research and development units, departments, and organizations, whether located in the private or the public sector. The choice of management style is based on the coding of the basic strategic choices of the research, whether incremental versus revolutionary and narrow versus broad in scope.

The joint development of new products occurs in interorganizational networks but science or, more specifically, particular areas of knowledge, move in a variety of networks creating communities of practitioners. Mohrman, Galbraith, and Monge provide a theory about this science or knowledge network. This also provides insights for the next part's topic of knowledge dynamics.

Not only do governments pay for the bulk of basic research, but they are also actively involved in the governance and coordination of much of this research. Georghiou's chapter examines the effects of the British policy of subjecting basic research to market competition in allowing both universities and private firms to compete for the same research money. He observes that there have been some unintended consequences to this policy, just as there have been to the state policy studied by Foster, Hildén, and Adler. And Kuhlmann and Shapira compare the governance styles of scientific research in Germany and the United States and illustrate the advantages of a European and American comparison, both as their topic and in their collaboration in their chapter.

The acceleration in production innovation and in state investments in basic research induced in economics a new way of thinking: evolutionary economics. The same result has occurred in the sociology of science: Rammert's contribution describes the evolution of the entire knowledge-producing system that focuses on the fundamental shift that necessitates new theories about innovation in knowledge dynamics. His examples cover a number of industrial sectors and scientific fields. The Rammert chapter is the last one in this part, and establishes the foundation for the next part on knowledge dynamics.

New agendas as suggested by the topic of knowledge dynamics

Because the industrial-innovation literature did not focus on research laboratories, especially in public research organizations, the problem of knowledge dynamics has not been studied as a problem. This does not mean that various topics that reflect the flow of knowledge have not become an important part of research, especially the idea-innovation chain (Kline and Rosenberg 1986; Hage and Hollingsworth 2000) and that of organizational learning (Nonaka and Takeuchi 1996). But the larger issue of knowledge trajectories has not been examined, including in the sociology of science. The definition of knowledge and its relationship to other major topics in the study of innovation is provided in the discussion of cross-cutting themes. One of the more critical implications of the recognition of the importance of knowledge, and also of the Rammert contribution, is the need to have a special part on knowledge dynamics.

Just as the organizational literature has ignored the role of science, it has also ignored the importance of design in the innovation process. One of the more interesting knowledge dynamics is the reciprocal generation of new design and new knowledge. Hatchuel, Weil, and Lemasson describe this process in their chapter, which challenges the organizational literature on innovation, and moves beyond it.

One of the more interesting ways in which research organizations and scientific communities are indirectly linked together is by the creation of generic research technologies—technologies that can be exploited in more than one scientific discipline. The importance of this knowledge dynamic is that it helps integrate the world of science and provides new ways for collective learning to occur. Shinn's contribution describes this kind of dynamic and explores the theoretical implications of generic technologies for the sociology of science.

The assumption that all technologies and scientific discoveries are good and desirable is implicit in much of the industrial-innovation literature. The chapter by Jolivet and Maurice attacks this assumption with an analysis of the European reaction to US firms' genetically modified seeds. Although Americans accepted this radical new technology quickly, Europeans have been more resistant. This very special kind of knowledge dynamic—different perceptions of the positive and negative aspects of a technology—is another ignored topic and needs to be addressed, especially now when there are considerable debates about the wisdom of pursuing certain kinds of research such as stem-cell research. This issue leads naturally into how specific research agendas are set or blocked in science. Van Lente's chapter provides one of the first examinations of how the control of rhetoric affects the choice of specific research arenas.

Concluding this part, and leading into the next, Finegold studies another critical aspect of knowledge dynamics, the role of deliberate changes in education programs in stimulating higher rates of innovation in industries or specific sectors of industry. His analysis compares the changes in educational policy in England, India, and Singapore.

New agendas as suggested by the topic of institutional change

A literature that was ignored in the study of industrial innovation was that on national systems of innovation. It is reviewed in Part I and also in the chapter by Casper. This literature emphasizes the *differences* between nations in Europe, the US, and Japan, explained by the ways in which their science and educational systems are structured and the consequences these have for rates of industrial innovation, especially industrial sectors at the nation-state level. The chapters by Kuhlmann and Shapira and by Finegold illustrate some of the basic differences in science and in education. Most of the existing literature on national systems of innovation, and that on industrial innovation, shares the same problem: it is largely static. Therefore, Part IV's focus is on the dynamic of institutional change, especially in its relationship to industrial innovation and scientific discoveries as well as government policies.

In the first chapter in this part, Hollingsworth presents a model of how the institutional context can influence scientific discoveries, including major breakthroughs. His work also provides the major debate that runs through this entire part: how much is institutional change simply a path dependency, and how much opportunity is there for organizational strategies or state policies designed to produce significant institutional change? This poses, in the most precise way possible, the theoretical question of the interrelationships between analytical levels.

One critical kind of knowledge dynamic is the creation of a new scientific paradigm. Van Waarden and Oosterwijk analyze how paradigmatic shifts in the telecommunications and pharmaceutical industries triggered structural evolution in these two industries towards greater knowledge specialization along the idea-innovation chain, which is defined in this chapter. This original research reports a comparative study of Austria, Finland, Germany, and the Netherlands. Studies of paradigmatic shifts are a very rich arena for future study, especially as scientific discoveries become more and more critical for industrial innovation. At the end of this contribution, the authors weigh the relative importance of the concept of the national system of innovation in the light of globalization, and especially of scientific research.

In the next contribution, by Hage, these two themes of path dependency and the impact of globalization are continued. The argument advanced in this chapter is that globalization and post-industrialization produced path dependency in Europe but not the US, again illustrating the importance of comparisons across the Atlantic. The specific institutional change examined is alteration in the dominant modes of coordination from vertical hierarchies to interorganizational networks.

This same theme of the dialectic between the national system of innovation, path dependency, and institutional change is continued in the chapter by Casper. He examines these issues in the biotech and the pharmaceutical industries in a comparison of the growth of small high-tech companies and their styles of product and process innovation in Britain, Sweden, and Germany. He finds some support for each perspective: there is an influence of the national system of innovation on path dependency, but there is also institutional change that departs from this.

In the last chapter in this part, Campbell presents yet another potential reconciliation between the path-dependency versus possible-departures-from-various-paths debate. Based on the rich experience of Eastern Europe and its attempts to produce dramatic institutional change in the form of markets and democracies, he proposes the idea of *bricolage* as a way of understanding how path dependency and institutional change can be understood together. Campbell's thesis is that in the process of institutional change, and especially radical change such as towards free markets or democracy, it is necessary to graft these institutional patterns on to existing rules within the society, and thus the process becomes one of *bricolage*, in the sense of trying what works.

Together these chapters provide a number of research agendas, not only for advancing the study of industrial innovation but also for the sociology of science, organizational learning, the study of networks, and research on the national systems of innovation. But perhaps the most important advance of these diverse literatures is a clear concept of knowledge as both a social force and as stock that has its own dynamic.

Since the book is an interdisciplinary effort of economists, political scientists, sociologists, management specialists, and historians of science, each specific contribution in one of these disciplines offers insight into the others. This is especially the case because of the focus on innovation and on knowledge production. It is hoped that this book represents an important step towards the creation of more multidisciplinary departments concerned with these themes.

As another step in this direction, each chapter, as well as representing in itself a new agenda of research, concludes with suggestions about future areas of research. These provide a powerful and complex agenda for building an integrated theory about innovation and knowledge. As a further step in the direction of building multidisciplinary knowledge, one can study in more depth the basic ways in which these contributions are integrated, which is our next topic.

The integration of the book

Earlier we suggested that the best way of producing an innovative book is to have a diversity of ideas. But these ideas also need to be carefully integrated. With so many new research agendas from different disciplines and different parts of the developed world, it might appear that there could be little cohesion to the contributions that are being made. This has indeed been a problem with many of the recent handbooks on innovation. What provides coherence and integration in this effort is a set of parallel themes—models, networks, coordination modes, state policy, and systemic change—which are discussed in each part of the book. Pervading our discussion are our three cross-cutting themes:

- knowledge as a foundation;
- social processes of evolution;
- the interaction of levels of analysis.

They are discussed in separate subsections following the discussion of the parallel themes.

Parallel themes

The coherence of a book can be measured by whether or not it is possible to construct an intellectual matrix that defines the contributions systematically by both section and theme. The extent of this book's coherence is provided in Table 1.2.

The introductions to each of the four parts acknowledge the great deal of empirical work, in several different disciplines, on the existing models of prediction or explanation. As exemplified above in our discussion of industrial innovation, each part of this book moves beyond the previous work in the organizational and management literature. And the integration of these varying perspectives provides a much deeper way of understanding organization and innovation. It is done in the book's *two* concluding chapters: as these chapters are particularly rich for managers who are interested in fostering creativity in their firms, we have a special concluding chapter that outlines the lessons for managers of R&D that can be drawn not only from the chapters from the first row of the matrix, but from all the contributions that have been written for this book.

As illustrated in the various articles on networks, the processes of knowledge evolution and growing specialization have made the networks a critical topic. Once one recognizes that the tendency for differentiation occurs along both the supply chain and the idea-innovation chain, as described in the cross-cutting theme of evolution, then the issue of integration becomes one of the major theoretical problems. But the content of the network is quite different in each of the four parts. This illustrates another way in which the contributions in one section can inform those in another, since one kind of linkage does not exclude the other. Again, by building across these four parts, a much richer view is obtained of how to measure and study the role of networks in

Table 1.2. Matrix structure of parallel themes

Intellectual connection	Industrial innovation	Scientific research	Knowledge dynamics	Institutional change
Models	Product versus process innovation *Damanpour and Aravind* Organizational learning? *Nonaka and Peltokorpi*	Design determinants of scientific ideas *Jordan*	Design and cognitive creation *Hatchuel, Weil, and Lemasson*	Institutions and path dependencies *Hollingsworth*
Networks	Interorganizational networks *Meeus and Faber*	Interpersonal relations and knowledge *Mohrman, Galbraith, and Monge*	Generic technologies and integration *Shinn*	Idea-innovation chains and institutions *Van Waarden and Oosterwijk*
Coordination modes	Markets and industrial innovation *Metcalfe*	Governance of science and innovation *Kuhlmann and Shapira*	Public opposition to innovation *Jolivet and Maurice*	Patterns of institutional and societal change *Hage*
State policies	The state and failed innovation *Foster, Hildén, and Adler*	Markets and scientific research *Georghiou*	Strategic turns in science policy *Van Lente*	The state and new industries *Casper*
Systemic change	Functions in NSI *Chaminade and Edquist*	Styles of knowledge regimes *Rammert*	Educational systems and innovation *Finegold*	*Bricolage* and institutional change *Campbell*

innovation and in knowledge creation. The theme of networks emerges in some of the other chapters as well. In particular, both the Rammert article on styles of knowledge regimes and the Hage chapter on patterns of institutional change have a considerable amount of material about the evolution towards interorganizational networks.

The third matrix row represents quite a different way in which organizations can be linked together: via 'coordination modes,' a term defined carefully in the contribution by Finegold, and in the introduction to the part on institutional change. Coordination modes vary considerably from market to state, and to lesser extents in the modes in between; again, because we have economists, sociologists, and political scientists, we have a much richer discussion of them than is typical in each of these disciplines. What is especially useful in the contributions in this matrix row is the debate about market versus non-market coordination modes. For example, the Kuhlmann and Shapira contribution provides an in-depth discussion of state coordination. A central idea in many of these discussions is the possibility that ideas, especially those that involve tacit knowledge, are more likely to be transmitted via various non-market coordination mechanisms: see the four chapters by Meeus and Faber; Mohrman, Galbraith, and Monge; Rammert; and Hage.

State policies continue the theme of non-market coordination, but with an explicit focus on policies designed to stimulate innovation, competition, or scientific research. We have contributions that illustrate both successes and failures of state policies. The fundamental issue is, when are state policies effective? The Foster, Hildén, and Adler chapter and the Georghiou contribution present some negative evidence; Casper provides some positive findings; Finegold's chapter provides examples of both. In the concluding chapter for academics, we return to the question of state-policy effectiveness, and attempt to provide some answers.

The contributions in the last parallel theme, that of systemic change, examine the problem of change from different angles that complement each other in various ways. One characteristic of systemic change can be its unplanned variety, illustrated in Rammert's and Hage's chapters, and the other its planned variety, represented in the contributions of Finegold and Campbell.

Each of these parallels provides a useful theoretical way of reading this book, depending upon readers' intellectual interests. Those concerned about the role of markets can read across the matrix row on coordination modes and also pursue similar topics, such as the role of public policy and of networks. Another form of integration is provided in the last two chapters: Norling guides managers of R&D concerned with stimulating more innovation or scientific discoveries; Hage and Meeus offer the beginnings of an attempt to construct a more complex theory of innovation that

unites these different research agendas and sections. Some of the contributions are relevant to more than one row or one section: for example, product and process innovation is discussed in the chapters by Rammert; van Waarden and Oosterwijk; Hage; and Casper; where two levels of analysis are combined—the organizational level and its macro-institutional context. There is the contribution by Jolivet and Maurice, which explores the problem of the acceptance of a radical product innovation (genetically modified crops).

And the Hatchuel, Weil, and Lemasson chapter explores the problem of industrial innovation internally from other than a structural perspective. Shinn and van Lente's contributions in Part III provide many illustrations for Part II, and also for the Hollingsworth chapter in Part IV.

Conversely, knowledge dynamics are reflected in the Mohrman, Galbraith, and Monge and Rammert contributions in Part II, and in the institutional-change chapters of van Waarden and Oosterwijk and Casper. As a final example, the problem of institutional change is involved in the Georghiou and Rammert chapters in Part II, in the contributions of Shinn and of Finegold to knowledge dynamics, and is reflected in the Meeus and Faber, and Foster, Hildén, and Adler chapters in Part I.

Three cross-cutting themes

But the five parallel themes are not the only ways in which integration has been achieved in this book. Equally, if not more, important are our three major cross-cutting themes of the foundation of knowledge, social processes of evolution, and the interpenetration of levels of analysis.

Why are these important? They form connecting links between the five disciplines represented in this book. Perhaps more critically, they are the basis upon which one can begin to construct a more viable theory of innovation and knowledge production.

For example, the cross-cutting theme of evolution describes a number of the structural causes that have made innovation and knowledge creation more critical to societies. The definition of knowledge and how it relates to innovation is the basic equation that connects many of the chapters in this book, and at the same time provides the basis for connection of new specialties that have merged recently. Finally, the interpenetration of analytical levels flows not only from the development of research at these disparate levels about innovation, but also the character of knowledge as a stock, which also exists at multiple levels: the individual, the community of researchers, the organization, the interorganizational network, the industrial sector or scientific discipline, and the nation state.

Knowledge as a foundation

Uniting the entire book is the theme of the role of knowledge. Yet this idea is an elusive one, difficult to define and about which there is much debate both among philosophers and social scientists. Part of the elusiveness stems from the considerable variety of scientific disciplines, their methods of research, and what they accept as evidence of what constitutes an established fact. Indeed, many scientists do not think that they are creating or building knowledge with their research. Even greater are the distinctions in the kinds of hands-on experiences associated with craft or artisan knowledge as opposed to those associated with the professional training of a physician or lawyer.

An important aspect of knowledge is its transmission via the process of collective learning or capacity building. What makes this aspect interesting is that it frequently involves *re*learning. It is this aspect that makes the transmission of knowledge somewhat difficult. Another reason for this difficulty is the importance of tacit—that is, uncodified—knowledge. A particularly rich exposition of the idea of tacit knowledge is found in the Rammert chapter. Therefore, one theme in the study of knowledge is the successful transmission as represented by relearning, not only at the level of the individual but, more importantly for the study of innovation, at the level of the collective.

Problematic, too, in defining knowledge are the philosophical debates between (among others) realist and idealist positions, which in the social sciences have centered on the concept of the social construction of knowledge, that is, the importance of power in the designation of what is taught, learned, and even what is accepted as a research finding. All of these divergences and debates make settling on a common definition difficult, if not impossible.

If knowledge is so difficult to define, why do we attempt to provide a definition? It is a problem over which most discussions of innovation and knowledge slide. To us, this does not seem appropriate. Better to attempt to provide a solution, even if it proves to be inadequate, because most of the chapters in this book refer, in one way or another, to a concept of knowledge and kindred ideas such as tacit knowledge, knowledge base, and organizational learning. Furthermore, the theory about the determinants of innovation, whether product, process/technological, or scientific discoveries, starts with some measure of the availability of a diverse set of capacities, which implies some kind of knowledge. And innovation, which means that something new has been added, implies the idea of some new knowledge, which in turn has feedback consequences on these capacities, as well as changing the larger institutional context. Given that we are interested in a definition of knowledge that relates to our problematic focus on product and process innovation, on scientific discoveries which are the essence of new knowledge, on learning and especially collective learning, or capacity building at both the individual and collective level, and institutional change, we propose the following definition.

Knowledge = the capacity to reproduce or to replicate findings, products and processes.

This definition may appear to be quite static, but changes in knowledge are defined by innovation in goods and services, scientific discoveries, and relearning. These changes add the dynamic quality, because they result in additions to the total stock of knowledge. It is worth repeating that this knowledge stock exists at multiple levels, ranging from the individual to the national state. The flows of knowledge reflect another aspect of change, and are discussed in Part III.

The static definition above calls attention to the importance of being able to replicate research results, which lies at the heart of what is accepted as established fact. In this context, Popper's (1959) idea of the ability to falsify an idea is an important contribution. Organizational ecologists describe organizations as having the capacity to reproduce the same product or provide the same service across time. In the work of Nelson and Winters (1982), this is referred to as 'having routines.' The whole issue of quality control illustrates how important the idea of reproducibility is. And the economists' idea of economies of scale is built on a similar notion. In health services, professionals attempt to replicate healing processes whether by drug or by surgery or some other kind of intervention. In educational services, professors attempt to reproduce a certain level of understanding or ability to reason critically in their students. As is apparent, this definition sidesteps the realist versus idealist debate, and can also be applied to individuals as well as collectives. But it is worth repeating that what interests us are changes in the stock of knowledge as reflected in innovations, which means changes in routines or learning new ways of doing things.

With this definition of knowledge we can construct a fundamental equation:

$$\text{Knowledge stock} + \text{collective learning} = \text{New knowledge or innovation.}$$

Furthermore, there are feedbacks from the creation of knowledge, resulting in innovation on the knowledge base, as reflected in various education programs and in various forms of collective learning, altering them in various ways. The really big question is whether or not the innovation also produces institutional change. This question forms one of the major debates in Part IV: are changes at the organizational level, including the creation of new organizational populations, independent of, and able to create major institutional change at, the institutional level; or are they inherently constrained by the national system of innovation and the basic norms and rules, including their mechanisms of enforcement?

A critical component of the equation is the issue of collective learning or relearning. Innovation requires the ability of a firm to recognize the value of new, external information, assimilate it, and translate it into the procurement and allocation of facilities, materials, components, and knowledge. For the firm, we want to understand what facilitates learning internally. The contributions by Nonaka and Peltokorpi, and by Hatchuel, Lemasson, and Weil explore this issue in original ways.

But internal learning is not the only source of innovation: the interaction with its environment determines a firm's access to a diversity of resources; the learning enables the firm to transform these resources into innovations. A number of the contributions indicate ways in which organizations can be connected to sources of tacit knowledge. Indeed the chapters in the Networks matrix row in Table 1.2 are rich in examples.

Learning is conceived as a process in which all kinds of knowledge are combined and recombined to form something new. Our equation above suggested that the knowledge base plus collective learning leads to new knowledge or innovation. What makes it collective is the interaction of this learning with some form of communication between people or organizations that possess different types of required knowledge. Basically, Lundvall's (1992) account of interactive learning clarifies:

- how technological and market dynamics pressurize firms to innovate their processes and products;

- how this innovation process impels firms to interact forward and backward in the production chain.

We have broadened his perspective to include the social processes of globalization, post-industrialization, and knowledge specialization, and have added the idea-innovation chain to that of the production chain. This leads us to our next theme, the social processes of evolution.

Social processes of evolution

The second of our cross-cutting themes concerns the various evolutionary forces creating pressures on both firms and governments to be more concerned about innovation and knowledge production.

The first and most obvious force is globalization. Globalization has, at minimum, three elements in its definition. The first is the commodity being traded, whether this is a good, a service, or an investment. What has become striking is the rapid movement of money through the world. The second and less obvious element is the number of different countries that can export products or services to any given country. Whereas only a few countries previously produced high-tech or knowledge-intensive products and processes, more and more countries are now involved. The third element is the consequences of globalization: the exporting of jobs overseas, especially from the developed to the developing countries; the creation of cross-national commodity chains, with rich countries keeping the high-skilled jobs and poor countries performing the labor; the increasing concentrations of wealth, both within countries and among them, as a consequence of these trends.[2]

A very special variant on globalization is the increasing number of countries that are investing in R&D, especially in basic research that provides the scientific discoveries so essential for continued industrial innovation, as indicated in the introduction to Part I. Perhaps the most extreme example, illustrated in the contributions of Finegold and of Casper, is the

attempt by many countries to build a biotechnology industry, and to provide the basic research in biology that is necessary for this. And in another sense in which products and services are produced globally, some countries have specialized in buying the rights to patents developed in other countries and then exploiting these patents effectively.

The second force is post-industrialization, defined as the growth in high-tech products and services that are increasingly provided in customized batches. As the labor force has become more educated, we have moved away from a mass society to one highly individualized, with many different niches in it. In the words of the economists, demand is becoming highly differentiated, requiring firms to respond to these varied demands. The explosion in the number of different models of the same product reflects this.

The third force is knowledge specialization, not just in product differentiation but more critically in the supply chain and in the idea-innovation chain. A definition of the latter is provided by the van Waarden and Oosterwijk contribution in Part IV, and examples of the former are given in the Meeus and Faber chapter in Part I. A major assumption is that, as knowledge grows, so do the cognitive limits of individuals and of organizations as represented by their core technology, resulting in an ever-increasing differentiation or specialization, so as to have enough depth in a specific competency. To continue to have innovation and collective learning requires that this diversity be integrated in some way. How to integrate across disciplines and organizations is a major issue. The process of specialization affects the supply chain as the new knowledge and technologies associated with new materials enter it, as well as the growing complexity of products, because of the development of new materials from research on nanotechnology and the addition of microchips to provide feedback and control. More products have multiple uses and add-ons necessitating much more complicated supply chains. One solution is the development of components, with different firms supplying different components.

As the amount of knowledge grows, it becomes more important for research organizations and firms to specialize in a specific part of the idea-innovation chain, whether this be in basic research, applied research, product development, manufacturing research, quality research, or market research. One of the unique aspects of this book is that it includes research findings about this new development in the contribution of van Waarden and Oosterwijk. In turn, these forces and processes make networks and non-market coordination modes critical because they can knit together, or integrate, the differentiated system, and networks of all kinds are one of the major parallel themes.

Why use these three processes as unifying themes? One of the central reasons is that they have caused the shift in competition from prices and quantity to newness and quality. Together these processes are generating pressures on governments, research organizations, and firms to adapt and to change their strategies and policies. Metcalfe's chapter describes the evolution in the economy, and Rammert's focuses on the evolution in the nature of knowledge regimes. But many of the policy papers can be seen as government responses to these forces. The chapter by Hage focuses on the consequences of these forces for the coordination modes in various countries, and specifically the movement towards interorganizational networks.

Consistent with the idea of knowledge specialization, considerable ferment exists in the world of science, and it is here that perhaps one observes the strongest tendencies towards evolution. Here are some of the discernible trends. Let us first consider universities, which are concerned not only with basic research but also education. Universities might be described as national knowledge stocks because they are concerned with absorbing knowledge worldwide and creating high-quality education programs to respond to industries' needs in times of expansion. There are several major trends emerging:

- growing links with application, increasingly blurring the borderline between

science and technology in frontier areas of research and the need for interdisciplinary approaches in complex problem solving, eliciting the growth of multidisciplinary research;
- a focus on generic technologies, which are either lacking in industrial R&D and have to be reinforced, or because industrial R&D is already ahead (mechatronics) and universities have to catch up;
- stronger university–industry interaction as a means of focusing and aligning basic research with industrial-knowledge demand (interorganizational partnerships in the Netherlands, and national research centres in the US are examples of this);
- the (re-)establishment of interface units, enhancing the possibilities for the business sector to access the internal capacity, skills, and know-how of university laboratories, thus reducing transaction costs of technology transfer;
- the establishment of joint research/technology development organizations. In many countries there is an increasing outsourcing of industrial R&D to universities.

This leads to our next cross-cutting theme, the importance of combining different levels of analysis.

The multiple levels of innovation

At the beginning of this chapter, we observed that one of the great advantages of having a mix of five different disciplines is that each of these has tended to find the causes of innovation, or the failure to innovate, in different variables and at disparate levels. Elsewhere, each of these levels tends to be studied in isolation, preventing us from coming to grips with which level is most critical. With multiple levels, we call attention to the general absence of studies of the macro-institutional level in organizational research on innovation, and even organizational research, more generally in the US. In contrast, the research on the national system of innovation and other macro-institutional perspectives is much

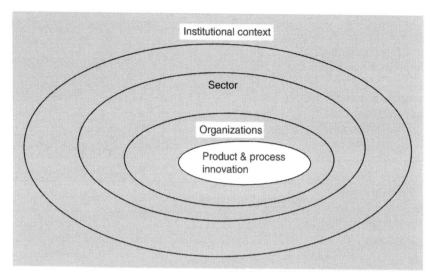

Fig. 1.1. A multi-layered perspective of innovation: is there co-evolution?

stronger in Europe, but they have tended to ignore concrete studies of organizations. Typically, when several levels are introduced into the analysis, one of the levels is reduced to a few variables, rather than being a robust example of what can be accomplished.

One reason to highlight four levels (see Figure 1.1), rather than just the organizational and macro-institutional levels, is to call attention to the importance of the industrial sector as a potential contingency factor that might explain when one level of analysis is more critical than another. But this diagram does not indicate that, at the level of the industrial sector or scientific discipline, there is a horizontal dimension, with one of the more common distinctions being between high-tech and low-tech sectors, which are discussed below. Much of the recent organizational theory (Hage 1980; Pavitt 1984) has emphasized the differences between economic sectors, suggesting that one needs to add contingencies at each level. In contrast, neither the literature on national systems of innovation nor the macro-institutional literature has made many distinctions by sector, except to observe that some societies perform better on measures of innovation in some sectors rather than others, as in the varieties-of-capitalism literature

reviewed in the Casper chapter. But this is at least the beginning of a movement towards the recognition of some contingencies by sector.

The most important avenue for future research in innovation is to examine each of the four levels as outlined in Figure 1.1 and determine which level is most critical, and under what circumstances or contingencies. To begin this process, the chapters of Rammert, Hollingsworth, van Waarden and Oosterwijk, Hage, and Casper contain multi-level analyses. In most instances, this reflects the combination of the organizational level with the national system of innovation (see the introduction to Part I for a definition) or macro-institutional level (see the introduction to Part IV for a definition). In some of these chapters, the industrial-sector level is included as well (for example, see van Waarden and Oosterwijk, and Hage chapters).

From an integrative perspective, we have several chapters that focus on the specific sector of biotechnology: those of Finegold, Hage, and Casper. Other chapters are detailed case studies of important industrial sectors: paper making in Foster, Hildén, and Adler's contribution, telecommunications in the van Waarden and Oosterwijk chapter. Certainly, if one accepts the distinction between high-tech industries and low-tech industries, as many do, there is a very different set of knowledge dynamics occurring in these sectors of society.

What has increased the importance of studying four levels in research on innovation is precisely the social processes of evolution that are forcing changes or adjustments upon the part of firms, governments, and the larger society. The question is whether or not all levels are actually changing and, if so, in a coordinated way. Some, like Freeman and Louçã (2001), refer to the four levels via the concept of co-evolution, meaning that there are simultaneous changes in the organization, science, sector, and institutional levels. But rather than make this assumption, it is preferable to treat it as a hypothesis: for example, is co-evolution more likely to occur in high-tech industries where radical innovation is common, and do the sector and even the institutional levels adjust accordingly, as Freeman has suggested?

The issue of co-evolution can be broken into two distinct ideas: first is co-variation at each level; second is whether the variation at each level is path dependent; if so, does the variation follow a path dependency, which is what is implied in the idea of evolution? As we have already observed, this is one of the major debates that runs through the entire fourth Part: the issue of evolutionary change via path dependency versus discontinuous change that does not appear to be evolutionary. In this context, various contingencies are suggested as to the relative independence of the organizational level from the national system of innovation.

We write 'co-evolution' in inverted commas to highlight the debates about the possibility for change at any of these levels. Some of these levels are more difficult to change than others. Indeed this is one of the major debates between, on the one hand, the organizational sociologists and policy makers, and the institutionalists on the other. The assumption made by all these individuals is that change is more difficult at the level of the society, but this may not always be the case. Do the science and educational systems respond quickly when they perceive demand for new areas of research and of education, as happened with

the expansion of the semiconductor and computer industries? In contrast, for high-volume, mass-production products, firms might be more resistant to meeting the challenges of globalization and post-industrialization than the macro-institutional sectors of science and education. Many large firms have shown more rigidity than the societies in which they are located, or at least parts of them. A final point worth mentioning is that the reasons for a lack of evolution are different at each of these levels, and need much more research.

Although one might remain sceptical about the principle of 'co-evolution,' there is another kind of trend across levels that is worth postulating: that is the growing trend towards connectivity across the different parts of society, reflected in the more complex interorganizational networks and, as knowledge dynamics unfold, the growing importance of linkages of various kinds, one reason why this is a section of the book. This does not mean that these linkages are necessarily very stable. Like knowledge, they can be fleeting, but their implications have yet to be understood and are the basis of an important line of research.

Summary

Our argument for the necessity of a new approach to the study of innovation has been based not only on its importance but also on the limitations in the existing literatures on management and organizational sociology, plus the merging together of a number of literatures. Consistent with what are standard themes in the study of innovation, we have argued that we need to have a diversity of ideas. This has been generated by selecting contributors from five disciplines and from multiple countries.

But rather than simply present twenty-one chapters, we have chosen to integrate them in an intellectual matrix that highlights how they articulate. The four Parts, industrial innovation, scientific research, knowledge dynamics, and institutional innovation, all work on our parallel themes of models, networks, co-ordination modes, policy studies, and systemic change. Many of the chapters are relevant to more than a single section or parallel theme, creating further integration. This matrix makes it easy for readers to work their way into the book according to their own preferences and interests. For managers of R&D who are interested in improving product and process innovation or scientific research, Norling has abstracted a number of practical principles from the various contributions.

If this were not enough integration, there are also three cross-cutting themes: knowledge as a foundation, social processes of evolution, and the interpenetration of the four levels. These cross-cutting themes form the basis of the beginnings of a much more complex and integrated theory about both innovation and knowledge, which is started in the concluding chapter. Our hope it that this theory, although not complete, will spur others to add to it, as well as focus their research in more decisive ways.

Notes

1. One cannot use the term 'sociology of knowledge' because this means the study of knowledge as a cultural product rather than as a stock of knowledge and of knowledge dynamics, which is the perspective in this book.
2. Other definitions of globalization are possible, including the idea of cultural globalization, which frequently takes the form of homogenization of culture. This is not the sense in which this concept is used here.

References

Cohen, W., and Levinthal, D. (1990). 'Absorptive Capacity: A New Perspective on Learning and Innovation.' *Administrative Science Quarterly*, 35: 128–52.

Damanpour, F. (1991). 'Organizational Innovation: A Meta-Analysis of Effects of Determinants and Moderators.' *Academy of Management Journal*, 34/3: 555–90.

European Innovation Scoreboard (EIS) (2003). **http://trendchart.cordis.lu/Scoreboard/ scoreboard.htm**

Freeman, C., and Louçã, F. (2001). *As Time Goes By: From the Industrial Revolution to the Information Revolution*. Oxford: Oxford University Press.

Hage, J. (1965). 'An Axiomatic Theory of Organizations.' *Administrative Science Quarterly*, 10: 289–320.

—— (1980). *Theories of Organizations*. New York: Wiley.

—— (1999). *Organizational Innovation (History of Management Thought)*. Dartmouth: Dartmouth Publishing Company.

—— and Aiken, M. (1967). 'Program Change and Organizational Properties: A Comparative Analysis.' *American Journal of Sociology*, 72: 503–19.

—— and Hollingsworth, J. R. (2000). 'A Strategy for the Analysis of Idea Innovation Networks and Institutions.' *Organization Studies*, 21: 971–1004.

Hall, P., and Soskice, D. (eds.) (2001). *Varieties of Capitalism*. Oxford: Oxford University Press.

Kline, S., and Rosenberg, N. (1986). 'An Overview of Innovation.' In L. A. Rosenberg (ed.), *The Positive Sum Strategy*. Washington, DC: National Academy of Sciences, 289.

Lundvall, B.-Å. (ed.) (1992). *National Systems of Innovation: Towards a Theory of Innovation and Interactive Learning*. London: Pinter.

Nelson, R. (ed.) (1993). *National Innovation Systems: A Comparative Analysis*. New York: Oxford University Press.

—— and Winters, S. (1982). *An Evolutionary Theory of Economic Change*. Cambridge, MA: Belknap Press.

Nonaka, I., and Takeuchi, H. (1995). *The Knowledge-Creating Company*. New York: Oxford University Press.

NSF. *Science and Research Indicators*. Found on **www.nsf.gov/statistics/indicators** for 2004, 2002, 2000, 1998, and 1996.

OECD (2001). *Innovative Clusters: Drivers of National Innovation Systems*. Paris.

Pavitt, K. (1984). 'Sectoral Patterns of Technical Change: Towards a Taxonomy and a Theory.' *Research Policy*, 13: 343–73.

Popper, K. (1959). *The Logic of Scientific Discovery* (trans. of *Logik der Forschung*, published in 1935). London: Hutchinson.

Zahra, S. A., and George, G. (2002). 'Absorptive Capacity: A Review, Reconceptualization and Extension.' *Academy of Management Review*, 27/2: 185–203.

Zammuto, R. F., and O'Connor, E. J. (1992). 'Gaining Advanced Manufacturing Technologies' Benefits: The Roles of Organization Design and Culture.' *Academy of Management Review*, 17: 701–28.

PART I

PRODUCT AND PROCESS INNOVATION

Introduction

Marius T. H. Meeus and Charles Edquist

Industrial innovation is considered fundamental to productivity growth and thereby to long-term socio-economic development in industrialized countries. According to growth accounting that emerged in the 1950s, the main source of growth in labor productivity is the 'residual,' which mainly consists of new products and processes based on the advancement of knowledge and technology. Increasing physical and human capital (through education) is seen as less important for increasing productivity. On the one hand, empirical research has revealed that innovation enhances the growth and survival of firms (Brouwer and Kleinknecht 1994; Archibugi and Pianta 1996; Audretsch 1995; Lawless and Anderson 1996; Metcalfe 1995). On the other hand, innovation is a very complex and risky process, with low success rates, and sometimes lethal effects. Innovations potentially disrupt and reform the organizational fabric, often in a fairly unpredictable and situation-specific way (Zammuto and O'Connor 1992; Dean and Snell 1991; Lundvall 1992; Leonard-Barton 1988; Dougherty and Hardy 1996).

After the cost cutting, downsizing, and re-engineering in the 1980s, both product and process innovation became levers in the late 1990s for companies to generate sustained competitive advantages. Intel, Nokia, Ericsson, Daimler-Chrysler, Microsoft, Du Pont, and many other multinational companies have generated continuous streams of innovations protecting their market positions. Yet there were as many examples of world-leading, market-dominating large firms like SSIH—the Swiss watch consortium—or IBM that were unable to respond to technological shifts such as the introduction of electronic watches and laptops. Christensen (1997) has shown that it was not their large competitors that had out-innovated them, but new entrants. This success paradox (Tushman *et al.* 1997) frames some of the questions in this part:

(*a*) What is innovation anyway, and how has it been conceptualized?
(*b*) What do we know about the antecedents of the rate of product and process innovation?
(*c*) How radical is the innovation?

Besides the variety in innovative performance of firms, there is the variety in innovative performance of economies at different levels: sectors, countries, and continents. This issue frames the second set of questions answered here:

(*d*) how do larger market environments, and
(*e*) how do regulatory and institutional environments impact on patterns of industrial innovation?

The innovation concept

Schumpeter described innovation as 'a historic and irreversible change in the way of doing things,' and 'creative destruction' (Schumpeter 1943). Innovations are here defined as new creations of economic significance, primarily carried out by firms. They include both product and process innovations. Product innovations are new or better products (or product varieties) being produced and sold; it is a question of *what* is produced. They include new material goods as well as new intangible services. Process innovations are new ways of producing goods and services; it is a matter of *how* existing products are produced. They may be technological or organizational. In this taxonomy, only goods and technological product innovations are material. The other categories are non-technological and intangible.[1] This taxonomy can be illustrated as in Table I.1.

Some product innovations are transformed into process innovations in a second incarnation. This concerns only investment products, not products intended for immediate consumption. For example, an industrial robot is a product when it is produced and a process when it is used in the production process. There are several other taxonomies of innovations and they can certainly be combined with each other. One may, for example, distinguish between:

(*a*) continuous small incremental changes;
(*b*) discontinuous radical innovations;
(*c*) massive shifts in some pervasive general purpose technology (GPT), sometimes called 'techno-economic paradigms' (Edquist and Riddell 2000).

Table I.1. A taxonomy of innovations

Types of Innovations			
Product Innovation		Process Innovation	
In goods	In services	Technological	Organizational

The idea is that some innovations change the entire order of things, making obsolete the old ways and perhaps sending entire businesses into the ditch of history. Other innovations, requiring only modest modifications of the old-world view (Van de Ven *et al.* 1999; Rogers 1995; Tidd *et al.* 2001), simply build on what is already present. An example of a radical or breakthrough innovation might be the first (marketed) design of an aircraft, the first integrated circuit, or the first development of penicillin. Examples of GPTs are information and communications technologies (ICTs), electricity, and the internal combustion engine. Such innovations might lead to the creation of brand-new industries. The incremental mode implies a more step-by-step approach of gradually improving existing products or processes. Damanpour and Aravind's chapter tries to find out which determinants of product and process innovation have been identified as having a consistent empirical effect.

Because their definitions ignore other aspects related to innovation processes, the distinction between what is incremental and what is radical has been reworked many times. To be more precise, it is useful to distinguish between three types of changes: changes at the level of the innovative product or process (technological characteristics, functions, quality); changes induced by the innovation at the level of the innovating agent (competencies, organizational structures, market position); changes induced by the innovation throughout the value chain—for example, for users' competencies, or supplier involvement. The dimensions of incremental and radical are mostly used to specify changes at the level of the product or process, but the incremental-radical continuum applies to the other aspects as well. Henderson and Clark (1990), Christensen (1992, 1997), and Afuah and Bahram (1995) developed this conceptualization. Each author showed, in different ways, how collaborative and competitive impacts and diffusion of innovations could be related to the innovation concept. Henderson and Clark (1990) and Christensen (1992, 1997) conceptualized innovation at the level of an artifact as changes in two dimensions: that is, linkage between core concepts and components on the one hand, and in reinforcement or replacement (= overturning) of core concepts on the other hand. 'Architectural innovation' is a rearrangement of the ways in which components relate to each other within a product's system design; the core concepts—the technological basis—are, at most, reinforced (for example, CPU-time is optimized in a computer). It is primarily a design- and production-driven innovation activity exemplified in Dell computers or Toyota in its supply-chain management. In the case of 'modular innovation,' the core concepts deployed in a component are overturned, while the product architecture is left unchanged: examples of this are read-write heads in disk drives, that were replaced first by ferrite technologies (in thin film heads), and later by magneto resistive heads.

'Incremental innovation' here refers to:

(a) improvements in component performance that build upon the established technological concept; or

(b) refinements in systems design that involve no significant changes in the technical relationships among components.

'Radical innovations' here involve both a new architecture and a new fundamental technological approach at the component level:

* mainframe—PC—laptop—palm;
* watermill—windmill;
* integrated steel production and minimills;
* standard spring-powered watch—the electric watch—the tuning watch—the quartz crystal watch.

The example of disk drives, taken from Christensen (1992, 1997), is helpful to understand both collaborative and competitive impacts of using this classification. In the disk drive industries, the drivers of performance improvement along the dimensions of performance most

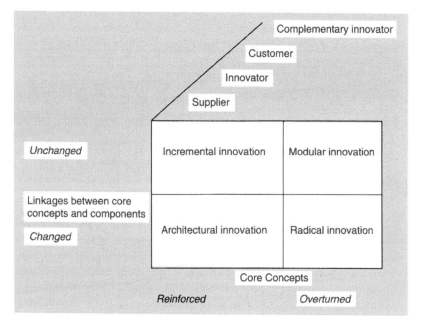

Fig. I.1. The hypercube of innovation

Note: The X and Y axes are the innovation-classifying factors. The Z axis is the innovation value-adding chain of key components: innovator, customer and supplier of complementary innovators.

Source: Afuah and Bahram 1995.

valued in established markets were generally new component technologies. Hence, market introduction of new components improving on main performance aspects was straightforward and easily sold to major customers. New architectural technologies, however, tended to redefine the product's functionality—the parameters by which system performance was assessed. Because of this, new architectures were generally deployed in new market applications. What seems to underlie the failure of established firms at points of architectural technology change in the history of disk drive industry is not failure to innovate in the laboratory, but failure to innovate in the market (see Figure I.1).

As is typical of architectural innovations, they employed proven component technologies and, at first, according to criteria highly valued in established markets, underperformed the dominant architectural technologies. The principal customers for the 14- and 8-inch architectures were the makers of mainframes and minicomputers, who applied total capacity and the speed of information storage and retrieval as performance criteria. The 5.25-inch drive architecture that emerged in 1980 was inferior to the 14- and 8-inch architecture along both dimensions (5.25: 5 MB storage, 160 ms access speed; larger drives' average storage 100–500 MB, speed 30 ms). For this reason, the established makers of mainframes and minicomputers ignored the new architecture. They continued to listen to their main customers, and this created a lock-in in the old architecture.

However, the emerging market of desktop PCs demanded this new functionality. It was based on performance dimensions that were new and overlooked because of the market segments serviced by the established firms. The firms that introduced the 5.25 architecture were new entrants to the industry. The point is that the substitution of new architecture for the old began long before the new technology became performance competitive. Christensen[2] gives a very nice summary of the way this competitive impact of architectural innovation was pulled by new user needs and unfolded between the late 1970s and 1995.

In the late 1970s, the market for disk drives consisted of makers of large mainframe computers. These customers demanded an aggressive improvement in capacity of more than 20 per cent a year, above the minimum required capacity of 300 MB. The leading and most innovative 14-inch drive makers (namely IBM, Memorex, EMM, and Ampex) competed vigorously, maintaining the industry's aggressive rate of R&D investment that had led to dramatic improvements in capacity and cost. During those years, a few start-ups developed 8-inch drives with less than 50 MB capacity, but only minicomputer start-up companies used them. Because these drives were easy to make, and because mainframe customers did not want them, profits margins and sales volume were extremely low.

New entrants struggled to find a viable market for these drives; mostly only minicomputer start-ups were interested in them. IBM and other established drive makers had to decide between two options: either they could divert scarce engineering and financial resources to this small new market and risk eroding their market share of the high-margin, high-growth 14-inch market; alternatively, they could wait until the market was big enough, and then invest aggressively to capture it. Unexpectedly, 8-inch drive makers sustained a capacity increase of more than 40 per cent a year. Their products soon met the needs of mainframe computer makers, while offering advantages intrinsic to a smaller disk, such as reduced vibration. Within four years, 8-inch drives had taken over the mainframe market. Although one-third of the 14-inch makers had introduced 8-inch models, with very competitive performance, every independent 14-inch drive maker had been driven out of the industry by the end of the 1980s. And of the seventeen disk drive companies existing in 1976, all but IBM had failed or had been acquired by 1995. The 8-inch manufacturers, however, were no wiser to the disruptive technology phenomenon, and found themselves fighting a losing battle several years later against the 5.25-inch drive.

The main inferences to be drawn from this example are:

(a) established firms could not anticipate the performance jumps in the 8-inch drives, so technological discontinuities remained the most important challenge in technology management;

(b) established firms remained top performers in radical component innovation, but they underestimated the competitive impacts of new architectures;

(c) the structure of established firms, combined with preferences of their lead users, made them bet on the wrong competences, and made them overlook new performance criteria.

Whereas Christensen emphasized the competitive impact of architectural innovation, Afuah and Bahram (1995) differentiated the focus of innovation in another way. They suggested that, in addition to probing the impact of an innovation on the innovator's own competences and assets, the innovator should also ask the question: 'What will my innovation do to the competence and products of my suppliers (original equipment manufacturers (OEM)), customers, end-user customers, and of key complementary innovators (software producers for IBM)?' What is, from the perspective of the innovators, a radical innovation may turn out to be an incremental innovation for the customer or the complementary innovator; what is, from the perspective of the innovator, incremental could be a radical innovation from the perspective of the customer, or the complementary innovator.

The impact of an innovation on the capabilities and assets of the other actors in the innovation chain is what largely determines the market success or failure of an innovation. Many complex high-technology products require that users invest time and money in learning how to operate and maintain the products. An innovation that destroys the knowledge that the customer has acquired is less likely to be adopted than one that enhances this knowledge and assets. The case of the electric car is a good example here. The Toyota Prius is a radical innovation to the car companies, to suppliers of key components like the power train, and to suppliers of the key complementary innovation—gasoline; to the customers, however, it is an incremental innovation. The DSK (Dvorak Simplified Keyboard) keyboard arranges the keyboard such that it allows 20–40 per cent faster typing than with the QWERTY keyboard; but this rival design implied enormous switching costs on behalf of the users, and would have such an impact on typing skills that adoption rates were minimal. The DSK keyboard was an architectural innovation in which core concepts and components had not changed; only the linkages had been changed.

These refinements in the innovation concept call for further reflections on two questions: How do radical innovations emerge? How is knowledge synthesized into new competencies in a firm while interacting with many partners? These issues are dealt with in the Nonaka and Peltokorpi chapter on the Toyota Prius case. The phenomenon of interorganizational linkages is addressed in the chapter by Meeus and Faber, which focuses on two questions: What effects do interorganizational relations have on the innovative behavior of firms? What induces the formation of interorganizational relations during innovation projects?

Comparing patterns of innovation: variety across the EU and the US, and within the EU

Many comparative studies have shown that patterns of innovation (Nelson 1993; Freeman and Soete 1997) within Europe and between Europe and the US are very different. What explains this uneven distribution of innovation?

The focus of this part of the book is on industrial innovation, and which factors are considered dominant in describing and explaining the occurrence and outcomes of innovation processes. Comparative analysis of performance indicators related to innovation in the European Union (EU) and the US shows remarkable differences (cf. Figure I.2). The European Innovation Scoreboard (EIS) 2003 explores in detail the

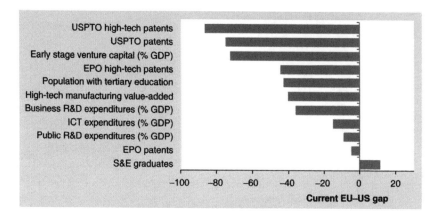

Fig. I.2. Current EU–US gap

Source: European Innovation Scoreboard 2003.

development of the EU/US gap for those indicators for which comparable data are available. As of last year, the US leads the EU in the majority of these indicators (10 out of 11).

The EIS (2003) reports that, at the current rates of change, none of the current EU/US gaps would be closed before 2010. Business R&D shows some weak signs of recovery but, since 2001, a new and increasing gap has appeared in public R&D (gross domestic expenditure on R&D minus business enterprise expenditure on R&D). Early-stage venture capital improves slowly, but the gap remains huge. As for human resources, the large gap in tertiary education persists. The EU weakness in education is further illustrated by the worrisome decline of the EU trend in lifelong learning (no comparable US data are available). The EU's only advance is in Science and Engineering (S&E) graduates. Only two indicators justify a more positive note: a very slow but noticeable, catching-up, value-adding process can be observed in high-tech manufacturing; a long-lasting catching-up process in ICT expenditures (EU/US gap cut by half since 1996).

Figure I.3 shows how the EU/US gap, measured from 1996 to 2002, evolves on the main innovation indicators. Variety in patterns of innovation is seen not only between the EU and the US. Figure I.4 shows that there is considerable variety in successful market introduction as a percentage of turnover in European countries. Italy is the leading country, followed by Spain, Ireland, and France, which all have performance scores above average. There is a large group of followers; among them are the UK, Germany, and Sweden.

At the level of innovation output, there is considerable variety within the EU. This also goes for innovation expenditures, one of the main input variables. What is especially interesting is that high innovation results—in

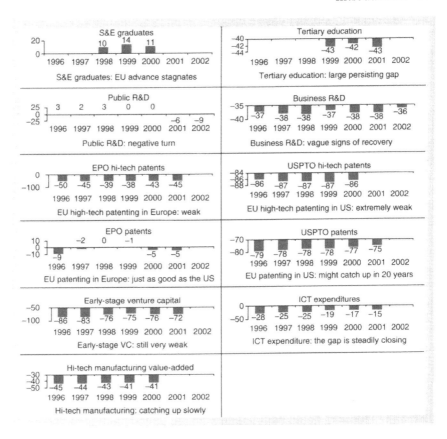

Fig. I.3. EU and US innovation indicators compared

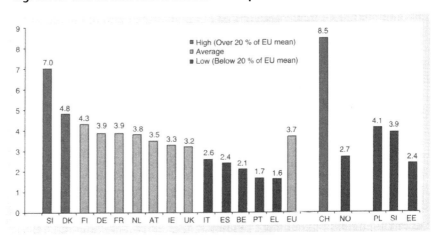

Fig. I.4. Sales due to products new to the market

Note: IT=Italy, ES=Spain, IE=Ireland, FR=France, FI=Finland, PL=Poland, DE=Germany, DK=Denmark, S=Sweden, NL=The Netherlands, UK=United Kingdom, AT=Austria, BE=Belgium, EU=European Union, IS=Iceland, NO=Norway, CH=Switzerland, EE=Estonia, SI=Slovenia, PT=Portugal, EL=Greece.

Source: European Innovation Scoreboard 2003.

terms of a high proportion of new-to-the-market products as a percentage of turnover—do not seem to be associated strongly with innovation expenditures of firms. Sweden has the highest innovation expenditures, but is outperformed in sales due to new-to-the-market products by seven other countries (cf. Figure I.4). This large variety in innovation indicators invites the question of which factors do determine innovation intensity, an issue that is taken up in the chapter by Damanpour and Aravind.

Innovation, market, and non-market factors: the innovation systems perspective

The variety in innovative inputs and outputs across countries and continents begs an explanation. In an innovation system, a broad number of market and non-market factors are brought together. The environment in which innovation emerges consists of many elements, often summarized in systems of innovation schemes (cf. Figure I.5). Several linkages connect the various players and subsystems in Figure I.5. Galli and Teubal (1997: 347) distinguish three types of linkages:

(*a*) market transactions, which involve backward and forward linkages as well as horizontal linkages;
(*b*) unilateral flows of funds, skills, and knowledge (embodied and disembodied) within and National System of Innovation as well as externally, between organizations and others located in other countries or NSIs;
(*c*) interactions, such as user-supplier networks.

These linkages are embedded in a wide variety of institutional arrangements, e.g. laws, norms, and traditions; regulations; policy-induced incentives and disincentives; specific allocation and decision-making mechanisms within organizations; cooperation agreements.

Innovation is highly contingent on historical circumstances, often captured as path dependency, and on the co-evolution of agents' behaviors and innovation. Institutional infrastructures and networks of research and innovation systems are historical products; they, in turn, continue to shape current innovation processes. In the past half-century, this area of society has been shaped by (national) state political interventions and private initiatives: political systems have developed research and innovation policy, in which they acted as catalysts, promoters, and regulators of innovation-related activities.

Since the 1970s, the triumph of high technologies has induced a broad spectrum of technology policy intervention measures in industrialized countries, and sparked off a technology race among them. In the same period, the spectrum of implemented instruments of research, technology, and innovation policy was widely differentiated, reflecting the scope of institutions and interests involved—from public funding of research organizations with various forms of financial incentives, to conducting research and experimental development in public or industrial research

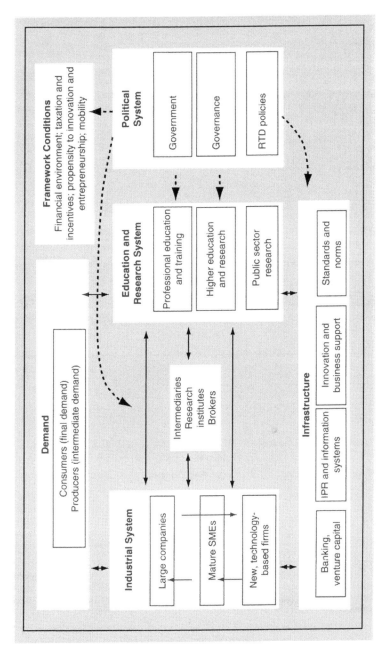

Fig. I.5. A system of innovation

Source: Kulmann and Arnold 2001.

laboratories, to the design of an innovation-oriented infrastructure, including the institutions, organizations, and mechanisms of technology transfer. In many European countries, these instruments have dominated research and technology policy for the last three decades.

Many different approaches to the relations between regulation, environment, and innovative firm behavior have evolved, especially in the field of environmental regulation. It has been argued that the deterministic nature of environmental laws establishing standards designed to control the material and energy outputs of society to the biophysical environment limits firms' strategic choices, and constrains their ability to innovate (Breyer 1982). More specifically, critics argue that the bureaucracy required to comply with environmental regulation restricts firms from pursuing cutting-edge technology. Managers of environmentally regulated firms argue that it is harder to innovate because regulations often change unexpectedly, and because regulators are unpredictable. This increased uncertainty motivates firms to de-emphasize risky strategies such as innovation. The net effect of these constraints is reduced innovation, which many argue puts environmentally regulated firms at a competitive disadvantage (Guttmann et al. 1992; Scherer and Ross 1990).

A competing argument is that, if viewed as an external jolt, environmental regulation can stimulate innovation within an organization (Marcus and Weber 1989; Meyer 1982). Such a jolt may appear disruptive and threatening to a firm, but it may be necessary to induce innovation (Schon 1971). In the absence of such stimuli, existing organizational practices are often not challenged, and members may resist innovation, fearing it will change the status quo (Van de Ven 1986). An example of the positive effect of environmental regulation on product innovation is found in chemical manufacturing. Faced with the rapidly approaching deadline for the worldwide phase-out of chlorofluorocarbons (CFCs), Imperial Chemical Industries (ICI), Du Pont, and Elf Atochem developed the technology required to produce CFC substitutes in record time, reduced from the industry norm of more than a decade to only five years (Weber 1993).

One can conclude that there are competing views on the link between (environmental) regulation and innovation (Porter and van der Linde 1995a, 1995b). Both positive and negative effects are reported and empirically confirmed. Of course this overview is not exhaustive and probably applies to many divergent contexts but, to say the least, it gives clear indications that we need better-specified theoretical models and data allowing us to test these competing views, as well as the factors mediating the effects of environmental regulation on environmental innovation.

These issues are addressed in the chapters of Foster et al. and of Metcalfe. Foster et al. discuss the issue of what the role of the state and related organizations can be in environmental innovation. They ask three questions:

(*a*) Is it generally possible, and under what conditions, can public interventions induce innovations?

(*b*) How does the intervention affect the economic and environmental performance of the firm?

(*c*) Can the induced innovations bring about beneficial societal effects?

Metcalfe deals with market features like competition and concentration and their impacts on innovation, adoption, and diffusion. He elaborates Schumpeterian ideas proposing that firms in concentrated markets have more incentives to innovate because they can more easily appropriate the returns from innovation. Scholars like Cohen and Levin (1989) and Baldwin *et al.* (2002) have qualified this proposal in different ways; for example:

(*a*) the firm's gains from innovation are greater in competitive than monopolistic industries;

(*b*) innovation is more intensive in the early stages of an industry's development when markets are less concentrated;

(*c*) large firms are more innovative in concentrated industries with high barriers to entry.

Furthermore, Metcalfe (1988) considers the relative importance of supply and demand. This is brought out most sharply when one considers the question of profitability as the incentive to the adoption and diffusion of a new technology. But profitability to whom? To the potential adopter or the potential producer (for innovations cannot be produced unless they can be profitably produced)? These and other issues are further elaborated in Metcalfe's chapter on markets and innovation.

Finally, Chaminade and Edquist discuss the institutional environment, the framework conditions and the infrastructure (cf. Figure I.5) governing innovation. They advance an alternative operationalization of the systems-of-innovation approach based on ten distinct activities that influence the development and diffusion of innovations. After a discussion of each activity, the field of innovation policy is entered, including reasons for public intervention in the innovation process and ways of identifying problems that should be subject to policy.

Notes

1. For further specifications of this taxonomy of innovations, see Edquist *et al.* (forthcoming: 10–17).
2. This summary (2004) can be found at this website: **www.christensen.com**

References

Afuah, A. N., and Bahram, N. (1995). 'The Hypercube of Innovation.' *Research Policy,* 24: 51–76.

Archibugi, D., and Pianta, M. (1996). 'Measuring Technological Change through Patents and Innovation Surveys.' *Technovation,* 16: 451–68.

Audretsch, D. B. (1995). 'Innovation, Growth and Survival.' *International Journal of Industrial Organization,* 13: 441–57.

Baldwin, J., Hanel, P., and Sabourin, D. (2002). 'Determinants of Innovative Activity in Canadian Manufacturing Firms.' In A. Kleinknecht and P. Mohnen (eds.), *Innovation and Firm Performance.* New York: Palgrave, 86–111.

Breyer, S. (1982). *Regulation and its Reform.* Cambridge, MA: Harvard University Press.

Brouwer, E., and Kleinknecht, A. H. (1994). 'Innovation in Dutch Industry and Service Sector.' (In Dutch: 'Innovatie in de Nederlandse industrie en dienstverlening. Een enquête-onderzoek.') The Hague: Economische Zaken.

Christensen, C. (1992). 'Exploring the Limits of the Technology S-curve, Part II. Architectura-Technologies.' *Production and Operations Management,* 1: 358–66.

—— (1997). *The Innovator's Dilemma.* Cambridge, MA: Harvard Business School Press.

Cohen, W. M., and Levin, R. C. (1989). 'Empirical Studies of Innovation and Market Structure.' In R. Schmalensee and R. D. Willig (eds.), *Handbook of Industrial Organization* II. Amsterdam: Elsevier Science Publishers, 1060–107.

Damanpour, F. (1991). 'Organizational Innovation: A Meta-Analysis of the Effects of Determinant and Moderators.' *Academy of Management Journal,* 34: 555–90.

—— (1996). 'Organizational Complexity and Innovation: Developing and Testing Multiple Contingency Models.' *Management Science,* 42/5: 693–716.

Dean Jr., J. W., and Snell, S. A. (1991). 'Integrated Manufacturing and Job Design: Moderating Effects of Organizational Inertia.' *Academy of Management Journal,* 34: 776–804.

Dougherty, D., and Hardy, C. (1996). 'Sustained Product Innovation in Large, Mature Organizations: Overcoming Innovation-to-Organization Problems.' *Academy of Management Journal,* 39: 1120–53.

Drazin, R., and Schoonhoven, C. B. (1996). 'Community, Population, and Organization Effects on Innovation: A Multilevel Perspective.' *Academy of Management Journal,* 39: 1065–83.

Edquist, C., and Riddell, C. (2000). 'The Role of Knowledge and Innovation for Economic Growth and Employment in the IT Era.' In K. Rubenson and H. Schuetze (eds.), *Transition to the Knowledge Society: Policies and Strategies for Individual Participation and Learning.* Vancouver: University of British Columbia, Institute for European Studies.

—— Hommen, L., and McKelvey, M. (forthcoming). *Innovation and Employment: Process versus Product Innovation.* Cheltenham: Edward Elgar.

European Innovation Scoreboard (EIS) 2003. **http://trendchart.cordis.lu/Scoreboard/scoreboard.htm**

Freeman, C., and Soete, L. (1997). *The Economics of Innovation.* London: Pinter.

Galli, R., and Teubal, M. (1997). 'Paradigmatic Shifts in National Innovation Systems.' In C. Edquist (ed.), *Systems of Innovation: Technologies, Institutions and Organizations.* London: Pinter, 343–70.

Guttmann, J. S., Sierck, A. W., and Friedland, D. M. (1992). 'The New Clean Air Act's Big Squeeze on America's Manufacturing Base.' *Business Horizons,* 35: 37–40.

Henderson, R. M., and Clark, K. B. (1990). 'Architectural Innovation: The Reconfiguration of Existing Product Technologies and the Failure of Established Firms.' *Administrative Science Quarterly,* 35: 9–30.

Kuhlmann, S., and Arnold, E. (2001). 'RCN in the Norwegian Research and Innovation System.' Background Report No. 12 in the Evaluation of the Research Council of Norway, Oslo: Royal Norwegian Ministry for Education, Research, and Church Affairs.

Lawless, M. W., and Anderson, P. C. (1996). 'Generational Technological Change: Effects of Innovation and Local Rivalry on Performance.' *Academy of Management Journal,* 39: 1185–217.

Leonard-Barton, D. (1988). 'Implementation as Mutual Adoption of Technology and Organization.' *Research Policy*, 17: 251–67.

Lundvall, B.-Å. (1992). 'User-Producer Relationships, National Systems of Innovation and Internationalisation.' In B.-Å. Lundvall (ed.), *National Systems of Innovation: Towards a Theory of Innovation and Interactive Learning*. London: Pinter, 45–67.

Marcus, A. A., and Weber, M. J. (1989). 'Externally Induced Innovation,' in A. H. Van de Ven, H. L. Angle, and M. S. Poole (eds.), *Research on the Management of Innovation: The Minnesota Studies*. New York: Harper and Row, 537–59.

Metcalfe, S. (1988). 'The Diffusion of Innovation.' In G. Dosi, C. Freeman, R. Nelson, G. Silverberg, and L. Soete (eds.), *Technical Change and Economic Theory*. London: Pinter.

—— (1995). 'The Economic Foundations of Technology: Equilibrium and Evolutionary Perspectives.' In P. Stoneman (ed.), *Handbook of the Economics of Innovation and Technological Change*. Cambridge: Basil Blackwell.

Meyer, A. D. (1982). 'Adapting to Environmental Jolts.' *Administrative Science Quarterly*, 27: 515–37.

Nelson, R. (ed.) (1993). *National Innovation Systems: A Comparative Analysis*. New York: Oxford University Press.

Nonaka, I. (1991). 'The Knowledge Creating Company.' *Harvard Business Review*, 69: 96–104.

—— Toyama, R., and Konno, N. (2000). 'SECI, Ba and Leadership: A Unified Model of Dynamic Knowledge Creation.' *Long Range Planning*, 33/1: 5–34.

Porter, M. E., and van der Linde, C. (1995a). 'Green and Competitive: Ending the Stalemate.' *Harvard Business Review*, September–October: 120–34.

—— (1995b). 'Toward a New Conception of the Environment–Competitiveness Relationship.' *Journal of Economic Perspectives*, 9: 97–118.

Rescher, N. (2003). *Epistemology: On the Scope and Limits of Knowledge*. Albany, NY: SUNY Press.

Rogers, E. M. (1995). *The Diffusion of Innovation*. New York: Free Press.

Scherer, F. M., and Ross, D. (1990). *Industrial Market Structure and Economic Performance*. Boston: Houghton Mifflin.

Schon, D. A. (1971). *Beyond the Stable State*. New York: Norton.

Schumpeter, J. (1943). *Capitalism, Socialism and Democracy*. New York: Harper.

Tidd, J., Bessant, J., and Pavitt, K. (2001). *Managing Innovation, Integrating Technological, Market and Organizational Change*. Chichester: J. Wiley and Sons.

Tushman, M. L., and Moore, W. L. (eds.) (1997). *Readings in the Management of Innovation*. Cambridge, MA: Ballinger Publishing Company.

—— and Nelson, R. R. (1990). 'Introduction: Technology, Organization and Innovation.' *Administrative Science Quarterly*, 35: 1–8.

—— Anderson, P. C., and O'Reilly, C. (1997). 'Technology Cycles, Innovation Streams, and Ambidextrous Organizations: Organizational Renewal through Innovation Streams and Strategic Change.' In M. L. Tushman and P. Anderson (eds.), *Managing Strategic Innovation and Change: A Collection of Readings*. Oxford: Oxford University Press, 3–23.

Van de Ven, A. H. (1986). 'Central Problems in the Management of Innovation.' *Management Science*, 32: 590–607.

—— and Angle, H. L. (1989). 'An Introduction to the Minnesota Innovation Research Program.' In A. Van de Ven, H. L. Angle, and M. Scott Poole (eds.), *Research on the Management of Innovation: The Minnesota Studies*. New York: Harper and Row, 637–63.

—— Polley, D. E., Garud, R., and Venkataraman, S. (1999). *The Innovation Journey*. Oxford: Oxford University Press.

Weber, J. (1993). 'Quick, Save the Ozone.' *Business Week*, 17 May, 78–9.

Zammuto, R. F., and O'Connor, F. J. (1992). 'Gaining Advanced Manufacturing Technologies' Benefits: The Roles of Organization Design and Culture.' *Academy of Management Review*, 17: 701–28.

2 Product and Process Innovations: A Review of Organizational and Environmental Determinants

Fariborz Damanpour and Deepa Aravind

Abstract

Empirical studies of the effects of organizational and environmental factors on product and process innovations are reviewed in this chapter. Our review shows that, for most determinants, the results across the studies are mixed and inconclusive. It also shows that most determinants do not differentiate between product and process innovations; when they do, the difference is more of degree than of direction of the effect. These findings suggest that there may be contextual or methodological conditions under which the effects of the determinants can vary, or that product and process innovations are complementary, not distinct. To help develop robust theories of product and process innovations in the future, we discuss the importance of investigating the influence of several moderators: industry type, radicalness of innovation, and phases of product life-cycle.

Introduction

Innovations in products and in processes have been studied since Schumpeter (1911) first distinguished them. Scholars have posited that these two kinds of technological change are central to the ability of firms to create competi-

tive advantage and to play an important role in economic growth (Jones and Tang 2000; Nelson and Winter 1982). Product and process innovations stimulate the growth and productivity not only of the firms that develop them, but also of other firms that adopt and use them. Thus their impact extends to the economic sector and thence to the nation and its international competitiveness and balance of trade (Meeus and Hage, this volume). The dichotomy between product and process innovations has been used to explain the business cycle (decline versus expansion), the product cycle (development versus maturity), employment, productivity, firm management, and appropriability and imitation (Archibugi et al. 1994). The importance of these two types of industrial innovation cannot be overstated.

According to Edquist et al. (2001), different patterns of diffusion of product and process innovations in different contexts are the result of different determining factors: some factors may influence one type of innovation and not another; and the importance of the effect of a specific factor may differ across types (Lunn 1986; Cohen and Levin 1989). Therefore, an understanding of the factors that lead to product and process innovations would be useful for understanding the possible economic consequences of these innovation types. As Cabagnols and Le Bas (2002: 114) state, 'If different types of innovation do not have the same economic consequences in terms of market share and/or profit rate, a better understanding of their determinants would help better understand the economic dynamics induced by technological change.'

Following the two hypotheses that are commonly attributed to Schumpeter—innovation increases with size of firm and with market concentration (Cohen and Levin 1989)—size and competition are the primary determinants of innovation to have been investigated in industrial organization research. However, reviews of the empirical studies associated with Schumpeter's two hypotheses have found the results inconclusive (Cohen and Levin 1989); inconsistent empirical results have inhibited the development of reliable theories of innovation; and lack of theory has motivated the application of econometric procedures to reduce estimation bias, but has not alleviated the problem of inconclusive results.

To explain sources of variation between the determinants and the innovative behavior of firms, sub-theories of innovation have been advanced for product and process innovations. For instance, the intensity of competition is more relevant for product than for process innovations (Kraft 1990); small firms tend to spend more resources on new products than on new processes (Fritsch and Meschede 2001); the availability of patent protection will have a stronger effect on product than process innovation (Cohen and Levin 1989). But a systematic review of the results of past research to identify the effects of salient environmental and organizational determinants on product and process innovations has not been conducted. Therefore, robust theories to guide future research on and investment in the two types of innovation efforts at firm or industry levels do not exist.

This book emphasizes the role of innovation types, whether industrial, scientific, social, or institutional (Meeus and Hage, this volume), addresses specific characteristics of each, and points out differences among them. This chapter focuses on two types of industrial innovation only. It reviews and integrates findings of the empirical studies of the determinants of product and process innovations to examine whether:

- product and process innovations are empirically distinguishable;
- a distinction between them would help remove the instability of the results of research on the effects of environmental and organizational determinants on innovation.

To avoid a biased selection of published studies, and to ease and facilitate continuation and extension of this review, we followed a systematic procedure in selecting the studies for this review. Appendix 1 describes it.

Our search showed that most empirical studies of innovation have focused on product innovations only. Because a main focus of our review is to distinguish the explanatory variables of product from process innovations, we consider empirical studies that examine the determinants of both innovation types. Seven organizational and five environmental factors from eighteen studies containing twenty-three independent samples are included in our review. These studies were conducted in North America (11 studies) and Europe (7), were published between 1983 and 2003, and were based on data collected mainly from firms in the manufacturing sector between the mid-1970s and mid-1990s. Table 2.1 shows the sources of the data, including country, industry, sample size, and data collection period for each study. Whereas reasons have been offered for the scarcity of empirical studies of the determinants of product and process innovations,[1] we presume that the information provided from the integration of the existing empirical studies, few as they might be, is superior to a single study or a traditional narrative review.

Determinants of product and process innovations

Our literature review for the selection of the relevant studies confirmed that scholars have paid insufficient attention to the analytical

Table 2.1. Data source in the studies

Study	Data source
Meisel and Lin (1983)	Data were collected for 1,026 business units from the PIMS database in the US in 1974.
Scherer (1983)	Data from 443 corporations mainly in manufacturing industry collected by the Federal Trade Commission's (FTC) Line of Business Survey in the US within 1974–6.
Ettlie *et al.* (1984)	Data were collected from 147 firms in the food-processing industry in the US in 1981. Follow-up interviews and survey were also conducted for a subset of these 147 firms.
Ettlie and Rubenstein (1987)	Data were collected from 348 manufacturing firms in the US during the years 1979–82.
Lunn (1987)	Data were collected for 302 US firms from Scherer's (1983, 1984) compilation of patent data (June 1976 to March 1977), COMPUSTAT, and the Bureau of the Census *Census of Manufacturing*.
Kotabe (1990)	Data were collected from 71 manufacturing subsidiaries of European and Japanese multinational firms operating in the US in the mid-1980s.
Kraft (1990)	Data were collected for 56 medium-sized firms in the metalworking industry in West Germany in 1979.
Coursey (1991)	Data were collected for 461 manufacturing firms (vast majority were small- or medium-sized) in the state of New York in the US in 1985.
Bertschek (1995)	Panel data for 1270 manufacturing firms in West Germany collected by the Ifo Institute, Munich, during the period 1984–8.
Cohen and Klepper (1996)	Data for 587 business units in the manufacturing industry collected by the FTC's Line of Business Program during the period 1974–6 in the US, and data developed by Scherer (1982, 1984) for patents granted from June 1976 to March 1977.
Arundel and Kabla (1998)	Data from 604 of Europe's largest R&D performing industrial firms collected in 1993 by MERIT in the Netherlands (data from firms in the European Union excluding France) and SESSI in France (data from firms in France).
Gopalakrishnan and Damanpour (2000)	Data were collected from 101 federally insured commercial banks in the US in 1994, and from the Sheshunoff database.
Martinez-Ros (2000)	Panel data for approximately 3,900 Spanish manufacturing firms from the *Survey of Business Strategies* (ESEE) collected over the period 1990–3.
Zahra *et al.* (2000)	Data were collected from 239 medium-size US manufacturing companies in 1994 using both mail surveys and secondary sources.
Fritsch and Meschede (2001)	Data were collected from approximately 1,800 manufacturing enterprises (including a relatively high share of small firms that do not belong to large multi-plant firms) in three regions in Germany during 1993–5.
Baldwin *et al.* (2002)	Data for approximately 1,600 manufacturing plants belonging to approximately 1,600 firms from the 1993 *Survey of Innovation and Advanced Technology* (SIAT) by Statistics Canada.
Cabagnols and Le Bas (2002)	Data for approximately 10,000 innovating firms during 1985–90, and 1650 during 1990–2, conducted by SESSI in France, which cover manufacturing firms, excluding the building and food industries.
Freel (2003)	Data were collected for 597 small- and medium-sized manufacturing firms as part of a *Survey of Enterprise in Scotland and Northern England* in the UK in 2001.

clarity of the definitions of product and process innovations (Archibugi *et al.* 1994). Researchers have used alternative definitions of these innovation types. Thus, before discussing their determinants, it is necessary to define product and process innovations.

Definitions of product and process innovations

Schumpeter defined product innovation as 'the introduction of a new good—that is, one with which consumers are not yet familiar—or a new quality of good' and process innovation as 'the introduction of a new method of production, that is, one not yet tested by experience in the branch of manufacture concerned [or] a new way of handling a commodity commercially' (Schumpeter 1911, as quoted in Archibugi *et al.* 1994: 7). Since then, there have been many publications, empirical as well as conceptual, on these innovation types in diverse fields. However, as two recent reviews of the definitions of product and process innovations suggest, conceptual differences between the two types are still not very clear, and classifications of innovations into the two types are not consistent (Archibugi *et al.* 1994; Edquist *et al.* 2001).

Despite incomplete conceptual clarity, we found common characteristics among the definitions of product and process innovations in the articles we reviewed: for example, authors acknowledge that product innovations result in product differentiation and/or an increase in product quality, whereas process innovations result in a decrease in the cost of production. The drivers of product innovations are mainly customer demand for new products and executives' desire to penetrate new markets; the drivers of process innovations are primarily reduction in delivery lead time, lowering of operational costs, and increase in flexibility (Boer and During 2001). Product innovations are products or services that are new to the market; process innovations are a firm's new ways of manufacturing existing or new products. That is, newness of product inno-

vations is defined at a more macro-level (market, industry), newness of process innovations at a more micro-level (firm, business unit).

At the level of the firm, innovation is often defined as the development and use of new ideas or behaviors. (A new idea could be a new product, service, production process, organizational structure, or administrative system.) Meeus and Edquist (this volume) divided product innovations into two categories—new goods and new services: new goods are material product innovations in manufacturing sectors; services are intangible, often consumed simultaneously to their production and satisfying non-physical needs of the user (Edquist *et al.* 2001). Meeus and Edquist (this volume) also divided process innovations into two categories—technological and organizational innovations: technological process innovations change the way products are produced by introducing change in technology (physical equipments, techniques, systems); organizational innovations are innovations in an organization's structure, strategy, and administrative processes (Damanpour 1987). The studies included in our sample focused mainly on innovation in goods and on technological process innovations. Therefore, we define product innovation as a new product or service introduced to meet an external user need, and process innovation as a new element introduced into a firm's production or service operation to produce a product or render a service (Knight 1967; Utterback and Abernathy 1975; Damanpour and Gopalakrishnan 2001).

Organizational determinants

Firm size, profit, capital intensity, diversification, exports, ownership, and technical knowledge resources are the organizational determinants that were used by more than one study in our sample.[2] The relationships between each of these variables and product and process innovations are shown in Table 2.2.[3] Appendix 2 provides the definitions and measures of these variables.

Table 2.2. Organizational determinants of product and process innovations

Determinants	Product innovation	Process innovation
Firm Size		
Meisel and Lin (1983)	Positive[b]	Positive[a]
Ettlie *et al.* (1984)	Positive[c]	Positive[c]
Ettlie and Rubenstein (1987)	Positive[c]	Positive[c]
Lunn (1987)		
High-technology industries	Positive[c]	Positive[c]
Low-technology industries	Positive[b]	Positive[c]
Kraft (1990)	NS	Positive[b]
Coursey (1991)	NS	Positive[c]
Bertschek (1995)	NS	Positive[b]
Cohen and Klepper (1996)	—	Positive[c]
Arundel and Kabla (1998)	Positive[b]	Positive[b]
Gopalakrishnan and Damanpour (2000)	Positive[c]	Positive[c]
Martinez-Ros (2000)	Positive[b]	Positive[c]
Zahra *et al.* (2000)	NS	NS
Fritsch and Meschede (2001)	Positive[c]	Positive[c]
Baldwin *et al.* (2002)		
100–499 employees	Positive[a]	Positive[b]
500 or more employees	Positive[c]	Positive[c]
Cabagnols and Le Bas (2002)		
1985–90 survey	Negative[c]	Negative[c]
1990–92 survey	Negative[c]	Negative[c]
Freel (2003)		
Supplier-dominated	Positive[b]	Positive[a]
Production-intensive	Positive[c]	NS
Science-based	Positive[a]	Positive[a]
Profit		
Meisel and Lin (1983)	NS	NS
Lunn (1987)		
High-technology industries	Positive[c]	Positive[c]
Low-technology industries	Positive[a]	NS
Kraft (1990)	NS	NS
Coursey (1991)	NS	NS
Zahra *et al.* (2000)	Positive[a]	NS
Capital intensity		
Lunn (1987)		
High-technology industries	—	Negative[a]
Low-technology industries	—	Positive[c]
Kraft (1990)	Positive[c]	NS
Martinez-Ros (2000)	Positive[c]	Positive[c]
Diversification		
Ettlie *et al.* (1984)	Positive[b]	NS

continues

Table 2.2. *Continued*

Determinants	Product innovation	Process innovation
Lunn (1987)		
High-technology industries	Positive[a]	Positive[c]
Low-technology industries	Positive[b]	Positive[b]
Zahra *et al.* (2000)	Positive[b]	Negative[b]
Cabagnols and Le Bas (2002)		
1985–90 survey	NS	Negative[c]
1990–2 survey	NS	Negative[c]
Exports		
Meisel and Lin (1983)	Positive[b]	Positive[b]
Kraft (1990)	NS	Negative[b]
Arundel and Kabla (1998)	Positive[b]	NS
Martinez-Ros (2000)	Positive[c]	Positive[c]
Zahra *et al.* (2000)	Negative[b]	Negative[b]
Ownership		
Kraft (1990)	Negative[c]	—
Martinez-Ros (2000)	NS	Negative[c]
Zahra *et al.* (2000)	Negative[c]	Negative[b]
Baldwin *et al.* (2002)	—	Positive[a]
Technical knowledge resources		
Ettlie *et al.* (1984)	Positive[c]	Positive[c]
Kraft (1990)	NS	—
Freel (2003)		
Supplier-dominated	NS	NS
Production-intensive	Positive[a]	NS
Science-based	NS	NS

Notes
- Statistical significance reported where available.
- The associations are mainly based on regression coefficients.
- Arundel and Kabla (1998), results for model 2 from Tables 2 and 4 are reported.
- Bertschek (1995), results from model 2, which controls for the multi-period data, are reported.
- Cabagnols and Le Bas (2002), results from Table 6.2 (pp. 144–6) are reported.
- Coursey (1991), for product innovation results from INNVO model (Table 2) and for process innovation results from PROCTECH model (Table 3) are reported.
- Ettlie *et al.* (1984), results for radical innovation from Table 3 are reported.
- Freel (2003), results from Table 2 and Table 3 for total sample are reported.
- Lunn (1987), results from Table 1 are reported.
- Zahra *et al.* (2000), results from Tables 3 and 4 are reported.

NS = Not Significant
[a] $p < .10$ [b] $p < .05$ [c] $p < .01$.

Firm size

Size of firm is by far the most-researched variable in our sample. Conceptually, the relationship between firm size and innovation is inconclusive (Baldwin *et al.* 2002; Cohen and Levin 1989; Nord and Tucker 1987): small firms seize discontinuous opportunities and innovate by introducing new products and new methods of production, by opening new markets and utilizing new sources of supply, and by reorganizing industries (Stevensen and Jarillo 1990, Utterback 1994); large firms have scientific knowledge and management expertise, production means, and other complementary assets, better access to capital, and some degree of monopoly power and, hence, would be more willing to risk investing in innovation (Afuah 2003). Empirically, meta-analytical reviews have reported a significant positive association between firm size and innovation (Damanpour 1992; Camison-Zornoza *et al.* 2004).[4]

Considering the relative effect of size on product and process innovations, most scholars posit that firm size would have a more positive association with process than with product innovations. Small firms tend to spend more resources on new products than on new processes because product innovations are perceived to be a better means of entry into a market than process innovations (Fritsch and Meschede 2001), and 'yield greater return from licensing and to spawn more rapid growth in output than process innovations' (Cohen and Klepper 1996: 233). Conversely, large firms benefit from investing in process innovations because they have a comparative advantage in exploiting their existing innovations in the market place (Cabagnols and Le Bas 2002; Cohen and Klepper 1996). This is in line with the results obtained by Scherer (1991), who found considerable evidence for the proposition that large firms devote more effort to process improvement than small firms.

Contrary to this theory, our review suggests that size has a positive association with both product and process innovations in ten studies (Table 2.2). The positive relation holds for firms in different size groups (Baldwin *et al.* 2002), in high- and low-technology industry groups (Lunn 1987), and in different industrial sectors (Freel 2003). This result coincides with the recent meta-analysis of Camison-Zornoza *et al.* (2004), which found non-significant differences between mean correlations of product and process innovations with firm size. In support of the theory, three studies reported that size has a more significant association with process innovations (Baldwin *et al.* 2002; Lunn 1987; Martinez-Ros 2000); three other studies found that size has no significance for product, but does have a positive effect on process innovations (Bertschek 1995; Coursey 1991; Kraft 1990). Therefore, the hypothesis that size is more advantageous for process than for product innovations is supported by several (but not the majority) of the studies.

Profit

Metcalfe (this volume) observes that high profitability in an evolutionary-efficient environment correlates positively with both product and process innovations. A firm's profit and cash flow reflect its internal financial capability and provide the resources needed for financing innovations (Cohen and Levin 1989; Kraft 1990). The studies in our sample did not propose any theory on the differential effects of firms' profitability on product versus process innovations. However, in general, organizational resources (time, cash flow, slack) are expected to facilitate all types of innovation. Innovation does not happen by itself; most innovations go beyond budgetary allocations and require special funding (Daft 2001). Financial resources would allow the organization to explore new ideas in advance of an actual need, to absorb failure, and to afford the cost of commercializing or implementing the innovation (Damanpour 1987). More resources would be needed if the innovation were more radical, complex, and risky, and also if it were more difficult to develop, commercialize, or implement.

Most studies in our sample reported that profit is not significantly associated with product and process innovations (Table 2.2). Lunn (1987) found that in the high-technology industries, where more resources are needed for innovation, profit positively affects both product and process innovations. However, in low-technology industries, he found a weaker effect on product and a non-significant effect on process innovations. It is interesting to note that the sample used by Kraft (1990) and Coursey (1991), who reported a non-significant effect on both innovation types, consisted of firms operating in mature industries such as the metalworking industry, construction materials, and agricultural goods, which, based on the definition given by Lunn (1987), can be considered as low technology. The findings of Zahra et al. (2000) are similar to Lunn's results for low-technology industries.

Overall, studies in our sample suggest that the association between profit and product innovation is mixed, and profit does not significantly influence process innovation.

Capital intensity

Capital intensity represents the intensity of physical capital that a firm has in its operations (Martinez-Ros 2000). Lunn (1987) studied the effect of capital intensity on process innovation only, and found a positive effect in the low-technology industries and a negative effect in the high-technology industries. This finding corresponds with the argument that the more capital-intensive processes in high-technology industries provide less room for process innovation because they are more automated and rigid (Martinez-Ros 2000).

On the other hand, Kraft (1990) found a positive effect of capital intensity on product innovation, and Martinez-Ros (2000) found positive effects on both product and process innovations. These results are in line with the argument that capital requirements act as a barrier to entry (Lunn 1987), and that 'the rents of innovation in more capital-intensive firms are less threatened as, in order to exploit the innovation, high investment in physical capital would be required' (Martinez-Ros 2000: 227).

Overall, while the results for product innovation are consistent (positive in two studies), they are inconsistent for process innovation (Table 2.2). Lunn (1987) posits that the argument for the effect of capital intensity on innovation applies better to process than to product innovations and does not include product innovation in his analysis. His finding for low-technology industries, however, is inconsistent with the finding of Kraft (1990), whose sample also consists of low-technology firms. The mixed results do not clarify whether firms with capital-intensive technologies would tend to innovate more in products or processes and suggest that better theory development and more research is needed.

Diversification

Nelson (1959) argued that because the result of research is unpredictable, diversified firms are more likely to be able to use new knowledge internally. Lunn (1987) presumed that diversification positively influences both product and process innovations and found positive associations with both in high- and low-technology industries (Table 2.2). Cabagnols and Le Bas (2002: 120), on the other hand, presume that the impact of diversification would be greater for product than for process innovations, because product innovation projects are riskier, and diversification may be a means of spreading risks over different projects. However, these authors report non-significant effects on product innovation and negative effects on process innovation (Cabagnols and Le Bas 2002). Contrary to both Lunn and Cabagnols and Le Bas, Ettlie et al. (1984) find a positive association with product and a non-significant association with process innovations, and Zahra et al. (2000) find a positive association with product and a negative association with process innovations.

Results for the influence of diversification on process innovation are mixed. For product innovations, the majority of studies found positive associations with diversification, which supports the assumption that product innovations are riskier than process innovation and diversified firms could more easily develop riskier product innovations. However, findings from other studies suggest that process innovations could be perceived to be riskier than product innovations. For instance, Gopalakrishnan et al. (1999: 159), in a study of product and process innovations in the banking industry, reported that bank executives consider process innovations to be significantly less autonomous, more complex, and more costly to implement than product innovations. Overall, the differential effect of diversification on product and process innovation (if any) is both conceptually and empirically undetermined.

Exports

Because the presence in foreign markets may require more technologically advanced products in order to remain competitive (Kraft 1990; Martinez-Ros 2000), export activity should favor innovation. Accordingly, Martinez-Ros (2000) and Meisel and Lin (1983) found that exports are positively associated with both product and process innovations (Table 2.2). However, according to Zimmermann (1987), exports could increase competitive pressure, and if Schumpeter's hypothesis on the negative effect of competition on innovation is valid, more intense competition should lead to exports having a negative effect on innovation. In accordance with this view, Zahra et al. (2000) found negative effects of exports on both product and process innovation. Kraft (1990) agrees with this view only if the market structure is measured on a domestic level; he measures market structure on the international level to investigate the impact of exports without the indirect influence of competition, and finds a non-significant effect of exports on product innovation and a negative effect on process innovation. But Arundel and Kabla (1998) find a reverse effect. In sum, results suggest that export activity has mixed effects on both product and process innovation and do not clearly differentiate between them.

Ownership

Ownership literature suggests that management-controlled firms are more risk averse than owner-controlled firms; hence, they would be less active in innovation (Fritz 1989). Zahra et al. (2000), on the other hand, argue that, because executives' wealth would become more dependent on the company's long-term performance, higher stock ownership gives executives better incentives to pursue innovation-related activities. The four studies in our sample compared external versus internal ownership at the country and firm levels. Baldwin et al. (2002) and Martinez-Ros (2000) used dummy variables that reflect foreign (external) versus domestic (internal) ownership of firms. Kraft (1990) used a dummy variable that represents management (internal) versus non-management (external) holdings of equity, and Zahra et al. (2000) used the percentage of total company stock held by the company's senior executives.[5] Although the results are not fully conclusive, they provide support for the argument of Zahra et al. (2000) that outside ownership negatively influences both product and process innovations.

Technical knowledge resources

This refers to the existence of technical groups, the proportion of technical employees, and employees' technical qualifications. Scholars of organizational innovation have argued that both the degree and the depth of knowledge resources facilitate innovation. Higher academic training, diversified backgrounds and skills, and the contrast and synthesis of different ideas and perspectives create a better understanding of new technical developments and processes and thus facilitate innovation (Dewar and Dutton 1986; Hage 1980; Kimberly and Evanisko 1981).

Results from the three studies included in Table 2.2 do not support these assertions. For example, even though Kraft (1990) expected a positive association of 'skill level of workforce' and 'training expenditures' with product innovations, he found non-significant effects of both measures. Ettlie *et al.* (1984), on the other hand, reported positive associations between knowledge resources and both product and process innovations. Freel (2003) employed the proportion of workforce classed as scientists and technologists and did not find significant effect on either type.

In summary, our review suggests: that size influences both innovation types positively; that profit does not significantly affect process innovation; that capital intensity, diversification, and exports more consistently influence product than process innovations; that the effects of ownership and technical knowledge resources are mixed and inconclusive. With the exception of firm size, which impacts process innovations more positively than product innovations in several studies, our review suggests that other determinants do not clearly differentiate between these innovation types.

Environmental determinants

The environmental variables examined by more than one study in our sample are competition, concentration, technological opportunity, appropriability conditions, and growth of demand. The relationships between these determinants and product and process innovations are shown in Table 2.3. Definitions and measures of these variables are shown in Appendix 2.

Competition and concentration

Competition and concentration represent market structure. A hypothesis attributed to Schumpeter posits that firms in concentrated markets would have more incentives to innovate because they can more easily appropriate the returns from innovation (Baldwin *et al.* 2002; Martinez-Ros 2000). Arrow (1962) and Scherer (1980), on the other hand, have argued that the gains from innovation are higher in competitive industries and that insulation from competition can cause bureaucratic inefficiencies that inhibit innovation (Baldwin

et al. 2002). Competition creates strong incentives to acquire knowledge and put it to productive use; innovation and competition are inseparable (Metcalfe this volume). Accordingly, scholars (Baldwin *et al.* 2002; Levin *et al.* 1985; Martinez-Ros 2000) have qualified the role of market structure on innovation in several ways: a firm's gains from innovation are greater in competitive than in monopolistic industries; innovation is more intensive in the early stages of an industry's development, when markets are less concentrated; large firms are more innovative in concentrated industries with high barriers to entry; once technological environment variables are added, there is a dramatic reduction in the observed impact of market structure on innovation.

A clear theory on the effects of competition or concentration on innovation has not yet emerged, but scholars have proposed that competition is more relevant to product than to process innovation (e.g. Kraft 1990). Under conditions of high competition, and if new products are not protected by patents, competitors would quickly reverse-engineer them. However, because process innovations are more internally driven and more easily kept secret, competitors cannot easily imitate them.

Considering the relation between competition and innovation, Baldwin *et al.* (2002) grouped firms according to the number of competitors. They found different results: in the case of fewer competitors, competition had positive and significant effects on product and process innovations; in the case of more competitors, the effects were non-significant. In general, three samples from two studies (Baldwin *et al.* 2002; Bertschek 1995) reported positive association between market competition and both product and process innovation (Table 2.3). However, four samples found non-significant relations with product innovations (Baldwin *et al.* 2002; Cabagnols and Le Bas 2002; Coursey 1991), and three samples reported negative relations with process innovations (Cabagnols and Le Bas 2002; Martinez-Ros 2000).

Considering the association between concentration and innovation, Lunn (1987), Meisel and Lin (1983), and Scherer (1983) found that

Table 2.3. Environmental determinants of product and process innovations

Determinants	Product innovation	Process innovation
Competition		
Coursey (1991)	NS	NS
Bertschek (1995)		
FDI	Positive[c]	Positive[c]
Imports	Positive[c]	Positive[c]
Martinez-Ros (2000)	Negative[c]	Negative[c]
Baldwin *et al.* (2002)		
6–20 competitors	Positive[c]	Positive[c]
Over 20 competitors	NS	NS
Cabagnols and Le Bas (2002)		
1985–90 survey	NS	Negative[c]
1990–2 survey	NS	Negative[c]
Concentration		
Meisel and Lin (1983)	Positive[b]	Positive[b]
Scherer (1983)	Positive[b]	Positive[b]
Lunn (1987)		
High-technology industries	NS	NS
Low-technology industries	Positive[b]	Positive[c]
Kraft (1990)	Positive[c]	NS
Technological opportunity		
Meisel and Lin (1983)	Positive[b]	Positive[b]
Lunn (1987)	Positive[c]	Positive[c]
Kotabe (1990)	NS	Positive[b]
Martinez-Ros (2000)	Positive[a]	Negative[c]
Zahra *et al.* (2000)	Positive[b]	Positive[b]
Baldwin *et al.* (2002)	NS	—
Cabagnols and Le Bas (2002)		
1990–2 survey (NSI)	NS	Negative[c]
Freel (2003)		
Supplier-dominated	NS	NS
Production-intensive	NS	NS
Science-based	Positive[a]	Positive[a]
Appropriability conditions		
Arundel and Kabla (1998)		
Patents	Positive[b]	Positive[b]
Secrecy	Negative[b]	Positive[a]
Baldwin *et al.* (2002)		
Patents	NS	—
Secrecy	—	NS
Cabagnols and Le Bas (2002)		
Patents	NS	Negative[a]
Secrecy	NS	NS

continues

Table 2.3. *Continued*

Determinants	Product innovation	Process innovation
Growth of demand		
Meisel and Lin (1983)	Positive[a]	NS
Lunn (1987)		
High-technology industries	Negative[c]	—
Low-technology industries	NS	—
Kotabe (1990)	Positive[b]	Positive[b]
Martinez-Ros (2000)	Positive[c]	Positive[c]
Cabagnols and Le Bas (2002)	NS	NS

Notes
- Statistical significance reported where available.
- The associations are mainly based on regression coefficients.
- Arundel and Kabla (1998), results for model 2 from Tables 2 and 4 are reported.
- Bertschek (1995), results from model 2, which controls for the multi-period data, are reported.
- Cabagnols and Le Bas (2002), results from Table 6.2 (128–30) are reported.
- Coursey (1991), results for product innovation from INNVO model (Table 2) and for process innovation from PROCTECH model (Table 3) are reported.
- Freel (2003), results for the total sample from Tables 2 and 3 are reported.
- Lunn (1987), results from Table 1 are reported; for technological opportunity, results encompass firms in both high- and low-technology industries.
- Martinez-Ros (2000), for competition, the results from Table 4 were inverted.

NS = Not Significant
[a] $p < .10$ [b] $p < .05$ [c] $p < .01$.

concentration affects both product and process innovations positively, although Lunn (1987) reported a positive effect only in the low-technology industries. Kraft (1990) found a positive effect on product innovations only. In general, from the four studies that examined the impact of concentration on product and process innovations, it appears that concentration: has a positive effect on both product and process innovations; it does not differently impact the two innovation types.

As Cabagnols and Le Bas (2002: 119) noted, whether market structure variables have a stronger positive impact on product than on process innovations (Scherer 1983), a more positive effect on process than on product innovations (Lunn 1986; Zimmermann 1987), or influences them equally remains an empirical question to be further investigated. However, when the findings of all studies that have examined competition and concentration in our sample are considered together (Table 2.3), they show that these factors influence product and process innovations in the same direction, and mostly positively. Therefore, although our review does not clarify the role of market structure on innovation, it does indicate that, contrary to expectation, in most studies, directions of the associations between the two market structure variables and innovation types are not opposite.

Technological opportunity

Technological opportunity reflects the influence of technology push which, according to Lunn (1987: 744), 'occurs when exogenous changes in scientific and engineering knowledge reduce the cost of new products and processes.' Greater technological opportunity would encourage innovation because the accumulated knowledge shared by firms due to spillovers or other effects reduces the cost of incorporating knowledge into new products and processes (Martinez-Ros 2000). Firms in technologically progressive industries would invest more in research and development and could have a higher-than-average propensity to patent (Lunn 1987). Under these conditions, technological opportunity would influence product innovations more positively than process innovations. However, when the product becomes more mature, competition shifts to production efficiency, and firms place more emphasis on process innovation, this effect may be reversed (Kotabe 1990).

The studies in Table 2.3 showed associations between technological opportunity and product innovation that were either positive (Freel 2003; Lunn 1987; Martinez-Ros 2000; Meisel and Lin 1983; Zahra *et al.* 2000) or non-significant (Baldwin *et al.* 2002; Cabagnols and Le Bas 2002; Freel 2003; Kotabe 1990); for process innovation, some also found negative effects. In general, the findings for both innovation types are inconclusive and a specific pattern for the influence of technological opportunity on product or process innovations does not emerge. However, the results from a study that controlled for sectors (Freel 2003) suggest that inter-industry differences might be the cause of inconclusive results: we will address this in the Discussion section.

Appropriability conditions

This refers to the extent to which a firm is able to capture returns from its innovations. If innovations were imitated easily, and firms could not benefit from them, they would be less motivated to innovate. Hence, firms use various forms of intellectual property protection, such as patents or secrecy, to protect their innovations from being copied (Baldwin *et al.* 2002). For product innovations, patents are considered to be an effective way of protection; for process innovations, secrecy is considered more effective than patents (Baldwin *et al.* 2002).

Three studies in our sample included patents and secrecy as ways of protecting a firm's product and process innovations. Arundel and Kabla (1998) found that patents positively influence both types of innovations, but secrecy positively affects process innovation and negatively affects product innovation. Baldwin *et al.* (2002) did not find significant effects on either type of innovation, and the only significant effect Cabagnols and Le Bas (2002) found was the negative effect of patents on process innovation. Thus, for both patents and secrecy the results from these three studies are inconsistent, which confirms Cohen's (1996) conclusion that 'although there is growing evidence of inter-industry differences in appropriability conditions, there is little empirical evidence as to the beneficial effect of these conditions on innovation activity across a wide range of industries' (as quoted in Baldwin *et al.* 2002: 94).

Growth of demand

Scholars have offered opposite views on the effects of market demand on product and process innovations. One view is that growth in demand would encourage firms to emphasize both innovation types (Kotabe 1990). A second view is that factors resulting in lower market uncertainty would favor product innovation, thus the volume of demand and its growth rate would have a stronger effect on product innovations than process innovations (Cabagnols and Le Bas 2002). According to a third view based on life-cycle theory, 'increasing demand should induce more process innovation, because the firm gains knowledge about demand and knows that it has developed a solution that can now be optimized' (Cabagnols and Le Bas 2002: 116). While the four studies that included this factor did not find consistent results of its effect on product and process innovations (Table 2.3), the findings of Kotabe (1990) and Martinez-Ros (2000) suggest that growth of demand positively affects both innovation types rather than favoring one type over the other.

In summary, although concentration and competition are expected to have opposite associations with innovation, they do not seem to distinguish between innovation types, because both have similar effects on product and process innovations in the majority of the studies. Technological opportunity influences product and process innovations differently, but the studies do not agree on the direction of influence. The results across studies for appropriability conditions and growth of demand are mixed and inconclusive.

Table 2.4 summarizes the findings of our review and compares them with the expected relations between the determinants and product and process innovations. In general, our review suggests that most determinants do not differentiate between product and process innovations. Lack of differences in the impact of various determinants on product and process innovations and lack of consistent findings for many of the factors suggest that:

- there may be contextual conditions under which the effects of the determinants on the innovation types vary; or
- product and process innovations may be complementary and not distinct.

We discuss these ideas below.

Discussion: Determinants of product and process innovations: are they context dependent?

Although our review of the empirical research suggests that organizational and environmental determinants of product and process innovations do not significantly differ, it is possible that, under certain conditions, one determinant would affect product innovation and, under others, it would affect process innovation. These conditions could include types of firm or industry (manufacturing versus service, high-technology versus low-technology, entrepreneurial versus mature), stages of product life-cycle (early versus late), economic conditions (prosperity versus depression), firm's role in the innovation process (developer versus adopter), and measure of innovation (patent versus R&D intensity). The data and methods used in the studies included in this review do not allow a systematic comparison of the effects of these moderators similar to those shown in Tables 2.2 and 2.3. Therefore, we discuss in this section the possible moderating effects of three salient factors—inter-industry differences, innovation radicalness, and product life-cycle—to suggest ideas for future research.[6]

Inter-industry differences

Scholars have acknowledged that firms in different industries differ in the degree to which they engage in innovative activity and have identified demand, technological opportunity, and appropriability as three main factors representing inter-industry differences for innovation (Cohen and Levin 1989). Sectoral differences in users' needs, sources of technology, and means of appropriating benefits also determine the technological trajectories of firms; that is, whether firms are supplier-dominated, scale-intensive, specialized-supplier, or science-based (Pavitt 1984).

The sources of innovation are different for firms in each of Pavitt's trajectories (Tidd *et al.* 2001). In supplier-dominated firms (agriculture, services, traditional manufacture), suppliers of machinery and other production inputs are the sources of technical change (Tidd *et al.* 2001). In scale-intensive firms (bulk materials, consumer durables, automobiles, and civil engineering), in-house design and production engineering departments, operating experience, and specialized suppliers are the main sources of technology. In specialized-supplier firms (machinery, instruments, and software), advanced users and the design and use of specialized inputs are the main sources of technology. Finally, in the science-based firms (electronics and chemicals), R&D and basic research are the main sources of technology. A recent addition (Tidd *et al.* 2001)

Table 2.4. Relations between determinants and product and process innovations

Determinant	Expected associations	Study's findings
Firm size	Size has a more positive association with process than with product innovations.	Studies generally reported that size positively influences both innovation types. While the majority (10 samples) found similar significance levels for both types, a sizeable minority (6 samples) found that size more significantly influences process innovations.
Profit	Financial resources facilitate innovation, but a theory on the resources' relative effect on innovation types has not been advanced.	Most studies reported that profit does not significantly influence process innovation; results for product innovation are mixed.
Capital intensity	High investment in physical capital helps innovation; relative effect on innovation types is unclear.	Capital intensity positively influences product innovation; results for process innovation are mixed.
Diversification	Diversification helps spreading risk of innovation; its impact is greater on riskier product innovation.	Findings more consistently suggest a positive effect on product innovation, but are mixed for process innovation.
Exports	Scholars agree that export increases competition, but do not agree on the effect of such competition on innovation. A clear theory does not exist.	Results are mixed and do not clearly differentiate between product and process innovations.
Ownership	Owner-controlled firms take more risk than management-controlled firms, but a theory on ownership's relative influence on innovation types does not exist.	Although not conclusive, results suggest that external ownership negatively influences product and process innovations.
Technical knowledge resources	Degree and depth of technical knowledge resources facilitate both innovation types.	Results are mixed and do not differentiate between product and process innovations.
Competition and concentration	Schumpeter proposed that firms in concentrated markets have more incentive to innovate, but others do not necessarily agree. Some scholars propose that competition more strongly relates to product than to process innovations.	Competition and concentration are expected to influence innovation in opposite directions. But results show that they either affect both innovation types positively (6 samples), or negatively (1 sample), or have no effect (3 samples); only 3 samples showed different effects. While findings do not clarify whether market structure positively or negatively influences innovation, they generally indicate that the direction and strength of its associations with product and process innovations are not different.
Technological opportunity	Technological opportunity more positively influences product than process innovations; however, this effect may reverse when the product becomes more mature.	The results for both innovation types are inconclusive, but suggest that inter-industry differences may be the cause of mixed results.

continues

Table 2.4. *Continued*

Determinant	Expected associations	Study's findings
Appropriability conditions	Whereas patents are an effective way of protection for product innovations, secrecy is an effective way of protection for process innovation.	For both patents and secrecy, the results are inconsistent, which suggests there is little empirical evidence as to the beneficial effect of patents or secrecy on product or process innovations.
Growth of demand	A clear theory on the effect of this factor on product and process innovations has not yet emerged.	Although the results are not consistent across all studies, findings from two studies suggest that growth of demand positively affects both product and process innovations.

to these four types of firms is the information-intensive firm (finance, retailing, publishing, travel), where in-house software and systems departments and suppliers are the main sources of technology. The role of suppliers, users, and manufacturers in each category is different, an argument also favored by von Hippel (1988). The innovative behavior of firms in Pavitt's sectors differs in the patterns of interaction among these actors, a point highlighted by Meeus and Faber (this volume) and Oerlemans *et al.* (1998). Further, Meeus *et al.* (2001) contend that patterns of interactive learning differ between sectors, some being more resource based, others affected more by the complexity of innovative activities.

Such inter-industry differences would imply that broad generalizations for all industries cannot be made regarding sets of factors that would determine whether product or process innovations are pursued. In fact, according to Pavitt (1990), a firm does not have completely free choice about whether or not to be product or process oriented, one of the main constraints being the nature of its accumulated technological competences, which is largely shaped by its core business, its principal activities, and its size. These constraints will determine the potential technological and market opportunities that the firm can exploit (Berry and Taggart 1994). Firms in certain industries (bio-pharmaceutical drugs, machinery) may focus

on product innovation; some (steel, chemical) may focus on process innovation; others (automobiles) may encourage and emphasize both types of innovation (Tidd *et al.* 2001). Thus, it is conceivable that firms in some industries emphasize product innovations and, in others, process innovations: for example, in petroleum refining, three-fourths of total R&D is dedicated to process innovations; in the pharmaceutical industry, the proportion is less than one-fourth (Cohen and Klepper 1996). But is it also conceivable that the determinants of product and process innovation and/or the strength of their impact on innovation types are different in each of these industries?

With two exceptions, the studies in our sample did not analyze inter-industry differences. Lunn (1987) distinguished between high- and low-technology industries. While Lunn found that size and diversification influenced product and process innovations positively in both industries, he also found differences between the two industries for three other determinants (Tables 2.3 and 2.4): capital intensity had a negative effect on process innovation in high-technology ($p < .10$) and a positive effect in low-technology industries ($p < .01$); concentration did not affect product or process innovations in high-technology industries, but had a positive effect on both in low-technology industry ($p < .05$); and growth of demand had a negative effect on product innovations in high-technology ($p < .01$) but no effect in low-technology industries.

Freel (2003) distinguished among three of Pavitt's sectors—supplier-dominated, production-intensive, and science-based—and found some differences for three determinants across the sectors. For example, he reported that size more strongly influenced product than process innovations in supplier-dominated firms, had a positive ($p < .01$) effect on product innovations but no effect on process innovations in production-intensive firms, and had equal influence on both innovations in science-based firms (Table 2.3).

Pavitt's sectors correspond more closely with manufacturing industries. Differences between manufacturing and service sectors can also moderate the effect of determinants on innovation. For example, meta-analytical reviews of associations between organizational factors and innovation have reported significant differences between manufacturing and service sectors (Damanpour 1991; Camison-Zornoza et al. 2004). None of the studies in our sample focused solely on the service sector to enable us to examine the differences between the determinants of product and process innovations in the two sectors. Evangelista (2000: 184) observed that we are a 'long way from having a satisfactory picture of the extent, role and nature of innovative activities in the service sector.' He contends that the distinction between product and process innovation in the service sector is more problematic because of certain peculiar features of service activities, including intangibility, simultaneous production and consumption, limited appropriability, and variability in customization and outcome (de Brentani and Cooper 1992; Evangelista 2000). Contrary to the manufacturing sector, where product innovations are emphasized, the service sector shows an orientation towards process innovation (Evangelista 2000).

In sum, the differences in the nature of the activities of manufacturing and service organizations could unequally affect the determinants of product and process innovations and the strength of their influence in each context. Therefore, we recommend that the design and data collection of future studies of the determinants of industrial innovations enable an analysis of inter-industry differences. The interaction between industry differences and innovation types may help explain inconclusive findings of research on the determinants of product and process innovations.

Innovation radicalness

This is an important attribute of innovation (Hage 1980), and the distinction between radical and incremental innovations is necessary for determining how innovation would influence a firm's effectiveness and competitiveness (Afuah 2003). Jordan (this volume) considers innovation radicalness as a pivotal dimension for developing R&D profiles in both science and technology. Radical innovations represent clear departures from existing practice; incremental innovations are minor improvements or adjustments in current technology (Dewar and Dutton 1986). Radical innovations would drastically influence a firm's competitiveness by making the existing technologies, products, or services obsolete or non-competitive; incremental innovations, although they might influence a firm's competitiveness, would not necessarily make the existing products or services obsolete or non-competitive (Afuah 2003). Radical innovations can create discontinuity in the product class and can potentially be 'competence-destroying', because the knowledge and skills required to exploit them are drastically different from those used for the existing product or processes (Tushman and Anderson 1986); incremental innovations are not 'competence-destroying', because they rely on existing knowledge and allow current product or process technologies to remain competitive.

Researchers have proposed differences between the determinants of radical and incremental innovations. For instance, technical knowledge resources and specialization would facilitate radical innovations more than incremental, whereas size and decentralization would affect incremental innovations more than radical (Dewar and Dutton 1986; Ettlie *et al.* 1984; Hage 1980). Jordan (this volume) argues that organizational design determinants of radical and incremental scientific innovations are not the same. Because product and process innovations can each be either radical or incremental, it is conceivable that the determinants of radical product innovation, incremental product innovation, radical process innovation, and incremental process innovation differ. Therefore, future studies should consider the moderating role of innovation radicalness for developing more robust theories of product and process innovations.

Product life-cycle

In both Abernathy and Utterback's (1978) product life-cycle model and Tushman and Anderson's (1986) cycles of technological change, patterns of development of product and process innovations differ over time: radical product innovations occur more frequently than process innovations in an industry's early life; after a dominant design emerges, radical process innovations occur more frequently; and finally, the rates of both product and process innovations slow down and become more balanced, and both innovations become less radical (Abernathy and Utterback 1978). Calantone *et al.* (1988) found support for patterns of product and process innovations according to the life-cycle model. McGahan and Silverman (2001: 1141), on the other hand, found 'no evidence of a shift from product to process innovations with industry maturity, and no evidence that leaders innovate less in mature industries than in non-mature industries.'

An explanation for inconsistent findings across the phases of product life-cycle could be due to sectoral differences. Barras (1986) argued that patterns of product and process innovations vary over different sectors. He explained that the three phases of the product cycle model advanced by Abernathy and Utterback (1978) applies to the production of goods embodying a new technology in the goods industries. In the user industries, which usually adopt the technology developed in goods industries, the cycle, which Barras termed the 'reverse product cycle,' operates in the opposite direction. That is, in the first phase, the technology is used to increase the efficiency of existing services; in the second phase, it is applied to improving the quality and effectiveness of services; and in the third phase, it assists in generating wholly transformed or new services (Barras 1986). Firms in goods industries are suppliers of innovation in the market; firms in user industries adopt the innovation.[7] The characteristics of suppliers and users, and the categories of variables that determine the success of an innovation in the market place versus its successful adoption in an organization, differ (Frambach 1993). Therefore, the determinants of product and process innovations in sectors with different patterns of product and process innovations might also differ. Such differences have not been empirically examined in the past; future research can contribute by filling this gap.

Product and process innovations: are they complementary rather than distinct?

Alternatively, the lack of differences in the impact of various determinants on product and process innovations found in this study could be explained by the two innovation types' being complementary. Tornatzky and Fleischer (1990: 20) consider product innovations as those that are *terminal* ('valuable in and of themselves') for their generators, and process innovations as those that are adopted as *instrumental* ('valuable essentially as a means toward some other outcome'). This viewpoint separates these innovation types based on sectors (Archibugi *et al* 1994); for example, an ATM is a product (terminal) innovation when produced in the goods industry, but when purchased by a bank in the user industry, as a part of bank's distribution system, it is a process (instrumental) innovation. Bhoovaraghavan *et al.* (1996) are critical of such distinctions and argue that, from a demand/consumer perspective, product and process innovations are not independent phenomena; they represent both ends of a continuum. According to these authors, the main question is to determine strategically 'which of these two should play the leading role in the launch of a new product that is successful in the marketplace' (Bhoovaraghavan *et al.* 1996: 234).

This view suggests that firms combine innovation types in a new way to contribute to their competitive advantage and help maintain or improve their performance (Roberts and Amit 2003). For example, Kraft (1990: 1029) contends that 'the implicit assumption that these two ways of technological advance are independent but determined by the same variables seems to be questionable.' Kotabe and Murray (1990: 402) stated that 'product and process innovative activities are intertwined such that continual improvement in manufacturing processes lead to product innovations.' Fritsch and Meschede (2001) also concluded that product and process innovations are interrelated: that is, product innovation could make corresponding process innovation necessary and process innovation could enable the firm to improve their product quality or to produce completely new products. New products both stimulate and result from new processes, and product innovation cannot take place without parallel process innovation (Tornatzky and Fleischer 1990; Voss 1994).

From this perspective, product and process innovations are complementary and firms should pursue both at the same time in order to derive full benefits (Capon *et al.* 1992; Damanpour and Gopalakrishnan 2001). In a study of US and European pharmaceutical companies, Pisano and Wheelwright (1995) found that companies in a variety of high-technology industries had gained tremendous advantages by treating process development as an integral part of product development. The congruent development of product and process innovations resulted in a smoother launch of new products, easier commercialization of complex products, and more rapid penetration of markets (Pisano and Wheelwright 1995). In his study of the market performance of Euro-

pean and Japanese multinationals, Kotabe (1990) found that the interaction of product and process innovations is a crucial determinant of market performance. Kotabe and Murray (1990: 403) concluded that 'when a high level of product innovation is backed simultaneously by a high level of manufacturing process innovations, such strategy will provide by far the strongest competitive advantage to the firm.'

These arguments and findings suggest that any emphasis on one type of innovation over the other type depends on the firm's strategic intent. For example, an organization's interest in quality control and re-engineering may motivate the organization to improve efficiencies and therefore emphasize process innovations over product innovations at a point in time. However, an organization may be motivated by increasing market share, winning customer loyalty, and staying ahead of competition, and therefore may focus more on product innovations (Damanpour and Gopalakrishnan 2001). Whereas many organizations, especially in high-technology industries, may find it difficult to be simultaneously strong in product and process innovations (Sen and Egelhoff 2000), those that are able to develop and manage the two together would perform better in the face of increasing global competition.

An examination of the lead or lag, or of the complementary patterns of development or adoption of product and process innovations over time, requires longitudinal data. Although the studies that rely on documentary sources of data (like those in our sample) usually collect data over a time interval, most pool the data for analysis. To examine the dynamics of product and process innovations, future studies would need to employ a longitudinal research design that allows both within- and cross-time analysis of data.

Conclusion

Both product and process innovations are important for the competitiveness of firms and the economic growth and advancement of society; thus, an understanding of what leads to each of these two types of industrial innovation is necessary. Our review points to certain gaps in the extant literature, such as a lack of sufficient empirical studies that:

- distinguish between product and process innovations;
- focus on specific firm or industry type;
- examine product and process innovations in the service sector;
- control for the phases of life-cycle;
- distinguish between radical and incremental product and process innovations.

While previous empirical studies considered direct effects of the determinants (Model A, Figure 2.1), future studies should include certain moderators in order to extend our understanding of the differences among and the dynamics of innovation types (Model B, Figure 2.1). The dearth of studies that deal with the determinants of product and process innovations in the service sector is specifically noteworthy, although service organizations and industries are a major economic force in many nations.

Most studies in our review used multi-industry data in their analysis. Pavitt (1986) and Porter (1985) argue that the pattern of technology development differs from industry to industry. The five technological trajectories developed by Pavitt and colleagues (Tidd *et al.* 2001), each with its own distinctive nature and sources of innovation, could provide a useful framework for examining the differences in the determinants of product and process innovation in different sectors. Because of the small number of studies that have examined determinants of innovation in individual industries, analysis of these trajectories is not possible at present. Future studies using single industry data analysis are needed to specify more accurately the dynamics of product and process innovations in specific industries, and to enable comparison of their determinants in different industries. For instance, Foster, Hildén, and Adler's (this volume) examination of the innovative behavior of pulp and paper

firms suggest that the perceived model of technology→innovation→productivity→competition, while suitable in some industries, may not be applicable to the paper industry.

Our review does not include the role of organizational structure and processes such as centralization, differentiation, culture, leadership, and communication. However, a meta-analytical review of the studies of organizational innovation that had included such factors showed that the differences between mean correlations of paired innovation types (product and process, administrative and technical, and radical and incremental) were not statistically significant at the 0.05 level (Damanpour 1991: 577). A recent meta-analysis of the combined moderating effects of innovation types, organizational types, and methods variables on the relationship between firm size and innovation confirmed the low moderating power of the three paired innovation types and concluded that types of innovation, in contrast to types of organization and measures of size, do not exert a significant effect on the size-innovation relationship (Camison-Zornoza *et al.* 2004: 350).

Lack of significant differences between the innovation types in these meta-analytic reviews, coupled with this study's findings that most environmental and organizational determinants do not differentiate between product and process innovations, suggest that in managing the process of technological innovations in organizations in general, and in allocating resources to innovations in product/services and methods of production in particular, corporate executives should focus more on the complementary relationships rather than the differences between innovation types. Research on effects of innovation types on firm performance supports this view. For example, Damanpour and Evan (1984) reported a stronger association between administrative and technical innovations in high- than in low-performance organizations. Capon *et al.* (1992) found that firms that emphasized both product and process development had the highest return on capital. Damanpour and Gopalakrishnan (2001) found that

Model A

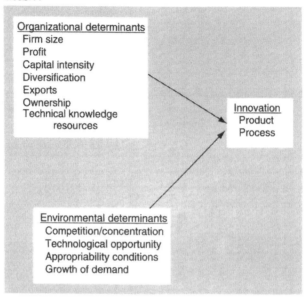

Model B

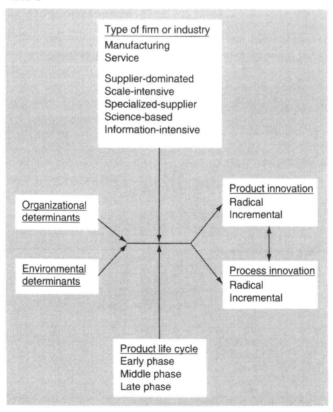

Fig. 2.1. Exisiting research model (Model A) and proposed model for future research (Model B)

high-performance organizations adopt product and process innovations more evenly than low-performance ones. A recent review of the relationship between innovation and organizational performance concluded that organizations that implement product innovations in conjunction with process innovations are most likely to achieve higher levels of performance (Walker 2004).

The managerial implication of these findings is that the synchronous development and implementation of product and process innovations could have positive performance consequences for the organization. Thus managers should avoid the tendency to focus on one type of innovation to the detriment of the other. They should always be cognizant of the importance of both, though they could focus more on product innovation during certain times and process innovation during other times. As stated earlier, when to focus on a specific type of innovation would depend primarily on the firm's sector and strategy.

This chapter has focused only on two types of industrial innovation, and has examined the role of several determinants. Parts II, III, and IV of this volume focus on other important innovation types, and discuss influences of additional factors. For example, considering industrial innovations, Foster et al. (this volume) focus on a factor not included in our review—the role of regulation in providing incentives and capacity for organizational innovation. Chaminade and Edquist (this volume) point out the importance of interactions between, and the hierarchy of, determinants. Meeus and Faber (this volume) focus on the role of interorganizational relations on the innovative behavior of firms. Despite its narrow focus, our review suggests that much still needs to be learned about product and process innovations and their determinants in different conditions. This points to opportunities that need to be explored and challenges that need to be met for developing robust theories of innovation, and to the continued fertility of this area for further research.

Appendix 1. Selection and inclusion of studies

We searched several electronic databases such as ABI/Inform, Econlit, Ei Compendex, Inspec, IEEE Xplore, and ACM Digital Library in May and June of 2003. Because we were interested in the studies of the determinants of product innovation and process innovation, we searched the title and abstract of the articles in these databases using keywords such as 'product innovation' and 'process innovation' with 'determinants,' 'predictors,' 'antecedents,' and 'factors.' We only considered peer-reviewed articles in English language. The above steps yielded approximately 350 articles.

We read the abstracts of the articles in this population. When the abstract was not sufficiently clear, we read the text. We excluded book reviews, conceptual articles, and empirical articles that were not directly related to the focus of the study; for example, articles on:

- determinants of other constructs including new product success (de la Fuente and Marin 1996; Vandenbosch and Dawar 2002; Cooper et al. 2001);[8]
- impact of innovation on employment or firm performance (e.g. Antonicci and Pianta 2002; Smolnyi 1998);
- development of econometric or mathematical models (e.g. Arend 1999; Thompson and Waldo 2000);
- determinants of product innovations only or process innovations only (e.g. Abratt and Lombard 1993; Fritz 1989; Herrmann 1997; Romano 1990) were excluded.

We thus narrowed our list of articles and thereafter examined the reference lists of these articles and identified additional publications. This process resulted in twenty-seven empirical papers that examined the determinants of both product and process innovation.

We coded these publications carefully and found that some have used the same datasets. When several articles were published from the same dataset, we included only one that contained more relevant data: for example, we included: Lunn (1987), but not Lunn (1986); and Kotabe (1990), but not Kotabe and Murray (1990) and Kotabe (1993). The articles used a variety of determinants; we included only the determinants that were considered by at least two authors and whose definitions corresponded with those provided in Appendix 2. If a study used multiple samples, we only reported the results from more than one sample if the samples were independent; for example, in Ettlie *et al.* (1984) the survey and interview studies overlapped, hence we included only the survey study. This process resulted in 23 samples from 18 empirical studies (16 journal articles and 2 book chapters) whose findings are reported in Tables 2.3 and 2.4.

Appendix 2.   Definitions and measures of variables

Product and process innovation

Product innovation is defined as new products or services introduced to meet an external user or market need, and process innovation is defined as new elements introduced into an organization's production or service operations to produce a product or render a service (Ettlie and Reza 1992; Knight 1967; Utterback and Abernathy 1975). Measures of these two types of innovation include product/process R&D expenditure (Fritsch and Meschede 2001; Meisel and Lin 1983), number of product/process patents (Lunn 1986; 1987), binary variables reflecting whether product and process innovations were introduced or not within a certain period (Bertschek 1995; Freel 2003), number of new products/processes introduced/adopted in the past few years (Ettlie *et al.* 1984), and factors derived from factor analysis of survey items (Zahra *et al.* 2000).

Organizational variables

Capital intensity represents the intensity of physical capital of a firm, and is used to differentiate production technologies (Martinez-Ros 2000). A high capital intensity implies a high degree of mechanization (Lunn 1987), processes that are more automated and rigid, and a less labor-intensive environment (Martinez-Ros 2000). It is measured as a ratio of sales to fixed assets (Martinez-Ros 2000) or as a ratio of the net value of a firm's plant and equipment to its employees (Lunn 1987).

Profit reflects the resources available for internal financing of innovations (Kraft 1990; Meisel and Lin 1983). It is measured by cash flow (Kraft 1990), income relative to sales (Lunn 1987; Meisel and Lin 1983), return on assets (Zahra *et al.* 2000), or executive perception of profits (Coursey 1991).

Diversification refers to the degree to which a firm is diversified in terms of the number of innovation projects it conducts (Cabagnols and Le Bas 2002; Lunn 1987). It is measured by concentration ratios of the industries of origin in which the firm has patents (Lunn 1987) or different branches in which a firm operates (Cabagnols and Le Bas 2002), sales outside the industry in which a firm operates (Ettlie *et al.* 1984), and factors from a multi-item scale (Zahra *et al.* 2000).

Exports. Firms with export activity would have presence in foreign markets (Martinez-Ros 2000). Exports are measured by a binary variable (1 if firm exports, 0 otherwise, Martinez-Ros 2000), by the share of sales delivered to other countries or markets (Meisel and Lin 1983; Kraft 1990; Arundel and Kabla 1998) and factors from a multi-item scale (Zahra *et al.* 2000).

Firm size. Several studies have measured size in terms of the number of employees (Baldwin *et al.* 2002; Ettlie *et al.* 1984; Fritsch and Meschede 2001; Kraft 1990; Martinez-Ros 2000; Zahra *et al.* 2000). It has also been measured as the firm's market share (Cabagnols and Le Bas 2002), sales (Arundel and Kabla 1998; Cohen and Klepper 1996; Lunn 1987), and total assets (Gopalakrishnan and Damanpour 2000). Bertschek (1995) has measured it as a ratio of firm's employment to industry employment.

Ownership. This variable compares external versus internal ownership (either to the firm or country). It is measured by a binary variable to reflect non-management versus management-controlled ownership (Kraft 1990), foreign-controlled versus domestic ownership (Baldwin *et al.* 2002; Martinez-Ros 2000), or the percentage of total company stock held by the company's senior executives (Zahra *et al.* 2000).

Technical knowledge resources refers to size, academic qualifications, or training of organizational members (Kraft 1990; Freel 2003). It is measured by the ratio of employees with an academic degree to all employees, the proportion of technical employees, and the presence of technical groups (Kraft 1990; Ettlie *et al.* 1984; Freel 2003).

Environmental variables

Appropriability conditions refers to the extent to which a firm is able to capture returns from its innovations. It is often operationalized by means of patents or trade secrets. Baldwin *et al.* (2002) used two binary variables, one for indicating use of patents and the other for indicating use of secrecy by the firm. Cabagnols and Le Bas (2002) measured it at the sector level by using indicators for measuring efficiency of patents and secrecy for protecting product innovations in comparison to process innovations. Arundel and Kabla (1998) used binary variables to measure the importance of patents and secrecy.

Competition refers to the intensity of market or technological competition (Martinez-Ros 2000). It has been measured by the number of competitors the firm faces (Baldwin *et al.* 2002), percentage of innovative firms in the sector (Cabagnols and Le Bas 2002), average gross profit market of the industry (Martinez-Ros 2000), the share of FDI and imports (Bertschek 1995), and an ordinal measure of the severity of foreign competition (Coursey 1991).

Concentration refers to the extent of market concentration, and reflects the market power of the firm (Lunn 1987). It has been measured in terms of the inverse of the number of main competitors of the firm operating in the major markets (Kraft 1990) and the industry four-firm concentration ratio (Lunn 1987; Meisel and Lin 1983; Scherer 1983).

Growth of demand for the product of the firm (Schmookler 1966) represents market power. It has been measured as the development of the market share of the firm over a period (Cabagnols and Le Bas 2002), the real growth rate of sales for a period (Lunn 1987; Meisel and Lin 1983), binary variables reflecting whether the market of the firm is in recession or not (Martinez-Ros 2000), and short-term versus long-term future market growth rate (Kotabe 1990).

Technological opportunity reflects the influences of technology push in the industry (Lunn 1987). It has been measured as the industry knowledge stock minus the firm R&D expenditure (Martinez-Ros 2000), a subjective measure indicating whether there have been or will be major technological changes in the products or in the methods of production (Kotabe 1990; Meisel and Lin 1983), a subjective measure indicating whether there were opportunities for innovation in the industry and for patenting innovations (Zahra *et al.* 2000), and firms' collaborative R&D with universities and other external R&D institutions (Baldwin *et al.* 2002; Freel 2003).

Notes

1. Empirical studies are scarce for several reasons. First, conceptually, what actually constitutes product and process innovations is a confused issue in the current literature (Bhoovaraghavan *et al.* 1996). Several researchers posit that it is difficult to differentiate between product and process innovations since they are inextricably interdependent (Pisano and Wheelright 1995). Many do not even define them, assuming terms are self-explanatory (Archibugi *et al.* 1994). Second, methodologically, 'the frequent use of proxy variables for innovative activity like R&D expenditures, employees working in R&D departments or total number of patents' makes the differentiation between product and process innovations difficult (Kraft 1990: 1029). Third, strategically, many firms, especially in the US, presume that their growth and profits come largely from new products; therefore, they put an undue emphasis on product innovations as a source of competitive advantage, and neglect process innovations (Kotabe 1990).

2. These determinants are typical in the studies of innovation by economists, which constitute the majority of the studies in our sample. Management researchers have examined additional organizational predictors of innovation, mainly structural variables (for example, centralization, specialization, professionalism) but also process, resources, and leadership variables. However, from this group of researchers, our search resulted only in four empirical studies that distinguished between product and process innovations (Ettlie *et al.* 1984; Ettlie and Rubenstein 1987; Zahra *et al.* 2000; Damanpour and Gopalakrishnan 2001).

3. Although we summarize the findings for each determinant and offer our interpretation, contrary to traditional narrative reviews, we report the direction and significance of the influence of the determinants on product and process innovations to enable the reader to make his or her interpretation of the results.

4. A U-shaped relationship between firm size and innovative activity has also been postulated, suggesting that both small and large firms can be innovative, depending on, for example, the technological opportunity of the industry in which the firms operate and appropriability

conditions. In some sectors, successful innovating firms will end up being very large; in other sectors, they end up being small, often with strong links to larger firms (Pavitt *et al.* 1987). In this review, effects of technological environment variables on the relationship between firm size and product and process innovations cannot be examined because a majority of studies have not included such moderating variables. We discuss these issues in the Discussion section.

5. We reversed the direction of Kraft's and Zahra *et al.*'s results to make their findings comparable with the other two studies that included ownership.

6. In addition to the effects of these factors, mixed and inconclusive research results can be due to divergence in the methods. Cohen and Levin (1989) in their review of antecedents of innovation point out inter-industry differences and measurement problems as sources of inconsistency. Camison-Zornoza *et al.* (2004) conclude that contradictory results of the size-innovation relationship are mainly due to operationalization of the variables, especially size. Although we agree that methodological differences play a crucial role, because our review involves twelve antecedents operationalized by variety of measures, we discuss the conceptual moderators only.

7. Since firms in the goods industry are manufacturing and those in the user industry are mainly service, Barras's model once again supports the importance of distinguishing between manufacturing and service sectors in the future studies of industrial innovations.

8. New product development (NPD) process and success have widely been studied and reviewed in marketing (e.g. Garcia and Calantone 2002). These studies were not included because they focus only on product innovation and do not include process innovation.

References

Abernathy, W. J., and Utterback, J. M. (1978). 'Patterns of Industrial Innovation.' *Technology Review*, June/July: 40–7.

Abratt, R., and Lombard, A. V. (1993). 'Determinants of Product Innovation in Specialty Chemical Companies.' *Industrial Marketing Management*, 22: 169–75.

Afuah, A. (2003). *Innovation Management*. New York: Oxford University Press.

Antonicci, T., and Pianta, M. (2002). 'Employment Effects of Product and Process Innovations in Europe.' *International Review of Applied Economics*, 16: 295–307.

Archibugi, D., Evangelista, R., and Simonetti, R. (1994). 'On the Definition and Measurement of Product and Process Innovations.' In Y. Shionoya and M. Perlman (eds.), *Innovation in Technology, Industries and Institutions: Studies in Schumpeterian Perspectives*, Ann Arbor: University of Michigan Press.

Arend, R. J. (1999). 'Emergence of Entrepreneurs Following Exogenous Technological Change.' *Strategic Management Journal*, 20: 31–47.

Arrow, K. (1962). 'Economic Welfare and the Allocation of Resources for Invention.' In R.R. Nelson (ed.), *The Rate and Direction of Inventive Activity*. Princeton: Princeton University Press.

Arundel, A. (2001). 'The Relative Effectiveness of Patents and Secrecy for Appropriation.' *Research Policy*, 30: 611–24.

—— and Kabla, I. (1998). 'What Percentage of Innovations is Patented? Empirical Estimates for European Firms.' *Research Policy*, 27: 127–41.

Baldwin, J., Hanel, P., and Sabourin, D. (2002). 'Determinants of Innovative Activity in Canadian Manufacturing Firms.' In A. Kleinknecht and P. Mohnen (eds.), *Innovation and Firm Performance*. New York: Palgrave, 86–111.

Barras, R. (1986). 'Towards a Theory of Innovation in Services.' *Research Policy*, 15: 161–73.

Berry, M. M. J., and Taggart, J. H. (1994). 'Managing Technology and Innovation: A Review.' *R&D Management*, 24: 341–53.

Bertschek, I. (1995). 'Product and Process Innovation as a Response to Increasing Imports and Foreign Direct Investment.' *Journal of Industrial Economics*, 43: 341–57.

Bhoovaraghavan, S., Vasudevan, A., and Chandran, R. (1996). 'Resolving the Process vs. Product Innovation Dilemma: A Consumer Choice Theoretic Approach.' *Management Science*, 42: 232–46.

Boer, H., and During, W. E. (2001). 'Innovation, What Innovation? A Comparison between Product, Process, and Organizational Innovation.' *International Journal of Technology Management*, 22: 83–107.

Cabagnols, A., and Le Bas, C. (2002). 'Differences in the Determinants of Product and Process Innovations: The French Case.' In A. Kleinknecht and P. Mohnen (eds.), *Innovation and Firm Performance*. New York: Palgrave, 112–49.

Calantone, R. J., di Benedetto, C. A., and Meloche, M. S. (1988). 'Strategies of Product and Process Innovation: A Loglinear Analysis.' *R&D Management*, 18: 13–21.

Camison-Zornoza, C., Lapiedra-Alcami, R., Segarra-Cipres, M., and Boronat-Navarro, M. (2004). 'A Meta-Analysis of Innovation and Organizational Size.' *Organization Studies*, 25: 331–61.

Capon, N., Farley, J. U., Lehmann, D. R., and Hulbert, J. M. (1992). 'Profiles of Product Innovators among Large US Manufacturers.' *Management Science*, 38: 157–69.

Cohen, W. M. (1996). 'Empirical Studies of Innovative Activity,' in P. Stoneman (ed.), *The Handbook of the Economics of Technological Change*. Oxford: Basil Blackwell, 182–264.

—— and Klepper, S. (1996). 'Size and the Nature of Innovation within Industries: The Case of Process and Product R&D.' *Review of Economics and Statistics*, 78/2: 232–43.

—— and Levin, R. C. (1989). 'Empirical Studies of Innovation and Market Structure.' In R. Schmalansee and R. D. Willing (eds.), *Handbook of Industrial Organization, II*. Elsevier.

Cooper, R., Edgett, S., and Kleinschmidt, E. (2001). 'Portfolio Management for New Product Development: Results of an Industry Practices Study.' *R&D Management*, 31: 361–80.

Coursey, D. H. (1991). 'Organizational Decline and Innovation in New York Manufacturing Firms: Different Effects for Different Innovations?' *Journal of Business and Economic Studies*, 1: 39–55.

Daft, R. L. (2001). *Organization Theory and Design*. Cincinnati: South-Western.

Damanpour, F. (1987). 'The Adoption of Technological, Administrative, and Ancillary Innovations: Impact of Organizational Factors.' *Journal of Management*, 13: 675–88.

—— (1991). 'Organizational Innovation: A Meta-analysis of Effects of Determinants and Moderators.' *Academy of Management Journal*, 34: 555–90.

—— (1992). 'Organizational Size and Innovation.' *Organization Studies*, 13: 375–402.

—— and Evan, W. M. (1984). 'Organizational Innovation and Performance: The Problem of Organizational Lag.' *Administrative Science Quarterly*, 29: 392–409.

—— and Gopalakrishnan, S. (2001). 'The Dynamics of the Adoption of Product and Process Innovations in Organizations.' *Journal of Management Studies*, 38: 45–65.

de Brentani, U., and Cooper, R. G. 1992. 'Developing Successful New Financial Services for Businesses.' *Industrial Marketing Management*, 21: 231–41.

de la Fuente, A., and Marin, J. (1996). 'Innovation, Bank Monitoring, and Endogenous Financial Development.' *Journal of Monetary Economics*, 38: 269–301.

Dewar, R. D., and Dutton, J. E. (1986). 'The Adoption of Radical and Incremental Innovations: An Empirical Analysis.' *Management Science*, 32: 1422–33.

Edquist, C., Hommen, L., and McKelvey, M. (2001). *Innovation and Employment: Process versus Product Innovation*. Cheltenham: Edward Elgar.

Ettlie, J. E., and Reza, E. M. (1992). 'Organizational Integration and Process Innovation.' *Academy of Management Journal*, 35: 795–827.

—— and Rubenstein, A. H. (1987). 'Firm Size and Product Innovation.' *Journal of Product Innovation Management*, 4: 89–108.

—— Bridges, W. P., and O'Keefe, R. D. (1984). 'Organization Strategy and Structural Differences for Radical versus Incremental Innovation.' *Management Science*, 30: 682–95.

Evangelista, R. (2000). 'Sectoral Patterns of Technological Change in Services.' *Economics of Innovation and New Technology*, 9: 183–221.

Frambach, R. T. (1993). 'An Integrated Model of Organizational Adoption and Diffusion of Innovation.' *European Journal of Marketing*, 27/5: 22–41.

Freel, M. S. (2003). 'Sectoral Patterns of Small Firm Innovation, Networking, and Proximity.' *Research Policy*, 32: 751–70.

Fritsch, M., and Meschede, M. (2001). 'Product Innovation, Process Innovation, and Size.' *Review of Industrial Organization*, 19: 335–50.

Fritz, W. (1989). 'Determinants of Product Innovation Activities.' *European Journal of Marketing*, 23: 32–43.

Garcia, R., and Calantone, R. (2002). 'A Critical Look at Technological Innovation Typology and Innovativeness Terminology: A Literature Review.' *Journal of Product Innovation Management*, 19: 110–32.

Gopalakrishnan, S., and Damanpour, F. (2000). 'The Impact of Organizational Context on Innovation Adoption in Commercial Banks.' *IEEE Transactions on Engineering Management*, 47: 14–25.

—— Bierly, P., and Kessler, E. H. (1999). 'Revisiting Product and Process Innovations Using a Knowledge-based Approach.' *Journal of High-Technology Management Research*, 10: 147–66.

Grunert, K. G., *et al.* (1997). 'A Framework for Analyzing Innovation in the Food Sector.' In B. Traill and K. G. Grunert (eds.), *Product and Process Innovation in the Food Industry*. London: Blackie Academic and Professional.

Hage, J. (1980). *Theories of Organizations*. New York: Wiley.

Herrmann, R. (1997). 'The Distribution of Product Innovations in the Food Industry: Economic Determinants and Empirical Tests for Germany.' *Agribusiness*, 13: 319–34.

Jones, O., and Tang, N. (2000). 'Innovation in Product and Process: The Implications for Techno-logical Strategy.' *International Journal of Manufacturing Technology and Management*, 1: 464–77.

Kimberly, J. R., and Evanisko, M. (1981). 'Organizational Innovation: The Influence of Individual, Organizational, and Contextual Factors on Hospital Adoption of Technological and Administra-tive Innovations.' *Academy of Management Journal*, 24: 679–713.

Knight, K. E. (1967). 'A Descriptive Model of the Intra-Firm Innovation Process.' *Journal of Business*, 40: 478–96.

Kotabe, M. (1990). 'Corporate Product Policy and Innovation Behavior of European and Japanese Multinationals: An Empirical Investigation.' *Journal of Marketing*, 54: 19–33.

—— (1993). 'Patterns and Technological Implications of Global Sourcing Strategies: A Study of European and Japanese Multinational Firms.' *Journal of International Marketing*, 1: 26–43.

—— and Murray, J. Y. (1990). 'Linking Product and Process Innovation and Modes of International Sourcing in Global Competition: A Case of Foreign Multinational Firms.' *Journal of International Business Studies*, 21: 383–408.

Kraft, K. (1990). 'Are Product- and Process-Innovations Independent of Each Other?' *Applied Economics*, 22: 1029–38.

Levin, R. C., Cohen, W., and Mowery, D. (1985). 'R&D Appropriability, Opportunity and Market Structure: New Evidence on the Schumpeterian Hypothesis.' *American Economic Review*, 75/2: 20–4.

Lunn, J. (1986). 'An Empirical Analysis of Process and Product Patenting: A Simultaneous Equation Framework.' *Journal of Industrial Economics*, 34: 319–30.

—— (1987). 'An Empirical Analysis of Firm Process and Product Patenting.' *Applied Economics*, 19: 743–51.

McGahan, A. M., and Silverman, B. S. (2001). 'How Does Innovative Activity Change as Industries Mature?' *International Journal of Industrial Organization*, 19: 1141–60.

Martinez-Ros, E. (2000). 'Explaining the Decisions to Carry Out Product and Process Innovations: The Spanish Case.' *Journal of High-Technology Management Research*, 10: 223–42.

Meeus, M. T. H., Oerlemans, L. A. G., and Hage, J. (2001). 'Sectoral Patterns of Interactive Learning: An Empirical Exploration of a Case in a Dutch Region.' *Technology Analysis and Strategic Management*, 13: 407–31.

Meisel, J. B., and Lin, S. A. Y. (1983). 'The Impact of Market Structure on the Firm's Allocation of Resources to Research and Development.' *Quarterly Review of Economics and Business*, 23: 28–43.

Nelson, R. R. (1959). 'The Simple Economics of Basic Scientific Research.' *Journal of Political Economy*, 1967: 297–306.

—— and Winter, S. G. (1982). *An Evolutionary Theory of Economic Change*. Cambridge, MA: Harvard University Press.

Nord, W. R., and Tucker, S. (1987). *Implementing Routine and Radical Innovations*. Lexington, MA: Lexington Books.

Oerlemans, L. A. G., Meeus, M. T. H., and Boekema, F. W. M. (1998). 'Do Networks Matter for Innovation? The Usefulness of the Economic Network Approach in Analyzing Innovation.' *Journal of Economic and Social Geography*, 89: 298–309.

Pavitt, K. (1984). 'Sectoral Patterns of Technical Change: Towards a Taxonomy and a Theory.' *Research Policy*, 13: 343–73.

—— (1986). 'Technology, Innovation, and Strategic Management.' In J. McGee and H. Thomas (eds.), *Strategic Management Research*. New York: John Wiley.

—— (1990). 'What We Know about the Strategic Management of Technology.' *California Management Review*, 32: 17–26.

—— Robson, M., and Townsend, J. (1987). 'The Size Distribution of Innovating Firms in the UK: 1945–83.' *Journal of Industrial Economics*, 35/3: 297–316.

Porter, M. E. (1985). *Competitive Advantage: Creating and Sustaining Superior Performance*. New York: Free Press.

Pisano, G. P., and Wheelwright, S. C. (1995). 'The New Logic of High-Tech R&D.' *Harvard Business Review*, Sept.–Oct.: 93–105.

Roberts, P. W., and Amit, R. (2003). 'The Dynamics of Innovative Activity and Competitive Advantage: The Case of Australian Retail Banking, 1981 to 1995.' *Organization Science*, 14: 107–22.

Romano, C. A. (1990). 'Identifying Factors which Influence Product Innovation: A Case Study Approach.' *Journal of Management Studies*, 27: 75–95.

Scherer, F. M. (1980). *Industrial Market Structure and Economic Performance*. Chicago: Rand McNally.

—— (1983). 'Concentration, Productivity, and R&D Change.' *Southern Economic Journal*, 50: 221–75.

—— (1991). 'Changing Perspectives on the Firm Size Problem.' In Z. J. Acs and D. B. Audretsch (eds.), *Innovation and Technological Change: An International Comparison*. Ann Arbor: University of Michigan Press, 24–38.

Schumpeter, J. A. [1911] (1934). *The Theory of Economic Development*. Cambridge, MA: Harvard U. Press.

Sen, F. K., and Egelhoff, W. G. (2000). 'Innovative Capabilities of a Firm and the Use of Technical Alliances.' *IEEE Transactions on Engineering Management*, 47: 174–83.

Smolnyi, W. (1998). 'Innovations, Prices and Employment.' *Journal of Industrial Economics*, 46: 359–81.

Stevenson, H. H., and Jarillo, J. C. (1990). 'A Paradigm of Entrepreneurship: Entrepreneurial Management.' *Strategic Management Journal*, 11(S): 17–27.

Tidd, J., Besant, J., and Pavitt, K. (2001). *Managing Innovation: Integrating Technology, Market, and Organizational Change*. New York: John Wiley (1st edn. 1997).

Thompson, P., and Waldo, D. (2000). 'Process versus Product Innovation: Do Consumption Data Contain any Information?' *Southern Economic Journal*, 67: 155–70.

Tornatzky, L. G., and Fleischer, M. (1990). *The Processes of Technological Innovation*. Lexington, MA: Lexington Books.

Tushman, M. L., and Anderson, P. (1986). 'Technological Discontinuities and Organizational Environments.' *Administrative Science Quarterly*, 31: 439–65.

Utterback, J. M. (1994). *Mastering the Dynamics of Innovation*. Boston: HBS Press.

—— and Abernathy, W. J. (1975). 'A Dynamic Model of Process and Product Innovation.' *Omega*, 3: 639–56.

Vandenbosch, M., and Dawar, N. (2002). 'Beyond Better Products: Capturing Value in Customer Interactions.' *MIT Sloan Management Review*, 43: 35–42.

von Hippel, E. (1988). *The Sources of Innovation*. New York: Oxford University Press.

Voss, C. A. (1994). 'Significant Issues for the Future of Product Innovation.' *Journal of Product Innovation Management*, 11: 460–3.

Walker, R. M (2004). 'Innovation and Organizational Performance: Evidence and a Research Agenda.' *AIM Research Working Paper*, Advanced Institute for Management Research, London.

—— Neubaum, D. O., and Huse, M. (2000). 'Entrepreneurship in Medium-Size Companies: Exploring the Effects of Ownership and Governance Systems.' *Journal of Management*, 26: 947–76.

Zimmermann, K. F. (1987). 'Trade and Dynamic Efficiency.' *Kyklos*, 40: 73–87.

3 Interorganizational Relations and Innovation: A Review and a Theoretical Extension

Marius T. H. Meeus and Jan Faber

Introduction

In the past decade, the amount of research on interorganizational relations (IOR) has grown significantly. Triggered by technological development, the deepening divisions of labor, and the specialization of firms, new metaphors are coined to summarize these changes in the industrialized countries; concepts like 'back-to-the-core business' or 'flexible production' have been created to label these structural changes. The network approach, rooted in anthropology and sociology, conceives of structure 'as patterns of specifiable relations joining social units—including both individual actors and collectives such as organizations' (Marsden 1990: 435). The growing attention to IOR in organization science is consistent with the growth of the phenomenon, shown in Figure 3.1.

Figure 3.2 reveals that, since the 1980s, a substantial percentage of the R&D partnerships have involved high-tech industries.

Obviously, the level of technological dynamics is associated with new organizational forms and, more specifically, all kinds of IOR. Their growth is also legitimized by an economic logic. Combs and Ketchen, Jr (1999: 884) report

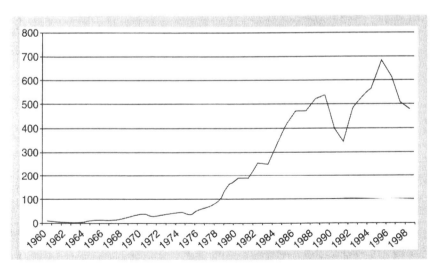

Fig. 3.1. The growth of newly established R&D partnerships (1960–1998)
Source: Hagedoorn 2002.

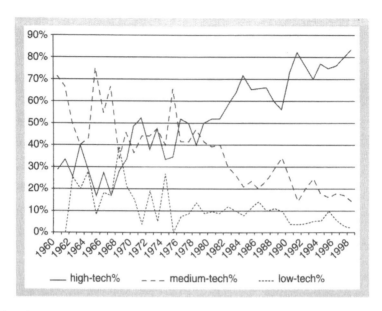

Fig. 3.2. The share (%) of high-tech, medium-tech, and low-tech industries in all newly established R&D partnerships (1960–1998)

Source: Hagedoorn 2002.

that firms low on resources successfully used inter-firm cooperation to surmount resource limitations, which enabled them to perform quite well in terms of return on assets (ROA). Yet this relation was heavily influenced by exchange conditions. If these pointed toward full ownership, inter-firm cooperation damaged performance severely. Mitchell and Singh (1996) showed that alliances raised the survival rates of organizations. Uzzi (1996) showed that apparel firms with strong ties to business groups had better chances of survival. In a sample of young biotech firms, Powell *et al.* (1996) found that those that formed many alliances had accelerated growth rates. We examine this link between the spread of new organizational forms and high-tech in two directions, asking the following questions: To what extent do IOR inspire innovative behavior in firms? And to what extent does innovative behavior encourage formation of IOR?

Our chapter has five parts: this introduction; a section that outlines a heuristic framework to specify the aspects of innovative behaviors, and the classification of interorganizational relations, as well as the bodies of literature we have reviewed; its third section describes first the review results of the consequences of interorganizational relations for innovative behavior, and then deals with the antecedents of interorganizational relations; the fourth section synthesizes the empirical findings in the third section into a framework that describes the mechanisms explaining the reciprocity of innovation and networks; the fifth section discusses the main findings in the light of some theoretical perspectives on IOR and formulates a research agenda.

Interorganizational relations and innovative behavior: conceptual issues and selection of literature

Conceptual issues

Innovation means 'the development and implementation of new ideas and knowledge into a socially and economically successful product, process or service innovation' (Van de Ven *et al.* 1999). 'Innovative behavior of firms' means all the behavior directly related to performing innovation; it can be defined at the macro-level of the total population of firms in a nation state, at the meso-level of sectors, or products, and at the micro-level of individual firms or business units within companies. More specifically, innovative behavior of firms means either the adoption or the generation of new technology, products, and/or processes. In this chapter we focus on the micro-level: implementing innovation; generating innovation by means of R&D; output changes (such as sales) due to innovation; knowledge support; innovation rate; innovative start-ups. In their definition of innovation quoted above, Van de Ven *et al.* talk of 'new' ideas, but newness of technologies in an absolute sense is not easy to discuss. The best example of newness is when patent applications are approved, and can be compared with existing patents.

The adoption and generation of innovative behavior involves a great diversity of inputs, actions, and outputs: R&D investments, a portfolio of R&D projects, innovative performance (granted patents, sales due to innovations), the setting up of structures for knowledge transfer, exchange, and acquisition. 'Adoption' means that a firm buys new technologies of manufacturers, which are implemented in the buyer's own company. 'Generation' of innovation and technology means that a company performs its own R&D, either informally or in organized R&D departments, and with specialized R&D personnel; these can be technical and also organizational, dealing with such things as applications for patent approval.

We use Gulati, Dialdin, and Wang's (2002) three dimensions of IOR. The first is the complete network, which encompasses all the relationships within a group of actors; networks can be described in terms of the level of centralization, which characterizes a pattern of relations from a network perspective. The second dimension is the pattern or configuration of distinct types of links: strong and weak links (interaction frequency), cohesive and bridging links, vertical (suppliers, buyers), horizontal (competitors), institutional and non-institutional links. The third dimension of IOR is the partner profile. To the links one can add such partner features as size, age, status, technological distance. We can also examine how central a focal unit might be to the network, measuring this against the total number of links that a firm has. Other types of centrality such as closeness and 'betweenness' centrality tell more about a firm's access to and control over information and other resource flows.

The key assumption analyzed here is the reciprocity of IOR and innovative behaviors: that innovative behaviors induce IOR, and IOR affect innovative behaviors. If IOR do indeed foster innovative behaviors, the likelihood of repeating this collaboration grows; hence we expect positive feedback loops between innovative behavior and IOR. Yet, IOR can also give disappointing innovation outcomes, which can cause negative feedback loops and lead to changes or terminations to courses of joint action.

Here we concentrate on IOR related to innovation: R&D collaboration; user-producer interaction and the learning processes involved; technology alliances; licensing agreements, etc. Institutional ties, such as those between industry and universities, are examined elsewhere in this book (see Part II). A further principle for us is that papers should take an explanatory approach: we therefore skipped the bulk of papers describing interorganizational structures of sectors, configurations of relations, and the like. In the selection process, we searched several bodies of literature. From the Innovation Studies journals we selected papers from between 1990 to 2003 on Research Policy, Technology Analysis and Strategic Management, and Technovation. The second body of literature comes from the organization and management science, in which IOR has had much attention for many years. We reviewed papers from 1990 to 2003 published in *Organization Studies, Organization Science, Management Science, Strategic Management Journal, Academy of Management Journal*, and *Administrative Science Quarterly*.

The literature review

Front-runners in developing a relational perspective on innovation

The relational perspective on innovation did not emerge recently. On the contrary, it has loomed in the social science and innovation studies literature of the late 1950s, the 1970s, 1980s, and 1990s. The seminal work of Coleman *et al.* (1957) was the first to reveal an effect of social networks on the adoption of new drugs among a group of physicians. Coleman *et al.* compared the network impact with the impact of personal characteristics (professional orientation) on adoption rates. Their tentative findings suggested that the network effect was the strongest: doctors who were more integrated introduced Gammanym considerably earlier than socially isolated doctors (1957: 267). Coleman and his collaborators also showed that the network effect is exhausted after a certain period of time, and that there might have been successive stages in the diffusion of this innovation through the community of doctors. The first networks to be operative as chains of influence appear to be those that connect the doctors in their professional relationships of advisers and discussion partners. Only then, it seems, does the friendship network become operative among those doctors who are influenced in their decisions more by colleagues they meet as friends than by those whom they consider advisers, or with whom they engage in discussion during their working hours. Finally, for those doctors not using the drug by about six months after its release, these networks seem completely inoperative as chains of influence. The social structure seems to have exhausted its effect: those doctors who have not responded to its influence by this time are apparently unresponsive to it.

In the 1970s and 1980s, Coleman's ideas were reanalyzed, especially by Burt (1980 and 1987). Burt's critique of Coleman set out to compare the mechanisms mediating between network structures/positions and diffusion. He claimed that the behavior of network members is affected both by the cohesion of a network and by the so-called 'structural equivalence' between actors. Burt defined 'structural equivalence' as two actors who have identical relations and hence jointly occupy a single position. In a topological sense, these actors are structurally equivalent. This allows an alternative, rival, and status-based explanation of network dynamics that complements the centre-periphery image in which the centrality of an actor is supposed to determine his actions. Status affects how a potential adopter evaluates the intangible advantages of adoption: structurally equivalent actors are mutual reference points, inducing imitation and hence competition for similar resources. The adoption of an innovation by a structurally equivalent network member signals its usefulness. Reanalyzing Coleman's data in his 1987 paper, Burt showed empirically that cohesion, in the aggregate, is a much weaker influence on innovation adoption than structural equivalence (1987: 1326). This does not mean that cohesion has no effects on expert perceptions; nevertheless, Burt claims that it is not cohesion but structural equivalence that creates the social pressure to adopt.

Authors like Teubal, Von Hippel, Lundvall, Edquist, and Pavitt write about research which has set the scene for a relational perspective on innovation. Some of them restrict themselves to conceptual and theoretical work (Lundvall 1992; Teubal 1976); others concentrate on empirical work (e.g. Von Hippel 1987; Pavitt 1984). In most of this research, the use of 'network' or 'interorganizational relation' has been more metaphorical than literal. Foremost, they stress the importance of interaction among several collective actors for the process of innovation.

In 1971, SPRU (Science and Technology Policy Research at Sussex University, England) (Freeman and Soete 1997) tested 200 measures explaining the patterns of success of innovation projects in chemicals and instruments. The single measure that discriminated most clearly between success and failure was 'user needs understood' (Freeman and Soete 1997). Teubal (1976) found the same 'market determinateness' in the Israeli medical electronics industry. In the seminal paper of Von Hippel (1976), empirical findings stressing the importance of sources for innovation external to the firm were presented. Of a total of 44 innovation projects in scientific instruments, 36 (81%) were user dominated. He found that it is the user who:

- perceives that an advance in instrumentation is required;
- invents the instrument;
- builds a prototype;
- proves the prototype by applying it;
- diffuses detailed information on the value of his invention.

Only when all of the above has transpired does the instrument manufacturer enter the innovation process. Typically, the manufacturer's contribution is then to:

- perform product engineering work on the user's device to improve its reliability, convenience of operation, etc.;
- manufacture, market, and sell the innovative product.

Interestingly, this user-dominated pattern appeared typical for more 'basic' innovations as well as for the minor and major improvement innovations. The user-dominated patterns described by Von Hippel also appeared to hold regardless of the size—and thus, presumably, of the internal R&D potential—of the commercializing company. Finally, Von Hippel observed that the pattern of a user-dominated innovation process appears to be true for companies who are established manufacturers of a given product line—manufacturers who ought to know about improvements needed in their present product line and ought to be working on them—as well as for the manufacturers for whom a given innovation represents their first entry into a new product line.

Pavitt (1984) elaborated the relational perspective on innovation by broadening the actor-set inside as well as outside the firm: he looked at linkages within the firm, stressing the role of internal departments, and between firms, stressing the role of suppliers, public R&D, etc. Remarkably, he did not mention the purchase and sales department, which links the firm to its suppliers and customers, as a source of innovation. Pavitt (1984: 354)

found that for supplier-dominated sectors (e.g. agriculture, housing, private services, traditional manufacture) the sources of innovative technology were suppliers, big users, and research extension services. For the scale-intensive sectors (e.g. bulk materials, assembly) he found that the production engineering department and (in-house) suppliers, as well as the R&D department, were the source of innovation. Innovations in the science-based industries (e.g. electronics/electrical, chemicals) originated in the R&D department, public science and production engineering, and in-house suppliers.

The empirical research of Nelson (1982) stressed the linkage between basic science and innovation. The strength of the linkage between firms and other technology-generating institutions in the US appeared to be strongly differentiated. From questioning research managers in 650 firms, it was found that all industries in the sample claimed a strong dependence on at least one field of basic or applied science, while a small number of industries—drugs, semiconductors, instruments—were very dependent on a single science. However, this did not mean that they had strong links with university-located research. In fact, only nine industries claimed close links with academic science. Over 40 per cent of the firms questioned claimed that suppliers of capital equipment and components were important sources of innovation inputs. Johnson (1992) reported that the Nordic Innovation Survey showed that customers are an important source of product-innovation ideas in Scandinavian firms. Universities and R&D institutions are also frequently mentioned.

Consequences of IOR for innovative behavior

Do networks impact innovation adoption/diffusion?

Johnson (1986) tested Burt's status effect in a community of fishermen along the North

Carolina coast engaged in a process of innovation of their equipment, in particular, the nets. Did fishermen quickly adopt an innovation after structurally equivalent (SE) fishermen adopted it? His findings tentatively lend support to Burt's hypothesis. Yet he also doubts the SE effect. For instance, the newly designed kicker plate spread almost randomly among members jointly occupying positions. One specification added by Johnson is specific fishing activities, because not every fisher needed this kicker plate. Because the measurement of structural equivalence does not take this into account, a main source of the structural equivalence is underspecified (1986: 354–5) and therefore loses its value as an explanation of the adoption of innovations.

Midgley *et al.* (1992–3) tested the relative effects of social cohesion versus structural equivalence in the setting of industrial diffusion processes. They found that the network structures of adopters can have important effects on the diffusion process. Their simulation results suggest that social cohesion is more significant early in the diffusion process, while structural equivalence is stronger after the first inflection point of the S-curve. Their survey results showed that both the structural equivalence and the social cohesion effect were operative during the diffusion of fax machines and PC networks.

Do networks foster innovative performance?

Hagedoorn and Schakenraad (1994) demonstrated a positive effect of entry into technology alliances on innovation rates. Ahuja (2000*a*) analyzed the interaction between direct and indirect links. He found that the number of direct links moderates the benefits of indirect links. Assuming that the focal firm can absorb the flow, embeddedness in a denser network, with greater numbers of direct and indirect links, would seem potentially beneficial. However, it turns out that the impact of indirect links decreases if a firm has more direct ones. Actors can only absorb and act on a lim-

ited number of links, and those with many direct links may be unable to profit from their indirect links, with their smaller contribution to an actor's knowledge base. A third issue discussed by Ahuja is which type of network structure is more beneficial: densely interconnected networks (cohesion theory) or structural-hole-rich networks (competition theory), or exclusive ties with a focal actor having few direct links and one partner with many direct links (exchange theory). Ahuja's findings show that having more structural holes in a firm's network, creating brokerage opportunities, has a negative impact on innovative activity. Ahuja's (2000*b*) longitudinal study of firms in the international chemical industry takes a structural point of view, and reports that higher numbers of both direct and indirect links have a positive impact on innovative output (as measured by patenting frequency), but that the benefits from indirect links are relatively low. The point is that, in many networks, indirect links have two counteracting functions that potentially nullify the net effect for the focal actor: on the one hand, they are resources that extend the actor's reach in the network and improve access to information; on the other hand, such links can also compete with the focal actor for this information.

Oerlemans *et al.* (1998) took an exchange-theory point of view and estimated the effects of information exchange with actors in innovation systems on the innovative performance of the focal firms. Its significance was measured as the average frequency of information and knowledge transfer between actors engaged in an innovation network. In general, they found that more frequent transfers with information brokers (trade organizations and consultancies) and so-called intermediary organizations (innovation centers and chambers of commerce) fostered innovative performance in firms. Innovative links with economic exchange partners (buyers and suppliers) turned out to have the most significant effects (ibid. 305). However the IOR impact proved to be highly contingent on sectoral technological dynamics. IOR between biotech start-ups concerned with innovation are basically created by the need to improve their image, their trustworthiness, and their embeddedness in external networks. Such relations among larger organizations are caused by technological complexity and uncertainty regarding complementary technology development (Baum *et al.* 2000).

Stuart (2000) investigates the relationship between intercorporate technology alliances and firms' innovation rates. He argues that alliances provide access, and therefore that the advantages which a focal firm derives from a portfolio of strategic coalitions depend upon the resource profiles of its alliance partners. In particular, large firms and those that possess leading-edge technological resources are posited to be the most valuable associates. The paper also argues that alliances are both pathways for the exchange of resources and signals that convey social status and recognition: when one of the firms in an alliance is a young or small organization or, more generally, an organization of equivalent quality, alliances can act as endorsements—they build public confidence in the organization's products and services and facilitate the firm's efforts to attract customers and other corporate partners. The findings from models of sales growth and innovation rates in a large sample of semiconductor producers confirm that organizations with large and innovative alliance partners perform better than otherwise comparable firms that lack such partners. Consistent with the status-transfer arguments, the findings also demonstrate that young and small firms benefit more from large and innovative strategic alliance partners than do old and large organizations.

Stuart *et al.* (1999) study how the interorganizational networks of young companies affect their ability to acquire the resources necessary for survival and growth. They propose that, faced with great uncertainty about the quality of young companies, third parties rely on the prominence of the affiliates of those companies to make judgments about their quality, and that young companies endorsed by prominent exchange partners will perform better than otherwise comparable ventures that lack

prominent associates. Results of an empirical examination of the rate of initial public offering (IPO) and the market capitalization at IPO of the members of a large sample of venture-capital-backed biotechnology firms show that privately held biotech firms with prominent strategic alliance partners and organizational equity investors go to IPO faster; they also earn greater valuations at IPO than firms that lack such connections. It is empirically demonstrated that much of the benefit of having prominent affiliates stems from the transfer of status that is an inherent by-product of IOR.

Powell *et al.* (1996) argue that because the knowledge base of an industry is both complex and expanding, and the sources of expertise are widely dispersed, the locus of innovation will be found in networks of learning rather than in individual firms. The large-scale reliance on interorganizational collaborations in the biotechnology industry reflects a fundamental and pervasive concern with access to knowledge. Powell *et al.* develop a network approach to organizational learning and derive firm-level, longitudinal hypotheses that link research and development alliances, experience with managing inter-firm relationships, network position, rates of growth, and portfolios of collaborative activities. These hypotheses were tested on a sample of dedicated biotechnology firms in the years 1990–4. Results from pooled, within-firm, time-series analyses support a learning view.

In sum, as can be observed from the findings, IOR offer serious benefits for the adoption of innovation and innovative performance. Obviously, networks enable the solution of specific problems associated with innovation: sharing risks, reducing uncertainties, getting access to external knowledge or information, and other resource bases facilitating learning and innovation. The general picture is that, in terms of interaction with users, embeddedness in the value chain is beneficial for innovative behavior and output. Degree centrality, network size and partner profiles all foster innovative output and knowledge exchange as well.

Table 3.1. Consequences of IOR for innovative behavior

Author(s)	IOR dimension	Type of innovative behavior	Effect + = positive − = negative
Powell et al. (1996)	Degree centrality	Firm growth	+
Shan et al. (1994)	Degree centrality	Start-ups	+ indirect, via size
	Size of the network	innovation-output	+
Stuart et al. (1999)	Partner profile	IPO rate/valuation of new ventures	+
Porter Liesbeskin et al. (1996)	To have social network exchanges	Integration, and knowledge support	+
Oerlemans et al. (1998)	Size of network	Innovation outcomes	+ in science-based sectors
Baum et al. (2000)	Size of the network Ties with rivals	Innovation outcomes	+
Stuart (2000)	Partner profile: resource profiles, status transfer	Firm growth and innovation rate	+ (esp. for small firms teaming up with large firms)
Baum et al. (2000)	Start-ups' alliance network efficiency (provide access to information and capabilities)	Innovative start-up R&D spending growth, dedicated R&D employees, patenting)	+ ns +
	Rival partner Biotech firms		all positive signs become negative
Ahuja (2000b)	Three aspects of a firm's ego network:	Innovation output (patents)	
	Direct ties +		+
	Indirect ties +		+ (impact of indirect ties is moderated by the number of a firm's direct ties)
	Structural holes +/−		−

Some comments

Although often citing each other, the authors of the empirical and theoretical literature on the user-producer interaction differ on three important points: the number and type of actors involved in the innovation process; the explanation of the strength of patterns of interaction; and the level of analysis.

Lundvall restricts his ideas on interactive learning to user–producer dyads, whereas in his discussion of national systems of inno-

vation he discerns a broader actor-set. The empirical literature also comes up with a broad variety of actors interacting in the innovation process. Pavitt (1984) and Von Hippel (1976) stress the role of suppliers. Although there is evidence that innovating firms cooperate with the knowledge infrastructure (Höglund and Persson 1987; Van Dierdonck 1990; Mitchell 1991), universities and providers of higher professional education do not play a role at the micro-level of interactive learning. This also applies to linkage amongst competitors (Von

Hippel 1987; Grabher 1991; Kleinknecht and Reijen 1992; Hagedoorn and Schakenraad 1994).

Second, there is the explanation of the strength of linkages. It is largely disregarded in the empirical literature, but Lundvall gives an implicit account for the occurrence of inter-action without a further empirical specifica-tion of it. In his view, every producer ought to have strong relations with each user (which is not the case according to Pavitt's findings). In Lundvall's theory, several mechanisms account for the formation and longevity of user-producer interaction. One mechanism is the complexity of the knowledge required for effec-tive interaction. As complexity rises, coopera-tion intensifies. Besides the technological dynamics and the changeability of user needs, both of which initiate innovation processes, the antidote to the neoclassical fallacy of opportunism—trust—is the main theoretical mechanism explaining network relations in the innovation process. But trust explains primarily the longevity of relations, not the strength or weakness of IOR. Finally, it is possible that, if knowledge flows become more complex, then actors—especially users—uncertain of the outcomes hesitate to continue such a relation.

Third, there is the issue of levels of analysis of innovation rates. Pavitt (1984) and Von Hippel (1976) on the one side and Lundvall (1992) on the other explain the patterns of interaction at different levels of analysis. Von Hippel's re-search is at the level of the development of artifacts. His research vividly illustrates the major contribution of users to the develop-ment of scientific instruments, while manufac-turers dominate the exploitation of these product innovations. Pavitt's research out-comes are clearly defined at the sectoral level, revealing differentiated patterns of interaction between several actors. Lundvall explains dif-ferent patterns of interaction at the level of firms, with the type and level of innovation. Combining these notions, we have at least two competing explanations for patterns of interaction: sectoral characteristics determine differences in patterns' interaction; or differ-ences in firm behavior determine different

patterns of interaction. Cohen and Levin (1989) and Freeman and Soete (1997) also stress that a different innovation rate can be explained bet-ter at the sectoral level than at the level of firms. The innovation rates in most papers from the nineties are at a pretty high aggregation level, measured as patenting and IOP rates.

A final remark is on measurement. Effects of IOR on innovative behavior such as status transfer, resource access, transfer of know-ledge, transaction costs are in general implied, not measured directly: in particular, alliances are measured via announcements, and nobody really checks whether these announcements have been implemented.

Antecedents of IOR

The second issue is the extent to which innova-tive behavior induces the formation and ex-pansion of IOR.

There are many theoretical papers that give a broad description of the antecedents of net-working; the empirical tests are restricted to some key factors. Unfortunately the hypothe-sized effects of these factors gained mixed em-pirical support. Oerlemans and Meeus (2001) explored some of the effects of exchange con-ditions on the likelihood of R&D collaboration as hypothesized by transaction cost econo-mists. Moderate uncertainty combined with high dependency and high frequency of know-ledge transfer indeed yields the highest per-centage of R&D cooperation between buyers and suppliers (Oerlemans and Meeus 2001: 84 and 88). Results from descriptive analysis (cross-tabulation) as well as a multivariate lo-gistic regression turned out to be consistent.

Singh's (1997) study's predictions are that businesses developing high-complexity tech-nologies face higher risks of failure than other businesses because of greater competency de-mands and higher organization costs. Further, alliances moderate such failure risks but pro-vide fewer survival benefits for businesses com-mercializing less complex technologies. Hypotheses were tested with longitudinal data from the US hospital software industry. High-complexity technology was associated

Table 3.2. Antecedents of IOR

Author(s)	Antecedent	IOR dimension	Type of effect
Greis *et al.* (1995)	Innovation barriers: funding availability, regulatory environment	External partnering	+
Gemünden and Heydebreck (1995)	Business strategy	R&D cooperation	+ + technological leader + customer-focused developer
Oerlemans and Meeus (2001)	Uncertainty (control), specificity, frequency	R&D collaboration	+ under moderate uncertainty
Kleinknecht and Reijen (1991)	Resource stock: R&D department	R&D cooperation	+
	Firm owns patents		+
	Buy license		+
	Applied for innovation grants		+
Singh (1997)	Technological complexity	Survival → Alliance announcements	+
Dutta and Weiss (1997)	Technological innovativeness	Size of partner network, compared to joint venturing	+ = more partnering (licensing, marketing agreements) than joint ventures
Stuart (1998)	Position in network (degree centrality) and partner prestige	Rate of alliance formation	+
Mowery *et al.* (1998)	Technological overlap	Inter-firm cooperation	+ partners in joint ventures have higher levels of technological overlap
Bidault *et al.* (1998)	Environmental pressures Social and industry norms	Adoption of early supplier involvement	+/ns there is an effect of regional origin
	Organizational choices		+
Ahuja (2000*a*)	Technical capital (patents)	Propensity to form technical links	+
	Commercial capital (assets)		+
	Social capital (degree centrality)		ns curvilinear
Tether (2002)	R&D	Innovation cooperation	+
	Investment in technology adoption		+
Orsenigo *et al.* (2001)	Technological change	Network dynamics as collaborative agreements over a 20-year period	?

with higher risk of failure, and alliances only partially moderated such risk.

The main inferences about the factors producing networks drawn from Table 3.2 are:

- business strategy, and environmental pressures forge R&D cooperation;
- the level of transaction costs as determined by exchange conditions induces R&D collaboration;
- resource deficits (funding availability and regulatory constraints), as well as above-average technological resource stocks (patents, in-licensing, innovation grants) foster R&D cooperation;
- technological change in a sector induces network dynamics, R&D leads to innovation cooperation, technological overlap spurs inter-firm cooperation, and technological complexity induces innovation;
- network position (measured as the degree to which it is central) is positively associated with collaboration.

The theoretical frameworks discussed so far (especially Gulati and Stuart) are dominated by a structural focus in which process and exchange aspects, as well as resource-based explanations, are left implicit, or only measured with proxies. Learning and information exchange and knowledge transfer are omitted. In our research, these were the key issues. We have focused more on the exchange and learning processes between innovator firms and their partners, and the different explanations advanced in the literature. We concentrate on a comparison of distinct dyads—broadly speaking, the buyer–supplier dyad, and the industry–university dyad (Meeus *et al.* 2001*a*; Meeus *et al.* 2001*b*) (results in Table 3.1). We found that the resources of (innovator) firms have a U-inverted relation with interactive learning. This relation is highly contingent on the type of partner and size of the firm. The linear effects of absorptive capacity were only supported for the interactive learning of larger firms in partnership with universities, whereas the linear effect of resource deficits did not receive any support. Mowery *et al.* (1998) also report limited effects of absorptive

capacity in the acquisition of capabilities through collaboration.

Why are interorganizational relations and innovative behavior reciprocally related?

From Lundvall's interactive learning, to our literature review

The starting point for this reciprocity idea was found in the work of Lundvall on interactive learning. In the innovation studies literature, especially that on the nature of the innovation process, knowledge and learning are considered the main factors in creating dyads, triads, networks, and clusters of innovation. Lundvall (1992) provides a theoretical perspective on user-producer interaction in innovation processes: 'A central activity in the system of innovation is learning, and learning is a social activity which involves interaction between people' (1992: 1). In Lundvall's theory, innovation is conceptualized as an informational commodity (Cohendet *et al.* 1993), and innovation profits are interpreted in a Schumpeterian way—as transitory. In this view, the acquisition and protection of information is essential in order to innovate and profit from the innovation. Lundvall perceives firms as knowledge-accumulating institutions. This is their *raison d'être*. Firms can more easily accumulate knowledge and utilize it than individuals can. Markets do not accumulate knowledge; they provide the invisible space connecting knowledgeable actors.

Lundvall's characterization of innovative firm behavior and cooperation is developed by taking the specific characteristics of the innovation process as its starting point and confronting it with the routine economic exchange process of commodities. Lundvall's theorizing on innovation as an interactive process is a departure from neoclassical assumptions about the behavior of firms. By definition, innovation is the creation of qualitatively different, new things and new knowledge. Therefore, agents involved in the creation and adoption of innovations cannot reasonably be assumed to know all the possible outcomes of their activities. This problem is aggravated by the fact that involved actors have different evaluations, preferences, and expectations of the outcomes of innovations during the development process (Van de Ven et al. 1999). So the bounded rationality of actors implies that agents behave differently and are not homogeneous in their behavioral rules. Rational calculation and decision making are severely constrained because, though inputs needed for an innovation can be estimated up to a certain level, outputs are difficult to forecast. Nobody knows beforehand how users will respond, or how user needs will change during the development process, things which will determine the heterogeneity of outcomes of innovation processes. Errors of estimation of future markets can go in either direction and they are often wild and inaccurate: the future markets for computers, polyethylene, and synthetic rubber were grossly underestimated, while nuclear power was vastly overestimated (Freeman and Soete 1997). This estimation problem with the trade-off between inputs and outputs makes innovation a complex process.

To be engaged in innovation demands mindsets and social norms other than the routine economic exchange. In the knowledge-intensive economy, behavioral norms other than those of rational, profit-maximizing, selfish economic actors are required. Economic actors will be involved more or less permanently in processes of interactive learning, sometimes demanding cooperation and sometimes the collective creation of complex new knowledge. Lundvall contends that interactive learning is seriously undermined if parties act exclusively from the viewpoint of calculation and maximizing profits. Interactive learning is based on discursive rationality more than on instrumental rationality and stresses sets of norms such as idle curiosity instead of efficiency, mutual respect instead of disrespect, and trust instead of opportunism.

Lundvall's theory of interactive learning is based on the idea of market failure when innovation occurs. Organized markets emerge because normal competition is hampered by interactions among organizations induced by innovation processes. The interaction of users and producers in the context of product innovations is based on communication about technological opportunities and user needs. Information about both technical opportunities and user needs enables firms to respond rapidly with innovations of product and processes. To exchange information more efficiently, a common code of communication is developed. It becomes increasingly costly to leave such a well-established relationship, and involves a loss of information capital. Organizations constituting the organized markets exchange qualitative information, and cooperate on the basis of trust and economic power. In the process of innovation, these specific organizational features are amplified such that users and producers develop durable and selective relationships from which an organized market emerges. Much of the discussion, seeing their emergence as an outcome of technological dynamics and associated levels of innovative activity, revolves around the problem of internalization and appropriability of knowledge. Innovations of all types demand knowledge transfer between suppliers of materials or components and the producer of the final product, without which the redesign of functions and qualities of any artifact is impossible. Cooperation in the definition stage, the development stage, or even introduction to user organizations is often used to appropriate that complex knowledge of users.

Our literature review results in two conclusions. For the relation of networks and innovation (displayed in Table 3.1) embeddedness in the value chain, in terms of interaction with users, is beneficial for innovative behavior and output. A degree of centrality, network size, and partner profiles all foster innovative output and knowledge exchange as well. The main conclusion on the antecedents of the formation of networks of IOR is that networks are: a function of environmental dynamics; business strategies; resource stocks of the collaborating organizations as well as of sectoral technological dynamics, technological complexity, and innovation performed by companies.

This implies that our research question can be answered positively: innovation and technological dynamics and the formation of networks are reciprocally related. The basic argument goes thus: the set of IOR of companies evolves in response to a firm's changing resource needs and acquisition challenges as its product and its R&D portfolio move through life-cycle stages. The environmental dynamics especially demand uncertainty, and technological dynamics motivate entrepreneurs to redefine business strategies pursuing innovation. The organizational form that is chosen should allow firms to acquire resources, and to share risks associated with innovation. As innovation means doing either different things or the same things in different ways, innovation always implies learning, exchange of knowledge, and, hence, intensified interaction of actors internal or external to the focal company. Both learning and interaction in their turn draw on the availability of resources that enable learning and interaction. To formalize our argument we use the following logical scheme:

$$I \rightarrow aA + aR \rightarrow N \rightarrow IP$$

Innovation (I) demands additional activities (aA) and additional resources (aR), which is made possible by the formation of networks (N) of IOR. Hence networks should foster innovative performance (IP). If this sequence of effects works out well, then this should allow for a new round of additional activities related to innovation and so on.

Why reciprocity, and when?

The positive answer to our research question leaves open some issues:

- how the reciprocity can be explained;
- under which contingencies these reciprocities are strengthened or weakened.

This brings us into the realm of speculation and theory development for how the networks of IOR evolved.

Two important mechanisms that explain the self-reinforcing effects of the innovation-network bond are described: performance feedback mechanisms and network-related rivalry. Several contingencies are taken into account: the type of innovation (radical or incremental), size/age of firms (start-up, early growth to mature, etc.); sector (high tech versus medium and low tech): the type of innovation systems network (highly centralized versus fragmented networks).

The performance feedback models describe behavioral consistency over time, and explain behavioral switches as a function of a decision rule in which social and historical aspiration levels are linked to actual performance. The performance feedback models consider learning as the product of experience by making the probability of changes conditional on the subject's history (Cyert and March 1963; Levitt and March 1988; Greve 1998). The mechanism that is key to producing future efficacious behavioral variation is a cognitive evaluation of experiences. Yet the sources for this cognitive evaluation can differ: the subject's own historical performance, or the performance of relevant and similar group members or competitors. Critical in this process is the availability of an aspiration level, defined as, 'the smallest outcome that would be deemed satisfactory by the decision-maker' (Schneider 1992; Kameda and Davis 1990). It is a result of a bounded rational decision-maker trying to simplify evaluation by transforming a continuous measure of performance into a discrete measure of success or failure (March and Simon 1958; March 1988). The aspiration level is the borderline between perceived success and failure and the starting point of doubt and conflict in the decision-making process (Lopes 1982; Schneider 1992).

Both the social and historical aspiration levels affect the probability of innovation (eventually in networks) by the functional form shown in Figure 3.1. The social aspiration level is based on the contemporary perform-ance of a sample group of subjects.

H1a When performance relative to the social aspiration level increases, the probability of innovation decreases.

H1b The decrease in the probability of innov-ation is greater for performance increases above the social aspiration level.

The historical aspiration level can be viewed as the result of an anchoring and adjustment pro-cess (Schneider 1992): last period's aspiration level is the anchor; this period's performance is the adjustment.

H2a When performance relative to the histor-ical aspiration level increases, the probability of innovation decreases.

H2b The decrease in the probability of innov-ation is greater for performance increases above the historical aspiration levels.

Other things being equal, the pattern dis-played in Figure 3.1 also applies to the organ-izational form(s) deployed to achieve certain aspiration levels. Hence, changes in organiza-tional forms—for example, from go-it-alone to cooperation—can be explained in a way simi-lar to the performance feedback effect. From what our review shows, it is plausible that in-novative performance is fostered by collabor-ation in the value chain. Hence we put 'Probability of innovation (in networks)' on the x-axis in Figure 3.3. This is meant to make clear that firms positioned on the left hand of

the assumed social/historical level of aspir-ation can switch their activities from routine products to innovation of products and from a go-it-alone mode to a collaborative mode. On the right hand of the social/historical aspiration levels, these switches are less probable.

Interestingly enough, this pattern of behav-ioral consistency has also been found in net-work behavior, where it was known as the 'repeated ties' or the 'experience' effect. In a longitudinal study, Gulati and Gargiulo (1999) investigated the process of alliance formation in three industries (new materials, industrial automation, automotive products). The dens-ity of direct alliances in prior time periods had a positive effect on the formation of new alli-ances, which suggests that successful collabor-ation is repeated due to the positive outcomes. Gulati (1999) reports that previous experience with alliances made entering new ones more likely, as did the network positions of firms. Firms that are centrally located in the alliance network—which means that they are members of a larger number of cliques (clique overlap centrality) and in closer reach of other actors in the network (closeness centrality)—are more likely to form new alliances.

This means that once a firm joins networks of IOR to facilitate innovation, the likelihood of being more successful is larger compared to firms that go it alone; and, given the aforemen-tioned performance feedbacks, we infer that

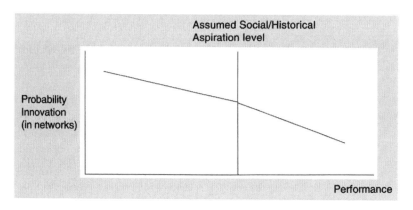

Fig. 3.3. Changing slope response
Source: Greve 1998.

there is an advantage to sticking to the network form. Of course, numerous contingencies will affect the self-reinforcing performance-feedback mechanism. One important limitation is that the mechanism can describe switches in innovation behavior, including the related organizational forms; but, without further specifications, this would only lead to a more centralized network in which the companies with central positions would gain network shares while the peripheral companies would incur losses.

So how is this process of network concentration stopped or redirected? Consider a situation in which a radical innovation offering disruptive technological breakthroughs is introduced in a market. Generator firms will often have had a change in some, even a majority, of main suppliers; they will also have had to look for buyers willing to incur losses in terms of sunk costs, and to experiment with the new equipment. This means that, at both the level of supplier and buyer, the network is fundamentally altered: the company that introduces the new equipment is thus entering a new strategic context in which performance patterns are set at distinct aspiration levels. In the case of incremental innovations, the process of network concentration will probably go on, because neither suppliers nor buyers will change drastically.

Another important contingency is whether the focal company is a start-up, an early-growth, or an established firm. Hite and Hesterly (2001) describe how start-up firms turn to their social relations, friends, and family—'identity-based networks'—to find the money and knowledge to respond to their resource challenges. When firms enter the early-growth stage, entrepreneurs can no longer rely solely on their personal networks; they must develop more calculative, intentional networks that complement their resource deficits. Thus they switch from strong ties to weak ties, and search for structural holes to build unique competences. In contrast, established firms act upon their network position and, depending on their main resource challenges, switch between weak ties and strong ties, often

pursuing control over the indirect ties of other firms.

A third contingency we include here is the type of sector. As we saw in Figure 3.2, the high-tech sector has a larger number of increasingly strong alliances and a network concentration that is probably much higher than the low- and medium-tech sectors. One of the interesting consequences is that CEOs spin their webs of IOR, and subsequently lose room for maneuver. They get locked in, and use 'bad news' to get rid of alliances, joint ventures, mergers, and acquisitions. The established firms tend towards collaboration with smaller, highly specialized suppliers, and this can radically alter their product portfolios. This is why centrality and indirect ties are important for larger, established firms.

A final contingency is the impact of the innovation-network bond on innovation systems. In European and other countries, there is a 'knowledge paradox.' Universities perform quite well in such things as international publications and patents, but R&D collaboration with industry is much less developed in the European Union (EU) than in the US. If one can talk of 'a US and an EU innovation system,' it seems that intellectual property rights hamper knowledge flows between companies in the EU system far more severely than in the US system. The systems differ greatly in knowledge sharing between the main knowledge producers (universities and larger public laboratories) and knowledge users/commercializers (the industry).

Main findings, future research

IOR have been studied from many different theoretical perspectives. Common to these perspectives is that organizations have exchanges with their individual environments; they also often inform—though mostly implicitly—the debate on the reciprocity of innovative behavior and IOR. The direct procurement of facilities, material, products, or revenue to ensure organizational survival has been an overriding reason for establishing them (Galaskiewicz 1985). This is a typical resource-dependence approach. Yet, given the option, firms would prefer not to establish IOR, because they can constrain their subsequent actions. Other approaches to IOR—decisions of whether to make it or buy it, to go it alone or cooperate, to adapt to environmental contingencies, the resource-based/dynamic capabilities perspective, and the transaction cost perspective—are typically geared to this issue. Seeking access to complementary capabilities and resources essential to the pursuit of their goals but currently in the control of other organizations or otherwise unavailable in their own resource base, organizations search for partners with whom to build cooperative links. Collaboration brings the partners uncertainties, and there are many sources of them. The two main ones are: the paucity of information about the reliability of the potential partner, and the difficulty in obtaining it. Both create a hurdle for any type of collaboration.

The hurdle is cleared in two ways: by allying with firms with whom the focal firm has had contacts for a long time; by choosing highly reputable partners. Safeguards that provide sanctions for unreliable behavior anticipate the problem of opportunism. Through collaboration, this interdependence and the associated uncertainties are managed. But several conditions have to be met in matters of similarity, power asymmetry, stability, the prior partnering experience, capabilities of partners, efficiency, and legitimacy.

For the performance effects of networking, one can say that there is ample support. Yet several of Ahuja's findings really put the effects of network size and structural holes into perspective. His work revealed that the benefits of the number of links, both direct and indirect, couldn't be understood in a simple linear manner: having more links is not better. Besides an informational advantage, a larger number of direct links implies that firms are confronted with their monitoring and utilization limitations, which reintroduces the resource-based and dynamic capabilities argument. Too many structural holes—which are supposed to improve the information brokerage position of the focal firm—in a network turns out to be detrimental for innovative activities. This implies that the resource base of a firm has a dual effect: a stronger resource base enables a firm to develop and maintain relations in networks but, since resource quantities are by definition limited, it also limits the number of links that can be utilized efficiently. The work of Coleman *et al.* also suggests that the network effect on innovation adoption is exhausted over time. The work of Burt, Johnson, Midgley, *et al.* shows that the membership of a network affects innovation adoption in distinct ways, with effects on status, as well as cohesion. So, this demands more specification of the network effect.

The findings on the antecedents of networking revealed that resource-based explanations gave mixed results on alliance formation as well as on learning. In both the resource deficits and the absorptive capacity arguments, our findings suggest the need to detail specific features of firms, the type of partners and exchange conditions in innovation networks. Our findings also suggest that the functional form of the effect (monotonic or non-monotonic) points to specific limits to the resource effects on network formation and learning.

'New' directions

What is too often ignored is that the complexity of activities performed by the focal firm determines its resource deficits, and the search direction for complementary resources, as well as the uncertainties involved. Although our findings only tentatively support these effects, we think that the analysis of the complexity effect is a very fruitful direction for future research. One can easily imagine how the complexity of a firm's innovative activities determines the search and partnering movement through a network. Such an approach could also be used for a comparison of the movements (horizontal/vertical) of innovator and non-innovator firms through their networks, and of the causes of such movements. In the innovation process, firms confront themselves with numerous limitations.

Innovation is a process in which uncertainties, information flows and interdependencies all rocket sky-high. This implies that the actor-sets organized with a focus on innovations have specific features. Formal arrangements do not control the exchange processes in innovation networks, and search processes are core activities there. Because resource problems and control problems are generally more explicit if a focal firm performs innovation projects, one would expect that the search for partners is a process that can be guided by the description of the resource deficits. Ahuja finds the issue of network size very important here, because it defines the set of potential partners. The smaller the set of potential partners, the larger the dependencies with this set of actors. Here a link with population ecology might be interesting.

In general, the measurement of governance mechanisms, resources, interdependence, control over resources and complementarity is rather opaque: because there is considerable overlap between these concepts, these imprecise measures cause many interpretation problems. The application of concepts denoting structural aspects of networks are often interpreted in terms of their processual/exchange implications. Take, for example, centrality: informational advantage or cliquishness means that a firm has network resources; another example is environmental dependence, which encompasses two sets of considerations: resource procurement and uncertainty reduction (Galaskiewicz 1985). We have never seen both dimensions clearly operationalized in a precise way. Grandori (1997) distinguished three types of interdependence: pooled, sequential, and intensified. We never saw these interdepend-

encies measured clearly. The same goes for governance mechanisms. Too often they are implied and not operationalized (Jones *et al.* 1997).

Another research agenda fitting the theme of learning and innovation has to do with knowledge flows and learning in networks. There is a large research domain labelled as 'science and technology indicators' in innovation studies. The research in this field purports to clarify international technological dynamics. In it one finds an enormous number of references related to the aforementioned theoretical and empirical work dealing with knowledge resources, technological capabilities, technological networks, spillovers, etc. However, the indicators used seldom or never measure 'real networks.' For instance, patent citations are considered to reflect a network. But is that the case? The citation of patents from different firms could as easily be interaction independent, and as such tell nothing about the 'real networks.' In this specific field, the most important renewal that has to take place is that attention shifts from networks measured by proxy to 'real networks.'

In future research, more attention should be paid to develop a clearer understanding of the empirical relations between the specific structural features of innovation networks and process aspects such as learning, partner search and choice, and resource transfer. The old Arrow learning curve approach is unsatisfactory here: it yields innovation output measures such as decreasing costs per unit produced, but black-boxes the actual learning process. We think that, in general, the learning perspective is an underanalyzed theme in network research.[1]

Notes

1. We have done some work on this issue already: L. A. G. Oerlemans, M. T. H. Meeus, and F. W. M. Boekema (2000), 'Innovation and Proximity: Theoretical Perspectives,' in M. B. Green and R. B. McNoughton (eds.), *Industrial Networks and Proximity*, Hampshire: Ashgate Publishing Limited, 17–46; L. A. G. Oerlemans, M. T. H. Meeus, and F. W. M. Boekema (2000), 'Learning, Innovation and Proximity: An Empirical Exploration of Learning. A Case Study,' in F. Boekema, K. Morgan, S. Bakkers, and R. Rutten (eds.), *Knowledge, Innovation and Economic*

Growth: The Theory and Practice of Learning Regions, Cheltenham: Edward Elgar, 137–64. In due course, two new papers of ours that dig deeper into the proximity issue will be published: L. A. G. Oerlemans, M. T. H. Meeus, and F. W. M. Boekema, 'On the Spatial Embeddedness of Innovation Networks: An Exploration of the Proximity Effect,' in *Book of Abstracts*, 40th Congress of the European Regional Science Association, Barcelona, August 2000: 125. A paper based on regional data will be published in the *Journal of Social and Economic Geography*, 2001/1.

References

Ahuja, G. (2000*a*). 'The Duality of Collaboration: Inducements and Opportunities in the Formation of Interfirm Linkages.' *Strategic Management Journal*, 21: 317–43.
—— (2000*b*). 'Collaboration Network, Structural Holes, and Innovation: A Longitudinal Study.' *Administrative Science Quarterly*, 45: 425–55.
Baum, J. A. C., Calabrese, T., and Silverman, B. S. (2000). 'Don't Go It Alone: Alliance Network Composition and Startups' Performance in Canadian Biotechnology.' *Strategic Management Journal*, 21: 267–94.
Bidault, F., Despres, C., and Butler, C. (1998). 'The Drivers of Cooperation between Buyers and Suppliers for Product Innovation.' *Research Policy*, 26: 719–32.
Burt, R. S. (1980). 'Innovation as a Structural Interest: Rethinking the Impact of Network Position on Innovation Adoption.' *Social Networks*, 2: 327–55.
—— (1987). 'Social Contagion and Innovation: Cohesion versus Structural Equivalence.' *American Journal of Sociology*, 92: 1287–335.
Buvik, A., and Gronhaug, K. (2000). 'Inter-Firm Dependence, Environmental Uncertainty and Vertical Co-ordination in Industrial Buyer-Seller Relationships.' *Omega*, 28: 445–54.
Cohen, W. M., and Levin, R. C. (1989). 'Empirical Studies of Innovation and Market Structure.' In R. Schmalensee and R. D. Willig (eds.), *Handbook of Industrial Organization* II. Amsterdam: Elsevier Science Publishers, 1060–107.
Cohendet, P., Héraud, J. A., and Zuscovitch, E. (1993). 'Technological Learning, Economic Networks and Innovation Appropriability.' In D. Foray and C. Freeman (eds.), *Technology and the Wealth of Nations: The Dynamics of Constructed Advantage*. London: Pinter Publishers, 66–76.
Coleman, J. S., Katz, E., and Menzel, H. (1957). 'The Diffusion of an Innovation among Physicians.' *Sociometry*, 20: 253–70.
Combs, J. G., and Ketchen Jr., D. J. (1999). 'Explaining Interfirm Cooperation and Performance: Toward a Reconciliation of Predictions from the Resource-Based View and Organizational Economics.' *Strategic Management Journal*, 20: 867–88.
Cyert, R. M., and March, J. G. (1963). *A Behavioural Theory of the Firm*. Englewood Cliffs, NJ: Prentice-Hall.
Czepiel, J. A. (1975). 'Patterns of Inter-organizational Communications and the Diffusion of a Major Technological Innovation in a Competitive Industrial Community.' *Academy of Management Journal*, 18/1: 6–24.
Doz, Y. L., Olk, P. M., and Smit Ring, P. (2000). 'Formation Processes of R&D Consortia: Which Path to Take? Where Does It Lead?' *Strategic Management Journal*, 21: 239–66.
Dutta, S., and Weiss, A. M. (1997). 'The Relationship between a Firm's Level of Technological Innovativeness and its Pattern of Partnership Agreements.' *Management Science*, 43: 343–56.
Edquist, C., and Lundvall, B.-A. (1993). 'Comparing the Danish and Swedish Systems of Innovation.' In R. Nelson (ed.), *National Innovation Systems: A Comparative Analysis*. New York: Oxford University Press, 265–99.
Freeman, C., and Soete, L. (1997). *The Economics of Innovation*. London: Pinter.
Galaskiewicz, J. (1985). 'Inter-organizational Relations.' *Annual Review of Sociology*, 11: 281–304.
Gemünden, H. G., and Heydebreck, P. (1995). 'The Influence of Business Strategies on Technological Network Activities.' *Research Policy*, 24: 831–49.

Grabher, G. (1991). 'Rebuilding Cathedrals in the Desert: New Patterns of Cooperation between Large and Small Firms in the Coal, Iron and Steel Complex of the German Ruhr Area.' In E. M. Bergman, G. Maier, and F. Tödtling (eds.), *Regions Reconsidered: Economic Networks, Innovation and Local Development in Industrialized Countries*. London: Mansell Publishing Limited, 59–78.

Grandori, A. (1997). 'An Organizational Assessment of Interfirm Coordination Modes.' *Organization Studies*, 18: 897–925.

Granovetter, M. (1973). 'The Strength of Weak Ties.' *American Journal of Sociology*, 78: 1360–80.

—— (1985). 'Economic Action and Social Structure: The Problem of Embeddedness.' *American Journal of Sociology*, 91: 481–510.

Greis, N. P., Dibner, M. D., and Bean, A. S. (1995). 'External Partnering as a Response to Innovation Barriers and Global Competition in Biotechnology.' *Research Policy*, 24: 609–30.

Greve, H. R. (1998). 'Performance, Aspirations, and Risky Organizational Change.' *Administrative Science Quarterly*, 43: 58–86.

Gulati, R. (1995). 'Social Structure and Alliance Formation: A Longitudinal Analysis.' *Administrative Science Quarterly*, 40: 619–52.

—— (1999). 'Network Location and Learning: The Influence of Network Resources and Firm Capabilities on Alliance Formation.' *Strategic Management Journal*, 20: 397–420.

—— and Gargiulo, M. (1999). 'Where Do Inter-organizational Networks Come From?' *American Journal of Sociology*, 103: 177–231.

—— Dialdin, D. A., and Wang, L. (2002). 'Organizational networks,' in J. A. C. Baum (ed.), *Companion to Organizations*. Oxford: Blackwell Publishers Ltd., 281–304.

Hagedoorn, J. (2002). 'Inter-Firm R&D Partnerships: An Overview of Major Trends and Patterns Since 1960.' *Research Policy*, 31/4: 477–92.

—— and Schakenraad, J. (1994). 'The Effect of Strategic Alliances on Company Performance.' *Strategic Management Journal*, 15: 291–309.

Hite, J. M., and Hesterley, W. S. (2001). 'The Evolution of Firm Networks: From Emergence to Early Growth of the Firm.' *Strategic Management Journal*, 22/3: 275–86.

Höglund, L., and Persson, O. (1987). 'Communication within a National R&D System: A Study of Iron and Steel in Sweden.' *Research Policy* 16: 29–37.

Johnson, B. (1992). 'Institutional Learning.' In B. A. Lundvall (ed.), *National Systems of Innovation: Towards a Theory of Innovation and Interactive Learning*. London: Pinter Publishers, 23–44.

Johnson, J. C. (1986). 'Social Networks and Innovation Adoption: A Look at Burt's Use of Structural Equivalence.' *Social Networks*, 8: 343–64.

Jones, C., Hesterly W. S., and Borgatti, S. P. (1997). 'A General Theory of Network Governance: Exchange Conditions and Social Mechanisms.' *Academy of Management Review*, 22: 911–45.

Kameda, T., and Davis, J. H. (1990). 'The Function of the Reference Point in Individual and Group Risk Decision Making.' *Organizational Behavior and Human Decision Processes*, 46/1: 55–76.

Kleinknecht, A., and Reijen, J. O. N. (1992). 'Why do Firms Cooperate on R&D? An Empirical Study.' *Research Policy*, 21: 347–60.

Levitt, B., and March, J. (1988). 'Organizational Learning.' In W. R. Scott and J. Blake (eds.), *Annual Review of Sociology*. Palo Alto, CA: Annual Reviews.

Lundvall, B.-Å. (1992). 'User-Producer Relationships, National Systems of Innovation and Internalization.' In B. A. Lundvall (ed.), *National Systems of Innovation: Towards a Theory of Innovation and Interactive Learning*. London: Pinter Publishers, 45–67.

March, J. (1988). 'Variable Risk Preferences and Adaptive Aspirations.' *Journal of Economic Behaviour and Organizations*, 9: 5–24.

—— and Simon, H. (1958). *Organizations*. New York: Wiley.

Marsden, P. (1990). 'Network Data and Measurement.' *Annual Review of Sociology* 1990: 435–63.

Meeus, M. T. H. (2000). 'Theory Formation in Innovation Sciences' [in Dutch, 'Theorie Ontwikkeling in de Innovatiewetenschappen: Een Bronnenonderzoek']. In W. de Nijs and S. Schruijer (eds.), *Ondernemen in de Wetenschap*. Apeldoorn: Garant: 203–16.

Meeus, M. T. H., and Oerlemans, L. A. G. (1993). 'Economic Network Research: A Methodological State of the Art.' In P. Beije, J. Groenwegen, and O. Nuys (eds.), *Networking in Dutch Industries*. Leuven/Apeldoorn: Garant.

—— Oerlemans, L. A. G., and Hage, J. (2001*a*). 'Patterns of Interactive Learning in a High-Tech Region. An Empirical Exploration of Complementary and Competing Perspectives.' *Organization Studies*, 22: 145–72.

—— —— —— (2001*b*). 'Sectoral Patterns of Interactive Learning: An Empirical Exploration of a Case in a Dutch Region.' Accepted May 2000 in *Technology Analysis and Strategic Management*, 13: 407–31.

—— —— —— (2004) 'Industry-Public Knowledge Infrastructure Interaction: Intra- and Inter-organizational Explanations of Interactive Learning.' *Industry and Innovation*, 11/4: 327–52.

Midgley, D. F., Morrison, P. D., and Roberts, J. H. (1992–3). 'The Effect of the Network Structure in Industrial Diffusion Processes.' *Research Policy*, 21: 533–52.

Mitchell, W. (1991). 'Using Academic Technology: Transfer Methods and Licensing Incidence in the Commercialization of American Diagnostic Imaging Equipment Research, 1945–1988.' *Research Policy*, 20: 203–16.

—— and Singh, K. (1996). 'Survival of Businesses Using Collaborative Relationships to Commercialize Complex Goods.' *Strategic Management Journal*, 17/3: 169–95.

Mowery, D. C., Oxley, J. E., and Silverman, B. S. (1998). 'Technological Overlap and Interfirm Cooperation: Implications for the Resource-Based View of the Firm.' *Research Policy*, 27: 507–23.

Nagarajan, A., and Mitchell, W. (1998). 'Evolutionary Diffusion: Internal and External Methods Used to Acquire Encompassing, Complementary, and Incremental Technological Changes in the Lithotripsy Industry.' *Strategic Management Journal*, 19: 1063–77.

Nelson, R. (1982). 'The Role of Knowledge in R&D Efficiency.' *Quarterly Journal of Economics*, 97: 453–70.

—— (ed.) (1993). *National Innovation Systems: A Comparative Analysis*. New York: Oxford University Press.

Oerlemans, L. A. G., and Meeus, M. T. H. (2001). 'R&D Cooperation in a Transaction Cost Perspective.' *Review of Industrial Organization*, 18: 77–90.

—— —— and Boekema, F. W. M. (1998). 'Do Networks Matter for Innovation? The Usefulness of the Economic Network Approach in Analyzing Innovation.' *Journal of Economic and Social Geography*, 89: 298–309.

—— —— —— (2001). 'On the Spatial Embeddedness of Innovation Networks: An Exploration of the Proximity Effect.' *Journal of Economic and Social Geography*, 92: 60–75.

Oliver, A. L., and Ebers, M. (1998). 'Networking Network Studies: An Analysis of Conceptual Configurations in the Study of Inter-Organizational Relationships.' *Organization Studies*, 19: 549–83.

Olk, P., and Young, C. (1997). 'Why Members Stay in or Leave an R&D Consortium: Performance and Conditions of Membership as Determinants of Continuity.' *Strategic Management Journal*, 18: 855–77.

Orsenigo, L., Pammolli, F., and Riccaboni, M. (2001). 'Technological Change and Network Dynamics: Lessons from the Pharmaceutical Industry.' *Research Policy*, 30: 485–508.

Pavitt, K. (1984). 'Sectoral Patterns of Technical Change: Towards a Taxonomy and a Theory.' *Research Policy*, 13: 343–73.

Porter Liebeskind, J., Oliver, A. L., Zucker, L., and Brewer, M. (1996) 'Social Networks, Learning, and Flexibility: Sourcing Scientific Knowledge in New Biotechnology Firms.' *Organization Science*, 7: 428–43.

Powell, W. W. (1996). 'Inter-organizational Collaboration in the Biotechnology Industry.' *Journal of Institutional and Theoretical Economics*, 152: 197–215.

—— Koput, K. W., and Smith-Doerr, L. (1996). 'Inter-organizational Collaboration and the Locus of Innovation: Networks of Learning in Biotechnology.' *Administrative Science Quarterly*, 41: 116–45.

—— —— Bowie, J. I., and Smith-Doerr, L. (2002). 'The Spatial Clustering of Science and Capital: Accounting for Biotech Firm-Venture Capital Relationships.' *Regional Studies*, 36: 291–305.

Robertson, S., and Gatignon, H. (1998). 'Technology Development Mode: A Transaction Cost Conceptualization.' *Strategic Management Journal*, 19: 515–31.

Schneider, S. L. (1992). 'Framing and Conflict: Aspiration Level Contingency, the Status Quo, and Current Theories of Risky Choice.' *Journal of Experimental Psychology: Learning, Memory, and Cognition*, 18/5: 1040–57.

Shan, W., Walker, G., and Kogut, B. (1994). 'Interfirm Cooperation and Startup Innovation in the Biotechnology Industry.' *Strategic Management Journal*, 15: 387–94.

Singh, K. (1997). 'The Impact of Technological Complexity and Interfirm Cooperation on Business Survival.' *Academy of Management Journal*, 40: 339–67.

—— and Mitchell, W. (1996). 'Precarious Collaboration: Business Survival after Partners Shut Down or Form New Partnerships.' *Strategic Management Journal*, Summer special issue 17: 99–115.

Stuart, T. (1998). 'Network Positions and Propensity to Collaborate: An Investigation of Strategic Alliance Formation in a High-Technology Industry.' *Administrative Science Quarterly*, 43: 668–98.

—— (2000). 'Inter-Organizational Alliances and the Performance of Firms: A Study of Growth and Innovation Rates in a High-Technology Industry.' *Strategic Management Journal*, 21: 791–811.

—— Hoang, H., and Hybels, R. C. (1999). 'Inter-Organizational Endorsements and the Performance of Entrepreneurial Ventures.' *Administrative Science Quarterly*, 44: 315–49.

Tether, B. S. (2002). 'Who Cooperates for Innovation, and Why: An Empirical Analyis.' *Research Policy*, 31: 947–67.

Teubal, M. (1976). 'Performance in Innovation in the Israeli Electronics Industry: A Case Study of Biomedical Electronics Instrumentation.' *Research Policy*, October: 354–79.

—— (1979). 'On User Needs and Need Determination: Aspects of the Theory of Technological Innovation.' In M. J. Baker (ed.), *Industrial Innovation: Technology, Policy, Diffusion*. London: Macmillan, 266–83.

Tsang, E. W. K. (2000). 'Transaction Cost and Resource-Bases Explanation of Joint Ventures: A Comparison and Synthesis.' *Organization Studies*, 21: 215–42.

Uzzi, B. (1996). 'The Sources and Consequences of Embeddedness for the Economic Performance of Organizations: The Network Effect.' *American Sociological Review*, 61: 674–98.

Van de Ven, A. H., Polley, D. E., Garud R., and Venkataraman, S. (1999). *The Innovation Journey*. New York: Oxford University Press.

van Dierdonck, R. (1990). 'University—Industry Relationships: How Does the Belgian Academic Community Feel about It?' *Research Policy*, 19: 551–66.

Von Hippel, E. (1976). 'The Dominant Role of Users in the Scientific Instrument Innovation Process.' *Research Policy*, 5: 212–39.

—— (1987). 'Cooperation between Rivals: Informal Know-how Trading.' *Research Policy*, 16: 291–302.

—— (1988). *The Sources of Innovation*. Cambridge: Cambridge University Press.

Wasserman, S., and Faust, K. (1994). *Social Network Analysis: Methods and Applications*. Cambridge: Cambridge University Press.

Williamson, O. E. (1999). 'Strategy Research: Governance and Competence Perspectives.' *Strategic Management Journal*, 20: 1087–108.

4 Knowledge-Based View of Radical Innovation: Toyota Prius Case

Ikujiro Nonaka and Vesa Peltokorpi

Introduction

Technological changes occur through non-linear cycles of incremental and radical innovations. In addition to incremental improvements on the existing technology and technological processes, occasional radical innovations emerge, with potentially devastating effects on the old technological and even economic systems (Schumpeter 1934). While potentially providing organizations with rare and inimitable sources of competitive advantage (Barney 1991), radical innovations are characterized by high risk and uncertainty. Because of the inherent uncertainty, companies tend to accumulate knowledge, techniques, and skills through incremental improvements.

The potential value of radical innovations for future technologies, products, and services has attracted wide interest among scholars (for example, Von Hippel 1988; Anderson and Tushman 1990; Damanpour 1991; Hage and Hollingsworth 2000; Van de Ven *et al.* 1999). Those studies identify important macro-level factors, but tend to perceive innovation as a *thing*. As a consequence, relatively little research examines innovation *processes*, defined as 'nonlinear cycle[s] of divergent and convergent activities' (Van de Ven *et al.* 1999: 16). Further focus on processes is important in order to identify the particulars that contribute to radical innovations as things. Knowledge-creation theory, with its dynamic process view, is used

in this chapter to explain radical innovation processes (Nonaka 1991, 1994; Nonaka and Takeuchi 1995; Nonaka and Toyama 2002).

Radical product innovations are products previously unavailable, products that improve performance significantly, or products removing some undesired quality (Hage and Hollingsworth 2000). One recent radical innovation of importance is the first commercialized hybrid electric vehicle, the Prius. The introduction of this vehicle in December 1997 marked a radical shift from more than a century of dominance by the internal combustion engine. In addition to introducing the revolutionary vehicle, Toyota Motor Corporation (hereinafter 'Toyota') made during the project radical improvements to new product-development efficiency. The development from an early prototype to mass-produced vehicles took only fifteen months. The case shows how idealistic projections and careful selection of people tie various knowledge domains into a dynamic coherence.

Four sections follow this introduction to the chapter: the basic tenets of the knowledge-creation theory are explained; the discussion moves from the *ba* (shared context in motion) to the SECI knowledge conversion process and the synthesis of the *ba* and the SECI; the Prius case is used to illustrate how the *ba* and the SECI act as a dynamic coherence in knowledge creation; the chapter closes with conclusions and managerial suggestions.

Context and process of knowledge emergence/creation

Knowledge-creation theory is based on related ontological and epistemological assumptions (Nonaka 1994; Nonaka and Takeuchi 1995). The time-space specific interactions in knowledge-creation are dialectical because the actors influence, and are influenced by, their surrounding reality. Two dialectical relations are germane to radical innovation: one between inner and outer dialectics, and the other between thought and action. Positivist management science can be said to ignore the context specificity and dialectic relationships because of the subject–object duality. As the world is frequently explained through structured immobility, little is known about knowledge-creation processes. In contrast, we describe the creation of knowledge as a dynamic phenomenon through the *ba* and the SECI process of knowledge conversion.

Context for knowledge emergence (ba)

A duality of objective structures and subjective consciousness exists in Anglo-American social theory (Brown 1978). These domains meet in organizations inhabited by subjective individuals. As objective structures shape human action in the structural approach, little focus is directed on the self-directing and dialectic human consciousness. But the human inner (processes internal to the individual) and outer (interactions between the individual and the external world) dialectics can instead be assumed to have an impact on the surrounding reality, which is why organizations do not always live up to their rational promise. An easy way out for many scholars is to explain organizations objectively, and to disregard intentionality, values, and other subjective human attributes. In contrast to objective positivism, the *ba* concept seeks to transcend some of this objectivity–subjectivity opposition (Nonaka and Konno 1998; Nonaka and Toyama 2002, 2003).

The paradox in objective theories is simultaneous explanation of structures suited to routine and non-routine tasks (Thompson 1967). Although scholars have sought to solve this puzzle over the years, objective explanations put humans into a secondary position. While positivist assumptions are based on a priori notions of human behavior, we claim, as do phenomenologists and pragmatists, that human thought and action are inherently subjective. Human actors can be explained through universal and idiosyncratic components (Hayek 1948), and have various, seemingly rational, means to reach their ends. Both means and ends are influenced by time-space in which humans draw on their values or ideals to determine appropriate means in context-specific action (Rescher 1987, 2003). The appropriate means to reach the goals is a matter of intelligent pursuit in which meanings are created by visualizing the projected act in the future (Schutz 1932). The objective structures may thus guide and create tendencies, rather than law-like causalities.

Organizations can also be described as intertwined collections of meaning structures (Nonaka and Konno 1998). By consciousness, individuals are united with the surrounding and unfolding reality. Understanding how reality 'works' is possible only through pure experience beyond the subject–object separation (Nishida 1970). New meanings at the individual level emerge from ontological nothingness or selflessness; the self itself is realized through the act of experiencing. That is, open *ba* generates new meanings. The highest form of knowing takes place from a point at which the knower and the known are one. Realization of the self and the surrounding reality as an absolute nothingness means that no relationship is exempt from the dialectic of coming to be and passing away. In the purest form, nothing exists in *ba* except relationships (Chia 2003). In presenting the individual as an effect of experience and relationships, Kitaro Nishida (1970) characterizes the individual self as a secondary social construction of reality. As the ability to achieve nothingness is a major cognitive achievement, and subject to individual variations, the self-transcending acts of pure experiencing and knowing by becoming can be facilitated through triangular phenomenological roles (Varela and Shear 1999). These roles, connecting inner self with the surrounding reality, are further important for consensual validation and nourishment of new ideas.

The intertwined inner and outer dialectics blur relations between individuals and their surrounding reality. Indeed, the boundaries between self and others, as well as firms and their environment, are essentially open. While the agency-structure interaction is explained in various ways, phenomenologists and structural sociologists agree that humans create knowledge in social interaction, through which they seek to fulfill their projects as embedded and intentional actors (Merleau-Ponty 1962; Giddens 1984). In the process, the surrounding reality both inhibits and provides opportunities to the consciously chosen projects. Humans have an embedded desire for projects that are both realistic but contain unreachable higher aspirations. There is hence a gap between aspiration and attainment. To quote Rescher (1987: 143): 'Optimal results are often attainable only by trying too much—by reaching beyond the limits of the possible. Man is a dual citizen of the realms of reality and possibility. He must live and labor in the one but strive toward the other.' Individuals, as self-promoting actors and parts of social collectives, seek to gain the essence of their surrounding reality through action-reflection dialectics (Polanyi 1952; Heidegger 1962; Schön 1983). The body acts as the medium between the mind and the environment (Merleau-Ponty 1962), which explains why some knowledge used in these projects is acquired through reflecting experiences in direct activities, and some of it occurs at a subconscious level.

The philosophical notion of *habitus* builds on structural assumptions towards shared knowledge (Husserl 1947; Merleau-Ponty 1962; Bourdieu 1985). In contrast to the structural dominance over humans, collective habits are developed over time through the dialectics between selfhood and otherness. *Habitus* can be understood as a kind of mind-less or unconscious orchestration of actions that does not presuppose agency and intentionality (Chia 2004). It is not about blind cultural programming of human behavior nor is it about the spontaneous creativity and intention of willful free agents. Rather, *habitus* as a set of internalized predispositions enables actors to cope with unexpected and changing situations by inducing non-deliberate responses which, while containing a degree of improvisation, reproduce regularities that make most human behavior 'expected' and 'rational.' However, social practice theorists, such as Bourdieu (1985), reject a linear and causal view of action. Instead of assuming that human actions follow structures, humans create space for interactions and the creation of meaning in the *ba* by their knowledge and intersubjectivity. In the supporting context, the interaction, respect for various ideas, and free exchange of tacit knowledge allows the new knowledge to emerge.

The role of an organization is to be a supporting mechanism for knowledge creation by connecting and energizing various *ba*. As tacit knowledge is located in various *ba*, managers create and foster free exchange of knowledge in, across, and between various *ba*. They also connect and rewire knowledge domains. Positioning is important: research estimates that 20 per cent of the actors hold about 80 percent of the network connections (Barbasi 2002). Managers can alternatively use hierarchies and formal power to connect *ba* at various organizational levels to form a greater *ba* (Nonaka *et al.* 2000a). It can further be claimed, as in recent research on transactive memory directories, that organizations can create structures enabling employees to use information technology to search and locate knowledge (Peltokorpi 2004). Also, physical places act as supporting contexts for the creation of knowledge (Nonaka and Konno 1998). The interfaces amongst the *ba* evolve through interaction, time, and intersubjectivity.

Firms connect, lead, and organize the interaction among the various *ba* through networks (Nonaka *et al.* 2000a). Through their small-world networks, people locate external collaborators with the right knowledge to speed up the product-development process. The small-world network properties enable the coexistence, even in a sparse network, of a high degree of clustering, and of short average path lengths to a wide range of nodes (Watts 2003). Once the structural view of networks is replaced with a relational view, networks act as both the outcome and initial condition of agency. People and firms feed into networks, energize them by their input, and, in turn, are enriched through the networks' content and activity. In contrast to the traditional concept of innovation networks (for example, Von Hippel 1988; Dyer and Nobeoka 2000), the relational network concept is not a way of presenting relations among fixed entities, but is about being itself (Merleau-Ponty 1962). What gives any object its properties as well as its ontology is a set of relations to other entities. This means that individuals, like organizations, are born, act, evolve, and cease to exist because of their networked part of the whole.

Knowledge-creation process (SECI)

Knowledge creation occurs through the dialectics of tacit and explicit knowledge (Nonaka 1994; Nonaka and Takeuchi 1995; Nonaka and Toyama 2002, 2003). Instead of adopting an objective approach to knowledge, the knowledge-creating theory draws from phenomenology, idealism, rationalism, and pragmatism. Knowledge is 'a dynamic human process of justifying personal belief toward the truth' (Nonaka and Takeuchi 1995: 58). The process takes place through the Socialization, Externalization, Combination, and Internalization modes (Figure 4.1).

Direct experiences during the socialization phase enable the accumulation of tacit knowledge. The emphasis on obtaining pure insights from experience embraces the phenomenological philosophy (Husserl 1931; Nishida 1970). The mind gives the accumulated experiences subjective personal meanings. Tacit knowledge is hard to externalize and explain, as there is more to know than can be explicitly communicated through language (Polanyi 1952). It is, therefore, important for actors to embrace contradictions rather than confront them (Nonaka and Toyama 2003; Varela and Shear 1999). The Structuration Theory links humans to the environment (Giddens 1984). Humans perform daily actions through

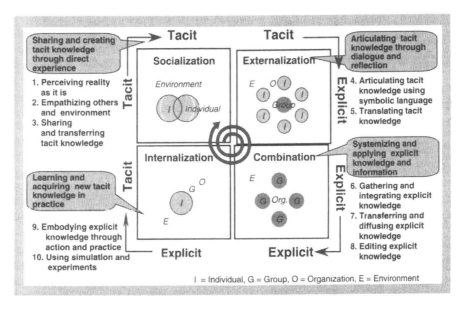

Fig. 4.1. SECI model

practical and discursive consciousness. Where the former refers to tacit stocks of knowledge from which humans unconsciously embrace environment, the latter describes conscious levels of knowing. Like tacit knowledge, practical consciousness reflects humans' being inside of the external world.

During the externalization phase, tacit knowledge is shared through metaphors, dialogues, analogies, and models. Metaphors can be described as the use of words and prototypes outside their conventional meanings. Dialogues can be distinguished from conversations: in conversations, people tender their case through logic, and try to change the opinions of other people; in contrast, dialogues, as 'a stream of meaning flow among and through a group of people,' are based on active listening and openness to changing opinions (Bohm 1990: 1). Important in the process is the diversity of knowledge, which increases the likelihood of the emergence of radical solutions (Hage 1999). Other enablers are intentionality (Searle 1983), love, care, trust (Nonaka and Takeuchi 1995), the embracing of paradoxes, and cultivation of opposite traits.

Firms systematize, validate, and crystallize the externalized tacit knowledge during the combination phase in more explicit forms for collective awareness and practical usage. Firms can, for example, break down concepts into smaller components to create systemic, explicit knowledge. These components are evaluated and validated rationally through their effectiveness. The process includes monitoring, testing, and refining to fit the created knowledge with the existing reality. The combination and distribution of knowledge can be facilitated by information technology, division of labor, and hierarchy.

New knowledge is acquired and utilized through experimentation and discipline during the internalization phase. This phase can be described as a transformational praxis in which knowledge is applied and used in practical situations, becoming a base for new routines (Nonaka and Toyama 2003; Nonaka et al. forthcoming). Explicit knowledge, such as product and service concepts, has to be actual-

ized through action, practice, and reflection so that it can really become owned (Nonaka and Toyama 2003). The consequent products work as a trigger to elicit tacit knowledge from customer demand and consumption patterns. External stimulus from customer reactions is thus reflected in the processes, starting a new spiral of knowledge creation.

The movement through the four modes of knowledge creation forms a spiral that becomes larger as it moves up the ontological levels (Nonaka and Toyama 2003). While the boundaries between the modes are claimed to be porous (Tsoukas 1996), growth of social knowledge occurs through the sequence of sensing ⇒ sharing ⇒ systematizing ⇒ utilizing. There is constant interaction between the different levels of analysis, because individuals interact with the environment, and new knowledge changes the societal rules and behavior. The accumulated knowledge, techniques, and skills set the agenda for further innovations that contribute to economic/societal change and the evolution of economic institutions.

Synthesis of context and process

A synthesis between context (*ba*) and process (SECI) enables knowledge creation. Interactions among the various *ba* are created by knowledge visions, which are value-driven articulations of an idealistic praxis for a social collective (Nonaka et al. forthcoming). Knowledge visions form a nexus among the past, present, and future, as the past has meaning only as a projection of the future. The future is not a determinate end but opens up a cascade of potentials. Middle managers bridge the top-management visions and the chaotic reality at the front line (Nonaka et al. 2000a). They take the roles of instructor, coach, mentor, and coordinator to encourage knowledge sharing, as well as to disseminate information.

While the visible hand of managerial control is occasionally important to speed up processes, the most effective part of power is rooted in language (Alvesson 1996). Managers have the opportunity, more than others, to give meanings to events, and in doing so contributing to the development of norms and values. Pettigrew (1977) describes this as the management of meaning, referring to a process of symbol construction and value use designed both to create legitimacy for one's own demands and to delegitimize the demands of others. Being able to define the reality of others, managers are powerful agents creating shared meanings, ideas, values, and reality through communication and the social construction of meaning.

Toyota Prius case[1]

Toyota has emerged as one of the world's leading automakers since its establishment in 1937. Global sales of Toyota and Lexus brands (combined with those of Daihatsu and Hino) totaled 6.78 million units in the fiscal year 2003. This is equivalent to one vehicle every six seconds. Besides 12 plants and 11 manufacturing subsidiaries and affiliates in Japan, Toyota has 51 manufacturing companies in 26 countries which produce Lexus- and Toyota-brand vehicles and components. As of March 2004, Toyota employs 264,000 people worldwide (on a consolidated basis), and markets vehicles in more than 140 countries. The automotive business, including sales finance, accounts for more than 90 per cent of the total sales, which came to a consolidated 17.29 trillion yen in the fiscal year ending March 2004. Toyota has not recorded an operating loss since it officially started to measure its profits in the 1940s.

Knowledge-based strategy

Although Toyota was, in many respects, successful in the early 1990s, the belief that there was no crisis waiting in the future was itself considered to be a potential crisis. Successes were seen to limit idealistic visioning of future possibilities. In order to avoid the curse of success, top management wanted to face future challenges at an advantage by developing core technology, path-breaking vehicles, and

new routines of product development for the 21st century. Without a path-breaking vehicle, it was felt that Toyota would be following a steady path of incremental innovation, rather than moving radically to a new level of existence.

In terms of the trends for future automobiles, top management had long been aware of the demand for cleaner air and greater fuel savings. This value-rational consideration was already explicit in the Earth Charter of 1992, which outlined goals to develop and market vehicles with the lowest emissions possible. Within the charter, the number one action guideline was 'always to be concerned about the environment.' President Fujio Cho made the following statement:

It bothers me when I'm told that, in the 100 years of automobile development, Japan has not contributed anything. Unfortunately, the starting lines were different, so nothing can be done about this. With respect to the environment, however, the starting line is the same for everyone. Toyota will make every effort so that we can hear that Japan's technology has contributed this much to the environment.

The global environmental concerns and consequent need for low-emission vehicles surfaced relatively late, giving Toyota an opportunity to break old technical systems with revolutionary, environmentally friendly technologies. The company had long worked to reduce emissions in internal combustion engines (ICE) and it was also one of the leading automakers in terms of fuel efficiency.

In reflecting on the Earth Charter, Toyota started to focus more seriously on alternative technologies in 1991. These efforts were stimulated partially by the Zero Emission Vehicle (ZEV) program implemented by the State of California in 1990. Managers realized quickly that car manufacturers need to respond to these increasingly environmental standards. While the company responded to the ZEV mandate with a battery powered RAV-4 in 1996, the car never became a commercial success: high costs and a low operating radius were unsolved problems with battery-powered vehicles. The Electronic Vehicle Development

Department consequently started a systematic study of the hybrid system, combining an engine and a motor. The current Prius chief engineer, Mr Masao Inoue, recalled:

Back in the early 1990s, when Toyota was developing a business case for hybrid technology, it was decided that the engineering program would need to be done almost entirely in-house. This meant that nearly every bit of design, engineering, parts production, and assembly would be done in-house. No partnerships. No contractors. No suppliers of major components or systems.

The in-house development was considered to enable the gaining of overall knowledge of the system and all included components. Top management was committed to providing the needed resources for the uncertain project. While risky, this approach allowed Toyota to develop an internal stock of rich tacit knowledge and reduce dependence on external partners during the early development phase. Embedding knowledge in a product could also yield more economic returns in the long run. It can further be assumed that external partners neither possessed the needed knowledge nor shared the same passion for the project. Toyota wanted to create an internal *ba* for the project and utilize external networked knowledge later, when smoothing the manufacturing processes and cutting down costs.

The origins of the Prius project

The origin of the Prius project can be traced back to initiatives made by a small study group that executive vice-president of R&D, Mr Yoshiro Kimbara, started in order to find alternative ways to increase future competitiveness in late 1993. The initiative had strong top-management support from the beginning. The general manager of the General Engineering Division, Mr Ritsuke Kuboshi, was soon selected to take over as leader of the initiative. He was formerly the chief engineer of Celica, with a reputation for being aggressive and strongly determined. Mr Kuboshi used his network to select ten middle managers for a small study group later called Generation 21st

Century (G21). They met once a week, on top of their daily work, to study the vehicle for the 21st century. As an indication of the vision and urgency, it was decided that the final output of the project should be a mass-produced, mass-marketed, fuel-efficient vehicle. Another mission was to search for a new way to do product development. They had only three months to complete the mission.

The high-level sponsorship increased commitment and urgency to create as comprehensive a plan as possible within the limited time period. As a guideline, they had only the implicit understanding that Toyota was seeking to move to the next level in automobile development. The fact that they were able to develop structured guidelines and a half-size blueprint indicates the level of commitment among employees. Toyota frequently uses projects to test the aptitude of potential hand-picked future managers. One such team member was Mr Sateshi Ogiso, who stayed with the hybrid car project until its launch in 1997. Although Mr Ogiso was only 32 years old, Mr Kuboshi gave him demanding responsibilities to cultivate his leadership skills. He was also made responsible for preparing the final report to the Toyota Motor Company Board.

The G21 group proposed to the Toyota Motor Company Board that the new vehicle should:

- realize a larger inner space by making the wheel base as long as possible;
- place the seats at a higher position to make it easier to get in and out of the car;
- make the car more aerodynamic by making the body height about 1,500 mm;
- improve fuel efficiency by 50 per cent in comparison to the vehicles in the same class (the target was 20 km per liter);
- use a small engine and efficient automatic transmission.

The reference vehicle for fuel efficiency was the Toyota Corolla. The report did not make any specific proposal to develop a hybrid car, partly because the development at that time was considered embryonic for mass production. Further, the G21 group was at this stage still largely influenced by the dominant paradigms of vehicle development. The next step was to develop a more detailed blueprint for the vehicle.

Hybrid vehicle development

Mr Takeshi Uchiyamada was selected to lead the second stage of the G21 project, in January 1994. The project mission was to create a global small car for the 21st century. Before the assignment, he was a test engineer, in charge of re-organizing R&D laboratories. These responsibilities made him knowledgeable about Toyota technologies, and able to locate people with the needed knowledge. Although a novice at leading a new-model development project, his technical knowledge, social networks, and relative inexperience were considered to be beneficial in developing a concept-breaking vehicle: the inexperience allowed him to evaluate critically and, if necessary, to break the established product-development routines and dominant thinking. And his social networks enabled him to locate the right people with the right knowledge.

The G21 project deviated in many respects from ordinary product-development projects at Toyota. The goal to develop a concept-breaking vehicle meant there was virtually no interaction with external parties. Attention to internal resources increased focus, commitment, and full utilization of knowledge resources. And the strong top-management support gave the team relatively free access to resources. According to Mr Uchiyamada, he was 'given a free hand, unbound by any of the usual corporate and engineering constraints; freedom from component sharing and commonalities, marketing considerations, and product hierarchy.' Moreover, the core members were solely dedicated to developing one end product. This enabled Mr Uchiyamada to focus on one task and build a self-sufficient team with all necessary capabilities. The carefully selected initial ten members with about ten years of experience came from eight technological areas, among them body, chassis, engine, drive system, and production technology. They were all in their in their

early thirties—old enough to have expertise, but young enough to be flexible.

Having a core team working solely for one product enabled the sharing of physical space. It was the first time at Toyota that a whole product-development project team had worked in one room. They installed two CAD systems and PCs in the room. The concentration of people increased commitment, knowledge exchange, and decision-making speed. In addition, the knowledge diversity and inter-functional interaction enabled team members to develop an overview of the whole project and the various challenges involved in real time. The experiences during the project were carefully documented and, after the hybrid vehicle project, sharing one big room in new-product development became a common practice at Toyota, because of the efficiency of knowledge combination.

Mr Uchiyamada was quick to identify factors that might slow the project down. He sought to increase commitment, interaction, and decision-making with such credos as:

- technology should be evaluated by everyone, regardless of specialty;
- one should think what is best for the product instead of representing one's own department's interests;
- one should not care about age or rank when discussing technologies.

These guidelines aimed to break down structural rigidities and promote openness for constructive criticism, finding solutions through cross-functional collaboration, and open sharing of ideas. He emphasized delaying final decisions in order to perceive problems from various angles, especially in the development of physical prototypes.

At the beginning of the project, the team followed the earlier guidelines for developing the twenty-first-century vehicle. The group envisioned a small sedan with seating for four adults that would focus on safety, looks, low pollution, and efficiency. A 1.3- or 1.5-liter direct injection engine was thought to be the answer for improved fuel efficiency. Top management rejected this first proposal due to its conventionality; the guidelines were consequently respecified to force the team to seek radical solutions beyond the existing technological boundaries, and, critically, top management made a decision to break organizational silos and forge creative linkages among various *ba*. It was for this reason that Executive Vice-President Wada established an idealistic goal of 100 per cent improvement in fuel efficiency. It was also emphasized that the goal of the project was a mass-produced vehicle.

The team set about the changes required by this rejection of a conventional engine in favor of the underdeveloped hybrid system. The first goal was to develop a concept car for the 1995 Tokyo Motor Show in late October—the team had about one year. The concept car was named Prius, the Latin word for 'before.' At that time, the leaders of the two projects— G21 and hybrid development—were separated in terms of goals and understanding. Initial interactions between the groups revealed problems in the fuel infrastructure: constant battery recharging was required in the hybrid technology. Toyota had developed the hybrid system at the Electronic Vehicle Department, but the now-urgent need to create a synergist hybrid technology for mass production meant there was also a need to build a bridge between technology development and manufacturing. To develop the production technologies for the hybrid system, Toyota created the Unit Production Technology Department.

In order to increase concentration and development speed, Mr Wada formed a project team to research and to analyze candidate hybrid technology for mass-produced vehicles. Because of the relatively long in-house experimentation, the team had many choices. According to Mr Uchiyamada: 'We had 80 research engineers and we got over 80 plans, including historical ones and those in the heads of our researchers.' Those eighty were first narrowed down by a newly developed analytical program to twenty, and then finally to four technologies. Each of these four was evaluated carefully through computer simulation. At the engineer management meeting, the final choice was made, based on efficiency, cost,

and business advantage. The approved technology was code-named 890T in mid-June 1995.

This hybrid system was still largely in the research phase, and the head of the hybrid project, Mr Takehisa Yaegashi, and his team of fifteen engineers worked day and night to build an engine for the concept car. They completed one in time for the Tokyo Motor Show. The concept car aroused wide interest at the motor show, but it was not ready for mass production. After the Motor Show, the Product Audit room, responsible for evaluating a car from the viewpoint of users, tested the car designed for mass production at the test course. On the first day, the car did not start. On the subsequent try, it moved only one meter. For fifty days, because of a basic engineering fault which caused logistical problems, the prototype Prius was not drivable: the engine and motor computers failed to communicate with each other. It finally moved about 100 meters under its own ICE/electric power in December 1995. Said Mr Uchiyamada, 'At that time, my first impression was "It moves!" rather than "It runs." '

The small event was one of the reasons that top management changed the target time for market introduction. The vehicle had originally been planned to be ready for the market at the end of 1999, which was a very short time period for designing, developing, and getting manufacturing ready for a complex product. It usually took an average of approximately four years for Toyota to develop a car; the ambitious goal for the Prius and its new technologies was fifteen months. However, when Mr Wada introduced a new plan for launch in 1998, the President, Mr Hiroshi Okuda, said, 'It is too late. Could it be one year earlier? It is important that we release it early. This car may change the future of the auto industry, not to mention the future of Toyota.' The Chairman, Mr Shoichiro Toyoda, and President Okuda insisted on the early launch in December 1997: another reason for it, according to Mr Uchiyamada, was the Kyoto Conference on global warming issues in 1997.

The change in market-introduction target time forced the team to find ways to cut the product-development cycle time. The challenge was to control the technological uncertainty, stimulate creative problem solving, and make sure that the time limits were not exceeded. Mr Uchiyamada decided to coordinate the whole project with three experienced members. One member was responsible for the hybrid system, another for the body, cost issues, and weight, with the remaining member responsible for regulatory issues, production planning, and marketing. Mr Uchiyamada explained, 'You have to keep the numbers down if you want to develop a vehicle quickly.' The intent was also to set parameters early and make key decisions as quickly as possible. The other key decision was to rely on extensive simultaneous engineering, which brought various groups together in product development. As the hybrid system was still in the research stage, both the research and the development were conducted at the same time.

Horizontal and cross-functional cooperation was critical in order to create a mutual understanding of the diverse technologies and the dependencies among them. In contrast with previous projects in which development teams submitted 'orders' to related departments to develop necessary components for new models, cross-functional cooperation during the hybrid project was intense and spontaneous. In addition, engineers had to possess and develop knowledge beyond their specialties due to the complex technological interrelations in the hybrid system. The efficiency of the hybrid system depends on the combination of all subsystems, and when problems occurred in the integrated system, engineers had to find the cause of the problem in close cooperation. All these functions sought to fulfill the vision of the criticality of speed.

The G21 was made an official product-development project and was renamed 'Product Planning Zi' in January 1996. In addition to the core members from G21, Mr Toshihiro Ohi joined Zi to oversee the entire process of commercialization, production technology, production, sales, and public relations. He was another heavyweight leader that was brought in at the right time to speed up the product launch. In addition to wide networks, he brought vital knowledge that was needed to commercialize the vehicle based on customer needs. Unlike other members, he had fifteen years of experience in product planning and commercialization for cars such as Tercel, Starlet, Corsa, and Sainos. Furthermore, because of the top priority of the project, Mr Uchiyamada was able to obtain the needed human resources. About 1,000 people eventually worked for the project at Toyota (most of them on a part-time basis).

In addition to the hybrid technology, the aerodynamics, maximum interior space, and ergonomics were all important to the vehicle's development. The small vehicle should be large enough inside to contain the hybrid technology and be comfortable to passengers. The vehicle was virtually built around passengers, as Mr Uchiyamada described:

I placed the driver not in a hypothetical automotive package, but in a comfortable seated posture mid-air. The front-seat passenger, likewise, was in a comfortable position with the right amount of space between him or her and the driver. Rear-seat passengers followed. And around the four floating people a package was created, with small outer dimensions and optimum interior space.

A competition was held to increase diversity in vehicle design in February 1996. The participants were from design centers in California, Europe, Tokyo, and Toyota City. Seven teams presented more than twenty designs. Five of them were chosen for further evaluation, with two reaching the final competition. They were from the Design Department at the Toyota headquarters and Calty Design Studio in California. The headquarters design was an extension of the existing models. In contrast, the Calty design was futuristic. The panel of judges consisted of Toyota executives and 100 employees. While the executives appreciated both designs, employees favored the futuristic design, and the winner was Calty's Mr Irwin Lui. The first-generation Prius was short and upright, distinctive among contemporary small sedans.

Battery joint venture

While most problems were eventually solved during the project, the persistent obstacle was battery technology. Because of the limited car size, the batteries needed to be small, but powerful enough for smooth functioning. They should also be reasonably priced and have a long life-cycle. Toyota did not have the in-house knowledge to tackle the problems involved, and formed a joint venture with Matsushita Battery Industrial Co. Ltd. (MBI) in December 1996. MBI was a logical partner: it had knowledge of developing batteries for electric-powered vehicles. The Electronic Vehicle Development Department had also cooperated earlier with MBI to develop a nickel hydrogen battery for RAV-4. In the joint venture, called Panasonic EV Energy Co. Ltd., the role for MBI was to produce battery cells, and Toyota's was to produce modules and holders.

The external cooperation brought with it the problems of developing the *ba* for smooth interaction. The early challenges in joint development and production were heat resistance and quality control. Because batteries were initially placed next to the engine, heat created problems. The early battery prototypes were heat sensitive and could not be used on hot days. The other challenge was to create a shared mindset on quality control: while MBI tolerated certain failure rates, tight quality control was required for hybrid car technology. Because of the early production problems, the first batteries produced less than one-third of the required performance.

The shared mindset came through movement of management-level employees, open dialogues, and shared understanding of the requirements needed for a mass-produced vehicle. When test drives took place in July and August 1997 in extreme conditions in Japan, Nevada, California, and New Zealand, engineers of Panasonic EV Energy participated to understand battery requirements in real conditions. These experiences in the actual context enabled engineers to detect problems quickly, and several modifications were made in the following months to prevent battery failures.

Product launch

The first media test-drive event was held at the Higashi-Fuji Proving Grounds, and the Prius was unveiled officially to the press in October 1997. The first Prius rolled out from the factory in December 1997. The world's first commercialized hybrid car was welcomed with wonder and surprise, and Toyota had to give three press conferences in Tokyo. The extensive media coverage, reasonable price, and general interest in hybrid technology resulted in a back-order of 3,500 hybrid vehicles, and the company quickly decided to raise production from 1,000 units to 2,000 units per month. The Prius received numerous awards, including Japanese Car of the Year and American Car of the Year, for its innovative product concept and technologies.

The in-house design and production proved to be beneficial for three reasons. First, it enabled the accumulation of knowledge, techniques, and skills: the first Prius included more than 300 patents. Because of the knowledge concentration, Toyota was able to develop and modify the hybrid system in a short time. According to Mr Takehisa Yaegashi, 'I hesitate to say we are much ahead of the others, but I do want to emphasize the difference between Toyota and the other companies ... We have a six-year advantage in mass-producing hybrid cars.' At the time of the launch of the first Prius, Toyota's closest rival, Honda, was estimated to be about three to four years behind Toyota in the late 1990s. Second, the company has been able to improve and simplify technology to reduce costs. According to Mr Inoue, 'We have been able to significantly reduce the cost of major hybrid components and sophisticated support systems through in-house R&D.' Third, Toyota has been able to use the created knowledge across all main developmental drives: alternative energy, such as compressed natural gas; diesel engines; gasoline engines; and electronic vehicles.

Second-generation Prius

At the New York Auto Show in 2003, the second-generation Prius demonstrated how quickly Toyota was able to make improvements in hybrid technology. In addition to increased attention to exterior design, the new model is based on a new main system called Toyota Hybrid System II (THS II), which has moved the hybrid technology to a new level. According to Mr Satoshi Ogiso, who had been with the Prius project from the beginning:

For the second-generation Prius, we focused on the promotion of widespread use and driving performance. No matter how good the environmental performance, if the system does not come into wide use, its benefits will not be felt by society. We completely abandoned the original concept of the Prius and started with a blank slate to create a new concept. Our biggest concern was whether we would be able to create a second-generation vehicle that could compare to the epoch-making first generation.

Despite only having a 1.5-liter engine, the hybrid system enables the new Prius to have acceleration that outpaces 2.0-liter-engine cars. The main focus in the development was placed on the high-voltage circuit between the battery and the generator, which increases the voltage to the motor provided by the battery. The battery still provides 21 kW, but it is smaller and less costly to produce than in the previous model.

When the second-generation Prius reached US showrooms in October 2003, dealers received 10,000 orders before the car was even available. The environmental performance of the new Prius has been praised in Europe, and it is subject to preferential tax treatment in fifteen countries, including France, the United Kingdom, and Germany. The new Prius has received several awards, such as 'Car of the Year' from the US magazine *Motor Trend*, 'Best Engineered Vehicle' from the Society of American Automotive Engineers, 'North American Car of the Year' from the US automotive journalists, and 'International Engine of the Year 2004' from international automotive journalists. In addition, support from users shows the extent to which Toyota has succeeded in raising consumer appreciation of hybrid vehicles.

Toyota is seeking to utilize hybrid technology in a wide range of vehicles. In the process, the horizontal deployment (*yoko-ten*) of ideas and technologies is being used extensively. The company has announced that, by summer 2005, it will start to sell a hybrid Lexus SUV, RX400h, whose V6 engine will deliver power rivaling that of a V8 engine, with the fuel efficiency of a compact car. The hybrid technology will be likely to be extended further to diesel engines in trucks and commercial vehicles.

Toyota is offering its technology to other automakers to create scale economies. Executive Vice President Akihiko Saito explains:

We believe that the proliferation of environmental technologies is essential. Proliferation is not something that Toyota can achieve on its own, and when considering the global environment, it is important that automakers from around the world work together. That is why we are considering disclosing technologies.

This has so far led to the signing of licensing agreements with Nissan and Ford. Several Japanese carmakers have launched a hybrid vehicle, or are seeking to launch one in the near future. The success of hybrid vehicles has not gone unnoticed in Europe, where several makers have announced market launches in 2003 and 2004. The patent activity on hybrid technology has accordingly increased rapidly in recent years. While most American carmakers have been skeptical about developing hybrid vehicles, General Motors announced in December 2002 that it plans to sell one million hybrid vehicles by 2010. The others are likely to follow.

Discussion

The flow of occurrences leading to the first commercialized hybrid vehicle evolved from policy statements and environmental concerns. While most automakers were hesitant to develop environmentally friendly technologies, the Toyota Motor Company Board had an idealistic vision to develop a twenty-first-century vehicle and product-development processes. This vision led to a creation of a small study team of ten middle managers in late 1993. Increased commitment brought in managers with wide experience and networks, and the new product-development project eventually tied together about 1,000 employees from various organizational layers in 1997. Important in the process was the right mix of people and selective top-management promotion of interaction and direction.

The G21 group responded to the ambitious goals by redefining processes and combining various knowledge domains, which enabled the fast organic growth from early policies to a mass-produced hybrid vehicle (Figure 4.2).

For example, the goal of 100 per cent improvement in fuel efficiency provided the needed push at the right time to break away from the dominant thinking in order to search for radical solutions; it also enabled the

rewiring of two dense, separated *ba* at Toyota and overcame silo barriers and human defense mechanisms. In order to promote urgency and direction, top management further began to emphasize that the future of the company largely depended on the completion and success of this single project. Project leaders conveyed this urgency to various *ba* by continually repeating and emphasizing the primacy of that underlying purpose. Experts with relevant knowledge and connections were included as the project evolved.

Efficient knowledge-creation processes within *ba* and among *ba* enabled the development of various subjective ideas into a concrete product. For example, Mr Uchiyamada visited research and development departments at the early stages of the project to create overall understanding of core technology and select the right people with relatively long experience. Desired competencies for team members were flexibility and deep understand-

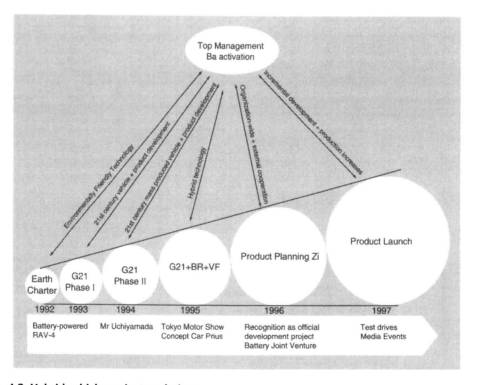

Fig. 4.2. Hybrid vehicle project evolution

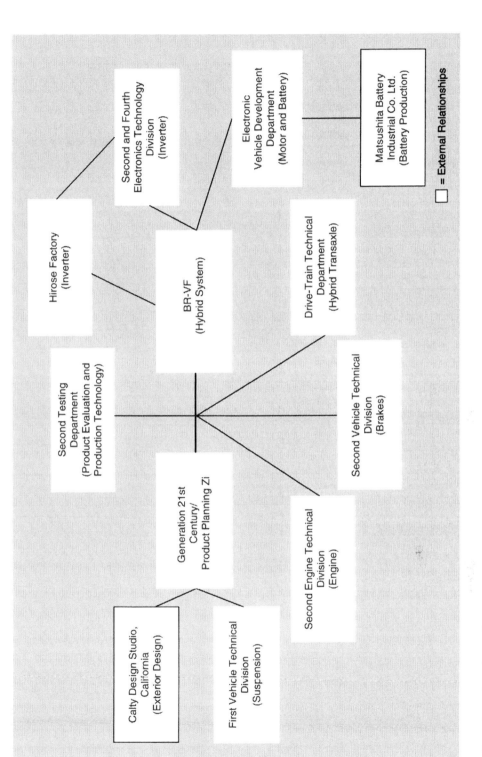

Fig. 4.3. Various *ba* at the Toyota Prius project

ing of interrelated technologies. This enabled fast and efficient knowledge exchange. The team members were quick to create the key concepts for the twenty-first-century vehicle, and ready to change their mindsets when their initial plan was rejected. Team members used information technology and visited other teams to share their embryonic concepts and ideas with relevant entities to get creative criticism and diverse ideas for improvements. They further documented their experiences in a 200-page book that has been used since to develop research and development processes at Toyota.

The urgency, the focus on a single product, and the strong top-management support enabled the core team to utilize various organizational resources quickly and efficiently, and to link various *ba* into a self-transcending and coherent fashion tied by a single overarching vision (Figure 4.3). The seemingly risky, but calculated, top-management goal was to create strong linkages and shared visions between the G21 group and the BR-VF hybrid system. While various *ba* were unified by a single vision, they were given autonomy to reach targets by self-organizing action. In addition, somewhat free utilization of organizational resources, the concentration on one product, and close cross-functional interaction enabled people and teams to bring their own experiences and contexts into *ba* to make the shared context a rich one.

In summary, the case shows that knowledge creation needs many, multi-level *ba*, connected to each other organically to form a greater *ba*. An organization is an organic configuration of *ba* in motion, where various *ba* form a fractal. Hence, leaders have to facilitate the differentiation and interweaving among a variety of seemingly distant and disconnected *ba*, and to synthesize the knowledge that emerges from the larger *ba*. An important enabler in the process was the accumulated tacit knowledge that was carefully connected and rewired from various organizational domains.

Conclusion

This chapter presents, through case analysis, a knowledge-based view of radical innovation. Instead of viewing innovation merely as a 'final' state, we describe innovation as a creative process. The knowledge-creating theory describes the process through the SECI and *ba*, in which subjective but rational human actors combine tacit and explicit knowledge in dialectic fashion. The context and processes are combined by vague and inspirational knowledge visions, which middle managers execute through intense social interaction. Their human connections enable them to link various *ba* as a large and systematic knowledge-creating system. While external networks enable firms to detect market possibilities, early phases of radical innovation are, provided that appropriate knowledge domains exist, conducted most efficiently within companies.

The development of the Prius was radical in at least three respects. First, it did not fit into any existing product line and it was designed to give Toyota a new image. Second, the vehicle used innovative technologies in its engine, motor, battery, and brakes, and in combining them into a hybrid system. Many of these new technologies were later applied to other products. Third, Toyota used a new method of product development to produce the Prius. While it usually took about four years for Toyota to develop a new model of an existing product, the Prius was developed in fifteen months. The increased speed in product development enabled Toyota to expand lines more rapidly and renew products more quickly. An interesting detail was the relative lack of external involvement in the product-development process.

A case study was used in this chapter to illustrate social processes leading to radical innovation in one company. While Toyota is a distinctive company with its own culture, practices, and history, case studies on radical innovation projects show that strong top-management vision and support, linkage of knowledge resources by middle managers, and strong knowledge fermentation, enable radical innovations, especially in Japanese organizations (Nonaka and Takeuchi 1995; Nonaka *et al.* forthcoming). In order to identify the mechanisms and differences between companies in radical innovation processes, future research could be conducted using both qualitative and quantitative methodologies for depth and breadth in understanding the phenomenon of radical innovation.

Note

1. The case is based partly on an unpublished work by Ryoko Toyama.

References

Alvesson, M. (1996). *Communication, Power and Organization.* Berlin: Walter de Gruyter.

Anderson, P., and Tushman, M. L. (1990). 'Technological Discontinuities and Dominant Designs: A Cyclical Model of Technological Change.' *Administrative Science Quarterly,* 35: 604–33.

Barbasi, A. (2002). *Linked: The New Science of Networks.* New York: Perseus Publishing.

Barney, J. (1991). 'Firm Resources and Sustained Competitive Advantage.' *Journal of Management Studies,* 17/1: 99–120.

Bohm, D. (1990). *David Bohm: On Dialogue.* David Bohm Seminars, Ojia.

Bourdieu, P. (1985). 'The Genesis of the Concepts of "Habitus" and "Field".' *Sociocriticism,* 1/2: 11–24.

Brown, R. H. (1978). 'Bureaucracy as Praxis: Toward a Political Phenomenology of Formal Organizations.' *Administrative Science Quarterly,* 23: 365–82.

Chia, R. (2003). 'From Knowledge-Creation to the Perfecting of Action: Tao, Basho and Pure Experience as the Ultimate Ground of Knowing.' *Human Relations,* 56/8: 953–81.

—— (2004). 'Strategy-as-Practice: Reflections on the Research Agenda.' *European Management Review,* 1: 29–34.

Damanpour, F. (1991). 'Organizational Innovation: A Meta-Analysis of Effects of Determinants and Moderators.' *Academy of Management Journal,* 34/3: 555–90.

Dyer, J. H., and Nobeoka, K. (2000). 'Creating and Managing a High-Performance Knowledge-Sharing Network: The Toyota Case.' *Strategic Management Journal,* 21: 345–67.

Giddens, A. (1984). *The Constitution of Society.* Berkeley: University of California Press.

Hage, J. (1999). *Organizational Innovation (History of Management Thought).* Dartmouth: Dartmouth Publishing Company.

—— and Hollingsworth, J. R. (2000). 'A Strategy for the Analysis of Idea Innovation Networks and Institutions.' *Organization Studies,* 21: 971–1004.

Hayek, F. A. (1948). *Individualism and Economic Order.* London: Routledge and Kegan Paul, Ltd.

Heidegger, M. (1962). *Being and Time.* New York: Harper and Row.

Husserl, E. (1931). *Ideas: General Introduction to Pure Phenomenology.* New York: Macmillan.

—— (1947). *Erfahrung und Urteil: Untersuchungen zur Genealogie der Logik.* Hamburg: Claassen and Goverts. (Eng. trans., *Experience and Judgment* by J. S. Churchill and K. Ameriks. London: Routledge and Kegan Paul, 1973.)

Lawson, T. (1997). *Economics and Reality.* New York: Routledge.

Merleau-Ponty, M. (1962). *Phenomenology of Perception.* London: Routledge.

Nishida, K. (1970). *Fundamental Problems in Philosophy: The World of Action, the Dialectical World.* Tokyo: Sophia University.

Nonaka, I. (1991). 'The Knowledge Creating Company.' *Harvard Business Review*, 69: 96–104.

—— (1994). 'A Dynamic Theory of Organizational Knowledge Creation.' *Organization Science*, 5/1: 14–37.

—— and Konno, N. (1998). 'The Concept of 'BA': Building a Foundation for Knowledge Creation.' *California Management Review*, 40/3: 40–54.

—— and Takeuchi, H. (1995). *The Knowledge-Creating Company*. New York: Oxford University Press.

—— and Toyama, R. (2003). 'The Knowledge-Creating Theory Revisited: Knowledge Creation as a Synthesizing Process.' *Knowledge Management Research and Practice*, 1/1: 2–10.

—— (2002). 'Firm as a Dialectic Being: Towards the Dynamic Theory of the Firm.' *Industrial and Corporate Change*, 11: 995–1019.

—— Peltokorpi, V., and Tomae, H. (forthcoming). 'Strategic Knowledge Creation: The Case of Hamamatsu Photonics.' *International Technology Management Journal*.

—— Toyama, R., and Konno, N. (2000*a*). 'SECI, *Ba* and Leadership: A Unified Model of Dynamic Knowledge Creation.' *Long Range Planning*, 33/1: 5–34.

—— —— and Nagata, A. (2000*b*). 'A Firm as a Knowledge Creating Entity: A New Perspective of the Theory of the Firm.' *Industrial and Corporate Change*, 9/1: 1–20.

Peltokorpi, V. (2004). 'Transactive Memory Directories in Small Work Units.' *Personnel Review*, 33/4: 446–67.

Pettigrew, A. M. (1977). 'Strategy Formulation as a Political Process.' *International Studies of Management and Organization*, 7: 78–97.

Polanyi, M. (1952). *Personal Knowledge*. Chicago: University of Chicago Press.

Rescher, N. (1987). *Ethical Idealism: An Inquiry into the Nature and Function of Ideals*. Berkeley: University of California Press.

—— (2003). *Epistemology: On the Scope and Limits of Knowledge*. Albany, NY: State University of New York Press.

Schön, D. A. (1983). *The Reflective Practitioner*. New York: Basic Books.

Schumpeter, J. A. (1934). *Theory of Economic Development*. trans. R. Opie, Cambridge, MA: Harvard University Press.

Schutz, A. (1932). *The Phenomenology of the Social World*. Evanston, Ill.: Northwestern University Press.

Searle, J. R. (1983). *Intentionality: An Essay in the Philosophy of Mind*. New York: Cambridge University Press.

Thompson, J. (1967). *Organizations in Action*. New York: McGraw Hill.

Tsoukas, H. (1996). 'The Firm as a Distributed Knowledge System: A Constructionist Approach.' *Strategic Management Journal*, 17, Winter Special Issue: 11–25.

Van de Ven, A. H., Angle, H. L., and Poole, M. S. (1999). *Research on the Management of Innovation: The Minnesota Studies*. New York: Oxford University Press.

Varela, F., and Shear, J. (1999). 'First-Person Accounts: Why, What, and How.' In F. Varela and J. Shear (eds.), *The View from Within: First Person Approaches to the Study of Consciousness*. Thorverton: Imprint Academic.

Von Hippel, E. A. (1988). *The Sources of Innovation*. Oxford: Oxford University Press.

Watts, D. J. (2003). *Six Degrees: The Science of a Connected Age*. New York: Norton.

5 Innovation, Competition, and Enterprise: Foundations for Economic Evolution in Learning Economies

J. Stanley Metcalfe

Introduction

This chapter reviews a triad of related ideas at the core of how an understanding of the connections between the growth of human capability and the growth of material standards of living are related. It outlines a framework in which this triad—innovation, enterprise, competition—and the development and growth of an economy are woven together in a fabric of change. The framework defines the restless nature of modern capitalism, its incessant capacity to transform itself from within in a continuous process of creative destruction. Capitalism can never be stationary. Its dynamic and transformative powers have marked effects across time and space. The chapter explores the foundations of this perspective; it justifies the label of 'restless' for the attributes of this economic system; it traces the consequences for the development of that economy and its connection with the continued growth in standards of living.

The metaphor of restlessness suggests a search for improvement and a sense of discomfort and uncertainty about how the economic order and the position of individuals within it will evolve. Foremost among these attributes are the connection between the triad and the growth of knowledge, the evolutionary nature of a competitive process grounded in innovative rivalry, and the complementary role of market and non-market processes in shaping innovation and the economic response to innovation. Among these non-market processes are those defined by innovation systems. As we witness at every level of aggregation in any modern economy, the institutional framework of modern capitalism is unique in its self-organization and self-transformation of economic activities. In short, a modern economy is incessantly evolving into a new economy; there is always an edge of modernity to the system so that, over extended periods of time, continuity can only be found in process and broad institutional form, not in morphology and activity. What is significant about the current phase of capitalism is that this process is spreading: an increasing number of economies are taking on the attributes of restless capitalism. China and India are leading examples, following the successes of S. Korea and the other 'tiger economies;' they will no doubt be followed by others on the edge of modernity, such as Brazil, or the transition economies of the former Soviet bloc (Amsden 1989, 2001). The emergence of new capitalisms can only accelerate change in the established Western economies, which will have to find new sources of comparative advantage to sustain their relatively high standards of living. That is to say, an innovative response will be required to the challenge of new international competition. But to say this is simply to recognize the force of history (Freeman and Louçã 2001; Landes 1998; Mokyr 1990, 2002).

It is manifestly the case that the economic world of 2004 in the US or Western Europe bears little resemblance to that of 1960 and even less to that of earlier times. As new products, made with new methods and new kinds of material and energy, have appeared and old ones disappeared, the entire pattern of production and the nature and location of resource allocation have changed. The consequent changes for the balance of urban and rural life have transformed these societies economically, socially, and politically in ways that no one could have foretold: patterns of resource allocation become radically different, the activities and economic ways of life of consecutive generations bear little resemblance to each other, and patterns of consumption include practices and purchases undreamt of by earlier generations. This is clearest in the increasing use of inanimate energy and the development of new material bases for production. As new professions and skills emerge, the division of labor changes dramatically; the share of labor directly engaged in agriculture declines to minuscule proportions, and the majority of the population comes to be classified as engaged in service activities. Even as recently as 1960, would many have imagined that the desktop computer would be almost as ubiquitous as the television in the households of a modern economy? That the occupation of computer programmer or software engineer would appear in a census of employment? Would anyone have foreseen the potential in global positioning systems for the logistical management of transport activity, or the prospect that disease might be defined at the level of an individual, as knowledge of genetics expands? Who would have foreseen that the film camera would have its market radically cut by the application of digital computer technology, or that this would enable the processing of images to move back into the home and away from specialized suppliers of processing and printing services? Who in the 1930s would have foreseen the role of the television in destroying the cinema industry, or would have imagined the effect of the refrigerator on patterns of household living? Few modern homes are lit by coal gas; not so in 1910. A negligible proportion of the population today works directly on the land; not so in 1870. Very few make the trip from Europe to New York or Australasia by ocean liner; not so in 1950. At the level of individual firms, household names come and go, as their underlying activities lose their competitive rationale. Only firms such as those involved in distribution and oil production, which retain a highly diversified character or supply inputs fundamental to the operation of the economy, show any sense of permanence. The record in this long-term perspective appears to be one of radical qualitative discontinuity, such that any comparison of a single economy over extended time is fraught with difficulty. But it takes place in the context of broad economic continuities, of which the most significant is the steady increase in per capita real income, at least insofar as we can trust these measures to reflect the underlying reality. For aggregate growth never happens without development, and structural transformation never occurs without innovation and the ongoing radical redevelopment of the economic structure: thus, economic change is always uneven within and between countries. The characteristic feature of such dimensions is their connection with new ways of living, new patterns of organization, and new ways of knowing. This was Schumpeter's great insight in his *Theory of Economic Development*, first published in 1912.

To understand this process, we need to understand the role of innovation as the transforming element in the economic picture. This—or rather, innovation in the context of an evolutionary competitive process—is the focus of this chapter, because innovation is the stimulus to and consequence of competition. This process is played out in multiple contexts whenever rival alternatives are subjected to selection according to their attributes and the questions asked of them by the selecting environment. More fundamentally still, as Loasby (2000, 2002) points out, in any modern economy there are multiple selection environments and ultimately they make choices between rival beliefs and ways of thinking and acting. We think of this most clearly when companies compete over different business conjectures within and between the rough classifications that we consider sectors of activity and market segments. The same processes are observed at regional, national, and international levels, as if the economic fabric is continually rewoven in new thread-counts, new colors, and new subtleties of composition.

But the more we increase the level of aggregation at which we represent this dynamic texture, the more we hide the evidence for innovation-based competition and development and the nature of the underlying dynamic processes. A macroeconomic approach cannot provide a basis for understanding the processes of innovation-based competition and adaptation in the allocation of resources. That understanding can only come through an analysis of innovation within firms, the supporting innovation processes, and the resolution of those innovations into changing patterns of resource utilization. As Schumpeter (1928: 378) put this fundamental point, 'what we unscientifically call economic progress means essentially putting productive resources to uses *hitherto untried in practice*, and withdrawing them from the uses they have served so far' (emphasis in original). What is at stake here is the central role of novelty generation in modern capitalism: that its market and non-market institutions seem designed not only to achieve self-organization of the activities currently known, but their self-transformation. The order or pattern of economic activity achieved by instituting processes like markets is transient and changes from within—it is both established and changed by the same processes. The central underlying reason behind this claim is not that capitalism is a knowledge-based system, for all economies are necessarily such, but that it is a particularly fertile knowledge-transforming system. Most fundamental of all, its instituted form is directed to the conduct and absorption of innovation.

Central to this view is the corollary that the economic world, in the sense used by Knight (1921) and Shackle (1972), must be uncertain: innovations are unforeseen events; they are based upon fallible conjectures and, consequently, not all innovations can be validated successfully in the market place. Those that are lucky conjectures may have an unpredictable economic life, and vary enormously in economic effect over time. The recent evolution of the mobile phone industry, or the camera industry, or the pharmaceutical industry provides countless examples of innovation-based competition in which there is unpredictability. This is obviously true with radical innovations, but it applies just as well to incremental innovations: their broad outlines may be foreseen by knowledgeable insiders, but as unforeseeable as with the most radical of ideas is which firms will introduce them, when they will appear, and in what precise form. Thus it was no doubt obvious to some that the combination of electronics and photography would redefine the camera, a possibility that the Polaroid Corporation apparently missed; but was it so obvious that the combination of camera and mobile phone was also a latent, real possibility? Since true innovations are singletons, this conclusion should come as no surprise. Uncertainty reflects partial knowledge, but it also makes possible access to new knowledge. Any knowledge-based system can only evolve if it is open; that is, if the solutions to prevailing problems serve to define unforeseen new problems. Closure means stagnation, an end to learning, an end to intelligent action, the denial of imagination, the end of history. This is why our understanding of modern capitalism is ultimately grounded in individual action, for it is in the minds of individuals that new problems are identified and posed. Progress is inherently micro, but its resolution is inherently systemic, and these systemic responses are hierarchical.

Three qualifications to this view follow. First, it does not mean that we are uninterested in aggregate measures of economic growth or their connection with rising levels of human welfare. Instead, the task is to understand macro-performance as the emergent consequence of a multitude of interdependent micro-processes. Macro-productivity growth, for example, is the outcome of countless changes in individual productive activities, each change of minimal consequence for the overall economy, amplified by processes of imitation, diffusion, and competitive growth into measurable, economy-wide effects (Field 2003). The system is constantly adapting to new opportunities and, in the process, defines new opportunities for further innovation and structural development.

Second, it is a restless system, because innovation is restless—and innovation is restless because knowledge is restless. Thus we shall argue that an economy of this knowledge-based kind can never be in equilibrium without destroying the foundations of the change processes on which its performance is premised. Only if knowledge were in equilibrium, whatever that may mean, could we expect modern capitalism to be in equilibrium. It turns out that the hypothetical stationary state is a thoroughly bad place to begin the study of the market economy, for once a system is in equilibrium it can only change through the imposition of forces not included in the list that accounted for the establishment of equilibrium. If we are to take seriously the role of knowledge and learning in transforming economic life, we have to recognize that much of that knowledge is created within the economic system and associated processes, and that this is the fundamental reason why an economy is never in equilibrium. It is always discovering from within reasons to change the prevailing allocation of resources (see Chaminade and Edquist, this volume) and each discovery is an addition to the total of knowledge contained and distributed across the economy. As Schumpeter told us many years ago, capitalism in equilibrium is, in this view, a contradiction in terms.

Third, since innovation entails the acquisition of new knowledge, we need to be clear what is meant by knowledge, and the processes by which it is generated and diffused. Knowledge has a unique property: it always and only

ever exists in the minds of individuals, and it is only in individual minds that new innovative concepts and thoughts can emerge. It is why we recognize the entrepreneur and the prize-winning scientist (they are different as individuals) and from it follows the fact that knowledge is always tacit; it is never codified as knowledge. What is articulated and codified is information. But information is only ever a public representation of individual knowledge; it is sometimes a virtually perfect representation, but in many significant cases it is not.

As Michael Polanyi (1962) expressed it, we know more than we can say and can say more than we can write. Since economic activity within firms and beyond depends on the ability of teams of individuals to coordinate their actions, it follows that processes must exist for correlating the knowledge of the individual members so that they understand and act in unison. For innovation, the internal organization and business plan of the firm are the primary means of coordinating the flow of information and correlating individual knowledge into the necessary hierarchy of understanding and actions. It may be helpful to conceive of the organization of a firm as an operator, a local network of interaction through which the knowledge of individual members of the firm is combined to collective effect (Leonard-Barton 1995). This spread of understanding in correlated minds is essentially a social process premised on human interaction. However, a chief consequence of information technology is that information can be communicated at a distance; this makes it possible for the firm to be included in wider, less personal networks, including the scientific and technological networks that communicate almost exclusively in written form. But to call these 'knowledge networks' is not helpful. They are information networks, perhaps better expressed as 'networks of understanding,' and their significance is in shaping what individuals in firms and other organizations transmit and receive as information. It is not that information is transmitted with error. It may be, but what matters is that information may legitimately be 'read' by recipient and transmitter in different ways. The interpretation of the message is in the different minds of the parties concerned (Arthur 2000). Indeed the growth of knowledge depends on this possibility of divergent interpretations of the information flux. All innovations, like scientific breakthroughs, are based on disagreement, on a different reading of information largely available in the public domain. Thus, the prior state of knowledge influences what is 'read' and what is 'expressed' and, as Rosenberg (1990), made clear, firms have to invest in their own understanding if they are to participate effectively in innovation information networks. This is why it is necessary for them to conduct their own R&D.[1]

Thus while information is a public good, in the sense of being usable indefinitely, it is not a free good; scarce mental capacity must always be engaged to convert it to and from private knowledge (Cohendet and Meyer-Kramer 2001; Witt 2003). Here we find one of the principle sources of variation in the innovation process: innovations are conceived in individual minds, and these minds differ. It needs only a moment's reflection to recognize that if all individuals held the same beliefs, there could be no growth of knowledge, no innovation, and thus no emergence of the beliefs in question in the first place. Idiosyncrasy, individuality, and imagination are the indispensable elements in the innovation process, and the way innovation policy is framed must recognize this fact; indeed, without them entrepreneurship would not be recognizable.

The obvious corollary for the policy or strategy process is that innovation cannot be planned from on high; it emerges from below. It follows that the institutions and organizations that manage the storage and exchange of information are thus crucial aspects of modern capitalism, and they are a mix of market and non-market arrangements. Nowhere is this perspective clearer than in the discussion of enterprise: enterprise depends upon believing that the economic world can be better organized, and on implementing those beliefs through innovation. Of course, many of the innovation conjectures are proved false and

are rejected in the market, but a sufficient number turn out to be correct, and it is on these that economic adaptation and the long-term growth in resource productivity ultimately depend.

Despite the primary role of disagreement and the entertainment of divergent expectations in economic progress, modern economic systems are not chaotic; they are highly structured and ordered, and that order comes from the working of the instituted, interdependent frameworks of market and non-market processes. Markets and other institutions produce order and stability in patterns of economic relations that can be replicated over time. Thus institutions play vital and complementary roles in framing the conditions for the growth and application of new knowledge without which standards of living could not have developed beyond the most rudimentary level. Adam Smith was the first economist of note to see this connection: his famous division of labor is a pattern of order contingent on the extent of the market and a process for acquiring new knowledge through specialized and focused learning processes that generates changes in the extent of the market (Richardson 1975). In the modern economy these learning processes have been highly refined, with the emergence of specific knowledge-generating activities not only in the R&D activities of larger firms but in the laboratories of universities and public research institutes. In turn, these activities depend on methods for instilling the individual capabilities through education and training that we summarize in the phrase 'human capital.' This 'knowledge sector' is, of course, embedded in the wider economy, being dependent on it for the resources it absorbs and for the directions in which funding takes new knowledge.

It is this paradoxical feature of knowledge-based capitalism that this chapter explores: through market and non-market types of instituting processes it is highly ordered and coordinated; these same ordering processes give rise to the opportunities that self-transform the existing order. Order is not equilibrium. In making this statement, we examine perspectives on competition, economic growth, and development and innovation that are fundamentally different from those associated with the economic thinking of the dominant orthodoxy. Order is transient, evolving rapidly in some directions, slowly in others; the structural changes that ensue are the context in which the growth of knowledge is stimulated to further redefine economic possibilities. The view we shall expand upon is inherently evolutionary in two senses: first, of cumulative unfolding; second, of change that is premised on processes of variation, selection, and innovation (Nelson and Winter 1982; Dosi 2000; Metcalfe 1998; Saviotti 1996; Mazzucato 2000; Witt 2003). Through the endogenous growth of knowledge, evolutionary processes transform established orders from within; they are the processes that have characterized humankind since earliest antiquity and the first agricultural settlements, and have culminated in the market capitalist economies of today. Innovations in the use of inanimate energy and materials are at the core of the innovation record, together with complementary changes in organization of productive activities. The growth of fundamental understanding of natural phenomena and human-made devices grew slowly at first; technology was thus for many centuries primarily empirical in its foundations and constrained in its diffusion across space and time by the limits on information technology that prevailed until the innovation of the printing press of 1453. This momentous technological discovery marked the beginning of the modern age: from then on, the growth of knowledge was rendered evolutionary. Now ideas could be stored more reliably; they could accumulate, they could be transmitted independently of the mobility of the originating mind and more rapidly to many minds; they could therefore stimulate the further growth of ideas in dissenting or reinforcing fashion.

Thus came into being the foundations of restless capitalism in which the market process harnessed the growth of knowledge to the pursuit of competitive advantage. However, this process is viable only if new beliefs can displace established beliefs, in enterprise as in science. Therefore, it is crucial that economic institutions permit the prevailing pattern of activity to be invaded by innovations. If the system had been evolutionarily stable, the great transformation in living standards over two millennia would not have occurred. Perhaps surprisingly, it is the ordering properties of the market process that stimulate and allow invasion; stability and challenge are inextricably linked. Thus restless capitalism is marked by evolutionary instability. Broad material progress follows, but at the cost of much distributed pain. Hopes are falsified, skills rendered obsolete, locations rendered uneconomic, communities disrupted. These costs of economic transition are not to be underestimated and must qualify any discussion of evolutionary progress (Witt 1998).

To summarize, all economies are knowledge based, a human attribute; they are all information based, a social attribute; consequently, they are all developing economies, an evolutionary attribute. But what is unique about modern market capitalism is the extent to which knowledge and information are harnessed to change the economic order from within. This is the core of Schumpeter's great legacy to the study of innovation—its cause and significance. In the rest of this chapter we provide a more detailed exegesis of these introductory remarks, beginning with innovation and enterprise, progressing through the accumulation of knowledge to competition and the connection with growth and development, and ending with the interaction between market and non-market arrangements in the innovation process.

Innovation and enterprise

One reason why innovation is such a problematic concept is the variety of forms that innovation can take (see Chaminade and Edquist, this volume). The familiar categories of product and process innovation extend to services as well as manufacture, to non-market organizations as well as market-oriented firms. They include innovations in organizations and market processes, and they are impossible to list in their entirety. Innovations that are defined within product or process architectures (Henderson and Clark 1990) or that are modular (Langlois and Robertson 1992) or that are competence destroying or competence enhancing (Tushman and Anderson 1986), are just the most prominent of recent attempts to classify innovations. Taxonomy is important, but in the current context there is merit in thinking not only in terms of what innovations are but of what characterizes innovation processes.

Innovation is, first and foremost, a matter of experimentation, normally but not only business experimentation, the economic trial of ideas that are intended to increase the profits or improve the market strength of a firm, or improve the performance of a non-profit organization such as a hospital or bureaucracy. In this regard, innovation is the principal way that a firm can acquire a competitive advantage over its business rivals. As a process of experimentation, a discovery process, the outcomes are necessarily uncertain; no firm can foresee whether rivals will produce better innovations; nor, even when all technical problems are solved, can it know in advance that consumers will pay a price and purchase a quantity that justifies the outlay of resources to generate a new or improved product or manufacturing process. As pointed out above, this is not a matter of calculable risk: probabilities cannot be formed for events that are unique, or that change the conditions under which future events occur. There is an inevitable penumbra of doubt that makes all innovations blind variations in practice, and the more the innovation deviates from established practice, the greater the fog of irresolution. Perhaps the fundamental point is that innovations are surprises, novelties, truly unexpected consequences of a particular kind of knowledge-based capitalism. This does not mean that innovation is irrational behavior: firms are presumed to innovate in ways that make the most of the opportunities and resources at their disposal. However, neither the

opportunities nor the resources available can be specified with precision in advance. Innovation is a question of dealing with the bounds on human decision-making; it is so much a matter of judgment, imagination, and guesswork, and the optimistic conjecturing of future possible economic worlds. Consequently, for example, innovative business or public strategies must be subject to the same penumbra of doubt about their effects on the innovation process; there will be unanticipated consequences of innovation policy and strategy, and great difficulty in evaluating cause and effect.

The second attribute of innovative activity is that it is embedded in the market process. Not only do firms innovate to generate market advantages over their perceived rivals, so that the functioning of markets shapes the return to innovation, but market processes influence the outcomes of innovation and the ability to innovate. Essential market process determinants of innovation activity are the way that users respond to an innovation and the ability of a firm to raise risk capital and acquire skilled labor and components necessary to an innovation. The fundamental test for successful innovation is not that it works, but that it is profitable *ex post*. This is a matter of market process. If markets are inefficient and distorted, this can only harm the innovation process; when incumbents and conservative users unduly control the relevant markets, the effect will be similar.

The third implication for the innovation process is that the systemic, emergent nature of group understanding leads directly to the basis of innovation systems (see Meeus and Faber, this volume). There is an increasingly elaborate division of labor in the generation of knowledge; to use an old economic concept, the division of innovative labor is becoming increasingly roundabout in nature. Since Adam Smith, scholars have recognized that the knowledge contained in any economy or organization is based on a division of mental specialism. It is not simply that the division of labor raises the productivity of the pinmaker; it also raises the productivity of the 'philosopher and man of speculation' and greatly augments the ability to generate knowledge in the process. When this division of labor is not contained within the firm, we have the conditions for an innovation system to emerge and the need to coordinate the various minds within that system. Innovation systems are the necessary consequence of this division of knowledge. They do not arise naturally; they have to be organized and are not to be taken for granted; from their perspective, their self-organization processes are a central concern. Innovation systems are, in Hayekian terms, a form of spontaneous order. Perhaps the most obvious characteristic of modern economies is the distributed nature of knowledge generation and the consequent distributedness of the resultant innovation processes across multiple organizations, multiple minds, and multiple kinds of knowledge (Coombs *et al.* 2003; Edquist *et al.* 2004). As a system, what matters are the natures of the component parts, the patterns of interconnection, and the drawing of the relevant boundaries. Each of these aspects forms a dimension of innovation practice.

Fourth and finally, it is helpful to group the factors that influence the ability to innovate into four broad categories: perceived opportunities, available resources, economic and other incentives, and the capabilities to manage the process. Innovation depends on the articulation of each of these elements and this largely defines the task of enterprise. Thus, for example, increasing the resources devoted to innovation is likely to run into rapidly diminishing returns if new market opportunities are not perceived, or if the management of innovation is weak and poorly connected with other activities in the firm. To understand the connection between innovation, competition, and their wider consequences, we need to place enterprise in its proper context, because the entrepreneurial function cannot be separated from the instituted structure of the economic system in which it is carried out. Its nature and consequences are embedded in the wider system of market and non-market economic institutions. The prevailing features

of a market economy produce a particular spectrum of entrepreneurial activities; in a different set of institutional arrangements—say of labor-managed firms, or of stakeholder capitalism—the entrepreneurial spectrum will take on a different hue, because those systems give different meanings and content to entrepreneurial activity and provide different incentive systems from shareholder capitalism (Adaman and Devine 2002). The rules for creating new business enterprises and for eliminating failing ones are particularly important in determining the enterprise characteristics of any economy.

What, then, are the instituted features of modern capitalism that create such a strong symbiosis between innovation, competition, and entrepreneurship? There are four. The first is the institution of the open market, in which every established business position is open to challenge, unless protected via a patent, copyright, or other limitation. If we see competition not as a state of affairs graded by the structure of the market but as a dynamic process of rivalry and struggle for a share of the market, then entrepreneurial activity is both necessary and sufficient to create competition. Provided the business idea is good enough and incumbents do not create sufficiently onerous, artificial barriers to entry, the general rule is that any market can be invaded. Indeed, competition authorities in the advanced economies spend a good deal of time preventing incumbent firms artificially closing off their markets to entry; and because any entrant incurs costs, there will usually be some compensating entrepreneurial advantage in the product design and quality, the method of production, or the scheme for distribution to customers that puts the incumbent at a disadvantage and which helps circumvent entry barriers. In this sense, entrepreneurship is pervasive because the idea of an open competitive market process is pervasive. A firm never quite knows where the threats to its existence will come from; frequently they come from such unanticipated directions that their significance is discounted until it is too late.

Second, markets play fundamental roles in relation to the incentives and rewards for entrepreneurial behavior. The prevailing market-based valuations of products and productive services allow the prospective entrepreneur to gauge the potential profitability of a new venture by virtue of its having to fit into the current pattern of activity. Market signals matter not only in the sense of encouraging the efficient use of existing business knowledge, the traditional argument in favor of the competitive organization of industry, but also in the deeper sense of guiding the competitive process of entrepreneurial change. Without market prices *and* quantities, no entrepreneur could judge that a business conjecture is potentially viable: he would be doubly blind, not knowing the plausibility of either the quantity conjectures or the value conjectures on which the plan depends. Markets generate this information and thus connect new beliefs with existing patterns of resource allocation. Even if the true margins of competition are initially misconceived and revealed in surprising ways *ex post*, all entrepreneurial conjectures compete with some existing activity. Notice that this remains true even for those radical entrepreneurial conjectures that introduce products or methods of production previously unheard of. Even these innovations must be conjectured to displace existing activities in consumers' expenditure and/or to utilize resources employable elsewhere in the economy. Existing market relations allow us to foresee the starting point for the entrepreneurial process, even if the radical reconfigurations of demand and reallocations of resources that flow from truly radical innovations cannot be envisaged.

Third, markets are instituted devices for generating low-cost access to consumers and productive services. Markets are both structures for indicating the terms on which resources are available and the channel to gain access to those resources. Open markets for skilled workers and for free capital, for example, are essential to an entrepreneurial economy; without them, the possibility of entrepreneurial behavior will be greatly circumscribed. Thus, there is a close correspondence between the institutions of the market

place and the spectrum of entrepreneurial behavior it engenders and supports. Consider, for example, the institution of the patent right. Patents provide important incentives to entrepreneurs because they protect a market opportunity for a circumscribed period. They protect against the narrow imitator who merely seeks to copy a novel idea and to free-ride on the imagination of others. But this protection is not absolute. In principle, any patent can be 'invented around;' indeed, the requirement that a patent be published indicates to potential inventors exactly what 'inventing around' would mean. The entrepreneur who bases a rival business on a different novel idea may thus destroy the economic basis of an established patent. Patents are an extremely clever institution; their protection is important but intentionally not unlimited, and the protection they afford is helpless in the face of other genuinely novel entrepreneurial actions. The crucial step is that patents come at a price for the inventor—the requirement to place information that is a fair representation of the invention in the public domain.

This takes us to the fourth aspect of the institutions of a market economy: the incentives they provide for entrepreneurship. Whether or not profits are the *primum mobile* for the entrepreneur, there can be no doubt that those profits are a necessary feature of such activity, and that their prospect is essential in the process of attracting risk capital to support conjectures for which there cannot be any basis in fact. Novelty may be its own reward, but is also the signal that what the successful entrepreneur does is economically superior to established competing activities. Abnormal profits, far from being an index of the absence of competition, are the very proof that competition is actively pursued, that resources are being reallocated. This is the crucial role that profits and losses play in the mobilization of changes in economic structure. By focusing on competitive equilibrium, we hide this from view.

A final, brief point relates to the wider significance of entrepreneurial activity in pointing to the particular mechanisms of economic change in modern capitalism. The fundamental issue here is that economic growth is never a steady advance, with all activities expanding at the same rate, as the prominent, aggregative theories of economic growth would have us accept. Thus, scale apart, one year is identical to the next, and it makes no difference whether growth is positive or negative. By contrast, we know that growth always follows on from development, from changing the economic structure quantitatively and qualitatively. Not only do activities change in relative importance, their absolute scale changes unevenly: while many grow absolutely, others decline absolutely; it is the diversity of growth experience that marks restless capitalism. The other side of the creative, entrepreneurial coin is that activities disappear from the economic scene—we cannot put resources to new uses without scaling down the old uses. In short, a theory of decline in economic activities is an integral part of any useful theory of economic growth. Intuitively, or at least with a reasonable knowledge of the history of the last two centuries, structural and qualitative change seem to be inseparable from the economic process. However, entrepreneurial activity is not simply about change, even in the general sense referred to here: entrepreneurial change refers only to change that arises from within the economic process, change that it is stimulated and enabled by the institutions of the modern market economy. Change of this kind is a non-equilibrium phenomenon; it cannot be understood by the methods of comparative statics or dynamics, for these always refer to the consequences of changes that arise from outside the economic system. We recognize again Schumpeter's insight that entrepreneurial-led change is based on a process of the internal self-transformation of the economic system. This process may be impossible to pin down in its details: it may be truly open-ended, historical, and entirely unpredictable in its effects. To understand the basis for this argument is an enormous challenge but, unless we make the effort, the role of the entrepreneur will remain elusive and, worse, marginal to economic thinking (Metcalfe 2004). Even more, we will never come to

understand the process of economic development, or why it is so unevenly distributed around the globe, and thus comprehend the reasons behind several of the major moral issues of our times.

Competition as an evolutionary process

We turn now to the role of competition and its dual relation with innovation processes. Competition links together innovation and the ongoing transformation of the economic system; it creates strong incentives to acquire practical knowledge and put it to productive use; it is thus the process on which depends the growth in productivity and standards of living that mark modern capitalism. However, this is not the static competition of the textbooks, the idea of competition as a state of equilibrium described by the number of competitors; rather, as Adam Smith well recognized (Richardson 1975), it is a process of changing economic structure premised on the differential behavior of rival firms. The contrast between competition as a state of market equilibrium characterized by a structure and competition as a process characterized by a rate and direction of change is scarcely new, but it is fundamental to drawing the connection between the growth of knowledge and economic progress.

Even leaving aside the fundamental contributions of Schumpeter and Hayek that competition is an entrepreneurial discovery process—that is to say, a process for learning how better to satisfy human need—other economists of note have indicated forcefully that all is not well with the idea of competition as a state of equilibrium. Thus J. M. Clark (1961), Brenner (1987), and Knight (1923) have explored the non-equilibrium aspects of the competitive process, arguing that competition is a process of rivalry and, for rivalry to be meaningful, the competitors need to be different. It is this recognition of the fundamental importance of variation that connects us immediately to the evolutionary foundations of competition, development, and the growth of knowledge.[2] Competition's character of a contest, a process of rivalry in conjecture between differentiated contestants, carries the implication that the outcome of the competitive process is open and unpredictable. An evolutionary perspective shows how these processes are connected to shifting patterns of resource utilization and the growth of productivity; they also help explain, for example, the presence of abnormal profits in the face of intense competition (Fisher *et al.* 1983).

To understand innovation-based competition, it is useful to divide the forces at work into two broad categories: variation and differentiation of behavior, and selection across those behaviors. In turn this leads to the central theorem of competitive capitalism—that more productive activities expand and absorb resources at the expense of less productive alternatives and, in the process, give rise to economic growth and development. This is the core of the line of analysis that, through innovation, connects the growth of knowledge to the growth in material welfare. The two principal sources of variation are innovation within the activities of established organizations and the entry of new organizations typically based on some innovative variation; the two principal forms of selection are competition within market environments and the demise of those activities that can no longer pass the test of economic viability. Thus innovation, entry, competitive selection, and exit are the markers for an active competitive process. As Hayek noted, 'to compete' is a verb, and verbs are action words; if nothing changes, there is no competition (Hayek 1948). Two important corollaries follow. The first is that competition involves decline as well as growth, the incorporation of new activities and the disappearance of former ones; we cannot understand economic growth and development without grasping that decline and disappearance occur simultaneously with expansion and entry, which surely is the point of creative destruction. Second, the process of competitive rivalry concentrates resources in the most competitive rivals and in the process destroys

the rivalry that drives competition: as evolutionary biologists put it, the process consumes its own fuel. Innovation is the way in which that variety is regenerated: it is not an optional extra to the analysis of competition, but essential to the maintenance of competition. Moreover, the rate and direction of innovation is inseparable from its market context, and the relation between market and non-market elements essential to the growth of knowledge. These processes operate within and between firms and industries, and within and between nation states; they are the forces that continually redefine and change the layout of the economic furniture.

The market contexts in which competition is played out may have multiple attributes. Differentiating market environments are the frequency with which competitive interactions occur, the degree to which the market participants are well informed about rival offers to buy or sell, the degree to which there are segmented niches, and the degree of regulation of the market process through law, administrative edict, and custom. Market processes in the narrow sense are always intertwined with wider regulatory frames that indeed may co-evolve with the development of the market.

As with all modern evolutionary approaches, the appropriate framework for a competitive process is based on the population method, for it is in populations that adaptation takes place in response to selection working on material provided by variation of relevant characteristics. A sketch of the underlying ideas will be useful at this stage. At the most primitive level, these populations are defined in terms of sets of activities—plants operated by firms that are competing for custom and resources in broadly similar ways. Each plant and firm combination has a set of competitive characteristics, those dimensions of its activity that are causally related to differential profitability, the growth of the capacity of the operating firm, and its differential innovative performance over time. These are the characteristics that underpin the competitive process and, when evaluated by the market environment, lead to the different levels of profitability associated with the operation of each plant. Thus what makes the different plants/firms part of the same population is their experience of the same processes of market evaluation. A firm that, for example, produces multiple products for different markets will be operating in different populations, and its overall performance will be an average of performance across the different market contexts. What is important to recognize here is that firms are not malleable entities capable of adapting instantaneously to any change in their environment. Rather, as organization theorists have frequently pointed out, they are subject to considerable inertia, so that the differences between them have a durable quality (Hannan and Freeman 1989).

For any firm, there are three broad groups of competitive characteristics, causal explanations of the multiple dimensions of firm behavior, to be brought into play:

- characteristics that causally influence the productive activity of the firm and determine its productive efficiency, its cost structure, and the quality of the products it produces. Some of these are grounded in matters of technology, but others are matters of organization, capability, and workplace culture—the internal rules of the game or the bundle of routines that determine how each individual member of the firm interacts with the others, together with the objectives motivating behavior at all levels (Nelson and Winter 1982);
- characteristics that causally influence the ability of the firm to expand its productive capacity in the population through processes of investment in physical and human capital. These relate to questions of motivation and ambition and the ability to manage change processes, as well as the ability to access the free capital to finance investment programs through retained profits of borrowing on the capital market;
- characteristics that causally influence the ability of the firm to innovate in terms of technology or organization and thus to alter the first set of characteristics above. All innovation presupposes a growth of

knowledge, and this will depend on the firm's ability to access external knowledge as well as the effort it devotes internally to support and manage innovation.

Under the rules of the capitalist game, the first cause of selection is profitability; this accounts for differential survival of plants and firms, the incentives to enter a market population, the differential growth of surviving plants and firms, and, in part at least, their differential innovative performance over time. Two causes underpin differential profitability at each point: the product quality and process efficiency-related characteristics of the rival firms, and the manner in which they are evaluated by the market environments for output and inputs in which any plant necessarily operates simultaneously. In an evolutionarily efficient market environment, similar characteristics are given similar values; thus, high profitability correlates positively with the production of better products by more efficient methods of production. If the capital and labor markets are efficient, high profitability correlates positively with the resources to expand more rapidly and take market and resource share away from less able rivals, while the least efficient operators are eliminated from the process. Finally, if markets for innovation work well, more creative firms are being founded that challenge the position of the incumbent market leaders.

Population analysis provides a rich approach to the dynamics of competition. Any individual firm is defined in these three dimensions and the characteristics in each dimension may not be independent, so creating multiple trade-offs between efficiency, investment, and innovation. Thus the ability to innovate may require internal processes that do not fit easily with other organizational rules and routines directed at ensuring the firm's efficiency (Nelson and Winter 1982; Meeus and Oerlemans 2000). As in all evolutionary argument, what matters is that the firms differ in at least one of the three dimensions, that there is variety in behavior on which selection can do its work. Even in this much simplified account, competition is multidimensional and cannot be reduced to behavior in the first set of characteristics. For example, it is not unknown for a firm to possess excellent technological or organizational capabilities that give it large cost advantages in its population, and yet for the owners of that firm to refuse to grow and invest to capitalize on these advantages. Similarly, there are firms that have excellent technology but fall behind through an unwillingness or inability to innovate as effectively as their rivals, eventually consigning themselves to history. It also follows that competitive advantage is not an intrinsic characteristic of any firm but an emergent consequence of the market populations in which it competes and of characteristics that differentiate it from the (changing) population of rivals. While the competitive characteristics of firms are often subject to considerable inertia, competitive advantages are typically transient, the inevitable consequence of the restless growth of knowledge that opens up the possibility of challenges to established activities.

As soon as we recognize that competition takes place in the three dimensions listed above, much of the record of business rivalry falls into view. So does the accuracy of Schumpeter's (1943) *mot juste* that it is a process of creative destruction. Moreover, the concept of the market environment involves much more than the role of product and factor markets in evaluating the current distribution of activity. It extends to the capital markets and the supply of finance for investment and innovation as well as the markets for skills so important in relation to innovation.

The wider-instituted context to competition and innovation

In this final section, we turn to a familiar theme in institutional economics: that markets are contingent on wider-instituted frameworks that govern the scope for human interaction—for example, in relation to the definition of property rights, the notion of

contract, and the rule of law. These instituted rules of the game are a fundamental basis for economic action and the allocation of claims on the product of the economic system. From this point of view, the order established in a market process depends on strong property rights in relation to the ownership and use of assets. Made clear less frequently is the corollary that the dynamics of the system depend on the weakness of property rights in terms of economic values. The owner of a business can be certain as to who has claims of a contractual or residual nature on the human and non-human assets employed, but there is no legally enforceable guarantee of the economic value of those assets as determined by the balance between revenue and cost streams. In a restless system, all such values are transient and reflect the fact that every economic activity is open to challenge from rivals with different approaches to the activity in question. Thus what is interesting about the institutions of capitalism is their openness to change, providing a context that creates incentives to discover better ways of using resources and makes it possible for innovation-based challenges to the established order. Property rights in inventions illustrate this point well. A patent is a right to control the exploitation of an invention, not a right to determine the value of that invention which is market contingent. Pure imitation can be prevented or redress sought in the courts, but this provides no protection at all from a rival invention based on different principles. In such a case, all the rivals can do is compete, which is one reason why the broader the scope of patents, the greater the risk to the competitive dynamic. Similarly, the rules about non-viability of economic activities are a crucial part of the instituted rules of market selection. The definition of a bankrupt entity, the rules of insolvency, and the administration of the associated assets are crucial aspects of the competitive rules of the game. These rules are often overridden if the business in question can be subsidized from private or public sources, but such subsidies usually work to reduce the efficacy of the evolutionary process and, if applied too broadly, can impose substantial long-run burdens on an economy. Markets matter in relation to enterprise and innovation in other ways too. In particular, the market for corporate control, the market in which business units are traded, is of fundamental importance to the idea of business experimentation and its connection with competition. The uncertainties of the innovation process mean that capabilities that have been mistakenly acquired need not become sunk costs, but can be transferred to a firm where their fit with the business strategy is better. Equally, a firm can use this market to acquire ready-assembled capabilities that would otherwise be beyond its competence to accumulate within its innovation time-frame. Thus the rules and costs associated with the trade in bundles of capability greatly influence the rate of business experimentation. The rules on mergers and acquisitions are another dimension of this same set of innovative influences.

These rules of the game imply that capitalism has evolved a set of institutions, the significance of which resides in their stimulation of the process of innovation-led competition. Usually enshrined in commercial law or practices of the public regulation of business, they underpin the experimental nature of restless capitalism. Thus competition authorities are important institutions in most modern economies; their role should not be interpreted in a narrow antitrust sense (in which regulation is interpreted as a way to impose perfect competition), but rather as procedures for keeping markets open and preventing incumbents from placing barriers to the invasion of their market positions. Nor, because there are inevitably strong incentives to limit the innovative capabilities of rivals, should this role be underestimated: its justification is found in the dynamics of the market process, not in the equilibrium analysis of market structure. Conversely, the most effective competition policy that can be designed is a pro-innovation policy that builds on the disaggregated, uncertain nature of the innovation process. This line of enquiry immediately opens up the fact that innovation processes within firms are aided and abetted by an external organization that provides information and other resources to facilitate innovation. Few if any modern firms innovate in isolation; rather they are embedded in relations with a broader texture of innovative agents in supply chains, user communities, universities, and other knowledge-generating, communicating, and storing organizations. In turn, this reflects the continual extension of the division of labor in the production of innovation-relevant knowledge: specialty after specialty is created; they are increasingly complementary elements in innovation; they become increasingly dissimilar and difficult to integrate (Coombs and Metcalfe 2000). Moreover, in relation to user or supplier-chain interactions in pursuit of innovation, these are inevitably grounded in the firm's market relations, so that innovation-related knowledge is created by the market process (see Chaminade and Edquist, and Meeus and Faber, this volume). Within the broader knowledge system, firms may play a unique role in relation to innovation. They are no longer the sole sources of innovation-relevant knowledge, but they remain the only organizations whose unique role is to combine the many disparate kinds of knowledge to practical innovative effect.

We cannot understand the competitive process in markets solely in terms of markets: this is the important contribution made by the literature on innovation systems (Freeman 1987; Lundvall 1992; Nelson 1993; Edquist 1999; Malerba 2004) that points to the complementary nature of market and non-market institutions and processes in innovation. This is not the place to explore this concept in detail, but two remarks are in order. First, it seems important to distinguish an innovation ecology from an innovation system. The former is the set of knowledge organizations and institutions that govern their activities in a given economy. Such an ecology has a strong national definition as well as subnational domains (in regions, for example) and is shaped by national characteristics in relation to law and polity. However, the ecology is not a system. A system requires component organizations to be drawn from the ecology and for those components to be connected in an innovative purpose. Systems are local, connected assemblages around the solution of innovation problems; they are dissolved or redefined as those problem sequences evolve. Thus what is significant about innovation systems beyond the first frontier is their emergent, transient nature, and it is a defining characteristic of restless capitalism that innovation systems are adaptive to innovation problems. The ecologies are, however, less transient: universities, research laboratories, as well as firms are typically far more durable than the innovation systems they contribute to, and necessarily so. Moreover, because many of these organizations must be organized and funded in the public sphere, there is a natural concern of innovation policy to sustain a rich national ecology out of which unknown innovation systems can emerge. In this light, policy for the formation of innovation systems becomes a policy to encourage collaboration and connection in the innovation process and to remove barriers to the self-organization of innovation systems.

The second remark is that, unlike innovation ecologies, innovation systems are increasingly not national: they have strong sectoral domains of definition and they are increasingly put together on an international scale. Thus the innovation systems of modern capitalism transcend national boundaries and are influenced by the policy and institutions of multiple ecologies. This is hardly surprising: fundamental research in science and technology has for long been based on international collaboration and, since 1945, international companies with research facilities in multiple economies have increased greatly in overall economic importance. But this is exactly the point: the innovation systems evolve to support the international nature of much modern business activity and so the search for innovation

advantage in different national markets leads to the formation of transnational innovation systems that support this competitive process.

Concluding remarks

Modern capitalism is a particular kind of knowledge-based economic system, one in which innovation, enterprise, and competition are connected through systems of complementary market and non-market instituted frameworks. These three processes are mutually defining, and together they form the connection between the growth of knowledge and the expansion of material welfare that defines a modern economy. In turn, the growth of innovative knowledge takes place within the market process and transforms the conduct of economic activity on a continuous basis, in an open-ended, unpredictable way. We have claimed that their dependence on selection variation processes that are interacting at multiple levels in an economy makes these processes essentially evolutionary in nature. Only if knowledge were to be in equilibrium could such evolutionary systems be in equilibrium. If this were ever to be so, the system would be irremediably stationary. That it is not—and history tells us so—testifies to the destabilizing institutions of modern capitalism through which market processes establish and simultaneously transform transient economic order. These processes of transformation are in turn shaped by the more widely instituted rules of the game, among the most important of which are the ecologies and systems that support the innovative activities of firms. In short, market and non-market forms are designed to underpin the restless nature of capitalism and to place the process of creative destruction at its core.

Notes

1. It is said that the British system of Industrial Co-operative Research Associations, set up primarily in fragmented industries, failed to raise innovation performance, precisely because their target

firms did not invest in acquiring their own capacity to understand the research and development carried out on their behalf.
2. Compare Joan Robinson's claim to the effect that the business view of competition largely means destroying competition in the economist's static sense (1954: 245). Schmalensee's (2000) assessment of the Microsoft case is along similar lines when he claims that competition in the software industry is 'a winner takes most' process in which strategies that do not exclude competition will not survive.

References

Adaman, F., and Devine, P. (2002). 'A Reconsideration of the Theory of Entrepreneurship: A Participatory Approach.' *Review of Political Economy*, 14: 329–55.

Amsden, A. H. (1989). *Asia's Next Giant: South Korea and Late Industrialization*. Oxford: Oxford University Press.

—— (2001). *The Rise of the Rest: Challenges to the West from Late-Industrializing Economies*. Oxford: Oxford University Press.

Arthur, W. B. (2000). 'Cognition: The Black Box of Economics.' In D. Colander (ed.), *The Complexity Vision and the Teaching of Economics*. Northampton, MA: Edward Elgar.

Brenner, R. (1987). *Rivalry: In Business, Science, among Nations*. Cambridge: Cambridge University Press.

Clark, J. M. (1961). *Competition as a Dynamic Process*. Washington, DC: Brookings Institute.

Cohendet, P., and Meyer-Kramer, F. (2001). 'The Theoretical and Policy Implications of Knowledge Codification.' *Research Policy*, 30: 1563–91.

Coombs, R., and Metcalfe, J. S. (2000). 'Organizing for Innovation: Co-ordinating Distributed Innovation Capabilities.' In N. Foss and V. Mahnke (eds.), *Competence, Governance and Entrepreneurship: Advances in Economic Strategy Research*. Oxford: Oxford University Press.

—— Harvey, M., and Tether, B. (2003). 'Analysing Distributed Innovation Processes.' *Industrial and Corporate Change*, 12/6: 1125–55.

Dosi, G. (2000). *Innovation, Organization and Economic Dynamics*. Cheltenham: Edward Elgar.

Edquist, C. (1999). *Systems of Innovation: Technologies, Institutions and Organisations*. London: Francis Pinter.

—— Malerba, F., Metcalfe, J. S., Montobbio, F., and Steinmueller, W. E. (2004). 'Sectoral Systems: Implications for European Technology Policy.' In F. Malerba (ed.), *Sectoral Systems of Innovation*. Cambridge: Cambridge University Press.

Field, A. (2003). 'The Most Technologically Progressive Decade of the Century.' *American Economic Review*, 93: 1399–413.

Fisher, F. M., McGowan, J. J., and Greenwood, J. E. (1983). *Folded, Spindled and Mutilated*. Boston: MIT Press.

Freeman, C. (1987). *Technology Policy and Economic Performance*. London: Pinter Publishers.

—— and Louçã, F. (2001). *As Time Goes By: From the Industrial Revolution to the Information Revolution*. Oxford: Oxford University Press.

Hannan, M. T., and Freeman, J. (1989). *Organizational Ecology*. Cambridge, MA: Harvard University Press.

Hayek, F. A. (1948). 'The Meaning of Competition.' In F. A. Hayek (ed.), *Individualism and Economic Order*. Chicago: Chicago University Press.

Henderson, R., and Clark, K. (1990). 'Architectural Innovation.' *Administrative Science Quarterly*. 35: 9–30.

Knight, F. (1921). *Risk, Uncertainty and Profit*. Boston: Houghton Mifflin.

—— (1923). 'The Ethics of Competition.' *Quarterly Journal of Economics*, 37: 579–624. Reprinted in F. Knight (1935), *The Ethics of Competition*.

Landes, D. (1998). *The Wealth and Poverty of Nations*. Boston: Little Brown.

Langlois, R. N., and Robertson, P. L. (1992). 'Networks and Innovation in a Modular System: Lessons from the Microcomputer and Stereo Component Industries.' *Research Policy*, 21: 297–313.

Reprinted in G. Raghu, A. Kuramaswamy, and R. N. Langlois (eds.), *Managing in the Modular Age: Architectures, Networks and Organizations*. Oxford; Blackwell.

Leonard-Barton, D. (1995). *Wellsprings of Knowledge: Building and Sustaining the Sources of Innovation*. Boston, MA: Harvard Business School Press.

Loasby, B. (2000). 'Market Institutions and Economic Evolution.' *Journal of Evolutionary Economics*, 10: 297–309.

—— (2002). 'The Evolution of Knowledge: Beyond the Biological Model.' *Research Policy*, 31: 1227–139.

Lundvall, B.-Å. (1992). *National Systems of Innovation: Towards a Theory of Innovation and Interactive Learning*. London: Francis Pinter.

Malerba, F. (2004). *Sectoral Systems of Innovation: Concepts, Issues and Analyses of Six Major Sectors in Europe*. Cambridge: Cambridge University Press.

Mazzucato, M. (2000). *Firm Size, Innovation and Market Structure*. Cheltenham: Edward Elgar.

Meeus, M. T. H., and Oerlemans, L. A. G. (2000). 'Firm Behavior and Innovative Performance: An Empirical Exploration of the Selection-Adaptation Debate.' *Research Policy*, 29: 41–58.

Metcalfe, J. S. (1998). *Evolutionary Economics and Creative Destruction*. London: Routledge.

—— (2004). 'The Entrepreneur and the Style of Modern Economics.' *Journal of Evolutionary Economics*, 14/2: 157–75.

Mokyr, J. (1990). *The Lever of Riches: Technological Creativity and Economic Progress*. Oxford: Oxford University Press.

—— (2002). *Gifts of Athena: Historical Origins of the Knowledge Economy*. Princeton: Princeton University Press.

Nelson, R. R. (ed.) (1993). *National Innovation Systems: A Comparative Analysis*. Oxford: Oxford University Press.

—— and Winter, S. (1982). *An Evolutionary Theory of Economic Change*. Boston: Belknap Press.

Polanyi, M. (1962). *Personal Knowledge: Towards a Post Critical Philosophy*. New York: Harper Torchbooks.

Richardson, G. B. (1975). 'Adam Smith on Competition and Increasing Returns.' In A. S. Skinner and T. Wilson (eds.), *Essays on Adam Smith*. Oxford: Clarendon Press.

Robinson, J. V. (1954). 'The Impossibility of Competition.' In E. H. Chamberlin (ed.), *Monopoly and Competition and their Regulation*. London: Macmillan.

Rosenberg, N. (1990). 'Why Do Firms Do Basic Research (With Their Own Money)?' *Research Policy*, 19: 165–74.

Saviotti, P. (1996). *Technological Evolution, Variety and the Economy*. Aldershot: Edward Elgar.

Schmalensee, R. (2000). 'Antitrust Issues in Schumpeterian Industries.' *American Economic Review*, 90/2: 180–3.

Schumpeter, J. A. (1912). (Eng. trans. 1961.) *The Theory of Economic Development*. New York: Oxford University Press [first Eng. trans. 1934].

—— (1928). 'The Instability of Capitalism.' *Economic Journal*, 38: 361–86.

—— (1943). *Capitalism, Socialism and Democracy*. London: George Allen and Unwin.

Shackle, G. L. S. (1972). *Epistemics and Economics*. Cambridge: Cambridge University Press.

Tushman, M. L., and Anderson, P. (1986). 'Technological Discontinuities and Organizational Environments.' *Administrative Science Quarterly*, 31: 439–65.

Witt, U. (1998). 'Imagination and Leadership: The Neglected Dimension of an Evolutionary Theory of the Firm.' *Journal of Economic Behaviour and Organisation*, 35: 161–77.

—— (2003). *The Evolving Economy: Essays on the Evolutionary Approach to Economics*. Cheltenham: Edward Elgar.

6 Can Regulations Induce Environmental Innovations? An Analysis of the Role of Regulations in the Pulp and Paper Industry in Selected Industrialized Countries

James Foster, Mikael Hildén, and Niclas Adler

The need for environmental innovations

The growth of industrial production has historically been associated with an increasing use of natural resources, with increasing emissions of pollutants to air, water, and soil, and with growing amounts of waste. To deal with these negative consequences of industrial production, environmental regulations have been introduced; but to truly decouple the link between growth of industrial production and increasing environmental degradation, it is obvious that process innovations are needed.[1]

From the point of view of public policy, three sets of interrelated questions emerge:

- What incentives drive process innovations, and how can regulations be designed to provide or strengthen those incentives? Is it simply a matter of increasing the stringency of the regulations, or are broader policy interventions needed?
- How do regulations affect long-term corporate market expectations, perceptions of risk and uncertainty, opportunity recognition and selection, and the competitive strategies of firms?

- How should environmental regulations that affect process innovations be evaluated? What factors should be taken into account in considering the effects of regulations on process innovations?

The first set of questions arises in any analysis of environmental policies. From the earliest period of developing environmental regulations in the US, the environmental economics literature has emphasized the argument that process innovation—rather than resource reallocation—is the key to effective solutions of environmental problems. Therefore, the creation of regulatory incentives for innovation has been seen as essential.[2] However, incentives to innovate can also be provided through direct R&D support or, indirectly, through the creation of markets for innovations; and the answers to the questions need to consider the interaction between different kinds of policy interventions.

Despite its potential for explaining both decisions and actions, the second set of questions, which focuses on how and why regulations affect businesses and their strategies, including innovation strategies, has hitherto received less attention (Fischer *et al.* 2003). The third set can be seen to raise some fundamental and practical issues linked to the two others; it also focuses on the dynamic nature of policy development.

In all attempts to answer the questions, it is essential to recognize the fundamental trade-offs among broad policy objectives that all public policies face. This means that policy makers face at least three key challenges: coherence of incentives across sectors; efficiency versus flexibility in regulations; and competition neutrality versus creation of competitive advantage. All of these may have consequences for process innovation incentives. For example, regulations aiming at protecting the environment, health, and safety may provide different incentives from those regulating labor markets or trade. From the point of view of the regulatory body, efficient regulations are those that are easy to implement and monitor, standardized, and predictable. Yet process innovations may require flexibility and adaptation to local circumstances, which quickly increases the administrative burden (see Foster 2000). Finally, there are strong political demands to create level playing fields to protect small firms, financially or otherwise disadvantaged firms, and to prevent the appearance of 'regulatory capture.' The price may be a removal of incentives for industry innovation.

To analyze the problems of encouraging process innovations with the help of regulations, we will argue that it is necessary to explore the possible mechanisms through which innovation may be effected and, in particular, the strategies and approaches that firms use in responding to the regulatory demands. For the empirical analysis of the relationship between innovations and regulations, we will use a comparative approach, and examine differences and similarities in the regulatory traditions between the US and selected European countries. As shown by Brickman *et al.* (1985), there are significant differences between the US and Europe in the general structure and function of regulations. Differences can also be found in the recognition of the relationship between regulations and innovations.[3]

We will specifically focus on the pulp and paper industry, which represents a mature industry that has nevertheless undergone some major restructuring in recent decades. Although the industry is still a major user of energy and raw material, the pulp and paper industry has on the whole been remarkably successful in reducing its emissions of polluting substances.

The chapter is organized as follows: in the second section, we explore the relationship between innovations, regulations, and the incentives for actors to innovate; in the third, we present an overview of the international pulp and paper industry; in the fourth section, we briefly review the environmental and economic performance of the pulp and paper industry; this review is then used to examine the relationship between regulations and innovations in the following two sections, leading to the general conclusions at the end.

The links between regulations, innovations, and business strategies

Mechanisms for encouraging innovations

This section identifies five basic mechanisms, as observed in literature and practice, through which regulations may affect incentives and the potential for innovations.

Regulations can create a barrier to incorporating particular technologies in either products or production processes. An example is the banning of the use of a particular substance, as in the prohibition of mercury-based slimicides in pulp and paper processes. They can also create barriers for expansion and growth within particular regions or for particular product lines, as in the US definition of 'non-attainment areas' that fail to meet air-quality standards. Within those areas, production expansion is subject to severe restrictions based on grandfathered emission rights or other restrictions of absolute emission levels. In some cases, the regulations create technology-specific barriers to expansion: thus, emission limits on incineration may block energy generation from waste.

Regulations can specify particular product qualities that can severely affect the use of certain production factors and thereby strongly direct innovation incentives: in Germany, standards require that finished paper contain a high percentage of recycled paper products; a similar example from the US transport sector is the requirement for ethanol-based gasoline in some parts of the country, which places significant cost and technical burdens on the oil refining and distribution system.

Regulations can be intentionally technology forcing, demanding major advances in existing products or processes, thereby creating new markets for innovations to satisfy standards that cannot be reached with commonly used technologies. For example, the environmental regulations have strongly encouraged the development of activated sludge technology for pulp and paper industries in Finland (Hildén *et al.* 2002). In the US transport sector, new diesel engine regulations have involved a number of industries in creating new markets for the development of new combustion technologies, fuel treatment technologies, and emission filters, traps, and catalysts. Similarly, California's ultra-low-emission automobile standards have created demand for a range of significant new technologies.

Regulations can demand the development of entirely new products or processes to substitute existing products and processes. The debate over chlorine-free pulp contained elements of this kind of regulation, and an example from the chemical industry, in response to clear evidence that CFCs were critically depleting stratospheric ozone, is their banning and replacement with new substances.

The first two mechanisms are based on a negative intervention; the last two are based on performance demands that explicitly require the creation of something new. The third can be seen either as a negative barrier or as a positive demand to develop new processes. Differences in institutionalizing regulatory regimes that affect the mechanisms are

also likely to affect the kind and degree of incentives that arise. Thus direct and detailed specification of a regulation at the legal level is likely to provide different incentives from a regulation that is based on a general legal framework specifying the rules for negotiating its detailed implementation.

Business strategies in the face of regulations

In trying to understand the possibilities of promoting innovations through regulations, of particular interest is the interplay between corporate strategy formation through opportunity recognition and selection on the one hand and, on the other, public environmental regulations. The basic, strategic, innovation-related reactions of firms to regulatory interventions include the following:

- compliance through acquiring new technology for pollution control. The standard example is the installation of end-of-pipe technology, thus creating a certain demand for innovations among suppliers in this field. The compliance enhances diffusion, but could also, if compliance costs are high, provide incentives for innovations and R&D spending;
- search for possibilities, either through technological innovations or through management innovations, to redesign production processes to reduce such needs as end-of-pipe pollution control: an example is the reduction of waste through reuse of waste products in the production process. This strategic reaction creates demands for innovations in-house, or in providers of production technology;
- closure or outsourcing of regulated products or production phases. An example is the withdrawal from particular markets. Outsourcing itself can, in some cases, be seen as a form of organizational innovation. Technological innovations may arise in the outsourced units;
- specific organizational innovation. By setting up joint ventures, trusts, etc., a firm

may be able to change its regulatory environment in such a way that liabilities are reduced, or that differences in regulatory regimes can be utilized: rules that apply to power plants may be partially different from those concerned with energy generation within a producing industry;

- active promotion of particular regulatory interventions in order to gain competitive advantage from the firm's own innovations. An example is the promotion of product standards for which the firm has already developed technology;
- active scanning of likely future regulatory interventions to gain competitive advantage, or innovative exploration of possibilities offered by existing regulations: the developing of technological solutions that clearly outperform existing regulatory demands is an example.

Skeptics of the possibility of encouraging innovations through regulations argue first that, though some businesses might not recognize competitive opportunities, this is not likely to be a systemic failure; it is also unlikely that government bureaucrats can identify beneficial competitive outcomes from regulation that decision-makers in firms cannot recognize.[4] However, these arguments curiously detach regulators and firms from the rest of society, and focus nearly exclusively on regulations that create barriers for expansion and growth within particular regions or for particular product lines, and on strategies for closure or outsourcing of regulated products or production phases, as described above. A different picture emerges when the broader societal context is introduced: for example, Vogel has argued that the British approach to environmental regulation has proved as effective as the American, though with lower expenditures and less political turmoil (Vogel 2003). Regulations are part of the process, not necessarily as absolute barriers, but as signals of broader societal concerns and demands for innovations to reduce those concerns. Related arguments have been put forward by Jasanoff (1993), who has stressed the role of procedures

in regulatory approaches. From this perspective, the observations of Bhat (2001) regarding innovations linked to regulations become comprehensible.

Regulations can clearly have positive innovation effects when they have created entirely new markets for pollution control. Examples include combinations of regulations that specify particular product qualities restricting the use of certain production factors, and regulations that create barriers for expansion and growth within particular regions, or for particular product lines. All of these require active scanning of likely future regulatory interventions to gain competitive advantage, or innovative exploration of possibilities offered by existing regulations leading to innovations in end-of-pipe technology.[5] A more difficult question is how costs incurred by regulation may affect innovation. On an empirical basis, Palmer *et al.* (1995) surveyed a range of firms subject to significant regulation. They found that most firms say that the net cost of regulation remains positive for them. But in an analysis of the Finnish pulp mills Hetemäki (1996) concluded that regulation had reduced the efficiency of pulp mills. One can argue that the costs have decreased the competitiveness of firms and their resources for innovation. However, the average costs or efficiency losses do not prove that there is no first-mover advantage in exploiting costly and risky environmental investment. This advantage can increase further when the dynamics of regulatory change are recognized. Provided that the regulations have a predictable course of development, first-mover advantages can increase with time. But if the regulatory environment is volatile and shifts its attention from one issue to another, the gains of first movers may dissipate and become a disadvantage. Thus a broad contextual framework is required for an analysis of the technological innovations and the economic performance of firms engaging in environmental performance strategies and/or overcompliance with existing regulation.[6]

One such broader contextual analysis of innovations and regulations has been carried out by the OECD, and one conclusion has been that 'strict regulations seem to have a particular detrimental effect on productivity the further the country is from the technology frontier' (Scarpetta and Tressel 2002). The implications are that regulatory reforms, especially those that liberalize entry, are very likely to spur investment. Environmental regulations based on product specifications that create entry barriers are thus likely to be particularly problematic from the point of view of providing positive incentives for innovations. The case of the pulp and paper industry will be used in the coming sections to elaborate further on the possible effects of regulation on firm strategies and process innovations.

The pulp and paper industry as a test case for innovation effects

The international pulp and paper industry offers a natural experiment to compare national regulatory regimes and the effects of regulatory diversity on the development, competitiveness, and innovativeness of pulp and paper companies. We will thus draw contrasts in its development, with particular emphasis on the industry in the US, Finland and Sweden, and Germany. In 1999, these countries produced about 55 per cent of the global wood-based pulp and 46 per cent of the global paper (FAO 2004). The differences between the countries (Table 6.1) provide a background against which process innovations can be viewed.

Basic, cross-national differences within the industry are that Finland and Sweden have paper industries that are based mainly on pristine fibre, with 5 and 17 per cent respectively of paper production being based on recovered paper, whereas German paper production is based up to 63 per cent on recovered paper. The US falls between the extremes with 38 per cent of paper production being based on recovered paper (Food and Agricultural Organization (FAO) 2001).

There are also clear differences in the position of the industry in the global production and consumption chain. In Finland and Sweden, pulp and, in particular, paper production is for export, with 85–90 per cent of production being exported. Imports of paper are insignificant. Of German paper production, about 50 per cent is exported, and slightly more than 50 per cent of consumption is imported. The US has a large domestic paper production, but its export is only about 10 per cent of production, and about 60 per cent of the total consumption is imported. Pulp production also shows structural differences. In Sweden and Finland, the largest part of the production (65 and 78 per cent respectively) of chemical pulp and more than 90 per cent of mechanical pulp is channelled into paper in the country itself. In Germany, domestic pulp production is very small, with imports dominating. The US produces significant amounts of pulp for domestic use, nearly 50 Mt, and also imports more than 6 Mt.[7]

The age and technological level of pulp and paper plants vary across countries. In Finland and Sweden practically all of the mills are modernized and have large capacities. The different plants are linked in large national and international networks by belonging to a few major firms. Pulp and paper plants are part of the same firm, and in several locations the mills form large integrated installations.

There are also significant differences in the relative importance of the pulp and paper industry in the studied countries. In all countries, the industry has played an important local role, as pulp and paper mills often have been significant installations around which communities have developed. In Finland, it was for a long time the very backbone and engine of the country's economic growth, and the major export industry; in Sweden, it was one among several important branches of industry; but in Germany and the US, the pulp and paper industry has mainly served a domestic market. The societal role of the industry is also reflected in the role given to the industry organizations: in Finland, the Finnish Forest Industries has been a major player in political initiatives, from developing legislation to deciding on monetary and labor policy; in Sweden, the role has been important, but not quite so dominant; in Germany and the US,

Table 6.1. Differences within the pulp and paper industry among the selected countries

Characteristics 2002	Finland	Germany	Sweden	US	Comment/reference
Production of paper and board (1,000 tonnes)	12,776	18,526	10,743	80,871	Data collected by VDP [a]
Export of paper and board (1,000 tonnes)	11,414	9,965	9,005	8,843	Data collected by VDP
Import of paper and board	366	663	9,651	16,135	Data collected by VDP
Mechanical pulp production (1,000 tonnes)	4,587	1,252	3,302	4,246	Data collected by VDP
Export/import of mechanical pulp (1,000 tonnes)	154/1	23/151	37/28	0/330	Data collected by VDP
Chemical pulp production (1,000 tonnes)	7,143	896	8,052	45,250	Data collected by VDP
Export/import of chemical pulp (1,000 tonnes)	1818/90	491/3856	3,089/273	5,006/4887	Data collected by VDP
Manufacture of pulp, paper, and paper products: Share (%) of value added in total manufacturing	16.1	2.4	10.6	3.8	Eurostat [b] US Census Bureau (2001), NAICS code 322
Industrial capacity	Predominantly large pulp (85 % > 50, 000 t/year) and paper mills (90 % > 50,000 t/year)	Both large and small mills; (55 % of pulp mills > 50,000 t/year; 40 % of paper mills > 50,000 t/year)	Predominantly large pulp (85 % > 50,000 t year) and paper mills, with some smaller (70 % > 50,000 t/year)	A large proportion of small establishments (32 % with less than 20 employees, 28 % with more than 100 employees)	EC (2001). US Census Bureau (2001), NAICS code 322
Age structure of industry	Predominantly new installations	Both new and (small) old installations	Predominantly new installations	Both new and old installations	Brännlund et al. (1996), EC (2001)

[a] VDP, German Pulp and Paper Association (2004).
[b] http://europa.eu.int/comm/eurostat/newcronos/suite/retrieve/en/theme4/sbs

the industry organizations have been one among many industrial branches, and unable to influence national policies to fit the needs of the industry.

Environmental and economic performance in the pulp and paper industry

Overall, the pulp and paper industry has been remarkably successful in increasing its environmental performance, and has managed to reduce the emissions of many pollutants to a small fraction of what they were two decades ago (see Table 6.2). This is true for both the US and the European industry: emissions, measured as biological oxygen demand (BOD), are now at the level of 1.3–1.4 kg/tonne production

in both the US and Europe; in the mid-1970s, they were more than six times higher. In leading countries such as Finland and Sweden, the BOD emissions are now below 1 kg/tonne of production.[8]

Other environmental variables, notably emissions of nutrients, chemical oxygen demand and chlorine to water, and acidifying substances to the air, show steady declines despite increasing production (Hildén *et al.* 2002; Naturvårdsverket 2003). Issues of recycling have also been high on the agenda (CEPI 2003) both in the US and Europe, particularly so in Germany.[9] The long-term figures for Finland show that the development has not been linear (Figure 6.1). A clear break occurred around 1970 and the same development can be seen in other countries.

The productivity of pulp and paper firms has increased worldwide, but there is large variation within the industry (Table 6.1). In

Table 6.2. Environmental characteristics for the pulp and paper industry in Finland, Germany, Sweden, and the US

	Finland	Germany	Sweden	US	Comments/ references
Current expenditure on env. protection as % of output (1999)	0.54 (above industry average)	0.53 (1998, below industry average)	0.58 (above industry average)	0.6 (above industry average)	Eurostat (2004) US Census Bureau (2002)
Environmental issues	Waste water and air emissions dominate, recycling recognized, biodiversity and other resource issues gaining importance	Recycling one of the dominant issues, water and air important, biodiversity resource issues as an argument in recycling	Waste water and air emissions dominate, recycling recognized, biodiversity and other resource issues gaining importance, somewhat more advanced than in Finland	Waste water, air, solid waste (recycling)	Eurostat and US Census Bureau (2002)

Source: Eurostat on-line data tables, available at http://europa.eu.int/comm/eurostat/newcronos/reference/display.do?screen=welcomeref&open=/intrse/sbs/enterpr&language=en&product=EU_MASTER_indus-U_MASTER_industry _trade_services_horizontal&root=EU_MASTER_industry_trade_services_horizontal&scrollto=0 (last visited 27 December 2004).

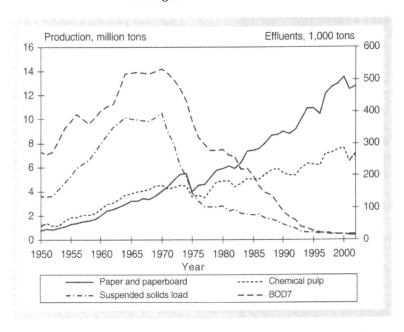

Fig. 6.1. Environmental performance of the Finnish pulp and paper industry with respect to the protection of waters

Source: Finnish Forest Industries Federation 2004*a*.

Finland and Sweden, the differences between individual installations are relatively small (Brännlund *et al.* 1996). In Finland, the industry association has concluded that 'the forest industry's machinery stock is relatively new and energy efficient at present' (Finnish Forest Industries Federation 2004*b*). In Germany and the US significant differences prevail, ranging from modern installations to antiquated production facilities. In the US there are still-active paper machines that were originally brought into service in the 1890s (McNutt 2003). The internationalization of the pulp and paper industry may gradually change the situation. Thus, the Finnish-Swedish company Stora Enso recently revealed that it will shut down two coated paper machines in the US. The move will take only 80,000 tonnes/year of capacity out of the market which, given that the average tonnage capacity of new machines in Finland and Sweden is more than 250,000 tonnes/year (European Commission 2001), is obviously not a large amount for two machines. The units affected include a paper

machine that is 63 years old and a second machine with 108 years of service (*Paperloop* 2000). This example illustrates the importance of eliminating relics in improving the environmental performance of the plant and the firm, and in increasing the average productivity of the firm.

The economic performance of the pulp and paper sector is that of a mature industry. In 2000, net sales of the global forest industry was approximately USD 450 billion. Paper and board account for 58 per cent, and the wood products industry for 38 per cent. The remaining 4 per cent is contributed by market pulp. Large, increasingly multinational companies produce modest profits, as shown by the 100 top companies analyzed by Pricewaterhouse-Coopers (Figure 6.2). This is also seen in the country statistics: in Finland, the average profit from 1994 to 2003, before extraordinary items and excluding capital gains, amounted to 7.0 per cent of net sales (Finnish Forest Industries Federation 2004*c*). As indicated by Figure 6.2, the US firms in the top 100 have

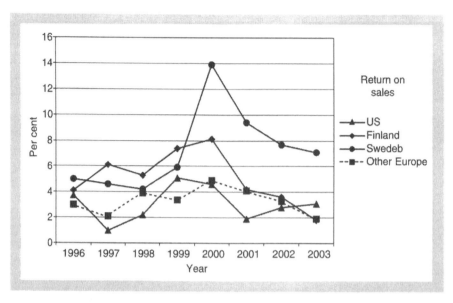

Fig. 6.2. Profit development for the top 100 companies. Upper panel return on investment employed; lower panel return on sales

Source: PricewaterhouseCoopers 2004.

had lower profits. This is also reflected in overall indices, which show that North American profits have been below those in Europe since the early 1990s (*Scandinavian Pulp and Paper Reports* 2005). In the US in general, there has been a declining rate of return on new investment since the early 1990s and declining profits (Center for Paper Business and Industry Studies 2003). Strong criticism has been directed at the US industry:

The industry is mature, capital intensive, extremely cyclical, seriously affected with failing performance and returns, monolithic and slow to change. Substantive assets are under-utilized and under-performing. Leadership seems largely to lack adequate vision, innovative thinking and a good solid understanding about the character of value. (McNutt 2000)

The role of regulation in inducing process innovations

Observations indicate that the state and environmental performance of the pulp and paper industry in the US is, on average, poorer than in the Nordic Countries and that rejuvenation of the pulp and paper industry has been faster in Finland and Sweden than in the US and Germany. What does this say about the possibilities of inducing innovations? Have the pulp and paper productivity gains and innovations been induced by regulations? Are they examples of the efforts of different companies in their search for competitive advantage through innovations? Or are they simply the result of differences in competitive and industrial structures in the studied countries?

The differences in productivity and environmental performance in a mature industry such as pulp and paper strongly suggest that societal factors have contributed significantly to the development. Technological factors alone are insufficient for explaining the changes that have occurred. Instead, the development has features that resemble the concept of technological momentum (Hughes 1994), and we argue that this momentum is the result of an interaction between societal policy processes with a developing technological base. Thus we are dealing with a momentum which is very different in each of the countries we have examined. This can be attributed to different socio-productive systems (Kemp 2000; Kemp *et al.* 2000), which determine, among other things, the networks among actors in the society. The significant differences among our countries' socio-productive systems are reflected in industrial, financial, tax, labor, and environmental policies: in Finland and Sweden, the pulp and paper industry has always had a strong say in the formulation of any policy that can affect the industry positively or negatively; in the US and Germany, other industries have had a stronger role in determining policies.

To what extent, then, can differences in regulatory policy or other efforts to induce innovations explain the observed differences between the pulp and paper industries in the studied countries? Compliance costs in Finland and Sweden are relatively low but certainly not insignificant, and provide some incentives for R&D spending, technology innovation, and other measures for environmental performance enhancement. But there must surely be other incentives, such as a reduction of the wider business risks associated with the environment, and gaining some competitive advantage, for improving environmental performance. Though the pattern is complex, this can be verified for the pulp and paper industry in Finland and Sweden.

Clearly, the Swedish and Finnish producers are heavily dependent on exports to Germany and other parts of Europe where the pulp and paper industry does not have a strong say in environmental policies, but can become the focus of stringent regulations, or draw public and political attention to environmental issues. The Nordic industries cannot take the risk of losing the game and, before the recent mergers of the industries, Swedish and Finnish paper and pulp exporters competed intensively on the European market. For the Nordic pulp and paper industries, environmental issues have become matters of business risk and strategy, which can be seen in corporate decisions to introduce environmental management systems (ISO 14001 and EMAS) and in extensive public reporting on the environment (see e.g. Finnish Forest Industries Federation 2004d). The development has been reinforced by an extraordinary openness in environmental issues in the Nordic societies, a high level of public environmental awareness, and significant public spending on the pulp and paper industry through education and R&D funding: all in all, this development has induced innovations. The contribution of the regulatory system has been important as a manifestation of a general and consistent policy line.[10] The significance of the regulatory system is that it has provided a clear signal on the direction of the development which, from a business strategy and risk perspective, is highly relevant (Hildén *et al.* 2002).

Although the pulp and paper industry in Sweden and Finland is characterized by generally bigger and newer mills than in the US and Germany, it would be wrong to assume that all innovations and productivity gains have been discovered and introduced in the Nordic countries, and only gradually diffused to other countries. As a further twist, there are examples of inventions that have originated in the Nordic countries but have reached the stage of an innovation in the US before being employed in the Nordic countries. This has been attributed to the lack of risk capital in Finland, especially before the merger of the biggest pulp and paper producers.[11]

In the US, more stringent regulation is a strong driver for investments in new, environmentally friendly technology. The incentive for action in 'beyond compliance' investments and improved corporate performance programs has not been the expectation of greater production efficiency, abatement cost reductions, or the discovery of competitive advantages; rather, the overriding consideration has been the need to manage business risks, with the dominant risks being the threat of ever more stringent regulation and the disruption of business, as well as the recognition of new and potential corporate liabilities.

This point is dramatically revealed in the behavior over the past ten years of Georgia

Pacific. Before the early 1990s, Georgia Pacific was, according to *Wall Street Journal* reports at the time, 'mocked as a poor corporate citizen' and was labeled by environmental interest groups as 'one of the nation's eight worst corporate polluters.' The firm viewed the non-compliance fines as simply another cost of doing business, and had routinely failed to comply with monthly pollutant discharge limits assigned by its water permits. But then the chaotic disruptions of the forest products industry in the Pacific Northwest occurred, with protection of the spotted owl eventually closing down nearly 80 per cent of the timber industry in the region. Events in the Northwest 'demonstrated to me that a confrontational, litigious solution was not going to work,' said A. D. Correll, Georgia Pacific's chief executive. 'The environmental groups are simply a fact of life, and we can no longer ignore them.'

Correll hired Lee Thomas, the former Environmental Protection Agency administrator, to run Georgia Pacific's environmental unit. As Thomas argued, 'Companies have decided to sidestep the 'environmental war' to make money. If we are going to be able to do business and make money, we've got to be prepared to accommodate the public's view, and special interest groups that represent that public.' Among other programs initiated by the company, the setting aside of tens of thousands of acres for nature conservancy was intended to give the company much freer rein to log elsewhere on its nearly six million US acres—and avoid a repeat of the chaos in the Pacific Northwest.

The abandoning of molecular chlorine for bleaching is also an example of innovations linked to business strategies and the possibilities of using regulatory gains. Swedish pulp and paper manufacturers were first movers in the chlorine-free market and, in contrast to Finnish producers, did not strongly oppose regulations concerning chlorine bleaching. However, the case also demonstrated significant differences towards this particular issue in the Swedish and Finnish society.

Rajotte (2003) suggests that the role played by markets in the innovation and diffusion of cleaner technologies was contingent on other factors:

Some firms were able to introduce and/or request totally-chlorine free (TCF) products and happily exploited public concern to their advantage, knowing the availability of chlorine-free technical means. In addition, latent green market demand could materialize through pioneer pulp firms identifying a niche.

Another example of the complex interplay between regulation, innovation, and environmental performance is provided by the rise of the recycled paper production. The business model that long guided the US industry was one that emphasized both vertical and horizontal integration of production and of products—from the forest to the broadest range of paper products, to the reuse of the waste paper. As well as being a model of remoteness from large urban areas, as the mills are typically located near the forest inputs and the finished paper converting plants near the customer, it is a model of large-scale industry. In Germany, where regulations effectively deny operation of craft pulp mills, and in China and other areas of Asia with limited, local forest inputs, the paper production relies on recycled paper. Recycled paper has long been with us, but as a poor relation of the large-scale, virgin pulpwood industry. Now, with recycled paper regulations and collection systems common to industrialized countries, waste paper has become the dominant paper trade item for the US and many other countries. Recycle mills, which avoid the worst polluting aspects of pulp mills, can be located near urban areas, can be smaller to serve a local market, and can dramatically reduce the burden on forestland.

Pressure from both environmental groups and public legislation has contributed to the increase in the recovery and recycling of used paper, and recycled paper today makes up over 40 per cent of the raw material for the world's paper mills. The wide variety of government regulations, subsidies, and so forth in the EU has made for large increases in the supply of used paper in Central Europe, and large reductions in prices for recycled paper. At the same time, there have been large increases in demand for products based on recycled fibres, partly because of the pressure that environmental organizations have exerted on newspaper and magazine publishers, and other paper users. Given increases in wood and production costs, the increasing supply of and demand for used paper has inspired some new strategies for the largest paper producers. Focusing on recycling mills brings benefits, both from the proximity to customers and from the cheap supplies of raw material in densely populated urban areas.

We can thus conclude that there are links between regulations and process innovations in the pulp and paper industry, but they are not straightforward. The key incentives are those that clearly relate to business strategies and products. This also means that the evaluation of regulations aiming at fostering process innovations must take into account the societal context and business environment, including the existence of pressure groups that exert their influence through the products and markets rather than through direct demands at the level of permit conditions for individual plants. The combined effects of the differences in the markets, business strategies, and the regulatory policies are capable of explaining the slower progress in environmental performance of US plants relative to those in the Nordic countries.

Regulations, investments, and productivity gains

From a perspective of interactions between regulations, business strategies, and society, the previous section revealed and explained differences in the environmental performance of the pulp and paper industry between the studied countries. As process innovations require significant investments, particularly in R&D, this section will examine differences in prerequisites for investments in the different countries.

The Nordic pulp and paper industry has invested more in R&D than the German one;

the figures for the US are somewhat contradictory (Table 6.3). According to NSF statistics, about 1 per cent of sales were invested in R&D. The number of R&D personnel relative to the total number of employees is lower than in Finland (and probably also than in Sweden, as most of the R&D costs are personnel costs). The total volume of R&D expenditure is not insignificant in the US, but it appears that much of the R&D for the industry is provided by 'outsourcing' to specialized R&D firms, meaning that internal R&D is not a source of potential competitive advantage for the different firms.

This suggests that it is necessary to include a broader set of factors that could potentially affect incentives for investments in R&D in the pulp and paper industry. Such additional factors are:

- the role of advantageous natural resource endowments;
- the existence of corporatist-industrial policies;
- the relative importance of the pulp and paper industry's economic importance to the country as a whole.

It is equally important to recognize disincentives for investments in the pulp and paper industry, because these are also likely to affect the possibilities for induced process innovation. Such disincentives are:

- global excess capacity;
- low and uncertain profitability;
- public policies (especially in the US) that maintain old, inefficient, and 'dirty' plants; counter-innovation incentives of 'locked-in' technology regulations;
- an extremely high capital investment intensity (in the US, the average annual investment per employee in the mid-1990s was $14,000 as compared to an average of $4,000 for all of manufacturing);
- given excess capacity and a large number of competitors, competition is based primarily on price, and margins are very low.

Under such circumstances, the most obvious and significant source of improved

Table 6.3. R&D in the pulp and paper industry in the studied countries: data from different sources

	Finland	Germany	Sweden	US
R&D relative to value added in 2001–2 in pulp and paper industry NACE code DE21 (Europe) and % of sales in SIC code 27 (US)	1.7 %	0.6 %	1.8 %	1 % of sales
Share of R&D employment in the number of persons employed in 2001–2 (%)	2.5 %	0.6 %	NA	1.8 %
R&D expenditure in pulp and paper (NACE 21)	86 million euro (2002)	79 million euro (2001)	126 million euro (2001)	1583(1998, million USD)

Note: Differences may exist in the definition of R&D.

Sources: Finland and Germany: Eurostat 2004. Annual detailed enterprise statistics on manufacturing 2002; Sweden: Svenska statistiska centralbyrån 2001; USA: US National Science Foundation 2004.

environmental, as well as economic, performance for the pulp and paper industry will not be driven by new technology, but rather the result of closing old or redundant production capacity. This is consistent with the circumstances in other mature industrial sectors, where it has been found that a majority of the increase in productivity over the past decade has resulted from the deaths and births of firms or plants, not from technological or other innovations (Bernard and Jensen 2005). Environmental regulations can clearly contribute to these kinds of disinvestments—factories are shut down when they can no longer meet regulations that tend to become more stringent over time. The environmental literature has largely focused on the efficiency of production resource input use: energy, raw materials, water, and so forth included in the concept of eco-efficiency.[12] However, the empirical data shows that of far greater importance—at least for the pulp and paper industry—is the efficiency of capital asset allocation.

The illustrations of the pulp and paper industry in the four countries also show that a lot of accessible and very promising technology is

enabling process innovations and environmental, as well as competitive gains. However, the win-win situation and the diffusion of technology necessitate clear incentives, and here the four countries show different incentives systems. A more stringent regulation in the 1990s inspired some new technology development and, more importantly, compelled broad-based diffusion of existing, advanced technology. As a result, there were some efficiency gains for certain producers. However, the return on capital data for the industry reveals that, on average, there has been a declining rate of return on new investment since the early 1990s, and many US and German companies have not adopted new technologies or process innovations.

What then, apart from a new round of more stringent regulations, could give incentives to the diffusion of the advanced technology that is currently available? The realistic answer is that, probably, there are no market or competitive incentives. If the industry is falling well behind the world technology frontier, this problem is aggravated. When a large part of the production facilities uses inferior technology, the degree of inferiority will increase with time. Thus it will take significant investments just to try to catch up with a receding target, let alone to gain advantages relative to that target. A reasonable proposition is that, as the gap between a firm's technology and the technology frontier widens, the incentives for innovation decline proportionate to the gap.

An illustration of this challenge is the statement made by George Weyerhaeuser, from Weyerhauser, US, in a recent interview. He was asked if US companies will be at a 'competitive disadvantage with their European and Asian counterparts because they are limiting their capital expenditures to historic low levels.' His answer was:

I think the North American industry is going to find itself at a disadvantage. ... For decades the industry has not earned a return on the capital we invested. So not just Wall Street, but our shareholders, our directors, now even our management teams are saying, 'No more.' The industry lost the right to grow. ... So we will end up having to compete with assets that are not as big as Asian assets, that are not as technically capable as what the competitors in Europe have. And we're going to find ourselves in a very, very tough competitive situation as a result. ... We're going to be competing against some big global monsters. And we're doing it with old equipment, old technology, and old cost structures. (Jensen 2000)

Many environmental advocates wish to see business opportunities in environmental performance. For example, a recent World Resources Institute report on the pulp and paper industry, dramatically titled 'Pure Profit,' said: 'Rich rewards are increasingly available to companies able to transform environmental concern into market opportunity or competitive advantage' (Repetto and Austin 2000b). This argument reflects the idea that important efficiency gains can be made through voluntary corporate actions to reduce pollution, and that the most stringent regulations are potentially the best for industry because they induce significant productivity gains and, thereby, important competitive gains in the aggregate. The report also details in a most interesting

way the differences in generally very sizeable financial losses that paper firms would confront with more stringent regulation: 'The same environmental standards are likely to have quite different impacts, individually and collectively, across companies in the industry.'[13] This argument is a very important counterpoint to the commonly expressed argument that more stringent regulations generally produce win-win outcomes for industry—an argument implying that all or most firms actually benefit from stringent regulation. In this study, however, more stringent regulation poses financial risk, and the concern is that financial markets should not recognize or penalize firms for that risk.

The concern is that, as indicated by price-earnings and price-to-book ratios, there are companies in a group that have quite similar valuations but differ substantially in their environmental risks. This suggests that markets have not fully assimilated companies' environmental risks. Similarly, analysts involved in credit ratings, and forming an overall judgment of a company's financial risks, may not yet have taken into consideration the potential outcomes from such environmental aspects for a company's earnings, cash flow, and balance sheets.

Regulations and regulatory policies will have to consider the capacity of different types of firms to improve environmental performance and gain from it. There are at least three concerns: low R&D spending; high investment requirements; and low margins. Given these conditions that affect all firms, it might seem that large firms would have an advantage over small firms in being able to fund and generate R&D and to finance major investments; and the high costs of increasingly stringent environmental regulations have reinforced other factors driving industry concentration. However, the large firms do not have an advantage in exploiting improved environmental performance in all aspects. They do have advantageous capacity for major investments, and have used that capacity to build new, very efficient, and clean paper mills. Yet, their emphasis may be on efficiency as opposed to flexibility. Paper-market analysts have commented at length on the risks in this approach: the new plants are highly optimized for particular products and to meet particular regulations; there is concern that flexibility in producing different types of products, or even variants on a given product, or meeting different regulatory standards, has been sacrificed. A current cost advantage has been acquired at the opportunity cost of limited ability to change course as new opportunities or new market challenges arise.

Conclusions

This chapter started by introducing three sets of questions about breaking the link between growth of industrial production and increasing environmental degradation, and to analyse the interplay between regulation, innovation, and economic and environmental performance. We argue that recent developments in the pulp and paper industry in the US, Germany, Sweden, and Finland are good illustrations of the complex and context-dependent set of answers to the questions raised.

The competitive incentives that drive process innovations are dependent upon each actor's current technological level and actual distance from the technological frontier in the industry. For the pulp and paper industry as a whole, technological innovations for end-of-pipe solutions to environmental problems have been strongly affected by regulatory concerns raised directly by authorities or indirectly through public concerns. More recently, process innovation has been encouraged in some parts of the industry by industry concentration, new product specialization, and process technologies associated with scale efficiencies. Further improvements in the environmental performance of the industry as a whole still demand significant diffusion of technology and further process innovations.

Regulations can and will affect long-term corporate market expectations, perceptions of risk and uncertainty, opportunity recognition and selection, and the competitive strategies of firms. Hence, process innovations can be induced, but not in a straightforward way. Strict technology-forcing environmental regulations may contribute to a restructuring of the industry, and may offer a particular segment of the industry a competitive advantage, as they have in the rise of the recycling industry in Germany. In these cases, the regulations may also generate some market-based incentives for process innovations, for example, through markets for high-quality recycled paper products, and provide firms that are able to exploit these with a competitive advantage. The regulations do not, however, induce significant process innovations on their own: they must be complemented by incentives for R&D in such other public policy areas as innovation, industry, and education. The leading role of the Nordic pulp and paper industries in the environmental field has not developed by chance, but through complex interplay between many societal features, of which the regulatory environment is only one. If this is not recognized, the dream of induced process innovation by public interventions is likely to remain a dream that may come true only occasionally, and under exceptional circumstances.

From a policy perspective, environmental regulations purposing to affect process innovations need be evaluated through their ability to identify, compensate for, complement, and leverage the multiple types of stimuli necessary. Hence, in evaluating the effects, it will be necessary to evaluate each regulatory initiative's interplay with other governing logics explaining decisions and actions in firms.

One way forward, as suggested by the lessons from the pulp and paper industry, seems to be through differentiated interventions and regulatory approaches that become integrated in the governing logic of each industry and targeted set of actors. Maybe, for example, regulations and regulatory approaches should not necessarily be built on coherence of incentives across sectors or competition neutrality. To achieve such interventions, policy makers may need to revisit some fundamental assumptions.

Notes

1. By 'process innovations' we mean changes in the actual production process or in end-of-pipe technologies that reduce effluents and emissions in separate treatments, or a combination of both.
2. Early examples of these arguments include: Orr (1976) and Kneese and Schultze (1975) who would contend that 'the most important criterion on which to judge environmental policies is the extent to which they spur new technology'.
3. The importance of innovations has been explicitly recognized only recently in the European regulatory context, and is still in many respects viewed as something external to the regulatory system. For example, Article 11 of the Council Directive 96/61/EC of 24 September 1996 concerning integrated pollution prevention and control (the IPPC Directive) only notes that 'Member States shall ensure that the competent authority follows or is informed of developments in best available techniques' (EC 1996). The whole Directive, which is the main legal basis for the environmental regulation of major industrial installations, does not include a single reference to innovations or incentives for innovations. In its recent communication on innovation policy, the Commission stresses, however, the links between innovations and environmental policies (EC 2003) In the discussion on chemicals regulation, innovation has also been an issue for a longer time: see, for example, Fleischer (1998).
4. An extensive critique of the Porter argument and a good summary of the economist's arguments critical of 'non-rational actor' models of corporate environmental behavior is found in Jaffe et al. (1995).
5. Taylor et al. (2003), 'In other words, government regulation created a market for scrubbing technologies. Market forces then drove innovation because the company that sold the best system had a competitive advantage.'
6. For example, **www.environmental-performance.org/index.php**; Kemp (1997). Repetto and Austin (2000a); Repetto and Austin (2000b).
7. All figures compiled by VPD, the German Pulp and Paper Association (2004).
8. US data from: American Forest and Paper Association (2000). European Data from CEPI (2003).
9. VPD, the German Pulp and Paper Association (2004).
10. Bhat (2001) finds empirical correlations that indicate positive effects of regulations on innovations.
11. Interview with developer of pulp and paper technology in Finland.
12. The World Business Council for Sustainable Development (2005) defines eco-efficiency as being achieved by the delivery of competitively priced goods and services that satisfy human needs and bring quality of life, while progressively reducing ecological impacts and resource intensity throughout the life-cycle, to a level at least in line with the Earth's estimated carrying capacity.
13. Repetto and Austin (2000b). The point is the authors attempt to identify the worst cases; they seek out those environmental issues which have in their words 'the greatest potential financial impacts' and the greatest 'potential adverse affect on competitive position.' The authors note that most firms readily claim they have limited environmental liabilities when, in fact, their potential liabilities could be very great, depending on regulatory actions.

References

American Forest and Paper Association (2000). *Environmental, Health and Safety Principles Verification Program*, Progress Report. **www.afandpa.org/Content/NavigationMenu/Environment_and_Recycling/Environment,_Health_and_Safety/Reports/EnvironmentalHealthSafety-ProgressReport.pdf** [22 October 2004].

Bernard, A. B., and Jensen, J. B. (2005). 'Firm Structure, Multinationals, and Manufacturing Plant Deaths.' Tuck School of Business at Dartmouth. Working Papers. **http://mba.tuck.dartmouth.edu/pages/faculty/andrew.bernard/deaths.pdf** [20 September 2005].

Bhat, V. N. (2001). 'Environmental Regulations and Innovation Activities: A View Point.' *International Journal of Environmental Studies, 58*: 741–8.

Brännlund, R., Hetemäki, L., Kriström, B., and Romstad, E. (1996). *Command and Control with a Gentle Hand: The Nordic Experience.* Sweden: Umeå University. Research Notes 115, Dept. of Forest Economics.

Brickman, R., Jasanoff, S., and Ilgen, T. (1985). *Controlling Chemicals: The Politics of Regulation in Europe and the United States.* New York: Cornell University Press.

Center for Paper Business and Industry Studies (2003). *State of the North American (& Maine) Pulp and Paper Industry: An Update and Outlook.* Orono, ME: Pulp and Paper Industry Foundation, Maine. **www.paperstudies.org/industry/030403_State_of_the_Industry_Maine.pdf** [21 October 2004].

CEPI (2003). *The European Paper Industry on the Road to Sustainable Development.* **www.cepi.org/files/sustreport03-141308A.pdf** [22 December 2004].

EC (1996). Council Directive 96/61/EC of 24 September 1996 concerning integrated pollution prevention and control (the IPPC Directive). *Official Journal* L 257, 10/10/1996: 26–40.

—— (2001). IPPC Reference Document on Best Available Techniques in the Pulp and Paper Industry 2001. Available at **http://eippcb.jrc.es/** [20 September 2005].

—— (2003). *Innovation Policy: Updating the Union's Approach in the Context of the Lisbon Strategy.* Communication from the Commission to the Council, the European Parliament, the European Economic and Social Committee and the Committee of the Regions, Brussels, 11.3.2003 COM(2003) 112 final.

Eurostat (2004). *Annual Detailed Enterprise Statistics on Manufacturing 2002.*

FAO (2001). *Recovered Paper Data 2001,* available at **www.fao.org/forestry/foris/webview/forestry2/index.jsp?siteId=3141&langId=1** [29 September 2003].

—— (2004). *Pulp and Paper Capacities 1999–2004,* available at **www.fao.org/forestry/foris/webview/forestry2/index.jsp?siteId=3141&langId=1** [29 September 2003].

Finnish Forest Industries Federation (2004*a*). *Forest, Environment and Industry: A Story of New Ways Forward.* **www.forestindustries.fi/files/julkaisut/pdf/8forest_environment.pdf** [22 September 2005].

—— (2004*b*). *Climate.* **http://english.forestindustries.fi/environment/climate/new.html** [21 October 2004].

—— (2004*c*). *Turnover.* **http://english.forestindustries.fi/finance/turnover/** [21 October 2004].

—— (2004*d*). *Environment.* **http://english.forestindustries.fi/environment/** [22 October 2004].

Fischer, C., Parry, I., and Pizer, W. (2003). 'Instrument Choice for Environmental Protection when Technological Innovation is Endogenous.' *Journal of Environmental Economics and Management, 45*: 523–45.

Fleischer, M. (1998). 'Key Issues Related to the Legislation of Chemicals in the EU.' In F. Leone and J. Hemmelskamp (eds.), 'The Impact of EU-Regulation on Innovation of European Industry.' Papers presented at the expert meeting on Regulation and Innovation, Seville, 18–19 January 1998. Available at **www.jrc.es/home/publications/publication.cfm?pub=26** [1 October 2003].

Foster, J. (2002). *The Dead Hand of Regulation,* MIT: Center for International Studies.

Hetemäki, L. (1996). 'Essays on the Impact of Pollution Control on a Firm: A Distance Function Approach.' *Finnish Forest Research Institute, Research Papers 609.*

Hildén, M., Lepola, J., Mickwitz, P., Mulders, A., Palosaari, M., Similä, J., Sjöblom, S., and Vedung, E. (2002). *Evaluation of Environmental Policy Instruments—A Case Study of the Finnish Pulp and Paper and Chemical Industries.* Monographs of the Boreal Environment Research, 21.

Hughes, T. P. (1994). 'Technological Momentum.' In M. R. Smith and L. Marx (eds.), *Does Technology Drive History?* Cambridge, MA: MIT Press, 101–14.

Jaffe, A., Peterson, S., Portney, P., and Stavins, R. (1995). 'Environmental Regulation and the Competitiveness of US Manufacturing: What Does the Evidence Tell Us?' *Journal of Economic Literature*, 33: 132–63.

Jasanoff, S. (1993). 'Procedural Choices in Regulatory Science.' *RISK—Issues in Health and Safety*, 4: 143–60.

Jensen, K. (2000). 'George Weyerhaeuser Shares Views on Industry and Technological Challenges.' *Pulp and Paper*, 27 March. **www.paperloop.com/db_area/archive/p_p_mag/2000/0003/ feat2.htm** [23 September 2005].

Kemp, R. (1997). *Environmental Policy and Technical Change: A Comparison of the Technological Impact of Policy Instruments*. Cheltenham: Edward Elgar.

—— (2000). 'Technology and Environmental Policy: Innovation Effects of Past Policies and Suggestions for Improvement.' *OECD Proceedings Innovation and the Environment*. Paris: OECD, 35–61.

—— Smith, K., and Becher, G. (2000). *How Should We Study the Relationship between Environmental Regulation and Innovation?* IPTS Publications. EUR 19827 EN. **www.jrc.es/home/pages/ detail.cfm?prs=567** [20 September 2005].

Kneese, A., and Schultze, C. (1978). *Pollution, Prices and Public Policy*. Washington, DC: Brookings Institution.

McNutt, J. A. (2000). 'Lessons from the Past: Other Industries Provide Impetus for Transformation.' *Pulp and Paper*, 74/3: 50.

—— (2003). 'The Paper Industry.' Presentation for the Sloan Workshop on Globalization. Center for Paper Business and Industry Studies (CPBIS), Georgia Institute of Technology and the Institute of Paper Science and Technology. **www.paperstudies.org/industry/020614_Paper_ Industry_Memo2.pdf** [20 September 2005].

Naturvårdsverket (2003). *Ingen övergödning: Underlagsrapport till fördjupad utvärdering av miljö- målsarbetet*. Swedish Environmental Protection Agency. **www.naturvardsverket.se/ index.php3?main=/dokument/mo/overvak.htm** [20 September 2005].

Orr, L. (1976). 'Incentive for Innovation as the Basis for Effluent Charge Strategy.' *American Economic Review*, 66: 441–7.

Palmer, K., Oates, W., and Portney, P. (1995). 'Tightening Environmental Standards: The Benefit-cost or the No-Cost Paradigm?' *Journal of Economic Perspectives*, 9: 119–32.

Paperloop (2000). 'Stora Enso Axes Two US Machines.' *Paperloop*, December 2000, p. 13. **www.paperloop.com/db_area/archive/ppi_mag/2000/0012/world.htm** [20 September 2005].

PricewaterhouseCoopers (2004). *Global Forest and Paper Industry Survey*. **www.pwc.com/ extweb/onlineforms.nsf/Notes?CreateDocument** [20 September 2005].

Rajotte, A. (2003). *Human Impacts on the Environment: Shifting Economic Growth towards Sustainable Modes of Development*. University of Jyväskylä. Available at **www.cc.jyu.fi/helsie/pdf/ rajotte** [30 September 2003].

Repetto, R., and Austin, D. (2000a). *Coming Clean: Corporate Disclosure of Financially Significant Environmental Risks*. Washington, DC: World Resources Institute.

—— —— (2000b) *Pure Profit: The Financial Implications of Environmental Performance*. Washington, DC: World Resources Institute. Available at **http://pubs.wri.org/pub-s_pdf.cfm?PubID=3026** [10 August 2005].

Scandinavian Pulp and Paper Reports (2005). SPPR Global Profitability Index. **www.sppr.no/ SPPR_Index.htm** [23 September 2005].

Scarpetta, S., and Tressel, T. (2002). 'Productivity and Convergence in a Panel of OECD Industries: Do Regulations and Institutions Matter?' Paris: OECD Economics Department Working Papers no. 342. Available at **www.olis.oecd.org/olis/2002doc.nsf/linkto/eco-wkp(2002)28** [15 August 2005].

Svenska statistiska centralbyrån (2001). Svenska företags FoU utgifter fördelad på produktgrupp 2001, **www.scb.se** [22 December 2004].

Taylor, M. R., Rubin, E. S., and Hounshell, D. A. (2003). 'Effect of Government Actions on Technological Innovation for SO_2 Control'. *Environmental Science and Technology*, 37: 4527–34.

US Census Bureau (2001). 1997 Economic Census *Manufacturing* Summary Series. **www.census.gov/prod/ec97/97m31s-gs.pdf** [27 December 2004].

—— (2002). 'Pollution Abatement Costs and Expenditures: 1999.' Issued November 2002. **www.census.gov/prod/2002pubs/ma200–99.pdf** [27 December 2004].

US National Science Foundation (2004). **www.nsf.gov/sbe/srs/iris/tables.cfm?pub_year=NSF%2001–305** [15 December 2004].

VDP, the German Pulp and Paper Association (2004). *Facts on Paper 2003*. Available at **www.vdp-online.de** [29 September 2004].

Vogel, D. (1986). *National Styles of Business Regulation: A Case Study of Environmental Policy*. Ithaca, NY: Cornell University Press. (Reprinted 2003, Beard Books.)

World Business Council for Sustainable Development (2005). *Ecoefficiency*. **www.wbcsd.org/templates/TemplateWBCSD5/layout.asp?type=p&MenuId=NzA&doOpen=1&ClickMenu=LeftMenu** [22 September 2005].

7   From Theory to Practice: The Use of the Systems of Innovation Approach in Innovation Policy

Cristina Chaminade and Charles Edquist

Introduction

Since the seminal work of Freeman (1987) on the Japanese national innovation system, the number of contributions to the systems of innovation approach at a national, sectoral, and regional level has grown (Lundvall 1992; Carlsson and Jacobsson 1993; Cooke, *et al.* 1997; Edquist 1997a; Edquist and Johnson 1997; Lundvall, Johnson, *et al.* 2002; Malerba 2004; Malerba and Orsenigo 1997; Nelson 1993).

The academic discussion started in the political sphere in the 1990s thanks to the Organization of Economic Cooperation and Development (OECD), which played a prominent role in promoting the use of the SI approach in the design and implementation of innovation policy in the OECD countries (Godin 2004). Among the diverse initiatives that took place in the OECD during the 1990s, the seven-year project on National Systems of Innovation (NSIs) (1995–2002) is of special relevance. The OECD had a great influence in the member countries and some of the governments soon adopted the innovation system approach in their innovation policy. However, as argued by Mytelka and Smith 2002, the SI approach has not been entirely successful in making the task of designing policy and proposing policy instruments easier.

This chapter proposes a way of dealing with such complex reality. By breaking down the operation of the SI into 'activities,' the role of the government and the interplay between private and public actors can be discussed, and specific recommendations on how and when public actors should intervene can be made. The point of departure of this chapter for the discussion of innovation policy is the 'generic' SI approach, as discussed briefly in its second section. This section also identifies the main components of the SI approach. The third section presents different approaches to classifying the activities in a SI; in the fourth section, the authors propose ten activities that capture the operation of an innovation system. The role of the public sector in each activity is then discussed, and a new research agenda is proposed; the final section draws some conclusions.

Systems of innovation[1]

There are almost as many definitions of SIs as authors, but most relate in some way to the definition of a system. According to Ingelstam (2002):

(a) a system consists of two kinds of *constituents*: there are, first, some kinds of *components* and, second, there are *relations* among them. The components and relations should form a coherent whole (which has properties different from the properties of the constituents);

(b) the system has a *function*—that is, it is performing or achieving something;

(c) it must be possible to discriminate between the system and the rest of the world; that is, it must be possible to identify the *boundaries* of the system. If we, for example, want to make empirical studies of specific systems, we must, of course, know their extension.[2]

A systemic approach is the point of departure for the literature on technological systems (Dosi 1982; Gille 1978; Hughes 1983; Rosenberg 1982), industrial systems (Hirschman 1958; Porter 1992), and innovation systems. Within this last group, and according to the level of analysis, it is possible to distinguish between (Edquist 1997):

- National Innovation Systems (Freeman 1987; Lundvall 1992; Nelson 1993);
- Regional Innovation Systems (Camagni 1991; Cooke *et al.* 1997; Braczyk *et al.* 1998; Cooke 2001; and Asheim and Isaksen 2002);
- sectoral and 'technological innovation systems' (Breschi and Malerba 1997; Carlsson 1995; Carlsson and Stankiewicz 1991; Malerba 2004).

For the purpose of the discussion here, we propose that an SI includes 'all-important economic, social, political, organizational, institutional and other factors that influence the development, diffusion and use of innovations' (Edquist 1997).[3] If all factors that influence innovation processes are not included in a definition, one has to argue which potential factors should be excluded—and why. This is quite difficult, since, at the present state of the art, we do not know the determinants of innovations systematically and in detail.

What are the components of an SI?

Organizations and institutions are often considered to be the main components of SIs, although it is not always clear what is meant by these terms. Let us, therefore, specify what

organizations and institutions mean here (Edquist 1997).

Organizations are 'formal structures that are consciously created and have an explicit purpose' (Edquist and Johnson 1997). They are 'players or actors.'[4] Some important organizations in SIs are firms (normally considered to be the most important organizations in SIs), universities, venture capital organizations, and public agencies responsible for innovation policy, competition policy, or drug regulation.

Institutions are 'sets of common habits, norms, routines, established practices, rules or laws that regulate the relations and interactions between individuals, groups and organizations,' (Edquist and Johnson 1997). They are the rules of the game. Examples of important institutions in SIs are patent laws, as well as rules and norms influencing the relations between universities and firms. Obviously, these definitions are of a Northian character (North 1990), discriminating between the rules of the game and the players in the game.

Which institutions and organizations are included within the boundaries of the system of innovation is a matter of discussion. Lundvall (1992) distinguishes between a narrow and a broad definition of an SI. The narrow one includes only the organizations and institutions involved in research activities (searching and exploring). This embraces universities, R&D departments in firms, and technological institutes. The broad definition, on the other hand, refers to all 'parts and aspects of the economic structure and the institutional set-up affecting learning as well as searching and exploring' Lundvall (1992: 12). This chapter adopts this broader perspective.

Implications of the SI approach for innovation policy

Innovation policy is public actions that influence innovation processes: that is, the development and diffusion of (product and process) innovations. The objectives of innovation policy are often economic ones, such as economic growth, productivity growth, increased employment and competitiveness. However, they may also be of a non-economic kind, such as cultural, social, environmental, or military. The objectives are determined in a political process, and not by researchers. They must, however, be specific and unambiguously formulated in relation to the current situation in the country and/or in comparison to other countries.

Understanding innovation as a complex interactive learning process has important implications for the design and implementation of any kind of policy to support innovation. It affects the focus of the policy, the instruments, and the rationale for public policy. This chapter will deal mainly with the first two issues, whilst the third will be discussed in detail in Chaminade and Edquist 2005.

The implications of the SI approach for public policy are better understood when its basic assumptions are compared to those of mainstream economics (Lipsey and Carlaw 1998; Smith 2000).

Knowledge, learning, and innovation in mainstream economics

One of the basic assumptions of neoclassical economic theory is perfect information: that is, all economic agents can maximize their profits because they have perfect information about the different options available to them. Knowledge is equal to information: that is, it is codified, generic, accessible at no cost, and easily adaptable to the firm's specific conditions.

These tacit assumptions about the properties of knowledge are included in the discussion about the process of invention. For Nelson (1959) and Arrow (1962), the knowledge emanating from research has some specific properties: uncertainty, unappropriability, and indivisibility (Smith 2000). 'Uncertainty' refers to the impossibility of knowing a priori the outcomes of the research process and the risk associated with it. 'Unappropriability' refers to firms' being unable fully to appropriate the benefits which derive from the invention. As

knowledge is information, freely accessible to all economic agents, this means that there is no incentive for the research activity. Finally, 'indivisibility' implies that there is a minimum scale of knowledge needed before any new knowledge can be created: that is, new knowledge is created on the basis of an existing pool of knowledge (inside or outside the firm). Therefore, it is difficult to separate what constitutes new knowledge from the knowledge that already exists.

For neoclassical economics, the innovation process is narrowed down to research (and invention). How to transform the results of the research activity into products or processes that can be traded in the market is a black box (Rosenberg 1982, 1994). For the neoclassical theorists, the process of innovation is a fixed sequence of phases, where some research efforts will automatically turn into new products.

These three characteristics of scientific knowledge (uncertainty, unappropriability, and indivisibility) will lead to an underinvestment in R&D activities. This constitutes the main rationale for public intervention in research activities. Policy makers have to intervene because of a *market failure*: private actors in the economies will systematically underinvest in R&D, not reaching the optimal allocation of resources for invention.

As argued by Smith (2000), the neoclassical approach, despite its many shortcomings, can be useful for understanding basic science, but it is very limited when trying to explain innovation activities, especially those with closer links to the market.

The policy implications that emerge from the market failure theory are, from a practical and specific point of view, not very helpful for policy makers. They are too blunt to provide much guidance. They do not indicate how large the subsidies or other interventions should be, or within which specific areas one should intervene. They say almost nothing about how to intervene: that is, which policy instruments should be used and the process through which they should be implemented. Standard economic theory is not of much help when it comes to formulating and implementing specific R&D and innovation policies. It only provides general policy implications: for example, that basic research should sometimes be subsidized (Edquist *et al.* 2004). The market failure approach is too abstract to be able to guide the design of specific innovation policies.

Knowledge, learning, and innovation in the SI approach

The general policy implications of the SI approach are different from those of standard economic theory.[5] This has to do with the fact that the characteristics of the two frameworks are very different. The SI approach shifts the focus away from actions at the level of individual and isolated units within the economy (firms, consumers) towards that of the collective underpinnings of innovation. It addresses the overall system that creates and distributes knowledge, rather than its individual components, and innovations are seen as the outcome of evolutionary processes within these systems.

The SI approach has its roots in evolutionary theory (Nelson and Winter 1982). Firms are a bundle of different capabilities and resources (Eisenhardt and Martin 2000; Grant 1996; Spender 1996) which they use to maximize their profit. Knowledge is not only information, but also tacit knowledge; it can be both general and specific and it is always costly. Knowledge can be specific to the firm or to the industry (Smith 2000).

The innovation process is interactive within the firms and among the different actors in the innovation system. At the level of the firm (Kline and Rosenberg 1986), innovation can take place in any part of the firm. Furthermore, Kline and Rosenberg argue that the process of mission-oriented research will be initiated only if the firm cannot find inside or outside the firm, the technical solution in the existing pools of knowledge (Kline and Rosenberg 1986: 291). The SI approach emphasizes the fact that firms do not innovate in isolation but with continuous interactions with the other actors in the system (at regional, sectoral, national, and supranational level).

The main focus of the SI approach is, therefore, the operation of the system and the complex interactions that take place among the different organizations and institutions in the system. Policy makers need to intervene in those areas where the system is not operating well. The policy rationale is based on systemic failures or problems rather than on market failures.

However, the notion of 'market failure' in mainstream economic theory implies a comparison between conditions in the real world and an ideal or optimal economic system. Hence, the notion of failure is associated with the existence of an optimum. However, innovation processes are path dependent over time, and it is not clear which path will be taken. They have evolutionary characteristics. We do not know whether the potentially best or optimal path is being exploited. The system never achieves equilibrium, and the notion of optimality is irrelevant in an innovation context. We cannot specify an ideal or optimal SI. Hence, comparisons between an existing system and an ideal or optimal system are not possible, and the notion of market failure loses its meaning and applicability. Not to lead thoughts in wrong directions, we therefore prefer to talk about systemic problems instead of systemic failures.

Systemic problems mentioned in the literature include (Smith 2000; Woolthuis, Lankhuizen, *et al.* 2005):

- infrastructure provision and investment, including the physical infrastructure (for example, IT, telecom, transport) and the scientific infrastructure (such as high-quality universities and research laboratories, technical institutes);
- transition problems—the difficulties that might arise when firms and other actors encounter technological problems or face changes in the prevailing technological paradigms that exceed their current capabilities;
- lock-in problems, derived from the socio-technological inertia, that might hamper

the emergence and dissemination of more efficient technologies;[6]

- hard and soft institutional problems, linked to formal rules (regulations, laws) as well as more tacit ones (such as social and political culture);
- network problems, which include problems derived from linkages too weak or too strong (blindness to what happens outside the network) in the SI;
- capability problems, linked to the transition problems, referring to the limited capabilities of firms, specially small and medium-size enterprises (SMEs), that might limit their capacity to adopt or produce new technologies over time.

It is obvious that not all these systemic problems can be solved by public intervention, and even in those cases where public intervention is expected, we know very little about how the intervention should take place.

How can we then identify 'problems' that should be subject to innovation policy? As argued earlier, we cannot compare an existing system with an ideal or optimal one (in order to identify a 'systemic problem'). This is contrary to most policy analysis, which basically compares existing situations with imaginary, supposedly optimal or ideal, ones.

What remains are empirical comparisons between different existing systems.[7] Comparison is a means for understanding what is good or bad, or what is a high or a low value for a variable in an SI. Pre-existing systems—national, regional, and sectoral—can be compared with currently existing ones. Or different currently existing systems can be compared with each other. These comparisons must be genuinely empirical and very detailed.[8] If so, they can identify problems that should be subject to policy intervention. Substantial analytical and methodological capabilities are needed to identify these problems.[9] This is what can be called benchmarking.

In order to be able to design appropriate innovation policy instruments, it is also necessary to know at least the most important causes of the problems identified. Not until they know these can policy makers know whether to influence or change organizations, or institutions, or the interactions between them—or something else. Therefore, an identification of a problem should be supplemented by an analysis of its causes as a part of the analytical basis for the design of an innovation policy.

In sum, understanding innovation as a systemic process has important implications for policy makers. The rationale for public intervention changes as well as the focus of that intervention. Under the SI perspective, policy makers need to address systemic problems. The design of an appropriate innovation policy based on the SI approach needs to start with a thorough analysis of the operation of the SI in focus. This is easier said than done. Scholars dealing with innovation systems have focused on the composition of the systems, in terms of institutions and organizations, as well as their measurement and comparison (Pavitt and Patel 1994). But we still know very little about the dynamics of SIs, or the activities within them. Some of the things we know are summarized in the next section.

Activities in the system of innovation: review of the literature

One way of analyzing SIs is to focus not only on their constituents but on what actually happens in the systems. At a general level, the main function—also known as 'overall function'—in SIs is to pursue innovation processes: that is, to develop and diffuse innovations. What we, from now on, call 'activities' in SIs are those factors that influence the development and diffusion of innovations.[10]

Although a system is normally considered to have a function, this was not addressed in a systematic manner in the early work on SIs. From the late 1990s, some contributions on functions or activities in innovation systems were published[11] (Galli and Teubal 1997;

Johnson and Jacobsson 2003; Liu and White 2001; Rickne 2000).[12]

As Table 7.1 shows, the variety of classifications is the result of the different research objectives and definitions of activities. It this sense, four approaches can be distinguished:

- innovation production process, looking at the different activities needed to turn an idea into a new product or process. Edquist (2004), Furman, Porter, *et al.* (2002), and Liu and White (2001) to a lesser extent, are examples of this approach;
- knowledge production process, focusing on how knowledge is created, transferred and exploited. There is here a strong emphasis on the channels and mechanisms for knowledge distribution. David and Foray (1994, 1995) and Johnson and Jacobsson (2003) follow this criterion. This is close to the Aalborg approach to innovation systems as learning systems and the emphasis placed on learning and knowledge dynamics in firms and networks (Lundvall, Johnson, *et al.* 2002);
- organizational performance, using the organizations as the starting point and identifying the activities of the different organizations that have an impact in the innovation system. Borrás (2004) would be an example of this approach;
- innovation policy, using innovation policy as a focal point, that is, what activities (and organizations) in the innovation system can be stimulated by public intervention. The OECD and other international organizations follow this approach. One point of criticism that can be expressed in relation to the OECD approach is that it considers only those activities that can be directly affected by public intervention. It ignores other activities in the system that are equally important, but whose links to innovation policy instruments are not so obvious.

We believe that a different approach is needed, one that starts with the relevant activities in the system of innovation and discusses,

for each of them, what is the division of labor between private and public actors in the performance of each activity.

This will provide policy makers with a new perspective on:

(a) what role they can play in stimulating different activities in the system of innovation;

(b) once the complex division of labor between public and private actors has been unfolded, what could be the appropriate instruments to do this;

(c) how to identify future research needs. This discussion will be taken forward in the next section.

Linking innovation activities in the system of innovation with innovation policy

We believe that it is important to study the activities in SIs—or causes/determinants of innovation processes—in a systematic manner. The hypothetical list of activities presented below is based upon the previous literature review and on our prior knowledge about innovation processes and their determinants. This list is provisional and will be subject to revision as our knowledge about determinants of innovation processes increases. On this basis, we argue that the activities listed below can be expected to be important in most SIs. The main activities in the system of innovation relate to the provision of knowledge inputs to the innovation process (1–2), the demand-side factors (3–4), the provision of constituents of SIs (5–7), and the provision of support services for innovating firms (8–10).

I. Provision of knowledge inputs to the innovation process

1. Provision of Research and Development (R&D) creating new knowledge, primarily in engineering, medicine and the natural sciences.

Table 7.1. Activities in systems of innovation

Author(s)	Definition of function or activity	Main criteria for classification	Breakdown of functions, activities or building blocks
Borrás (2004)	Activities of the different organizations in the system of innovation affecting innovation performance	Role of institutions in the system of innovation	Five generic functions are identified: to reduce uncertainty; to manage conflict and cooperation; to provide incentives, to build competences and to define the boundaries of the system. Ten specific functions in the system of innovation are listed: 1. production of knowledge 2. diffusion of knowledge 3. appropriation of knowledge 4. regulation of labor markets 5. financing innovation 6. alignment of actors 7. guidance of innovators 8. reduction of technological diversity 9. reduction of risk 10. control of knowledge use.
David and Foray (1994, 1995)	Factors affecting the knowledge distribution power of an SI	Knowledge distribution processes organized according to the relationship between organizations	1. Distribution of knowledge (DoK) among universities, research organizations, and industry 2. DoK within a market and between suppliers and users 3. Reuse and recombination of knowledge 4. DoK among decentralized R&D projects 5. Dual technological development of civilian and military technologies
Edquist (2004)	Factors that influence the development and diffusion of innovation	Determinants of the innovation process	1. Knowledge inputs to the innovation process 2. Demand-side factors 3. Provision of constituents in SIs 4. Support services for innovating firms

continues

Table 7.1. Continued

Author(s)	Definition of function or activity	Main criteria for classification	Breakdown of functions, activities or building blocks
Furman, Porter, et al. (2002)	Building blocks required to produce and commercial-ize a flow of technologies new to the world over the long term	Determinants of national innovative capacity	1. Strong innovation infrastructure 2. Strong innovation environments (incl. input conditions, demand conditions, related and supporting industries, and context for firm strategy and rivalry) 3. Linkages between 1 and 2
Galli and Teubal (1997)	Factors affecting the pro-duction and diffusion of innovations	Activities according to type of organization (hard or soft)	Hard functions 1. R&D 2. Supply of scientific and technical services to third parties Soft functions 3. Diffusion of information, knowledge, and technology to bridging organizations 4. Policymaking by government offices 5. Design and implementation of institutions 6. Diffusion/divulgation of scientific cultures 7. Professional coordination through academies, prof. associations.
Johnson and Jacobsson (2003)	Factors that affect the knowledge production processes	Knowledge production processes that can be influenced by public policy	1. Creation of new knowledge 2. Guidance of the research process 3. Provision of resources 4. Generation of knowledge economies 5. Dissemination of market information

Liu and White (2001)	Factors that influence the development, diffusion, and use of technological innovation	Knowledge production process	1. Research 2. Implementation 3. End use 4. Linkage 5. Education
OECD (2002a)	Core blocks in the system of innovation to be considered in a comprehensive innovation policy approach	Innovation Policy	1. Enhancing firm innovative capacities (capacity building) 2. Exploiting Power of markets 3. Securing Investment in knowledge 4. Promoting the commercialization of publicly funded research 5. Promoting cluster development 6. Promoting internationally open networks

2. Competence building (provision of education and training, creation of human capital, production and reproduction of skills, individual learning) in the labor force to be used in innovation and R&D activities.

II. Provision of markets—demand-side factors

3. Formation of new product markets.

4. Articulation of quality requirements emanating from the demand side with regard to new products.

III. Provision of constituents for IS

5. Creating and changing organizations needed for the development of new fields of innovation, for example, enhancing entrepreneurship to create new firms and intrapreneurship to diversify existing firms, creating new research organizations, policy agencies, etc.

6. Provision (creation, change, abolition) of institutions—for example, IPR laws, tax laws, environment and safety regulations, R&D investment routines, etc—that influence innovating organizations and innovation processes by providing incentives or obstacles to innovation.

7. Networking through markets and other mechanisms, including interactive learning between different organizations (potentially) involved in the innovation processes. This implies integrating new knowledge elements developed in different spheres of the SI and coming from outside with elements already available in the innovating firms.

IV. Support services for innovation firms

8. Incubating activities, for example, providing access to facilities, administrative support, etc. for new innovating efforts.

9. Financing of innovation processes and other activities that can facilitate commercialization of knowledge and its adoption.

10. Provision of consultancy services of relevance for innovation processes,

for example, technology transfer, commercial information, and legal advice.

Here we are placing greater emphasis on activities than much of the early work on SIs. Nonetheless, this emphasis does not mean that we can disregard or neglect the components of SIs and the relations among them. Organizations or individuals perform the activities, institutions provide incentives and obstacles influencing these activities. To understand and explain innovation processes, we need to address the relations between activities and components, as well as among different kinds of components.[13]

We believe that understanding the dynamics of each of these activities can be a useful departure point for identifying the role of the government in stimulating the innovation system and the division of labor between public and private actors.

Provision of knowledge inputs to the innovation process

Provision of R&D

R&D is an important basis for some innovations, particularly radical ones in engineering, medicine, and the natural sciences. Such R&D has traditionally been an activity partly financed and carried out by public agencies. This applies to basic R&D, but also to more applied kinds of R&D in some countries. This publicly performed R&D is carried out in universities and in public research organizations. NSIs can differ significantly with regard to the balance between these two kinds of organizations. In Sweden, less than 5 per cent of all R&D is carried out in public research organizations. In Norway, this figure is more than 20 per cent. Public organizations carrying out R&D are also governed or influenced by different institutional rules in different national systems.

However, a considerable part of the R&D in some countries is financed and carried out by the private sector, primarily firms.[14] In 1999, the proportion of all firm-financed R&D in the OECD countries ranged from 21 per cent

(Portugal) to 72 per cent (Japan) (OECD 2002b). Such data may be a way of distinguishing between different types of NSIs. In most NSIs in the world today, little R&D is carried out and most of this is performed in public organizations. Most of these countries are poor or medium-income countries. Those few countries that do a lot of R&D are all rich, and much of their R&D is carried out by private organizations. This includes some large countries, such as the United States (US) and Japan, but also some small and medium-sized countries such as Sweden, Switzerland, and South Korea. There are also some rich countries that do little R&D, for example, Denmark and Norway.

Because innovation processes are evolutionary and path dependent, there is the danger of negative lock-ins, that is, trajectories of innovation that lead to inferior technologies resulting in low growth and decreasing employment. Potentially superior innovation trajectories may not take off and the generation of diversity may be reduced or blocked. In such situations, governments may favor experimentation and use R&D subsidies to support possible alternatives to the winning technologies (Edquist et al. 2004).[15]

Therefore, public organizations can influence R&D activity in different ways, from direct investment and performance through public universities and research centres to stimulating alternative technologies via R&D subsidies. However, much research is needed to understand the relationship between R&D, innovation, productivity growth, the role of R&D in innovation in different sectors, and the impact of different instruments on the propensity of the firms to invest in R&D.

Competence building

The concept of competence building is usually linked to the qualification of human resources. However, it involves other processes and activities related to the capacity to create, absorb, and exploit knowledge.

Here we follow the definition of Lundvall, Johnson, et al. (2002) of competence building that includes: 'formal education and training, the labor market dynamics and the organization of knowledge creation and learning within firms and in networks' (Lundvall, Johnson, et al. 2002).

Education and training of importance for innovation processes (and R&D) are primarily provided by public organizations (schools, universities, training institutes) in most countries. However, some competence building is done in or by firms through learning-by-doing, learning-by-using, or learning-by-interacting. Competence building leads to creation of human capital accumulated in the heads of people: that is, it is a matter of individual learning, the result of which is controlled by individuals.[16]

The organizational and institutional contexts of competence building vary considerably among NSIs. There are particularly significant differences between the systems in the English-speaking countries and continental Europe. However, scholars and policy makers lack good comparative measures on the scope and structure of such differences. There is little systematic knowledge about the ways in which the organization of education and training influences the development and diffusion of innovations. Since labor, including skilled labor, is the least mobile production factor, domestic systems for competence building remain among the most enduringly national of elements of NSIs.

Competence building should not only be limited to human capital. Organizations have competences that exceed those of the employees. Human capital is hired by the company but is always owned by the individual. However, there are ways by which the firm can capture individual knowledge and transform it into organizational knowledge. The organization of the processes of knowledge creation and learning within the firm and in networks are also part of the competence-building activity. Those processes have received attention from the scholars only very recently (Chaminade 2003; Edvinsson and Malone 1997; Guthrie and Petty 2000; Nooteboom 2004; Sanchez, Chaminade, et al. 2000; Tsekouras and Roussos 2005) and many questions remained unanswered.

The role of the government in the timely provision of qualified human resources is clear, although the division of labor between private and public actors is still under debate. However, the situation is very different when we come to components of competence building such as knowledge and learning dynamics. We know very little about knowledge dynamics in firms and in networks. Evidence is based on cases; these can seldom be compared and the evidence is not enough to make generalizations. Little can be said about the role of government supporting these processes, although some attempts have been made (European Commission 2003; OECD 1999). It remains an issue to be further developed.

Formation of new markets and articulation of quality requirements— demand-side factors[17]

In the very early stages of the development of new fields of innovation, there is uncertainty whether a market exists or not. An illustrative example was the belief that the total computer market amounted to four or six computers in the 1950s. Eventually markets develop spontaneously.

One example of market creation is in the area of inventions. The creation of intellectual property rights through the institution of a patent law gives a temporary monopoly to the patent owner. This makes the selling and buying of technical knowledge easier.[18] Public policy makers can also enhance the creation of markets by supporting legal security or the formation of trust.

Another example of public support to market creation is the creation of standards. For example, the Nordic Mobile Telephony Standard (NMT 450) created by the Nordic telecomunication offices (PTTs) in the 1970s and 1980s— when they were state-owned monopolies—was crucial for the development of mobile telephony in the Nordic countries. This made it possible for the private firms to develop mobile systems (Edquist 2003).

In some cases, the instrument of public innovation procurement has been important for market formation. Public innovation procurement is the public buying of technologies and systems, which did not exist at the time. This has been—and is—an important instrument in the defence material sector in all countries. It has also been important in infrastructure development (telecom, trains, etc.) in many countries.[19]

There may also be public subsidies intended to enhance adoption of innovations. One example is subsidies that exist in many countries for electricity produced by windmills.

The provision of new markets is often linked to the articulation of quality requirements, which may be regarded as another activity of the SI. Articulation of quality requirements emanating from the demand side with regard to new products is important for product development in most SIs. It is an important activity, enhancing innovation and influencing processes of innovation in certain directions. Most of this activity is performed spontaneously by demanding customers in SIs. It is a result of interactive learning between innovating firms and their customers. In investigations of collaboration between organizations in their pursuit of innovation such collaboration is one of the most frequent.

Quality requirements can also be a consequence of public action, for example, regulation in the fields of health, safety, and the environment, or the development of technical standards. Public innovation procurement normally includes a functional specification of the product or system wanted, and this certainly means demand articulation that influences product development significantly.

But we know very little about the formation of new markets and the articulation of quality requirements. Instruments such as public procurement, regulation, or subsidies can influence these activities, but further discussion is needed on the adequate division of labor between public and private actors.

Provision of constituents

Creation and change of organizations

As pointed out above, organizations are normally considered to be one of the main components in systems of innovation. Entry and exit of organizations, as well as change of incumbent organizations, is therefore naturally an important activity constituting a part of the change of systems of innovation as such.

Creation and change of organizations for the development and diffusion of innovations is partly a matter of spontaneous firm-creation (through entrepreneurship) and diversification of existing firms (through intrapreneurship). However, public action can facilitate such private activities by simplifying the rules of the game and by creating appropriate tax laws. New R&D organizations and innovation policy agencies can also be created through political decisions.

One important role of policy is to enhance the entry and survival of new firms by facilitating and supporting entrepreneurship. As compared to incumbents, new entrants are characterized by different capabilities, and they may be the socio-economic carriers of innovations. They bring new ideas, products, and processes. Hence, governments should create an environment favorable to the entry of new firms and the growth of successful small and medium-sized firms. Survival and growth of firms often require continuous (or at least multiple) innovation, particularly in high-tech sectors of production.

Enhancement of entrepreneurship and intrapreneurship is a way of supporting changes in the production structure in the direction of new products. There are actually three mechanisms by which the production structure can change through the addition of new products:

(1) existing firms may diversify into new products (examples are found in Japan and South Korea);
(2) new firms in new product areas may grow rapidly (the US provides an example);
(3) foreign firms may invest in new product areas in the country (Ireland is an example).

To add new products to the existing bundle of products is important, since the demand for new products often grows more rapidly than for old ones—with accompanying job creation and economic growth. New products are also often characterized by high productivity growth.

Governments should therefore create opportunities and incentives for changes in the production structure. Policy issues in this context concern how policy makers can help develop alternative patterns of learning and innovation, and nurture emerging sectoral systems of innovation.

In any system of innovation, it is important, from a policy point of view, to study whether the existing organizations are appropriate for promoting innovation. How should organizations be changed or engineered to induce innovation? This dynamic perspective on organizations is crucial in the SI approach, in both theory and practice. Creation, destruction, and change of organizations were very important in the development strategies of the successful Asian economies and they are crucial in the ongoing transformation of Eastern Europe. Hence, organizational changes seem to be particularly important in situations of rapid structural change which, in turn, is linked to building the capacity to deal with changes.

Interactive learning, networking, and knowledge integration

We pointed out above that relations among components are a basic constituent of systems. Interactive learning is a basis for competence building. The SI approach emphasizes interdependence and non-linearity. This is based on the understanding that firms normally do not innovate in isolation but interact with other organizations through complex relations that are often characterized by reciprocity and feedback mechanisms in several loops. Innovation processes are not only influenced by the components of the systems, but also by the relations between them. This captures the non-linear features of innovation processes and is one of the most important characteristics of the SI approach.

The interactive nature of much learning and innovation implies that this interaction should be targeted much more directly than is normally the case in innovation policy today.[20] Innovation policy should not only focus on the components of the systems, but also—and perhaps primarily—on the relations among them. Relations between organizations may occur through markets but also through other mechanisms. This implies integrating new knowledge elements developed in different spheres of the SI and coming from outside with elements already available in the innovating firms.

Most interaction between organizations involved in innovation processes occurs spontaneously when there is a need. The activity of (re)combining knowledge—from any source—into product and process innovations is largely carried out by private firms. They often collaborate with other firms, but sometimes universities and public research organizations are also involved. The long-term innovative performance of firms in science-based industries is strongly dependent upon the interactions between firms and universities and research organizations. If they are not spontaneously operating smoothly enough, these interactions should be facilitated by means of policy. Here formal institutions are important, as we will see in the next subsection.

The relations between universities and public research organizations on the one hand and firms on the other are coordinated only to a limited degree by markets. This linkage activity is addressed (by policy) in different ways, to different extents in different NSIs, and sometimes not at all. Incubators, technology parks, public venture capital organizations—to be discussed in later subsections—may also be important in similar ways. This means that the public sector may create organizations to

facilitate innovation. At the same time, however, it may create the rules and laws that govern these organizations and their relations to private ones—that is, create institutions (Edquist *et al.* 2004).

Creation and change of institutions

As shown above, institutions are normally considered to be the second main component (in addition to organizations) in SIs. The creation, abolition, and change of institutions are activities crucial to the maintenance of SIs' dynamism.

Important institutions in systems of innovation are intellectual property rights (IPR) laws, technical standards, tax laws, environment and safety regulations, R&D investment routines, firm-specific rules and norms, etc.; these influence innovating organizations and innovation processes by providing incentives or obstacles for organizations and individuals to innovate.

IPR laws are considered to be important as a means of creating incentives to invest in knowledge creation and innovation (and, as we have seen, they are leading to the creation of markets). Tax laws are also often considered to influence innovation processes. An important question here is which kinds (and levels) of taxes become obstacles or facilitators of innovation (and entrepreneurship).

We have already mentioned the important role of institutions in facilitating the interaction between organizations in the previous subsection. Governments may, for example, support collaborative centers and programs, remove barriers to cooperation, and facilitate the mobility of skilled personnel between different kinds of organizations. This might include the creation or change of institutional rules that govern the relations between universities and firms, such as the one in Sweden stating that university professors shall perform a 'third task' in addition to teaching and doing research: that is, interact with the society surrounding the university, including firms (Edquist *et al.* 2004).

Some kinds of institutions are created by public agencies. They are often formal (codified) ones. Others develop spontaneously over history without public involvement. There are institutions that influence firms and there are institutions that operate inside firms.[21]

Those formal institutions that are created by public agencies are policy instruments. Public innovation policy is largely a matter of formulating the rules of the game that will facilitate innovation processes. These rules might have nothing to do with markets, or they might be intended to create markets or make the operation of markets more efficient.

Just as in the case of organizations, it is important, from a policy point of view, to study whether the existing institutions are appropriate for promoting innovation and to ask the same question of how institutions should be changed or engineered to induce innovation. Here, too, the evolution and design of new institutions were very important in the development strategies of the successful Asian economies as well as in the ongoing transformation of Eastern Europe. Hence, institutional (as well as organizational) changes are particularly important in situations of rapid structural change.

Support services for innovative firms

Incubation

Incubating activities include such things as provision of access to facilities and administrative support for new innovating efforts. We know very little about how incubating activities emerge in the SI. Incubating activities have been carried out in science parks to facilitate commercialization of knowledge in recent decades. That this activity has become partly public has to do with the uncertainty characterizing early stages of the development of new products, which means that markets do not operate well in this respect.

However, innovations are also emerging in existing firms through incremental innovation and when they diversify into new product areas. In those cases, the innovating firms normally provide incubation themselves. There is a need to understand better the conditions under which incubation needs to be a public

activity and when it should be left to the private initiative.

Financing

Financing of innovation processes is necessary for the commercialization of knowledge into innovations and their diffusion. Financing of innovation is primarily done by private organizations within innovating firms, through stock exchanges, by venture capital organizations, or through individuals (business angels). Again, however, financing is sometimes—for example in the form of seed capital—provided by public organizations in many countries, including the US.

As in all public interventions, financing should only be provided publicly when firms and markets do not spontaneously perform this activity (for example, when uncertainty is too large). But the question is not just when the public sector should finance innovation activities but also how: that is, what should be the instruments and what should be the appropriate balance between public and private funding in a particular SI.

Provision of consultancy services

Consultancy services are very often of importance for innovation processes. Those of relevance for innovation processes are, for example, technology transfer, commercial information, and legal advice. They are primarily carried out by private organizations. If they are large and rich in competence in various fields, the innovating firms themselves may do this in cases where the innovations are created by diversification processes. They may also be provided by specialized consultancy firms both in such cases and in cases where a new firm is established around the innovation.

Specialized consultancy firms are normally classified as Knowledge Intensive Business Services (KIBS), a service sector that is growing rapidly. KIBS firms provide services in the field of computer hardware and software, other technical services, management, marketing, patenting, legal advice, accounting, etc.

But there are certain cases (groups of SMEs, mature sectors) where these services are also provided by public authorities, either directly or by their acting as broker between firms and service providers. Examples of these can be found in regional public agencies.

Once again, the discussion of the division of labor between public and private actors needs to be supported by more evidence of the systemic problems that give reason for public intervention.

Conclusions and future research agenda

This chapter has placed a great emphasis on activities that operate in SIs. However, this emphasis does not mean that we can neglect the components of SIs and the relations among them. *Organizations* or individuals perform the activities and *institutions* provide incentives and obstacles. We believe that the analysis of innovation systems proposed here can fruitfully be used for innovation policy purposes, and that the activities that influence innovation processes in the systems are a useful point of entry in the policy analysis. Thereafter, one can identify the organizations performing the activities and see that there is not a one-to-one relation between them, but that a certain kind of organization can perform more than one activity and that many activities can be carried out by more than one category of organization.

A similar exercise can be carried out for innovation policy: we can analyze the division of labor between private and public organizations with regard to the performance of each of the activities in innovation systems, investigate whether these activities are performed by private or public organizations, and whether this division of labour is motivated or not.

The policy discussion at each point is focused upon changes in the division of labor between the private and the public spheres or upon changes in those activities already carried out by the public agencies. This includes adding new public policy activities as well as terminating others. Terminating activities carried out by public organizations are not the least important!

However, the discussion of the division of labor between private and public organizations is burdened by our lack of knowledge on several issues that should be part of our future research agenda in innovation policy and systems of innovation:

(*a*) How is each of the activities related to the propensity to innovate in the system of innovation?

(*b*) Which institutional rules are governing each activity?

(*c*) What is the role of private and public actors for each activity; and how has it evolved over time?

(*d*) What are the differences between countries in these respects?

Notes

1. This section is partly based on Edquist (2004).
2. Only in exceptional cases is the system closed in the sense that it has nothing to do with the rest of the world (or because it encompasses the whole world).
3. A more detailed discussion of the different definitions of national systems of innovations can be found in Edquist (2004).
4. Although there are kinds of actors other than organizations—for example individuals—the terms 'organizations' and 'actors' are used interchangeably in this chapter.
5. The rest of this subsection is based upon Edquist (2001) and Edquist *et al.* (2004).

6. One clear example of lock-in is fossil energy (Smith 2000). The productive system is so dependent on fossil energy that it is preventing the expansion of new forms of energy (such as solar, eolic, etc.).

7. One may also compare existing systems with 'target systems,' that is, systems that have characteristics that are wanted by someone. Such target systems must not, however, be confused with ideal or optimal systems.

8. To carry out such comparisons is one of the objectives in a current project, which compares the national systems of innovation in Norway, Sweden, Finland, Denmark, the Netherlands, Ireland, Singapore, Hong Kong, Taiwan, and South Korea. It will be published as Edquist and Hommen (2004).

9. Such capabilities are also needed to design policies that can mitigate the problems.

10. Examples of activities are R&D as a means of the development of economically relevant knowledge that can provide a basis for innovations, or the financing of the commercialization of such knowledge, that is, its transformation into innovations. The activities in SIs are the same as the determinants of the main function. An alternative term to 'activities' could have been 'subfunctions.' We chose 'activities' in order to avoid the connotation of 'functionalism' or 'functional analysis' as practiced in sociology, which focuses on the consequences of a phenomenon rather than on its causes, which are in focus here (Edquist 2004).

11. We have broadened our analysis to include not only those contributions specifically dealing with activities in the NSI but also those that discuss the determinants of the innovation capacity or learning competences of a national system of innovation as both are relevant for the discussion on policy issues.

12. This work is summarized in Edquist (2004).

13. These relations are addressed in Edquist (2004).

14. There are also public financial support schemes to stimulate firms to perform R&D. One example is tax credits for R&D.

15. Another policy instrument that can be used for the same purpose is public innovation procurement.

16. There is also organizational learning, the result of which is controlled or owned by firms and other organizations. Organizational learning leads to the accumulation of 'structural capital,' a knowledge-related asset controlled by firms (as distinct from 'human capital'). An example is patents. Organizations can also accumulate knowledge thanks to their ability to combine knowledge bases of individuals. Organizations have an interest in transforming individual knowledge into organizational knowledge.

17. In the discussion here, we have chosen to discuss the two activities related to the demand side that were mentioned in the beginning of the fourth section in one subsection. They could as well have been discussed in separate subsections.

18. Paradoxically, then, a monopoly is created by law, in order to create a market for knowledge: that is, to make it possible to trade in knowledge. This has to do with the peculiar characteristics of knowledge as a product or commodity. It is hard for a buyer to know the price of knowledge, since you do not know what it is before the transaction. (If you know what it is, you do not want to pay for it.) In addition, knowledge is not worn out when used—unlike other products.

19. Public innovation procurement is analysed in Edquist et al. (2000).

20. Interactive learning has been studied empirically by Lundvall (1992) and Meeus and Oerlemans (2001).

21. For taxonomies of institutions see Edquist and Johnson (1997).

References

Arrow, K. (1962). 'Economic Welfare and the Allocation of Resources for Invention.' In R. Nelson (ed.), *The Rate and Direction of Inventive Activity*. Princeton: Princeton University Press, 609–29.

Asheim, B., and Isaksen, A. (2002). 'Regional Innovation Systems: The Integration of 'Sticky' and Global 'Ubiquitous' Knowledge.' *Journal of Technology Transfer*, 27: 77–86.

Borrás, S. (2004). 'System of Innovation: Theory and the European Union.' *Science and Public Policy*, 31/6: 425–33.

Braczyk, H. J., Cooke, P., and Heindenreich, M. (eds.) (1998). *Regional Innovation Systems: The Role of Governance in a Globalised World*. London: UCL.

Breschi, S., and Malerba, F. (1997). 'Sectoral Innovation Systems: Technological Regimes, Schumpeterian Dynamics and Spatial Boundaries.' In C. Edquist (ed.), *Systems of Innovation: Technologies, Institutions and Organizations*. London: Pinter, 130–56.

Camagni, R. (ed.) (1991). *Innovation Networks: Spatial Perpectives*. London: Belhaven Press.

Carlsson, B. (ed.) (1995). *Technological Systems and Economic Performance: The Case of Factory Automation*. Dordrecht: Kluwer Academic Publishers.

—— and Jacobsson, S. (1993). 'Technological Systems and Economic Performance: The Diffusion of Factory Automation in Sweden.' In D. Foray and C. Freeman (eds.), *Technology and the Wealth of Nations*. London: Pinter.

—— and Stankiewicz, R. (1991). 'On the Nature and Composition of Technological Systems.' *Journal of Evolutionary Economics*, 1/2: 93–119.

Chaminade, C. (2003). 'How IC Management Can Make Your Company More Innovative.' In N. I. Fund (ed.), *How to Develop and Monitor Your Company's Intellectual Capital*. Oslo: Nordic Industrial Fund.

—— and Edquist, C. (2005). 'Rationales for Innovation Policies: Theories and Criteria for Public Innovation.' CIRCLE Discussion Paper Series. Lund: Center for Innovation Research and Competence in the Learning Economy.

Cooke, P. (2001). 'Regional Innovation Systems, Clusters, and the Knowledge Economy.' *Industrial and Corporate Change*, 10/4: 945–74.

—— Gomez Uranga, M., and Etxebarria, G. (1997). 'Regional Systems of Innovation: Institutional and Organizational Dimensions.' *Research Policy*, 26: 475–91.

David, P., and Foray, D. (1994). *Accessing and Expanding the Science and Technology Knowledge Base: A Conceptual Framework for Comparing National Profiles in Systems of Learning and Innovation*. Paris: OECD.

—— —— (1995). 'Interactions in Knowledge Systems: Foundations, Policy Implications and Empirical Methods.' *STI Review*, 16: 14–68.

Dosi, G. (1982). 'Technological Paradigms and Technological Trajectories.' *Research Policy*, 11/3: 147–62.

Edquist, C. (1997a). 'Systems of Innovation Approaches: Their Emergence and Characteristics.' In C. Edquist (ed.), *Systems of Innovation: Technologies, Institutions and Organizations*. London: Pinter.

—— (ed.) (1997b). *Systems of Innovation: Technologies, Institutions and Organizations*. London: Pinter.

—— (2001). 'Innovation Policy: A Systemic Approach.' In B-Å. Lundvall and D. Archibugi (eds.), *Major Socio-Economic Trends and European Innovation Policy*. Oxford: Oxford University Press.

—— (2003). 'The Fixed Internet and Mobile Telecommunications Sectoral System of Innovation: Equipment, Access and Content.' In C. Edquist (ed.), *The Internet and Mobile Telecommunications System of Innovation: Developments in Equipment, Access and Content*. Cheltenham: Edward Elgar.

—— (2004). 'Systems of Innovation: Perspectives and Challenges.' In J. Fagerberg (ed.), *The Oxford Handbook of Innovation*. Oxford: Oxford University Press, 181–208.

—— and Hommen, L. (2004). 'Comparative Framework for, and Proposed Structure of, the Studies of National Innovation Systems in Ten Small Countries.' Lund University, Working Paper.

—— —— (2005). 'Small Economy Innovation Systems: Comparing Globalisation, Change and Policy in Asia and Europe.' Cheltenham: Edward Elgar.

—— and Johnson, B. (1997). 'Institutions and Organizations in Systems of Innovation.' In C. Edquist (ed.), *Systems of Innovation: Technologies, Institutions and Organizations*. London: Pinter, 33.

—— Ericsson. M-L., and Sjögren, H. (2000). 'Collaboration in Product Innovation in the East Gothia Regional Innovation System.' *Enterprise and Innovation Management Studies*, 1.

—— *et al.* (2004). 'Sectoral Systems: Implications for European Innovation Policy.' In F. Malerba (ed.), *Sectoral Systems of Innovation: Concepts, Issues and Analysis of Six Major Sectors in Europe*. Cambridge: Cambridge University Press.

Edvinsson, L., and Malone, M. (1997). *Intellectual Capital: Realizing Your Company's True Value by Finding Its Hidden Brainpower*. New York: Harper Collins.

Eisenhardt, K., and Martin, J. (2000). 'Dynamic Capabilities: What Are They?' *Strategic Management Journal*, 21/9: 1105–21.

European Commission (2003). Study on the Measurement of Intangible Assets and Associated Accounting Practices. Available at **http://europa.eu.int/comm/enterprise/services/business_ services/documents/ studies /intangiblesstudy.pdf.%20APRIL% 202003**

Freeman, C. (1987). *Technology Policy and Economic Performance: Lessons from Japan*. London: Pinter.

—— (2002). 'Continental, National and Sub-National Innovation Systems: Complementarity and Economic Growth.' *Research Policy*, 31: 191–211.

Furman, J., Porter, M., *et al.* (2002). 'The Determinants of National Innovative Capacity.' *Research Policy*, 31: 899–933.

Galli, R., and Teubal, M. (1997). 'Paradigmatic Shifts in National Innovation Systems.' In C. Edquist (ed.), *Systems of Innovation: Technologies, Institutions and Organization*. London: Pinter.

Gille, B. (1978). *Histoire des techniques*. París: Gallimard.

Godin, B. (2004). 'The New Economy: What the Concept Owes to the OECD.' *Research Policy*, 33/5: 679–90.

Grant, R. (1996). 'Toward a Knowledge-Based Theory of the Firm.' *Strategic Management Journal*, 17 (Winter special issue): 109–22.

Guthrie, J., and Petty, R. (2000). 'Intellectual Capital Literature Review.' *Journal of Intellectual Capital*, 1/2: 155–76.

Hirschman, A. (1958). *The Strategy of Economic Development*. New Haven: Yale University Press.

Hobday, M. (2004). 'Firm-Level Innovation Models: Perspectives on Research in Developed and Developing Countries.' *Technology Analysis and Strategic Management*, 38.

Hughes, T. P. (1983). *Networks of Power: Electrification in Western Society 1880–1930*. Baltimore: Johns Hopkins.

Ingelstam, L. (2002). 'System—Att Tänka Över Samhälle Ach Teknik ('Systems—to Reflect over Society and Technology'—in Swedish).' *Statens Energimyndighetents förlag*. Eskilstuna 2002.

Johnson, A., and Jacobsson, S. (2003). 'The Emergence of a Growth Industry: A Comparative Analysis of the German, Dutch and Swedish Wind Turbine Industries.' In J. S. Metcalfe and U. Canter (eds.), *Transformation and Development: Schumpeterian Perspectives*. Heidelberg: Physica/Springer.

Kline, S., and Rosenberg, N. (1986). 'An Overview of Innovation.' In L. A. Rosenberg (ed.), *The Positive Sum Strategy*. Washington, DC: National Academy of Sciences, 289.

Lipsey, R., and Carlaw, K. (1998). 'A Structuralist Assessment of Technology Policies: Taking Schumpeter Seriously on Policy.' Industry Canada Research Publications Program, 123.

Liu, X., and White, S. (2001). 'Comparing Innovation Systems: A Framework and Application to China's Trasitional Context.' *Research Policy*, 30/7: 1091–114.

Lundvall, B.-Å. (ed.) (1992). *National Systems of Innovation: Towards a Theory of Innovation and Interactive Learning*. London: Pinter.

—— Johnson, B. *et al.* (2002). 'National Systems of Production, Innovation and Competence Building.' *Research Policy*, 31: 213–31.

Mahdi, S. (2003). 'Search Strategy in Product Innovation Process: Theory and Evidence from the Evolution of Agrochemical Lead Discovery Process.' *Industrial and Corporate Change*, 12/2: 235–70.

Malerba, F. (ed.) (2004). *Sectoral Systems of Innovation: Concepts, Issues and Analyses of Six Major Sectors in Europe*. Cambridge: Cambridge University Press.

Malerba, F. and Orsenigo, L. (1997). 'Technological Regimes and Sectoral Patterns of Innovation Activities.' *Industrial and Corporate Change*, 6: 83–117.

Meeus, M. T. H., Oerlemans, L. A. G., and Hage, J. (2001). 'Patterns of Interactive Learning in a High-Tech Region: An Empirical Exploration of Complementary and Competing Perspectives.' *Organization Studies*, 22/1: 145–72.

Mytelka, L. K., and Smith, K. (2002). 'Policy Learning and Innovation Theory: An Interactive and Co-Evolving Process.' *Research Policy*, 31.

Nelson, R. (1959). 'The Simple Economics of Basic Scientific Research.' In N. Rosenberg (ed.), *The Economics of Technological Change*. Harmondsworth: Penguin Books: 478.

—— (ed.) (1993). *National Innovation Systems: A Comparative Analysis*. New York: Oxford University Press.

—— and Winter, S. (1982). *An Evolutionary Theory of Economic Change*. Cambridge, MA: Harvard University Press.

Nooteboom, B. (2004). *Inter-Firm Collaboration, Learning and Networks: An Integrated Approach*. London: Routledge.

North, D. C. (1990). *Institutions, Institutional Change and Economic Performance*. Cambridge: Cambridge University Press.

OECD (1999). *Measuring and Reporting Intellectual Capital: Experience, Issues and Prospects*. Amsterdam: OECD.

—— (2002a). *Dynamising National Innovation Systems*. Paris: OECD.

—— (2002b). *OECD Main Science and Technology Indicators*. Paris: OECD.

Pavitt, K., and Patel, P. (1994). 'National Innovation Systems: Why They Are Important and How They Might Be Measured and Compared.' *Economics of Innovation and New Technology* 3/1: 77–95.

Porter, M. (1992). *The Competitive Advantage of Nations*. London: Macmillan.

Rickne, A. (2000). 'New-Technology Based Firms and Industrial Dynamics: Evidence from the Technological Systems of Biomaterials in Sweden, Ohio and Massachussetts.' Department of Industrial Dynamics, Chalmers University of Technology.

Rosenberg, N. (1982). *Inside the Black Box*. Cambridge: Cambridge University Press.

—— (1982). 'Technological Interdependence in the American Economy.' In N. Rosenberg (ed.), *Inside the Black Box*. Cambridge: Cambridge University Press.

—— (1994). *Exploring the Black Box: Technology, Economics and History*. Cambridge: Cambridge University Press.

Sanchez, M. P., Chaminade, C. *et al.* (2000). 'Management of Intangibles: An Attempt to Build a Theory.' *Journal of Intellectual Capital*, 1/4: 312–27.

Smith, K. (2000). 'Innovation as a Systemic Phenomenon: Rethinking the Role of Policy.' *Enterprise and Innovation Management Studies*, 1/1: 73–102.

Spender, J. C. (1996). 'Making Knowledge the Basis of a Dynamic Theory of the Firm.' *Strategic Management Journal*, 17 (Winter special issue): 45–62.

Tsekouras, G., and Roussos, G. (2005). 'Learning Networks and Service-Oriented Architectures.' In D. Schwartz (ed.), 'Encyclopedia of Knowledge Management.'

Woolthuis, R. K., Lankhuizen, M., *et al.* (2005). 'A System Failure Framework for Innovation Policy Design.' *Technovation*, 25: 609–19.

PART II

SCIENTIFIC RESEARCH: NEW FRAMEWORKS

Introduction

Jerald Hage

Research on industrial innovation has focused on neither studies of industrial research laboratories, nor the more general, nebulous phenomena of science; yet scientific research lies at the heart of a vibrant economy, a secure society, and the system of knowledge production. More and more product innovations, even those in the traditional sectors, are based on advances in science. This is obvious in the case of such high-tech industries as semiconductors, pharmaceuticals, aircraft, etc.; but is also true for such traditional industries as mattress manufacturing, light bulbs, and many simple home products. A major reason for the changes in many of these industries is the development of new materials. Often based on nanotechnology that allows for the creation of materials with very special optical, electrical, and physical properties, they are important in the many mundane products in the home: examples range from construction materials made of new plastic to more scientifically designed mattresses, to say nothing of stain-proof pants.

Everyone stresses the importance of scientific research for the economy. And since 9/11, there has been a surge in research in matters of defense: on how to detect anthrax in letters, weapons hidden in luggage, viruses in the air, etc. The problem of how to protect various installations such as dams, bridges, skyscrapers, subways, schools, hospitals, and government buildings from terrorist attacks will require a whole new set of protective devices. Although this is applied research, much of it requires advances in basic scientific research as well.

Despite the large amount of literature on industrial innovation (as indicated in the bibliographies in the first part), and despite many case studies in both the history of science and in science and technology evaluations, there has been surprisingly little accumulation of research findings on scientific research. In this part on research agendas, we explain why this is so and suggest new frameworks. The major argument is that too much attention has been focused on descriptive detail and counts of papers, patents, and even citations, and not enough on developing general variables that would discuss how much of a scientific advance had been made. This part provides a framework that corrects for this deficiency.

Justification for a separate section on scientific innovation rests on two intellectual arguments about how thinking about science and technology needs to be changed. The first proposed change is to develop a new vocabulary, not only for making science and technological evaluations, but to examine the nature of research projects that produce scientific advances. The second proposed change is to develop frameworks for understanding knowledge trajectories and collective learning, to parallel the work on organizational learning.

Each of these points needs to be briefly developed. The intellectual development of the literature on industrial innovation started because there were general variables for describing the nature of organizations that had high rates of innovation (see Hage 1965; Damanpour 1991; Zammuto and O'Connor 1992; Hage 1999). With this, it became possible to test hypotheses in different kinds of organizations and accumulate a set of findings across time. The same kind of framework is needed for studies of scientific research projects and is included in the first contribution in this part. Drawing parallels wherever possible with the work on product and process innovation is perhaps the most important intellectual reason for this new agenda and way of thinking in the sociology and history of science literatures. Two contributions in this book provide models of how this can be done: one is for contemporary research and thus the sociology of science (see Jordan in this part); the other is for historical research and thus the history of science (see Hollingsworth in Part IV).

At the same time, precisely because the content of industrial products is quite different from the nature of knowledge, parallels between industrial innovation and scientific research can easily break down. To study adequately how knowledge transforms itself over time, propagates through space, and precipitates collective learning requires a different kind of framework from the one used for the study of interorganizational relationships. The contributions of Mohrman, Galbraith, and Monge (in this part) and of Shinn (in the next) provide examples of how to approach this issue.

The advantages of combining different disciplines and both American and European perspectives are especially evident in this part. With two Americans and three Europeans, the former concentrating on the meso-level of organizations and management, the latter on the macro-level, the five contributors reflect a diversity of disciplinary specialties as well. Together, they provide a more complete and rich perspective in their suggested frameworks for studying science and technology than would be possible otherwise.

Frameworks for new research agendas

The interrelationship between scientific research and industrial innovation requires a slight shift in emphasis in the national system of innovation, described by Chaminade and Edquist in the previous section as a 'knowledge production system.' In each arena in the idea-innovation network, as defined by Kline and Rosenberg (1986) and described in the contribution by van Waarden and Oosterwijk in Part IV, knowledge is created: it is, however, a different kind of knowledge. The arenas are basic science knowledge, applied science knowledge, prototype or model development knowledge, manufacturing knowledge, quality-control knowledge, and consumer or marketing knowledge. This knowledge can be in the form of ideas or tools/techniques and also involves communities of practice. In each area, there is a problem of transition: that is, will the basic science knowledge be transformed into applied science knowledge? Nor is this a linear process. New knowledge in any one of these arenas creates opportunities for the creation of new knowledge in any one of the others. Although this section concentrates primarily on new basic science and applied science knowledge, it is important to recognize the different arenas in the knowledge production system.

Defining the object of research on scientific advances

Despite the importance of new discoveries and new research technologies, science and scientific innovation has been much less studied than product and service innovation. The literature on R&D management has been much more concentrated on industrial research laboratories (Hauser and Zettelmeyer 1997; see Jordan *et al.* 2003 for an annotated bibliography; Kim and Wilemon 2003; Saleh and Wang 1993; Schuman *et al.* 1995; Thomas and McMillan 2001). While some studies have examined scientific laboratories (Brown 1997; Coccia 2001; Grimaldi and von Tunzelmann 2002; Sadeh *et al.* 2000), they have not looked at the degree of radicalness in the scientific advance. Making up for this lacuna provides another justification for a separate section devoted to this topic.

In the shift to the study of scientific research there are three separate problems to be addressed:

(*a*) What are the kinds of outcomes of scientific research?
(*b*) What are the dimensions for describing these outcomes?
(*c*) What is the appropriate unit of analysis?

One of the most striking differences between scientific research and industrial innovation is that multiple outcomes from the research can be seen. Jordan in the first contribution suggests that there are the following:

- scientific advances as in new ideas, concepts, and theories;
- new research tools;
- new capabilities in the human capital of the researchers;
- collective learning as knowledge diffuses.

Given this book's unifying theme of knowledge, the definition of science is not limited to advancing knowledge for knowledge's sake, or of gaining certainty by replicating the findings, but includes these other aspects as well. This listing moves considerably beyond the framework of Larédo and Mustar (2000) who suggested contrasting research organizations on the basis of what they called their 'profiles' in terms of their functional contributions to society. Here Jordan is emphasizing the importance of quantifying these outcomes in some way, which leads to the next, critical point, the dimensions for doing so.

Jordan proposes two general variables for dimensions: the degree of radicalness and the scale of the research. These dimensions can be employed to describe either the outputs or the strategic choices and thus the intent of the research. One of the dimensions, incremental versus radical, is also used by Hollingsworth in his contribution in Part IV. Both contributions indicate how this could be measured. The main point is that, by emphasizing the radicalness of the scientific advance, research on the evaluation of science and technology (S&T) can then begin to accumulate sorely needed findings across quite disparate areas of substantive research.

One story untold even in the history and sociology of science is the critical importance of what might be called 'big science:' that is, the huge investments needed to study particular aspects of the physical environment. (Hagstrom wrote a book on this topic in 1965 but big science has largely disappeared from the research agenda.) Consider all the costs of obtaining measures of the wind patterns worldwide to make predictions thirty days in advance; or the cost of oceanographic ships to study water currents to assess the consequences of global warming; or radio telescopes for research on black holes and the beginnings of the universe. Nor are expensive research technologies restricted to the physical and biological sciences. Various surveys, such as the general social survey in the US, or the innovation survey in Europe, represent attempts to provide expensive data for others to analyze. The role of research technologies has largely been ignored; yet it is a vital topic that needs much more attention than it has received. Again, several of the contributions in this book call attention to how research technologies can be usefully examined.

The two suggested approaches have different advantages for someone interested in studying scientific research. By emphasizing two dimensions, and also by indicating the multiple ways in which they can be measured, as well as the idea of multiple outcomes, Jordan provides a framework not only for S&T evaluation but for the sociology of science. By emphasizing prizes, Hollingsworth provides a contrasting framework for the historical study of radical scientific advances or, as he labels it, innovation.

The argument of both contributions is that rather than study papers or patents, one needs to quantify the extent of the advance represented in the paper or patent. A single paper was used to propose the special theory of relativity and the model of DNA. Clearly, these were extraordinary contributions to science. The same problem exists at the level of patents: the patent for transistors represented a far more radical advance than many other patents that are registered. One might assume that citation counts might do this, but there is often not a one-to-one correspondence between the number of citations and the radicalness of the advance in science, akin to radical innovation in Part I.

Once one starts to measure the radicalness of advance, the first, obvious way to see how thinking is changed is to draw parallels with the literature on radical innovation in Part I. But this is not a simple borrowing from the industrial innovation literature. Jordan's second dimension of scale allows one to begin to rethink the literature on industrial innovation and observe that not enough attention has been paid to large systemic innovations such as new cars, planes, and the Internet. The contribution by van Waarden and Oosterjick in Part IV describes radical innovation in the telecommunications system and how it was altered as a consequence.

Once one can measure particular kinds of outputs along specific dimensions, there still remains the issue at what level this should be accomplished. Both the Jordan and the Hollingsworth contributions are concerned with what might be called the lowest common dominator, namely the research project. But they are equally concerned with the context of the research organization as well. It is this multi-layered approach that makes both of their frameworks particularly rich for advancing the study of scientific research. Again, this can have a positive benefit for industrial innovation research which, rather than studying the multiple levels, has tended to concentrate only on the innovative organization and, one might add, a single model of it, namely the organic structure (see Hage 1965; Damanpour 1991; Zammuto and O'Connor 1992; Hage 1999). Indeed, this is one of the reasons why studies of industrial research within firms have not received much attention until recently.

The focus on the research project and, beyond this, the research organization rather than the brilliant or creative scientist, represents a shift for the sociology and history of science which has emphasized the individual rather than the team. No one would think today that product innovation is the work of a single individual, and recent research has demonstrated how the discoveries of Edison and of Ford were dependent upon teams of individuals and networks of innovation. This same collective perspective is sorely lacking in the study of scientific research.

Frameworks for the determinants of the degree of radicalness in the scientific research advance

Once decided on the measures of what is a radical advance, the next logical question is, what are the determinants of these advances at either the organization/project or the institutional level? Again, the contributions of Jordan and of Hollingsworth provide interesting and somewhat contrasting models of determinants. Hollingsworth emphasizes the constraints of the national system of innovation; Jordan describes multiple models for examining both projects and research organizations. Hollingsworth has a more probabilistic view of radical scientific advances or innovation than does Jordan. Finally, Jordan stresses more the extra-organizational linkages in two of her four generic types of research profiles. This is another needed corrective to the industrial innovation literature, which has traditionally emphasized the organic model and internal cross-functional teams rather than external linkages (Hage 1999).

Regardless, both provide rich research agendas about how to study the management of research projects, laboratories, and their organizations. This leads naturally into the testing of hypotheses about how the management and leadership of research projects and organizations can influence the kinds of outcomes that are obtained. This means not only the gradual accumulation of knowledge about the production of knowledge, but also begins to provide governments with ways of intervening in their research organizations so as to facilitate more radical advances if so desired.

Both of these models of determinants have emerged from extensive inductive and deductive research. Jordan has been studying technical progress in the scientific research projects of the Department of Energy, while Hollingsworth has been focusing on biomedical research in the US and Britain. Thus, their frameworks are well grounded in what might be called the reality of scientific research.

Frameworks for measuring knowledge trajectories and collective learning

It is fascinating to compare the nature of the linkages illustrated in the various contributions in Part I on product and service innovation and this one on scientific advances as measures of innovation. In the Meeus and Faber contribution, interorganizational networks are concrete entities; in the Mohrman *et al.* contribution, we have a community of practice generated by flows of knowledge. The reason for these differences stems partly from the goals of these specific studies. In the former, the objective is to ask whether interorganizational networks contribute more to product innovation than if each organization were to conduct its research on its own. In the latter instance, the objective is to plot how knowledge propagates across space, adding value to public investments in basic research.

Mohrman, Galbraith, and Monge see basic science as an ecosystem that operates as a self-organizing community of communities. Communities (of professional practice) exist at multiple levels of analysis and cut across traditional organizational boundaries. They arise to address particular problems and scientific questions, and re-form over time as the state of knowledge in an area advances, particular pathways fail, and/or the nature of the problems that participants find interesting changes. Basic research communities are knowledge networks.

The framework provided in the Mohrman, Galbraith, and Monge contribution allows one to measure the extent of the knowledge community and also the trajectory of knowledge and how this can change over time. As Jordan observed in her contribution, one of the important outcomes of scientific research is both the development of new capabilities and also collective learning. The Mohrman, Galbraith, and Monge contribution allows us to address this problem directly.

Shinn's contribution provides another approach to measuring the amount of collective learning that has occurred. He observes how some technologies become generic ones with applications across a number of disciplinary fields, creating in turn another kind of community of practice, one based on a common technology. Subsequently, this has led to the formation of new professional associations concerned with the specific research technology.

Frameworks for studying the problem of coordination of scientific research

The book's introduction highlighted the very special problems associated with the flows of knowledge. One suggestion was that markets may not be the best coordination modes for facilitating the transfer of knowledge from one part of the scientific subsystem to another. The contribution by Georghiou demonstrates that, in the case of Britain and basic scientific research, markets are not effective. Here is a set of ideas that now can be tested in other countries, where similar experiments of introducing market competition are unfolding.

Although a number of trends in the scientific subsystem were mentioned, their consequences for increased competition within this subsystem were not highlighted. In the past, industrial research laboratories in the electrical products and chemical industries could afford to take a long time to develop their radical new innovations. But, nowadays, businesses experience:

(a) competition over speed in the development of new products and technologies;

(b) the geographic spread of competition with the increasing globalization of scientific and technological research;

(c) the increasing concentration of research in some very large firms because of mergers and acquisitions across international boundaries;

(d) at the same time, increasing pressure for shareholder value, meaning that investments in R&D are being evaluated for both productivity and development time. Time-to-market is now a critical concept, and, particularly in the automobile, industry organizations report their reductions in the amount of time needed to develop new models.

Despite all these pressures, there has been a paradoxical decline in number and size of corporate laboratories performing basic research. Instead, there is greater reliance upon universities and public research laboratories to perform this function. Observe that this is breaking the pre-existing link between basic and applied research that existed in companies like General Electric or Siemens. But the corporate laboratories that remain are now encouraged to outsource some of their research and, at the same time, compete for research funds with universities and public research laboratories.

Public laboratories have had their missions questioned by legislatures increasingly concerned about accountability. These laboratories have also been encouraged towards diversification in funding sources: that is, to look for funding from the private sector as indicative of their merit. Public-sector reform has led to the privatization of some research laboratories. Again, this trend heightens competition.

Frameworks for studying the production of knowledge

Another practical reason for separating research that is clearly scientific from industrial research is the amount of ferment in the scientific subsystem, a comment already made in the book's introduction. Because of globalization, post-industrialization, and the continued growth in knowledge and its consequences for specialization in both the supply chain and the idea-innovation chain, the process of knowledge production has changed a great deal recently. These processes are also producing increased interdisciplinary and collaborative research that blurs the functions of government, industry, and academic research. These issues are considerably elaborated in the work of Rammert in this part and in van Waarden and Oostervijk's contribution in the last part on institutional change. These contributions reflect attempts to build a new theory of knowledge production, one that considers the problem of the dynamics of knowledge flows.

Both of these frameworks indicate that, as one way of providing non-market coordination, there has been an increase in the number of inter-organizational relationships. The Meeus and Faber contribution in Part I and the Hage contribution in Part IV provide a considerable amount of evidence for this. The reasons why this shift is occurring are provided in van Waarden and Oostervijk's contribution: that is, growing specialization is creating fragmentation along the idea-innovation chain/network; and interorganizational relationships as a mode of coordination are a mechanism for dealing with this (also see the argument in Hage and Hollingsworth 2000). In their contribution, Oosterwijk and van Waarden document both an increase in market competition but also an increase in non-market modes of coordination along the idea-innovation chain.

Parallel to these increases in market competition and non-market coordination modes, part as cause and part as consequence, has been the growing complexity of the governance structure of the scientific subsystem, which is documented in the contribution by Kuhlmann and Shapira. The contrasts between Germany and the US document both general trends and different styles of governance. Although competition is increasing, there is evidence of another trend, 'the triple helix' of three major research-performing sectors (government, university, and industry) where many more partnerships between these different sectors are being developed.

Finally, all of this has implications for what are called knowledge regimes: that is, the organization of knowledge itself. The contribution by Rammert documents the movement away from functional specialization, which is what was described above, to what he calls fragmented diversity. Again, the contribution of Oosterwijk and van Waarden provides an illustration in two high-tech industries.

As Nelson and Rosenberg (1993) highlight, there exist some theoretical and empirical gaps that make ambiguous many of the findings concerning the causal relationships between organizational determinants and innovation outcomes at the national state level. We believe there is a need for a model of knowledge production to help clarify the precise way in which different kinds of organizations have specialized along the idea-innovation chain/network connect (or not), and what consequences this has for the innovation of the national system of innovation. We return to this problem in the concluding chapter.

References

Brown, E. (1997). 'Measuring Performance at the Army Research Laboratory: The Performance Evaluation Construct.' *Journal of Technology Transfer*, 22: 21–6.

Coccia, M. (2001). 'A Tool for Measuring Performance in Research Organizations.' In D. Kocaogho and T. Anderson (eds.), *Technology Management in the Knowledge Era*. Piscatawey, NJ: IEEE Operations Center, 160–8.

Damanpour, F. (1991). 'Organizational Innovation: A Meta-Analysis of Effects of Determinants and Moderators.' *Academy of Management Journal*, 34/3: 555–90.

Grimaldi, R., and von Tunzelmann, N. (2002). 'Assessing Collaborative, Pre-Competitive R&D Projects: The Case of the UK Link Scheme.' *R&D Management*, 32: 165–73.

Hage, J. (1965). 'An Axiomatic Theory of Organizations.' *Administrative Science Quarterly*, 10: 289–320.

—— (1999). *Organizational Innovation (History of Management Thought)*. Dartmouth: Dartmouth Publishing Company.

—— and Hollingsworth, J. R. (2000). 'A Strategy for the Analysis of Idea Innovation Networks and Institutions.' *Organization Studies*, 21: 971–1004.

Hagstrom, W. O. (1965). *The Scientific Community*. New York: Basic Books.

Hauser, J., and Zettelmeyer, F. (1997). 'Metrics to Evaluate R, D, and E.' *Research Technology Management*, 40: 32–8.

Jordan, G. B., Streit, L. D., and Matiasek, J. (2003). 'Attributes in the Research Environment that Foster Excellent Research: An Annotated Bibliography.' SAND Report 2003–0132. Albuquerque, NM: Sandia National Laboratories.

Kim, J., and Wilemon, D. (2003). 'Sources and Assessment of Complexity in NPD Projects.' *R&D Management*, 33: 15–30.

Kline, S., and Rosenberg, N. (1986). 'An Overview of Innovation.' In L. A. Rosenberg (ed.), *The Positive Sum Strategy*. Washington, DC: National Academy of Sciences, 289.

Larédo, P., and Mustar, P. (2000). 'Laboratory Activity Profiles: An Exploratory Approach.' *Scientometrics*, 47: 515–39.

Nelson, R., and Rosenberg, N. (1993). 'Technical Innovation and National Systems.' In R. Nelson (ed.), *National Systems of Innovation: A Comparative Analysis*. Oxford: Oxford University Press, 3–21.

Sadeh, A., Dvir, D., and Shenhar, A. (2000). 'The Role of Contract Type in the Success of R&D Defence Projects under Increasing Uncertainty.' *Project Management Journal*, 31/3: 14–22.

Saleh, S. D., and Wang, C. K. (1993). 'The Management of Innovation: Strategy, Structure, and Organizational Climate.' *IEEE Transactions on Engineering Management*, 40/1: 14–21.

Schuman, P. A., Ransley, L., and Prestwood, C. L. (1995). 'Measuring R&D Performance.' *Industrial Research Institute*, May–June: 45–54.

Thomas, P., and McMillan, G. S. (2001). 'Using Science and Technology Indicators to Manage R&D as a Business.' *Engineering Management Journal*, 13/3: 9–14.

Zammuto, R. F., and O'Connor, E. J. (1992). 'Gaining Advanced Manufacturing Technologies' Benefits: The Roles of Organization Design and Culture.' *Academy of Management Review*, 17: 701–28.

8 Factors Influencing Advances in Basic and Applied Research: Variation due to Diversity in Research Profiles

Gretchen B. Jordan[1]

Introduction

Despite the increasing importance of basic and applied research to the economy and to national defense, despite increased calls worldwide for accountability and demonstration of results, particularly for publicly funded research, and despite the increasing diversity of research laboratories, there is no adequate theory about what factors influence advances in science, or how the different structures and strategies of research laboratories or units influence those advances. There is not even agreement on how to describe the strategies or intended outcomes, or how to measure the outcomes of these science advances. Some refer to our current situation as the 'globalization learning economy' (Archibugi and Lundvall 2001). In addition to globalization, there has been increasing specialization in knowledge production and in organizations, bringing with it the need for interdisciplinary work and collaboration.

Such a theory could build on contingency theory, which suggests that organizations can have different strategies and structures, and that they perform best if strategy and structure are aligned. We suggest here that, given the increasing diversity in research structures and strategies, a contingency theory for science management is needed in order to discuss the factors that influence scientific advances, and

to assess research and research advances appropriately.

Both Crow and Bozeman (1998) and Larédo and Mustar (2000) have suggested that the current classification of research projects does not capture their essential differences, but their solutions do not provide a theory about the diversity of structures and strategies involved in scientific and technological research. Innovation theory stemming from Burns and Stalker (1961) similarly is inadequate in that it typically focuses on the entire organization and, we would suggest, one organizational model (the organic organization, characterized as having few authority levels, informal, decentralized, and extensive communication), rather than the specific units where different types of research are conducted (Zammuto and O'Connor 1992). In the industrial innovation literature, there is only one study that examines the structure and performance of research laboratories as such (Hull 1988). Read (2000) has suggested that, in general, there is not an adequate theory of innovation that explicates clearly the organizational contingencies and how these might vary. We would suggest that the monolithic treatment of research, technology, and product development (R&D) and, within that, basic and applied research stems from the lack of studies of diverse research projects, especially those involving basic and applied scientific research. Further, this monolithic treatment limits the opportunities to describe the different outcomes of research projects in ways that can be generalized across a number of projects with similar aims (Bozeman 2004). Performance is measured as discrete outputs, such as the development of a new algorithm or sensor, rather than the contribution of either of these to increased accuracy of predicting wind speed in hurricanes, or the soundness of the science underlying weather prediction more generally.

Although we are beginning to see many studies of research laboratories and organizations (see bibliographic review of Jordan *et al.* 2003 and, more specifically, Auditor General of Canada 1999; Brown 1997; Ellis 1997; Joly and Mangematin 1996; Menke 1997; Szakonyi 1994a, 1994b), the fact remains that none of these studies has focused on the measurement of incremental versus radical innovation, and then assessed which factors maximized these outcomes. Furthermore, most of these studies focus on new product development in industrial laboratories; few involve an important arena relative to innovation, namely basic science. And while there is now a new and growing literature on projects (Kim and Wilemon 2003; McDermott and O'Connor 2002; Shenhar 1998, 2001; Thamhain 2003), most of the emphasis here has been on developing dimensions of complexity or uncertainty at the project/product level, rather than an overall theory of diversity of scientific research projects. However, since complexity and uncertainty are critical elements that must be included, the concerns of these papers can inform the development of a satisfactory theory about research diversity.

There are three reasons why a theory about the diversity of any and all organizations is not sufficient for R&D, or for basic and applied research. First, the management of a large number of researchers is very different from the typical management issues involved in contemporary firms or public bureaucracies (Clarke 2002). Among the differences is the implied motivation of the researchers, which may be more about curiosity than money. Also, it is generally understood that the time-frames

for successful outcomes can be quite long and uncertain, and tasks are not routine or repetitive. Second, research is increasingly separated into distinct units: more and more basic research is being conducted either outside the firm (Hage and Hollingsworth 2000; Lundvall 1992; Meeus and Faber, forthcoming; Oosterwijk and van Waarden 2003) or in special kinds of interorganizational relationships, such as research consortia and global alliances. Finally, as with, for example, nanotechnology, scientific research is increasingly organized into highly fluid research projects, making the issue of how these projects should be grouped and managed more and more problematic.

Even though a theory for the diversity of research projects is different from the theory of organizations, it can be informed by the latter's suggestions of basic sources of tension in the management of multiple researchers engaged in a variety of projects. Proposed here as a first step in building a theory of research diversity is a Research Profiles Framework: it has been developed from a combination of review of the innovation and R&D management literature and lessons learned in focus groups with scientists and engineers. In particular, it builds on insights from the Competing Values Framework (Cameron and Quinn 1999) which, in turn, grew out of contingency theory (Hage 1980; Lawrence and Lorsch 1967; Mintzberg 1979; Perrow 1967). In particular, the older theories from the 1960s emphasized contingencies of strategy and structure as related to their environmental context. Although the proposed framework was developed independently of the growing literature on projects (Kim and Wilemon 2003; McDermott and O'Connor 2002; Shenhar 1998, 2001; Thamhain 2003), this literature complements the Research Profiles Framework in its diversity of research projects, and can be synthesized with it.

In this chapter, the Research Profiles Framework is put forward as a developing theoretical framework for understanding the diversity of research projects and organizations with multiple projects, and how this affects scientific advances. The first section indicates the various reasons for a separate framework for research projects. The following sections describe, and provide the justification for, the choice of specific strategic and structural contingencies that define distinctive profiles of research activity. Then managerial problems in the generic research profiles are described. Prior to concluding, implications for managers of organizations with multiple research profiles are discussed: these include the possibility of using the Research Profiles Framework as an organizing principle, beyond specialties and problem areas, for creating, within a research organization, groups of research projects that have specially trained managers.

Scope and terminology

The primary focus of this framework is the project, but implications for managers of the larger organization within which projects usually sit cannot be avoided. The term 'project' is used here to encompass a set of activities in a knowledge area that are focused on a particular question or problem area at a particular time. A research project may be an entirely separate organization, or a project within a firm or mission agency, such as the US Navy or NASA, or a research consortium. Even within organizations where research is basically the only activity, research activities are increasingly organized in projects and teams. Researchers are frequently engaged in multiple projects, raising the question of what is the best way of grouping these projects. Furthermore, the fluidity of the project's personnel across time also raises questions about the best way of grouping projects into administrative units of a higher order. Given this diversity, for our definition of 'research organization' we use Westley's definition of an organization as a 'series of interlocking routings, habituated action patterns that bring the same people around the same activities in the same time and places' (Westley 1990: 339).

Although we think that the Research Profiles Framework will apply equally well to technol-

ogy and product development, this chapter is focused on strategy, structure, and management of basic and applied research, not on technology and product innovation. This separation of science from technology is deemed impossible by some, but most recognize a blurry distinction between science and technology (Faulkner 1994). Some technologies are strongly science related. Similarly, some science is strongly technology related. Further, the non-linear models of R&D all show interactions and flows between research and technology (Kline and Rosenberg 1986; Hage and Hollingsworth 2000). A few in the social studies of science, starting with Bruno Latour (1987), speak of 'technoscience,' and see science and technology as a single methodology, where the research process and what is deemed a legitimate scientific or technological advance from that process are socially constructed and political in nature. There is a benefit, however, in distinguishing between science and technology, even while recognizing that, in the process of creating either scientific or technological advance, the two are often tied together. After reviewing past and current literatures, Faulkner concludes that there are nuances, and that these are useful because they serve as vague umbrella terms that roughly define limits. The limits are useful precisely because, she says, the socio-technical organization of science is different from technology (with technology being more hierarchical), as are the purpose or orientation, and the cognitive and epistemological features (Faulkner 1994).

In the Research Profiles Framework we describe here, we define the advances in science more broadly than others would, but in ways that have emerged since 1995 as requirements for documenting the outcomes of science increased. The National Science Foundation (NSF) has described its outcomes in terms of ideas, tools, and people. People can be individuals or 'communities of practice.' Of course, transitions from advances in science to further research or application in a problem area are also considered important outcomes of the research. These five areas were agreed upon at a workshop on knowledge benefits sponsored by the US Department of Energy (Lee *et al.* 2003). Thus our definition of the results of science is not limited to advancing knowledge for the sake of knowledge or potential application, or of gaining certainty by replicating the findings. In addition to work that advances knowledge of science (understanding of phenomena), there are advances that are new research technologies, instruments, or procedures, including new ways of organizing how research is accomplished. (See the chapter by Terry Shinn for examples of these.) Thus 'advance' includes advances in theories, laws, and general principles, measurement tools, operating principles, or in properties of materials. As mentioned, there are also advances in knowledge and skills as these are embodied in people (human capital), and in the growth of communities of practice around a discipline or problem area. Outcomes are also marked as transitions to further work, such as occurs when others extend what is known, or enter a field, as they have done with biomaterials, or when a concept or research tool is moved into commercial development.

Why a theory for profiles of research projects?

Increasingly, research is being pursued in a range of organizations of all sizes and shapes, from large, high-budget research laboratories to small research facilities. At one extreme are the many small high-tech research companies (Powell 1998), as represented by biotechnology start-ups, where applied research and product development are the major activities. At the other extreme, also in the private sector, are the chemical, pharmaceutical, and semiconductor industries, all of which have very large research projects, sometimes in a central headquarters, sometimes in multiple units clustered by product line or in a country. But it is in the public sector, which has been frequently ignored in the literature, where more and more basic and applied research is typically con-

ducted. Here, as in the private sector, extremes exist: they range from the many small research projects funded by the National Institutes of Health (NIH) or the NSF to the mission-oriented research conducted at the large Department of Energy (DOE) national laboratories at Livermore, Los Alamos, Chicago, and elsewhere. Parallels to these exist in Europe, with the European Organization for Nuclear Research (CERN) perhaps the most prominent example.

We suggest four reasons why we need a theory about the diversity of research and how this research is managed. First, science is too important for industrial innovation and meeting public goals, such as health, to be ignored. Second, knowledge production has changed, and current frameworks do not reflect those changes. Third, more study is needed of the management of large-scale research. Fourth, pressures to assess and demonstrate progress and outcomes using a one-size-fits-all approach and current measures can put perverse incentives into the system, as well as understate the value of the science advance. We will explain each of these in turn.

First, scientific research, and especially basic research, is increasingly becoming the basis for success in industrial innovation and achieving public goals. The importance of basic research is obvious in the case of the pharmaceutical, semiconductor, and other large high-tech industries, in which to be on the cutting edge necessitates exploratory research. Less obvious, but also increasingly important, is the need to make products neutral relative to their environment—that is, to reduce products' negative health and safety consequences for individuals, as well as their consumption of energy and scarce resources. This has amplified the importance of basic and applied research in, for example, material sciences. Basic sciences are also now seen as contributors to national goals such as health and a competitive economy. Indeed, the terrorist attacks of 11 September 2001 have made national security as important as the health of the economy: this has spurred applied research into, and technological development of, a variety of sensors for detection of biological and chemical weapons, as well as other kinds of anti-terrorist technologies.

Second, the Burns and Stalker (1961) organic model is obsolete because, in both the private and the public sectors, the world of research has changed dramatically. There are now more varieties of innovation than the simple incremental vs. radical innovation distinction: for instance, the hypercube innovation model of Afuah and Bahram (1995) has drawn attention to more complex views of types of innovation, including the architectural notion of innovation, though it does not include a discussion of the different organizational models needed to produce these kinds of innovation. More generally, there has been a movement away from the linear model of research towards a chain-link model (Kline and Rosenberg 1986) and the differentiation of the idea-innovation network—that is, the movement back and forth between manufacturing research, basic research, applied research, quality research, product development, and marketing research (Hage and Hollingsworth 2000). Given this, neither functional arrangement by managerial specialties nor a product arrangement appears appropriate for describing research activities, particularly with the change to multiple projects of short to medium duration (Davis and Lawrence 1977; Cleland 1984).

The dimension of external relationships, essential to understanding both the strategy and structure of science (and R&D), is not explicit in the organic model; we suggest that it should be. Most complex projects frequently involve multiple scientific and engineering disciplines that change across time because, as new problems present themselves, new kinds of expertise become essential. Indeed the complexity of many of these projects is far beyond what is usually considered in the organizational literature (but for an interesting exception on the complexity of new product development see Kim and Wilemon 2003). Separate organizations are now specializing in one or two of these areas of research (Hage and Hollingsworth 2000; Oosterwijk and van Waarden 2003). As companies specialize more and more

in product research, basic and even applied research has been moving out of the firm. This shift toward specialization accompanied by increased intra- and interorganizational collaboration has led to a whole series of new ideas about the organization of science, including the triple helix (Etzkowitz and Leydesdorff 1997), new systems of knowledge production (Gibbons 1994), and distributed innovation processes (Rammert 2003). See the chapter by Rammert for a discussion of functional specialization and fragmental distribution of research.

Third, also missing in the analysis of innovation is an understanding of large-scale technical systems (Mayntz and Hughes 1988), such as the electrical, railroad, and telephone systems, large-scale scientific research such as global climate change, and large-scale projects about the physics of space. The White House Office of Science and Technology is interested in how to choose and manage big science projects. Shenhar (2001) has recently included project scope in his discussion of engineering projects, using the language of systems. But further research is needed to determine whether the distinctions in science between a concept, a hypothesis, a theory, and a paradigm (which roughly parallel Shenhar's dimension of concept, creation of a new component, creation of a new system involving the component and, finally, a system of systems) can be related to the scale and size of basic and applied research. It appears the same distinctions can be made. For instance, in a quick glance at the research conducted at the US National Oceanic and Atmospheric Administration (NOAA), we see that there is work on both the science and the technology involved in the development of a new cloud sensor, research on a suite of sensors that measure a variety of properties, research on the oceanographic system, and, of course, the system of systems (ocean, atmosphere, solar, etc.) as reflected in the science of weather forecasting.

Fourth, following the lead of industry, which wants to know the return on investment for R&D activities, there has been a major shift toward assessing the progress and outcomes of all public research programs, part of a larger movement toward increasing public accountability. A number of countries, including Japan and France, have created national committees for research evaluation, or offices within ministries with this objective. Within the US, the Government Performance and Results Act of 1993 (GPRA) and the Office of Management and Budget's Program Assessment Rating Tool (PART) require most public agencies, including research organizations, to develop and implement performance measures and evaluate the quality, relevance, and outcomes of R&D that they fund. Measurement always perturbs a system, and research is no different. That you get what you measure is true. Since it is easier to measure tangibles, current assessment tends to be on discrete results such as papers, awards, and citations. Ideally there would be more general variables that described the progress of science toward ten- and fifteen-year goals. These general variables would be related to the dimensions of a diverse set of outcomes, and the process of measurement would also differ. Geisler (2000) concludes that current metrics miss the temporal dimension between scientific outputs and technological accomplishments, and that motivations for measurement differ. The short-term criteria for control are often in conflict with the long-term criteria for judging value and strategy realization.

Choosing strategic and structural dimension

What theoretical characteristics might one desire in a framework of research project diversity? Ideally, a framework should encompass dimensions that tap into the fundamental dilemmas, tensions, and problems of conducting research. Since our whole interest is in making meaningful distinctions, we want to isolate multiple dimensions of both strategy and structure. These dimensions should be most appropriate for describing differences in research outcomes and tasks or activities. Furthermore, a theoretical concern is to

connect these choices as much as possible with the existing literature.

To arrive at the dimensions for strategy and structure, we conducted an inductively based exploratory study to determine the critical factors facing the research worker and manager. This is what Quinn and Rohrbaugh did when developing the Competing Values Framework, though they focused on the 'cognitive structure of the organizational theorist' (1983: 365). We sought to gain a good understanding of the cognitive structure of the participants in the research process. As a result of the exploratory focus groups, representing a diversity of project types, we identified aspects of organizational structure and management practices specific to research projects, which we then used to design a survey instrument for assessing an research organizational environment (Jordan 2005; Jordan et al. 2003a). The findings of the exploratory study and of subsequent surveys suggest that the diversity of research organizations can be sorted according to two primary dimensions for strategy and two for structure, which in turn are highly related to two other structural dimensions.

For scientific research, the task environment is the knowledge world or the 'state of the art'—that is, how much is known in a problem area, and what is considered to be an important scientific concern or requirement. The first strategic choice reflects to what extent, given the concern or requirement, an advance will be attempted, or, if measured after the fact, has occurred in this state of the art. This strategic choice connects to the research on innovation (Hage 1999), where the distinction is frequently made between incremental or evolutionary vs. radical or revolutionary advances. To this we would add the more recent developments in the suggestion of a hypercube of innovation (Afuah and Bahram 1995) regarding the existence of modular and architectural innovations, and how radical the change is for various parties involved (the innovator, the user, those supporting innovators and suppliers to the innovator).

We define as 'revolutionary' or 'radical' research that is fundamentally new, or that makes a significant advance in the science involved. 'Significance,' in addition to the amount of change from the current state of the art, includes the centrality of the advance to the field or problem area, and the difficulty of achieving that advance, which is related to the number of the unknowns involved. Radicalness is a continuum; the distance of the advance compared to the state of the art would be determined by peers. The hypercube theory for new products reminds us that, for science to be successful, it must be absorbed and supported. If new science falls into a science trajectory and is relatively easily assimilated, this must be need-sustaining and, though it could be radical, it is not as radical as it would be if it created a new need. Explicating modular and architectural scientific advances could also be helpful in considering the influence of the existing technical environment on the level of advance in a particular area.

Not all radical advances are seen as that at the time, and some radical advances happen without planning for them. Managers can and do, however, set out to make radical advances, and want to understand how to improve the chances that one will occur within some predetermined time-frame. But because strategies or intentions represent aspiration levels, errors are possible on both ends of the evolutionary vs. revolutionary continuum: not setting the bar high enough, and setting the bar too high. The organization must decide whether to pursue new science and new areas of application, or exploit those that exist. The differences are large in terms of risk, size, and timing of the payoff from the research. This is one of two dilemmas related to strategy and the intended outcomes of the research.

A second dimension of research strategy, which also gives rise to a dilemma, is the scale, in terms of breadth, depth, and reach, of the outcomes of the research. For this dimension, we have to consider the scale of outcome in relation to some predetermined period of time, because the scale of impact may not be recognized as large for decades. In addition to being of obvious interest to stakeholders, scale of impact is important because it affects decisions about the resources required to undertake the research successfully. For science, we suggest there are at least four characteristics that determine the scale of the research undertaking and outcome:

- one is the number of variables and iterations of experiments involved, related to knowledge outputs and to requirements for people-hours or equipment to collect and process test data;
- a second is the extent of coverage of conditions for these variables, such as temperature range or geographic area;
- a third is the extent to which the conditions examined are extreme—conditions usually requiring special instruments or physical structures, or locations that are expensive—and the outcomes therefore of great interest;
- a fourth is the number of disciplines, fields, or problem areas that are affected by the advance.

Scale is a dimension commonly used in science. A 1994 National Academies of Science (NAS) report states that large projects are required for problems that can be pursued only by using large, complex facilities and platforms, extensive campaigns, or multipoint observations (NAS 1994): in all of these areas, there must be one project that is responsible for managing the combined activities. Scale is dependent in part on the cognitive definition of the scope of the scientific research: many sciences—for example, the earth sciences—are inherently systemic and, in these instances, studying a component or part of the problem is not easily accomplished in isolation, and can indeed give false information. Also, more scientific problems are perceived to be complicated both because of external influences that must be taken into account, and internal processes that must be modeled more or less well to improve the quality of the prediction (Boesman 1997; Kodama 1992; Miller and Morris 1999). Almost by definition, new understanding of a system and its multiple relationships affects a wide array of subsystems and components. The relative scale of the research outcomes thus leads to another strategic choice: the dilemma of choosing a scale for a research project that is considered value for money and one that covers enough of a problem area to give correct answers. If large problems can be divided into components, or if there are avenues for accessing and feeding a larger system, project scale may remain smaller.

External influences on intended outcomes

Although we have employed the word 'choice' about an organization's research strategy, there still remains the question of how freely managers can select particular strategies. The choices may be dictated to them by the environment, such as agendas of funding agencies, or control exerted by the state, or certain crises that demand attention (DiMaggio and Powell 1983). Choices are also bounded by resource availability and the state of the art in the area of science.

For example, to recognize, absorb, and take advantage of new knowledge and technologies generated elsewhere, there may be a need for investment in incremental work to build a knowledge and competency base. This has happened in parts of the world in the field of biotechnology. Projected shortages of physicists or engineers, or building a capability in biomaterials are other examples of agenda setting. Preferences and funding for research areas change as social values, demographics, and pressing problems change: thus we now notice increased funds for cancer research, geriatrics research, and science seeking to understand how to safeguard the stockpile of nuclear weapons using simulation techniques. Environmental concerns have pressured both public research laboratories and private companies to focus on such radical research agendas as hydrogen energy, the non-polluting car, and the 'green warship.' Pressing needs for immediate improvements in national security after the events of 9/11 might mean greater emphasis

on incremental advances to ensure the quick transfer of current technology into security applications. Furthermore, the culture surrounding the peer panel review process creates a bias toward what is often termed 'normal science,' that is, toward incremental advances in knowledge (Braun 1998).

There are also considerations related to the science itself and its related market. There is a discussion of markets in the chapter by Luke Georghiou, but not much has been written about this elsewhere that we are aware of. An area of research needed is to see if there are parallels in science with the discussions of strategy in the literature on technology and product development. Balachandra and Friar (1997), looking at project selection for innovation, suggest that three key areas of context (contingencies) need to be examined, and that they will have different relative importance. They are:

- type of innovation (radical vs. incremental) which the Research Profiles Framework covers;
- technology (high to low);
- market (new or existing).

Pathways in science, in contrast to technology and product development, have not been studied, apart from speaking of emerging fields and mature science.

Characterizing four distinctive research strategy profiles

Four distinctive kinds of strategic focuses are generated by the intended outcomes of the relative emphasis on the revolutionary or radical aspects of the scientific discovery, and the scale of the advance from the research project. They can be labeled with terms used in the Competing Values Framework (DeGraff and Lawrence 2002):

- *Be New*, which is the combination of radical and small scale;
- *Be First*, which is the combination of radical and large scale;

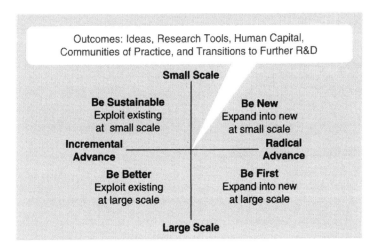

Fig. 8.1. Research profiles: strategic dilemmas and outcomes in scientific research

- *Be Better*, which is the combination of incremental and large scale; and
- *Be Sustainable*, which is the combination of incremental and small scale (see Figure 8.1).

In the technology realm, these four profiles are easy to see as the phases of technology development and maturation. The *Be New* profile is the creative period during which many revolutionary and perhaps radical new ideas are generated and tested by individuals or small groups. After proof of concept, there may be a decision to move forward with large-scale development efforts in order to *Be First* in producing a new product. Once the product is commercialized, there is a phase where the strategy is to *Be Better*, and to improve or differentiate the product and the way in which it is produced through incremental R&D, often on a large scale. To *Be Sustainable*, a mature product may require some incremental R&D. This profile can also be seen as the beginning of new product development, where mastering the current knowledge and capabilities in an existing area of research is a precursor to exploratory research on related new product concepts.

The dynamics of moving across four outcome profiles for science are not quite so familiar, but that is the subject of this chapter.

We will provide examples in each of the four areas of outcomes: ideas, research tools, human capital, and communities of practice. There are exceptions to these profiles, of course, and projects may not fall into just one profile.

Be New. The combination of small scale with revolutionary discovery is typically found in the development of a new concept, or theory, or field of research, as in the first research on the genetic model (Kohier 1979) or on DNA (Watson and Crick 1953). Organizations rely on this type of small-scale research to provide new ideas that may proceed to research at a larger scale of data collection or application development. The discovery could be a new research instrument or technique, such as a micro-fluidic drop ejector for very small platforms. An individual or small group becomes knowledgeable in this new concept or tool, and as others begin to work in the area, a new community of practice would emerge. People would perhaps refer to 'an emerging field.'

Be First. The combination of larger scale with revolutionary or radical advances is 'big science' that attempts to do something never done before. The project goal might be to be first with a whole new system, such as the research to map the human genome, or to build a new scientific instrument that will allow otherwise impossible research to be done. Fusion energy research is another example, as is the physics of space. In the latter two cases, tests are done under extreme conditions and, because of the difficulty of the questions addressed, all three cases require expensive measurement instruments or sustained funding over many years. In technology, the development of the Apollo rocket is an example from the past; a present one is the work on the international space station. People must capture the new knowledge and learn how to build and operate new instruments. Communities of practice must produce and absorb this knowledge. In the case of large expensive projects, key stakeholders must also know enough about the science and its promise to agree to the investment.

Be Better. In contrast, the combination of less radical and large scale can be an improvement in a more mature field or problem area. An example is the systematic collection of data from multiple points to study aspects of the environment such as air quality. Another is research in materials and improving simulation methods in order to improve the science behind stockpiling atomic weapons, so that these arsenals are safeguarded against the ravages of time and other dangers. Likewise, NOAA is concerned with improvements in the amount of time for warnings of impending violent weather. In the social sciences, examples are the slow improvements in the data collection methods associated with national surveys, such as the Census, or the General Social Survey. In the area of research tools, the operation and upgrading of large-scientific-user facilities fits this profile. Large-scale efforts to build human capital in a discipline, such as support of physics or mathematics graduate students, is also a possible intended outcome, as is the informal network these graduates form when they enter the workforce.

Be Sustainable. The last combination—less radical research with small scale—involves sustaining or extending existing science concepts, tools, human capital, or networks. A general example would include research to extend a concept to an application not far removed, a specific one to modify a vacuum chamber for

a different set of experiments. Many of the research projects that are funded by the NIH or the NSF, while doing excellent and challenging research, fit the general example because they are not aimed at radical advances: a second specific example might be attempts to improve the capability of weather detection systems, such as cloud sensors; in terms of human capital goals, coursework to keep technical staff current in their fields or train university students in research fits this profile, as does investment to bring different communities of practice together to determine new directions or build excitement around an existing problem area.

Characterizing four distinctive research structures

As we will discuss in more detail in this section, the strategic dimensions of radicalness and scale suggest structural dimensions and tensions that are important to research. Radicalness is related to complexity and to organizational autonomy: the more radical the aim of the research, the more complex the problem and the team, including external parties, needed to address that aim successfully. Scale is related to the resources and the amount of autonomy required for the research. Given the axiom of contingency theory that structure should follow strategy, research outcomes will be determined in part by how managers handle these structural tensions.

First, we will examine the dimensions of project size and research autonomy. To borrow from contingency theory: from the dimension of scale flows the obvious dimension of size of the project; from size flow a number of consequences that create tensions about coordination and control mechanisms; these inevitably impact project autonomy. We define autonomy as the extent to which research direction and technical decisions are made by researchers, rather than coordinated for them. As we discussed earlier, project size is often a reflection of the scale of the intended outcome. Research project size includes monetary costs, the number of researchers, and the amount and variety of equipment involved. Scale is relative, of course, and what is large in one subfield or group of organizations may not be large in another. In dollar terms, three typical project sizes are annual budgets of under $1 million, between $1 and $10 million, and over $10 million. Obviously, the range of over $10 million is enormous, and can reach into the billions when it involves a new energy system, such as one based on hydrogen. Some would classify the cost of a project as an indicator of the radicalness of the innovation (McDermott and O'Connor 2002); but because some projects of this nature are of low cost and others high, we would argue that this should be kept quite separate from measures of revolutionary breakthroughs in science or in technology.

Research projects of a large size and scale require a significant investment in management control and coordination, though this would probably still be less control than would be true for non-technical, repetitive work. Large projects typically need a substantial support staff and systems for required services: accounting, human resources, and libraries. Larger teams need leaders who can allocate resources and maintain communication and focus among team members. Management also helps to define and communicate clear goals and strategies, and align groups with them. Indeed, the success of a research project often depends on management correctly positioning the research to fulfill a need or fill a niche. In this manner, it is essential for managers to be technically competent and able to orchestrate new ideas through the organization. Overall, large projects have a unifying planning process that makes it possible to set specific scientific goals and to track progress against those. Organizational structures required for large projects are similar to the mechanical (as opposed to organic), bureaucratic structures found in the chemical and electrical product research laboratories in terms of hierarchy, coordination and control, rewards for administrative ability, and interdependence among organizational subunits.

However, too much coordination can stifle creativity. Shenhar (2001) provides a vivid description of how managers in this profile attempt to bureaucratize the coordination of the disparate parts of the research.

Research projects of small size do not require a great deal of management control, oversight, or support. The researchers themselves largely determine the research direction of these small projects, and individuals have a great deal of autonomy to make technical decisions. The independence and autonomy of academics are well known, as is the fact that researchers are motivated as much by recognition of their work and the intrinsic pleasure of directing and doing their research as they are by extrinsic rewards. However, there is evidence that researchers do best with some pressures to deliver (Pelz and Andrews 1976). It is possible that, left alone, the research will not continue to be cutting edge. When it is also important to have a critical mass of projects in an area, and to have research that is related to the larger organization's mission, a manager also worries about how much autonomy to allow small projects in setting the research direction. Another tension is being able to meet corporate requirements such as safety, security, and accountability for these small projects with minimum burden on researchers' time. At the level of the larger organization, there can be many of these small, often one-of-a-kind projects. Managers cannot look across these disparate projects easily; thus, the organizational system requires agility and flexibility.

The other dimension of structure, namely the complexity of research as represented by the variety of scientific and engineering disciplines involved, is essentially a measure of the amount of expertise required to accomplish the research project. This could be the expertise and perspectives of an individual as well as a group. At one end of the spectrum are ongoing 'specialized' teams with the specific expertise known to be required to complete the task. At the other end are highly complex and diverse research teams. Like size and research autonomy, complexity is equally well established in the literature—in this case, the literature on innovation (Hage 1965 and 1980), and technology (Perrow 1967; Shenhar 2001), and the literature on creativity (DeGraff and Lawrence 2002) and science more generally (Hagstrom 1965). The dimension of complexity, which is multifaceted, has also emerged in the new project management literature (Kim and Wilemon 2003). Indeed, this is one of the single most important findings in the innovation literature (Damanpour 1991; Hage 1999). Innovation is more likely to occur when there is a complex division of labor knowledge or experience mindsets (if the knowledge sets are integrated).

There are areas where the notion of complexity appears: the idea of Pasteur's quadrant (Stokes 1995) is bringing together the perspectives of advances of knowledge in both science and an application area; we often hear that discovery occurs on the margins of disciplines. Recognizing the absence of the studies of complexity relative to scientific research, and funded by several countries including NSF in the United States, Hollingsworth and Hage launched a large-scale historical study of radical innovation in biomedicine. Preliminary results indicate that, at the level of the research project, the relationship between complexity and radical innovation holds without qualification. Studies of multiple institutions found that when diverse groups of researchers were integrated, they were more likely to make significant contributions, as measured by winning Nobel and other prizes (Hollingsworth 2002). For example, groups often included both a biologist and a physician. Diversity can also be the mix of researchers and technologists. Our research at Sandia National Laboratories has found that many research projects have representation from six or more departments. One of the more interesting aspects of these complex research projects is the nature of the equipment that must be used to develop new ideas in science. Because typically this equipment did not previously exist, or could not be purchased, the research unit had to build these measurement instruments themselves to demonstrate the correctness of some new concept or idea. This requires an-

other dimension of knowledge, namely the presence of technical specialties that have the capability for pursuing the development of new instrumentation.

This structural dimension also includes whether this expertise is within the project's own organization, or whether organizations external to the project's are involved. Thus an external focus within the research unit becomes a second aspect of complexity. This notion is the basis for the growing literature on interorganizational networks, but it is not part of existing contingency theory (Lawrence and Lorsch 1967; Cameron and Quinn 1999). Complexity is essentially a measure of the knowledge pool of the research project and, even in large organizations such as the national laboratories or mission agencies, it is possible that not all of the necessary skills and attributes are to be found in the same research unit or organization. Therefore, as the complexity of research grows, there is usually an increasing need to search for expertise or information outside of the organization. This need was one reason behind the emergence of the interorganizational network literature (Alter and Hage 1993; Dussauge and Garrette 1999; Doz and Hamel 1998; Hagedoorn 1993; Harbison and Pekar 1998; Häkansson 1990; Inkpen and Dinur 1998; Jarillo 1993; Kogut *et al.* 1993; Lundvall 1993; O'Doherty 1995; Powell 1998; Van de Ven and Polley 1992).

We note that there is a certain irony in this. Those projects that are already more complex because of their revolutionary strategy are precisely the ones that are most likely to recognize the need for other pools of knowledge. As Zammuto and O'Connor (1992) argued in explaining the adoption of the radical process technology of flexible manufacturing, this flows from their aspirations, and thus makes the strategic choice so critical. Meeus and Faber (forthcoming) observe this to be true on the interorganizational side of the structure as well.

Managers face tensions related to complexity and organizational autonomy. As Nooteboom (1999) has observed, radical innovation is created by increasing cognitive distance; but as it increases, the tendency is for people to communicate less. The managerial problem is to develop a number of mechanisms to encourage the sharing of tacit knowledge (Nonaka and Takeuchi 1995). We would suggest that, within the context of science and smaller research projects that address only a component of a larger system or area, this requires more than the classical mechanisms suggested by Lawrence and Lorsch (1967). Similarly, interactions that bring in resources beyond those available to the project, often from outside the research unit/organization, have both benefits and costs. The first and most obvious cost is the time and effort involved in the knowledge search process. Where is the needed expertise located? The second is the loss of project or organizational autonomy when bringing in external parties. Whether they bring funds, people, equipment, or ideas to the table, they will want a share in the decision-making and the outcomes of the research.

The two dimensions of size/research autonomy and complexity/external ties (see Figure 8.2) yield four distinctive structural profiles of research projects.

- *Be New*: small, autonomous, complex research projects with interorganizational ties.
- *Be First*: large, coordinated, complex research projects with interorganizational ties.

- *Be Better*: large, coordinated, specialized research projects with organizational autonomy.
- *Be Sustainable*: small, autonomous, specialized research projects with organizational autonomy.

Managerial challenges, strategy, and structure

The central idea in contingency theory is that structure must follow from strategy. Thus the two basic dimensions of strategy dictate how the research tasks should be structured: if the combinations described above are not selected, then there is a mis-fit, which will be followed by related poor performance in papers, patents, citations, and other measures of scientific progress.

Before introducing more specific management challenges related to the four profiles, here is a summary of the strategic dilemmas and structural tensions that managers face.

- One strategic dilemma is the setting of the aspiration level: not too low to achieve radical advance, and not too high for existing capabilities and opportunities in an area.

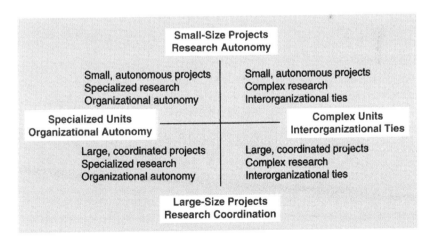

Fig. 8.2. Research profiles: structural tensions in scientific research

- A second strategic dilemma is the choice of a scale of research: too small or too large for the number of variables, systemic relationships, situations covered, and areas or field affected.
- A structural tension that follows from strategy is the complexity of the project: the more that managers choose to pursue a radical breakthrough in science, the more diverse the perspectives and knowledge sets needing to be integrated in the research unit.
- A structural tension follows from the degree of complexity: coordination tension in the complex research projects, because of collaboration with external organizations, where more specialized research projects can maintain organizational autonomy.
- Another structural tension that follows from strategy is the size of the project in terms of funds, people, and tools. As the scale of the research broadens, the resources needed grow.
- A structural tension related to the dimension of project size is how much autonomy can be allowed, with more coordination necessary for large projects than for small.

These strategic dilemmas describe the characteristics of intended science advances and the related structural tensions that help determine if the project will reach its intended outcomes; in addition, this DOE study has identified twelve areas mentioned by researchers and their managers as necessary for them to do excellent research (Figure 8.3) (Jordan *et al.* 2003*a*, 2003*b*; Jordan and Streit 2003). By 'excellent research' they meant research that is innovative, stands the test of time, and transitions to further research or helps solve a real-world problem. All of these aspects of the research environment are important, but some are more important for some research profiles than others (Jordan *et al.* 2005). For example, the uncertain nature of research means that time to think and explore is required but, if the aim of the research is a radical breakthrough, more time for exploration is needed.

The twelve areas of key management challenges that also determine advances in basic and applied research are as follows.

Aspects more important for research that strives for radical advances:

(1) Encourage exploration and risk taking: includes ensuring that researchers have time to think and explore, resources and freedom to pursue new ideas, and a systematic way (even if that is researcher action) of identifying new projects, partnerships, and opportunities.
(2) Integrate ideas internally and externally: includes internal cross-fertilization of

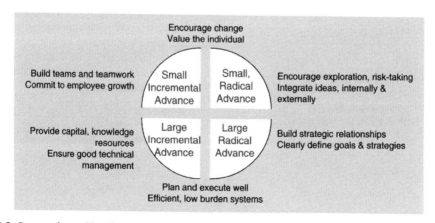

Fig. 8.3. Research profiles: key management challenges

technical ideas, external collaborations and interactions, and an integrating and relevant research portfolio.

(3) Build strategic relationships: includes maintaining good relationships with research sponsors, building and maintaining a reputation for excellence in research, and having a relationship with the larger organization such that senior managers are champions for basic and applied research.

(4) Clear project goals and strategies: includes defining and communicating a clear research vision and strategy (often more specific for larger projects), having sufficient and stable research funding, and investing in future, as well as current, capabilities.

Aspects more important for research that strives for less radical advances:

(5) Build teams and teamwork: includes having a climate of cooperation rather than competition within the project and organization, good internal project communication, and managers who add value to projects.

(6) Commitment to employee growth: includes a good process for technical career advancement, good educational and professional development opportunities, and high-quality staff internal to the project's organization.

(7) Excellent capital and knowledge resources: includes in-house, near-state-of-the-art research equipment and a good physical work environment, good research competencies, and resources to pay competitive salaries and benefits.

(8) Good technical management: includes managers who are both technically informed and decisive, a process for giving rewards and recognition that researchers respond to, and allocation of internal research funds perceived by researchers to be reasonable and fair.

Aspects more important for small-scale projects:

(9) Encourage change at the project level: includes autonomy in decision-making about research, a commitment to critical thinking about research direction and progress, and maintaining a sense of challenge and enthusiasm.

(10) Value the individual: includes the optimal use of each person's skills, having respect for people, and an atmosphere of trust and integrity.

Aspects more important for larger-scale projects:

(11) Plan and execute projects well: includes, in addition to planning and executing well, having measures of technical progress appropriate for the project, so that performance is driven the right way, and organization-wide measures of progress that are appropriate for the project's profile.

(12) Efficient, low-burden internal support systems: includes user-friendly laboratory services such as a library, good laboratory systems such as accounting and security at the lowest possible cost in dollars and time, and overall overhead rates low enough for research projects to be cost competitive.

For each of these twelve areas, the DOE study has a list of sub-areas of researcher concerns. Questions in the DOE Research Environment Survey (Jordan *et al.* 2003a) probe for, given the characteristics of their work, the percentage of time these aspects are true in the researchers' work environment. Continuing research will inquire about the specific mechanisms and skills that managers can apply in each of these areas. The expectation is that these will vary according to the profile of the research, and that managers will want to recognize the differences in the nature of the work and adjust their management practices accordingly. Here are just a few examples.

One aspect of project planning and execution is having appropriate measures of project success. There are a number of challenges for managers seeking to recognize differences in how different kinds of research should be assessed (NAS 1999). The benefits of review and of strategic relationships are openness to external events and strong feedback loops that keep the research on the cutting edge or on target for satisfying a specific need. Some of the mechanisms for keeping abreast of the literature and in planning research—interlocking spheres of information through such means as external review, advisory committees, and the involvement of the science and technology community—can provide both the stability and the stimulation necessary for making science advances, particularly revolutionary advances. The expertise of people involved in the reviews, both formal and informal, will differ depending on the profile, as will the questions asked, the criteria for assessment, and data considered. Another question is how often projects should be reviewed. Overlapping oversight currently has some projects at some of the DOE national laboratories feeling that they are over-reviewed, taking time away from research without sufficient additional benefit to offset the costs.

Another challenge is to improve technical management within the research unit: it requires skill in administration and management of people as well as knowledge and skills in the research area; the actual expertise may lie in separate departments not under the control of the project manager; and finding technical managers for larger projects is likely to be particularly difficult. Indeed, it is this kind of profile that led to the creation of matrix authority structures and, as is well known, these have not always functioned well, because researchers find it difficult to serve two masters, discipline and project (Davis and Lawrence 1977).

Another managerial problem is having an optimal mix of staff on small projects. It could be that the research unit is so small that technologists and assistants are not part of the team, and the researcher has to do many tasks, such as setting up samples and filing reports, that do not make optimum use of his or her skills. A manager can alleviate some of this by providing shared technologist and administrative resources. Paradoxically, another problem of the small specialized research profile is that many of the researchers may have multiple projects—perhaps too many. As Pelz and Andrews (1976) demonstrated many years ago, a few were beneficial for creativity, but the benefits dropped rapidly with an increase in numbers.

A final example is that of researchers in small projects who are typically given greater levels of autonomy. Management challenges here are the costs for the research organization of unrestricted research; if there is too much autonomy, researchers may not make any connection with others with whom they share similar interests. This is a problem in some American universities, where researchers in the same discipline, but located in different departments, do not interact. Such solitary tendencies of researchers in small, less radical research projects make encouraging teamwork a managerial problem. In the other three profiles, teamwork is more or less forced upon the researchers because of either the scale or the radicalness of the objective; but that is not to say that there are not problems with the integration of those teams.

Organizational implications of research profiles

We have described the Research Profiles Framework for projects but, as we have mentioned, there are obvious and important implications for organizations that are groups of projects, because these decisions are not made just at the project level. The outcomes of the research depend upon a match between strategy and structure. Three implications are discussed briefly here. How can an organization use the Research Profiles Framework to examine its portfolio of projects? How does an organization structure itself to manage groups of projects that have very different intended outcomes? Finally, what are the implications for measuring the performance of basic and applied research?

An advantage of the Research Profiles Framework is that it can make explicit the composition of a portfolio of projects along multiple dimensions, and show this diversity in one picture that stimulates discussion of trade-offs among small and large-scale projects, and radical and more specialized research projects, within a particular area of research. Given the current state of the art in an area, the resources available, the importance of moving forward in this area, and the scientific and technical opportunities to do so, what should the current mix be? Are there enough projects aimed at radical breakthroughs? Is now the time for large data-collection efforts that require large projects? Are projects transitioning from one profile to another, such as from extending an internal knowledge and skill base to then move forward to working on cutting-edge research in that area? A portfolio can get out of balance. For example, a recent National Academy of Sciences study of the US space physics program concluded that, while big science was uniquely suited for accomplishing some scientific goals, 'these projects have also been accompanied by implementation delays ... and the sapping of the base-funded programs' (National Research Council 1994). The recommendation was that future portfolio assessment should specifically address the balance between large and small projects.

What are the implications for the structure of organizations? Given these alternative research profiles, the Research Profiles Framework suggests that, in addition to grouping disciplines and specialties by their content and by the nature of the problem that they are working on, another basic organizing distinction within a research organization be made on the basis of the research profile. For example, within a problem area such as Pulsed Power Sciences, all of the research projects associated with the *Be New* profile could be in the same group, with a manager trained in how to manage that profile. There would be a different manager for those responsible for the day-to-day operation of the experimental facility. The profile managers and the Pulsed Power Sciences manager would meet together to look across the profiles at the performance as a whole. The evident differences in the nature

of the work, in the structural tensions, and the managerial challenges, mean that the managerial skills needed in each profile are also quite different: so different structures and management practices would be in place in the different groups. At a minimum, managers could be made sensitive through training to the differences in the profiles.

Just as Ciba Vision and other 'ambidextrous' organizations have recognized the need for selecting different managers for different kinds of business activities (O'Reilly and Tushman 2004), the same principle can be applied to these four kinds of research profiles. In effect, we have expanded existing contingency theory beyond the distinction between organic and mechanical to include the dimension of complexity and interorganizational ties. The Research Profiles Framework approach is a return to the basic insight of Chandler (1962) in his famous work on the relationship between strategy and structure. In his study of Dupont, he observed that such aspects as the technology of production and the marketing varied by product line. Basically, the recognition of distinctive research profiles is making the same argument, except that we are suggesting that it is the differences in the strategic choices of evolutionary versus revolutionary and small scale versus large scale that entail different structuring of the research projects, as well as distinctive managerial problems.

Turning to research assessment, the worldwide movement toward performance-based management of science and technology programs aggravates the problems of not recognizing diversity and the part that each research profile plays in the larger scheme of advancing scientific and technological knowledge. It can lead to the undervaluing of incremental innovation, setting up reward systems that reward occasional, serendipitous accomplishments but which, at the same time, eliminate research that might provide radical advances because someone insists that there be measures of tangible outcomes now. The Research Profiles Framework can help set performance expectations for the different profiles. The dimensions also suggest general variables to measure—the radicalness and scale, as determined by peers, of the research progress and outcomes. Within a problem domain, there are continuous variables related to the domain and the knowledge; they can be assessed on the dimensions of the research profiles, in addition to the traditional measures of papers, awards, citations, discrete milestones met, and single-word ratings, such as 'outstanding,' from peer reviewers.

Such measures as a per cent increase in the accuracy of prediction of ozone levels, the reliability of estimates, the speed with which a measurement is achieved, or how important a science advance about ozone is to geophysical science are all examples of these continuous measures. Real progress in ozone research, for example, could then be measured by the per cent gain in each of these various attributes, rather than just a single milestone, such as development and implementation of a new algorithm. The basic argument is that the use of general measures of scientific or technical progress or, at a minimum, estimates of the amount of technical progress likely with certain kinds of research (for example, radical versus incremental), makes it possible to better determine the critical pathways of progress, both within and across research projects, as well as larger research units.

Conclusions

We have argued that, in the study of innovation, learning, and macro-institutional change, the study of the management of basic and applied science projects is too important a topic to be ignored. Basic and applied research is important to the success of the economy and the strengthening of national security. However, there is increasing diversity in research that is not captured in current theory on how it should be managed to achieve scientific advances. In addition to radical and incremental research outcomes, there are other types of innovation, such as architectural and modular. The view of the production of research has

changed from the linear model to the chain-link or idea-innovation-chain model. Because collaboration is the norm, the old distinctions between research done in industry, government, and academia are blurred. With increasing knowledge specialization, a major question is how to access and share tacit knowledge across the specialties. A theory that describes this diversity is needed. Such a theory can build on contingency theory, which suggests that different organizations can have different strategies and structures, and that their performance will be best if strategy and structure are aligned.

The Research Profiles Framework proposed here is a first step in developing a theory of diversity in the management of research. The Research Profiles Framework extends the Burns and Stalker organic/mechanistic model, with its tensions between flexibility and control, to include the structural tensions between having interorganizational ties or organizational autonomy. The framework links tensions in the nature and structure of the research work to tensions in two other arenas: the strategies of desired outcomes and specific management challenges. Dimensions of nature of task and structure are small-size, autonomous vs. large-size, coordinated projects, and complex projects with interorganizational ties vs. specialized projects with organizational autonomy. These dimensions are linked to dimensions of hoped-for outcomes in some specified period of time, which are small vs. large scale of impact, and the radicalness of the outcome in terms of the degree to which it advances the state of the art and is central to the area of research. As has been indicated in our discussion of each of the four profiles, in addition to whether or not the organizational structure matches the research strategy, specific management styles will also determine whether or not the desired scientific and organizational learning and advances—be they finding new fields or exploiting existing competencies—actually take place. In each instance, different obstacles to advances present themselves. When the objective is radical innovation, then usually the problem is how to integrate diverse research teams (and consider impact with up- and downstream actors). As more and more of the research occurs in an interorganizational setting, this presents an obstacle to how much advance occurs. In contrast, when the objective is incremental innovation, the problem is frequently low levels of intra-organizational integration and aspiration, especially in small research projects where the researcher has considerable autonomy.

The four profiles resulting from the Research Profiles Framework not only delineate these issues effectively but also provide some interesting insights into how to arrange research projects, how to assess their progress, and how to analyze a portfolio of projects within a research organization. In addition to the usual criteria of scientific and engineering specialty and program focus, we suggest that projects in larger research organizations can be grouped on the basis of their research profile. Once divisional structures are established, managers can specialize in the kind of profile for which they have expertise in handling the dilemmas, tensions, and key management challenges. Furthermore, with this organizing

device, research organizations can assess progress along general variables that are the profile dimensions and, within a given area of research or research problem, assess their balance across profiles. Thus, we suggest that the Research Profiles Framework has the potential to provide solutions to a number of basic issues in the management of basic and applied research, while recognizing fundamental managerial challenges as well.

Throughout, we have suggested a number of areas where further research can be done. The DOE Study agenda is to build a store of knowledge from collected data and observations at the project and laboratory levels and, from this, to validate the Research Profiles Framework, define new measures of progress and value, and elucidate links between particular management actions, achievement of desired outomes within the planning horizon, and characteristics of the research profile. There are obvious linkages with the study of organizational learning, organizational change, and interorganizational networks. This is an exciting area of research, and one that is increasingly recognized as important to the study of innovation.

Note

1. This research has been performed under contract DE-AC04-94AL85000 with Sandia National Laboratories. Sandia is operated by Sandia Corporation, a subsidiary of Lockheed Martin Corporation. This work is funded by the US Department of Energy Office of Science, and has been done in collaboration with the Center for Innovation at the University of Maryland, Dr Jerald Hage and Jonathan Mote. The opinions expressed are those of the author, not the US Department of Energy or Sandia National Laboratories. For comments or more information contact the author at **gbjorda@sandia.gov** or 202-314-3040.

References

Afuah, A. N., and Bahram, N. (1995). 'The Hypercube of Innovation.' *Research Policy*, 24: 51–76.

Alter, C., and Hage, J. (1993). *Organizations Working Together*. Newbury Park, CA: Sage Publications.

Archibugi, D., and Lundvall, B.-A. (2001). *The Globalizing Learning Economy*. Oxford: Oxford University Press.

Auditor General of Canada (1999). 'Attributes of Well-Managed Research Organizations.' In *1999 Report of the Auditor General of Canada*: ch. 23.

Balachandra, R., and Friar, J. H. (1997). 'Factors for Success in R&D Projects and New Product Innovation: A Contextual Framework.' *IEEE Transactions on Engineering Management*, 44/3: 276–87.

Boesman, W. C. (1997). 'Analysis of Ten Selected Science and Technology Policy Studies.' *Working Paper*, 97–836 SPR. Washington, DC: Congressional Research Service.

Bozeman, B. (undated). *Public Value Mapping of Science Outcomes: Theory and Method*. A monograph found at **www.cspo.org/home/cspoideas/know_flows/Rock-Vol2-1.PDF** 2004.

Braun, D. (1998). 'The Role of Funding Agencies in the Cognitive Development of Science.' *Research Policy*, 27: 807–21.

Brown, A. (1997). 'Measuring Performance at the Army Research Laboratory: The Performance Evaluation Construct.' *Journal of Technology Transfer*, 22: 21–6.

Burns, T., and Stalker, G. M. (1961). *The Management of Innovation*. London: Tavistock.

Cameron, K. S., and Quinn, R. E. (1999). *Diagnosing and Changing Organizational Culture: Based on the Competing Values Framework*. Reading, MA: Addison-Wesley.

Chandler, A. (1962). *Strategy and Structures*. Cambridge, MA: MIT Press.

Clarke, T. E. (2002). 'Unique Features of an R&D Work Environment and Research Scientists and Engineers.' *Knowledge, Technology and Policy*, 15/3: 58–69.

Cleland, D. I. (1984). *Matrix Management Systems*. New York: Van Nostrand Reinhold Company.

Conner, K. R., and Prahalad, C. K. (1996). 'A Resource-Based Theory of the Firm: Knowledge versus Opportunism.' *Organization Science*, 7/5: 477–502.

Crow, M., and Bozeman, B. (1998). *Limited by Design: R&D Laboratories in the U.S. National Innovation System*. New York: Columbia University Press.

Damanpour, F. (1991). 'Organizational Innovation: A Meta-Analysis of Effects of Determinants and Moderators.' *Academy of Management Journal*, 34: 555–90.

Davis, S. M., and Lawrence, P. R. (1977). *Matrix*. Reading, MA: Addison-Wesley.

DeGraff, J., and Lawrence, K. (2002). *Creativity at Work: Developing the Right Practices to Make Innovation Happen*. San Francisco: Jossey-Bass.

DiMaggio, P., and Powell, W. (1983). 'The Iron Cage Revisited: Institutional Isomorphism and Collective Rationality in Organizational Fields.' *American Sociological Review*, 48: 147–60.

Doz, Y. L., and Hamel, G. (1998). *Alliance Advantage: The Art of Creating Value through Partnering*. Boston: Harvard Business School Press.

Dussauge, P., and Garrette, B. (1999). *Cooperative Strategy: Competing Successfully through Strategic Alliances*. Chichester, NY: Wiley.

Ellis, D. (1997). 'Modeling the Information-Seeking Patterns of Engineers and Research Scientists in an Industrial Environment.' *Journal of Documentation*, 53/4: 384–403.

Etzkowitz, H., and Leydesdorff, L. (eds.) (1997). *Universities and the Global Knowledge Economy: A Triple Helix of University-Industry-Government Relations*. London: Cassel Academic.

Faulkner, W. (1994). 'Conceptualizing Knowledge Used in Innovation: A Second Look at the Science-Technology Distinction and Industrial Innovation.' *Science, Technology, and Human Values*, 19/4: 425–58.

Geisler, E. (2000). *The Metrics of Science and Technology*. Westport, CT: Quorum Books.

Gibbons, M. (ed.) (1994). *The New Production of Knowledge*. London: Sage.

Hage, J. (1965). 'An Axiomatic Theory of Organizations.' *Administrative Science Quarterly*, 8: 289–20.

—— (1999) 'Organizational innovation and organizational change.' *Annual Review of Sociology*, 25: 597–622.

—— (1980). *Theories of Organizations: Form, Process, and Transformation*. New York: Wiley.

—— and Dewar, R. (1973). 'Elite Values vs. Organization Structure in Predicting Innovation. *Administrative Science Quarterly*, 18: 279–90.

—— and Hollingsworth, R. (2000). 'A Strategy for Analysis of Idea Innovation Networks and Institutions.' *Organization Studies*, 21: 971–1004.

Hagedoorn, J. (1993). 'Strategic Technology Alliances and Modes of Cooperation in High-Technology Industries.' In Grabher, G. (ed.), *The Embedded Firm: On the Socioeconomics of Industrial Networks*. London: Routledge, 116–38.

Hagstrom, W. O. (1965). *The Scientific Community*. New York: Basic Books.

Håkansson, H. (1990). 'Technological Collaboration in Industrial Networks.' *European Management Journal*, 8: 371–9.

Harbison, J. R., and Pekar, P. P. (1998). *Smart Alliances: A Practical Guide to Repeatable Success*. San Francisco: Jossey Bass.

Hollingsworth, J. R. (2002). 'Institutionalizing Excellence in Biomedical Research: The Case of Rockefeller University.' In D. Stapleton (ed.), *Essays on the History of Rockefeller University*. New York: Rockefeller University Press.

—— Hage J. T., and Hollingsworth, E. J. (forthcoming). *The Search for Excellence: Organizations, Institutions, and Major Discoveries in Biomedical Science*. New York: Cambridge University Press.

Hull, F. (1988). 'Inventions from R&D: Organizational Designs for Efficient Research Performance.' *Sociology*, 22/3: 393–415.

Inkpen, A. C., and Dinur, A. (1998). 'Knowledge Management Processes and International Joint Ventures.' *Organization Science*, 9: 454–68.

Jarillo, J. C. (1993). *Strategic Networks: Creating the Borderless Organization*. Oxford: Butterworth-Heinemann.

Joly, P. B., and Mangematin, V. (1996). 'Profile of Public Laboratories, Industrial Partnerships and Organization of R&D: The Dynamics of Industrial Relationships in a Large Research Organization.' *Research Policy*, 25/6: 901–22.

Jordan, G. B. (2005). 'What is Important to R&D Workers.' *Research Technology Management*, 48/3: 23–32.

—— and Streit, L. D. (2003). 'Recognizing the Competing Values in Science and Technology Organizations: Implications for Evaluation.' In P. Shapira and S. Kuhlmann (eds.), *Learning from Science and Technology Policy Evaluation*. Cheltenham: Edward Elgar.

—— —— and Binkley, J. S. (2003a). 'Assessing and Improving the Effectiveness of National Research Laboratories.' *IEEE Transactions in Engineering Management*, 50/2: 228–35.

—— —— and Matiasek, J. (2003b). 'Attributes in the Research Environment that Foster Excellent Research: an Annotated Bibliography. SAND 2003–0132. Albuquerque, NM: Sandia National Laboratories.

—— Hage, J. T., Mote, J. E., and Hepler, B. (2005). 'Investigating Differences among R&D Projects and Implications for Management.' *R&D Management*, 35/5: 501–11.

Kim, J., and Wilemon, D. (2003). 'Sources and Assessment of Complexity in NPD Projects.' *R&D Management*, 33/1: 16–31.

Kline, S., and Rosenberg, N. (1986). 'An Overview of Innovation.' In R. Landau and N. Rosenberg (eds.), *The Positive Sum Strategy*. Washington, DC: National Academy Press, 227–305.

Kodama, F. (1992). 'Technology and the New R&D.' *Harvard Business Review*, July–August: 70–8.

Kogut, B., Shan, W., and Walker, G. (1993). 'Knowledge in the Network and the Network as Knowledge: The Structuring of New Industries.' In G. Grabher (ed.), *The Embedded Firm: On the Socioeconomics of Industrial Networks*. London: Routledge, 67–94.

Kohier Jr., R. E. (1979). 'Warren Weaver and the Rockefeller Foundation Program in Molecular Biology: A Case Study in the Management of Science.' In N. Reingold (ed.), *The Sciences in the American Context: New Perspectives*. Washington, DC: Smithsonian Institute Press, 249–93.

Larédo, P., and Mustar, P. (2000). 'Laboratory Activity Profiles: An Exploratory Approach.' *Scientometrics*, 47: 515–39.

Latour, B. (1987). *Science in Action: How to Follow Scientists and Engineers through Society*. Milton Keynes: Open University Press.

Lawrence, P., and Lorsch, J. (1967). *Organizations and Environment*. Boston: Harvard Business School.

Lee, R., Jordan G. B., Leiby, P., Owens, B., and Wolf, J. L. (2003). 'Estimating the Benefits of Government-Sponsored Energy R&D.' *Research Evaluation*, 12/3: 189–95.

Lundvall, B.-Å. (1992). *National Systems of Innovation: Towards a Theory of Innovation and Interactive Learning*. London: Pinter.

—— (1993). 'Explaining Interfirm Cooperation and Innovation Limits of the Transaction-Cost Approach.' In G. Grabher (ed.), *The Embedded Firm: On the Socioeconomics of Industrial Networks*. London: Routledge, 52–64.

McDermott, C., and O'Connor, G. C. (2002). 'Managing Radical Innovation: An Overview of Emergent Strategy Issues.' *Journal of Product Innovation Management*, 19/6: 424–39.

Mayntz, R., and Hughes, T. P. (eds.) (1988). *The Development of Large Technical Systems*. Boulder, CO: Westview Press.

Meeus, M., and Faber, J. (forthcoming). 'Networks of Inter-organizational Relations and Innovation.' In J. Hage (ed.), *Innovation, Knowledge Dynamics and Institutional Change*.

Menke, M. M. (1997). 'Essentials of R&D Strategic Excellence.' *Research Technology Management*, 40/5: 42–7.

Miller, W. L., and Morris, L. (1999). *Fourth Generation R&D: Managing Knowledge, Technology, and Innovation*. New York: Wiley.

Mintzberg, H. (1979). *The Structuring of Organizations*. Englewood Cliffs, NJ: Prentice-Hall.

Mockler, R. J. (1999). *Multinational Strategic Alliances*. Chichester, NY: Wiley.

National Academy of Sciences (1999). *Evaluating Federal Research Programs: Research and the Government Performance and Results Act*. Committee on Science, Engineering, and Public Policy. Washington, DC: National Academy Press.

National Research Council (1994). *A Space Physics Paradox: Why Has Increased Funding Been Accompanied by Decreased Effectiveness in the Conduct of Space Physics Research?* Washington, DC: National Academy Press.

Nonaka, I., and Takeuchi, H. (1995). *The Knowledge-Creating Company: How Japanese Companies Create the Dynamics of Innovation*. New York: Oxford University Press.

Nooteboom, B. (1999). 'The Dynamic Efficiency of Networks in Interfirm Networks.' In A. Grandori (ed.), *Organization and Industrial Competitiveness*. London: Routledge, 91–119.

O'Doherty, D. (ed.) (1995). *Globalisation, Networking, and Small Firm Innovation*. London: Graham and Trotman.

O'Reilly III, C. A., and Tushman, M. L. (2004). 'The Ambidextrous Organization.' *Harvard Business Review*, April: 74–81.

—— —— (1996). 'Ambidextrous Organization: Managing Evolutionary and Revolutionary Change.' *California Management Review*, 38/4: 8–30.

Oosterwijk, H., and van Waarden, F. (2003). 'The Architecture of the Idea-Innovation Chain.' Unpublished paper, School of Public Policy, Utrecht University.

Pelz, D. C., and Andrews, F. M. (1976). *Scientists in Organizations: Productive Climates for Research and Development*. Ann Arbor: University of Michigan Press.

Perrow, C. (1967). 'A Framework for the Comparative Analysis of Organizations.' *American Sociological Review*, 79: 686–704.

Powell, W. W. (1998). 'Learning from Collaboration: Knowledge and Networks in the Biotechnology and Pharmaceutical Industries.' *California Management Review*, 40: 228–41.

Quinn, R. E., and Rohrbaugh, J. (1983). 'A Spatial Model of Effectiveness Criteria: Towards a Competing Values Approach to Organizational Analysis.' *Management Science*, 29: 363–77.

Rammert, W. (2003). 'Two Paradoxes of Fragmented Knowledge Production: Combining Heterogeneous and Cultivating Non-Explicit Knowledge.' Paper presented at the annual conference of SASE, Aix-en-Provence, France.

Read, A. (2000). 'Determinants of Successful Organisational Innovation: A Review of Current Research.' *Journal of Management Practice*, 3: 95–119.

Shenhar, A. J. (1998). 'From Theory to Practice: Toward a Typology of Project-Management Styles.' *IEEE Transactions on Engineering Management*, 45/1: 33–48.

—— (2001). 'One Size Does Not Fit All Projects: Exploring Classical Contingency Domains.' *Management Science*, 47/3: 394–414.

Stokes, D. E. (1997). *Pasteur's Quadrant: Basic Science and Technological Innovation*. Washington, DC: Brookings Institution Press.

Szakonyi, R. (1994*a*). 'Measuring R&D Effectiveness I.' *Research Technology Management*, 37/2: 27–32.

—— (1994*b*). 'Measuring R&D Effectiveness II.' *Research Technology Management*, 37/3: 44–56.

Thamhain, H. J. (2003). 'Managing Innovative R&D Teams.' *R&D Management*, 33/3: 297–311.

Van de Ven, A. H., and Polley, D. (1992). 'Learning While Innovating.' *Organization Science*, 3/1: 32–57.

Watson, J., and Crick, F. (1953). 'A Structure for Deoxyribose Nucleic Acid.' *Nature*, 171/4356: 737–8.

Westley, F. R. (1990). 'Middle Managers and Strategy: Microdynamics of Inclusion.' *Strategic Management Journal*, 11: 337–51.

Zammuto, R., and O'Connor, E. (1992). 'Gaining Advanced Manufacturing Technology Benefits: The Role of Organizational Design and Culture.' *Academy of Management Review*, 17: 701–28.

9 Network Attributes Impacting the Generation and Flow of Knowledge within and from the Basic Science Community

Susan A. Mohrman, Jay R. Galbraith, and Peter Monge

Introduction

Though its goals and focus are often politically contested (Stokes 1997), public funding of basic science research has become accepted policy in many parts of the world. Even as corporate research and development (R&D) has increasingly acquired a short-term focus (Tucker and Sampat 1998), public support for basic science research has increased steadily in both the European Union (European Commission 1994, 1998, 2002) and the United States (US Government 2003). Such investment in science is made with the expectation that society will benefit from the research, particularly in areas of security, health, and environmental sustainability (for example, see European Commission 1994, 1998, 2002; AIP 1999; Lee *et al.* 2003). It is also widely believed that competitiveness in the knowledge economy depends on generating scientific knowledge to underpin continuing innovation and knowledge leadership (Brooks 1986; Stokes 1997). But government-funded research has been too ineffectively organized to diffuse the knowledge it produces and to stimulate its use in technology innovation (Branscomb 1992; Pinelli and Barclay 1998). In the US, there have been calls for publicly funded research to focus on knowledge transfer, increased leverage of knowledge, and the consequent extension of the value of the knowledge that is generated (Pinelli and Bar-

clay 1998; Branscomb 1992; Alic 1993), thus enhancing national capability. Yet little academic work has examined how the organization of basic research affects the value that is yielded (an exception is Jordan, this volume).

Our intent is to develop a theory-based framework to examine how organization affects the value that flows from basic science research. We view the basic science research arena as an ecosystem with finite resources and multiple populations that co-evolve as they establish competitive and cooperative relationships in their attempts to thrive (Aldrich 1999; Astley 1985; Monge and Contractor 2003). Despite the importance of government agencies, foundations, and corporations in providing resources and influencing the direction of research, we argue that the basic science ecosystem operates as a self-organizing community of communities (Kauffman 1993, 1995; Tuomi 2002): they exist at multiple levels of analysis and cut across traditional organizational boundaries; they form to address particular problems and questions; and they re-form through time as the state of knowledge in an area advances, particular pathways fail, and/or the nature of the problems that participants find interesting changes.

Basic research communities are knowledge networks—networks in which knowledge is a key resource that flows between the network nodes. Monge and Contractor (2003) advocate a three-tiered approach to network analysis:

(*a*) theoretical analysis of the networks under consideration, and their decomposition into their potential multi-level components explained by those theories;

(*b*) examination of the attributes of the nodes;

(*c*) examination of the multiple (multiplex) relations in the network, including dynamic relations over time.

Monge and Contractor argue that the veracity of the theory may then be tested by examining whether the theoretically predicted connections occur more frequently than would be expected by chance.

Drawing on the literature on networks in general, and on networks in R&D in particular, this chapter provides a theoretical analysis of basic science research networks and lays out a research agenda for exploring whether this approach explains what features of the network generate a flow of value, and what organizing approaches can foster the self-organizing activities that enable knowledge to move down and across the value stream. By 'value,' we mean the usefulness of knowledge to others in the community and/or the stream that flows from basic research through to product and process innovation and commercialization of new products and services, a flow which can occur laterally among members of the basic science research community, or vertically through levels of application (Hage and Hollingsworth 2000).

The chapter will be organized as follows: First, we discuss the nature of the work of science, and present attributes that we believe to be key determinants of forms of basic science networks. Second, we present our view that basic science research occurs in a co-evolving ecological community. We discuss the various populations and elements that constitute the ecosystem, the resources and value that flow through and out of it, the dynamic and self-organizing nature of its networks, and the exogenous forces that operate on them.

Throughout the chapter, we suggest a research agenda aimed at testing the descriptive accuracy and explanatory power of our framework, and at refining understanding of basic science research networks and the way in which they contribute value. In particular, we are concerned with their contribution to society's capacity for innovation.

The work of science

Four aspects of scientific work are germane to understanding basic science's networks. Each aspect underscores science's relational nature, and enables us to consider its dynamics and outcomes from a network perspective.

Science research is a collective endeavor

Science is inherently a social activity and a collective task (Kantorovich 1993). Our perspective is that knowledge is contextual and relational—people construct knowledge as they interact in a social context, and this knowledge in turn influences their behaviors, perceptions, and cognitions (Berger and Luckmann 1966). For example, through its education, training, and research, the molecular biologist community has developed ways of understanding and approaches to studying the human cell that differ from those held by the communities of chemists or genomists who study the cell. Pelz and Andrews (1966) found that high-performing scientists, although self-directed by their own ideas, interacted vigorously with colleagues; Polanyi (1962) has referred to the 'invisible college' of scientists who, in pursuing their own research interests and focuses and applying their own personal judgment, are in fact cooperating as members of a closely knit organization, continually heedful of the knowledge that is being generated by others in their field. Throughout the world and over time, this invisible college spreads and organizes around a set of problems and questions to be solved. Root-Bernstein (1989) describes the scientist as working within the flow of activities.

He describes individual scientists as conducting activities that are nestled between an evolving stock of existing knowledge and the problem or question at hand. Thus viewed, the conduct of science is inherently relational.

The empirical study of networks in R&D settings has focused on how knowledge actually flows and becomes accessible throughout the network. Its transfer is affected by its type (codified or tacit), the type of network link (strong or weak), the network structure (sparse or dense), and the person's or institution's position (central or peripheral) in the network (Hansen 1999; Powell 1998). Network patterns effective for the transfer of knowledge have been found to differ depending on whether the research task involves fundamental research, development, and/or technical support (Allen *et al.* 1980; Katz and Tushman 1979). As we will discuss later, the effective flow of knowledge through networks is, at least in part, an organization design problem.

The creation and leverage of knowledge

The creation and leverage of knowledge is enabled by two generic processes: knowledge sharing and knowledge combination (Schumpeter 1934; Nahapiet and Ghoshal 1998). Scientific advances have often been attributed to exposure to new information and/or perspectives, which may be achieved through 'smart foraging' (Perkins 1992), well-adapted search (Koestler 1964), exploration rather than exploitation (March 1991), taking advantage of serendipity and variation, and/or through reasoning by analogy or combination that may be triggered by being exposed to more than one problem at a time (Kantorovich 1993). One way that such broad exposure can be achieved is through sharing and collaboration among colleagues, each of whom has a unique set of experiences and knowledge.

Novel solutions are particularly likely to emerge if different knowledge bases are combined. The pooling of multiple and distinct knowledge resources to create new knowledge has theoretical underpinnings in Ashby's (1956, 1962) work on 'requisite variety:' for survival,

the internal diversity of a self-regulating system should match the diversity of the environment. Requisite variety's ensuring that multiple knowledge domains be represented can provide a robust basis for learning and innovation: it ensures multiple perspectives, and can facilitate the innovative process by enabling novel associations and linkages (Cohen and Levinthal 1990). Stokes (1997) has pointed out the role that 'cross-overs' play in the innovation process. For example, scientists trained in nuclear rather than solid states physics led the development of radar and semiconductors. This had less to do with the intellectual content of the problem than with their knowledge of the sophisticated electronic circuits and instrumentation and systems thinking required for studying very complex physics problems (p. 103). On the other hand, the extent of cross-field versus within-field interaction required to advance knowledge can vary tremendously, even within the same discipline. In tracking and relating a decade of publications in two subfields of condensed matter physics research, Hicks (1992) found that one of the two fields, spin-glass, depended heavily on cross-discipline communication and combination. This contrasted with research in superfluid helium three, in which deep knowledge of a particular field, and of the instrumentation that evolved in concert with the theory and empirical work, meant that knowledge was created by a very focused, single-discipline community concentrated in a few institutions that housed the needed technology. Advancing knowledge in superfluid helium three depended on the person-to-person exchange of tacit, complex, and highly specialized knowledge. Researchers in spin-glass, on the other hand, depended on being exposed to a variety of literature at a more general level, and were less dependent on person-to-person exchange.

The single- versus multiple-discipline requirements for knowledge flows has also been elaborated in the R&D literature, particularly the literature dealing with pharmaceutical development. The work of Henderson (1994) and Powell (1998) demonstrates how institutions have designed themselves to capture deep-discipline external knowledge and then integrate it into the institution's problem-relevant knowledge base. Pharmaceutical discovery units encourage their scientists to publish and remain part of external discipline communities. These scientists then work together in cross-community teams to develop new chemical entities that have value in treating diseases. Biotech firms often have alliance managers whose task is to capture knowledge from collaborations and integrate it into the firm. Again, the flow of knowledge both into and across the organization, and the sharing and combining of knowledge to generate new knowledge and innovations become a problem of organization design.

Understanding and application

Science is often carried out in service of and in close relation to application. In *Pasteur's Quadrant*, Donald Stokes (1997) debunks the notion that there is always a linear path from basic science research through the development of technology to application—a notion that has underpinned science policy in the US for many decades. In fact, much science is simultaneously oriented toward fundamental understanding and application. Stokes's image (p. 88) is of fundamental research and application progressing along dual, upward, loosely coupled trajectories. Each trajectory is at times strongly influenced by the other, with use-inspired basic science research often in the linking role. He points out that pooling of multiple knowledge bases often enables this interplay between mission-oriented work and basic science: for example, physicists trained in nuclear physics were key to the development of molecular biology, a field that is almost completely use-inspired. He offers this as a good example of bringing together broad and deep scientific capabilities to meet societal needs.

Schon (1963) referred to such appropriation of knowledge from different disciplines as the displacement of concepts, and argued that it is a major path through which individual creativity is transformed into innovation and becomes part of social and professional practice. Tuomi (2002) argues that recombinatorial innovation—innovation that relies on the combination of previously disparate knowledge bases and technologies—requires relational change at the network level: people need to get access to knowledge resources previously unknown to them or inaccessible. He argues that, along with cultural norms that do not value or invite novel interpretations and strict definitions of professional identity, rigid organizational and network architectures can impede the mobility required for the recombination of knowledge. In understanding the flow of value from basic science research, one can look laterally at the combination of knowledge across basic science disciplines (in a sense, one discipline becomes a locus of application of the knowledge of another discipline) and vertically along the value stream toward technological and, eventually, commercial application.

Pelz and Andrews (1966) found that the highest-performing scientists were interested in both application and pure science. In his study of Nobel laureates, Hurley (1997) found that a sense of purpose is an important factor motivating many creative scientists. He quotes one of the laureates in his study, Simon van der Meer, a Dutch physicist who worked on stochastic cooling and the accumulation of antiprotons: 'I would agree that an important factor in a creative scientist is an interest in attacking existing theories. The main interest, however, is in looking at something with a purpose—being able to do something that you need for something else' (p. 63). Whether the potential application is to other problems of basic sciences or downstream to the development of technology to address social needs and wants, carrying out science with an application in mind requires a connection or relation to the world of applications. For example, in a study of applied mathematicians' networks of interaction (Mohrman *et al.* 2004), university mathematicians often reported that, to advance their mathematics, it is imperative to find links to scientific and engineering applications that enable a test of their theories. Thus, a primary linkage supporting both the sharing and combination of knowledge is the nexus between applications and science: to understand the network linkages through which knowledge flows to others who can apply it and through which the application needs become known to the basic science researcher is vital, and the concern of policy makers with the societal benefits of basic research investment makes it especially so.

Relational quality

The increasingly large-scale nature of science also underscores its relational quality. Requirements for enabling technology and information-processing capability have grown tremendously (for example, Office of Science 2003). Scientists are using tools of immense capacity, which are transforming the conduct and methods of science and the epistemologies of scientists (Brockman 2002; Shinn, this volume). Huge installations of plant and equipment, such as the particle accelerators that are found at CERN or SLAC (Stanford Linear Accelerator Center), are often needed to expand knowledge. The Human Genome Project in the US alone consumed over $3 billion, and entailed involvement from over sixteen major science centers in many countries, multiple governments, and commercial firms. Organizational capabilities were critical to the success of this project (Sulston and Ferry 2002; Lambright 2002). Even the conduct of a smaller, focused research program can require high levels of organizational skill to manage complex sets of interrelated research steps and projects. As told by Hurley (1997), Baruch Blumberg's Nobel Prize-winning research, which discovered an antigen associated with viral hepatitis and followed through to the development of a hepatitis vaccine, entailed six streams of work; each of these involved the establishment of a 'small organization,' including many cross-cutting mini-teams in different countries and organizational settings. Similarly,

in his study of the science funded by the Basic Energy Sciences (BES) of the US Department of Energy, Bozeman (1999) found a transition from the funding of relatively small-scale, individual-investigator, disciplinary-oriented projects to almost exclusively multidisciplinary research team activities. The complexity of the enterprise underscores both the difficulty and importance of the linkages that allow knowledge to be shared, reused, and combined, and the organizational challenges that underpin these capabilities.

The scientific enterprise as a co-evolving ecological community

Because of its heavily relational character and its dependence on a limited supply of resources, basic science research can be profitably examined as a co-evolving ecological system. Studies of community ecologies operating under co-evolutionary principles examine how networks of organizational populations and other entities interact with each other in environmental niches to acquire scarce resources (Monge and Contractor 2003). Through time, new forms and relationships emerge (Carroll and Hannan 2000; Ruef 2000).

Aldrich (1999) defined an organizational community as a set of co-evolving organizational populations joined by commensalistic and symbiotic ties through their orientation toward a common technology, normative order, or legal-regulatory regime (see also Greve 2002; Hunt and Aldrich 1998; Rao 2002). By 'commensalistic' ties is meant relations among members of the same or similar populations that are both cooperative and competitive. In cooperative relations, populations seek to collaborate with each other for mutual advantage (which is why these are sometimes called mutual ties; see Hawley 1986). In competitive relations, populations engage each other in an attempt to acquire what belongs to the other, with the intent of driving the other populations out of existence, or at least subsuming them under their own aegis (Aldrich 1999). Symbiotic ties exist when members of different populations (that is, different species or organizational forms) relate to each other in ways that are mutually beneficial. As Porac (1994) says, 'communities are enacted by people who are involved in multidimensional relationships' (p. 452). This means that scientific knowledge communities are composed of diverse, multiplex networks (Kogut 2000), in an ecological system in which they compete for scarce resources such as grants, journal space, graduate students, and expert talent; all the while, they relate to others who are generating relevant knowledge and pursuing either complementary and/or competing theoretical and empirical approaches.

As Donald Campbell (1960) pointed out nearly half a century ago, sociocultural knowledge and knowledge communities grow and are transformed on the basis of three basic sociocultural evolutionary principles: variation, selection, and retention (March 1994).

Variations

Variations are introduced into knowledge communities in two forms. The first is blind, random, or unplanned variations (Romanelli 1999); the second is purposive variations, those innovations that people consciously attempt to generate (Madsen *et al.* 1999). Both kinds are important to knowledge creation. Variation occurs naturally through individual scientists' pursuit of their own interests guided by their theoretical beliefs as they pursue related topics of investigation (Hurley 1997). Even in the context of 'big science' that demands overall coordination, variation is achieved because scientists are pursuing their individual approaches and exercising investigatory freedom within the overall context of directed research goals. Hurley (p. 95) points to the Human Genome Project as an example of the purposeful creation of variation by housing different activities in different centers, relying on the application and discovery of new approaches and techniques within each.

Selection mechanisms

Selection mechanisms are evaluative processes designed to choose better alternatives and reject poorer ones. One common selection process is interorganizational or interpopulation imitation, where members of one community copy the practices of similar populations or communities (Miner and Raghavan 1999)—a process made possible only if a population has access to the knowledge of the other. As a self-correcting process, science is very much in the business of selecting the best possible variations. While the scientific method itself yields 'results' that can select or deselect various theories and methods, peer review processes both for funding and publication, and other forums for collegial exchange and feedback point to the relational nature of the selection process.

Retention mechanisms

Retention mechanisms are those activities and practices that embed selected variations into organizations, populations, and communities. Nelson and Winter (1982) describe in considerable detail how selected variations become routinized and, therefore, standard practices. (See also, Feldman 2000, for how routines also lead to continuous change.) In scientific knowledge communities, variations can be routinized in standard laboratory procedures, textbook knowledge, norms and values for refereeing articles and books, computer programs for data analysis, the hardware and software components of various enabling technologies, and a host of other accepted-as-correct ways of doing scientific work. As Anderson (1999) points out, 'Campbell (1965) showed how evolution through variation, selection, and retention can occur in a variety of ways, such as the selective survival of social structures, selective borrowing and imitation, selective promotion of individuals who propagate some variation, or selective repetition of behaviors that seem to be associated with success' (p. 137).

The extent and balance of competitive, mutual, and symbiotic relationships vary over the life-cycle of the populations that comprise a community (Astley and Fombrun 1987; Hannan and Freeman 1977). A new population often occupies an open environmental space or niche where abundant resources exist. However, as more organizations enter the population, and as more populations become active in the ecosystem, the demand for resources exceeds the supply, and competition for these resources ensues (Astley and Fombrun 1987). As organizations begin to compete, they also form the beginnings of interrelationships; this makes apparent the need for the community and leads to its creation (Astley 1985).

Knowledge entrepreneurs or 'champions' (Ingram 2002) play a central role in the generation of knowledge. Like other entrepreneurs, knowledge champions typically see new opportunities to develop new knowledge communities around intellectual or scientific innovations (Aldrich 1999). This requires the development of both knowledge and communication networks: knowledge networks are the connections among 'who knows what' in the community, thereby connecting the various elements of the entire knowledge domain; communication networks represent 'who talks with whom,' thereby providing channels for information and flow of knowledge among community members. In science research, entrepreneurs often create these knowledge and communication networks by convening conferences on new approaches to ideas, organizing convention sessions on cutting-edge topics, editing special issues of journals on emerging ideas, sponsoring new lines of intellectual work, encouraging the development of new scientific tools, and generally taking the lead in promoting new knowledge and innovative ideas. Though not in the scientific knowledge community per se, Anderson (1999) shows how the entrepreneurial skill of venture capitalists introduces considerable variation into the community of high-technology firms.

Community networks provide 'an ecological context for changes in knowledge' (that is, knowledge evolution, Schultz 2003; see also March *et al.* 2000) because 'changes in some parts of the knowledge structure tend to induce changes in other, related, or similar parts' (p. 440). When new connections are established in the network, disparate pieces of knowledge need to be reconciled. Similarly, when novel circumstances occur in the community environment, knowledge stored in organizational routines needs to be modified to adapt to the new circumstances. 'From this knowledge ecology perspective, knowledge evolves through interactions between new knowledge and prior, related knowledge' (Schulz 2003: 441; see also Burgelman 1991).

Knowledge communities typically form around knowledge technologies (Hunt and Aldrich 1998). These are particular intellectual frameworks for thinking about intellectual and scientific problems. Often they include real physical technologies, like gene splicers or linear accelerators. Sometimes they are ideological frameworks, such as critical or interpretive perspectives, that indicate what should be legitimate objects of study and appropriate ways to study them.

Viewing basic science research as a co-evolving ecological community raises some interesting research questions that revolve around what kinds of links and dynamics in the ecosystem work to accelerate or retard the flow of value and the rate of knowledge leverage and generation in the ecosystem as a whole. These include the following:

(a) What changes in the overall configuration of the network result in changes in the amount of sharing and combining of knowledge that occur, and the rate at which knowledge is leveraged, reused, tested, and generated?

(b) How do competitive, symbiotic, and commensalistic ties influence the leverage and advance of knowledge, and the overall value created by basic science research in the ecosystem?

(c) How do the various mechanisms of variation, selection, and retention operate to accelerate or retard knowledge advance and leverage?

The elements of the ecological community

Concepts similar to those in the literature of the co-evolving ecological community have been empirically observed and described in the context of science and technology development. In the BES studies, Bozeman (1999) used the term 'knowledge value collective' in his observation of sets of individuals connected to one another by their use of a particular body of information for a particular type of application. To refer to a more tightly concentrated set of individuals from multiple institutions who contribute resources and interact with each other directly in pursuit of a knowledge goal he used the term 'knowledge value alliance.' Brown and Duguid (1991, 2000) call these communities of practice. Echoing the theoretical framework of the co-evolutionary theorists, Bozeman noticed that such alliances emerge fairly soon after the first research efforts on some scientific or technological problem take place; they grow in size and complexity; then interest wanes, or new alliances sprout from the old ones as new research directions emerge. In describing the evolution of the Internet and other recent communication technologies, Tuomi (2002) shows how new directions are often crafted both from the combination of knowledge from multiple communities of knowledge and from the increasing division of labor within a well-known field. Both of these dynamics have the capacity to spawn new fields. (See Shinn, this volume, for a discussion of the history of thought regarding how these two processes result in the formation of new disciplines.)

Traditional academic approaches to knowledge creation have focused on firm-based knowledge processes (Levitt and March 1988). As Lee and Cole (2003) point out, there are several problems with this approach. First, knowledge is seen as a private commodity which the firm owns and must protect in order to profit from its intellectual capital. Second, knowledge is created under norms of authority and power that require extensive supervision and other hierarchical forms of control. Third, participants in the innovation process are typically restricted to employees of the firm, which severely limits the knowledge-creation resources.

Lee and Cole (2003) argue for a contrasting community-based model of knowledge creation. This approach seems much more suited to understanding the progress of science, including its contribution to societal and commercial innovation. Here, knowledge is openly generated by the community of interested participants rather than by the members of a closed firm. 'Knowledge is public but can be owned by members who contribute to it as long as they share it' (p. 635). Knowledge is made available to all members of the community and to anyone else who is interested in it. Typically, members of the community are distributed geographically and therefore need a variety of communication mechanisms to engage the innovation process.

These authors provide a case study of the Linux open source community to elucidate the factors that lead to innovative knowledge generation. They identify three principles that govern these processes. The first is a set of norms for knowledge creation that center on public sharing and critical review of each contribution; this leads to a self-correcting system in which all interested parties take responsibility for developing the best knowledge possible. Second, the compensation and rewards for successful contribution center on status and prestige rather than more traditional forms of monetary compensation; as Lee and Cole (2003) state, 'Norms of knowledge creation are different in how intellectual property rights can be assigned in novel ways that promote trust, ensure the sharing of knowledge, and reveal error' (p. 642). Third, the community is built on a (digital) communication system that facilitates speedy, efficient, and accurate sharing of information to everyone who needs or wants to know, not just privileged knowledge elite. In his account of the Linux community, Tuomi (2002) reports the same core principles, and also emphasizes the self-organizing character of this community of communities, and its ability to simultaneously coordinate, control, and promote local innovation. He emphasizes the criticality of the simultaneous unfolding of the social and technical systems, and of this self-organizing system's capability of aligning its social and technical architectures.

Hunt and Aldrich (1998) studied the World Wide Web commercial community. They identified seventeen different populations, including browser and search engine developers, web page designers, standards-setting bodies, and government regulatory agencies. Their analysis led to a model of community evolution that focused on '(1) the importance of technological innovation, (2) the central role of entrepreneurial activities, and (3) the dependence of community development on the establishment of legitimacy' (p. 277). Similarly, in observing how the development first of ARPANET, then of the World Wide Web, and finally of the Linux Operating System evolved, Tuomi (2002) described a dynamic set of specialized communities that focused on the development of specific enabling technologies and/or functionalities that addressed their own interests, needs, and problems (see also Castells 2002). He describes these dynamic sets of communities as having a fractal character: the evolving division of labor can create communities that spin off further communities. A similar dynamic was noted by Mohrman et al. (2004) in their study of mathematicians and computational scientists researching large-scale simulation methodologies. Communities that were loosely interrelating and dynamic formed to address mathematical and computational challenges surfaced by various applications. According to Tuomi, two different innovation dynamics can be created as new communities spin off: if the communities reinforce the existing division of labor and reproduce the functional differentiation of the social system, incremental changes in its stocks of knowledge are most likely to occur; if knowledge resources are combined and a new division of labor results, the new community that is formed may constitute the formation of new knowledge specialties, new combinations of the disciplines, new frameworks and meanings

required to address new problems, new paths and solutions. Thus the dynamic configuration and evolution of communities comprise an ecosystem of communities of practice, each with its social stocks of knowledge. At this ecosystem level, the speed of recombination of knowledge domains has increased and some believe that it is now the dominant space for innovative work (Tuomi 2002; Stokes 1997).

These network forces are evident in research in the relatively new field of biotechnology. Knowledge network connections are particularly important in industries where there is rapid technological development, knowledge is complex, and expertise is distributed around many organizations (Powell et al. 1996; Owen-Smith and Powell 2004). Although some researchers have focused on formal organizational linkages, Liebeskind and her colleagues (1996) documented how two new biotech firms used social networks to source scientific knowledge almost exclusively through individual-level exchanges of knowledge and research collaborations that involved no formal organizational market agreements. In other work, it has become apparent that, as the industry has moved in the direction of maturity and as resources become more constrained (Powell et al. 2002), the nature of the relationships in the biotechnology ecosystem has changed to include more linkages focused on the commercialization of the technology. The innovation process depends on creating a social (organizational) architecture that allows both connection to the deep discipline knowledge that flows from discipline communities and also combination of knowledge from the multiple disciplines and functions working in drug development and commercialization.

Such communities share some common features. First, they are made up of researchers, users, and sometimes other stakeholders. Both Tuomi (2002) and Bozeman (1999), for example, support the position that the meaning and value of knowledge emerges only in its use, and that users are a necessary part of networks of innovation. Second, their members come from a diverse set of organizational locations. Bozeman's knowledge value collectives, for example, may include researchers from various national laboratories, multiple universities, funding agencies, corporations, and diverse users in various organizational settings. These knowledge networks are distributed repositories of knowledge elements from a larger knowledge domain, tied together by knowledge linkages within and among organizations. Bozeman did not find formal projects or other administrative units particularly helpful in understanding the flow of knowledge. We found the same in our study of applied mathematics researchers: it was impossible to identify stable structures because the work was characterized by multiple overlapping and dynamic collaborations loosely connected by a common understanding of the field of knowledge being advanced and the problems needing to be solved for that purpose (Mohrman *et al.* 2004). And Liebeskind *et al.* (1996) found that formal organizational relationships do not capture the flow of basic knowledge. Third, both Bozeman and Tuomi describe dynamic communities in which individuals come and go, move from center to periphery, and sometimes spin off to form a new community. The same dynamism is apparent at the organizational level in the biotechnology ecological community (Powell *et al.* 2002).

Formal membership designations, such as 'organizational,' 'departmental,' or 'project memberships,' do not readily describe many of the network elements that contribute to the creation and flow of basic science value—these processes do not necessarily map onto formal 'permanent' organizations at any level (Monge and Contractor 2003). As is true in the invisible college, communities form around problems. They exist at multiple levels of inclusion, as overlapping and fluid sets: individuals can belong to multiple communities and link to others, can join or leave according to the problems to be solved, the tasks to be accomplished and their knowledge, levels of interests, resources, and commitments (Hollingshead *et al.* 2001). The nature of the network at any point in time depends fundamentally on the nature of the work that is being done (Hicks 1992; Tuomi 2002; Bozeman 1999).

As we consider the composition of the basic science research ecosystem, some important research questions relating to the mixed motives of different populations and at the different levels of analysis become evident.

(a) How does the value yielded by a firm-based model of science and innovation research differ from that of a community-based model? When considering the contribution of basic science research to the innovation potential of the society, are these two models synergistic?

(b) How do linkages across various kinds of populations in the ecosystem constrain or enable the flow of research knowledge to potential applications? For example, under what conditions do linkages between research centers and laboratories, universities, and corporations provide to companies a flow of knowledge that can inform innovation? Should these linkages be direct or indirect, weak or strong, in sparse or dense networks?

(c) How are overall levels of knowledge generation and of innovation affected by the alignment or tensions of the motives of agents at different levels: the funding agencies and policy makers that operate on the system as a whole, the organizations that act to foster their own health and survival in the ecosystem, and individuals who operate in dynamic knowledge communities to solve problems and pursue interests?

(d) What network linkages best support the creation of knowledge and enhance the likelihood that value will be created and leveraged from new knowledge communities formed through combination or differentiation of existing ones?

The resources

Three types of resources are particularly germane to a scientific ecological community:

knowledge, finances, and people. The flow of knowledge through the network, making it available to be used for multiple purposes and combined to yield novel frameworks and solutions, and new meaning, is the essence of the flow of value through the ecosystem. It is also essential to the advance of scientific knowledge. To understand fully how investment in research yields value, one must look beyond the individual program or project and see it as embedded in an overall network of activities of many loosely or closely tied communities. These connections enable knowledge to be re-used or applied in a synergistic knowledge creation that builds on shared tacit and explicit knowledge and novel combinatorial solutions: a breakthrough achieved at one node in the network has been influenced by the research carried out in another location and, in turn, has the potential for influence elsewhere.

However, knowledge does not flow simply because people see themselves as part of a knowledge community. The populations of the ecosystem compete for financial resources, which influences the flow of knowledge. It has long been known, for example, that scientists who work in corporate research laboratories are constrained by whether and when they can publish their results. Similarly, scientists working in national security areas of federal laboratories may be subject to complete or partial constraint. Although several studies (e.g. Blumenthal *et al.* 1986) have found that faculty working in industry-supported biotechnology publish and patent at higher rates than their peers who do not have such funding, they are also more likely to report that their research has become a trade secret, and that commercial considerations influence their choice of projects (Blumenthal *et al.* 1986).

One key challenge concerning the funding of basic research is that there are many more ideas worth investigation than there are funds to support the research (Stokes 1997). The resource shortage influences the flow of knowledge by affecting the focus of research as well as the kinds of collaborations that occur. As in all crowded ecosystems, different kinds of alliances and relationships are formed, as re-searchers and institutions band together to get the resources needed to survive. For example, the funding of centers often catalyzes the formation of coalitions or communities that vie for center funding. This may lead to multidiscipline and multifunctional communities that might otherwise not have emerged, as well as to linkages across the value stream that promote technology diffusion (Crow and Bozeman 1998). Calls for proposals that designate certain constituency involvements shape the membership of the would-be communities that try for the funding, and thus the knowledge combinations that are formed, as well as the problems that are addressed. For example, the call for proposals for the 6th Framework Programme (European Commission 2002) stipulates that the Framework provide only partial funding and that the applicants must find complementary sources of funding for their projects. It further stipulates that certain funds are available only to consortia of applicants, and that formal consortia agreements are required of the recipients. The dependence of these network arrangements on funding has been demonstrated: some funding-induced coalitions have rapidly dissolved when funding is no longer available. The experience of the US National Science Foundation (NSF) with their prestigious Engineering Research Centers is an example. These centers were established to conduct basic, long-term research, while linking in companies that could ultimately benefit from such research and carrying out an educational mission of developing new scientists and engineers. Companies participating willingly, and reporting that they received value from these centers as long as they were being funded by NSF, often showed dwindling interest in membership when asked to assume a greater share of the funding of the research (Feller *et al.* 2002).

The third resource that moves through the basic science ecosystem is people. Their movement between organizations within communities is particularly interesting. People are repositories for much tacit knowledge, and they carry their knowledge from one locale to another. Unless private firms are involved, the fluid and boundaryless nature of basic science collaborations means that this often happens without their disruption. People can continue to derive knowledge from their participation in multidirectional and cross-boundary collaborations, bringing to each the knowledge they derive from the others, enabling the development of new knowledge and the leverage of existing knowledge, and to some extent shaping the human capital, the social capital, and the intellectual capital that result (Nahapiet and Ghoshal 1998). But movement of people to new institutions may also mean reduced ease in the sharing of tacit knowledge, altered priorities, and perhaps changes in individuals' centrality in various communities.

Thus, the flow of value through the basic research network must be viewed in broader terms than that of the knowledge that is shared and combined. The development of human capital and knowledge communities (research capability) is one of the societal values derived from basic science research (Jordan, this volume) and one of the purposes of the public funding of such research (Bozeman 1999). Human capital and vigorous knowledge communities are considered part of a nation's technical and innovative capabilities. An additional benefit of the movement of people between communities is the social capital that is developed—the trust, goodwill, and familiarity that enables people to call on one another for knowledge resources, to extend their mutual contacts to access knowledge beyond their personal networks (see, for example, Mohrman and Galbraith 2005).

Generative mechanisms that affect the value produced by the basic science research ecosystem often operate by influencing the patterns of flow of these three resources: knowledge, financial resources, and people. Interesting research questions include:

(a) To what extent, and how, are the flow and leverage of knowledge and the generation of new knowledge and innovation related to the amount and flow of financial resources in the basic science research ecosystem? What network patterns are required if greater financial input is to result in greater yield of value?

(b) How do the patterns of flow of financial resources (the network viewed as a flow of funds) relate to the creation of human and social capital required to generate and leverage knowledge and create value?

(c) Does funding that is concentrated on highly productive, mature knowledge communities yield more value over time than funding that is spread over a more varied set of communities and/or that is concentrated on new knowledge communities?

The emergence of scientific ecological communities

It has become clear from the above discussion that basic science communities emerge and, to a large extent, self-organize. Kontopolous (1993) has generated a framework for understanding how systems emerge. He posits models of emergence that range on a continuum from highly reductionist, where the system is an aggregate of changes in independent elemental parts, to where the mechanisms of emergence are highly holistic and hierarchical, where lower levels of the system are determined by higher or more macro levels. Basic science communities fall into the middle and most complex position, referred to as 'heterarchy' (McKelvey 1997; Kontopolous 1993), in which there are multiple orders determined by multiple levels with many linkages and directions of influence. Although funding agencies, for example, yield massive influence through their choice of content focus and the organizational forms they support and encourage, they are but one influence that shapes the flow of knowledge value in the basic science system. Corporations, markets, legislators, professional societies, and the scientists themselves through their interests, friendships, collaborations, and breakthroughs continually operate to shape the ecosystem. Each of these may be considered as different but interacting populations in the community. Thus, harnessing basic science research to support national innovation and/or other agendas requires understanding the generative mechanisms that influence behavior within a complex system. It can neither be controlled from the top nor managed through the encouragement of independent, high-performing units. Nor can it emerge spontaneously from the bottom up, with little nurturing or assistance. Rather, it takes a combination of top-down, bottom-up, internal, and external mechanisms to generate a vibrant, value-producing scientific community.

Although institutional and national patterns of funding and policy may shape and influence the knowledge networks of basic science, these remain fundamentally self-organizing networks. Commenting on the self-organizing character of the Linux community, for example, Monge and Contractor point out that the success of its form in delivering value rests on its micro-foundations—the motivations of individual humans that choose freely to contribute—as well as on macro-foundations—the social and political structures that channel these contributions to a collective end (2003: 322). Even within a large research program centered in one organization such as a national laboratory or center, collaborations snake around the world. Researchers determine whom they talk to, with whom they share and combine knowledge, where they will seek out new knowledge, what conferences they will attend, and what proposals they will pursue. The components of the system are self-generative: both the people and the knowledge self-create and self-renew. The vigor of the generation, flow, and leverage of knowledge within and out of the basic research community depends on this ongoing self-organization; it depends on the strength of connections within and across communities.

The knowledge emanating from basic science research creates value when someone else learns it and applies it to their problems—when it is incorporated into social practice (Tuomi 2002). The Pelz and Andrews studies show that scientists naturally create value when they interact with one another and pass on research results. In this study, the highest-performing scientists, the best value creators, are the best interactors. However, this self-organized flow of knowledge usually follows the path of least resistance. It flows easily within 'thought collectives' (Fleck 1979)—communities that share a stock of knowledge, including ways of interacting, methodologies, theories, and social structures. The knowledge transferred confirms existing models and theories and flows within the community. But when the knowledge constitutes a breakthrough or radical innovation, the flow follows the path of most resistance. The knowledge is often tacit and complex; when flowing between communities, it entails shifts in meaning and may require changes in methods and social structure.

The flow or transfer of complex knowledge across communities may take place in a two-or-more step process. Following research on public opinion, Allen (1977) and his colleagues found a two-step process of knowledge transfer through 'opinion leaders' in laboratories; they called them 'gatekeepers.' These boundary spanners read more literature and had more external contacts, patents, and publications; they were plugged into various external communities; they became aware of new knowledge, learned it, and then integrated it into the product development projects of the laboratory. The nature of the work makes a great deal of difference in the amount of communication that is required and the extent to which a gatekeeper model is preferred to multiple collaborative contacts (Allen, *et al.* 1980; Katz and Tushman 1979). These researchers found that, in basic research, strong ties within projects and widespread communication with external colleagues are positively associated with performance: these patterns lead to collaborations that enable effective transfer of knowledge. In the more effective product-development projects, on the other hand, contact between the research function and the product-development functions of the firm tended to be handled by gatekeepers.

A two-step process for transfer between research and downstream functions was articulated by Morton (1967), the former head of AT&T's Bell Laboratories. Morton's process was part of his 'Barriers and Bonds' theory of innovation. He said that if two activities took place in the same organization, there was a bond between them: they would have similar goals, language, and so on; if they were in different organizations there was a barrier. If two activities took place at the same location, there would be a bond between them: people would meet face-to-face and develop relationships; if the activities took place at different locations, then it was a barrier. His belief was that it was necessary to have one bond and one barrier to get successful innovation. If two activities took place in the same organization at the same place, transfer of knowledge would be easy but there would be little new innovation to transfer. If two activities took place in different organizations in different places, innovations would happen but it would be extremely difficult to transfer the knowledge across two barriers. The theory developed from his experience at Bell Labs, where he found that if the Bell Labs at Murray Hill, NJ developed a new technology in switching, it was almost impossible to transfer this knowledge to Western Electric's switching division at Columbus, OH. So he created a Bell Labs satellite unit at each Western Electric (AT&T's manufacturing subsidiary—now Lucent) site. The Bell Labs at Murray Hill, NJ transferred the switching technology to Bell Labs in Columbus, OH. (Thus, the knowledge was transferred across one barrier and one bond.) Then from the Bell Labs Columbus satellite unit, the knowledge was transferred to the Western Electric Labs in Columbus. (Again the transfer took place across one bond and one barrier.)

The Bell Labs experience in transferring knowledge through barriers and bonds is instructive. The scientists at the Bell Labs found their satellites to be difficult to deal with: they were always demanding changes to the technology. An internal review of transfers of technology found that the changes suggested by

the satellites resulted in necessary modifications to integrate the new technology into the Western Electric business model. The satellites were declared a success.

The Bell Labs' experience is instructive since it shows that the value of a technology to product development is latent until a user unlocks the value through the work or collaboration to create a business model (Chesbrough 2003). The two-step process also appears in an example from research on mental health practices. The observation that university-based research results are often not adopted by mental health practitioners led to the design and creation of Cuyahoga County Community Mental Health Research Institute (CCCMHRI), a partnership between Case-Western Reserve University and the Mental Health Board in Cleveland (Biegel *et al.* 2001). Led by co-directors from the Board and the University, the Institute conducted research and modified practices to be cost-effective and consistent with the managed care environment. The assessment of the Institute, like the one for Bell Labs satellites, was a positive one. The two-step process modified the knowledge to fit their equivalent of a business model and was learned and integrated into the board's delivery practices.

Centers such as NSF's Engineering Research Centers or DOE's Scientific Discovery through Advanced Computing (SciDAC) Centers are the public-sector equivalent of these satellites. Although activities within these centers are largely self-organizing, starting with the applications processes by participants who want to create one, the centers are designed so that multiple communities contribute talent; this talent collaborates to generate and transfer knowledge to users who will create value. Technology transfer is less likely if centers are not involved (Crow and Bozeman 1998). Nor, if it is based on assumptions of one-way transfer of knowledge (Pinelli and Barclay 1998; Bikson *et al.* 1987), is it ensured by simply setting up centers and/or creating intermediary roles such as technology transfer offices. Studies of the dissemination of the knowledge created by government-funded research studies find that approaches that encourage knowledge utilization after the research has been completed are generally ineffective. Pinelli and Barclay conclude that linkages between applications and basic science researchers must occur during the idea-generation phase of the innovation process, rather than as a linear process. For this to happen, they argue, government funding agencies must better understand and attend to the flow of knowledge and to establishing policies and practices that encourage early connection and dissemination.

The design of centers, satellites, and other approaches to linking across the value chain is an important research issue in the creation of value. Henderson (1994) provides some insight into how these institutions can be designed, how they operate to acquire knowledge from different communities, and then integrate it to create new knowledge and new value. She and her colleagues examined the transition by pharmaceutical companies from an intuitive drug discovery approach dominated by synthetic chemists to one of science-based research through a structure change to highly specialized discovery units populated by multiple science specialties: synthetic chemists, pharmacologists, animal biologists, biochemists, molecular biologists, physiologists, analytical chemists, computer scientists, molecular kinetics specialists, and so on. Top academic scientists were recruited to lead the specialist departments. These discovery units search the knowledge of many new and old science communities for new ideas. Firms successful in making this transition encouraged links enabling new knowledge to be sought and brought into the firm. The new specialists were measured and rewarded for their publication records and standing in their professional communities. Access to state-of-the-art discipline knowledge was part of the challenge of creating value through the firm's research. The other was integrating it with the knowledge of other communities in order to create new chemical entities (NCEs). The most successful firms organized by therapeutic areas as well as by scientific specialty in a matrix structure. They were able to create a balanced distribution of power across these new departments, through a leadership team that debated proposals in a peer-review-like manner, and collectively allocated resources and set priorities. Members of the multiple knowledge communities focused on and shared their knowledge about application areas such as oncology, cardiology, and respiratory diseases. When a promising molecule was discovered, they organized into cross-functional teams to create an NCE. The research discovery activities in the low-performing companies were run by a laboratory director who made the decisions, and continued to be dominated by the views of the synthetic chemists. The best value-creators integrated cross-community knowledge at three levels: the leaders integrated knowledge across communities and therapeutic areas when setting priorities; the therapeutic areas integrated knowledge across communities when generating proposals; cross-functional teams integrated knowledge when working to create an NCE.

The successful firms were 25 per cent more productive in creating NCEs. They created discovery organizations with the following characteristics: the talent of the unit was the members of their discipline-based research communities and networks; they shared knowledge and then combined it in their discovery units which were designed for integration; the knowledge was integrated in focused application units to create NCEs, which were then transferred to the development units to create new molecules to treat diseases. The successful discovery units were designed to access new scientific knowledge from the relevant scientific communities and integrate this knowledge to create value. When knowledge acquisition required external collaboration, these firms formed alliances with other firms. The knowledge from these alliances was captured through scientists who participated in the collaboration and through 'alliance managers' whose task it was to see that knowledge is learned and distributed within the firms (Powell 1998). Again, a two-step process was used to transfer knowledge. This knowledge flowed in part through the time-honored self-organizing processes within communities, and then was integrated in structures that were designed to combine cross-community knowledge.

From this example we get a sense for the multiple levels of analysis that interplay in the yielding of value, particularly innovation, from basic research. One key focus is the self-organizing processes of scientists who form collaborations and share knowledge both within and across knowledge communities, in the process potentially yielding new knowledge communities and fields of study. This includes both the within and cross-organization linkages along the value stream, enabling the reciprocation between science and application. At another level of analysis are the structural units—departments, projects, business units—that administratively shape behavior and networks as well as the barriers and incentives that are built into their design and operation. Societal agents such as funding agencies, scientific associations, and journals shape the overall agenda and enforce it through funding and review processes. We have singled out the importance of centers, institutes, satellites, and research departments designed to fill in structural holes in the value network. To intervene to some purpose in the dynamics of the network, and to shape institutional and organizational practices that encourage the flow of value, the impact of these varying agents at multiple embedded and cross-cutting levels of analysis must be understood. To understand better what kinds of interventions will yield the greatest value, a research agenda might examine such questions as:

(a) In the cycle of evolution of particular basic science knowledge domains, when are connections to application most important, and how should these connections change over time for maximum translation of knowledge into innovative accomplishments?

(b) What are the policy and managerial measures that can be taken to generate a greater flow of value through the investment in basic science research? If such research occurs largely within self-organizing communities, what measures will cause the ongoing self-renewal activities in the overall ecosystem to be heedful of the ways in which increased value can be created and focused?

(c) For institutions in linkage positions in the networks that form the value chains that, eventually, deliver value and yield innovation, what design features and network patterns are important?

Conclusion and implications

We have made the case that the metaphor of an ecosystem helps us to understand more fully how the value generated by basic science research contributes to innovation, or to any mission. Science research is a heavily relational activity: it is made up of dynamic knowledge networks embedded in an ecosystem composed of co-evolving populations and communities. Focusing on linear flows between particular laboratories and downstream users captures only one small part of the system, and does not address its overall innovative capability. Concentrating on the role of formal organizations in this process may result in failing to see where knowledge is actually produced and leveraged, and how it does or does not flow into use.

Idea-innovation networks (Hage and Hollingsworth 2000) consist of both horizontal links across an arena such as basic research, and vertical links along the value chain. The horizontal linkages of basic science research reach into invisible colleges and multifaceted collaborations, made up of intricate networks with a variety of linkages. The link to product development or other applications may require a series of two-step processes. Because time and energy are limited, links to application may occur at the expense of time spent generating new scientific knowledge. Strengthening these links may require the creation of special structures that focus on them, such as satellites, institutes, and centers (Crow and

Bozeman 1998) with funding for the participation of scientists in application-oriented activities; alternatively, they may be built to draw application perspectives into the activities of basic science research communities. To grasp fully how value is created through basic science research, we must understand the dynamics of knowledge networks and the ecological field in which they operate.

Research methodologies must be capable of examining complex, multi-level phenomena and applying multiple theories to explain the generative forces that simultaneously affect behavior in the network (Monge and Contractor 2003). At the institutional level, the flow of value is shaped, enabled, and constrained by policy and funding decisions. Likewise, organization-level decisions about what relationships to formalize, fund, and encourage influence the relationships in the community. The activities of individuals and of formal and informal communities of practice are integral. Network organizations are organized around and created out of complex webs of exchange and dependency relations among multiple organizations, individuals, and communities. Among the many causal dynamics simultaneously at work are theories of economic self-interest and resource dependency, of homophily (that people will connect to those who are similar along key dimensions), of weak and strong ties, and of tacit and codified information. Each theory will describe some, but not all, of the behavior within the network; each theory's perspective will yield understanding of the basic science research networks that can contribute to application and to how, through managerial and policy decisions and actions, to intervene purposefully in the ecosystem to increase the value flow.

Viewing basic science research as an ecosystem in which value lies in the overall knowledge that is produced and applied provides a framework for making science policy decisions based on an understanding of such a system's complex dynamics. Eventually, it might be possible to model those dynamics and predict the impact of various policy interventions more accurately. Clearly, science policies impact the variation and selection processes in the knowledge system; they thus may be a force that constrains or enlarges the domain of investigations, and/or fosters continuity or radical new directions. Additionally, through collaboration requirements built into the calls for research proposals, and through the funding of various kinds of cross- and within-community institutes, funding agencies can affect the linkages in the basic science networks and promote knowledge combination and sharing, and connection to application. Funding patterns also impact the development of human and social capital. By concentrating funds on high-performing teams and knowledge champions, greater knowledge may be yielded in the short term, but by funding young scientists with novel approaches, variety may be encouraged, new knowledge breakthroughs may occur, and a broader and deeper pool of human capital may result. As more is learned about the knowledge dynamics of the basic science ecosystem, its dynamic communities and networks, and the impact of various patterns of linkages on knowledge leverage and innovation, this knowledge can inform science policy.

The vast majority of organizational research about innovation has focused on the private sector, and the within- and cross-firm approaches that foster increased innovation. The recent literature examining the networks of the biotechnology industry and alliance behavior has drawn attention to the rich ecological community in which science- and technology-based firms exist, to the importance of the firm's network linkages, and to the complex interplay between public and private elements of the ecosystem. To develop science and technology policies and managerial and organizational approaches that will increase the value derived from the large investment in fundamental research, it is necessary to understand both how such networks operate and the role of basic science research in innovation. This chapter has aimed at a deeper understanding of the multi-level network dynamics that constitute the conduct of basic science research, and its connections to the innovative capacity of society.

References

Aldrich, H. (1999). *Organizations Evolving*. Thousand Oaks, CA: Sage.

Alic, J. A. (1993). 'Technical Knowledge and Technology Diffusion: New Issues for U.S. Government Policy.' *Technology Analysis and Strategic Management*, 5: 369–83.

Allen, T. (1977). *Managing the Flow of Technology*. Cambridge, MA: MIT Press.

—— Lee, D., and Tushman, M. (1980). 'R&D Performance as a Function of Internal Communication, Project Management, and the Nature of the Work.' *IEEE Transactions on Engineering Management*, EM-27/1: 2–12.

American Institute of Physics (1999). *Bulletin of Science Policy News,* 145/7 October.

Anderson, P. (1999). 'Venture Capital Dynamics and the Creation of Variation through Entrepreneurship.' In J. A. C. Baum and B. McKelvey (eds.), *Variations in Organization Science: In Honor of Donald T. Campbell*. Thousand Oaks, CA: Sage, 155–68.

Ashby, W. R. (1956). *Introduction to Cybernetics*, New York: Wiley.

—— (1962). 'Principles of the Self-Organizing System.' In H. von Foerster and G. W. Zopf (eds.), *Principles of Self-Organization*. New York: Pergamon Press.

Astley, W. G. (1985). 'The Two Ecologies: Population and Community Perspectives on Organizational Evolution.' *Administrative Science Quarterly*, 30: 224–41.

—— and Fombrun, C. J. (1987). 'Organizational Communities: An Ecological Perspective.' *Research in the Sociology of Organizations*, 5: 163–85.

Berger, P. L., and Luckmann, P. (1966). *The Social Construction of Reality*. Garden City, NY: Doubleday.

Biegel, D. E., Johnsen, J. A., and Shafran, R. (2001). 'The Cuyahoga County Community Mental Health Research Institute.' *Research on Social Work Practice*, May, 11/3: 390–403.

Bikson, T. K., Quint, B. E., and Johnson, L. L. (1987). *Scientific and Technical Information Transfer: Issues and Options*. Washington, DC: National Science Foundation.

Blumenthal, D., Gluck, G., Louis, K. S., Stoto, M. A., and Wise, D. (1986). 'University-Industry Research Relationships in Biotechnology: Implications for the University.' *Science*, New Series, 323/4756: 1361–6.

Bozeman, B. (1999). *The Research Value Mapping Project: Qualitative-Quantitative Case Studies of Research Projects Funded by the Office of Basic Energy Sciences*. Final Report to the Office of Basic Energy Sciences, Department of Energy.

—— and Rogers, J. (2001). 'Strategic Management of Government-Sponsored R&D Portfolios.' *Environment and Planning C: Government and Government and Policy*. 19/3: 412–42.

Branscomb, L. M. (1992). 'U.S. Scientific and Technical Information Policy in the Context of a Diffusion-Oriented National Technology Policy.' *Government Publications Review*, 19: 469–82.

—— (1999). 'The False Dichotomy: Scientific Creativity and Utility.' *Issues in Science and Technology*. Fall: 66–72.

Brockman, J. (2002). Introduction to *The Next Fifty Years: Science in the First Half of the Twenty-First Century*. New York: Random House, xiii–xv.

Brooks, H. (1986). 'National Science Policy and Technological Innovation.' In R. Landau and N. Rosenberg (eds.), *The Positive Sum Strategy: Harnessing Technology for Economic Growth*. Washington, DC: National Academy Press, 119–68.

Brown, J. S., and Duguid, P. (1991). 'Organizational Learning and Communities of Practice: Toward a Unified View of Working, Learning and Innovation.' *Organization Science,* 2: 40–57.

—— —— (2000). *The Social Life of Information*. Boston, MA: Harvard Business School Press.

—— —— (2001). 'Knowledge and Organization: A Social-Practice Perspective.' *Organization Science,* 12: 198–213.

Burgelman, R. A. (1991). 'Intraorganizational Ecology of Strategy Making and Organizational Adaptation: Theory and Field Research.' *Organization Science,* 2: 239–62.

Campbell, D. T. (1960). 'Blind Variation and Selective Retention in Creative Thought as in Other Knowledge Processes.' *Psychological Review*, 67: 380–400.

Carroll, G. R., and Hannan, M. T. (2000). *The Demography of Corporations and Industries*. Princeton: Princeton University Press.

Castells, M. (2002). *The Internet Galaxy: Reflections on the Internet, Business, and Society*. New York: Oxford University Press.

Chesborough, H. W. (2003). *Open Innovation*. Boston: Harvard Business School Press.

Cohen, W. M., and Levinthal, D. A. (1990). 'Absorptive Capacity: A New Perspective on Learning and Innovation.' *Administrative Science Quarterly*, 35/1: 128–52.

Crow, M., and Bozeman, B. (1998). *Limited by Design: R&D Laboratories in the U.S. National Innovation System*. New York: Columbia University Press.

Dalpe, R. (2003). 'Interaction between Public Research Organizations and Industry in Biotechnology.' *Managerial and Decision Economics*, 24: 171–85.

European Commission (1994). *The 4th Framework Programme in Brief*. European Commission: December.

—— (1998). *The 5th Framework Programme in Brief*. European Commission: December.

—— (2002). *The 6th Framework Programme in Brief*. European Commission: December.

Feldman, M. S. (2000). 'Organizational Routines as a Source of Continuous Change.' *Organization Science*, 11: 611–29.

Feller, I., Ailes, C. P., and Roessner, J. D. (2002). 'Impacts of Research Universities on Technological Innovation in Industry: Evidence from Engineering Research Centers.' *Research Policy*, 31: 457–74.

Fleck, L. (1979). *Genesis and Development of a Scientific Fact*. Chicago: The University of Chicago Press.

Greve, H. R. (2002). 'Interorganizational Evolution.' In J. A. C. Baum (ed.), *The Blackwell Companion to Organizations*. Malden, MA: Blackwell Publishers.

Gulati, R., Nohria, N., and Zaheer, A. (2000). *Strategic Management Journal*, 21: 203–15.

Hage, J., & Hollingsworth, R. (2000). 'A Strategy for the Analysis of Idea Innovation Networks and Institutions.' *Organization Studies*, 21/5: 971–1004.

Hansen, M. T. (1999). 'The Search-Transfer Problem.' *Administrative Science Quarterly*, 44: 82–111.

—— and Freeman, J. (1977). 'The Population Ecology of Organizations.' *American Journal of Sociology*, 82: 929–84.

Hawley, A. (1986). *Human Ecology: A Theoretical Essay*. Chicago: University of Chicago Press.

Henderson, R. (1994). 'The Evolution of Integration Capability.' *Industrial and Corporate Change*, 3/3: 607–30.

Hicks, D. (1992). 'Instrumentation, Interdisciplinary Knowledge, and Research Performance in Spin Glass and Superfluid Helium Three.' *Science, Technology and Human Values*, 17/2: 180–204.

Hollingshead, A. B., Fulk, J., and Monge, P. (2001). 'Fostering Intranet Knowledge-Sharing: An Integration of Transactive Memory and Public Goods Approaches.' In S. Keisler and P. Hines (eds.), *Distributed Work: New Research on Working across Distance using Technology*. Cambridge, MA: MIT Press, 235–55.

Hunt, C. S., and Aldrich, H. E. (1998). 'The Second Ecology: Creation and Evolution of Organizational Communities.' *Research in Organizational Behavior*, 20: 267–301.

Hurley, J. (1997). *Organisation and Scientific Discovery*. Chichester: John Wiley and Sons.

Ingram, P. (2002). 'Interorganizational Learning.' In J. A. C. Baum (ed.), *The Blackwell Companion to Organizations*. Oxford: Blackwell, 642–63.

Ireland, R. D., Hitt, M. A., and Vaidyanath, D. (2002). 'Alliance Management as a Source of Competitive Advantage.' *Journal of Management*, 28/3: 413–46.

Kantorovich, A. (1993). *Scientific Discovery*. Albany: State University of New York Press.

Kash, D. E. (1992). 'Innovation Policy.' In S. Okamura, K. Murakami, and I. Nonaka (eds.), *Proceedings of the Second NISTEP International Conference on Science and Technology Policy Research*. Tokyo: Mita Press, 139–48.

Katz, R., and Tushman, M. (1979). 'Communication Patterns, Project Performance, and Task Characteristics: An Empirical Evaluation and Integration in an R&D Setting.' *Organizational Behavior and Human Performance*, 23: 139–62.

Kauffman, S. A. (1993). *The Origins of Order: Self-Organizing and Selection in Evolution*. New York: Oxford University Press.

—— (1995). *At Home in the Universe: The Search for the Laws of Self-Organization and Complexity*. New York: Oxford University Press.

Kealey, T. (1996). 'Technical Knowledge and Technology Diffusion: New Issues for U.S. Government Policy.' *Technology Analysis and Strategic Management*, 5: 369–83.

Koestler, A. (1964). *The Act of Creation* (repr. 1978). London: Pan Books.

Kogut, B. (2000). 'The Network as Knowledge: Generative Rules and the Emergence of Structure.' *Strategic Management Journal*, 21/3: 405–25.

Kontopoulos, K. M. (1993). *The Logics of Social Structure*. New York: Cambridge University Press.

Lambright, W. H. (2002). 'Managing 'Big Science': A Case Study of the Human Genome Project.' *New Ways to Manage Series*: the Pricewaterhouse Coopers Endowment for the Business of Government, March.

Lave, J., and Wenger, E. (1991). *Situated Learning: Legitimate Peripheral Participation*. Cambridge: Cambridge University Press.

Lee, G. E., and Cole, R. E. (2003). 'From a Firm-Based to a Community-Based Model of Knowledge Creation: The Case of the Linux Kernel Development.' *Organization Science*, 14: 633–49.

Lee, R., Jordan, G., Leiby, P., Owens, B., and Wolf, J. (2003). *Estimating the Benefits of Government-Sponsored Energy R&D*. Oak Ridge National Laboratory.

Levitt, B., and March, J. G. (1988). 'Organizational Learning.' *Annual Review of Sociology*, 14: 319–40.

Liebeskind, J. P., Oliver, A. L., Zucker, L., and Brewer, M. (1996). 'Social Networks, Learning, and Flexibility: Sourcing Scientific Knowledge in New Biotechnology Firms.' *Organization Science*, 7/4: 428–43.

McKelvey, B. (1997). 'Quasi-natural Organization Science.' *Organization Science*, 8: 352–80.

Madsen, T. L., Mosakowski, E., and Zaheer, S. (1999). 'Static and Dynamic Variation and Firm Outcomes.' In J. A. C. Baum and B. McKelvey (eds.), *Variations in Organization Science: In Honor of Donald T. Campbell*. Thousand Oaks, CA: Sage, 213–36.

March, J. G. (1991). 'Exploration and Exploitation in Organizational Learning.' *Organization Science*, 2: 71–87.

—— (1994). 'The Evolution of Evolution.' In J. A. C. Baum and J. V. Singh (eds.), *Evolutionary Dynamics of Organizations*. New York: Oxford University Press, 39–49.

—— Schulz, M., and Zhou, X. (2000). *The Dynamics of Rules: Change in Written Organizational Codes*. Stanford, CA: Stanford University Press.

Meeus, M. M., Oerlemans, L., and Hage, J. T. (2001). 'Patterns of Interactive Learning in a High-tech Region.' *Organization Studies*, 22: 145–72.

Miner, A. S., and Raghavan, S. V. (1999). 'Interorganizational Imitation: A Hidden Engine of Selection.' In J. A. C. Baum and B. McKelvey (eds.), *Variations in Organization Science: In Honor of Donald T. Campbell*. Thousand Oaks, CA: Sage, 35–62.

Mohrman, S. M., and Galbraith, J. R. (2005). *Dynamics of the Adaptive Mesh Refinement (AMR) Network: The Organizational and Managerial Factors that Contribute to the Stream of Value from the Basic Research Funding of the Office of Science*. Technical Report Contract DE-AC02-04ER30321, The Center for Effective Organizations, University of Southern California.

—— —— and Monge, P. R. (2004). *Network Attributes Influencing the Generation and Flow of Knowledge within and from the Basic Research Community*. Technical Report Contract 84882. The Center for Effective Organizations, University of Southern California.

Monge, P. R., and Contractor, N. S. (2003). *Theories of Communication Networks*. Oxford: Oxford University Press.

Morton, J. (1967). 'A Systems Approach to the Innovation Process.' *Business Horizons*, 10/2: 27–36.

Nahapiet, J., and Ghoshal, S. (1998). 'Social Capital, Intellectual Capital and the Organizational Advantage.' *Academy of Management Review*, 23/2: 218–24.

Nelson, R. R., and Winter, S. G. (1982). *An Evolutionary Theory of Economic Change*. Cambridge, MA: Belknap Press of Harvard University Press.

Office of Science (2003). *A Science-Based Case for Large-Scale Simulation*.

Owen-Smith, J., and Powell, W. W. (2004). 'Knowledge Networks as Channels and Conduits: The Effects of Spillovers in the Boston Biotechnology Community.' *Organization Science*, 15/1: 5–21.

Pelz, D. C., and Andrews, F. M. (1966). *Scientists in Organizations*. New York: John Wiley Press.

Perkins, D. (1992). 'The Topography of Invention.' In R. W. Weber and D. Perkins (eds.), *Inventive Minds*. Oxford: Oxford University Press.

Pinelli, T. E., and Barclay, R. O. (1998). 'Maximizing the Results of Federally Funded Research and Development through Knowledge Management: A Strategic Imperative for Improving U.S. Competitiveness.' *Government Information Quarterly*, 15/2: 157–72.

Polanyi, M. (1962). 'The Republic of Science.' *Minerva*, 1: 54–73.

Porac, J. (1994). 'On the Concept of 'Organizational Community.' ' In J. A. C. Baum and J. V. Singh (eds.), *Evolutionary Dynamics of Organizations*. New York: Oxford University Press, 451–6.

Powell, W. (1998). 'Learning from Collaboration.' *California Management Review*, 40/3: 228–40.

—— Koput, K., and Smith-Doerr, L. (1996). 'Interorganizational Collaboration and the Locus of Innovation: Networks of Learning in Biology.' *Administrative Science Quarterly*, 41/1: 116–45.

—— White, D. R., Koput, K., and Owen-Smith, J. (2002). 'Network Dynamics and Field Evolution: The Growth of Interorganizational Collaboration in the Life Sciences.' A Santa Fe Institute publication, December.

Rao, H. (2002). 'Interorganizational Ecology.' In J. A. C. Baum (ed.), *The Blackwell Companion to Organizations*. Malden, MA: Blackwell Publishers, 541–56.

Romanelli, E. (1999). 'Blind (But Not Unconditioned) Variation: Problems of Copying in Sociocultural Evolution.' In J. A. C. Baum and B. McKelvey (eds.), *Variations in Organization Science: In Honor of Donald T. Campbell*. Thousand Oaks, CA: Sage, 79–91.

Root-Bernstein, R. (1989). *Discovering*. Cambridge, MA: Harvard University Press.

Ruef, M. (2000). 'The Emergence of Organizational Forms: A Community Ecology Approach.' *American Journal of Sociology*, 106: 658–714.

Schon, D. (1963). *Invention and Evolution of Ideas*. London: Social Science Paperbacks.

Schultz, M. (2003). 'Pathways of Relevance: Exploring Inflows of Knowledge into Subunits of Multinational Corporations.' *Organization Science*, 14: 440–59.

Schumpeter, J. A. (1934; repr. 1962). *The Theory of Economic Development: An Inquiry into Profits, Capital, Credit, Interest and the Business Cycle*. Cambridge, MA: Harvard University Press.

Stokes, D. (1997). *Pasteur's Quadrant: Basic Science and Technological Innovation*. Washington, DC: The Brookings Institution.

Sulston, J., and Ferry, G. (2002). 'The Common Threat: Science Politics, Ethics and the Human Genome.' European Commission 2002.

Thomke, S., von Hippel, E., and Franke, R. (1998). 'Models of Experimentation: An Innovation Process—and Competitive—Variable.' *Research Policy*, 27: 315–32.

Tucker, C., and Sampat, B. (1998). 'Laboratory-Based Innovation in the American National Innovation System.' In M. Crow and B. Bozeman (eds.), *Limited by Design: R&D Laboratories in the U.S. National Innovation System*. New York: Columbia University Press.

Tuomi, I. (2002). *Networks of Innovation*. Oxford: Oxford University Press.

US Government (2003). Historical Tables. Budget of the United States Government: Fiscal Year 2004. Washington, DC: US Government Printing Office.

Innovation, Learning, and Macro-institutional Change: The Limits of the Market Model as an Organizing Principle for Research Systems

Luke Georghiou

Introduction

Current conventional wisdom is that research and innovation policies should be designed on the basis of the 'systems of innovation' approach (Freeman 1987; Lundvall 1992). This aims to go beyond the familiar market-failure rationales for public policy that have dominated thinking since the 1980s, and that have their origins in the seminal work of Arrow (1962). Rather than focus on the public-good nature of science and the issue of spillovers, the 'systems failure' approach is focused upon the importance of institutions and the linkages between them (Metcalfe 1995; Smith 2000). However, much more attention has been paid to some parts of the system than to others, and little to the overall dynamics. To consider the three main types of research-performing institutions: numerous studies have addressed changes in universities and, to a lesser extent, those in corporate R&D; a much smaller literature has examined changes in government laboratories (Crow and Bozeman 1998; Boden et al. 2004). During the past two decades, these institutional types have been subject to substantial changes in their operating environment and to reform in their own structures. One way in which to summarize these changes is to say that research performers have deliber-ately been more exposed to market forces and the incentive structures that accompany these. It would be an oversimplification to describe this process as one of opening up fully to mar-ket forces: in many cases, the structures that now exist are not the direct result of markets but rather the attempts of policy makers to reproduce in the design of organizations, or in their funding environment, some aspects of market conditions. Such elements as competition for contracts, pressure to raise funding from new sources, changes in govern-ance and performance measurement have been introduced, but the extent to which they constitute a market is one that we shall explore below.

The chapter seeks to examine the conse-quences of operating research policy for the provision of scientific services on the basis of a competitive market and, within that, the role of public research organizations. While several of the trends are found in all major research-performing nations, many of the changes here will be exemplified by reference to changes in UK policy since the 1980s. In essence, it will be argued that these changes have created condi-tions in which the three major research-per-forming sectors have moved away from an initial clear division of labor to a situation where they now are frequently in competition for the same work. The positive and negative consequences of this convergence are explored in the chapter. Policies for improving linkages within the innovation system are reconsidered in the light of these considerations.

Research-performing sectors

First we take the developments in the three main research-performing sectors: corporate R&D, government laboratories, and universities.

Corporate R&D

The environment for business R&D has been characterized by globalization: its mergers and acquisitions have increased pressure for shareholder value, leading to demands for increased productivity and throughput in R&D. There has been a general decline in the number and size of corporate laboratories, with budgets now predominantly held by operating divisions. This has caused the bulk of research to be focused on current business problems. Recent work has shown that leading-edge companies have now moved to a relationship between research and business which is better characterized as a partnership, not least because of the need to extend technological horizons beyond those of the immediate market. One consequence is a rapid increase in outsourcing, from both the private R&D service sector and from universities. Howells (1997) suggests a move, over the last decade, from around 5 per cent of business R&D outsourced to between 10 and 15 per cent. However, the corporate labs, in response to having to compete for company funds, have themselves also tended to seek external work. The environment for knowledge generation is now better characterized as an ecology rather than a single firm (Coombs and Georghiou 2002). The histories of two corporate laboratories serve to illustrate these trends.

BT Exact

The first example is BT Exact, the name given since 2001 to the former laboratories of British Telecom, historically the national telecommunications provider, and still the dominant player in the UK market. Today, BT Exact describes itself as a communications- solutions provider in the global market. These laboratories had their origins in the 1930s; their most recent previous existence was as the corporate research laboratories of the company. A strong technological inheritance includes a patent portfolio of over 2,000 and a workforce of 3,000 technologists (a figure now doubled by a recent merger with the firm's IT supplier). At one level, the core activity is the same, providing research-based services for BT and its customers. However, much has also changed. The shift of BT to being a service company has naturally led to much closer research relationships with customers, but business is being sought beyond that group. Targets as ambitious as 40 per cent of revenue have been set for external earnings, and the laboratory is a profit center rather than a cost center. Accompanying this has come an opening up of its site to a range of technology businesses and collaborative ventures with universities. For example, Chimera, Essex University's Institute for Socio-Technical Innovation and Research, is one of several university outposts located on the company's site at Adastral Park. Thus, as the laboratory becomes a business with its own customer relationships and collaboration networks, we see a gradual disconnection from the parent company.

Roke Manor Research

The second case, Roke Manor Research, was founded in 1956 as a laboratory working on military communications for the Plessey Company. From the mid-1980s, it expanded from its defense-related origins to work on topics for the commercial telecommunications market, and GSM cellular telephony. Following the takeover of Plessey by the German electronics company Siemens and the UK's GEC in 1990, Siemens took part-ownership of Roke Manor

Research in 1990. In 1991, GEC sold its 50 per cent shareholding in Roke Manor Research, and the company became wholly owned by Siemens. However, during that transition the laboratory became a business, today with 431 employees, carrying out contract research and development for communications, networks, and electronic sensors. In 2002/3 Siemens accounted for 44 per cent of its custom, government 47 per cent, and commercial work 9 per cent. The commercial element had been higher but the fallback in telecommunications and growing opportunities in defense and security have changed the balance. Roke Manor invests 7 per cent of turnover in its own R&D to support the business. In a more pronounced fashion than the previous example, this laboratory has evolved to being a business that trades in research services.

Government laboratories

Public laboratories have frequently faced an expiry of the validity of their original missions; this has driven them in the direction of diversification. Some of the largest in this category have been the laboratories formerly responsible for the development of civil nuclear power, where the mission either transferred to the public sector or became irrelevant following a national political decision to withdraw from that technology. A wide variety of missions remain common in the public sector including:

- pure scientific research to support the national science base (for example, research in astronomy and particle physics);
- public information services (e.g. meteorology, air quality reports);
- support for regulation and legislation, either nationally or in international agreements (e.g. analysis, forensics, environmental impacts, health and safety);
- support for procurement (notably in defense equipment);
- services for industry or agriculture (e.g. standards, measurement and calibration, technology transfer/extension);

- support for policy (e.g. scientific advice on public health) (Boden et al. 2001).

However, the assumption that government should also be responsible for the provision of the services that it consumes has been challenged. The general trend of public-sector reform or 'new public management' (Dunleavy and Hood 1994) set in train a sequence of commercialization, agencification, and, in many cases, privatization of government laboratories. The result in the UK has been a variety of institutional formats, with many of the most prominent laboratories transformed into contract research organizations in which their original work for government is only a part of the portfolio of work for which they compete. A recent cross-European study has shown that while privatization is a relatively rare event outside the UK, reduced core funding from government, and the consequent need to raise funds from contracts, have led to similar management changes. While some large national institutions have hit hard times, others flourish and the sector as a whole is dynamic, with around half of all European laboratories being founded or radically restructured in the past twenty years (PREST 1993). This situation is in marked contrast to that in the US where, despite some initiatives aimed at commercialization, the basic institutional structures have remained essentially unchanged for several decades (Crow and Bozeman 1998). The extent to which such labs have changed varies to a large degree in the UK. Again we may consider two examples (drawn from PREST 1993).

Building Research Establishment

The Building Research Establishment (BRE) began its existence in 1921 as a centrally funded government laboratory in the area of construction. From the outset, it was concerned with standards and guidelines for practitioners, long-term testing, and advisory services for industry, as well as more fundamental research. Its mission was to serve the users of buildings (and not the construction industry). Thus, it has always had the character or mission of an applied research institute. It became a recognized leader in its field, with a worldwide reputation for its excellence, expertise, and impartiality. In 1988 it ceased to be funded directly by parliamentary vote, instead being funded through contracts with its parent ministry, the Department of the Environment (DoE). From 1990 it became an Executive Agency and gained a Chief Executive, some financial autonomy, and was expected to negotiate specified programmes of work with the Department. The Agency was expected to win a minimum of 10 per cent and a maximum of 15 per cent of its income from private work. From 1992, its mission shifted as the DoE acquired an industry-sponsorship role and, in turn, sought the research it supported to have more emphasis upon this.

Full privatization came in 1997; it was pursued with politically driven haste in order to be completed before the General Election. Privatization was through sale by competitive tender, but the successful bid was led by the management and based upon ownership by the Foundation for the Built Environment, a not-for-profit company of a legal form known as a 'company limited by guarantee.' The Foundation's members are firms, professional bodies, and other organizations across the construction industry and building users, including some universities with built environment research groups. The governing structure aims to prevent any single interest group having undue influence, and thus to preserve BRE's independence and impartiality. BRE now describes itself as a research-based consultancy. Its income is around £30 million per year, and this has remained stable since privatization. Government work has dropped from 95 per cent to 60 per cent of income since privatization. A five-year guarantee of government work, given on privatization, expired in 2002. Since then, BRE has found it easier to bid for government contracts outside the protected area of Construction Sponsorship. It is likely that the proportion will decline further, perhaps to 50 per cent, but probably no lower.

Income from non-government sources has grown significantly, split more or less evenly between problem-solving/consultancy work and testing/certification work. BRE exists in a complex system where competitors are frequent collaborators and, sometimes, customers, as are those affected by its regulatory work for the government.

One major concern is for the national facilities that BRE traditionally operated. Some existing long-term experiments are being terminated to allow the sale of land, and it is not clear why the government should be motivated to provide facilities for a private organization. To a lesser extent, renewal of the knowledge base is also a problem, particularly as staff turnover is now 17 per cent per year, a strong contrast from the lifetime-career model once normal in the civil service.

In sum, BRE is now a seller of science-based services. (Many of them are not so much research as application of existing knowledge.) These services are provided to a market far more diverse than in BRE's previous existence as a government laboratory.

AEA Technology

AEA Technology plc is a private-sector company listed on the London Stock Exchange. It was formerly part of the United Kingdom Atomic Energy Authority, a public corporation created by an Act of Parliament in 1954. Its purpose was to oversee the United Kingdom atomic energy project. In September 1996, the AEA was privatized and the name changed to AEA Technology (AEAT). At the time of privatization, the organization still bore many of the hallmarks of its nuclear inheritance. Three-quarters of its business was in the UK; some 54 per cent of its £253m turnover was nuclear related; its largest single customer was its former parent, the 'new' UKAEA. In its last financial year as a state-owned company, 60 per cent of sales were to the public sector or to the UK government itself. Today, AEAT describes itself as one of the world's leading innovation businesses in science and engineering. The business focuses on five key areas: technology-based products; specialized science; environmental management; improving the efficiency of industrial plant; risk assessment and safety management. In each area, AEAT offer services, products, consultancy, software, and technology transfer. In the financial year 2000–1, turnover was £378.7 million.

Since privatization, AEAT has maintained its scientific capacity and expanded its social science capability, which has grown and is now significant; focusing the activity of the AEAT group has been a major strategic aim. Focus for the future of AEAT is now on just two businesses, rail and environment. Funding from UK central government has fallen dramatically in the nuclear sector, but income for environmental work has grown. AEAT has no core government funding; all its income from this customer is won by competitive tender. Nor is AEAT the monopoly supplier to UKAEA. There were no guaranteed contract arrangements for AEAT when it was privatized.

Since privatization, the AEAT connection with academic institutions has weakened, and participation in public research funding schemes is much reduced. In 1986, AEAT had 150 cooperative PhD students; by 2003, it had only three. AEAT employs 2,500 graduates; these are identified as the qualified staff and their number has risen since privatization, mainly as a result of acquisitions. Since commercialization, the profile has become younger; a number have left and moved into academia, to UKAEA or into other public-sector organizations; several moved from nuclear and retrained to work in environment and other sectors. One example of priority change is in the publishing activity of AEAT: as part of company policy, the number of publications since privatization has decreased greatly. Publishing, even in academic journals, is now seen as a marketing activity, of significance only if relevant to the areas of current business. The commercial pressures are considered to be too great to allow time to write papers for publication.

This case marks the most complete transition to the private sector of any former government laboratory. As a result, AEAT's responses to market signals have led it far from its original knowledge base.

Universities

If we turn to the third research-performing sector, universities, the trends are well documented. The massification of higher education, the squeezing of overburdened academics' time, the growing demands of accountability systems, and the costs of keeping up with the research front, rising with the increasing sophistication of equipment, all put pressure on research. At a time when there is a general recognition of its increasing economic significance, a further barrier is created by the increasing discrepancy between the disciplinary organization of universities and the interdisciplinary framework of research. A combination of insufficient public funding and substantial policy pressure has led to a strong growth in the proportion of income coming from industry and from other contract sources. In the UK, funding of university research by industry rose 53 per cent between 1995/6 and 2000/1 to a point where it constituted 11.7 per cent of external research income (that is, excluding that proportion of general university funding attributed to research—in the UK's case, the 'QR' funds allocated through the Research Assessment exercise). This was, however, heavily concentrated in a few institutions, and this peak fell to 10.3 per cent in 2001/2. Public funding, increasingly linked to co-financing by industry, is calibrated by these inputs.

The interface with government laboratories has been complicated by the integration of some of these into universities. For example, since 1996 the Southampton Oceanography Centre at the University of Southampton has operated a joint venture between the University and the Natural Environment Research Council, which relocated its Institute of Oceanographic Sciences and Research Vessel Services there, with scientific synergy a major aim. Similarly the Natural Resources Institute, which carries out research, consultancy, and training in areas relevant to overseas development, has become part of the University of Greenwich.

Pressure is also in place to raise income from commercialization. In terms of its economic weight, this is an exaggerated phenomenon: intellectual property income is growing, but it is small, the licensing income being around 1.4 per cent of external research income (double the proportion in the previous year but half of that in the US). Around half of UK universities own spin-offs, but many of these are also economically insignificant. According to the most recent Higher Education Business Interaction Survey (Department for Employment and Learning 2004), total revenues from intellectual property exceeded the costs of protection by a ratio of 2.6: 1; benefits were skewed towards a small number of universities; and 62 institutions spent more on protection than they received in revenue.

As a growing proportion of universities put in place an infrastructure to secure intellectual property, institutions find themselves in a situation where they are competing for funding from industry but, at the same time, competing with industry in sectors where they attempt to launch firms or provide research services. There are also concerns about the effects of excessive patenting on the practice of research itself (Nelson 2004). Government policy has strongly promoted commercialization of research, financially and institutionally, though an initial preoccupation with the numbers of spin-offs as a performance indicator has now been replaced by a more realistic expectation, particularly since the downturn in the stock market.

These pressures are creating consequences for the organization of research. A constant refrain from industry has been for university research to be concentrated in 'Centers of Excellence.' This is quite logical from a firm's perspective: it reduces the search and transaction costs in dealing with academic collaborators. Indeed, there has been a tendency towards 'broadband' relationships where companies sponsor larger centers and also furnish their scientific and managerial training requirements from a much more limited range of (usually elite) institutions (Howells et al. 1998). The tendency to concentration has found an echo, both from the academics themselves seeking economies of scale in equipment and other resources needed to compete internationally, and from governments seeing it as a means to ensure international excellence when resources are scarce. The European Union has also moved in this direction, replacing a good deal of its project funding with much larger 'Networks of Excellence' with the explicit aim of restructuring science.

University of Manchester

Thus far, aggregation into larger units has been limited in the university sector. Some mergers have taken place, with many of London's smaller postgraduate medical research institutions joining colleges of London University. To date, the only fully-fledged coming together of research-intensive universities has been that between the Victoria University of Manchester and the University of Manchester Institute of Science and Technology, which from 1 October 2004 was reconstituted as a new university, the University of Manchester. The two institutions had the advantage of contiguous campuses and a long history of shared services and collaboration, but the rationale for the change is presented far more as a response to the increased competitiveness of research and the need to achieve critical mass if world-class standards are to be achieved. A further rationale exists in the more efficient use of teaching resources, offering students a wider range of course combinations. But it is in the area of research that the real challenges exist. Entirely new structures have been designed, with much larger units and the elimination of most committees. Visioning over a ten-year period and strategic planning are seen as key instruments to achieve change. 'Exemplary knowledge and technology transfer' form one element of this vision. Recognizing the absence of a large endowment base (the lack of which is usual for UK universities), the need to leverage intellectual property and knowledge transfer are stressed as income generators, with ambitious targets to grow commercialization activities. This growth comes from a base that is already relatively strong in the predecessor institutions, both of which had holding companies dealing with licensing and spin-offs, though on different models. While it is the case that most research-based universities in the UK are seeking increased revenue from commercialization and industrial research contracts, and certainly the case that Manchester has other goals and aspirations relating to world-class research and teaching excellence, the prominence given to these market-related activities also should be noted.

Convergence and its consequences

The combination of these trends in the three sectors has been summarized as a phenomenon of convergence (Georghiou 1998): research performers move from having differentiated funding sources and roles to all being, to some extent, in competition as sellers in a market for contract research—the 'contract research space' (Figure 10.1). This was, of course, not an empty space to start with: private contract research providers and not-for-profit associations also exist. So the questions are of how the contract research space is instituted and regulated, to what extent it is a mar-

ket in information and if so, what kind of market it is.

While this chapter is principally concerned with the suppliers of research in the contract research space, it is worth considering the buyers at this point. If contract research is defined as research in which the agenda is set by the buyer and defined in a contract for its provision by another party, the UK market may be seen to have several different procurement models in operation. The situation that seems perhaps the most straightforward is where a firm or government department issues a call for tender for a research project to meet a particular commercial or regulatory need. But many assumptions are hidden here. For example, how was the need and hence the call for tender specified? Is it set in the context of an ongoing research agenda developed by the suppliers? Is there true competition in the market, either in terms of capability or in terms of charging a commercial rate? Are the buyers able to articulate the need in sufficient detail that the research choices to be made are also specified, or is this a function of the expertise of the bidders? When the research is done, how will the results be presented and used? Does the buyer need research expertise to interpret the results and translate them to action? Projects are generally not in themselves solutions

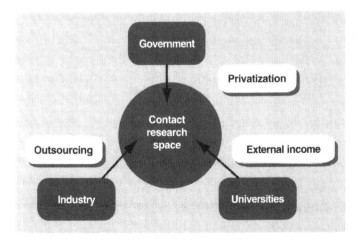

Fig. 10.1. Convergence into contract research space
Source: Georghiou 1998.

to problems, in which case, how will results from different sources be aggregated to provide those solutions?

European competition regulations demand that procurement be conducted in a transparent and competitive fashion, with, for example, projects above a financial threshold having to be advertised in the *Official Journal of the European Communities*. In fact, previous regulations for public procurement have been found to inhibit necessary contact between buyer and seller in complex technological situations, resulting in a new Directive which makes provision for a 'competitive dialogue' in which the needs of the buyer can be articulated (OJ 2004).

A key issue in a market of this kind is that of intellectual property. While at first sight it might seem to be a statement of the obvious that the buyer of contract research would own the rights to the results, this is not always the case; even where it is, there are areas of contention or variance. Separating the specific knowledge generated under contract from the pre-existing knowledge base of the organization (known usually as background) is one area of potential contention. Moreover, universities normally seek rights to use results for further research, and indeed may be in breach of their charters if they do not (though delays in publication to allow for patenting in the UK and/or European system are generally accepted). Government departments may impose particular IP conditions upon their suppliers, but increasingly the trend is also towards encouraging those suppliers to exploit the knowledge themselves. For universities, such arrangements had been in place since the 1980s (in a situation analogous to the US Bayh–Dole Act), but more recent reforms have sought to give ownership of IP to government laboratories, with the presumption that the opportunity to generate revenue for themselves and their employees would act as an incentive for them to commercialize their work (Baker 1999).

Elements of contract research may also be seen in some of the more traditional sources of research funding, including grant funding to universities from research councils. A substantial proportion of this funding is dedicated to collaborative programs of various kinds, in which the needs of a partner firm or of a sponsor ministry affect or even define the research agenda. Much of the European Union's Framework Programme is similarly dedicated to addressing industrial competitiveness (in which case, the European Commission is a proxy client aiming to represent the needs of firms in that sector) or policy needs in areas such as health and environment. Needs are articulated through published 'work programs' that provide an indicative guide for project proposals. Selection is by merit review. Interestingly, while for academics these programs represent work at the more applied end of the spectrum that they occupy, for contract research laboratories of a more traditional nature they create a rare opportunity to perform strategic research which strengthens their capability to deliver contract research for clients.

In its role as a purchaser of research, industry can also produce varied conditions. At one extreme, universities and laboratories are seen simply as lower-cost suppliers for technical services; at the other, the industrial input is fundamentally one of sponsorship, whereby resources are put into a priority field, usually through a center of excellence, without any attempt to direct the work, in the expectation that radical discoveries will ensue. In the broader debate in the UK, the main concern in respect to industrial R&D is in terms of the willingness of firms to spend money on it, and, in turn, on outsourced or sponsored research. While business R&D followed the international trend of increasing in real terms since the 1980s, its share of GDP has fallen to 1.9 per cent, driven by cuts in defense expenditure during the 1980s and by slow growth in many sectors. With the exceptions of the pharmaceutical and aerospace sectors, there are few major UK firms spending substantial amounts, though foreign affiliates partly make up the gap, and often have explicit strategies for working with the science base.

In summary, the contract research space contains a wide variety of contractual agreements between the main purchasers of research and their suppliers. These vary in terms of the degree of competition involved in the award of the contract, the control exerted upon the content of the research, and the arrangements for subsequent exploitation of the knowledge produced. Against this backdrop, we may now explore the phenomenon of convergence in more detail. In principle, as shown in Table 10.1, it can be seen to have both positive and negative consequences:

Competition in and for the market

Efficiency gains rest on the presumption that a greater number of players able to compete for research contracts achieves cost reduction. There is no systematic evidence to date of cost reduction resulting from competition. However, at least one laboratory (the National Physical Laboratory, initially in a near-monopoly position with a monopsonistic client for its metrology research, and still only facing competition at the margins) has claimed that it can deliver more science for the same money to its government customer because of the savings in overheads and administration arising from not having to follow civil service practices in areas such as staffing (Wallard 2001). But the same source, after noting that 80–5 per cent of the laboratory's income still comes from public sources, states:

Indeed it would have been surprising if the laboratory had been able to generate substantial new contract income because of its overhead costs and competition from contract research from the less expensive University and Research and Technology Organization (RTOs) competitors which work in similar areas. The Laboratory's policy has been to avoid such competition and only to seek work which is entirely consistent with its core mission. (p. 206)

The overall picture is complicated by generally increased research productivity through improved equipment and techniques (for example, cost saving through automation of areas such as gene sequencing, and through the use of modelling in place of expensive apparatus). Further complication, at least in the academic sector, comes from proven under-pricing and cross-subsidy of research. A government review concluded that:

There is now compelling evidence of a significant funding gap in higher education science and engineering research (which makes up the great majority of the UK SEB [Science and Engineering Budget]) which points to the present level of output being unsustainable in the medium to longer term. (HM Treasury 2002: 18–19)

It goes on to argue that no institutions have, thus far, become insolvent because they are cross-subsidizing research from other earned income such as overseas student fees, but that this situation was unsustainable. Furthermore, underfunding was being maintained through large-scale underinvestment in the physical infrastructure for research.

This raises the question of whether research can be a contestable market when some of the

Table 10.1. Consequences of Convergence

Positive	Negative
Efficiency through competition	Overcrowding in the contract sector
Closure of uncompetitive performers	Loss of coverage and variety
Cross-subsidy to original mission	Cross-subsidy or movement from original mission
Contestable scientific advice	Compromised scientific advice
	Loss of externalities
	Purchasers have no intrinsic interest to secure supply
	Investment difficult in face of uncertainty and large capital requirement

players are not motivated by profit. Opening a part of the activity to competition also runs the risk of creaming off the more lucrative consultancy opportunities, weakening the ability of players to perform longer-term research activities.

A market of this kind depends upon closure of uncompetitive performers. The standard competitive model has threat of closure as a powerful stimulus to raise efficiency. However, the public or mixed ownership of many institutions and, in particular, the position of universities creates a strong limitation. Even leaving research in favor of a teaching-only role is unlikely because of regional or local pressures to maintain the image of a local knowledge economy. To date, an apparently competitive structure for UK universities has been underwritten by bail-outs of financially failing institutions, though individual departments (often in core subjects such as Chemistry) are closed with increasing frequency if they are deemed to fail in terms of research or inadequate student numbers. Guaranteed government contracts initially maintained the presence of privatized labs, but their scale and specialization also creates both a major barrier to entry for competitors and a sense that most would have to be maintained by some means if they encountered financial difficulties.

Moving to a more systemic perspective leads to the issue of whether there is overcrowding and loss of variety among research performers. It is clear that the processes driving universities, government labs, and industrial labs to contract research space are largely distinct, and do not take account of this systemic effect. There has to be some concern that suppliers of research-based information are likely to converge on the opportunities for finance and, in consequence, also in their underlying capabilities. Short-term gains in performance could be made at the expense of long-term ability to adapt to new and unforeseen problems.

Much depends upon the behavior of what are usually called 'intelligent customers.' Without some sophistication on their part, the customer-contractor relationship may inhibit internalization of externalities, as value-for-money contracting may not allow a price high enough to maintain long-term capabilities. The case studies showed some evidence of a shift to technical services and consultancy at the expense of research, and certainly away from broader scientific activities. The operation of a market in information requires a degree of stability in supply arrangements, but too narrow a commercial attitude of customers may prevent development of stable market relationships.

Scientific advice to government

Looking at the potential benefit of contestable scientific advice, the market model can be presented as a way to ensure plurality of advice in the public domain and an escape from capture of advice within a single ministry acting as regulator and promoter. However, building long-term commercial relationships limits the ability of the supplier to give unpalatable advice. Furthermore, through their business relationships with a regulated sector, suppliers may have independent interests that they would be reluctant to jeopardize.

A particular challenge for the 'intelligent customer' is maintaining capability to respond to a possibly unforeseen catastrophic event. The BSE crisis is one example; a major fire in the London underground a few years ago is another. The government needs experts at short notice to investigate the causes of a crisis and to recommend a course of action. The problem is that the characteristic of being unforeseen means that it is unlikely that specific contracts for relevant research would have been procured. Hence, there would be no incentive for the research providers to maintain the necessary capability. Only a longer-range research program could ensure that those engaged in it would have the knowledge and skills to adapt to the new situation. In some cases this can be furnished by universities: in the government's successful strategy to eradicate a recent foot-and-mouth epidemic, key players came from universities to work in a task force dedicated to the problem. However, in other cases, such as the fire mentioned above, the appropriate expertise was only on hand because of the ongoing activity of the relatively unreformed Health and Safety Laboratories of the Health and Safety Executive. The model is one of accumulation of knowledge rather than its production to order.

Commercialization and investment

At the core of university motivation for contract research and commercialization is generation of revenue to spend on core activities—in effect, cross-subsidy. As we have seen, skewed returns mean that the majority lose money, or get a low return on costs of IPR protection. Taking money out of commercialization also ignores the financial needs of that activity: they can be weakened by inadequate investment and exposed to competitive pressure from those not engaged in cross-subsidy. A second-generation strategy is emerging among some of the more experienced institutions: the revenue-earning limits of commercialization are being recognized, and a rationale of commercialization as a service to the economy and the community is emerging instead. The potential high returns for the lucky few are, in effect, a lottery which cannot be entered without a strategy for intellectual property protection, and unless expectations have been substantially moderated. Government policy has also moved away from a central focus on spin-offs to a more balanced view of business–university links. The new consensus is well summarized in the influential Lambert Review—an inquiry into business-university links commissioned by the Treasury from Richard Lambert, former editor of the *Financial Times* (HM Treasury 2003). This stated in its summary:

The Review expresses concern that universities may be setting too high a price on their IP. Public funding for basic research, and for the development of technology transfer offices, is intended to benefit the economy as a whole rather than to create significant new sources of revenue for the universities. (p. 4)

and

... there has been too much emphasis on developing university spinouts, a good number of which may not prove to be sustainable, and not enough on licensing technology to industry. (p. 5)

Finally, the effect of uncertainty on capital investment may be considered. Development of supply capability depends upon substantial investments in the generation of knowledge, including, in many cases, the large facilities that provided the initial rationale for public involvement. Given the uncertainty of R&D as a business, a lack of collateral for ideas makes raising capital difficult. In short, a very competitive research market will make it difficult to raise funds for investment in equipment and intangibles when the public-sector legacy is exhausted: customers will absorb specific costs, but not generic or overhead items. Remedial action has taken place in the university sector, but it is an ongoing issue.

Consequences for collaboration

What are the consequences of these trends and issues for collaboration between the sectors? Collaboration in research and innovation has been empirically shown to depend mainly upon complementarity, whereby partners seek competencies and characteristics that they do not possess (Guy *et al.* 1991). Convergence means that similar organizations find it more difficult to cooperate. Examples of areas of conflict include:

(*a*) Disputes between industry and universities over ownership of intellectual property: *The Lambert Review* proposed that universities should own IP where the research was publicly funded but did not recommend specific legislation to enforce this. The Confederation of British industry reacted negatively to this, recommending instead that IP ownership and rights in the company's own area of interest should vest with the company, with the expectation that the university would then be permitted to exploit the IP in other applications and even work with other companies where it is agreed that there is no competition.

(b) Disputes between government laboratories and industry over the role of the former in managing a mixture of commissioning work on behalf of government, contracting in the market, and commercialization of intellectual property. Documented examples are harder to come by because of the sensitivities involved, but the case of the largest laboratory, the former Defence Evaluation and Research Agency (DERA) now privatized as QinetiQ, was commented on in a Parliamentary Report. A joint report by the House of Commons Select Committees of Defence and of Trade and Industry noted that companies had three main concerns about DERA: whether enough of its research was contracted out to industry, whether it duplicated industry's own research, and the availability of DERA's research to companies (House of Commons 1995). Since privatization, concerns have continued to be raised as QinetiQ remains the main source of advice on procurement. In a memorandum to the Defence Select Committee, a statement by QinetiQ indicated the tensions that have to be managed:

There are occasions when the MoD sees it as in its interest for QinetiQ to engage with Defence Companies in technology transfer before or during a procurement contract for which QinetiQ is separately giving technical advice. In these circumstances the company establishes separate teams with robust firewalls between to serve the two or more different customers (in the case, for instance, where there are several competitors). (House of Commons 2003)

In addition to the above two cases, competition has grown between universities and government or recently privatized laboratories, at present mainly in the direction of universities bidding for work that was formerly the exclusive preserve of the laboratory.

Against this, new forms of collaboration have emerged. There is an increasing role for former public laboratories, as intermediaries between academia and industry are able to use their consulting skills to translate and apply knowledge in problem-solving mode for small and medium-sized firms. There is

also an emergence of hybrid organizations such as company-sponsored laboratories on campuses and university laboratories on commercial premises (as illustrated in the example of BTExact).

Conclusions

What conclusions may be drawn from the sum of these changes, which could be described as an experiment in the operation of research through a market model, in the UK system over two decades? One conclusion is methodological—that the health of a research system should be judged by its total capabilities, not simply by the capabilities of its components. It is clear that reforms in one part of the system have effects upon other parts which are not necessarily foreseen or accounted for by those undertaking them.

In terms of implications for public policy, the principle that seems to emerge is the need for policies that create conditions in which all parts of the system are fully networked, but preserve specialized functions. This must go beyond rhetoric and ensure that incentive structures are fully aligned with this objective. Research funders, in particular, need long-term strategies that take account of the need to maintain the capability of their supply institutions. The market model offers insights and benefits, but can also cause key concerns to be overlooked. A wise policymaker should expect the emergence of a market to be a starting point for the design of ways to evolve a research system, and never a solution in itself.

Using the framework of convergence, it is possible to build a simple policy typology whereby research and innovation policies may be categorized into three groups: support for core mission, support for convergence, and support for networking.

Policies in support of the core mission for the Science Base and/or universities generally involve finance for research which does not meet the definition of contract research—in other words, funding that allows the research performer to pursue its own objectives. This includes general university funding, responsive-mode Research Council grants, and studentships/fellowships. For industry, such core policies include R&D subsidy, fiscal incentives for R&D or innovation, competence enhancement, and public procurement. Finally, for public laboratories, such policies include core funding for research programmes (probably jointly defined with the client ministry) and investment in facilities.

Policies in support of convergence for the science base are characterized by efforts to make organizations more capable of raising funds from market sources or, more narrowly, sources that previously were dedicated to other types of research performer. These include promotion of universities owning companies and/or IPR, and of selling technical services. For industry, they include floating corporate laboratories as businesses and corporate universities. For public laboratories, they include promotion of commercialization, and selling research and technical services to industry.

If we take it as a given that the research performers are now in a situation where they are increasingly likely to be in complex relationships both competitive and collaborative, with other performers of all types, then there is a need for policies that increase the level of self-organization of the research system. Such policies mainly address systemic elements and, in particular, networking. They include support for developing an interface culture through very professional liaison and intellectual property offices. Universities need to be encouraged to work for industry in contract and collaborative research, but their role in such projects should be one of knowledge generation and validation—in other words, to provide capabilities that companies do not have themselves. Firms should, in turn, be encouraged to stretch their technological developments. Similarly, public laboratories can work with companies to commercialize their results. The use of technology foresight can create a shared space in which new, future-oriented visions can be built as a pre-require-

ment for the formulation of new networks (Georghiou 1996). The longer-term perspectives involved can alleviate competitive pressures enough to allow cooperative positions to be established. It is probably no coincidence that the UK entered and maintained a commitment to foresight at the same time as the emergence of the market model.

As a final consideration, it is worth considering the extent to which the UK's experience is specific, or whether the trends discussed, and some of their positive and negative consequences, are more general. The first issue is one of whether convergence is a reality elsewhere. For corporate R&D, the picture certainly holds for Europe, and there is evidence of similar pressures in Japan. Mowery has noted a reduction in central corporate R&D in the USA, with firms 'externalizing' their operations through a number of alternative collaborative arrangements (Mowery 2001).

An international pattern may be seen for public-sector laboratories. Though the phenomenon of privatization on this scale is largely a British one, all of the other pressures faced by labs are generic in Europe; the key one for convergence is the continuing pressure to increase the proportion of earnings from the market rather than from traditional government sources. In Japan, reform of public laboratories has partly followed the UK pattern: they are being given the status of Independent Administrative Agencies and required to raise a proportion of their income from industry and government contracts. The same status has been given to National Universities in Japan, and there is certainly strong policy pressure for them to engage in commercialization activities.

The situation is less clear for US universities. Commercialization and research contracting are well established and even used as a model, but they remain at relatively low levels in terms of overall research income. It is possible that declining funding for State Universities may propel those with the right capabilities in this direction. In the laboratory sector, Bozeman and Dietz (2001) noted a succession of technology transfer initiatives, but without large-scale effects to date. While those authors saw the true value of DOE national laboratories deriving from the core scientific and technological work pertaining to their historic and statutory missions, they did not expect this to be recognized, seeing instead 'an environment that entails high uncertainty and competition.'

Overall it may be concluded that the market model has transformed the research system, and that the transformation has led to both positive and negative effects. Unfortunately, a lack of systemic evaluation means that the balance cannot be calculated at present. Indeed, despite the twenty-year time scale of the reforms, some of the effects may not be manifested for an even longer period because of the very long lead times involved in the renewal of both infrastructure and human capital in science. At this stage, probably the best policy strategy for science and innovation is an adaptive one that seeks to make the system work better by exploiting positively the differences between actors, and seeking to maintain the variety that increases the total capability of the system.

References

Arrow, K. (1962). 'Economic Welfare and the Allocation of Resources for Invention.' In R. Nelson (ed.), *The Rate and Direction of Inventive Activity*. Princeton: Princeton University Press.

Baker, J. (1999). 'Creating Knowledge, Creating Wealth, Realising the Economic Potential of Public Sector Research Establishments.' A Report to the Minister for Science and the Financial Secretary to the Treasury. HM Treasury, August.

Boden, R., Cox, D., Georghiou, L., and Barker, K. (2001). 'Administrative Reform of United Kingdom Government Research Establishments: Case Studies of New Organizational Forms.' In D. Cox, P. Gummett, and K. Barker (eds.), *Government Laboratories: Transition and Transformation*. NATO Science Series. Amsterdam: IOS Press.

—— —— Nedeva, M., and Barker, K. (2004). *Scrutinising Science: The Changing UK Government of Science*. Basingstoke: Palgrave Macmillan.

—— and Dietz, J. S. (2001). 'Research Policy in the United States: Civilian Technology Programs, Defense Technology and the Deployment of the National Laboratories.' In P. Larédo and P. Mustar (eds.), *Research and Innovation Policies in the New Global Economy: An International Comparative Analysis*. Cheltenham: Edward Elgar.

Coombs, R., and Georghiou, L. (2002). 'A New Industrial Ecology.' *Science*, 296: 471.

Crow, M., and Bozeman, B. (1998). *Limited by Design: R&D Laboratories in the US National Innovation System*. New York: Columbia University-Press.

Department for Employment and Learning, Higher Education Funding Council for England, Higher Education Funding Council for Wales, Scottish Higher Education Funding Council, Office of Science and Technology (2004). *Higher Education: Business Interaction Survey 2001/02*.

Dunleavy, P., and Hood, C. (1994). 'From Old Public-Administration to New Public Management.' *Public Money and Management*, 14/3: 9–16.

Freeman, C. (1987). *Technology Policy and Economic Performance: Lessons from Japan*. London: Pinter Publishers.

Georghiou, L. (1996). 'The UK Technology Foresight Programme.' *Futures*, 28/4: 359–77.

—— (1998). 'Science, Technology and Innovation Policy for the XXIst Century.' Opening address to PREST 20th Anniversary Conference *Science and Public Policy*, April 1998.

Guy, K., Georghiou, L., Quintas, P., Cameron, H., Hobday, M., and Ray, T. (1991). *Evaluation of the Alvey Programme for Advanced Information Technology*. London: HMSO, 49.

Her Majesty's Treasury (2002). *Cross-Cutting Review of Science and Research: Final Report*. March.

—— (2003). *Lambert Review of Business-University Collaboration: Final Report*. London: HMSO.

House of Commons (1995). *Aspects of Defence Procurement and Industrial Policy: First Report of the Defence Committee and Trade and Industry Committee*. HMSO: London.

—— (2003). Memorandum submitted by QinetiQ (15 January 2003): *Minutes of Evidence Taken before the Defence Committee*, Tuesday, 21 January 2003, Paragraph 19.

Howells, J. (1997). 'Research and Technology Outsourcing.' Centre for Research on Innovation and Competition (CRIC) Discussion Paper 6, University of Manchester, UK. See **http://les1. man. ac.uk/cric/papers.htm**

—— Nedeva, M., and Georghiou, L. (1998). *Industry-Academic Links in the UK*. Higher Education Funding Council for England, December 1998.

Jaffe, A. B. (1996). *Economic Analysis of Research Spillovers: Implications for the Advanced Technology Program*. NIST, US Department of Commerce Technology Administration. NIST GCR: 97–708.

Lundvall, B.-Å. (ed.) (1992). *National Systems of Innovation: Towards a Theory of Innovation and Interactive Learning*. London: Pinter.

Metcalfe, J. S. (1995). 'The Economic Foundations of Technology Policy: Equilibrium and Evolutionary Perspectives.' In P. Stoneman (ed.), *Handbook of the Economics of Innovation and Technological Change*. Oxford: Blackwell Handbooks in Economics.

Mowery, D. C. (2001). 'The United States National Innovation System after the Cold War.' In P. Larédo and P. Mustar (eds.), *Research and Innovation Policies in the New Global Economy: An International Comparative Analysis*. Cheltenham: Edward Elgar.

Nelson, R. R. (2004). 'The Market Economy and the Scientific Commons.' *Research Policy*, 33: 455–71.

Official Journal (OJ) of the European Union (2004). Directive 2004/18/EC of the European Parliament and of the Council of 31 March 2004 on the coordination of procedures for the award of public works contracts, public supply contracts, and public service contracts, Article 29.

PREST *et al.* (1993). *The Public Research System in Europe: a Comparative Analysis of Public, Semi-Public and Recently Privatized Research Centres*. Report to European Commission. **www.cordis.lu/ indicators/publications.htm**

Roke Manor Research Business and Technology Review **www.roke.co.uk/download/brochures/ Annual_ Review_2003.pdf**

Smith, K. (2000). 'Innovation as a Systemic Phenomenon: Rethinking the Role of Policy.' *Enterprise and Innovation Management Studies*, 1/1: 73–102.

Wallard, A. (2001). 'Successful Contractorisation: The Experience of the National Physical Laboratory.' In D. Cox, P. Gummett, and K. Barker (eds.), *Government Laboratories: Transition and Transformation*. NATO Science Series. Amsterdam: IOS Press.

11 How is Innovation Influenced by Science and Technology Policy Governance? Transatlantic Comparisons

Stefan Kuhlmann and Philip Shapira

Introduction

What influence does the political governance of an innovation system have on innovation performance? In the science and technology policy field, this is one of those perpetually important questions, simple to ask, yet not so simple to answer, particularly since the process of answering inevitably raises further issues that need to be explained or investigated. In this chapter, we attempt to answer the question about the relationship between innovation governance and performance first by understanding practice through theory, then by testing that theory with comparative evidence. We address our key question in four steps:

(a) exploring the role of public policy in the context of a heuristic concept of systems of innovation;

(b) conceptualizing and discussing the political governance of systems of innovation, and probing how governance might influence patterns of innovation performance;

(c) examining four cases—two each from Germany and the United States—with a focus on the relationships between governance, policy, and innovation outcomes;

(d) comparing these cases and drawing conclusions.

A heuristic concept of systems of innovation

The concept of 'systems of innovation' was first introduced to explain differences in the competitiveness and innovativeness of economies (Freeman 1987; Lundvall 1992; Edquist 1997). It was recognized that patterns of scientific and technological specialization and related cultures and norms (each rooted in historical origins, in particular industrial, research, and governmental institutions, and in inter-institutional relationships) crucially affected the ability of academic and economic actors to produce, and of policymakers to support, successful innovations. Comparative empirical studies have used this approach on national and regional scales, as well as at the level of individual technological or sectoral developments (Nelson 1993; Cooke 2001; Malerba 2004).

We regard the concept of innovation systems not so much as a 'final' theory but as a heuristic aid that guides us in developing and testing further ideas about innovation precursors and processes. An innovation system, in our understanding, encompasses the broad array of institutions and relationships involved in scientific research, the accumulation and diffusion of knowledge, education and training, technology development, and the development and distribution of new products and processes. Among the integral components of innovation systems are regulatory bodies (standards, norms, laws) and public and private investments in supporting infrastructures. Innovation systems extend over schools, universities, research institutions (education and science system), industrial enterprises (economic system), the politico-administrative and intermediary authorities (political system) as well as the formal and informal networks of the actors of these institutions. Innovation systems are distinctive; their competitive scientific, educational, and technological profiles and strengths develop only slowly, based on deep-rooted exchange relationships among the institutions of science and technology, industry, and the political system. As 'hybrid systems' (Kuhlmann 1999), innovation systems cut across, or are linked into, other societal areas, for example, education or business entrepreneurship, and they are critical in the processes of modernization and economic development (OECD 1999).

The hybrid institutional infrastructures and networks of innovation systems did not come into existence spontaneously: in the past 150 years, this area of society has been intensely influenced by government interventions, mainly by the nation state. National political systems, themselves differentiated, developed science, research, technology, innovation, and other policy activities, in which they acted as catalysts, promoters, and regulators of elements of innovation systems which were emerging in many places: the establishment of technical colleges and universities with industrial or economic development missions in France, Germany, and the US demonstrates this. In the twentieth century, the innovation systems of the industrialized countries co-evolved with their national political systems, and have firmly established country-specific characteristics. It is because of this close interweaving with respective country political systems that one speaks of 'national' innovation systems.

In the last few years, however, the discussion has grown more complex, to include regional and sectoral innovation systems and their roles (Cooke 2001; Malerba 2004). Underpinning this debate has been the remarkable growth of subnational innovation policy initiatives: in the US, almost every state government now has science, technology, and innovation policies, as do regions in Europe and elsewhere. In Europe, there has been the further development of supra-national innovation policies and programs, for example, through the European Union's Framework Programs and the promotion of a European Research Area.

The political governance of systems of innovation

As initiatives to foster innovation have expanded, with policies implemented at different levels of government and with varied goals, important issues arise as to how such increasingly complex systems of innovation are governed. Clearly, policy-making is seldom a matter of top-down decision-making and straightforward execution. Rather, policy decisions are negotiated in multi-actor arenas and networks that may stretch over several politico-administrative systems at different levels (Marin and Mayntz 1991). Indeed, we suggest that in an innovation-systems framework, policy making should be modeled as a process of competition, networking, and attempts at consensus-building between heterogeneous (corporatist) actors representing different societal subsystems. Typical perspectives of key actor groups represented in the science and technology policy arena are those of:

1. Science. The science system is represented by universities (researchers, administrators), non-university public research institutes (basic or applied research), or related professional associations like science councils (depending on their role in the national research system). Science actors are typically interested in such things as funding for research institutions and projects, the development of

scientific knowledge and careers (publications, professional leadership), the growth of disciplines or thematic areas, and the training of young researchers.

2 Industry. The world of technology- or knowledge-seeking industrial companies is represented by research laboratories of large enterprises or by industrial research associations, and (less frequently in policy arenas) by individual smaller companies. Typical interests are the exchange of pre-competitive technological knowledge, the creation of new knowledge through research cooperation with other companies or public institutes, the joint development of technical norms and standards, the appropriation of new knowledge (patents), and the realization of new products and processes in markets (successful innovations).

3. Other societal actors. Numerous relatively well-organized interest groups and non-governmental organizations (NGOs) are engaged in innovation policy arenas, for example, in environmental or life-sciences research. These groups are often very heterogeneous in purposes and perspectives: some actively seek to promote certain kinds of research or technological outcomes; others are directly opposed to the same.

4. The politico-administrative system. This is represented in science and technology policy arenas by elected legislators, ministers or departmental secretaries, and by related governmental institutions and bureaucracies; each is differently constructed and empowered on national, regional (for example, German *Länder*; the US states), and transnational (European Parliament; Council of Ministers; EU Commission) levels. Politico-administrative systems are themselves characterized by diverse actor perspectives and interests: for example, parliamentarians who adopt positions of strong interest groups (for example, industry or NGOs); or science and technology policy administrators who may seek to further the positions of their own agencies (not necessarily those of the broader scientific community).

What room for maneuver do such actor groups in the science and technology arena have? From a conceptual perspective, one may employ the notion of actor-centered institutionalism (see Mayntz and Scharpf 1995; Scharpf 2000). Here, institutions are defined as sets of rules providing the actors with reliable expectations and social sense. However, Mayntz and Scharpf model reliability only with respect to regulatory, not to normative or cognitive, aspects. To explore the dynamics of new actor constellations and institutional settings, it is necessary to analyze 'soft' forms of social rules, not yet frozen into codified regulations. The identification of the dynamics and impacts of experiences embodied in institutions may help in understanding the strategies of corporatist and other organized actors in the governance of research and innovation systems, whereby governance is understood as a kind of evolutionary social order. Our concept goes beyond the 'triple-helix' characterization of university-industry-government relationships (Etzkowitz and Leydesdorff 2000), which implies more concordance, and less conflict, negotiation, and reconciliation in goals than we can observe.

To sum up, negotiating actors, with their different responsibilities (for example, policymakers define programs, allocate budgets; researchers define themes, purchase equipment; industry looks for competitive advantages), pursue different, partly contradictory, interests; they represent different stakeholders' perspectives, and construct different perceptions of 'reality' or institutional 'frames' (for example, Callon 1992; Schön and Rein 1994). Established power structures and the shape of arenas may vary considerably between nation states (or regions). Normally, national, regional, or transnational governmental authorities in multi-actor arenas of innovation policy play important but not dominant roles. In many cases they perform the function of a mediator, facilitating alignment between stakeholders, equipped with a 'shadow of hierarchy' (Scharpf 1993), rather than operating as a top-down steering power. Eventually, 'successful' policy-making means compromising through 're-framing' stakeholders' perspectives and joint production of agreement.

We suggest that we can operationalize this model by examining a series of key variables which characterize the political governance of innovation systems. These include:

- the institutional setting (concentration vs. fragmentation; hierarchy vs. self-control);
- the arena (including specific interrelationship of public and private actors and their strengths and weaknesses);
- mechanisms for priority-setting in public and private organizations;
- allocation procedures for funding and other resources (including the role of evaluation, foresight, and other methods);
- regulatory regimes (laws, standards, norms, corporate governance); and
- organizational orientation (including organizational cultures, belief systems, and taken-for-granted rules, as well as formal mission statements and organizational structures).

We hypothesize that political governance, in the above sense, has a decisive influence on an innovation system's performance. Patterns of performance can be characterized by a variety of measures, including scientific specialization and productivity, technological competencies, dominant sectors and clusters, market characteristics, entrepreneurship, and innovation outputs (including patents, new products or processes, or trade marks).

In this chapter, we explore our hypothesis through examination and comparison of both broader trends and specific cases in the German and US innovation systems. Both of these countries have large, mature, and complex research and innovation systems. The US is the world's biggest national spender of funds on research and development, while Germany is the world's fourth-biggest (measured by purchasing-power parity). Normalized by gross expenditures on research and development (GERD) as a share of gross domestic product (GDP), the US and Germany are fairly close together, ranking 6th and 8th among OECD members, with GERD/GDP at 2.7 per cent and 2.5 per cent respectively (2002 data). However, business performs relatively more R&D in the US than Germany, while the reverse is true for government. Also, while overall densities of patenting (per million population) are not too dissimilar, there are proportionately more high-technology patents (and higher levels of venture capital) in the US when compared with Germany (see Table 11.1).

Table 11.1. Selected Innovation Measures: US and Germany

Measure	US	Germany	Ratio US / Germany
S&E graduates (% of 20–9 years age class)	10.2	8.0	1.3
Population with tertiary education (% of 25–64 years age class)	37.2	22.3	1.7
Business R&D performance (% of GDP)	2.10	1.76	1.2
Higher education R&D performance (% of GDP)	0.40	0.40	1.0
Government R&D performance (% of GDP)	0.70	0.33	0.6
Manufacturing value added in high-tech sectors (%)	11.9	23.0	0.5
ICT expenditures (% of GDP)	8.2	6.9	1.2
EPO + USTPO patent applications (per million pop.)	492.3	457.3	1.1
EPO + USTPO high-tech patent applications (per million pop.)	148.9	65.2	2.3
Early stage venture capital (% of GDP)	0.22	0.04	5.2

Sources: EU Innovation Trend Chart; OECD, for most recent year (2001, 2002).

This data confirms that our task of associating governance systems with innovation outputs will not be easy or straightforward. While we see relatively similar levels of investment in R&D (as a percentage of GDP), there are differences in the composition and performance of R&D, complementary investments in human capital, information technologies, and financial pools for innovation. There is an apparent difference in innovation outcomes (the share of high-technology patents in all patent applications). In trying to explain this, many historical and contemporary economic and industrial factors are going to be relevant. We recognize this readily and openly, but argue, too, that the characteristics and dynamics of governance also need to be distinguished and traced in terms of their own relative influence on innovation outcomes. We seek to do this by analyzing trends and cases within the German and US innovation systems.

The German innovation system: moving from differentiation to innovation

In Germany, the governance of science and research is largely determined by the federal system, in which political responsibilities are divided between the central government and the states (*Länder*). The states are principally responsible for science and education, thus they finance the largest part of the university budgets (see Figure 11.1).

Overall, the German national innovation system is well developed: in 2000, about 480,000 people (full-time equivalent) were employed in R&D; the total expenditures for R&D amounted to approximately €50.1 billion or 2.45 per cent of the GDP (BMBF statistics 2002). The research infrastructure can be characterized as systematically differentiated:

- Industry carries out the greatest part of German R&D. With 306,700 R&D employees, industry invested about €33.6 billion in R&D in 1999 (or about 67 per cent of the total), mostly in applied research and experimental development.
- The second-largest share of R&D expenditure (€8.1 billion) falls to 345 higher education institutions (101,500 R&D employees), including 116 universities. They concentrate on basic and long-term, application-oriented research, for the most part financed by the federal states and the German Research Association (DFG),

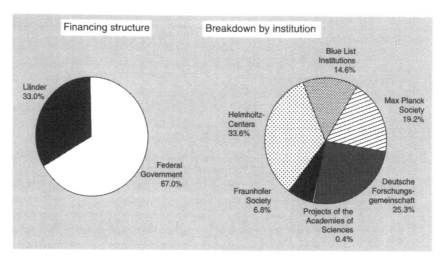

Fig. 11.1. Joint research funding by German Federal and Länder governments in 2001
Source: BMBF 2002.

a government-funded, but largely independent, research sponsor.

- The 15 centers of the federal Helmholtz Association of National Research Centers (21,500 R&D employees; €2.1 billion) focus on research that is long term and considered risky or entails high costs (plants/facilities) and large research teams.
- The institutes of the Max Planck Society (9,200 R&D employees, €1.0 billion), a research organization which has its origins in the Kaiser-Wilhelm-Gesellschaft established in 1911, concentrate on selected fields of basic research in the natural sciences and humanities. They focus on those research areas in which a significant knowledge and development potential is assumed, while not yet anchored in university research, either because of their interdisciplinary character or because of the resources required.
- The institutes of the Fraunhofer Society (9,000 R&D employees, €730 million) aim to promote the practical utilization of scientific knowledge via long-term application-oriented and applied research. The Fraunhofer Society principally carries out contract research, financed partly by industry and partly by government bodies.
- The G.W. Leibniz Science Association (WGL) is an umbrella organization of non-university research institutes supported by the federal government and the federal states (c.10,000 employees, €811 million).
- Finally, the research institutions of the Confederation of Industrial Research Associations (AIF) conduct applied research and experimental development for the sector-specific needs of industrial enterprises. Their research palette, financed partly by public means and partly by industry, is especially geared to small and medium-sized enterprises which are organized in industry-sectoral research associations.

In 2000, the federal government spent a total of €8.4 billion on research and development.

The spectrum of instruments available to German research and technology policy is broad, including institutional support for research facilities (roughly half of federal spending), various financial incentives (these are programs costing, roughly, the other half of federal expenses) for research and experimental development in public or industrial research laboratories, and the creation of an innovation-oriented infrastructure, including the institutions and mechanisms of technology transfer. These instruments characterize, by and large, the practice of research and innovation policy in the Federal Republic of Germany since the 1970s.

It is an open question whether the considerable institutional differentiation of the German system is a reflection of an advanced functional division of complex work or just fragmentation. The political governance, in any case, is characterized by the absence of a dominant actor. The Bund-Länder-Konferenz (the Federal Government and the States Conference on Education Planning and Research Promotion) may serve as an example: for decades it has been facilitating an alignment of procedures, quality criteria, etc., for science and education between the federal states and the national government; at the same time, the in-built need for the production of consensus has brought along a dangerous propensity to institutional conservatism, hampering, for example, a modernization of the university system, repeatedly and urgently called for by experts.

Furthermore, in Germany, public research, technology, and innovation policies are no longer exclusively in the hands of the various national authorities: increasingly, national initiatives are supplemented by, or are even competing with, regional innovation policies or transnational programs, in particular the activities of the European Union (see Kuhlmann and Edler 2003). At the same time, industrial innovation increasingly occurs within international networks.

The industrial R&D and innovation activities in Germany are focused more on various fields of mechanical engineering and less on micro-electronics, information technology, and biotechnology. Indeed, several R&D-intensive sectors have seen weakening in their trade positions, with the exception of the automotive sector (Figure 11.2). The German profile is almost the opposite of the American one. Only in the area of chemistry do the two countries have similar structures (Schmoch *et al.* 1997).

Germany's science and innovation performance still continues at a high level. Germany's innovators focus on the world market, mirrored in a steep increase in the number of patent applications relevant to the world market (Grupp *et al.* 2003). In the 1990s, the country was able to maintain its position as the second most important net technology exporter. Since 2001, Germany has edged slightly ahead of Japan as an exporter of R&D-intensive goods. In the 1990s, the German foreign-trade success was increasingly based on the automobile sector. In the traditionally strong chemical and machinery industries, however, Germany lost some of its ability to compete in the world market in the late 1990s, and imports increased. In the past decade, the dominant sectors in worldwide

technology trade were pharmaceuticals and the products of information and communications technology: their shares in global industrial trade have almost doubled. Germany joined the international trend here, but in a modest way. More recently, companies based in Germany's Eastern *Länder* have contributed to the increase in technology exports.

Education and science provide the basis for technological performance and innovation. In the science field, Germany was able to increase its share in worldwide publications during the 1990s by 1.5 per cent to a level of about 9 per cent today. Advances were made also in terms of quality: Germany's scientists publish an above-average number of articles in journals with an international readership, and their research, on the whole, is well cited. The research communities' scientific and technical qualifications are important for innovation performance. Since the 1990s, however, shortages are developing in both the secondary and the tertiary sector. The number of training contracts for technical occupations provided under the German dual vocational training system has been declining.

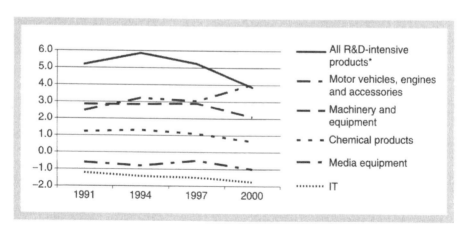

Fig. 11.2. Contribution by R&D-intensive goods to Germany's foreign trade balance 1991–2000

Note: A positive value indicates that this sector contributes to moving the foreign trade balance into a surplus.

* Including non-allocable complete manufacturing facilities, etc.

Source: Grupp *et al.* 2003.

Case Studies

Case Study I: Biotechnology innovation policy in Germany

Political and economic context

In Europe, since the 1990s (in the US, earlier), one can find an increasing clustering of biotechnological research and commercialization activities in a few prominent regions—a phenomenon that is of growing importance for national competitive advantage in biotechnology. Typical of regional biotechnology clusters are excellent research institutions and qualified scientists; a high density of start-up firms; the provision of venture capital, technology-transfer institutions, and local science parks; and close (and often informal) collaboration structures between scientists, firms, banks, and local government (Dolata 2003; Enzing *et al.* 1999).

Before the 1990s, Germany's national biotechnology effort was focused on large industrial companies like Hoechst, Bayer, or BASF; policymakers 'seemed to accept the view that German competitiveness in biotechnology could be achieved without creating new forms of corporate enterprise, such as the American start-up firms, or more systematic university–industry linkages' (Jasanoff 1985: 30; Dolata 2003). The situation changed significantly when the German government initiated the BioRegio program in 1995, in the context of the ambitious goal to become Europe's leading biotechnology nation by 2000. BioRegio started as a contest to stimulate the creation of biotechnology clusters and the commercialization of scientific knowledge (Kaiser 2003*a*). Seventeen regions entering the contest had to demonstrate that they could establish a working and interacting infrastructure for the commercialization of biotechnology. The Federal Ministry for Education and Research (BMBF), responsible for the program, tried to strengthen embryonic locations. The winners of the competition—the bioregions around Cologne (BioRegio Rheinland), Heidelberg (BioRegio Rhein–Neckar–Dreieck), and

Munich—had already been favored in the 1980s through the establishment of national genetic research centers. The BioRegio program supported 57 regional R&D projects between 1996 and 2000 and invested a total of €72 million. At the same time, encouraged by the BioRegio program, a biotech cluster emerged in Munich, with about 120 biotech and pharmaceutical companies. Here, other regional initiatives played a role. The Bavarian state government participated more readily than most other state governments in the federal-state consultations which took place prior to the BioRegio contest; the state also started initiatives to upgrade the research infrastructure and the provision of risk capital at the regional level (Kaiser 2003*a*).

Institutional setting, policy arena, and decision-making

In Germany, national innovation policy played an important catalytic role in the emergence of regional biotechnology clusters. Regionalization has not led to a loss of influence of national policies. Instead, this is a 'guided' regionalization, stimulated and coordinated first of all by national policies—and, of course, underpinned by relevant efforts of regional authorities and actors (Dolata 2003). Significantly, the BioRegio program stimulated the commercialization of biotechnological research. Provision of funds was not the primary factor; rather, this was the establishment of a network structure in different local clusters involving relevant private and public actors. This fostered more fluid and self-organized collaborations within industry and between industry and academia (Dolata 2003)—with and without state aid.

The development of the German biotechnology sector was further aided by public stimulation of venture capital for the expansion of high-technology industries, thus mitigating the traditional bank-centered financial system in Germany (Kaiser 2003*a*). In Germany, the ratio between private and public risk capital within the biotechnology industry today is 1 to 0.8, higher than in the US or Britain. On the other hand, not all of the governing components of the innovation system are aligned to support sustainable growth: Germany's complex and inflexible decision-making procedures in higher education still make it difficult to quickly respond to new labor-force needs, and the German biotechnology industry has fields where qualified personnel are not available, for example, in bioinformatics.

Regulatory regimes

In 1990, an Embryo Protection Law was enacted: it prohibited researchers from using embryos for genetic experiments; meanwhile, a Genetic Engineering Law set legal standards for the authorization of genetic engineering laboratories and production facilities, and regulated field trials with genetically modified organisms. The latter was revised in 1993 in order to reduce administrative hurdles for the authorization of biotechnological research and production. Compared to US standards, the law remained more restrictive, especially by requiring public participation in administrative authorization procedures. In 1995, new European procedures allowed community-wide authorization for medical products; it was granted on the basis of a scientific evaluation by a newly established European Agency for the Evaluation of Medicinal Products. Nevertheless, the overall regulatory regime still has a strong regional dimension in federally organized Germany. Since states have the authority to enforce biotechnology-related regulations, the degree to which enforcement agencies are supportive of the industry differs from state to state, according to the political attitude of the respective government (Kaiser 2003a). The state government of Bavaria, for example, introduced a legislative proposal to amend the federal genetic engineering law in November 2000; this was to speed up the implementation of a European Directive that further deregulated the biotechnology sector, especially through the simplification and shortening of administrative procedures for operating research and production facilities (Kaiser 2003a).

Innovation performance

Today, the Munich Biotech Cluster has the highest density of corporate actors in the life-sciences industry in Germany and is second (behind London) in Europe. In 2000, the Munich-based life-sciences industry employed about 12,200 people (Kaiser 2003a). Of the twenty largest pharmaceutical companies in the world, five are represented in Munich, and the number of small and medium-sized biotech companies has increased dynamically between 1996 and 2000 from 36 to 101. The cluster's knowledge base is fed by several high-level research institutes (for example, the Max Planck Institutes for Biochemistry and Neurobiology, the Fraunhofer Institute for Process Engineering and Packaging, the Bavarian Institute for Soil and Plant Production) and universities (the clinical center of the University of Munich, the Technical University of Munich, University of Applied Sciences Weihenstephan (Kaiser 2003a)). New opportunities for biotech start-ups are especially evident in the development of new R&D technologies for the pharmaceutical industry. Having developed rapidly in recent years following BMBF's initial stimulation, young German biotech firms show strength in the development of new platform technologies. In the context of a new division of labor in pharmaceutics, small R&D-intensive companies are developing active substances, especially in the first phases of the R&D process. In the future, the innovative strength of major international pharmaceutical companies may be less critical for Germany's long-term competitive strength than the interaction between them and the small, highly specialized firms (Legler et al. 2001).

Case Study II: Helmholtz Association

Political and economic context

The fifteen national research centers of the Helmholtz Association perform long-term oriented research aiming at the preservation and improvement of the foundations of human life; in functional terms, the Helmholtz mission is to some extent compatible with that of the US national laboratories. Helmholtz research entails high costs (plants/facilities) and large research teams (Hohn and Schimank 1990), mainly funded by the German federal government (90 per cent) and co-funded by several federal states (10 per cent). In the past years, the centers have transformed their fields of activity significantly, and today they cover six core fields: energy, earth and environment, health, key technologies, structure of matter, transport and space.

A few years ago, on behalf of the federal government and of the states, the German Science Council (Wissenschaftsrat) organized a 'system level evaluation' of the Helmholtz Association, assessing the mission and the performance of the Association and its centers (Wissenschaftsrat 2001). The evaluation report criticized the centers' heterogeneous performance profile, rooted in a lack of incentives and of networking across all levels of the Association. The report recommended new performance-oriented governance and funding structures, replacing the former system of center-oriented institutional (block) funding.

Institutional setting, policy arena, and decision-making

In 2001, as a reaction to the critical evaluation, the Association gave itself a new strategic orientation aimed at making more efficient use of its scientific resources for Germany's long-term competitive strength. Today, the Helmholtz Association is a registered association; the fifteen Research Centers are legally independent bodies. A full-time President is responsible for implementing a newly established, program-oriented funding system, and has a small fund to speed up the reform process (annual budget totaling €25m).

The new program-oriented funding—the core of the reform—aims at:

- focusing the scientific work along research programs across institutional and disciplinary borders; and

- the introduction of performance-based research funding.

Previously, the sponsors (federal government and states) gave funding directly to the centers. Today financial resources are paid increasingly through defined scientific programs rather than to the centers. The research programs are organized around the Association's six major research fields.

The central decision-making bodies at the Helmholtz Association form an assembly: it is made up of internal members of the Association, and a Senate of external members. The members of the annual general meeting are the directors of the associated Helmholtz centers; the members of the Senate are representatives of federal government and federal states, as well as representatives of science and research, business and industry, and other research organizations. The Senate commissions the evaluation of research programs by independent, internationally acknowledged experts and receives their review reports. These evaluations serve as a basis for the funding recommendations which the Senate makes to the association's financial sponsors—that is federal government and federal states—on how much support funding the individual research programs and core topic areas will receive.

The two research fields of health and of transport and space, involving a total of ten programs, were the first to be reviewed in 2002. One hundred scientists took part in the review, half of them from abroad. As a result, the research fields received funding totaling €500m in 2003. The fields of energy and of earth and environment were reviewed in 2003, while those of structure of matter and of key technologies have been evaluated in 2004.

Innovation performance

Since their founding in the 1950s and 1960s, the Helmholtz centers have had a strong focus on long-term and basic technological research, and the idea of the autonomy of science has been deeply rooted in the identity of their staff. Since the mid-1990s, the centers have been exposed to political pressure to shift their orientation further down the innovation chain. As a consequence, there are now organizational and cultural tensions, even frictions, with respect to the centers' mission. They are still expected to perform long-term, problem-oriented research. For decades the centers' budget was almost entirely covered by public institutional funding. Only since the late 1990s has the share of external project funding grown (modestly). Helmholtz has managed to get some project funding through the European Union's programs, though the system evaluation criticized that only in a few cases did Helmholtz act as an international coordinator of collaborative projects (Wissenschaftsrat 2001). The evaluation report was also critical about the scientific-publication output of several of the centers, and their attractiveness for international visiting researchers. With respect to Helmholtz's innovation performance, 3,300 collaborations were reported in 1998, of which 60 per cent generated some income. But the evaluation stated that collaboration with industry was quite unevenly distributed across the centers; the same applied to patent applications and licenses. Figure 11.3 illustrates a general trend in all German non-university research organizations—that of increasing patent applications, with Helmholtz and Fraunhofer (far smaller than Helmholtz) in

leading positions, even before the new governance was introduced in Helmholtz.

The new system aims to stimulate a further increase in Helmholtz's innovation output; in 2004 Helmholtz's patent applications amounted to 500; there was a €12 million income from licenses, and some 30 start-up companies were established. Within the association, nevertheless, the fifteen centers enjoy a high degree of strategic autonomy: it remains to be seen whether and to what extent they will adapt their internal governance with respect to innovation performance.

The US innovation system: networked innovation structures

The United States is the largest of all the national systems of innovation, certainly when measured by total R&D investment. About 45 per cent of global R&D spending occurs in the US. In 2002, having grown strongly throughout the 1990s, experiencing an annual average real growth rate of 5.8 per cent between 1994 and 2000, total US R&D spending (private and public) was an estimated $292 billion. More recently, R&D spending has been affected by economic slowdown, but still managed to grow by 2.4 per cent between 2001 and 2002. Overall, R&D spending as a ratio of GDP

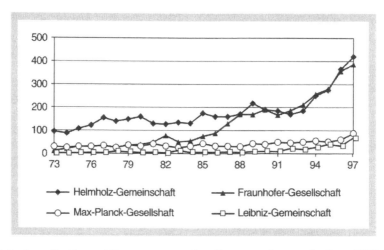

Fig. 11.3. Patent applications of German non-university research organizations 1973–1997
Source: BMBF 1999.

reached an estimated 2.8 per cent in 2002, up from 2.3 per cent in 1981 (NSF 2003).

Industry is the biggest source of national R&D funding in the US—$190 billion in 2002, equivalent to 66 per cent of the total (see Table 11.2), which is slightly above the German level. The federal government supplies most of the rest of R&D funding ($81 billion in 2002 or 28 per cent of the total), with universities, states, and non-profit institutions funding a further $17 billion (6 per cent). In recent decades, the industry share of US R&D funding has increased (7.6 per cent annual real growth 1994–8), with the share provided by non-manufacturing industries growing particularly rapidly. Conversely, the share of US R&D funding provided by the federal government has declined, from 67 per cent in 1964 and 47 per cent in 1988 to 28 per cent in 2002. Within the federal R&D budget, the three largest allocations are to defense, health, and space respectively at 55 per cent, 23 per cent, and 11 per cent of the 2002 total (NSF 2002). There have been significant shifts in civilian (i.e. non-defense) federal research priorities: the concentration was on space research in the 1960s, energy research in the 1970s and early 1980s; this shifted to a major focus on health-related and life-sciences research in the 1990s through to the present (American Association for the Advancement of Science (AAAS) 2004).

The performers of R&D in the US are diverse, involving a variety of private, public, academic, and other non-profit institutions. Judging by allocation of funding (2002), industry itself performs about 70 per cent of US R&D. Federal government departments and laboratories undertake about 9 per cent of R&D. Nonprofit institutions and externally administered, federally funded R&D centers account for a further 8 per cent. Universities and colleges undertake some 13 per cent of US R&D.

In addition to being large, the US innovation system is highly decentralized. The checks-and-balances system of government distributes decision-making power among executive, legislative, and judicial branches, with federal agencies themselves often having considerable leverage in determining research priorities. State governments are also increasingly involved in science and technology policies, with spending in this area by the states growing to perhaps $5 billion annually (roughly equivalent to a European framework program in scale, although organized from the bottom up, with fifty different and unconnected state governments making decisions). In the US system, the governance of education is highly fragmented, with states and localities (not the federal government) responsible for education, but also with a combination of public and private institutions. In particular, state governments and their university systems oversee state institutions of higher education, including land-grant universities and other major public research-performing universities.

Table 11.2. Shares of national R&D expenditures, by source of funds, performing sector, and character of work, in the United States 2002 (%)

	Source of funds	Use of funds
Industry	66	70
Federal government	28	9
Universities	3	13
Other non-profit	3	4
Federally funded research and development centers	—	4

Note: Figures are rounded to nearest whole number. National R&D expenditures were an estimated $276 billion in 2002.

Source: National Science Foundation 2004, *Science Indicators*, Figure 4.2.

This decentralized, often fragmented system notwithstanding, it is clear that the US innovation system as a whole has undergone major paradigm shifts (see Tassey 1992; Crow 1994; Galli and Teubal 1997). Following the end of the Second World War, there was an emphasis on building up the capabilities of national laboratories, universities, and corporate research centers and promoting R&D for such government-oriented missions as defense, energy, space exploration, health, and agriculture. In the 1960s, a focus on technology transfer emerged, with the aim of promoting greater civilian spin-off from mission-driven public R&D. Now, a third shift is underway, although one that itself has taken an unexpected turn. In the 1990s, prompted by increased global economic competition and the end of the cold war, US policy became more attentive to explicit civilian commercialization goals, especially in the areas of life sciences, electronic commerce, and new fields such as nanotechnology. Since 2001, there has been a further turn into new defense concepts and homeland security (now associated with a $1 billion research budget). However, despite these recent changes in emphasis, there remains a consistent accent on new implementation concepts in the innovation system—eschewing traditional linear 'pipeline' models of technology 'push' and 'pull' for more complex and iterative perspectives on the technology development and diffusion process (Branscomb and Florida 1998). In particular, new patterns of industry collaboration and commercialization are being promoted, through industry consortia, university–industry linkages, and public-private partnerships.

In understanding this reorientation of innovation concepts, there is a tendency to focus on changes at the federal level. However, there is another important strand to the story, namely the states' role in establishing new policy concepts and frameworks (see also Shapira 2001). During periods when the federal government has been reluctant to promote civilian technology explicitly, states have often done so under the guise of economic development.

State universities and engineering experiment stations (and their successors) have long been engaged in collaborative, pragmatic relationships with local industries. States have sought to promote economic development by establishing regional technology clusters and alliances, and have considerably increased their investments in technology policies, including university/non-profit research centers, joint industry-university research partnerships, direct financing grants, incubators, and other programs using research and technology for economic development (Coburn and Berglund 1995; Berglund 1998). The relevance of the states is not simply that their budgets have increased, but also that in America's decentralized federal system, the states are, in their own right, a countervailing source of policy intervention and experimentation.

Case Study III: A new innovation mission: university research commercialization in the US

Political and economic context

The Carnegie Foundation for the Advancement of Teaching (2001) identifies over 3,900 institutions of higher education in the US in eighteen categories ranging from doctoral-degree granting institutes to a variety of baccalaureate and specialized-degree-awarding colleges. More than 15 million students attended these institutions (1998 data). Within this diverse set, Carnegie further recognizes 261 doctoral/research universities, of which just under two-thirds are public universities and just over one-third are private. While faculty members at many institutions undertake research, it is these doctoral/research universities—and particularly a subset of about 150 designated by Carnegie as 'extensive doctoral research' universities—which are at the core of the US university research enterprise. R&D expenditures at US colleges and universities totaled about $30 billion in 2000 (NSF 2002); more than four-fifths of university R&D is undertaken by extensive doctoral research universities (Carnegie Foundation 2001: Table 13).

The most rapid growth in the relative scale of university research took place in earlier decades. In 1960, US universities performed 5.1 per cent of all US R&D. The university share grew to 10.4 per cent by 1980 and to 11.4 per cent by 2000 (data in NSF 2002: Appendix Table 4.3.) Thus, in the 1960s and 1970s, university research greatly increased its share of national R&D performance in the US, but growth slackened in more recent years. However, from the 1980s through to the present, the key change is in the scope of what universities do. In addition to long-held missions of research, teaching, and public service, leading US research universities are today increasingly engaged as knowledge hubs seeking to link with industry, commercialize research, foster new technology ventures, and anchor regional innovation complexes.

Institutional setting, policy arena, and decision-making

In adopting an innovation mission and, perhaps more important, in defining the character of that mission, the institutional setting and governance of US universities is critical. There is central control neither of education nor of research organization and funding. Rather, in the system of governance, responsibility is decentralized and diffused. Higher education is a responsibility of states (for public higher education), while in all states there are both public (state) and private universities and colleges. Individual university systems and institutions have flexibility in development of research initiatives and the design of curriculum (with accreditation devolved to a variety of non-profit discipline and regional organizations). Teaching expenditures are financed primarily through tuition and state allocations (for public universities). Conversely, research funding for universities derives significantly from the federal government: in 2000, the federal government sponsored 58 per cent of research expenditures at US universities, with 7.3 per cent from states, 7.7 per cent from industry, and the balance from internal sources or foundations (NSF 2002). Multiple federal agencies and

programs sponsor university research, with the biggest sponsors (in 2001) being the National Institutes of Health (NIH $10.7b), NSF ($2.6b), DOD ($1.5b), NASA ($0.8b), and DOE ($0.7b). Despite massive recent growth in NIH life-sciences funding, the federal government share of academic R&D has actually declined (it comprised 67 per cent in 1980), encouraging universities to seek other funding sources. The share of academic R&D that is industry funded has grown. (It was 4.1 per cent in 1980; today, the industry share is nearly twice this.) University engagement with industry and regional economic development is not new in the US. In the late nineteenth century, the system of US land grant universities was established with explicit aims to promote industry and agriculture. Some private universities, such as MIT, were established around the same time with local industrial support. Throughout the early twentieth century, university–industry linkages grew in the US, notably in engineering and applied sciences, with collaborations between university and corporate researchers (Mowery et al. 2001). The early 1970s saw the emergence of Industry–University Cooperative Research Centers (sponsored by NSF). However, in the 1980s, to encourage research commercialization, and to harness the resources of universities not only for the development of new scientific knowledge, but for technological and industrial competitiveness, there was a series of changes in the policy framework and environment. In 1980, the landmark Bay–Dole Act promoted the commercialization of federally funded R&D and allowed universities to obtain and transfer (to corporations) patents and licenses derived from federally sponsored research. Further federal initiatives followed, including the Small Business Innovation Research Program (1982), National Cooperative Research Act (1984), and NSF's Engineering Research Centers (1985), through to the doubling of the NIH budget (1998–2003), all of which provided incentives to universities and their faculty to focus research on emerging technologies that could be commercialized. Similarly, at the state level, numerous states in the last two-to-three decades have initiated programs to foster university-driven regional innovation, including Pennsylvania's Ben Franklin program, Ohio's Thomas Edison program, and the Georgia Research Alliance. Universities themselves, often influenced by stylized models of Stanford University and MIT, have also promoted institutional change from within, developing new research mechanisms and incentives, technology licensing offices, faculty commercialization programs, incubators, and royalty agreements. And, while core goals of scholarship and teaching remain, there has been increasing motivation among (and incentives for) faculty to engage in venture start-ups and university–industry research collaborations.

These developments have been accelerated by the decentralized, fragmented, multi-actor systems within which American universities operate. In what has been called the state-level 'laboratory of democracy,' to experiment with, and improve upon, strategies to promote university-driven regional innovation (Osborn 1988), states have competed against, and learned from, one another. To promote innovation today, states have moved from limited linkages between academic science and economic development to the institutionalized partnerships between academic and business (Plosila 2004). Federal support—through changes in the regulatory framework, incentives for research-industry collaboration, and early funding of emerging new areas (such as in life sciences or nanotechnologies)—has further encouraged state and private university institutions to engage more deeply in the development and transfer of research. For example, a current NSF program, Partnerships for Innovation, provides resources for universities, businesses, and communities to collaborate in commercializing research.

Innovation performance

Are there measurable innovation outputs from the adoption of new innovation missions by US research universities? In 1982, US universities and colleges were awarded 464 patents; in 2000, universities were granted 3,598 patents. The number of institutions receiving

patents has more than doubled over this time period (75 in 1982, 180 in 1998). Growth in university patenting was particularly strong in the period 1981 through 1997; there was a dip in university patenting in 2000, although this seemed to have stabilized by 2001 (Figure 11.4). University licensing revenues from patents totaled $1.24 billion in 2000 (Association of University Technology Managers (AUTM) 2003). Growth in university patenting has occurred in both public and private institutions. There is evidence to justify the view that reforms such as Bay–Dole have encouraged increased patenting by American universities, although it has been noted that other factors are also at work (shifts in research fields, university policies on disclosure, see Mowery *et al.* 2001). Jaffe (2000) finds that patent intensity (patents per million inflation-adjusted R&D dollars) increased by a factor of three between 1980 and 1999, although he also suggests that the growth in the number of university patents not cited in their first five years signifies a decline in patent quality. University technology officers report over 2,000 commercial products between 1998 and 2002, and the formation of over 4,300 companies associated with university licensed technologies between 1980 and 2002, of which about one-half remain in business. Company start-up rates increased from about 240 a year in 1994 to 450 in 2002.

However, there remains some uncertainty about the scale and value of the innovation outputs associated with university research. Studies which have examined the relationships between university R&D spending and formation of new firms are not definitive (see, for example, Kirchoff *et al.*, 2002), while there are few robust analyses of the influence of policy initiatives such as university technology parks or technology incubators on regional innovation. Additionally, there is concern about university allocation of private rights to the results of publicly sponsored research. Universities generally argue that exclusive licensing is essential if companies are to make the additional investments needed to bring promising research to the market. However, others note that more than one-third of university

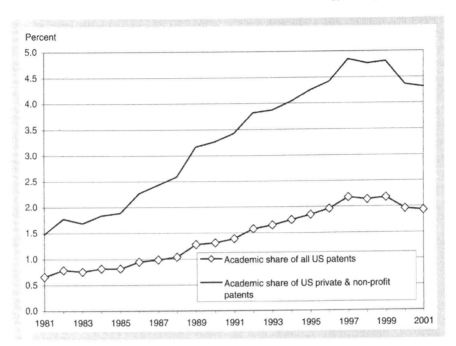

Fig. 11.4. Trends in share of university patents in the US patents 1981–2001

Source: NSF 2004, Figure 5-46 source data.

exclusive licenses are granted to large corporations; that these arrangements go against traditions and ethics of university openness, and may lead to the blocking of valuable innovations in certain cases (Shulman 2003).

Arguably, the context for the changing scope of what American universities do has been provided by broad developments in the economy and polity as a whole: shifts in production systems, the rise of science and technology-based industries, and developments in the way knowledge is generated, used, and valued. In addition, specific features and changes in the governance and expectations of research systems, and in the role of academic institutions in the US at federal and state levels, have stimulated the enlargement of the university mission to more fully encompass innovation. The decentralized, open, and flexible character of the US university system allowed American academic institutions to become early movers (compared with most other developed countries) in adopting innovation missions. The

evidence to date indicates some success in the performance of this new innovation mission, although not without growth in debate and challenge about the balance of public and private costs and benefits accrued.

Case Study IV: Biotechnology development and policy in the United States[1]

Political and economic context

The scientific breakthroughs that fostered modern biotechnology occurred on both sides of the Atlantic, as well as in other parts of the world. However, the early biotechnology industry grew more rapidly in the US than elsewhere. By the late 1970s, nearly 100 biotechnology companies had emerged in the US, with concentrations in northern California and Massachusetts (Swann *et al.* 1998). Despite considerable national policy activity in Europe and other developed countries to boost life sciences and the biotechnology industry, the US remains at the forefront of biotechnology research and commercialization.

There has been no single, consistent national policy to develop biotechnology in the US; even if it could be agreed upon, and despite calls for more coherent policies, the innovation system is too decentralized to pursue the implementation of such an approach. Rather, over time, and within a framework of interaction among different stakeholders and actors (including universities, scientists, research sponsors, large pharmaceutical companies, small dedicated biotechnology firms, venture capitalists, and government agencies), multiple national and local initiatives have been pursued. They have advanced the US biotechnology sector directly. Policies have conflicted, and there has certainly been duplication. Nonetheless, the cumulative result has been massive research support to develop the science as well as the emergence of an institutional and regulatory environment which has facilitated research-industry collaboration, the mobility of scientific and entrepreneurial human capital, and incentives for commercialization.

As might be expected, the industry has not developed uniformly. Three-quarters of large biotechnology firms and start-ups are located in just nine (of fifty-one) metropolitan areas. The early leaders—San Francisco and Boston— remain prominent. New York and Philadelphia have built biotechnology strengths on historic pharmaceutical concentrations. The other leading centers are Washington–Baltimore, Los Angeles, and the more recent entrants of San Diego, Seattle, and North Carolina's Research Triangle (Cortright and Mayer 2002). The success of these metropolitan areas has drawn attention to regional innovation clusters and prompted other state governments and cities to develop their own life-sciences and biotechnology strategies. Such regional efforts to encourage the diffusion of the industry are often well funded (for example, Florida recently pledged more than $300 million to attract a bioresearch center to the state).

Institutional setting, policy arena, and decision-making

Many factors contributed to the initial growth of US biotechnology enterprise. Universities emerged as sources of knowledge and innovation for biotechnology companies, and helped to precipitate regional clustering (Jaffe 1986). Biotechnology is characterized by the high degree of complex and tacit knowledge embedded in researchers. Consequently, the relationship between researchers and entrepreneurs was not hands-off: indeed, the early scientists responsible for advances in biotechnology were also central to initial commercialization. Start-up firms such as Genentech, Biogen, and Hybritech were founded by such scientists. The location of these firms was determined by that of the scientists working in university laboratories (Audretsch 2001).

Perhaps in contrast to European counterparts, most leading US research universities were very willing to encourage biotechnology research commercialization. Policy changes to allow universities to claim (and commercialize) intellectual property resulting from government funded research played a role, including the 1980 Bay–Dole Act. US universities and regions also drew on experiences from prior rounds of university-industry commercialization in other high-technology industries. Efforts to commercialize research in high-technology locations like the San Francisco Bay Area or Boston could attract seasoned venture capitalists. Since most biotechnology start-ups did not possess any material assets, they relied on the reputation of 'star scientists' to convince venture capitalists involved in companies or serving on scientific boards (Zucker et al. 1998). Genentech made the first biotech initial public offering (IPO) in 1980, and its success in raising capital set the trend for other young biotech companies to secure finance.

Recent decades have seen growth in finance for biotechnology research and commercial start-ups. Federal life-sciences research funding (basic and applied, excluding development and R&D facilities) rose (in 2004 constant dollars) from under $6b in 1970 to about $14b in 1997, then grew dramatically by 2004 to over $29b led by the doubling of research funding at NIH (AAAS 2004). By discipline, the share of life sciences in all federal research funding has jumped from 29 per cent in 1970 to 54 per cent in 2004, with much of the increase going to the biomedical sciences (AAAS 2004). State governments and private philanthropic foundations (such as the Howard Hughes Medical Institute) have also developed significant biotechnology-related research funding programs. Venture capital investment in US biotechnology was about $3.6 billion in 2003, down from the 2000 peak of $4.3 billion, but more than double the 1998 figure of $1.5 billion (data from Price-WaterhouseCoopers 2004).

Regulatory regime

The regulatory framework for the US biotechnology sector is complex and fragmented: it reflects the character of the innovation system and the overlapping influences of different institutions, actors, and decision-making systems. At least four major (and intersecting) regulatory 'regimes' can be identified: science, commercial, labor, and user.

- The governance of the science regime involves actors that include research sponsors, universities, scientists and professional associations, journals and review panels, federal and state agencies, and ethicists, religious, and other non-governmental associations.
- The commercial regime, for intellectual property, patenting, commercialization and business development, involves different combinations of university, governmental, and corporate actors, including technology transfer offices, patent regulators, securities, financial and tax regulators, investment companies, and the stock market.
- The labor regime involves public- and private-sector employers, educational institutions, trade associations, and government who are involved in the co-production of external and internal

labor market rules and customs (including those related to hiring, career development, firing, and mobility).
- The user regime encompasses federal and state agencies concerned with health, drug, food, and agricultural regulation, the medical system, insurers, and other non-governmental organizations.

Legislators and the judiciary are involved in all four regulatory regimes.

Actions and decisions in all of these regulatory systems have influenced innovation processes and outcomes: in the science regulatory sphere, concerns have been raised about ethics, conflicts of interest, and scientific honesty, and there has been an expansion of institutional review mechanisms to increase oversight of scientific research; in labor regulation, the sector has benefited from labor flexibility (for example, between research institutions and companies) and attracting talented foreign-born scientists. But recent immigration restrictions may make the US less attractive to foreign students and researchers. The regulatory climate affects not only the character and pace of innovation, but also its location. Compared with Germany (Kaiser 2003b), public opinion and the political environment in the US were generally supportive of research in biotechnology during the early decades of development, although some localities (in California, Massachusetts, and elsewhere) imposed additional regulations. More recently, regulatory climates have changed, with increasing controversy at the federal level about certain kinds of biotechnology research, leading to limitations on using human stem cells in federally funded research (NIH 2004). However, new state-funded (as well as private) initiatives are seeking to overcome this limitation, led by the successful passage in November 2004 of California's Proposition 71 (Stem Cell Research and Cures Initiative). Over the next ten years, California will invest $3 billion raised through tax-exempt bonds to support human stem-cell research. It has been suggested that this will attract new researchers and scientific entrepreneurs to the state

(Philipkoski 2004), as well as stimulate other US states to develop similar funding schemes.

Innovation performance

One key measure of output, particularly in the view of state policy-makers, but also at the federal level, is business development. The US biotechnology industry has grown significantly from a small base in the 1970s, although estimates of its current size vary. An Ernst and Young study (cited in US Biotechnology Industry Organization 2004) identified over 1,460 dedicated US biotechnology companies (including nearly 320 public companies) in 2002, with 194,600 employees, revenues of $29.6 million, and research expenditures of $20.5 million. Another study, using a broader definition of the biosciences sector—including drugs and pharmaceuticals, medical devices and equipment, agricultural feedstock and chemicals, and research and testing (excluding academic and hospital bioscience research)—reports US employment of 885,000 in 2003 (Battelle 2004). The true scale of the US biotechnology enterprise is probably between these two estimates (for example, there is biotechnology employment beyond members of the industry association, but not all the employment reported in the second study is biotech related). Nonetheless, all estimates confirm a significant growth in employment and enterprise activity in US biotechnology over the past two decades.

Other innovation output measures, such as patents or biotechnology drug approvals, also show long-run growth. US biotechnology patents grew sharply from 2,160 in 1989 to over 7,800 in 1998; similarly, new biotechnology drug and vaccine approvals went from 6 in 1989 to 25 in 1998. Recently, these measures show some hiatus: biotechnology patenting has remained at a constant, but flat, level through to 2002 (most recent data), with new biotechnology drug and vaccine approvals growing more slowly to 37 in 2003.

The cycle time from R&D to commercialization in biotechnology is lengthy (particularly in biomedical areas, with stringent federal testing rules); so many potential products are still in the pipeline. Still, slower recent growth in these innovation output indicators stands in contrast to the sustained increase in biomedical and drug research investment over the past decades. Arguably, prospects for rapid advances in certain biomedical areas (for example, results from human genome sequencing) were overestimated; at the same time, reflecting perhaps their 'scientific' heritage, companies may have been slow in adopting efficient methods to increase the effectiveness of available scientific advances and their own R&D, although this is now said to be changing (see Hall 2003).

The case shows that the relationships between public and private inputs and innovation outputs are not necessarily achieved directly or swiftly. But it does demonstrate a series of relationships between governance and innovation consequences. A decentralized policy system, institutional flexibility, and multi-actor regulation have stimulated several regional biotechnology innovation clusters in locations which have strong research capabilities and parallel commercialization infrastructures. It has also fostered a sector with great diversity, yet also facing uncertainty about the direction and impact of regulation, the availability of human capital (domestic and foreign), and future research support. At the same time, the very flexibility of the system of governance allows new leadership to implement fresh approaches to problems rapidly, as in the case of California and stem-cell research funding.

Insights from the cases: relating governance to innovation performance

What do these four cases tell us about the relationships between political governance of innovation systems and innovation performance? In the German biotechnology case, the emergence of bioregional policies and institutional reforms appears to be creating conditions where the landscape of innovation is changing. While institutional character and agenda-setting systems remain as neutral or contradictory, allocation procedures, regulation, and cultural orientation are now more supportive. New small biotechnology companies have emerged, and regional clusters (as in the case of Munich) are forming. On the other hand, some unchanged national governance characteristics hamper the dynamics, for example, in the area of biotech-related higher education. Germany's institutionalized joint-decision procedures (*Politikverflechtung*), requiring horizontal policy coordination between the federal states, as well as vertical policy coordination for the education system and with the national level, mean that reforms are unlikely to happen early enough to react to immediate needs (Kaiser 2003b).

The reform of the Helmholtz research centers is much more problematic, perhaps because, in part, the reform is occurring primarily within a publicly orientated system of governance. The institutional setting, priority mechanisms, and resource allocation procedures are now more favorable to new research and commercialization strategies. But within the quite autonomous fifteen centers, the cultural orientations and 'self-regulation' systems of institute personnel may be slower to change.

In the US university case, what at first might seem to be a weakness in governance—the decentralization of management to states and private institutions who might not be inter-ested or able to invest in knowledge-driven research—actually turns out to be a great strength when high levels of federal and industrial research funding are made available to the system. A group of powerful research universities has emerged: these institutions have aggressively pursued the pathways opened up by regulatory changes concerning intellectual property and new economic development and commercialization missions, to establish hubs of innovation.

The US biotechnology case further demonstrates that decentralized governance systems with multiple actors (but without the need to develop consistent coordination) can encourage innovation, especially if supported by significant research and commercialization infrastructure. The role of regional clusters within the broader innovation system was highlighted. Decentralized governance also brings with it uncertainty and weaknesses in planning, but at the same time fosters multiple sources of leadership that can help to overcome challenges.

In all four cases, we find evidence that the characteristics of governance influence the ways in which specific innovation systems evolve, and the outcomes that are produced. Table 11.3 displays a summary of the role of various governance variables with respect to the innovation performance of the US and Germany.

The decentralized yet interconnected structure and governance of the US innovation systems have encouraged flexibility, responsiveness

Table 11.3. Relevance of governance variables for innovation performance in US and Germany

Governance variable	Relevance for innovation performance	
	US	Germany
Institutional setting	federal, flexible	federal, segmented
Character of arenas	manifold, competition oriented	manifold, consensus oriented
Means of priority setting	diverse	fragmented
Allocation procedures	competition oriented	increasingly competition oriented
Regulation	supportive	increasingly supportive
Cultural orientation	supportive	increasingly supportive

to change, tolerance of research risk, and the ability to be an early mover in making organizational changes to embrace emerging technologies (for example, new interdisciplinary research centers) or mission needs (for example, reforms and incentives to encourage research commercialization). It is proven as a system that can consistently develop major research breakthroughs and dynamic new technology sectors and regions, although it is weaker in addressing challenges of technology diffusion and modernization in existing industries and older regions. The German innovation system emerged as one that was more strongly segmented into functional organizations and disciplines. This system is strong scientifically, and is well organized to develop technology in production and process technology fields, but the division of research institutions and functions at first has seemed slow to take up the challenges of the organizational change necessary for new strategies of innovation. More recently, the pace of change has picked up, and new strategies (such as BioRegio) are beginning to bear fruit. Results from other recent efforts (such as shifting of patent rights to universities and away from individual professors) remain to be seen.

Thus, in both countries, what may once have been more linear or segmented systems of research and development are evolving into networked systems of institutions, researchers, companies, and other stakeholders. Policies developed through multi-actor innovation governance systems have influenced these changes, with impacts for innovation performance, although with much variation. Yet, while there are similarities in current paths between the US and Germany, the starting points are different, as are the relative influences of governance actors. Thus, it may well be that while the systems in each country move in parallel for a while or even cross, they do not necessarily converge. Also, while innovation outputs in both systems are increased, the particular characteristics of these outputs are also likely to continue to differ, reflecting contrasts in mediating institutional and policy systems.

Note

1. This section draws on background research by Ajay Bhaskarabhatla.

References

American Association for the Advancement of Science (2004). R&D Budget and Policy Program, Federal Support of Research by Discipline, FY 1970–2004, American Association for the Advancement of Science. Retrieved November 17, 2004, from **www.aaas.org/spp/rd**

Association of University Technology Managers (2003). *AUTM Licensing Survey*: FY 2002. Northbrook, IL.

Audretsch, D. B. (2001). 'The Role of Small Firms in U.S. Biotechnology Clusters.' *Small Business Economics*, 17: 3–15.

Battelle Technology Partnership Practice and SSTI (2004). *Laboratories of Innovation: State Bioscience Initiatives—2004*. Columbus, OH: Battelle Memorial Institute.

Berglund, D. (1998). *State Funding for Cooperative Technology Programs*. Columbus, OH: Battelle Memorial Institute.

Biotechnology Industry Organization (2004). *Editors' and Reporters' Guide to Biotechnology*. Washington, DC.

BMBF (Bundesministerium für Bildung und Forschung) (1999). *Zur technologischen Leistungsfähigkeit*. Bonn: BMBF.

—— (2002). *Facts and Figures Research 2002*, Bonn: BMBF.

Branscomb, L., and Florida, R. (1998). 'Challenges to Technology Policy.' In L. Branscomb and J. Keller (eds.), *Investing in Innovation: Creating a Research and Innovation Policy that Works.* Cambridge, MA: MIT Press.

Callon, M. (1992). 'The Dynamics of Techno-Economic Networks.' In R. Coombs, P. Saviotti, and V. Walsh (eds.), *Technological Change and Company Strategies: Economic and Sociological Perspectives.* London: Academic Press, 72–102.

Carnegie Foundation for the Advancement of Teaching (2001). *Carnegie Classification of Institutions of Higher Education,* 2000 Edition. Menlo Park, CA.

Coburn, C., and Berglund, D. (1995). *Partnerships: A Compendium of State and Federal Cooperative Technology Programs.* Columbus, OH: Battelle Memorial Institute.

Cooke, P. (2001). 'Regional Innovation Systems, Clusters, and the Knowledge Economy.' *Industrial and Corporate Change,* 10/4: 945–74.

Cortright, J., and Mayer, H. (2002). *Signs of Life: The Growth of Biotechnology Centers in the U.S.* Washington, DC: Brookings Institution.

Crow, M. (1994). 'Science and Technology Policy in the US: Trading in the 1950 Model.' *Science and Public Policy,* August.

Dolata, U. (2003). 'International Innovative Activities, National Technology Competition and European Integration Efforts.' In J. Edler, S. Kuhlmann, and M. Behrens (eds.), *Changing Governance of Research and Technology Policy: The European Research Area.* Cheltenham: Edward Elgar, 271–89.

Edquist, C. (ed.) (1997). *Systems of Innovation: Technologies, Institutions and Organisations.* London: Pinter.

Enzing, C., Reiss, T., Schmidt, H., *et al.* (1999). *Inventory of Public Biotechnology R&D Programmes in Europe,* Volume 1: Analytical Report, Luxembourg: Office for Official Publications of the European Union (Doc. EUR 18886/1 EN).

Etzkowitz, H., and Leydesdorff, L. (2000). 'The Dynamics of Innovation: From National Systems and "Mode-2" to a Triple Helix of University-Industry-Government Relations.' *Research Policy,* 29: 109–23.

Freeman, C. (1987). *Technology Policy and Economic Performance: Lessons from Japan.* London: Pinter.

Galli, R., and Teubal, M. (1997). 'Paradigmatic Shifts in National Innovation Systems.' In C. Edquist (ed.), *Systems of Innovation: Technologies, Institutions, and Organisations.* Pinter: London.

Grupp, H., Legler, H., Gehrke, B., and Breitschopf, B. (2003). *Germany's Technological Performance 2002.* Summary. Report written on behalf of the Federal Ministry of Education and Research (BMBF) by a group of institutes coordinated by Fraunhofer-Institut für Systemtechnik und Innovationsforschung. Karlsruhe: Niedersächsisches Institut für Wirtschaftsforschung, Hanover: Institut für Wirtschaftspolitik und Wirtschaftsforschung der Universität Karlsruhe. Bonn: (BMBF).

Hall, S. (2003). 'Revitalizing Drug Discovery,' *Technology Review,* 30 September.

Hohn, H-W., and Schimank, U. (1990). *Konflikte und Gleichgewichte im Forschungssystem: Akteurkonstellationen und Entwicklungspfade in der Staatlich Finanzierten Außeruniversitären Forschung.* Frankfurt: Campus.

Jaffe, A. (1986). 'Technological Opportunity and Spillovers of R&D: Evidence from Firms' Patents, Profits and Market Value.' *American Economic Review,* 76/5: 984–1001.

—— (2000). 'The US Patent System in Transition: Policy Innovation and the Innovation Process.' *Research Policy,* 29: 531–7.

Jasanoff, S. (1985). 'Technological Innovation in a Corporatist State: The Case of Biotechnology in the Federal Republic of Germany.' *Research Policy,* 14: 23–38.

Kaiser, R. (2003a). 'Innovation Policy in a Multi-level Governance System: The Changing Institutional Environment for the Establishment of Science-based Industries.' In J. Edler, S. Kuhlmann, and M. Behrens (eds.), *Changing Governance of Research and Technology Policy: The European Research Area.* Cheltenham: Edward Elgar, 290–310.

—— (2003b). 'Multilevel Science Policy and Regional Innovation: The Case of the Munich Cluster for Pharmaceutical Biotechnology.' *European Planning Studies,* 11: 7.

Kirchoff, B., Armington, C., Hasan, I., and Newbert, S. (2002). *The Influence of R&D Expenditure on New Firm Formation and Economic Growth*. Research report for Office of Economic Research, Small Business Administration, Washington, DC.

Kuhlmann, S. (1999). 'Politisches System und Innovationssystem in "postnationalen" Arenen.' In K. Grimmer, S. Kuhlmann, and F. Meyer-Krahmer (eds.), *Innovationspolitik in Globalisierten Arenen*. Leverkusen (Leske+Budrich), 9–37.

—— and Edler, J. (2003). 'Governance of Technology and Innovation Policies in Europe: Investigating Future Scenarios.' In P. Conceição, M. Heitor, and G. Sirilli (guest eds.), *Technological Forecasting and Social Change*, special issue *Innovation Systems and Policies* (forthcoming).

Legler, H., Licht, G., and Egeln, J. (2001). *Germany's Technological Performance*. Summary Report 2000, Mannheim *et al.* (mimeo).

Lundvall, B.-Å. (ed.) (1992). *National Systems of Innovation: Towards a Theory of Innovation and Interactive Learning*. London: Pinter.

Malerba, F. (ed.) (2004). *Sectoral Systems of Innovation: Concepts, Issues and Analyses of Six Major Sectors in Europe*. Cambridge: Cambridge University Press.

Marin, B., and Mayntz, R. (eds.) (1991). *Policy Networks*. Boulder, CO: Westview.

Mayntz, R., and Scharpf, F. W. (1995). 'Der Ansatz des Akteurzentrierten Institutionalismus.' In R. Mayntz and F. W. Scharpf (eds.), *Gesellschaftliche Selbstregelung und Politische Steuerung*. Frankfurt: Campus, 39–72.

Mowery, D., Nelson, R., Sampat, B., and Ziedonis, A. (2001). 'The Growth of Patenting and Licensing by U.S. Universities: An Assessment of the Effects of the Bay–Dole Act of 1980.' *Research Policy*, 30: 99–11.

National Science Foundation (2002). *Science and Engineering Indicators 2002*. Arlington, VA: National Science Board.

—— (2003). *Slowing R&D Growth Expected in 2002*. NSF 03–307, Arlington, VA: National Science Board.

—— (2004). *Science and Engineering Indicators 2004*. Arlington, VA: National Science Board.

Nelson, R. (ed.) (1993). *National Innovation Systems: A Comparative Analysis*. New York: Oxford University Press.

NIH (2004). Funding of Research on Human Embryonic Stem Cells. National Institutes of Health. Retrieved November 15, 2004, from **http://stemcells.nih.gov/research/ registry/eligibility-Criteria. asp**

OECD (1999). *Managing National Innovation Systems*. Paris: OECD.

Osborn, D. (1988). *Laboratories of Democracy*. Boston, MA: Harvard Business School Press.

Philipkoski, K. (2004). 'The Stem Cell Gold Rush.' *Wired*, 4 November. **www.wired.com/news/ medtech/0,1286,65588,00. html**

Plosila, W. (2004). 'State Science- and Technology-based Economic Development Policy: History, Trends and Developments, Future Directions.' *Economic Development Quarterly*, 18/2: 113–26.

PricewaterhouseCoopers (2004). MoneyTree Survey, retrieved from **www.pwcmoneytree.com** 17 November 2004.

Scharpf, F. W. (2000). *Interaktionsformen: Akteurzentrierter Institutionalismus in der Politikforschung*. Opladen: Leske and Budrich.

—— (ed.) (1993). *Games and Hierarchies and Networks*. Boulder, CO: Westview.

Schmoch, U., Abramson, N. H., Encarnacao, J., and Reid, P. (1997). *Technology Transfer Systems in the United States and Germany: Lessons and Perspectives*. Binational Panel on Technology Transfer Systems in the United States and Germany. Washington, DC: National Academy Press.

Schön, D., and Rein, M. (1994). *Frame Reflection: Toward the Resolution of Intractable Policy Controversies*. New York: BasicBooks.

Shapira, P. (2001). 'US Manufacturing Extension Partnerships: Technology Policy Reinvented?' *Research Policy*, 30: 977–92.

Shulman, S. (2003). 'Big Ivory Takes License.' *Technology Review*, 7 April.

Swann, G. M. P., Prevezer, M., and Stout, D. K. (1998). *The Dynamics of Industrial Clustering*. Oxford: Oxford University Press.

Tassey, G. (1992). *Technology Infrastructure and Competitive Position*. Norwell, MA: Kluwer.

US Biotechnology Industry Organization (2004). *Editors' and Reporters' Guide to Biotechnology.* Washington, DC: Biotechnology Industry Organization.

Wissenschaftsrat (2001). *Systemevaluation der HGF: Stellungnahme des Wissenschaftsrates zur Hermann von Helmholtz-Gemeinschaft Deutscher Forschungszentren.* Cologne.

Zucker, L., Darby, M., and Brewer, M. (1998). 'Intellectual Human Capital and the Birth of U.S. Biotechnology Enterprises.' *American Economic Review,* 88/1: 290–306.

Two Styles of Knowing and Knowledge Regimes: Between 'Explicitation' and 'Exploration' under Conditions of Functional Specialization or Fragmental Distribution[1]

Werner Rammert

Innovation and knowledge changes: knowledge regimes and styles of knowing

Changes in knowledge are the driving force behind modern innovation. They propel scientific innovation, industrial innovation, and the institutional system of innovation; they involve both the regime of knowledge production and the style of knowing.

Innovation depends on many forms of knowledge production in various institutional settings. Scientific innovation in the academic field is as important as technological innovation in industry: new knowledge matters not only when scientific discoveries are made or technological projects are pushed forward but also when industrial products are improved, when services and markets are created, and when political funding programs are launched. These various forms of knowledge production, with their different rules and different resources, have to be coordinated into a system of innovation. A 'knowledge regime' is the specific way in which the differentiation between the scientific, industrial, and political dynamics of innovation is shaped, and in which the interplay among them is institutionalized. If one can identify new patterns of coordination both within and among the academic, industrial, and political fields of knowledge production, and if these patterns can be condensed into a coherent set of rules of the game, a new knowledge regime can be said to have emerged.

Innovation also depends on the kinds of knowledge that are produced and on the styles of knowing that are cultivated. Modern society is famous for its processes of formal rationalization, codification, and 'scientification' of traditional knowledge. New types of explicit scientific and codified knowledge are preferred to diffuse and diverse kinds of knowing: ascertained laws to observed regularities in science; tested procedures to engineering routines in technology; fixed patents to ascribed engineering capabilities in law; calculated business plans to diffuse commercial intuitions in economy. Underlying the traditional theories of science, and increasingly relevant, is how different kinds of knowledge connect to one another. Studies of both scientific and industrial innovation emphasize the critical role of tacit and non-explicit knowing, of knowledge that is incorporated in bodies, brains, and technologies, and of transdisciplinary expertise. Consequently, the issue of knowledge styles is on the agenda of innovation research.

Throughout history, we have experienced some great knowledge changes: the transition from oral to written knowledge in Ancient Greece, when knowledge regimes based on religious doctrines and theoretical reasoning took precedence over those based on mythical narration and magical skills; the media revolution from written to printed knowledge, which turned secret and local knowing into public and universal knowledge. This latter change gave rise to modern knowledge regimes, which substituted formal rationality for routine and tradition, scientific knowledge for experience. Changes in knowledge regimes bring changes in the kind of knowledge favored, and in how the production, distribution, and use of knowledge are institutionalized.

Today, we again face significant knowledge changes. A review of the literature of science and technology studies shows that the production of scientific knowledge seems to be in a process of fermentation. Since the nineteenth century, it has mostly been institutionalized as a relatively autonomous activity of the modern research and teaching universities. Disciplines, scientific communities, and faculties have been differentiated according to internal criteria that subscribed to the idea of the unity of science. This type of disciplinary institutionalization favored complementary specialization and created an enormous increase in scientific productivity. But it showed great resistance to external challenges, such as cooperation with other disciplines in order to solve transdisciplinary problems, or to coproduce technological innovations with economic and political partners. New patterns of scientific knowledge production, such as mixed epistemic cultures around laboratories, clustered networks of researchers around great problems of mankind, discipline-crossing communities around generic research technologies, and

interorganizational communities of practice, evolved inside and outside the academic field. These kinds of 'post-normal science' (Funtowics and Ravetz 1993) developed alongside the canonical system of closed disciplines and universally oriented scientific communities. This chapter examines the patterns labeled sometimes as 'fragmentation,' sometimes as 'disunity of science,' and sometimes as transdisciplinary 'mode 2' of scientific knowledge production, to see whether they have a common denominator, so that we can speak of the emergence of a new knowledge regime. It argues that, as an institutional response to the heterogeneity caused by overspecialization, a regime of 'fragmental distribution' emerges in some fields of knowledge production. The fragmental knowledge regime does not replace the dominant regime of functional specialization, but works alongside it.

The literature of the socio-economics of innovation and organization sociology gives the impression that extensive knowledge changes are taking place in the field of industrial innovation: the specialization of economic and academic organizations is partly fading; firms are turning themselves into knowledge-producing units, and universities into knowledge-exploiting units; in discovery, invention, innovation, and diffusion, the concepts of functional division and sequential organization that constitute a national system and a standard course of innovation are being partly given up in favor of concepts of distributed innovative processes (see, for example, Coombs et al. 2003) amongst a greater variety of organizations, geographically dispersed, but operating at the same time. This chapter argues that the increasing number of networks of innovation characterized by interactions between heterogeneous organizations is a particular mode of coordination, and an adequate institutional answer to the problems of a fragmental distribution of innovative activities.

This chapter, insofar as it offers a modest conceptual solution to the problems and claims of theories such as postmodernism, the knowledge society or reflexive modernization, also relates to the macro-sociological debates on the changes of modern society. It argues that we can distinguish between two types of knowledge regimes: a functional regime and a fragmental one. These respond differently to the consequences of functional differentiation: the one continues specialization and standard integration, the other takes advantage of fragmentation and loose coupling. It also argues that we can distinguish between two styles of knowing: one that favors explicit knowledge over tacit, another that explores the relations between these two kinds of knowledge. It is an open and empirical question how far and in which fields of the innovation system the changes correspond with one another. The rise of a fragmental knowledge regime and of an explorative style of knowing that can be deduced from the patterns described in the literatures may signal critical macro-institutional changes (see Campbell, this volume).

The next section reviews these three bodies of literature from the perspectives of whether and where distinct patterns of knowledge production can be detected. The section aims to construct coherent patterns of rules that may indicate the coexistence of two knowledge regimes.

The following section focuses on the relation between explicit knowledge and tacit knowing. It analyzes this distinction between the 'known' and the 'knowing' (Dewey and Bentley 1949) and 'explicit' and 'tacit' knowledge (Polanyi 1966) to see the way in which the reference to (or reflection or awareness of) tacit knowledge constitutes a different style of knowing.

The final section draws two lines of research from the conceptual considerations and the empirical studies presented: to identify distinct patterns of innovation paths and their changes, we need more comparisons between individual innovation biographies in different technological and industrial fields; and we need finer-grained analyses of innovative constellations of scientific, technological, and institutional concepts. To be able to choose specific, well-adapted styles of management and consistent strategies for institutional policies, science managers, funding institutions, and politicians need to know about the different patterns of knowledge production, of the distinct styles of knowing, and of knowledge regimes.

Changing patterns of knowledge production: between 'functional specialization' and 'fragmental distribution'

Relations between types of social differentiation and regimes of knowledge production

Societies and social systems can be analyzed from the two complementary perspectives of diversity and coordination. Since Karl Marx and Emil Durkheim, sociologists have pointed out the interrelatedness of various kinds of division of labor and mechanisms of social coordination, of types of social differentiation and regimes of social integration. If we want to examine the question of a fundamental knowledge change, as supposed by adherents of postmodernism or by proponents of the coming of a 'knowledge society,' then we have to look for changes in the division of knowledge production and for changes of institutional and organizational patterns.

Modern society differs from traditional society in the emergence and predominance of a new type of social differentiation, characterized by patterns of functional specialization of spheres, values, and institutions. This type of differentiation splits society into complementary parts, organized around functions that are different, but of the same status. Spheres of action and sets of orientations, such as the economic, the political, and the scientific subsystems, are separated horizontally from each other; they differ according to their specialized contribution to the reproduction of society; they increase their efficiency by generating a self-referential orientation that follows its own code and purifies it from other influences; they achieve a relative autonomy from outside interventions by establishing a system of self-organization. As the functions are indispensable, and cannot be substituted by another subsystem, all functionally specialized systems are basically equally important. In modern society, a functionally specialized system of horizontal division ('segmentation') prevails over the vertical system ('stratification') as the dominant pattern of differentiation. The production and use of economic, governmental, and scientific knowledge is institutionalized according to this type of division, and the knowledge processes are concentrated in the specialized spheres of society: thus, scientific knowledge production is concentrated in the academic sphere of universities and research institutes, gains high institutional autonomy and self-governance, can be characterized by the focus on basic science, and is mainly organized along scientific disciplines; industrial knowledge production takes place in specialized research and development departments and in industrial research institutions, is strongly oriented towards applied sciences, and follows the lines of technological trajectories and industrial fields.

This kind of knowledge specialization creates problems of coordination and coupling between the separate subsystems. Markets of patents and licenses, and establishing research and development departments in big corporations become the predominant means of coordination between science and industry. When the state aims at the enforcement of military power, or at the acceleration of economic innovation, it recruits consultants from science, industry, established mission-oriented laboratories, and domain-oriented governmental research to bring together the separate knowledge processes. The strategy influences the patterns of coordination to maintain the lines of specialization of scientific disciplines, industrial branches, and policy domains. The whole system of specialized knowledge production is treated as if it were integrated by a standard model of sequential innovation. This oft-described model attributes the functional parts of the innovation process to the separate

institutional fields. It functions as a productive fiction, because the different contributions can be combined by the actors, who follow the sequential order of discovery, invention, and innovation. It could be said that this regime of complementary and specialized knowledge production emerged as an institutional answer to the functional type of social differentiation.

But the increase in knowledge specialization, the globalization of knowledge production, and the acceleration of the pace of innovation have led to unexpected problems of synchronization and mutual adjustment: when a function or some rules are adapted from one institution and integrated into another, functional specialization lines between scientific, industrial, and political institutions are sometimes crossed; functional divisions between scientific disciplines, industrial branches, and political domains lose their exclusive character in some fields; the sequence of innovation processes gets more and more disrupted by parallel processes of scientific and technological innovation; in some places, the standard model of innovation is broken into pieces, and the dispersed fragments are then put together into a heterogeneous patchwork of innovation. As the following sections will demonstrate, all these changes show a common pattern that indicates the emergence of a fragmental type of social differentiation and an affiliated fragmental regime of distributed knowledge production.

A fragmental type of social differentiation splits a heterogeneous whole into parts that are of the same kind, but of a different status. For example, regional innovation networks almost always include the same mixture of elements: scientific, economic, and political actors and institutions. But some innovation networks, such as the Silicon Valley network of microelectronics and software industry, or the Baden-Württemberg network of mechanical engineering and car production, set the benchmarks; others are imitators and followers in the global competition. The differentiation of scientific disciplines is rooted in well-defined and theory-bound fields of research. Each of them enjoys the same highly regarded status of

Table 12.1. Types of societal differentiation

	Functional	Fragmental
Type of division	separated + horizontal	combined + heterogeneous
Means of coordination	market; science	networks; mixed cultures

certain knowledge that we call scientific truth. But when the number of mission-oriented research projects and research programs increases, the lines of disciplinary knowledge production are crossed; several pieces of disciplinary knowledge, many methods of scientific and technological knowledge production, and different kinds of knowing (such as building laboratories or patenting products and procedures) must be mixed. Then a new type of fragmented knowledge production, rooted in the same combinations of heterogeneous knowledge fragments, emerges. Its quality can no longer be assessed by the peers of one discipline, but needs heterogeneous expert groups and mixed epistemic cultures from across the classic disciplines.

This fragmental type of differentiation differs radically from the functional one: separation is given up in favor of heterogeneity and reflexivity. Functionally specialized institutions and pure scientific disciplines do remain fundamental factors for knowledge production, but sometimes in the background; they lose their prerogative to act as the dominant pattern or personnel on the stage of innovation; heterogeneous innovation networks, transdisciplinary expert groups and mixed epistemic cultures take over prominent roles. This chapter will demonstrate that a new knowledge regime of fragmental distribution is rising alongside the regime of functional specialization, in close relation to the fragmental type of social differentiation.

Scientific innovation: between disciplinary research and distributed knowledge production

We can see the institutionalization of modern science in the establishment of scientific disciplines, affiliated scientific communities and academic organizations such as institutes, laboratories, faculties and universities, all following the same order of specialization. The disciplinary division of knowledge production around theoretical core programs prepared most of the ground for scientific growth; the unity of disciplinary research and teaching enabled the accumulation and integration of scientific research results. This can be seen as an epistemological emancipation: disciplinary research was freed from the earlier dependence on theology and philosophy; the hierarchy of faculties was dissolved; the stratified pattern of pre-modern science was replaced by a functional pattern of disciplinary specialization; there was an increase in the number of disciplines, subdisciplines, and specialties. As a consequence, the edges of disciplines became frayed; the overall idea of the unity of science (which philosophers of science had developed into a philosophical program to keep exploding scientific disciplines together when the real unity became threatened) broke down. The more that knowledge specialization progressed, the more new research fields emerged; they came up either at the periphery of traditional disciplines or between them. Occasionally they were integrated into established disciplines; more commonly, they turned into a new kind of discipline characterized by a looser coupling of disciplinary fragments. Computer science (or informatics, as this interdisciplinary research field is called in France and Germany) derives from such different subdisciplines as formal logics, applied mathematics, cognitive psychology, computer linguistics and automata construction, but it never gained the paradigmatic status of an integrated discipline as did, for example, plasma physics as part of physics. If we look at the fields of artificial intelligence or robotics, we can no longer call them a closed and unified discipline or a functionally specialized part of a discipline. They are neither parts of mathematics nor of mechanical engineering; their status remains that of a loosely coupled transdisciplinary field composed of specialties. The knowledge production is distributed over heterogeneous fields of research. If they succeed in developing common practices of borderline activities in so called 'trading zones' between disciplines, and if they share a pidgin kind of communication, these fields of distributed knowledge production gain stability (Galison 1996; Meister 2002).

Scientific communities, closely connected with the concept of scientific disciplines, underlie the same patterns of change that result from the ongoing specialization and multiplication. Teaching, education, and graduation usually follow the lines of the disciplines; faculties, departments, and universities are mainly organized according to disciplinary specialization. But when we actually look into some research laboratories, mission-oriented projects and mixed research teams are increasingly preferred to the disciplinary division of knowledge production. We still find designations such as 'physics' or 'mathematics' or 'chemistry' or 'biology' department, reminiscent of educational backgrounds. But all empirical descriptions of laboratories agree that either the broader concept of 'epistemological cultures' (Knorr Cetina 1999; Traweek 1988) or the narrower concept of 'experimental cultures' (Rheinberger 1997) is a more meaningful delineation of the reality of distributed knowledge production than 'disciplinary scientific community.'

The modern university seems to be transforming itself in a similar way. The growth of disciplines can no longer be organized under the umbrellas of faculties. Faculties have often turned into formal organizational units that hold together very heterogeneous departments and institutes. Universities are no longer able to house all disciplines and faculties. They too follow the pattern of specialization: they strengthen particular profiles and core competences and seek for alliances with other universities. This trend started early, when teaching and research universities, business and medical schools and technological institutes were established as distinct units of the academic system. Nowadays, a greater diversity of units offers a smaller number of disciplines, bound together under a particular mission: the Hertie School of Governance combines Political Science, Management Science, and Law in Berlin; the Keck School of Applied Life Sciences combines Biology, Engineering, Innovation Management and Ethics in Claremont, Southern California. The teaching starts to be separated from the disciplinary curricula; modularization begins to break the traditional curriculum into interchangeable pieces; multiple forms of reintegration (such as Master Studies in computer-aided façade design in architecture, or special educational programs for the breeding of experimental animals in biology) become possible; curricula often relinquish the idea of a specialization based on a disciplinary identity in favor of a flexible combination of divergent knowledge pieces and competences; the unity of teaching is fragmented; the universities, with their teaching units, gradually become 'multiversities' (at first, Kerr 1963) with a selected set of disciplines instead of the whole universe of the sciences and humanities (see Schimank and Stölting 2001).

A further consequence of the multiplication and specialization of research fields is a change in the pattern of interdisciplinary cooperation. The traditional pattern of scientific cooperation was hierarchical: the research subject was defined primarily from the perspective of one leading discipline; the other disciplines took the auxiliary roles, like statistics for biology or instrument making for nuclear physics. This strongly stratified pattern within and between disciplines has now been replaced by weaker forms of intra- and interdisciplinary cooperation. Based on a functional division of knowledge production, each field of a discipline contributes a well-defined part to a visible and shared objective. Even established disciplines, like physics, chemistry, or geology, develop a pattern of internal disciplinary specialization that shows little communication between the fields: high energy physics, for example, is internally subdivided into theoretical, experimental, and instrumentation micro-cultures with a low intensity of exchange between them, as Peter Galison (1997) has demonstrated. This functional division sometimes dissolves, such as when, because the research subject is still in the making, a clear object or aim is missing, or when these processes are situated between the engaged disciplines: in artificial intelligence, for example, one can always find leading ideas and concepts that coordinate the distributed research activities, but they change every five to ten years and, with them, the dominating disciplinary fields—from the engineering sciences in the early cybernetic phase, to cognitive sciences during the classic phase of the physical symbol systems approach, to the brain sciences under the paradigm of non-symbolic artificial intelligence (see Gardner 1985). Artificial intelligence has developed beyond the functional pattern of specialized disciplinary contributions: cooperation is more fluid and more heterogeneous than in the established disciplines. One can distinguish it as a particular pattern—the fragmental pattern of interdisciplinary cooperation in a field organized like a patchwork in progress, the focal points of which are still changing.

Academic research has long been seen as mainly shaped by theoretical developments, and the cognitive and organizational development of scientific disciplines as following a congruent pattern (see Whitley 1984; Fuchs 1992). Even as the research instruments were obviously growing in size or complexity (as in high energy physics or simulation techniques), their development was seen as being in line with the disciplinary pattern of theory-building. But there are also many examples of the fact that scientific instruments are not only transferred from one to another field, but that they have triggered new research fields across the established disciplinary boundaries, such as the electronic microscope that pushed nanotechnology, or have even created a new quasi-disciplinary research field, like the com-

puter that enabled computer sciences. Further examination of the crossing of disciplinary boundaries yields many more examples of migrating instruments and transdisciplinary communities. 'Generic technologies' (Shinn and Joerges 2002) push patterns of disciplinary specialization in the direction of new combinations of disciplinary fragments. Nanotechnology can be defined as such a transdisciplinary field of heterogeneous knowledge pieces and competencies created by the capacity of the new microscope to move tiny molecular particles. It is neither a subdiscipline or specialty of physics, nor a mere part of mechanical or chemical engineering: it includes parts, procedures, and pieces from all these fields; they are not integrated like the branches of one disciplinary tree, but they follow a pattern like the knots and roots of a rhizome that nourishes the distributed knowledge production.

Such criss-crossing and recombining seem also to open academic research to interdisciplinary cooperation and to external definitions of the research subjects. The complexity of research subjects is greatly increasing in some fields because of the expansion of areas of interest for many disciplinary and external actors: the physics of efficient energy production is turning nowadays into a transdisciplinary, combined effort to design sustainable technical systems of energy circulation; the narrow perspective of weather forecasting that is a specialty of the physics of thermo-dynamics is now widened to the enlarged perspective of complex climate research, which combines such heterogeneous fields as paleo-climatology, computer simulation and others (Stehr and von Storch 1999).

The expansion of computerized work and computer-mediated communication changes the subjects of disciplinary research and, even more, the conditions of scientific work and communication itself: ergonomics comes out of a complementary combination of physiological, psychological, and sociological aspects of workplace design; with the rise of computerization and new technological media came a new kind of engineering named software engineering. Such subjects then turn into more hybrid subjects that ask for a new transdisciplinary approach: human-computer-interaction studies, based on both activity theory and the theory of cognition (see Chaiklin and Lave 1998; Engström and Middleton 1996); high-tech workplace studies that follow an anthropological approach to frame all fragmented aspects (Button 1993; Star 1996); socionics that uses sociological models to design software agents and architectures of intelligent information systems and that changes sociological theory-building by making technical agents and human actors into a hybrid subject (Malsch 1998, 2001; Burkhard and Rammert 2000). The influences of computerization and the Internet on the reorganization of scientific disciplines and communities are still an open research question (for some distinct perspectives, see Merz 2002). These change processes underlie the question of whether they give rise to a new pattern of heterogeneous cooperation between diverse and distributed research fields, and under what conditions they strengthen the functional specialization and integration of academic research.

We can summarize that, in many research fields, the subjects of research are becoming more complex in their elements and relations, more heterogeneous in their disciplinary perspectives, and more hybrid in their focus. Critics might object that these developments are not really a new phenomenon: it seems to be the normal process for engineering and applied sciences. Even if we accepted this objection, we could argue that the complementary specialization among basic sciences and applied or technological sciences often dissolves. Fundamental research can be found in technological fields—a Nobel Prize awarded for the invention of the electronic microscope—and technological and practical aspects are closely intermingled with basic scientific advances, as in molecular biology. The terms 'technosciences' and 'high technologies' (Latour 1987; Rammert 1992) were coined to grasp this crossing of two distinct knowledge styles and research cultures. It seems obvious that a new pattern of interdisciplinarily distributed knowledge production has developed alongside the established pattern of disciplinary and complementary interdisciplinary academic research. The crossing of disciplinary borders sometimes takes on a new quality, and the heterogeneous cooperation is beyond the usual interdisciplinary cooperation. The new knowledge situation can better be described as being in a state of loosely coupled distribution rather than of functional or even hierarchical integration. It is far beyond the 'finalization' phase of completed and perfected disciplines (see Böhme et al. 1976). It looks more like the patchwork of disciplinary knowledge fragments.

Industrial innovation: between complementary divisions and distributed innovation processes

Since Schumpeter, economic enterprise has been acknowledged as the core of technological innovation. But the research on technological innovation has taught that the firm is only one important locus, along with communities of practitioners and technological systems (see Constant II 1987), and that there are changes in the patterns of corporate innovation in relation to technological regimes, industrial structures, and supporting institutions (see Nelson 1998). Both the socio-economics of innovation and the industrial organization literature have produced a rich stock of knowledge about the interrelationships between technological, organizational, and institutional developments. I shall review a few results from the perspective of whether they indicate significant changes of the patterns of knowledge production.

In the beginning, scholars of the economics of innovation focused on the question of how far the technological innovation depends on scientific advances in the basic sciences. A controversy between adherents of the 'technology-push' (Schmookler 1966) and the 'demand-pull' (Nelson 1959) approach came up which was later on dissolved by making distinctions between radical and incremental innovations (Abernathy and Utterback 1978) and between traditional and 'science-based' industries. The chemical and the electro-mechanical industry became the paradigms for the organization of science-based industrial innovation (Noble 1977). They introduced the industrial laboratory and the R&D department into the enterprise. R&D was integrated into the divisional structure of the big corporation: its function was to maintain a narrow relationship with scientific advances and the scientific community, and to translate scientific discoveries and technological options into new products and production processes.

This internalization of scientific knowledge production established patterns of functional specialization inside and outside the firms: their organizational structure was characterized by an internal differentiation between R&D, production, marketing, and other departments corresponding to the complex environment of the firm (see Lawrence and Lorsch 1967); the institutional structure of the environment was designed as a standard sequence of special innovation processes, from basic scientific discoveries to technological innovation, to broad diffusion of the product. The whole innovation process was organized as a sequence of separable stages linked by relatively minor transitions to allow for adjustments between stages. The prevailing pattern of the specialized functions can be characterized as a 'linear sequential coupling' (Van de Ven 1988: 111 ff.).

When the challenges of the knowledge production grow, and when the diversity of knowledge resources increases, firms change their dominant strategy. Because they cannot afford to produce all the necessary knowledge themselves, they start a process of externalization of the knowledge production. They concentrate their activities on their 'core competencies,' a set of differentiated skills, complementary assets, and routines that provide the basis for a firm's capacities and sustainable advantage in a particular business (see Prahalad and Hamel 1990).

The multiplication of relations with knowledge producers outside the firm counteracts the reduction of the sets of knowledge production inside. Relations of one kind are maintained with the externalized knowledge producers; together, they sometimes build a kind of 'virtual organization' which offers products and services as a combined system. Relations of a second kind are the punctual and temporary cooperation with competitors in 'alliances.' A third kind is the intensification of the user–producer interaction (see Hakansson 1987; Lundvall 1988) in order to launch user-induced innovations with success. All these changes of the patterns of innovation are systematically collected under the new term 'networks of innovators' (Freeman 1991).

Since the 1980s, all kinds of networks have been created: joint ventures, joint R&D agreements, technology exchange agreements, common licensing, research associations, and computer databases. Their forerunners were the early research associations of the leather and color industries in the 1890s. Formal networks reach back to the time-limited cooperations between the petrol corporations in the 1930s, and in the chemical and the synthetic rubber industry. The 'network form of organization' (Powell 1990) takes on a new quality when it changes from a type of 'strategic network' (Sydow 1992) between firms for keeping control to a type of 'idea-innovation-network' (Hage and Hollingsworth 2000) of firms and other organizations to share knowledge. The latter network is more combinatory than stratified, and is better described as a 'loosely coupled system' (see Weick 1976). The 'linear sequential coupling' pattern is replaced in some industries by the pattern of 'simultaneous coupling' (Van de Ven 1988). This second regime of a 'distributed innovation process' (von Hippel 1988) can be observed mainly in the new industrial branches of high technologies such as electronics, information technologies, new materials, and biotechnologies: many small and medium-sized firms create networks of knowledge production with one another and with some big corporations.

Industrial innovation is distributed over firms' internal and external places of knowledge production, over the user-firms, producer-firms, small start-up firms and big incumbent corporations; it is also distributed over heterogeneous institutions, like science, the economy, and the state. It is pushed and pulled by a highly diverse spectrum of actors: university departments, governmental research institutes, risk capitalists. The boundaries between scientific innovation and industrial innovation are blurring, especially in the high-technology and new economy sectors. That is why the next section deals with the change of the patterns on the higher (macro-) level of heterogeneous networks of innovation.

Regime innovation: between a specialized and standardized system of coordination and a fragmented and fluid order of interactive networking

Crossing the boundaries between specialized systems is one sign of a new knowledge regime. The functional differentiation between scientific innovation and industrial innovation blurs and it is no longer sufficient to focus on the change from disciplinary to interdisciplinary patterns of scientific innovation. Industrial laboratories and research firms have developed into an integral part of a research system that cannot be restricted to the academic sector of scientific knowledge production. At some interfaces, the industry–university relations cross the boundaries between economy and science; it is even claimed that a kind of a 'triple-helix model' of relations among industry, university and government, interlocking science, economy and politics, evolves (Etzkowitz and Leydesdorff 1997).

However, the analysis of industrial innovation demands a view of organizational patterns and interorganizational relations between firms and industries that is broader than the firm and inter-firm view. To concentrate on 'technical systems' (defined as networks of agents in an economical industrial arena (Carlson and Stankiewicz 1995)) or on exclusively 'industrial networks' (Imai and Yamasaki 1992) is to take too narrow a view. The concept of 'large technological systems' (Hughes 1987) crosses the borders of the industrial arena and interweaves scientific, industrial, and political agents and artifacts. The concept of 'user-producer-interactions' (Lundvall 1988) has been broadened to regional and 'national systems of innovation' (Nelson 1983; Edquist 1993); these include actors and institutions from the educational, research, legal, administrative, and economic subsystems. A similar change can be confirmed for the change of network concepts: from strategic networks, industrial networks or functionally structured idea-innovation networks (see Hage and Hollingsworth 2000) to heterogeneous networks of innovation that include the really mixed spectrum of many and diverse academic, economic, and governmental agencies (Powell et al. 1996; Rammert 1997).

Looking at the overall and distributed innovation processes from such a broadened perspective, one realizes how a growing diversity of academic, industrial, and governmental organizations are linked to one another in a distinctive pattern. The institutions of basic scientific research, like some Max Planck Institutes or the GMD Institute of Mathematics and Data Processing in Germany, are transformed into more market-oriented institutions. Universities are influenced to develop and differentiate into research universities, professional schools, or regional higher-education schools. The rise of the 'entrepreneurial university' (Etzkowitz 2002) that follows the example of Harvard, MIT, and Stanford demonstrates the success of a hybrid type of institution that follows the logics of both the academic and the economic system. A similar confluence of formerly separated systems can be observed in the sphere of industrial organizations: the organizational population not only shows a higher diversity of types, but some start-up firms and some big corporations turn into knowledge-creating companies (Nonaka and Takeuchi 1995) that take over some norms of academic research and innovative work (Hirschhorn 1988).

Building bridges between the separate systems is a second significant criterion. When the parts of the innovation system are functionally differentiated, then the specific tasks and orientations are organized in sets that are homogeneous but separate: scientific research in disciplinary communities with publications as the product, industrial innovation in enterprises and technological communities with patents or new products as output. Bridging the systems requires a differentiation of further transfer subsystems, the establishment of standardized interfaces and a long-term perspective of coordination. The coupling of the separate, but homogeneous, units follows a linear and reverse linear mode. Such kinds of standardized coordination between the separate systems can be found in the fields of mechanical engineering and chemistry, which have a long experience of institutionalization.

But there are different developments in fields like microelectronics, software technology or biotechnology, where the units are becoming more and more heterogeneous. Universities supplement their traditional research and teaching tasks with managerial competences, to apply for patents, to found start-up firms, and to sell educational programs. Business enterprises develop scientific education and research competencies alongside their management and venture competencies. The innovation system then consists of fragmented parts which are coupled, more opportunistically than systematically, and under a shorter temporal perspective. The interfaces could be described as 'floating' and look more like trading posts, where exchanges are performed based on trust relationships. The exchanges are neither spot encounters, as with markets, nor strongly coupled, as in corporations. They are longer term (or medium term?) and loosely coupled, as in interactive networks. The style of bridging between the heterogeneous units follows an interactive mode of innovation. The arm's-length relations between separate systems are now replaced by an enacted web of weak links between heterogeneous units.

Breeding different regimes of knowledge production is the institutional answer for these processes of knowledge specialization and fragmentation. They are based on two styles: functional integration and interactive coordination. How can the formation of different regimes be described? In the science and technology studies literature, the distinction between a 'mode 1' and a 'mode 2' of scientific knowledge production is heavily debated (Gibbons *et al.* 1994). 'Disciplinary' and 'transdisciplinary' knowledge production, 'decontextualized science' and 'context-sensitive science' are distinguished. Some research fields are claimed to be shaped by segregated scientific communities, others by 'integrated arenas' constituted by heterogeneous participants. The separation of scientific and public debates is in some fields replaced by a 'hybrid agora' and by 'transaction spaces' between science and politics. Homogeneous disciplinary advisers give way to 'socially distributed expertise' (Nowotny *et al.* 2001). If these changes of pattern are interpreted not as a linear transformation, but as two possible styles of knowledge regimes, then both styles can be found in some research fields of classical scientific disciplines, such as nuclear physics or molecular biology, as well as in new interdisciplinary research fields like environmental science or risk analysis. The attribution of a mode 2-type depends on the intensity and scale of heterogeneous contexts which are thrown into interaction with the scientific knowledge production.

Networking among homogeneous units and networking among heterogeneous parts are two distinct modes of coordination. From the industrial economy and organizational sociology literature, one can learn that strategic networks among firms such as the road construction, insurance, or film industries are quite different from 'hybrid networks' (Teubner 2002) of innovation, which interweave agents from heterogeneous institutional fields, like university biology departments, the National Health Institute, patent lawyers and private companies. In the new economy, especially in the high-technology sector, the interaction between developers, producers, and users is transformed from a highly standardized relation between separate systems of an academic and a commercial world into a 'hybrid regime' where the different standards and values have been intermingled (Owen-Smith 2003: 1100). Because they include more heterogeneity and offer more open access than corporate systems, networks of business groups and 'authoritatively grounded' networks, like those in South Korea and Japan (Hamilton and Feenstra 1998), are different from heterogeneous networks of 'industrial districts' and 'market-oriented voluntaristic' networks, like those in Taiwan and China. 'Hybrid networks' describes a third method of allocation: it is neither a contract relationship between agents on a market, nor an organizational relationship between members and the management of a hierarchy; it is a particular type of networking relation, beyond both and based mainly on trust (Powell 1998*a*).

Networks of innovation are built both to exploit common but distributed resources and, at the same time, to explore knowledge spaces with a common interest, but from diverse perspectives. 'Exploitation-focused' networks tend to develop hierarchical, exclusive, and corporate structures, whereas 'exploration-focused' networks are always in the making, interactively learning and open to new agents. In chemistry or in car production, the strategic networks are strongly controlled by big corporations and the whole system of innovation is integrated with the complementary functions of research, development, patenting, testing, production, and quality control. In biotechnology or information technologies, heterogeneous networks of innovation are emerging; they show a high intensity of interaction and mutual learning among generally equal but heterogeneous agents. The agents of knowledge production are multiplied; hybrid formations prevail; the networking shows a more interactive, heterogeneous, and heterarchical form than in the corporate networks. Is this change restricted to the ageing of industries and technologies—from young industries and technologies with ill-defined problems and low codification of knowledge to established ones with highly defined trajectories and knowledge stocks—or is it more general and inclusive of the whole social system of production and innovation?

The acceleration of the pace of knowledge production is a significant general trend. It has limited the coordinative capacities of the linear-sequential mode of integration. The standard model produced certainty and coherence by connecting the stages of inception, invention, and innovation consecutively. But in the mode of parallel-interactive coordination, the different activities, from the idea creation to the marketing, are fragmented and distributed over many different arenas (see Hage and Hollingsworth 2000; Coombs *et al.* 2003). That means that they are performed simultaneously and dispersedly, as in parallel computing processes. Problems of continuity and synchronization arise because of the differences in pace. Innovation networks with permanent and parallel interactions are the institutional answer to these problems (Rammert 2000).

To sum up, we can distinguish between two styles of knowledge regimes: a regime of further functional specialization and a regime of fragmental distribution. They shape the boundaries, linkages, and units of the knowledge, industrial, and innovation systems differently. In the debates about the changes of modern society, we can now rediscover some of the features. Adherents of postmodernism point to the 'dissolving of boundaries,' the 'fragmentation' of social units (Baudrillard 2004) or the 'patchwork character' of institutions (Lyotard 1984), but analytical and empirical specifications are missing. The proponents of a theory of 'reflexive modernization' (Beck *et al.* 1994) have been more successful in the identification of some processes of 'de-differentiation' and 'heterogenization' in many fields of modern society, which is characterized by continuous functional specialization and the unintended reflexes on its problematical consequences. But this chapter argues that one can distinguish a new style of knowledge regime which is related to a fragmental mode of social differentiation. However, an epochal change from one type of society to another is not generally supposed. It is just one possible hypothesis that the consequences of functional specialization generally lead to a new

Table 12.2. Regimes and levels of knowledge production

Levels	Regimes	
	Functional specialization	Fragmental distribution
Interaction	disciplinary communities specialized disciplines + and subdisciplines	communities of practice heterogeneous expert groups
Organization	formal, internally specialized orgs., for example, faculties, divisions, firm alliances	interdisciplinary research research institutes virtual enterprises
Society	strategic networks policy networks specialized + complementary institutional regimes	heterogeneous interactive networks of collaboration and innovation experimental regimes of institutional learning

knowledge regime throughout the society. Other hypotheses may state that it is the character of scientific disciplines or the nature of the technologies that make the difference; others may emphasize the temporal differences between technologies in an early stage of ferment and variation and in a normal state of dominant design and stabilization (Tushman and Rosenkopf 1992); yet others could look for institutional paths of innovation in certain industries and in particular countries to explain different regimes (Hall and Soskice 2001). In time, we will be able to draw a map that shows in which areas and at what times one of the two styles of regimes is predominant.

Between two styles of knowing: processing explicit knowledge or exploring the relation between explicit and tacit knowing

Changing relations between explicit knowledge and tacit knowing: the 'circle of explicitation'

When the modes of knowledge production are changing, an obvious question is whether the kind of knowledge that is preferred in society

also undergoes a significant transformation. Following Max Weber and Werner Sombart, modern societies are characterized by a transformation of knowledge: of traditional into scientific-rational knowledge, and of tacit into conscious knowledge. Modernization means to make more and more explicit the inherent relations between means and ends, and to reorganize the activities under the imperatives of progress, accountability, and efficiency. Explicit knowledge, like written and codified laws, profit-and-loss account, or scientific explication, is preferred above all kinds of non-explicit knowledge, such as traditions of jurisdiction, economic intuition, technical experience, and practical rules. Formalization, codification, and scientific explication mark the cognitive aspect of modern society; functional specialization marks the institutional aspect.

As digital computing revolutionizes all functions of information processing, the production, distribution, and reproduction of information, and the processes of the explication and formalization of knowledge are intensifying. All kinds of knowledge are now gradually transformed into formalized information, into electronic databases and into expert systems: in the case of research knowledge, the formulas and the textbook knowledge are transformed into computerized data processing and electronic archives; in the case of production knowledge, all relevant data, like parts, prices, and parameters of a product, and the temporal and spatial position in the production process are represented in information, construction, and management systems. Computer programs coordinate thousands of operations and hundreds of variations in order to create a car with exactly those specifications for which a particular buyer has asked. All pieces of knowledge that can be made explicit by being written down or, more narrowly, condensed to a formal rule and represented by an algorithm, can be handled with the help of the computer.

The stock of explicit knowledge will be augmented during the digital revolution on a scale similar to the print revolution. The print revolution increased the mass of explicit knowledge by the printing of books, circulation of periodicals and accumulation in archives, and enabled the specialization of new genres of knowledge and of more scientific disciplines; will the digital revolution merely continue this process of explicitation and codification, or will it also lead to comparable changes in the ways of knowing, as the print revolution did?

Before we can answer this question, some terms have to be clarified. Usually, knowledge is seen as a substance, something that is embodied in books, brains, patent formulas, or computer programs. In this book, we define knowledge as the capacity to reproduce or to replicate findings, products, and processes (see Introduction). This capacity is influenced both by people who know and by the media that store and process information. We use the term 'knowing' to stress that knowledge emerges out of the interaction between the knower and the known (see Dewey and Bentley 1949). Neither collecting books, nor navigating through digital archives amounts to really knowing something. To draw knowledge from explicit knowledge such as a printed formula or a programmed expert system, one needs non-explicit knowledge—to be able to read and understand, to translate it into effective action, or to learn from the interaction. Without a rooted relationship in tacit knowing there exists no explicit knowledge. 'Hence all

knowledge is either tacit or rooted in tacit knowledge. A wholly explicit knowledge is unthinkable' (Polanyi 1969: 144).

Usually, explicit knowledge and tacit knowledge are conceived of as two expressions of one and the same kind of knowledge that can be transformed into the other (by, for example, 'externalization,' see Nonaka and Takeuchi 1995). The relation between them then takes the form of a zero-sum game: the more knowledge is made explicit, the more tacit knowledge diminishes. But this juxtaposition contradicts the logical argument of Polanyi and the empirical reality of knowledge differentiation. If each new piece of explicit knowledge implies a new tacit dimension of knowing—for example, how to read the formal terms, or how to interpret the program-processed results—then the dimensions of explicit knowledge and tacit knowing must be different. They are two kinds or modes of knowing that cannot be substituted one for the other. If the process of knowledge specialization creates more and more domains of knowledge, then at the same time it increases the realm of not knowing something explicitly. One aspect of this phenomenon is discussed as the concept of 'Not-Knowing' ('Nicht-Wissen'), a kind of structural ignorance that emerges out of the hitherto unknown relations that come up with every new piece of knowledge (Luhmann 1992; Beck 1992; Smithson 1993). Another aspect is the realm of the unarticulated knowledge (see Cowan *et al.* 2000). With knowledge specialization, this special kind of tacit knowing is growing, because more and more sectors of society and categories of people act on the basis of rules and routines, explicit knowledge of which only exists in the domains of highly specialized experts.

In a double sense, the relevance of tacit knowing is increasing under conditions of knowledge differentiation. On the one hand, the production of new knowledge always produces new methods, machines, and media, the handling and interpretation of which requires tacit knowing. It is a question of time and money whether some parts of this non-explicit knowledge will be articulated and even codified. (Other parts of implicit experience cannot be made explicit.) On the other hand, the fragmentation of the different knowledge domains reduces the population explicitly knowing the particular code books, and increases the need to rely on tacit knowing. If we conceive of tacit knowing and explicit knowledge as distinct but interdependent dimensions, then we discover a paradoxical relation between them: the more that knowledge is made explicit by processes of codification and computing, the greater the increase in the importance of non-explicit knowledge to appropriate and integrate it. I call this paradoxical relation the 'circle of explicitation:' making explicit the hitherto unarticulated produces new tacit dimensions of knowing and, at the same time, increases the importance of non-explicit knowledge.

This fundamental, paradoxical circle of explicitation cannot be resolved but, to cope, societies have developed different styles of knowing. It is misleading to distinguish societies only by the dominant kind of knowledge, such as traditional and tacit knowledge in pre-modern societies, and rational and explicit in modern societies. It is the relation between these kinds of knowledge that creates the style of knowing. A functional specialization regime prefers the abstraction, formalization, and universal codification of knowledge; to facilitate the internal processing and the exchange between the different spheres of action, knowledge is likely to be transformed into standardized packages of information; this style of knowing follows the mechanical and universal model of information processing. A fragmental distribution regime prefers situational and associative knowledge to connect the modularized and heterogeneous pieces of knowledge and enable learning; this style of knowing follows the organic model of cultivating knowledge, and explores the fluid relations between explicit and tacit knowing. For example, the explicit fundamental knowledge of molecular biologists can be only successfully transferred to bio-engineering and business when it is combined with tacit knowing of how to organize a laboratory and how to

evaluate the chances of different claims. Both styles have in common that they encompass both, the tacit and the explicit kind of knowledge. But they differ in the way that they relate the two: the *style of explicitation* is based on excessive explanation and exploitation of codified knowledge, whereas the *style of exploration* trusts more in the tacit circulation and informal integration of the implicit and explicit knowledge.

Unraveling the critical importance of tacit knowing and the technical roots of scientific innovation

Science is the most prominent endeavor that presupposes, produces, and uses explicit knowledge. Premises and prerequisites have to be made explicit, propositions and conclusions have to be formulated in a precise and explicit way, and methods and instruments have to be operated according to explicit rules in order to receive methodically controlled knowledge that can be replicated universally. Science seems to be the holy empire of explicitness.

From the social studies of science literature we have, however, learnt that even scientific knowledge is based on non-explicit knowledge. It was the Polish physician and pioneer of science studies Ludwik Fleck (1935) who demonstrated that scientific statements and interpretations of empirical observations, as well as expected effects of instruments, are deeply embedded in an unarticulated and shared frame of what might be called a 'thought collective' that he called 'thought style'. It is this group-bound, gestalt-oriented, and incorporated kind of knowledge that the chemist and philosopher of science Michael Polanyi later defined as 'tacit knowledge' and, with reference to that the physicist and historian of science Thomas Kuhn, coined the famous term 'paradigm.' This statement about the rooting of scientific knowledge in tacit knowing does not mean that science has always remained an art, as some postmodern thinkers assume, but that the achievements of modern science are necessarily interwoven with non-explicit knowledge. It is Polanyi who expresses the paradoxical relation between implicit and explicit knowledge more sharply: 'Any attempt to gain control of thought by explicit rules is self-contradictory, systematically misleading and culturally destructive' (Polanyi 1969: 156). He states that the process of formalization of any knowledge that excludes any element of implicit knowledge destroys itself. As a consequence, one has to consider the relation between the two kinds of knowing, namely how it is organized and how it is changing over time.

Scientific innovation takes place via the interference of all kinds of knowledge, of formalized theoretical and implicit practical knowledge, of articulated, instrumentally incorporated and habitually embodied knowledge. It is not only a result of explicitation and information processing. The sociologist of scientific knowledge Harry M. Collins (1974) has already demonstrated that published physical and technological knowledge is not sufficient if one wants to replicate a path-breaking experiment with success. At least one person who has been a member of the scientific research group, or who has shared the practices of the group for a while as a visitor, is needed. Collins's TEA Laser Set study emphasizes the importance of shared collective experiences and incorporated knowledge for the production and reproduction of scientific knowledge. Mathematics is surely the scientific discipline with the highest degree of explicitness. Even here, the proving of calculative procedures and the formal examination of proofs with and without the help of computers are based on different kinds of background knowledge that cannot be precisely articulated and made completely explicit (Heintz 2000: 175) After laboratory study in a mathematical research institute, the sociologist of science Bettina Heintz could demonstrate the existence of distinct mathematical cultures or thought styles which differ according to whether they accept computer-based procedures of proving or not. Some mathematicians accept only human, not computer-based, proof because, despite the latter's being based

on an explicit computer program, one cannot reconstruct and control all of its operations and instructions.

It is not only the computer that has changed the knowledge style in science. From the beginning, modern sciences have been closely connected with technical instruments. Scientific disciplinary knowledge was usually separated from the conditions of its production, untainted by human and technical interferences, and made explicit as a disciplinary code. Historians of science have criticized this idealistic view of a merely cognitive and completely codified disciplinary knowledge based on the style of perfect explicitation. They emphasize the whole system of knowledge production that includes the material side and the social practices of closure and coherence building. If we acknowledge that scientific knowledge is a mixture of both—of universal, calculative knowledge and of context-specific practical design, codified design parameters and implicit hands-on experience (Vincenti 1990)—then a style of exploring the relations between the two can be distinguished from the style of establishing and exploiting the explicit. As the variety, size, and complexity of technical instruments have been steadily increasing—think of the complicated system of accelerators, detectors and calculation programs in high energy physics, or of the loosely coupled worldwide network of observation stations with various documentation techniques, simulation-models and the link between supercomputers in fields like climatology and oceanography—classical disciplines turn into multidisciplinary fields of research. The more the scientific disciplines and subdisciplines are fragmented and reorganized in this way, and the more they are interwoven with technical instruments, the more the above-described explorative style of knowing favors innovation.

Some of these so-called research technologies develop towards 'generic technologies' (Joerges and Shinn 2001) that establish new fields of knowledge and innovation alongside and between the established scientific disciplines. The spectrometer, the electronic microscope, and computer simulation are examples of this type of generic technology, around which particular communities of knowledge arose (see Shinn in this volume; and Shinn 1997) and circulated both explicit and codified knowledge via special journals and tacit knowing via technical experience.

When the experience of a research technology is combined with the highly explicit knowledge of scientific disciplines, another kind of exploration takes place. I recall the development of the computer sciences when such heterogeneous competencies as those of fundamental mathematics and formal logics were brought into close interaction with those of practical deciphering and telegraphy engineering (Heims 1980; Rammert 1992). In the case of the great human genome project, Craig Venter deviated from the strategy that would have been an example of the style of explicitation—of making explicit one genome structure after the other: instead, he followed an intuitive strategy of searching for patterns by a complicated computer program, an example of the style of exploring the relations between the explicit and the tacit knowledge.

Summarizing, we can say that two styles of knowing exist side by side. The explicitation style endeavors to establish a disciplinary code and to enlarge the core knowledge; specialized knowledge is aggregated and integrated into the fundamental scientific code of the discipline. When this explicit knowledge base has gained a particular grade of unification and perfection, then the knowledge can be practically applied during the so-called phase of 'finalization' (Böhme *et al.* 1976). Nuclear physics, plasma physics, synthetical chemistry, molecular biology, or neoclassical economy are research fields in which this style of knowledge production is dominant.

If we accept the description of the actual knowledge as changing towards a more heterogeneous and distributed knowledge production, and if we acknowledge the increasing use of a growing variety of technical instruments, then we may discover another style of knowing. It is more sensitive to the relations between explicit and non-explicit aspects of knowing, but it does not only rely on the fundamental and formal processes of explicitation; it produces coherence between heterogeneous participants by processes of enculturation that create communities of knowing by interaction, learning, and soft theorizing. That means that, instead of axiomatic theories that only allow formal integration of knowledge pieces and follow a limited logic of algorithmic information-processing, scenarios, and simulation models enable a tacit integration. There also remains a strong tendency to make explicit more and more of the unarticulated knowing, and to codify the heterogeneous knowledge fragments; but the explicitation will be limited by time and costs (see Cowan *et al.* 2000). As processes of fragmentation and instrumentation continue to augment the spheres of non-explicit knowing, the search for heuristic strategies will expand, and strong codification will be more and more restricted to only the most relevant cases. The style of exploring the relations between the explicit and the tacit dimension of knowledge fosters a strategy of 'satisficing' (Simon 1954) that concentrates on the tuning between the different kinds of knowing.

Overcoming the limits of explicit and rational knowledge by tacit knowing and trustful cooperation between firms in industrial innovation

The modern economy is said to be based on rational choices between goods of which the values and costs can be made explicit. But, especially in the field of innovation and technological choices, it is evident that decisions follow plausible rules of thumb and organizational routines more than explicit rational calculations (see Nelson and Winter 1982). Numbers are definitely the basis of book-keeping and controlling in the modern firm, but knowledge is more than number-crunching: it is the capacity to relate these numbers to other numbers and to interpret their relevance in relation to earlier experiences and future constellations.

Under the regime of fragmental distribution, the conditions of temporal planning and of clear cost calculation worsen: the usual uncertainty of economic decision-making is multiplied; a complex 'circle of uncertainties' (Rammert 2002: 177) that limits the explicit calculation of risks, returns, and benefits is created. Firms wanting to invest in an innovation do not know whether:

- they can get access to the relevant scientific information;
- they can select the relevant technological information out of the rising flood of information;
- they have the capacity to process and convert it into useful knowledge;
- the innovation process will come up with a technologically feasible product;
- this product can be produced economically;
- a new market can be established;
- users will accept the product and tolerate its unintended consequences;
- the developer will get a fair return on investments and risks;
- property rights are sufficiently protected;
- the product will meet the compatibility requirements of technical standards and legal norms.

The standard solutions of how to cope with uncertainty, such as establishing close relationships with a faculty or a disciplinary research institute, or hiring R&D people with a sound disciplinary background, are limited by fragmentation processes in the fields of scientific knowledge production. If you are not sure which of the many diverse competencies you need, then you have to make links with many different research fields and institutions. It is no longer just a question of buying the knowledge and being the first to know; the complexity of 'distributed innovation processes' (Coombs et al. 2003) requires new strategies of building alliances with complementary or even with competitive firms to share the knowledge and the risks.

In the industrial economics literature, we find two significant changes in this area: from

an information approach to a knowledge approach, and from knowledge-using to knowledge-creation. The information approach conceives of the firm as a 'response to information-related problems' (Fransman 1998: 149). Information is defined as closed sets of data which can be appropriated and processed by firms. It is a common belief of the views known as 'principal agent' and 'transaction costs' views that firms organize innovation on the basis of asymmetrical information, and thereby economize on 'bounded rationality' and opportunism. The knowledge approach, however, emphasizes the problem of uncertainty and the fact that information is always incomplete and has to be interpreted. As Freeman (1991: 501) says: 'The problem of innovation is to process and convert information from diverse sources into useful knowledge about designing, making and selling new products and processes.' It conceptualizes the firm as a 'repository of knowledge,' meaning a collection of routines, organizational experiences and dynamic capabilities, as evolutionary and institutional economics have proposed. The dynamic capability of a firm consists of unique organization skills; their replication is limited because they are tacit in nature, not codified, or embodied (Teece and Pisano 1998: 206). They must be built because they cannot be bought. If one crosses the 'ease of replication' (easy—hard) dimension with the dimension of 'intellectual property rights' (loose—tight) (Teece and Pisano 1998: 207), one can distinguish between weak, moderate, and strong 'appropriability regimes' in firms: weak appropriability regimes favor the augmentation of small start-up firms, whereas strong regimes reinforce the dominance of big, established firms.

The second shift—from the concept of knowledge-using firms to the concept of knowledge-creating firms (Nonaka and Takeuchi 1995)—seems to be more radical and relevant than the first one. It theorizes about the difference of tacit and explicit knowledge and the spiral processes of knowledge changes. It emphasizes the relations and interactions that produce and reproduce different kinds of knowledge sets. 'Systemic knowledge assets' develop out of the connection of packaged explicit knowledge (Nonaka *et al.* 2000: 15). There is no doubt that this process of 'combination' belongs to the knowledge style of explicitation. If tacit knowledge is shared by common experience (what is called 'socialization'), then 'experiential knowledge assets' rise. 'Conceptual knowledge assets' are articulated in the process of 'externalization,' and 'routine knowledge sets' mark the opposite process of 'internalization' when explicit knowledge is turned into embedded and embodied routines. One can object that the externalization of tacit knowledge is not so easy, and that not all modes of tacit knowing can be made explicit (Clegg and Ray 2003), but the shift from a concept of the firm as an 'information-processing entity' to a 'dynamic configuration of knowledge creation' (Nonaka *et al.* 2000: 17) is path breaking.

This concept can be expanded to interorganizational relations between firms. Under conditions of fragmental knowledge production, firms must incorporate a diversity of sources of learning to raise their potential rate of innovation. A traditional innovation strategy of firms was to aim to integrate diverse technologies by building large-scale plants and big corporations. This diversification and integration strategy, successful during the inter-war and early post-war period, became exhausted in the 1970s. Up-scaling the corporation was replaced by building corporate international networks. Since the late 1980s, the formation of 'a more complex integrated and interactive network for the generation of new competence' could be observed at 166 firms in the US and Europe (Cantwell and Piscitello 2000: 26). These international networks resemble heterarchies of learning more than hierarchies of power. As knowledge and learning processes are distributed, new bodies of knowledge are generated mainly by the encounter and interaction of two or more previously existing bodies of organizational knowledge. They are assured by inter-firm technological agreements, as demonstrated by an analysis of the automobile robotics sector (Lazaric and Marengo 2000: 56).

There are many ways in which the different modes of tacit knowing and technological learning take place. First is the continual exchange of knowledge within 'technological communities' (Constant 1987; Rosenkopf and Tushman 1998). Technological communities are parts of a professional community, such as the society of electrical engineering or of chemistry. They communicate explicit knowledge by influencing educational programs, and non-explicit knowledge by meetings and committees. A second way can be seen in the informal help between particular 'communities of practice' (Wenger 1998, 2000). Communities of practice are usually bound to a local community and its practices. These 'networks of practice' (Brown and Duguid 2000: 28) are not working together at one place, but they build a virtual guild that shares similar practices and indirect communications. A third, more general, way is the organizational or interorganizational practice of doing something together—coordinated activities without explicit communication. The firm, the organization, or the interorganizational network are made up of diverse communities with different practices and interpretative systems that, in addition to their local practices, share collaborative practices. Innovation is then a systemic process that involves linking the inventive knowledge of diverse communities into something robust and rounded enough to enter the market place (ibid. 26).

One can relate different grades of explicitation and of explorative experience to different kinds of organizations and different types of technology or phases of technology development. Since Max Weber, the bureaucratic organization has been defined by written rules and explicit procedures of rational action, especially by the criterion of formal membership, as a prototype of formal organizations. They have the highest degree of explicitation and the lowest degree of tolerance for informal knowledge and practical experiences. They are well designed for the processing of clear-cut information and homogeneous knowledge domains, such as financial control, or mass production of a technological product, or a service that is stable and not too complex. Big corporations have developed a divisional structure and an internal differentiation between domains of routine and domains of innovation. They still show a high degree of explicitation, but the differentiation and dynamics of the environment enforces lateral communication and informal knowledge processes (see Hage 1999). In low-tech industries, the central focus of knowledge generation is on the organization of practical knowledge (Hirsch-Kreinsen *et al.* 2003); in high-tech industries, other strategies gain importance, like coordinating the diverse sources of knowledge and balancing the relation between explicitation and explorative experience.

Networks of cooperation between firms are created when the share and the diversity of non-explicit knowledge increases in comparison to homogeneous and explicit knowledge. We distinguish three types of networks. Networks of small firms are formed in order to put a more complex technology or service on the global market; they link their particular organizational knowledge, explicit and non-explicit, with other firms which have a complementary capability. An example from the car industry is the supplier firms which join to build a 'virtual organization' (Davidow and Malone 1992). Networks of small firms, like start-ups, and big, established firms develop when small firms need more structured knowledge for the production and distribution of a new product and the big firms want to appropriate the explorative capacity and the tacit knowledge of start-ups: cooperation between small biotech firms and big pharmaceutical corporations are prime examples. Networks of big corporations take the form of a time-limited and product-oriented alliance to share their knowledge bases and to develop a particular new product while limiting the risk and the costs of a radical innovation that requires a high rate of non-explicit knowing: joint ventures in the telecommunication sector for the appliance of the UMTS-technology are an example of this strategy of combining the diverse technological capabilities.

Thus, we can see that the different kinds of non-explicit knowledge grow under the regime of fragmental distribution (see also Rammert 2004). One needs additional practical knowledge to embody heterogeneous knowledge in functioning technology and to combine different products and services to make a complex technological system. More time, money, and personnel are needed to span the boundaries between technological domains and diverse industrial traditions, as well as greater space for the explorative combination of explicit and non-explicit knowledge and for the 'generative dance between organizational knowledge and organizational knowing' (Cook and Brown 1999: 381).

The rising relevance of tacit rules and trust relationships in heterogeneous networks of organization and under conditions of distributed governance

Modern states are based on explicit constitutional rules and administrative laws. For a long time, they were considered the central agency of regulation. Governments developed specialized ministerial administrations which defined the explicit legal frameworks of financial, economic, or science and technology policies. But policy studies demonstrated that a formulated political program or a legal regulatory framework could only gain the intended results if the conditions of its implementation, especially the interests, the knowledge, and experiences of the heterogeneous collective actors in the field, are known and acknowledged (Mayntz 1993). New forms of governance were tested, like corporate governance, that divided governance between state authorities and private associations (see Hollingsworth *et al.* 1994). Under a regime of functional specialization, this kind of oligopoly among the relevant collective actors allowed concerted actions and regulations, because the different stocks of domain knowledge were communicated.

The knowledge condition becomes much more difficult when we move towards a regime of fragmented distribution. The numbers and diversity of actors who participate in the innovation process is growing. The knowledge that is needed to formulate effective and consensual regulations is more radically distributed over many fields of expertise. The knowledge base that was once certain becomes quickly obsolete in high-technology fields. The new knowledge has many locations in a variety of communities and organizations, and cannot be easily transformed into packaged and portable information. Under these fragmented conditions, a coordination mechanism is required that achieves both: it must maintain the diversity of actors and their knowledge perspectives; at the same time, it has to create a culture of trust and cooperation. A kind of distributed governance that refers both to explicit rules that frame the collaboration and to an implicit cultural model or 'hidden curriculum' (Rammert 2002: 180) that constitutes an informal platform of collective learning arises among heterogeneous actors. The new institutional answers to this problem are procedures of mediation that bring together actors with different and even dissenting knowledge perspectives, and interactive networks of innovation that pool the heterogeneous knowledge capabilities and follow a mixed model of formal contracts and tacitly confirmed routines.

This model of mixed kinds of knowing and of distributed governance in a fragmented knowledge regime can be observed at each level. On the micro-level, new situations of 'distributed co-operation' between heterogeneous people (see Hutchins 1996, 1998; Strübing et al. 2004) and situations of 'distributed action' between hybrid agencies, like machines, programs, and people (see Rammert 2003; Rammert and Schulz-Schaeffer 2002) arise, especially in high-tech workplaces. They replace the Fordist model of work organization that is mainly shaped by a sequential and overspecialized functional division of work planning and performance. On this level, governance is assured by an overlapping of the diversity of skills and a low grade of formalization of knowledge, by a balanced mixture of explicit standards and tacit routines, by explicit 'boundary

objects' and other more implicit borderline activities which mediate and translate between the different knowledge cultures (Galison 1997; Star and Griesemer 1989).

On the meso-level of organization, when the knowledge conditions get more fragmented, and when knowledge-changes raise the level of uncertainty, mechanical and divisional forms that rely on a high grade of explicitation and formalization dissolve and change into modes of organizing that are organic, and design, process, and project-oriented (see Hage 1999). When the knowledge acquisition by purchase, license, or internal research is restricted, then both firms and government agencies have to join heterogeneous networks of innovation in order to participate in the collective knowledge creation and diffusion processes.

On the macro level of society and its subsystems, these heterogeneous networks of collaboration and innovation take over the role of distributed governance (Powell 1998b; Callon 1992; Hage and Hollingsworth 2000). They unite scientific, economic, and political actors and their different codes of action and knowing, such as research universities and start-up firms, nonprofit organizations and venture capitalists, law firms and consultants, and so on. As Powell (1998b: 231) says: 'Heterogeneity and interdependence are greater spurs to collective action than homogeneity and discipline.' When collaboration and collective learning across institutional boundaries seem to become the critical factor in the competition, then it is ineffective to formalize all decisions and approvals and to make all kinds of knowledge explicit. In many fields it is sufficient to develop routines of cooperation and informal channels of communication. These forms of collective learning can evolve to formal and hybrid networks where the many independent actors blend into the unity of a collective actor. Or they can develop as 'informal networks' (OECD 2001: 8) that are especially capable of incorporating tacit knowledge into their learning processes. They have less need of formal procedures and explicit treaties, being instead built on informal meetings and relations of mutual trust.

Finally, the considerations and findings of this section demonstrate that explicit knowledge will not lose its relevance or even diminish under the fragmental knowledge regime; nor will all tacit knowledge be made explicit. There is no zero-sum game between the two kinds of knowledge. Both can grow and gain more relevance at the same time. It is a question of relating the two kinds of knowledge that shapes two styles of knowledge regimes.

When a knowledge-based society is confronted with the rise of material complexity, when it has to cope with a growing discontinuity in the course of innovation, and when it has to integrate an increasing diversity of actors and perspectives, there is a strong tendency to raise the level of explicitation in all fields. Additionally, computer technologies and the progress of telecommunications strengthen the tendency to make knowledge more explicit. But, as was demonstrated for the scientific, the industrial, and the regime innovation, neither the fundamental, the financial, nor the temporal limits of explicitation disappear. On the contrary, there is an emerging strong ten-

dency to take care of all kinds of non-explicit knowledge: the unknown, the uncodified, and the uncodifiable tacit knowledge. The rising sphere of the unknown is acknowledged by a higher consciousness of risk, by organizational forms of sharing it, and by methods of preparing for risks by scenario- and simulation-techniques. More and more fields of uncodified knowledge are accepted and organized by routines and routing paths of collective learning. The tacit knowledge that cannot be codified is mobilized in hybrid work situations, in informal networks of practice, and in heterogeneous networks of innovation and learning. All these forms of non-explicit knowledge gain a high relevance under a fragmented knowledge regime. The explorative knowledge style that is characterized by an experimental balance between the tacit and the explicit knowledge becomes a necessary condition for:

- the creation and diffusion of scientific facts and technological artifacts in a growing multidisciplinary landscape;
- pushing technological innovation in times of high uncertainty;

Table 12.3. Styles of knowing

	'Explicitation'	'Exploration'
Features	abstraction formalization universal codification standard packages of information	embodiment association situated, localization modularized patchwork of heterogeneous knowledge
Models	mechanical + universal model of information processing	organic + practical model of knowledge cultivating
Methods	acquisition decision-making purchase and licensing	inquiry waging learning
Emphasis	excessive explanation exploitation of codified knowledge	tacit circulation + integration experimental balance between the implicit and the explicit

- distributed governance of innovation processes when the central authority of the state is fading, and when the certainty, disposability and transferability of knowledge across organizational and institutional boundaries are restricted.

Fragmental diversity and explorative learning in heterogeneous networks of innovation

From the debates about the coming of post-industrial, postmodern, reflexive, modern, and knowledge-based society, we know a lot about the rough features of the society we are living in. But we should develop more precise concepts of the organizing principles and epistemic styles that produce different types of social differentiation and various patterns of institutionalization. Two coexisting types of social differentiation have been distinguished, and two coexisting styles of knowledge regimes identified: the knowledge regime of fragmental distribution neither substitutes for nor follows the regime of functional specialization. One first research task is to analyze the particular distribution of these patterns over the basic institutions of a society: the national systems of innovation, the industrial branches, the technological domains, and the scientific disciplines. The aim of this kind of research is to receive a fine-grained picture of the whole landscape of innovation from the perspective of dominant regimes and styles. For politicians and practitioners of innovation, it may be important to know the answers to these questions: does this kind of innovation flourish better when locally concentrated or dispersed; when temporally sequential and synchronized, or in a simultaneous and instantaneous jazz-like concert?

This chapter has assumed that there is also a close relationship between the institutional and the epistemic dimension of the two knowledge regimes, and develops the concept of knowledge styles beyond the traditional distinction between tacit and explicit knowledge. It has argued that one can distinguish three kinds of knowing; tacit, implicit, and explicit knowing. It has also been argued that knowledge styles are characterized by the emphasis on a certain kind of knowledge, and by the kind of connection they make between the types of knowing. It has discussed in detail two knowledge styles, and loosely related them to the knowledge regimes: the knowledge style of explicitation that favors an excessive strategy of explanation and codification, and the knowledge style of exploration that trusts more in the experimental balance between the explicit and the implicit. It is a second research task to analyze the advantages and disadvantages of both these knowledge styles under different conditions (such as a high diversity of actors and perspectives versus a standard range of plurality, or high uncertainty versus routines of risk management) and in particular situations (such as research and radical innovation versus production and improvement). The aim of this chapter's kind of research is to develop an analysis to help identify different knowledge styles and their critical implications for the innovativeness of organizations and systems of innovation. It will be very important for innovative organizations' future designs and decisions to know more about the right balance between different kinds of knowledge: for example, about the limits and cost of excessive explicitation, about the methods of cultivating the different kinds of non-explicit knowledge, and about the organizational forms that allow the crossing of different kinds of knowledge to establish a productive balance between explicit knowledge and tacit knowing.

Economic analyses of knowledge and innovation usually follow an approach too narrow to grasp the interrelations and the dynamics of the distributed innovation processes. Quantitative analyses present leading factors and develop rough indicators of the systems of innovation. But they, too, miss the qualitative differences of local situations and institutional constellations. There are two ways to overcome the qualitative blindness. A first line of research elaborates on the approaches to the dynamics and journeys of innovation towards a finer-grained approach of comparative innovation biographies. Its aim is to chart the various paths taken by innovations more closely than studies which search for typical phases of technological development. The focus of interest concerns critical events, typical sequences, and hybrid constellations that include actors, agencies, and artifacts. The creation and the co-production of particular paths of scientific and technological innovation (see Garud and Carnoe 2001) are more important than the path dependencies. Closely interrelated with this line of research is a second; this complements the longitudinal view of the biographical analysis with a cross-sectional view of constellations of actors and artifacts, of rules and resources. It aims at a comparative analysis of innovation regimes under different aspects, for example 'technological regimes' (Stankiewics 2000), 'organizational regimes,' 'innovation regimes' (Hage 2003), 'appropriability regimes' (Teece and Pisano 1998), and knowledge regimes. It elaborates the new institutional approaches towards a multi-level analysis (see Hollingsworth 1999) that includes the analysis of communities and constellations on the levels of interaction, organization, and society.

The main interest of both kinds of research is to discover and to discriminate typical patterns of change in the distributed processes of knowledge production. When we are able to spell out the set of rules that constitute particular regimes, then we can answer the questions of which kind of organization, which style of knowing, and which form of intermediary association can be rated as a more adequate institutional arrangement than another. This kind of research goes beyond economic approaches of maximization, or management approaches of searching for a 'champion' or identifying a best practice by benchmarking evaluations. It supposes that there are many good practices, and that quality and effectiveness depend on particular situations and constellations which, for example, allow collective learning for a longer period rather than concentrating on exploitation of knowledge in an actual situation. This kind of approach has highly important implications for innovation management and innovation policy: it frees managers and politicians from following a general pattern, and opens spaces for diversity. Scientific disciplines and transdisciplinary research fields require different strategies: conventional technologies and complex technological configurations do not follow the same logic of development; traditional industrial branches and heterogeneous networks of collaboration need different policies and legal frameworks. Therefore, we have to distinguish between styles of knowledge regimes and develop adequate policies for each one. If the conditions of fragmental distribution are expanding, politics and management have to try to maintain both the creative diversity of actors, opinions, and perspectives as well as the institutionalization of codes, cultural models, and procedures that enable processes of collective learning. And if the need for the explicitation, codification, and computerization of knowledge continues to be strong, politicians and managers would be well advised to pay heed to the explorative style of knowing that brings different kinds of knowledge in balance with one another.

Note

1. The author thanks Jerry Hage, Martin Meister, Terry Shinn, and Ingo Schulz-Schaeffer for critical comments and constructive suggestions.

References

Abernathy, W., and Utterback, J. (1978). 'Patterns of Industrial Innovation.' *Technology Review*: 41–7.

Baudrillard, J. (2004). *Fragments: Conversations with Francois L'Yvonnet*. London: Routledge.

Beck, U. (1992). *Risk Society: Towards a New Modernity*. London: Sage.

—— Giddens, A., and Lash, S. (1994). *Reflexive Modernization*. Cambridge: Polity Press.

Böhme, G., van den Daele, W., and Weingart, P. (1976). 'Finalization in Science.' *Social Science Information*, 15: 307–30.

Brown, J. S., and Duguid, P. (2000). 'Mysteries of the Region: Knowledge Dynamics in Silicon Valley.' In C-M. Lee *et al.* (eds.), *The Silicon Valley Edge*. Stanford, CA: Stanford University Press: 16–39.

Burkhard, H-D., and Rammert, W. (2000). 'Integration kooperationsfähiger Agenten in komplexen Organisationen: Möglichkeiten und Grenzen der Gestaltung hybrider offener Systeme.' Working Paper TUTS-WP-1-2000, Technical University Berlin.

Button, G. H. (1993). *Technology in Working Order: Studies of Work, Interaction, and Technology*. London: Routledge.

Callon, M. (1992). 'The Dynamics of Techno-Economic Networks.' In R. Combs, P. Saviotti, and U. Welsh (eds), *Technological Change and Company Strategies*. London: Academic Press, 72–102.

Cantwell, J., and Piscitello, L. (2000). 'Accumulating Technological Competence: Its Changing Impact on Corporate Diversification and Internationalization.' *Industrial and Corporate Change*, 9/1: 21–51.

Carlson, B., and Stankiewicz, R. (1995). 'On the Nature, Function and Composition of Technological Systems.' In B. Carlson (ed.), *Technological Systems and Economic Performance: The Case of Factory Automation*. Dordrecht: Kluwer Academic Publishers, 21–56.

Chaiklin, S., and Lave, J. (eds) (1998). *Understanding Practice: Perspectives on Activity and Context*. Cambridge: Cambridge University Press.

Clegg, S., and Ray, T. (2003). 'Power, Rules of the Game and the Limits to Knowledge Management: Lessons from Japan and Anglo-Saxon Alarms.' *Prometheus*, 21/1: 23–40.

Collins, H. M. (1974). 'The TEA Set: Tacit Knowledge and Scientific Networks.' *Science Studies*, 4: 165–86.

Constant II, E. W. (1987). 'The Social Locus of Technological Practice: Community, System, or Organization?' In W. E. Bijker, T. P. Hughes, and T. Pinch (eds). *The Social Construction of Technological Systems*. Cambridge, MA: MIT Press, 223–42.

Cook, S., and Brown, J. S. (1999). 'Bridging Epistemologies: The Generative Dance between Organizational Knowledge and Organizational Knowing.' *Organization Studies*, 10/4: 381–400.

Coombs, R., Harvey, M., and Tether, B. S. (2003). 'Analyzing Distributed Processes of Provision and Innovation.' *Industrial and Corporate Change*, 12/6: 1125–55.

Cowan, R., David, P. A., and Foray, D. (2000). 'The Explicit Economics of Knowledge Codification and Tacitness.' *Industrial and Corporate Change*, 9: 211–53.

Davidow, W., and Malone, M. S. (1992). *The Virtual Corporation*. New York: Harper Collins.

Dewey, J., and Bentley, A. (1949). *Knowing and the Known*. Boston: Beacon Press.

Edquist, C. (ed.) (1993). *Systems of Innovation: Technologies, Institutions and Organizations*. London: Pinter.

Engström, Y., and Middleton, D. (eds.) (1996). *Cognition and Communication at Work*. Cambridge: Cambridge University Press.

Etzkowitz, H. (2002). *MIT and the Rise of Entrepreneurial Science*. London: Routledge.

—— and Leydesdorff, L. (eds.) (1997). *Universities and the Global Knowledge Economy: A Triple Helix of University-Industry-Government Relations*. London: Pinter.

Fleck, L. (1935). *Genesis and Development of a Scientific Fact*. Chicago: University of Chicago Press.

Fransman, M. (1998). 'Information, Knowledge, Vision and Theories of the Firm.' In G. Dosi, D. Teece, and J. Chitry (eds.), *Technology, Organization and Competitiveness*. Oxford: Oxford University Press, 147–91.

Freeman, C. (1991). 'Networks of Innovators: A Synthesis of Research Issues.' *Research Policy*, 20/5: 499–514.

Fuchs, S. (1992). *The Professional Quest for Truth: A Social Theory of Science and Knowledge*. Albany: State University of New York Press.

Funtowicz, S., and Ravetz, J. (1993). 'Science for the Post-Normal Age.' *Futures*, September: 739–55.

Galison, P. (1996). 'Computer Simulations and the Trading Zone.' In P. Galison and D. Stump (eds.), *The Disunity of Science: Boundaries, Contexts, and Power.* Stanford, CA: Stanford University Press, 118–57.

—— (1997). *Image and Logic: A Material Culture of Microphysics*. Chicago: University of Chicago Press.

Gardner, A. (1985). *The Mind's New Science*. New York: Basic Books.

Garud, R., and Karnoe, P. (eds) (2001). *Path Dependence and Creation*. Mahwah, NJ: Lawrence.

Gibbons, M., Limoges, C., Nowotny, H., Schwartzman, S., Scott, P., and Trow, P. (1994). *The New Production of Knowledge: The Dynamics of Science and Research in Contemporary Societies*. London: Sage.

Hage, J. (1999). 'Organizational Innovation and Organizational Change.' *Annual Review of Sociology*, 25: 597–622.

—— (2003). 'A Contingency Theory of Innovation Regime and Appropriate Institutional Concept.' Unpublished Paper. International Conference, INNOVERSITY, at the Technical University of Berlin.

—— and Hollingsworth, R. (2000). 'A Strategy for Analysis of Idea Innovation Networks and Illustrations.' *Organization Studies*, 21: 971–1004.

Hakansson, H. (ed.) (1987). *Industrial Technological Development: A Network Approach*. London: Croom Helm.

Hall, P., and Soskice, D. (2001). *Varieties of Capitalism*. Oxford: Oxford University Press.

Hamilton, G. G., and Feenstra, R. C. (1998). 'Varieties of Hierarchies and Markets: An Introduction.' In G. Dosi, D. Teece, and J. Chitry (eds.), *Technology, Organization and Competitiveness*. Oxford: Oxford University Press: 105–45.

Heims, S. (1980). *John von Neumann and Norbert Wiener: From Mathematics to the Technologies of Life and Death*. Cambridge, MA: Harvard University Press.

Heintz, B. (2000). *Die Innenwelt der Mathematik: Zur Kultur und Praxis einer beweisenden Disziplin*. Vienna: Springer.

Hirschhorn, L. (1988). *The Workplace Within*. Cambridge, MA: MIT Press.

Hirsch-Kreinsen, H., Jacobson, D., Laestadius, S., and Smith, H. (2003). 'Low-Tech Industries and the Knowledge Economy: State of the Art and Research Challenges.' Working Paper No. 1. University of Dortmund.

Hollingsworth, R. (1999). 'Doing Institutional Analysis: Implications for the Study of Innovations.' Unpublished paper. Madison: University of Wisconsin.

—— Schmitter, P., and Streeck, W. (eds.) (1994). *Governing Capitalist Economies*. New York: Oxford University Press.

Hughes, T. P. (1987). 'The Evolution of Large Technological Systems.' In W. E. Bijker, T. P. Hughes, and T. J. Pinch (eds.), *The Social Construction of Technological Systems*. Cambridge: MIT Press, 51–81.

Hutchins, E. (1996). *Cognition in the Wild*. Cambridge, MA: MIT Press.

—— (1998). 'Learning to Navigate: Understanding Practice. Perspectives on Activity and Context.' In S. Chaiklin and J. Lave (eds.), *Understanding Practice: Perspectives on Activity and Context*. Cambridge: Cambridge University Press.

Imai, K., and Yamasaki, A. (1992). *Dynamics of the Japanese Industrial System from a Schumpeterian Perspective*. SJC-R Working Paper Series.

Joerges, B., and Shinn, T. (2001). *Instrumentation: Between Science, State, and Industry*. Dordrecht: Kluwer.

Kerr, C. (1963). *The Uses of University*. Cambridge, MA: Cambridge University Press.

Knorr Cetina, K. (1999). *Epistemic Cultures*. Cambridge: Harvard University Press.

Latour, B. (1987). *Science in Action: How to Follow Scientists and Engineers through Society.* Cambridge, MA: Harvard University Press.

Lazaric, N., and Marengo, L. (2000). 'Towards a Characterization of Assets and Knowledge Created in Technological Agreements: Some Evidence from the Automobile Robotics Sector.' *Industrial and Corporate Change,* 9/1: 53–86.

Lawrence, P. R., and Lorsch, J. W. (1967). *Organization and Environment: Managing Differentiation and Integration.* Boston. Division of Research, Harvard Business School.

Luhmann, N. (1992). 'Ökologie des Nicht-Wissens.' In N. Luhmann (ed.), *Beobachtungen der Moderne.* Opladen: Westdeutscher Verlag, 149–220.

Lundvall, B.-A. (1988). 'Innovation as an Interactive Process: From User-Producer Interaction to the National System of Innovation.' In G. Dosi, C. Freeman, R. Nelson, G. Silverberg, and L. Soete (eds.), *Technical Change and Economic Theory.* London: Pinter, 349–69.

Lyotard, J.-F. (1984). *The Postmodern Condition.* Minneapolis: University of Minneapolis Press.

Malsch, T. (ed.) (1998). *Sozionik: Soziologische Ansichten zur künstlichen Sozialität.* Berlin: Sigma.

—— (2001). 'Naming the Unnamable: Socionics or the Sociological Turn of/to Distributed Artificial Intelligence.' *Autonomous Agents and Multi-Agent Systems,* 4: 155–86.

Mayntz, R. (1993). 'Networks, Issues, and Games: Multiorganizational Interactions in the Restructuring of a National Research System.' In F. W. Scharpf (ed.), *Games in Hierarchies and Networks.* Frankfurt am Main: Campus, 189–209.

Meister, M. (2002). 'Grenzzonenaktivitäten: Formen einer Schwachen Handlungsbeteiligung der Artefakte.' In W. Rammert and I. Schulz-Schaeffer (eds.), *Können Maschinen handeln? Soziologische Beiträge zum Verhältnis von Mensch und Technik.* Frankfurt am Main: Campus, 189–222.

Merz, M. (2002). 'Kontrolle—Widerstand—Ermächtigung: Wie Simulationssoftware Physiker konfiguriert.' In W. Rammert and I. Schulz-Schaeffer (eds.), *Können Maschinen handeln?* Frankfurt am Main: Campus, 267–90.

Nelson, R. R. (1959). 'The Simple Economics of Basic Scientific Research.' *Journal of Political Economy,* 67/3: 297–306.

—— (ed.) (1993). *National Innovation Systems: A Comparative Study.* Oxford: Oxford University Press.

—— (1998). 'The Co-Evolution of Technology: Industrial Structure, and Supporting Institutions.' In G. Dosi, D. Teece, and J. Chitry (eds.), *Technology, Organization, and Competitiveness.* Oxford: Oxford University Press, 319–35.

—— and Winter, S. (1982). *An Evolutionary Theory of Economic Change.* Cambridge, MA: Belknap Press for Harvard University Press.

Noble, D. F. (1977). *America By Design: Science, Technology and the Rise of Corporate Capitalism.* New York: Alfred A. Knopf.

Nonaka, I., and Takeuchi, H. (1995). *The Knowledge-Creating Company: How Japanese Companies Create the Dynamics of Innovations.* New York: Oxford University Press.

—— Toyama, R., and Nagata, A. (2000). 'A Firm as a Knowledge-Creating Entity: A New Perspective on the Theory of the Firm.' *Industrial and Corporate Change,* 9: 1–20.

Nowotny, H., Scott, P., and Gibbons, M. (2001). *Re-Thinking Science: Knowledge and the Public in an Age of Uncertainty.* Cambridge: Polity Press.

OECD (2001). *Innovative Networks: Co-operation in National Innovation Systems.* Paris: OECD Office of Publication.

Owen-Smith, J. (2003). 'From Separate Systems to a Hybrid Order: Accumulative Advantage across Public and Private Science at Research One Universities.' *Research Policy,* 32: 1081–104.

Polanyi, M. (1966). *The Tacit Dimension.* Garden City, NY: Doubleday.

—— (1969). *Knowing and Being: Essays by Michael Polanyi,* ed. M. Grene. Chicago: University of Chicago Press.

Powell, W. W. (1990). 'Neither Market, nor Hierarchy: Network Forms of Organization.' *Research in Organization Behavior,* 12: 295–336.

Powell, W. W. (1998a). 'Trust-Based Forms of Governance.' In R. M. Kramer and T. R. Tyler (eds.), *Trust in Organizations.* Thousand Oaks, CA: Sage, 51–67.

Powell, W. W. (1998b). 'Learning from Collaboration: Knowledge and Networks in the Biotechnology and Pharmaceutical Industries.' *California Management Review*, 40/3: 228–40.

—— Koput, K., and Smith-Doerr, L. (1996). 'Interorganizational Collaboration and the Locus of Innovation: Networks of Learning in Biotechnology.' *Administrative Science Quarterly*, 41/1: 116–45.

Prahalad, C. K., and Hamel, G. (1990). 'The Core Competence of the Corporation.' *Harvard Business Review*, May–June: 79–91.

Rammert, W. (1992). 'From Mechanical Engineering to Information Engineering: Phenomenology and the Social Roots of an Emerging Type of Technology.' In M. Dierkes and U. Hoffmann (eds.), *New Technology at the Outset: Social Forces in the Shaping of Technological Innovations*. Frankfurt am Main: Campus, 193–205.

—— (1997). 'Innovation im Netz: Neue Zeiten für technische Innovationen. Heterogen Verteilt und Interaktiv Vernetzt.' *Soziale Welt*, 48/4: 397–416.

—— (2000). 'Ritardando and Accelerando in Reflexive Innovation, or How Networks Synchronise the Tempi of Technological Innovation.' Working Paper TUTS-WP-7-2000, Technical University Berlin.

—— (2002). 'The Cultural Shaping of Technologies and the Politics of Technodiversity.' In K. H. Sörensen and R. Williams (eds.), *Shaping Technology, Guiding Policy: Concepts, Spaces and Tools*. Cheltenham: Edward Elgar Publishing: 173–94.

—— (2003). 'Technik in Aktion: Verteiltes Handeln in Soziotechnischen Konstellationen.' In T. Christaller and J. Wehner (eds.), *Autonome Maschinen*. Wiesbaden: Westdeutscher Verlag, 289–315.

—— (2004). 'The Rising Relevance of Non-Explicit Knowledge under a New Regime of Knowledge Production.' In N. Stehr (ed.), *The Governance of Knowledge*. New Brunswick: Transaction Publishers, 85–102.

—— and Schulz-Schaeffer, I. (eds.) (2002). *Können Maschinen handeln? Soziologische Beiträge zum Verhältnis von Mensch und Technik*. Frankfurt am Main: Campus.

Rheinberger, H.-J. (1997). *Toward a History of Epistemic Things: Synthesizing Proteins in the Test Tube*. Stanford, CA: Stanford University Press.

Rosenkopf, L., and Tushman, M. L. (1998). 'The Coevolution of Community Networks and Technology: Lessons from the Flight and Simulation Industry.' *Industrial and Corporate Change*, 7: 311–46.

Schimank, U., and Stölting, E. (eds.) (2001). *Die Krise der Universitäten*. Leviathan Sonderheft 20. Wiesbaden: Westdeutscher Verlag.

Schmookler, J. (1966). *Invention and Economic Growth*. Cambridge: Cambridge University Press.

Shinn, T. (1997). 'Crossing Boundaries: The Emergence of Research-Technology Communities.' In H. Etzkowitz and L. Leydesdorff (eds.), *Universities and the Global Knowledge Economy*. London: Pinter, 85–96.

—— (2001). 'The Research-Technology Matrix: German Origins, 1860–1900.' In B. Joerges and T. Shinn (eds.), *Instrumentation between Science, State and Industry*. Dordrecht: Kluwer Academic, 69–95.

—— and Joerges, B. (2002). 'The Transverse Science and Technology Culture, Dynamics and Roles of Research-Technology.' *Social Science Information*, 41/2: 207–51.

Simon, H. (1954). 'A Behavioural Theory of Rational Choice.' *Quarterly Journal of Economics*, 69: 99–118.

Smithson, M. (1993). 'Ignorance and Science: Dilemmas, Perspectives and Prospects.' *Knowledge: Creation, Diffusion, Utilization*, 15: 133–56

Stankiewics, R. (2000). 'The Concept of Design Spaces.' In G. Ziman (ed.), *Technological Innovation as an Evolutionary Process*. Cambridge: Cambridge University Press, 234–47.

Star, S. L. (1996). 'Working Together: Symbolic Interactionism, Activity Theory, and Information Systems.' In Y. Engeström and D. Middleton (eds.), *Cognition and Communication at Work*. Cambridge: Cambridge University Press, 296–318.

Star, S. L., and Griesemer, J. R. (1989). 'Institutional Ecology: 'Translations' and Coherence. Amateurs and Professionals in Berkeley's Museum of Vertebrate Zoology, 1907–1939.' *Social Studies of Science*, 19: 387–420.

Stehr, N., and von Storch, H. (1999). *Klima—Wetter—Mensch*. Munich: Beck.

Strübing, J., Schulz-Schaeffer, I., Meister, M., and Gläser, J. (eds.) (2004). *Kooperation im Niemandsland: Neue Perspektiven auf Zusammenarbeit in Wissenschaft und Technik*. Opladen: Leske and Budrich.

Sydow, J. (1992). *Strategische Netzwerke: Evaluation und Organisation*. Wiesbaden: Gabler.

Teece, G., and Pisano, G. (1998). 'The Dynamic Capabilities of Firms: An Introduction.' In G. Dosi, D. Teece, and J. Chitry (eds.), *Technology, Organization, and Competitiveness*. Oxford: Oxford University Press, 193–212.

Teubner, G. (2002). 'Hybrid Laws: Constitutionalizing Private Governance Networks.' In R. Kagan, K. Winston, and M. Krygier (eds.), *Legality and Community*. New York: Rowman and Littlefield.

Traweek, S. (1988). *Beamtimes and Lifetimes: The World of High Energy Physics*. Cambridge, MA: Harvard University Press.

Tushman, M., and Rosenkopf, L. (1992). 'Organizational Determinants of Technological Change: Towards a Sociology of Technological Evolution.' *Research in Organization Behavior*, 14: 311–47.

Van de Ven, A. H. (1988). 'Central Problems in the Management of Innovation.' In M. L. Tushman and W. L. Moore (eds.), *Readings in the Management of Innovation* (2nd eds.). Cambridge: Ballinger Publishing, 261–74.

Vincenti, W. (1990). *What Engineers Know and How Thev Know It*. Baltimore: Johns Hopkins University Press.

Von Hippel, E. (1988). *The Sources of Innovation*. New York: Oxford University Press.

Weick, K. (1976). 'Educational Organizations as Loosely Coupled Systems.' *Administrative Science Quarterly*, 21: 1–19.

Wenger, E. (1998). *Communities of Practice: Learning, Meaning, and Identity*. New York: Cambridge University Press.

Whitley, R. (1984). *The Intellectual and Social Organization of the Sciences*. Oxford: Clarendon.

PART III

KNOWLEDGE DYNAMICS IN CONTEXT

Introduction

Harro van Lente and Susan A. Mohrman

This part focuses on the generation and use of knowledge in and across contexts, especially on how knowledge trajectories are shaped and lead to innovation. It considers the core theme of this book from a different perspective: the creative, indispensable, multifaceted, and multi-level relationships between knowledge, innovation, and institutional change. Understanding their dynamics thus requires a variety of approaches, and so the chapters in this part represent an array of disciplines and levels of analysis. Where Part I focused on firms and industries—traditionally a very important setting for innovation and institutional change—and the second on the scientific system, we now will review and investigate the problem of innovation and institutional change as manifested in the problem of the dynamics of knowledge—its generation, interpretation, and use—within and across organizational, societal, and institutional boundaries.

That we live in a 'knowledge society' seems uncontested, although the meaning and the implications of such a notion often remain obscure. It is clear, though, that knowledge dynamics are key to many studies of innovation, and that attaining a full understanding of them requires going beyond a narrow focus on particular organizations and levels of analysis such as firms, universities, projects, and/or teams. So this part investigates knowledge dynamics in a broader context: it pays attention to the micro-dynamics that have to do with how people, teams, and organizations learn and create knowledge; it also studies the macro-contexts that help shape these dynamics, such as communities of practice, networks of practice, invisible colleges, institutional frameworks, and cultural values and norms. Knowledge trajectories result from simultaneous and interacting dynamics at both the micro and macro levels.

A brief reflection on several key explanatory frameworks for knowledge dynamics is warranted. There are rich theoretical and empirical traditions dealing with the theme of knowledge dynamics, ranging from concerns about R&D management to cultural shifts related to new forms of knowledge practices. Three themes stand out in this work on how knowledge is generated and used: power, sense-making, and future orientation.

The notion of power stands out in post-Marxist and critical theories such as those of Habermas (1971) and Foucault (1977). A main message here is that knowledge production is not innocent: it tends to align with the asymmetric division of power. Habermas analyzes how dominant knowledge paradigms are connected to power; Foucault stresses in his historical and discourse studies how social science is the other side of the coin of social power. Although these views are based on a very different intellectual framework from the mainstream management literature, concepts such as 'path dependency' (Cohen and Levinthal 1990), 'dominant logics' (Nelson and Winter 1982), and 'translation' (Latour 1987) put a lens to the power of embedded capabilities, governance processes, and 'thought-worlds' (Fleck 1935, 1979) that constrain dialogue, radical or discontinuous innovation, and even the production of new knowledge. Management research, for example, has shown that core rigidities and competency traps may evolve when individuals attempt to preserve the status quo and limit new insights (Levitt and March 1988; Leonard-Barton 1992). Because of standards for admission and gatekeeper roles, even communities of practice may become impediments to learning (Wenger 1998).

Knowledge is never to be taken for granted: it always needs a context of beliefs and action in order to be meaningful. This is a second major theme in knowledge's rich tradition. Simon's (1969) notion of bounded rationality is a useful start for such an understanding. It criticizes the idea that actors can or will decide on the basis of complete information, and will process this information in a rational way. Instead, actors (firms, people) do not use optimization procedures in their decision-making, but rely on 'satisficing'. This concept stresses the weight of context in the use of information—and of situated learning. Weick's (1979, 1995) notions of sense-making imply a step further: knowledge cannot exist as such, because it always needs a context in order to have any significance. Sense-making describes the mechanisms by which knowledge is appropriated and made meaningful. At the same time, sense-making recreates the cognitive and social schemes by which people operate.

The chapter's third theme is that knowledge is deeply related to lines of action: it draws from and builds on ongoing practices and it informs future actions. Using the definition provided in the introduction to this book, knowledge can be defined as 'the capacity to reproduce or replicate findings, products, or processes.' Yet, especially in the case of innovation, knowledge refers to future possibilities. Knowledge sheds light on the next steps, on the potential of new directions, and about new questions. The close link between the past, the present, and the action that will create the future is stressed in the notion of path dependency, as well as in practice theory (Bourdieu 1990; Lave 1988), structuration theory (Giddens 1993), and activity theory (Vygotsky 1978; Engström 1999), all of which are grounded in the ongoing practical activities of human agents in particular historical, cultural, and institutional contexts. The members of a community are both constrained by their context and operating to change it—to create new possibilities, knowledge, and innovation. This is consistent with the definition in the introduction to this book: that new knowledge or innovation is a function of an existing stock of knowledge plus collective learning.

The chapters in Part III will shed light on each of these themes that are deeply connected to knowledge dynamics and the resulting trajectories of knowledge. Arguing from quite different theoretical traditions, and focusing on phenomena at quite different levels of analysis, these chapters have in common the belief that knowledge moves in directions that are shaped by the past and that embody the aspirations and actions of actors—directions that are informed by existing knowledge and shaped by scenarios of the future. These scenarios are created by actors and require the unfolding of new knowledge.

Knowledge dynamics is a broad concept: it includes the activities of single research groups, design choices in new product development, the unfolding of new scientific disciplines, and the social learning that takes place when radical innovations are introduced to the market (Jasanoff *et al.* 1995). Discussions of knowledge dynamics, therefore, generally focus on specific contexts that are chosen because of their capacity to influence knowledge trajectories. In this part, five key settings have been selected:

- the design context;
- the context of scientific disciplines;
- the public context, where new technologies are contested and put on trial;
- the (science) policy context;
- the context of higher education.

These contexts are mutually interacting and, as mentioned in the introductory chapter, are believed by some theorists to co-evolve. Yet the actions and influence of each on knowledge dynamics must be understood in order to fully understand knowledge trajectories and their impact on innovation.

First, Hatchuel, Lemasson, and Weil provide an analysis of the knowledge dynamics in design processes. They argue that design processes are not the expression of some pre-given knowledge in material and organizational form, but are to be seen as knowledge projects themselves. The design of artifacts and of their intended usage is at the same time the design of implied knowledge. Design is seen as the dual generation of concepts and knowledge, and design strategies regulate knowledge creation, dynamic work, and organizational designs. The greater the distance between concepts and knowledge, the more likely that radical innovation and new lineages of product will result from design activities. Viewed in this way, innovativeness is a design capability.

Shinn's chapter investigates the knowledge dynamics in the context of emerging disciplines. Shinn locates his contribution in a perennial debate about the role of differentiation or integration in the dynamics of new disciplines. While centrifugal forces of modern specialization and task divisions affect the science business, Shinn also argues emphatically that, without countervailing integration, the science project would soon come to a halt. There is a special integrating role for what he calls 'generic technologies'—instruments developed to meet the requirements of particular lines of research that could not otherwise be pursued, and which come to be used in multiple research fields. By making possible research activities not otherwise within reach, these technologies influence the trajectory of knowledge in multiple fields. And in what becomes a mutually reinforcing knowledge dynamic, the experience of using the generic technology to address problems in multiple disciplines impacts the trajectory of knowledge about it. Shinn argues that the dynamics of the emergence and evolution of these generic technologies are closely related to innovation.

Jolivet and Maurice examine the mobilization of public opinion for or against specific innovations. The example they investigate is the striking difference between the US and Europe in the acceptance of genetically modified food. It is well known that all kinds of concerns in Europe rigorously restricted the production and use of GMO food such as corn and soy, whereas developments in the US proceeded quickly and were not publicly salient. Jolivet and Maurice's analytical framework is societal, which combines a cognitive approach to learning with a comparative method of institutions. Important institutional differences are to be found in the strength of a trusted public authority such as the FDA and the historical background of agriculture.

Van Lente's chapter, examining the issue of rhetoric and the mobilization of scientific opinion for or against certain areas of research, looks at the strategic turn in science policy and the role that rhetoric plays in shaping science strategy. In his analysis, the perceived potential of scientific disciplines is what increasingly leads knowledge dynamics. Anticipation has become a striking characteristic of the business of science: a scientific claim is a claim about the future importance of a finding, to be tested by scientific contesters; it is a claim, so to speak, about future truth. This anticipatory nature of science is reinforced by the policy need to choose between areas of investigation and the dominant tendency to base decisions on expected future economic gains of these areas. This, on the one hand, makes policymakers susceptible to claims of future usefulness of a research area, and, on the other, enforces the mobilization strategies of scientists to come up with claims about future benefits.

In the Part's final chapter, Finegold continues the comparative focus in discussing knowledge dynamics in the context of societal education systems. He analyzes programs in Europe, the US, and East Asia, and points out the mutual influence of innovation and educational institutions. He argues that patterns of education yield human capital with limitations, propensities, and capabilities for different kinds of innovation. Also, as new products and processes are created, educational programs to teach how to operate and repair these innovations develop, especially when the volume and technical complexity of the products is such that it requires extended training. For example, training in automobile and electronics repair and in computer programming are now important technical programs found in many community colleges in the US; in the healthcare sector, many new kinds of para-professional degree programs that are closely associated with new occupations have been created. In this instance, we observe the feedback process of innovation: the knowledge base is organized into occupations created and sustained in part through innovation in education and training programs. One of the major insights in the Finegold contribution is the importance of the population level of organizations and its characteristics. In particular, he investigates the deliberate attempt to create a whole new community of learning built around the biotech industry in Singapore. These programs involve the teaching of new disciplines in higher education, but they also exist at other levels of the educational system, and include the establishment of research parks and programs to attract the human capital required to seed the community. As he demonstrates, the government in Singapore has created a learning environment at the collective-population level of organizations.

Table III.1 illustrates how power, sense-making, and orientation towards the future are manifested in the perspectives on knowledge dynamics in these five chapters. The chapters do not set out to address these themes explicitly; they are implicit in each chapter's treatment of assumptions about knowledge. All frameworks imply contention that influences the trajectories along which knowledge will unfold—at stake is whether the knowledge dynamics will yield knowledge advances that are radical versus incremental. In all these frameworks, collective sense must be made of the usefulness and desirability of alternatives in order both to guide knowledge-creating activities and to attach meaning to the knowledge. In all these chapters, an innovation's interpretation is seen to be made in the context of aspirations for the future: innovations are rejected or advanced based on how the future is envisioned and on beliefs about how the innovation will affect that future.

Table III.1. Three major knowledge themes as they appear in Part III

Context	Power	Sense-making issues	Future/next steps
Hatchuel: design	Concepts express intended influence and new directions.	Distance between concept and knowledge drives collective processes of new knowledge production.	A design projects the future use.
Shinn: disciplines	Differentiation vs integration: knowledge advances in the struggle between discipline-driven and problem-driven search activities.	Researchers assess the usefulness of the application of generic technologies within and across boundaries.	Generic technologies have the potential to enable new fields of investigation and to integrate across disciplines.
van Lente: science policy	Discourse and policy are dominated by contending promises.	Promises create urgency to act and guidance to research.	Science policy is part of interlocked promises that propel the directions of scientific inquiry.
Jolivet and Maurice: public opinion	State influenced by power of public opinion and mobilization of interest groups.	Stakeholders make sense of the nature and desirability of new technological options.	Accepting or rejecting new technologies is based on constructed scenarios about their impact.
Finegold: education systems	The state, the public, and private stakeholders in education influence the form and purposes of education.	Policy-makers and stakeholders make sense of purpose of education—and of who should be educated and what should be taught.	Education policy proceeds based on a vision of desired societal capabilities, including innovation.

All chapters also point to the special role of non-market mechanisms of coordination that influence innovation. These include conceptual break-throughs that enable new ways of organizing knowledge and work, prob-lems that can only be solved through the invention of new tools, rhetoric, and civil action that aligns divergent stakeholders and institutions, and actions by the state that realign the knowledge dynamics in a society. This part addresses the dynamics of institutional change and society's innova-tive capabilities because they are integrally intertwined with knowledge dynamics. These chapters lead naturally to the last part of the book, which directly tackles these critical societal-level phenomena.

References

Bogner, W. C., and Thomas, H. (1994). 'Core Competence and Competitive Advantage: A Model and Illustrative Evidence from the Pharmaceutical Industry.' In G. Gamel and A. Heene (eds.), *Competence-based Competition*. Chichester: Wiley.

Bourdieu, P. (1990). *The Logic of Practice*. Stanford, CA: Stanford University Press.

Cohen, W. M., and Levinthal, D. A. (1990). 'Absorptive Capacity: A New Perspective on Learning and Innovation.' *Administrative Science Quarterly*, 35: 128–52.

Engström, Y. (1999). 'Activity Theory and Individual and Social Transformation.' In Y. Engeström, R. Miettinen, and R. L. Punamäki (eds.), *Perspectives on Activity Theory*. Cambridge: Cambridge University Press.

Fleck, L. (1935/1979). *Genesis and Development of a Scientific Fact*, ed. T. J. Trenn and R. K. Merton. Chicago: University of Chicago Press (first published in German, 1935).

Foucault, M. (1977). *Discipline and Punish*. New York: Pantheon.

Giddens, A. (1993). *New Rules of Sociological Method*. Stanford, CA: Stanford University Press.

Habermas, J. (1971). *Knowledge and Human Interests*. Boston: Beacon Press.

Jasanoff, S., Markle, G. E., *et al*. (eds.) (1995). *Handbook of Science and Technology Studies*. London: Sage.

Latour, B. (1987). *Science in Action: How to Follow Scientists and Engineers through Society*. Cambridge, MA: Harvard University Press.

Lave, J. (1988). *Cognition in Practice: Mind, Mathematics and Culture in Everyday Life*. Cambridge: Cambridge University Press.

Leonard-Barton, D. (1992). 'Core Capabilities and Core Rigidities: A Paradox in Managing New Product Development.' *Strategic Management Journal*, 13 (Summer Special Issue): 111–25.

Levitt, B., and March J. (1988). 'Organizational Learning.' *Annual Review of Sociology*, 14: 314–40. Palo Alto, CA: Annual Reviews.

Nelson, R., and Winter, S. (1982). *An Evolutionary Theory of Economic Change*. Cambridge, MA: The Bellhop Press of Harvard University Press.

Simon, H. A. (1969). *The Sciences of the Artificial*. Cambridge, MA: MIT Press.

Vygotsky, L. S. (1978). *Mind in Society: The Development of Higher Psychological Processes*. Cambridge, MA: Harvard University Press.

Weick, K. E. (1979). *The Social Psychology of Organizing*. New York: Random House.

—— (1995). *Sensemaking in Organizations*. Thousand Oaks, CA: Sage.

Wenger, E. (1998). *Communities of Practice: Learning, Meaning, and Identity*. Cambridge: Cambridge University Press.

13 Building Innovation Capabilities: The Development of Design-Oriented Organizations

Armand Hatchuel, Pascal Lemasson, and Benoit Weil

Summary

Much has been written about innovation. Yet we still do not have a good understanding of how to develop innovation capabilities in organizations. In this chapter, we claim that this weakness of contemporary organizational research on innovation comes from the failure to recognize *design theory* as a prerequisite to such an understanding. We base this claim on two detailed qualitative studies of companies that showed the rare capability to innovate repeatedly over several decades. By using existing knowledge to create new products and existing products to create new knowledge, these companies constantly stimulated innovative design. A recent definition of design theory—that design is the dual generation of concepts (innovations) and knowledge (competencies)—captures the logic of innovation; design strategies are regulations central to both knowledge-creation and for a dynamic division of work (organizations, networks, partnerships). The definition also supports a new contingency approach to organizations, which fits with the cases presented, and predicts the emergence of 'design-oriented organizations' (DO2) that cannot be reduced to a functional matrix or network structure, and can be identified only through their design-based model of growth.

Introduction: a design-based approach to innovation

In this chapter, we develop our hypothesis that many of the weaknesses of contemporary organization's theories of innovation come from the failure to recognize design theory as a way to understand and operationalize innovation capability. (It is also true that the central role of design activities in any innovation process is not well recognized.) Classic organizational theory was mainly developed in companies when and where the scope and influence of innovation activities were less important than in contemporary companies. The classic organizational determinants of innovation which have been studied in the literature are specialization, functional differentiation, professionalism, formalization, centralization, managerial attitude towards change, managerial tenure, and technical knowledge resources (Kline and Rosenberg 1985; Damanpour 1991). All these variables belong to economics, or to the structural theory of bureaucracy. Their limited relevance to the study of companies where innovation and R&D are central forces has already been suggested (Allen 1977; Hage 1999; Buderi 2000). More recently, the development of project-based organizations (Lundin and Midler 1999; Lenfle and Midler 2000) or network leadership (Gawer and Cusumano 2002) has been reported as an important management trend aiming to foster the development of new products. The terminology naming this evolutionary development is not from the language of organizational structure: the word 'project' belongs to the vocabulary of architecture. During the life of a project, several organizational structures which are shaped by design choices and strategies (for instance, competing design teams) can be implemented or generated. Thus the development of project-based organizations signals the need to enrich the organizational language with concepts from design theory or practice. The following are first claims in support of the chapter's main hypothesis:

- empirical research shows that design activities are the core regulating process of innovative companies and their major innovation capability. Hence, we can expect that highly innovative companies will have developed specific design strategies and competencies.
- design can be fully captured neither by a 'structure-conduct performance' paradigm, nor within an 'organizational learning' perspective. To describe design requires a specific 'model of thought' (Simon 1979). We present an approach where concepts and knowledge have to be both distinguished and connected (Hatchuel and Weil 2003). This model allows us to interpret the power of the design strategies observed in innovative companies.
- design theory also offers a new contingency approach to organizations based on the specific descriptors of design work: it signals the emergence of design-oriented organizations.

Overview of the chapter

The first section of this chapter briefly surveys the innovation literature's main findings, in which we underline the multiple implicit uses of design metaphors or concepts. The second section discusses two case studies of companies that have been repeatedly innovative over at least three decades.[1] Both cases highlight the design dynamics underlying the companies' powerful capabilities to innovate. In the third section, we introduce elements of design theory: we define design as a collective activity aiming to expand concepts and knowledge. We use this framework to revisit our empirical material and the theory of classic organizational structures. Finally, we discuss what we call DO2, which seem highly adapted to contemporary competition through innovation.

Innovation as a design capability

From organic structures to creative behavior

Organizational research on innovation was initially developed through the structure-conduct-performance paradigm. Authors tried to find the type of structure or structural variables that favor and support innovation (Kline and Rosenberg 1985). The organic structure was early suggested as better adapted to innovation by Burns and Stalker (1968), and this finding has been repeatedly confirmed in the literature (Damanpour 1991). However, the concept of organic structure presents theoretical limitations. Organic structures are described as complex organizations with lower levels of formalization and centralization. But what does 'organic' mean, and how can we use it to describe a 'structure'? Is there a difference between an organization that continuously adapts its structures and an 'organic' one? This concept is ambiguous; it is more rigorous for research purposes to speak of an 'organic behavior' that allows organizations to generate evolving structures. Even assuming that it is well established that organic behavior is robustly correlated to innovation, one cannot say that all forms of organic behavior generate innovation. Therefore, we need to be more specific: the proposition is true only for the type of organic behavior that produces new knowledge and new ideas: that is, creative behavior. Therefore, we can reformulate the research question and ask: what are the determinants of creative behavior within organizations? This is, in our view, a fruitful way to depart from the traditional inquiry of looking for the best structures.

Creative behavior can be inhibited by structural rigidity and high formalization, as has been constantly confirmed in the literature. But only the negative can be proved: structures can limit the capacity to innovate, but they cannot create it. Research needs a complementary hypothesis. If creative behavior cannot be obtained just by weakening structural rules, we claim that it has to be a design capability—that is, a set of design strategies, design rules, and design cultures—that form the competence to innovate. To support this claim, we briefly survey the implicit use of design metaphors or principles in the innovation literature.

Design metaphors in the literature

Much of the innovation literature emphasizes learning processes, knowledge management, absorptive capacities, and networks. It is beyond the scope of this chapter to survey all these trends. However, all these findings underline characteristic features of design processes or design metaphors. It is widely recognized that innovation needs dynamic networks of participants (clients can also be useful participants) who contribute to the transformation of the initial ideas and make them more concrete and viable (Van de Ven et al. 1999). Innovation is also associated with complex learning in uncertain contexts (Wheelwright and Clark 1992; Jonash and Sommerlatte 1999; MacCormack and Verganti 2000), going from tacit to explicit and specific knowledge (Nonaka and Takeuchi 1995; Von Krogh et al. 2000) and requiring strong experimentalism (Leonard-Barton 1995). All these observations are common features of any collective design process, be it artistic, archi-

tectural, or involving engineering work. And when it is deemed that innovation has to be oriented by some 'mappings,' 'guiding patterns,' or 'framings' (Van de Ven et al. 1999), such terms are metaphors of a design strategy, a design rule, or brief (a classic practice in industrial design).

Design rules and design cultures as innovation capabilities

Other authors directly link design rules and design cultures to creative behavior. In a special issue of *Organization Science* edited by Karl Weick, jazz improvisation was used as a metaphor for organized autonomy and creativity (Barrett 1998). Jazz improvisation is an individual design work (improvisation), yet embedded in a large set of design rules (concerning theme, harmony, rhythm) and a structured organizational process (the band, the rehearsals, the instruments). All these elements allow the jazz soloist to be innovative within pre-established limits. Jazz players can freely improvise provided that they do not attempt to transform a piece of jazz into a piece of classical music. Hence, the design rules and culture of jazz are a good example of an innovation capability.

Revisiting complex work division

Innovation has been related to the concept of 'complex division of labor' (Hage 1999). For Hage, complex division of labor is central to innovation processes because it 'refers to the intellectual-or problem-solving, or learning capacities of the organization, to say nothing about the creative capacities' (Hage 1999: 605). Our line of thought is close to this perspective. Yet Hage's formulation keeps innovation research within a structuralist paradigm. Instead, we consider that complex division of labor is the consequence of design capabilities. Even within a seemingly stable design-based structure (for example, a unit of mechanics and a unit of acoustics in an R&D department), an innovative process will generate new issues, new tasks, and new problems which will be allocated to the units according to design rules and trade-offs. This also means that design rules will change the boundaries, competencies, and behavior of these units, even if they broadly keep their technical orientation. In short, the division of design work is not the same concept as the division of routinized—or pre-designed—work.

Let us summarize the state of the art of research about innovation:

- innovation processes cannot be defined by a set of structures, but by evolving design actions that determine a dynamic sequence of processes, knowledge productions, and organizational relations;
- the innovation process is always described with metaphors that belong to the design tradition: architecture, mapping, framing, patterns. Yet, in spite of these recurrent references to design, no link between organization theory and design theory has been considered.

This gap explains, in our view, the limited influence of the innovation and learning literature on management practice. Left with no clear operational principles, no clear meaning of what is good mapping, framing, or networking, many companies have equated the development of innovation capabilities with the implementation of project management (Lundin and Midler 1999), with improved R&D or with knowledge management. These proposals seem easier to implement than concepts like the 'innovation journey' (Van de Ven et al. 1999), the 'absorptive capacity' (Cohen and Levinthal 1990), or 'technology integration' (Iansiti and West 1997).

In this chapter we establish that introducing design theory solves (or at least clarifies) the old organizational enigmas concerning innovation and, by combining design strategies and organizational principles, offers accurate managerial principles for developing innovation capability. Yet such claims have to be grounded on appropriate empirical observations. Thus, in the next section we present insights from two companies selected because they have developed innovation capabilities over several decades, in a persistent, successful, and sustained way.

The design logic of innovative companies: two case studies

Research methodology

Our research program (Hatchuel et al. 2001; Chapel 1997) followed these principles:

- we targeted companies that showed a clear capability to innovate repeatedly over long periods of time;
- we did not focus our research on isolated innovations, nor on quantitative rates of innovation, because these classic forms of data are not appropriate to identify capabilities to innovate or how these capabilities have been developed by the company;
- we analyzed the history of the products, the history of their skills and know-how, and the evolution of the organizational behavior of the company;
- to capture the long processes of generation that produced innovative design capabilities and specific organizational behavior, we combined 'deep collaborative research' (Hatchuel 2001) over three or four years with historical research.

What are our main findings? These companies struggled to design, simultaneously, lineages (a notion that will be discussed below) of products and of competencies. This produced an interesting growth model supported by design strategies: incremental product innovations were used to develop a radically new competency, or radically new products were designed using many well-established competencies (Chapel 1997; Hatchuel and Lemasson 1999). Hence, specific design strategies were the operational support and implementation of the innovation capability. These design strategies determined the pacing of innovations (Eisenhardt and Brown 1998) and they nurtured the growth and competitiveness of the company.

Behind a seemingly evolutionary process (Burgelman and Rosenbloom 1989), we could identify visible and repeated strategies. In this chapter, we illustrate these findings with two contrasting cases: Tefal, a fifty-year-old company devoted to kitchen and home equipment, and Sekurit Saint-Gobain, a century-old company specializing in the making of glass for the car industry.

Case Study I: Tefal 1974–1997: Innovative design strategies as the managerial core of a company

Tefal is a company belonging to the French group, SEB. For at least two decades, this company showed very successful and innovative growth (Chapel 1997). During the 1960s, it became famous for its Teflon-coated pans. Thereafter, it showed a sustained ability to develop innovative products in the very competitive sector of domestic goods for cooking and the home. While many of its competitors tried to win market shares by developing production units in low-wage countries, Tefal chose to innovate permanently and to create new businesses with high-value-added products allowing high profit rates. For at least one decade, Tefal was one of the most profitable companies in its sector. This strategy of 'repeated innovations' (Chapel 1997) emerged at an early stage in Tefal's history. The unlimited opportunities created by Teflon coating was one of the impetuses that led the managers of Tefal to organize an increasingly collective process of innovation that was quite different from classic research departments or engineering design units. It is not possible for us to give the entire history of Tefal, but we can summarize the distinctive features of its innovation process, which combined design strategies and organizing principles over a period of two decades:

(a) *A high-level innovation committee.* Regularly held and headed by Tefal's CEO, this committee was charged with launching new ideas and concepts and monitoring their maturation process. It was composed of all the functional departments and of all the 'innovation teams' (see below). One of its main roles was to identify, through new products, what were called 'innovation fields': that is, a set of concepts that designate a new area of product development and/or competency development. Hence, the core design strategies of the company were both centralized and widely discussed.

(b) *Innovation teams.* An innovation team was composed of a product engineer and a marketing specialist; together they studied the new innovation field and concepts launched by the committee. This team of two had to integrate the new concepts into suggestions for new products (or a new family of products) by activating all the functional units of the company and/or external competencies. This team cannot be simply defined as 'cross-functional', because the two members behaved differently from the way a classic product engineer or a marketing specialist would behave alone. The team had both to define concepts and to organize their exploration. Nor does this description fit the classic distinction between 'lightweight' and 'heavyweight' project leaders (Clark and Wheelwright 1992). Their influence and mobilizing capacity was not pre-defined, but appeared largely dependent on the content and acceptance of the developments they proposed.

(c) *A culture of collective design through 'prototyping' discussion.* The role of the innovation committee in the treatment and evolution of new concepts was essential. A concept that seemed not very promising could suddenly, after a meeting, become attractive and receive legitimacy and priority from all the departments. The 'innovation groups' used to go as fast as possible in transforming a concept into mock-ups or prototypes that could be discussed, criticized, and improved by the committee. The political arena of the committee created a logic of 'rapid experimentation' (Leonard-Barton 1995) which aimed not only to validate technical details but also to stimulate the strategic discussions of the concepts and to mobilize the multiple forms of expertise available in the committee. In some cases, the committee could behave as an early design team.

Product lineages and design strategy: innovating within self-imposed dominant designs.

The logic of 'repeated innovations' needs both stability and change. New designs appeared interesting if they created important learning opportunities with reduced economic risk. In the literature, a 'dominant design' usually means the design standard selected by the market. But from another point of view, a dominant design can be a voluntary design strategy, i.e. selected combinations of self-imposed design choices and their related competencies; a strategy that offers long-term and large-scale potential for product developments. In Tefal's case, identifying such design strategies was a well-established process that had been learned by the managers during the early years of the company. The first development of Teflon-coated pans was reinterpreted neither as a technology-push nor as a market-pull product. It was perceived as a 'design strategy,' generating a potentially wide array of cooking instruments of the same lineage: that is, several generations of products linked to the same innovation field that could support new values as well as changes in cooking habits and gender status in contemporary societies. A product lineage should not be confused with a product line : the latter describes an existing variety of products having common attributes. The former is a design strategy that potentially allows the development of several generations of product lines; these are not built on a group of attributes but on a group of competencies. Recent Teflon-coated pans use completely different parts and materials from the old ones, yet they belong to the same lineage. Following this design strategy, hundreds of

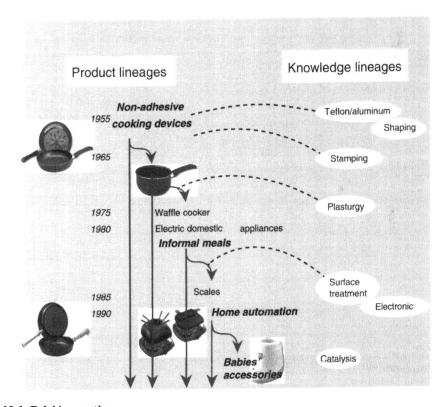

Fig. 13.1. Tefal innovations

products were developed. Moreover, consciously formulating voluntary design strategies helped to identify deviant ones that could offer alternatives and open the way to new product lines. In addition to the Teflon-coating, at the beginning of the 1980s, an electric heating system and a specific plastic structure were being developed for a waffle cooker. This deviant strategy was explored step by step and led to the opening up of a new lineage and product division as successful as the old lineage of Teflon-coated pans (see Figure 13.1 for a schematic representation of the product lineages).

Knowledge exchange and reuse across teams and product lineages

The flow of newly emerging products was also a vehicle for exploring new knowledge and new competencies. Not all concepts ended up as products, but several skills acquired for a mature concept were put to use in another context. For instance, the experience gained in a less successful electronic device became an important input into a very innovative improvement in home weighing-scales. A thorough longitudinal study of the Tefal products shows that such transfers have been systematically explored (Chapel 1997). One of the striking effects of this process is that the old Teflon-coated aluminum pan is still a successful lineage precisely because it benefited from technical competencies acquired in new series of products. Paradoxically, the capability to spin off knowledge is favored by the design logic of product lineages. This combination of design principles and organization principles has been collectively learned over time. A standard organizational study of Tefal would not have elucidated this capability: the names of the departments are classic, the product committee exists in many companies; the specific behavior of the innovation teams is not easy to identify by interviews. From an outsider's point of view, the only strong sign of a specific innovation capability was the amazing development of hundreds of innovative products during decades.

Case Study II: Saint-Gobain Sekurit: Rebuilding R&D as an innovation capability

This second case (Lemasson 2001) differs from the first one in many respects, yet it also confirms our main hypotheses. The history of Tefal showed an expansion of the organization around its innovation process. Saint-Gobain Sekurit went through a sequence of R&D models that corresponded to different regimes of innovation and therefore to different design strategies. During the last decade, it shifted from a classic R&D organization to an innovation regime similar to Tefal's, yet on a wider scale and with more complex products.

1965–1995: Innovating within a dominant design

Saint-Gobain Sekurit is the European leader in car glazing. Until the mid-1990s, car glazing followed a typical dominant design: the 3D shape dictated the main performance criteria (the more curved, the better). Other standard specifications, such as durability, strength, optical quality, were carried over from product to product without modifications, and this corresponded to the usual logic of the relationship between the car designer and the glazing unit designer. This design strategy was fully consistent with a classic and efficient R&D structure: a powerful research department developed advanced knowledge, while a development department warranted that each product met its specifications. This organization limited and shaped the innovation capability in two ways: any new skill had to be consistent with the complex and sophisticated knowledge specific to the glass used in the car industry; and any new specification had to be consistent with all the car constraints and with the current dominant design.

1995–2000: Getting into intensive innovation

After some years of learning, this conservative strategy was changed with substantial results. Saint-Gobain Sekurit is today putting a lot of new products on the market and, in spite of being in a slow-growing business, has been hitting an impressive 8 per cent growth in annual turnover for the last two years. All the products have new functions largely different from the old dominant design, and they require new skills. The overall logic of the design activities has changed dramatically. The R&D structure was transformed through the birth of an 'innovation laboratory' that included the research laboratories. This innovation laboratory had two main missions: to submit new questions to the research laboratories (whenever they can be solved by scientific procedures) and to provide the development departments with innovative types of products (such as athermic windshields) with reduced risk. Thus the innovation laboratory had both to trigger research and to work with the customers to establish new specifications and new skills validated by prototypes (such as levels of energy transmission through the windshield, sputtered nanometric transparent metal layers). This innovation laboratory had an unusual agenda compared to other R&D models (Roussel *et al.* 1991; Myers and Rosenbloom 1996): first on the list was the design of prototypes and new products that were too innovative to be treated directly by the development department; second was the exploration of generic functions in cars (such as thermal comfort or communication) that could lead to new glass products; third was the management of meetings and working groups with design experts from the car makers. Such a new capability emerged in an interesting way.

The emergence of an innovative design strategy

Detailed investigation showed no organizational revolution, nor formulation of a new strategic vision. It had been a step-by-step learning process, led by new product concepts and new customer requirements, that propelled new design strategies (see Figure 13.2 and Table 13.1 for a brief overview of the evolution).

An important event occurred when the research laboratory was suddenly asked to change from its customary study of scientific phenomena to take charge of an emergency innovative project concerning the design of athermic windshields. This first move was successful and was followed rapidly by others. Each of these projects brought the study of a new function and new specifications to car glazing. Though not classic research work, these projects needed the skills (know-how, instruments) of research people. The new style of work was well received by the researchers who perceived that such innovative projects enacted a new logic of competition through innovation.

With the steady introduction of innovative projects, the same thing happened with newly

Table 13.1. ZAF Organization at Saint-Gobain Sekurit 1995–2001

1995 Research lab structure	**Departments**: organic chemistry, inorganic chemistry, physics, optics, technology Work organization: several small research 'projects' for each researcher.
1998 Project organization	**Matrix structure**: same departments + a small number of NPD projects Project names: athermic windshield for Renault, multiconnection for BMW, defrosting coating for Audi.
2000 Lineage organization	**Emergence of 'competence cells'**: connection cell, athermic cell (in charge of monitoring several projects and creating tools and knowledge for future NPD projects).
2001 Innovation and design-oriented organization	**Emergence of 'exploratory projects'**: low cost, fast exploration of an innovation area (fictive example: making use of active glass surface for communicating with other car drivers).

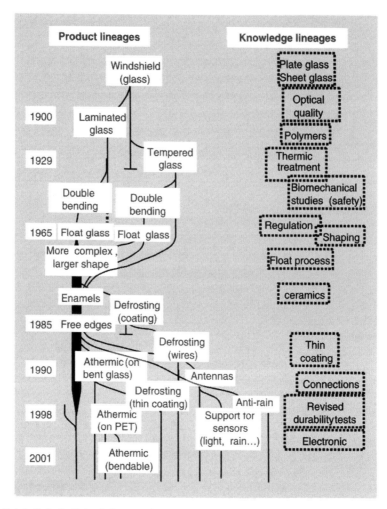

Fig. 13.2. Saint-Gobain Sekurit Innovations

required and acquired skills. All this was interpreted as a different strategic issue: the old commodity of car glazing was becoming a multifunctional product with a large number of new possible lineages. Successful projects also reused newly acquired knowledge but not necessarily in the parent project. Gradually, the laboratory was organized according to functional lineages of projects: groups of projects requiring the same design logic and a common core of skills. 'Lineage leaders' were appointed to take charge of the growth of one function and/or one technology through series of projects, be it product projects or compe-

tency projects. For instance, it was decided to design an athermic line involving several interdependent projects and technologies for athermic windshields, calling for research studies in other laboratories. This regularly provided the development department with new reliable product families to be offered to the car maker.

This was not the last step. With multifunctional glazing, growth would come from a strategic development of new sources of value. This meant defining a new concept of car glazing. A '3D-shape in glass' was the concept behind the old dominant design; 'a car-glazing with a common mix of functions' corresponded to the concept of the lineage organization. Now, a new phase emerged when designers began to think of car glazing as an 'isolating-communicating membrane.' Because it revealed new, unexplored design areas, this new concept triggered innovation. New, small teams were launched to explore the rapid filling in of these 'missing' aspects with low-cost and efficient prototypes and 'exploratory partnerships' (Segrestin 2003) with car makers. Finally, this case shows how adopting new design dynamics generates new innovative behavior.

From cases studies to theoretical issues: a critical view of R&D

In spite of their contrasting contexts, these two case studies evoke the same theoretical remarks. Let us stress that, in our research rationale, the goal of the case studies is not to validate any universal link between variables. We aim to identify, both empirically and theoretically, a new model of organizational behavior and collective thinking, a model that could have been predicted theoretically if we had had the good theory.

Design strategy as a guide for dynamic organizing

The Tefal model was created from the unexpected bringing together of a very old and common tool, the pan, with one of the most modern forms of material: Teflon polymer. The design of a pan was ageless and straightforward, but the properties given by Teflon coating to the pan opened a new set of values and reopened the business issue. A new universe of goods had opened, provided that the organization supported its exploration. In this unknown world, design strategies became the guidelines for organizational changes. Design strategies shaped the organic behavior of

the first teams and the generation of new competencies. These competencies, once established, stabilized profitable products, skills, and divisions. Then new design strategies triggered the whole process again, while benefiting from the past knowledge produced, and so forth. This theoretical scheme requires several conditions to become reality: obviously, it needs customers that are able and willing to pay for the continuous innovation process; it also requires that some rigid dominant design is not maintained by a leading coalition of customers and competitors.

Similar innovative behavior within different organizations

Comparing these two cases, we notice that only recently did the old and big company, Saint-Gobain Sekurit, enter a competitive environment similar to Tefal's. When this happened, the design of a windscreen became not a product issue but a strategic issue; likewise the design of pans became strategic when it incorporated the Teflon properties. Therefore, in recent years, the organizational logic of the two companies became closer, not in their structure (there is no pure research department at Tefal) but in their fundamental behavior and design strategy. In both cases, innovation is now the result of a systematic, intentional, repeated, paced, and conceptually shaped design strategy. This strategy triggered the core metabolism of the firm because it gave rules and meaning to sustained innovative behavior.

Design strategies and the shaping of collective behavior

Tefal and Sekurit may be considered examples of a new model of organization, but we still cannot describe such a model in terms of structural principles or stable processes. And the idea of the 'innovation laboratory' lacks any value if we do not specify the type of behavior that is generated in this laboratory. So what is missing in classic organization theory that explains this model? The basic language of organization theory has been shaped by a production paradigm that comes partly from the economic tradition: organizations are described by sets of tasks that have to be performed by agents. Yet the organizing work needed to define and generate these tasks (their design) is not taken into account. Finally, without the help of design theory, our understanding of innovation and our capacity to detect and identify new forms of organization have been hampered. Thus in the next section, to capture all these findings within a consistent conceptual framework, we introduce elements of design theory.

From design theory to innovative capabilities: towards design-oriented organizations

Introducing design theory: the concept-knowledge approach

Why is it difficult to define design theoretically? We all know that architects design buildings, engineers design technical equipment, graphic designers design shapes and symbols. Yet, this does not tell us what design is and how it operates. Usually, design is described as a sequence of tasks to be performed: establishing requirements, defining alternatives, validating and selecting solutions. But such language only describes design planning. Instead, the aim of a design theory is to describe the logic of design, and its input and output.

Design is not representing things, but 'presenting' and expanding concepts

The content of design is difficult to capture because we think about things as if they exist, and as if the problem were to represent them correctly. Moreover, when we experience design situations, we tend to reflect on them not in terms of design theory but in terms of problem-solving or decision theory (Simon 1979, Hatchuel 2002). For example, many of us

who have organized a party will only remember the hard choices we had to face. We rarely think about the whole process as one of design.

To improve our description of design operations, we use elements of a design theory called Concept-Knowledge theory or C-K theory (Hatchuel 2002). These will directly offer a theoretical support to our findings about the operations of innovation within organizations. Let us consider the example of organizing with a group of friends a particularly desirable kind of party (we will call it a 'smart party'). Saying that we want to design a 'smart party,' we abandon the logic of representation. We may use knowledge coming from our memory of past parties. The 'smart party' we want to design may look like these old ones, but what is essential to design is that this 'smart party' does not already exist. Let us call the concept 'the smart party that we want to organize with the properties we desire.' This concept is a comprehensible start for a design team. 'Design' implies operations by which different embodiments of this party concept will be tried out. One of these outputs may be a well-defined and innovative party; but the design operations will also produce various concepts of parties that will never be realized and several pieces of new knowledge will be generated during the process. Thus, design begins with the presentation of concepts (they can be verbal or non-verbal—a drawing, a short melody), starting points that make further design work possible. In organizational terms, design concepts do not directly describe tasks, but are a prerequisite to their definition.

Design expands through different 'design spaces'

The core operation of design is the progressive definition of concepts through an expansion of their formulation. For instance, we can expand 'a smart party' by adding the property of being a 'fancy-dress party' provided that 'fancy-dress' is understandable: that is, it belongs to existing knowledge. Whenever the party is declared as a fancy-dress party, the world of fancy-dress costumes may be settled as a design space (Hatchuel *et al.* 2005): a space that can be investigated, evaluated, and where one can discuss which kind of fancy dress is wanted. Which one can be found easily? Should party members invent their own costumes, or should we select a dress code? A design space also provides conditions for the investigation and generation of new knowledge: for instance, a visit to a specialized fancy-dress shop where new concept expansions could be found—perhaps, the combination of a fancy-dress ball with some other entertaining event. This highlights a central feature of a design strategy: the selection, expansion, and modification of design spaces. For example, industrial designers often expand the first design space of a two-dimensional drawing ('a rough') into a 3D solid object ('a mock-up') which introduces new dimensions like aesthetics or assembly possibilities. Hence, design spaces allow the simultaneous expansion of concepts and knowledge; but they follow different paths, depending on the design spaces that are available or strategically investigated by the organization. This also means that co-evolution within design activities (Lewin and Volberda 1999) is central and manageable.

We have presented such a theory of design in more detail in other papers (Hatchuel 2002: Hatchuel and Weil 2003). However, these introductory elements are sufficient to clarify the link between design strategies and an innovation capability.

Design strategies: a central innovation capability

The preceding theoretical elements allow us to establish three important results which are strongly consistent with our empirical finding that innovation capabilities are anchored in the design competencies of a company:

- design strategies stimulate the knowledge metabolism;
- classic organizational factors are poor descriptors of design;

- we can revisit classic organization theory with the concept of C-K distance.

Design strategies stimulate the knowledge metabolism of the organization

Design activates existing knowledge (stimulates existing resources and uses them at relevant places and moments) and helps to generate new ones (transforms and creates resources): it controls the metabolism of knowledge (and learning) in the organization. We use this biological metaphor to insist on the transformational aspects of knowledge. One central mechanism of this metabolism is the use of concepts acting as 'understandable fictions.' Still, concepts do not fully control the metabolism of knowledge, nor do they determine it. They only provide an entry to a design strategy that can be used as a management tool to influence the knowledge metabolism. Let us use our same example and consider a company offering the service of organizing fancy-dress parties for clients. We can contrast two types of design strategies corresponding to two different knowledge metabolisms.

Oriented exploration: a fixed set of design spaces

The company offers fancy-dress parties that are defined by a fixed sequence of design spaces (the client is asked to choose from some established list of schemes and variables to shape the desired party). These schemes will limit and orient the knowledge metabolism. And a stable division of skills and knowledge areas can be sustained. Learning (including research) and innovation are still possible, provided that they only explore these design spaces. Using a classic typology, this is not only knowledge exploitation but also oriented exploration of new knowledge. The company is not a mechanistic bureaucracy: what has been routinized is not the detail of operations but the design rules that generate the definition of these operations. Jazz or soccer are good examples of such form of routinized design spaces.

Creative design: an open set of design space

In this second case, the type of fancy-dress party offered to the clients is no more defined by a fixed list of schemes. The knowledge metabolism is widely open and non-routinized. A stabilized division of knowledge areas will be difficult to maintain; learning can change the design spaces in unexpected ways, and even the initial concept of fancy-dress could be re-generated. Innovation is now intensive and can occur at all levels of the processes.

Obviously these are extreme ideal-types, and our cases studies suggest that innovative companies combine both design strategies.

The Tefal case shows a striking example of how non-routinized metabolism has been used to launch new lineages of products, and thus to foster new knowledge. Yet, the task of the high-level innovation committee was precisely to manage the tensions between the two design strategies and to cross-fertilize their outputs. This explains:

- how old products like pans are still innovative and profitable using knowledge developed by younger families of products like weighing-scales;
- how new families of products can benefit from old families of products that host them like surrogate mothers.

The Saint-Gobain Sekurit case shows how an R&D department is transformed when design strategy stimulates non-routinized metabolism. It underlines the conceptual work done on car glazing to allow for numerous knowledge expansions. It also highlights the role of the client in such transformations. Non-routinized metabolism is almost impossible if clients are not willing to accept and share the design strategy and work: such innovative capabilities have also to be supported by at least some of the suppliers in the network.

Classic organizational factors are poor descriptors of design strategies

A second important result is that the main organizational factors or determinants used in the literature to explain innovation (Damanpour 1991) cannot capture the preceding design strategies:

Organic behavior

This can be observed in both cases, but the content, purpose, and target of this behavior are not the same in each case. In the Tefal case, we notice a strong exchange of knowledge across different lines of products; in other companies, the organic behavior may exist only within product families. Hence, it is not the organic character of some behavior that counts, but in what and when it occurs.

The number of specialties

This is not relevant to analyze design strategies. What defines a non-routinized metabolism is not the number of knowledge areas but the intensity of the renewal and expansion logic of these specialties. In both case studies, we can observe periods of rapid increase of new specialties and periods of relative stability: obviously, innovation capability is linked to this ability to reshape and create specialties.

Degree of formalization

This variable is defined as the degree of *task* specification. Obviously the concept can be misleading when it is applied to design strategies. Design tasks can be specified in great detail without determining the concepts (products) and knowledge areas produced. This means that a high specification of concepts is not a feature of routinized or non-routinized metabolism: for example, highly, complex equipment requires extensive conceptual specifications (functions, contexts, usages), and yet one can find both types of design strategies in this sector. What should be considered is the logic of concept expansion, the variety of design spaces considered, the knowledge lineages explored.

Centralization

In design, this is not necessarily a source of bureaucratic behavior. If centralization reinforces a logic of routinized metabolism, it will diminish the innovation capability; if centralization allows conceptual expansion and renewal of design strategies, like in luxury goods where famous designers have a personal control on all products, then it can favor the rate and scope of innovations.

To put it briefly, classic organizational variables lose any operational meaning and power of predictability when we characterize design processes. Therefore, when studying the innovative capability of organizations, the main contingent variables have to be good descriptors of the design process. However, innovative capability is not simple creativity, nor research capacity, nor technological capacity, nor good teamwork, nor good networking. It is an organizational capability focused on design strategy and design work, a capability in which development and learning is ruled by the logics and intensity of design.

Design descriptors for a contingent analysis of innovative organizations: the concept-knowledge distance

Following our theoretical framework, to capture the design process we have to trace the activation of concepts by knowledge and vice versa. This suggests that the intensity and amount of the design work that has to be done could be related to a concept-knowledge distance (C-K distance). The bigger a C-K distance at the beginning of a new project, the more non-routinized metabolism is needed, the higher the potential for innovation, and the more design has to be managed and guided by a strategy. Let us stress that the C-K distance is a cognitive judgment, not a countable measure. It can only be evaluated by a group of designers and it is dependent on their own knowledge. Quantitative proxies could be investigated *ex post* (like the number and variance of design strategies that have been explored), but this is still a research issue. What can be observed is that, in a context of routinized metabolism, designers usually think that they can find an acceptable way, and this means a small C-K distance. In non-routinized

metabolism, it will be quite the contrary (a large C-K distance) as designers will face concepts that can be expanded through a large number of design spaces and they will experience an important lack of knowledge. Therefore, C-K distance could be at least considered as an ordinal scale of design intensity. But it also offers a new contingency perspective which captures the innovation capability of organizations.

Revisiting classical structures with C-K distance as a contingent variable

Bureaucratic innovation

The classical bureaucracy is described as a routinized and standardized organization. This concept describes organizations where design work is limited and with no innovation capability. Yet this is an extreme ideal-type. It is more realistic to see bureaucracies as organizations where some capabilities of innovation exist, yet where the acceptable C-K distance in design is obviously small. This does not mean that the knowledge and competencies of the company are limited, but that new concepts can be introduced only if they are readily interpreted and designed within existing knowledge or within very limited expansions of it. Team sports, orchestras (including jazz, dance, and theater) are organizations of such type, where infinite variations of behavior are possible within a strong set of rules. These rules control the design capability, and therefore the innovative potential within narrow boundaries.

The functional structure as a stabilized innovation capability

The functional structure appeared with the development of expert departments that had to suggest new work rules to operational units. The functional division of work is generally structured by product areas (brands, systems, components, parts) or knowledge areas (processes, markets, professional skills). Actually, 'functions' would be better described as 'specialized design departments' that produce design rules within constrained design and controlled knowledge metabolism. The functional organization corresponds to an innovation capability that has to cope with increased C-K distance in the projects of the company. Yet, this innovation capability has to be regulated by the structure of the functions. One function can be very innovative provided that it doesn't interfere too strongly with the other functions: for instance, car glazing can be very innovative without creating constraints on the car engine. The functional structures allow more than component innovations; they may even accept architectural innovations in products whenever the function areas can expand their skills without deep restructuring. It is the type of organization where the structural variables have certainly the strongest impact on the innovation capability. The limit of this capability appears when innovation needs the creation of new, and therefore unknown, functions or changes the functional division of work (Henderson and Clark 1990).

Projects and matrix organizations: unstable equilibria

The matrix system is the first organizational model that distinguishes between knowledge-oriented units (the functions) and concept-oriented units (the projects, or divisions). This scheme responds to the necessity of increasing C-K distance in current design work in order to cope with more competitive contexts, changing requirements and fast knowledge renewal. Thus, in spite of their name, the 'functions' in a functional structure cannot be similar to 'functions' in a matrix. In the functional structure, the mission of functions is to design operations. In the matrix, both functions and projects (or programs) claim to design work rules. Moreover, projects can face new concepts (products, services, systems) that need knowledge which is not available in the existing functions. This leads to external knowledge acquisition, a method for knowledge expansion that, with the intensification of innovation, became, not surprisingly, very popular in management. For this same reason,

matrix organizations tend to be unstable: they combine a mix of routinized and non-routinized design that is not managed as such. Whenever innovation becomes more radical or more intensive, all projects must face higher C-K distances, and tensions will appear at top management level and between the projects and functions (Olilla *et al.* 1998). Innovation will increase, but it takes a wild and costly form: 'skunk work,' conflicting projects, high mortality of new ideas. It is a central conjecture induced by our findings that the observed instability of matrix organizations (Katz and Allen 1985) paved the way to new forms of organizations. We have observed such new forms in our case studies, and their existence can be theoretically predicted within our contingency perspective.

Beyond matrix: (DO2)

Design-oriented organizations (DO2) are not matrix structures: we enter the world of DO2 (Hatchuel and Weil 1998) precisely when two types of situations occur repeatedly within a competitive context:

- new bodies of knowledge are identified and could be explored, yet nobody clearly knows which concepts could be developed and which might lead to new products (for example, new materials, new communication standards);
- new product or service concepts are identified but nobody knows which body of knowledge will be necessary to transform the concept in acceptable products (for example, environmental or safety services).

In such contexts, there are no organizational principles that can guide effective work and adequate work division. Classic structural entries are missing: standards, outputs, skills are not yet defined because they are the output of the design work. Mutual adjustment, the principle of 'adhocracies,' has no clear operational meaning. Who is going to adjust to whom, if there is no network? What can be the logic of collective action in such contexts? It needs the emergence of actors and collective work to be able to establish design strategies, principles, rules, and to manage the non-routinized metabolism associated with these strategies. Therefore, as in Tefal and in the new R&D of Saint-Gobain Sekurit, one can predict that a great deal of managerial attention and skill will be devoted to the following activities which are the core of the innovative power of the company:

- generate 'innovation fields' (Hatchuel *et al.* 2001)—that is, potential spaces of development with high C-K distance—and manage the expansion processes of concepts and knowledge as a core competency;
- maintain product lineages using repeated designs that can benefit from the outputs of the innovation fields, and use known products to explore the new concepts and knowledge;
- monitor the expansion processes of concepts and knowledge by creating coordinated work groups intended to treat these two different logics of exploration equally;
- detect all opportunities of fast-to-market products which can serve as vehicles for the control of the customer value of the expansion processes;
- use the design process as a support for personal development and skill building for the greatest possible number of personnel.

These activities are the operational landmarks and managerial logics of DO2 and this, in our view, explains at least one part of the mystery and enigma of repeatedly innovative companies.[2]

Conclusion: capabilities of innovation as the development of collective design-thinking

In this chapter we have explored a new approach to the innovation capability of organizations. We have grounded this approach on empirical and theoretical material.

Empirical material

We chose to study neither innovations, nor rates of innovations, but the behavior of companies that have been innovating successfully during long period of times. We have paid detailed attention to the history of their products and to the type of knowledge (technical or not) that they have developed through their innovations. This material led us to focus on design work and thinking as central processes for the formation of these innovation capabilities. We have also observed that design work, knowledge expansions, and innovation shape and stimulate the metabolism of knowledge in organizations.

Theoretical material

To interpret these observations, we introduced elements of a design theory based on the concept-knowledge distinction and expansions. This framework allows a new operationalization of the innovation capability and offers descriptors of the design process. The same framework brings a new contingent perspective on organization theory. We have briefly revisited bureaucracies, functional organizations, and matrix structures and we interpret them as attempts to cope with increased C-K distances in the design process. We also identified the existence of organizations that are better adapted to manage situations with a high C-K distance: these organizations that we call design-oriented organizations have the capability to manage design strategies and to support non-routinized knowledge metabolism.

One central result of this research is that these organizations cannot be uniquely defined or identified with classic structural variables (functions, matrix, networks). The price to pay for the understanding of these organizations is *to depart from classic structural theory and to mobilize design theory as a basic and legitimate analytical framework* (or determinant) of the organizational phenomena. Therefore, we combined organization theory with design theory, and this approach offers a fresh and operational perspective on the development of innovation capabilities in contemporary contexts. Finally, this reminds us of an old and forgotten issue in organization theory: it is not possible to define work division (inside and outside the company) independently of some design strategy. If work division controls design, we get the structuralist view of organizations and markets; if design controls work division, we have the functional view; if design and work division interact and foster knowledge expansions, skill development, and growth (be it internal or external), we enter the world of intensively innovative organizations.

Notes

1. These cases are part of a larger empirical investigation that cannot be fully described here.
2. Recent work on Edison's laboratories show that they clearly presented all the features of DO2 (Millard 1990).

References

Allen, T. J. (1997). 'Architecture and Communication among Product Development Engineers.' Working Paper, MIT, Sloan School of Management, the International Center for Research on the Management of Technology, September 1997.

Barrett F. J. (1998). 'Creativity and Improvisation in Jazz and Organizations: Implications for Organizational Learning.' *Organization Science*, 9/5, September–October: 605–22.

Buderi, R. (2000). *Engines of Tomorrow: How The World's Best Companies Are Using Their Research Labs to Win the Future*. New York: Simon and Schuster.

Burgelman, R. A., and Rosenbloom, R. S. (1989). 'Technology Strategy: An Evolutionary Process Perspective.' *Research on Technological Innovation, Management and Policy*, 4.

Burns,T., and Stalker, G. M. (1968). *The Management of Innovation*. London: Tavistock Publications.

Chapel, V. (1997). 'La Croissance par l'innovation intensive: de la dynamique d'apprentissage à la révélation d'un modèle industriel, le cas Téfal.' Paris: École des Mines de Paris.

Clark, K. B., and Fujimoto, T. (1991). *Product Development Performance: Strategy, Organization and Management in the World Auto Industry*. Boston: Harvard Business School Press.

—— and Wheelwright, S. C. (1992). *Revolutionizing Product Development*. New York: The Free Press.

Cohen, W. M., and Levinthal, D. A. (1990). 'Absorptive Capacity: A New Perspective on Learning and Innovation.' *Administrative Science Quarterly*, 35: 128–52.

Damanpour, F. (1991). 'Organizational Innovation: A Meta Analysis of Determinants and Moderators.' *Academy of Management Journal*, 34/3: 555–90.

Eisenhardt, K. M., and Brown, S. L. (1998). 'Time Pacing: Competing in Markets that Won't Stand Still.' *Harvard Business Review*, March–April: 59–69.

Gawer, A., and Cusumano, M. A. (2002). *Platform Leadership: How Intel, Microsoft, and Cisco Drive Industry Innovation*. Boston: Harvard Business School Press.

Hage, J. T. (1999). 'Organizational Innovation and Organizational Change.' *Annual Review of Sociology*, 25: 597–622.

Hatchuel, A. (2001). 'The Two pillars of New Management Research.' *British Journal of Management*, 12: 33–9

—— (2002). 'Towards Design Theory and Expandable Rationality.' *Journal of Management and Governance*, 5: 3–4.

—— and LeMasson, P. (1999). 'Firm Growth and Repeated Innovation.' European Meeting on Applied Evolutionary Economics, Grenoble, 7–9 June.

—— and Weil, B. (1999). 'Design-Oriented Organizations: Towards a Unified Theory of Design Activities.' 6th International Product Development Management Conference, Churchill College,

Cambridge, UK, 5–6 July 1999, pp. 1–28.

—— & Weil, B. (2003). 'A New Approach to Innovative Design : An Introduction to C-K Theory.' *Proceedings of ICED Conference*, Stockholm, August 2003.

—— Lemasson P., and Weil, B. (2001). 'From R&D to R-I-D: Design Strategies and the Management of "Innovation Fields." ' *Proceedings of the IPDM Conference*, EIASM, 2001, Entschedde, Netherlands.

—— Weil, B., and Lemasson, P. (2005). 'The Development of Science-Based Products: Managing by Design Spaces.' *Innovation and Creativity Management Journal*, 14/4: 345–55.

Henderson, R. M., and Clark, K. B. (1990). 'Architectural Innovation: The Reconfiguration of Existing Product Technologies and the Failure of Established Firms.' *Administrative Science Quarterly*, 35: 9–30.

Iansiti, M., and West, J. (1997). 'Technology Integration: Turning Great Research into Great Products.' *Harvard Business Review*, May–June: 69–79.

Jonash, R. S., and Sommerlatte, T. (1999). *The Innovation Premium: How Next Generation Companies Are Achieving Peak Performance and Profitability*. Reading, MA: Perseus Books.

Katz, R., and Allen, T. J. (1985). 'Project Performance and the Locus of Influence in the R&D Matrix.' *Academy of Management Journal*, 28/1: 67–87.

Kline, S., and Rosenberg, N. (1986). 'An Overview of Innovation.' In R. Landau and N. Rosenberg (eds.), *The Positive Sum Strategy: Harnessing Technology for Economic Growth*. Washington, DC: National Academy Press, 275–305.

Lemasson, P. (2001). 'De la R&D à la R-I-D: modélisation des fonctions de conception et nouvelles organisations de la R&D.' PhD dissertation, École des Mines de Paris.

Lenfle, S., and Midler, C. (2000). 'Managing Innovative Projects in Upstream Industries: The Case of a French Steel Group.' In P.-J. Benghozi, F. Charue-Duboc, and C. Midler (eds.), *Innovation-Based Competition and Design Systems Dynamics: Lessons from French Innovative Firms and Organizational Issues for the Next Decade*. Paris: L'Harmattan, 193–217.

Leonard-Barton, D. (1995). *Wellsprings of Knowledge*. Boston: Harvard Business School Press.

Levinthal, D. A., and Warglien, M. (1998). 'Landscape Design: Designing for Local Action in Complex Worlds.' *Organization Science* 9/5, September–October: 342–55.

Lewin, A. Y., and Volberda, H. W. (1999). 'Prolegomena on Co-evolution: A Framework for Research on Strategy and New Organizational Forms.' *Organization Science*, 10/5, September–October: 519–34.

Lundin, R., and Midler, C. (eds.) (1999). *Projects as Arenas for Policy and Learning*. Boston: Kluwer Academic Publishers.

MacCormack, A., and Verganti, R. (2000). 'Managing the Sources of Uncertainty: Matching Process and Context in New Product Development.' International Product Development Management Conference, Leuven, Belgium, EIASM, Katholieke Universiteit Leuven: 347–68.

Millard, A. (1990). *Edison and the Business of Innovation*. Baltimore: The Johns Hopkins University Press.

Miller, W. L., and Morris, L. (1999). *Fourth Generation R&D: Managing Knowledge, Technology, and Innovation*. New York: John Wiley and Sons.

Myers, M. B., and Rosenbloom, R. S. (1996). 'Rethinking the Role of Industrial Research.' In R. S. Rosenbloom and W. J. Spencer (eds.), *Engines of Innovation, U.S. Industrial Research at the End of an Era*. Boston: Harvard Business School Press, 209–28.

Nonaka, I., and Takeuchi, H. (1995). *The Knowledge Creating Company: How Japanese Companies Create the Dynamics of Innovation*. New York: Oxford University Press.

Olilla, S., Norrgren, F., and Schaller, J. (1998). 'Political Skills in Leading Product Development Projects.' 5th International Product Development Management Conference, Como, Italy, EIASM.

Roussel, P. A., Saad, K. N., and Erickson, T. J. (1991). *Third Generation R&D: Managing the Link to Corporate Strategy*. Boston: Harvard Business School Press.

Segrestin, B. (2003). 'The Management of Exploratory Partnerships.' EGOS presentation July 2003. (See also PhD dissertation, École des Mines de Paris, 2003.)

Simon, A. H. (1979, 1989). *Models of Thought*, 1 and 2. New Haven: Yale University Press.

Van de Ven, A., Polley, D. E., Garud, R., and Venkataraman, S. (1999). *The Innovation Journey.* New York: Oxford University Press,

Von Krogh, G., Ichijo, K., and Nonaka, I. (2000). *Enabling Knowledge Creation: How to Unlock the Mystery of Tacit Knowledge and Release the Power of Innovation.* New York: Oxford University Press.

Wheelwright, S. C., and Clark, K. B. (1992). *Revolutionizing Product Development, Quantum Leaps in Speed, Efficiency, and Quality.* New York: Macmillan.

14 New Sources of Radical Innovation: Research Technologies, Transversality, and Distributed Learning in a Post-industrial Order

Terry Shinn[1]

Introduction

This chapter connects literature in the history and sociology of science and technology to studies in organizational sociology of innovation. The focus is the emergence of a new technical/organizational/epistemological environment specifically propitious to the growth of radical innovation. What exactly is 'radical innovation'? Radical innovation may usefully be defined as the generation of novel products or/and processes with simultaneous reference to a vertical and a horizontal axis. The vertical axis refers to the raw materials, components, and production operations involved in the manufacture/production of products, and the horizontal axis refers to the variety and range of end-user products and services which the innovation directly fits or strongly affects. Items such as the electric motor (1840), automatic switching mechanisms (1880), transistors (1950), and the microprocessor and computer are instances of this breed of radical innovation, as they figure centrally in the components and the processes that are involved in manufacturing a huge stretch of unrelated products, and as they themselves constitute end-user products or comprise components in many finished products. Other instances of incontestably radical

innovations include the ultracentrifuge (1930), rumbatron (1945), Fourier transform spectroscopy (1960), the laser (1960), and the C++ object-oriented simulation language (1985). Radical technologies such as these are general-purpose technologies. This category of innovation has long fuelled, and continues to fuel, vigorous and sustained economic growth, and it is clearly implicated in significant changes of life, both for the better and for the worse!

The characteristics and underlying dynamics of this form of radical innovation will be explored in this chapter, with particular attention to its impact on organizational structure and operations, sites and forms of creativity, and learning paths and processes.

The definition of radical innovation proposed here diverges from more routinely employed understandings on two grounds. One frequent usage of the term 'radical innovation' is framed in terms of markets and profits. In this view, a radical innovation is one that extends a market, generates new markets, and yields elevated profits. Another understanding of 'radical innovation' is coupled to the notion of extreme novelty. According to this understanding, an entity that is entirely unprecedented is a radical innovation. Both of these perspectives are one-dimensional, and both deal with very limited and superficial features of innovation. The potential worth of the concept of radical innovation recommended in this chapter lies first of all in its potential to see radical innovation as necessarily integrating both technological and market forces. Second, the proposed definition opens the way to the identification and characterization of the complex components and dynamic interactions encapsulated in radical innovation.

The discussion that follows deals mainly with one expression of radical innovation, a historically recent and increasingly dominant form of radical innovation known as 'research technology.' Research-technology-driven innovation is fundamentally transverse, as it spills over into and affects a vast scope of intellectual, technical, and economic domains that lie far from the nexus of origin. It hence serves as a vehicle that traverses the usual boundaries that define and protect established lines of mental or material work—scientific disciplines, organizational departments, technological spheres, and other expressions of habitus and language. In so doing, research technologies defy classical divisions of labor. They allow specialty occupations and organization their customary autonomy, and also richly contribute to them through the introduction of new material resources. They similarly contribute to the organizational capacities of existing groups. This is achieved because the research technology adopted and appropriately adapted by particular groups carries with it an intrinsic technical logic and language which is everywhere necessarily imported, thereby furnishing organizations an important tool for transverse communication. The inherent language of the innovation becomes a kind of lingua franca spoken by all of its users. This is fundamental, since it reduces the often negative effects of an intellectual, organizational, and economic order that is increasingly crippled by hyper-differentiation and fragmentation.

Much sociology of innovation and of science and technology has explored change with reference to the concepts of integration and differentiation—or, more accurately, integration

versus differentiation. Classical sociology of science often equates cognitive growth with processes of fresh waves of intellectual and organizational differentiation. For their part, numerous sociologists of industrial change have striven to derive a formula that would permit escape from the paradox of occupational and functional differentiation versus integration. Differentiation is frequently identified with the mobilization of resources and specialization of skills required for effective and efficient pursuit of a standardized task, such as manufacturing. By contrast, industrial research is often equated with the integration of highly dispersed ideas, material elements, and expertise that lie well beyond standard organizational boundaries. But managerial control over industrial research groups calls for limitations on assertive integrating initiatives. Constraints and the imposition of boundaries result in incipient differentiation which, although efficacious in routine reproduction of goods, nevertheless proves detrimental to the generation of industrial novelties.

In this chapter it will be demonstrated that inside research-technology integration and differentiation do not function as opposites, but instead constitute complementary sides of the same coin. An appreciation of this subtle, and historically recent, situation is a key component to an understanding of the dynamics of radical innovation.

Research technology first arose in Germany in the last decades of the nineteenth century, in response to the military demands associated with nation building, the technical and organizational exigencies of rapid economic growth rooted in the technologies of the second industrial revolution, and, finally, in response to an increase in the number and complexity of scientific disciplines, and to the institutionalization of research as a differentiated cognitive and technical function. Research technologies quickly spread to France and to Great Britain in the 1920s and to the US in the 1930s. In the 1940s and 1950s, research-technology-driven radical innovation initiatives took root in the USSR and Japan. A concatenation of factors and forces conjoined to emergent disposition in science, state, and industry thus framed the rise of research-technology ventures, and the last five decades have witnessed not merely the perpetuation of research-technology ventures, but indeed their multiplication and emergence as an often crucial source of radical innovation.

Research technologies are characterized by three features. First, they entail 'generic' devices. These often take the form of scientific instrument or methodology whose purpose is detection, measurement, or control. Generic instruments express some fundamental instrument principle. This permits the research technology to be general, open-ended, and flexible. A research technology is thus not created to solve a well-defined, local, and short-term technical problem. It is instead associated with fundamental research into instrument theory and the potential of said theory to find expression in as yet unspecified spheres. It is the generic feature of research technologies like the transistor, computer, chemical engineering, Fourier transform spectroscopy, and the ultracentrifuge that has allowed them to be re-embedded into particular application markets by highly specific occupation-based engineering endeavours.

Second, much research-technology work is carried out in an 'interstitial' arena. While research technologists may be associated with a principal organization, such as a university, a firm, a research agency or state technical service, they nevertheless tend to move from organization to organization—he who works for everyone is the bondsman of no one. Through occupying the spaces between dominant organizations, research technologists thereby enjoy an environment where opportunities and resources are maximized. The interstitial arena also functions as a platform that fosters multiple selective, intermittent, and temporary boundary crossing. Although research technologists pursue the design and production of their generic instruments in an interstitial arena, their workflow calls for strong interaction with specialized occupational/professional groups in science, state, or industry at two junctures. They cross countless occupational boundaries during their quest for generic instrument ideas and for necessary technical information or components. Boundary crossing similarly takes place when research technologists look outward from the interstitial arena while validating their devices through determining their relevance and applicability in specific technical settings. During this phase, research technologists hence consult and advise the staff of particular disciplines, companies, and state technical services on how a generic device or methodology might be adapted to specific end-user needs. Herewith, the differentiated profile of specialty groups is not challenged or encroached on by research technology and, reciprocally, neither is the integrative dynamic of research technology sanctioned by differentiationist logic. Integration and differentiation operate as if they are two sides of the same coin, a new configuration and concept in the pursuit of innovation.

The third characteristic of research technology is its strong connection with 'metrologies.' Generic devices usually contribute to precision. They introduce greater levels of accuracy, establish new units of measurement, norms, and standards, or express or utilize precision (digitalization versus the erstwhile analogical system) in some unprecedented fashion. There is thus a link here between metrologies and radical innovation. These metrologies constitute the foundation of the afore mentioned lingua franca, the (language) of research technology that promotes trans-occupational/professional, disciplinary, and function-based communication. The metrology-ground lingua franca permits transverse intelligibility. This lingua franca may incorporate elements of vocabulary and terminology, image-based representations, abstract formal expressions, or even a historically original paradigm.

The radical innovation of research technologies cannot be encapsulated in a formula, and it furthermore largely escapes the logic of frame working structure. The radical innovation of research technology is grown, a product of a dynamic meshing of forces of integration and differentiation, where the interface is often improvised, yet in a framework of adaptive structure.

The radical element of innovation derives from the generic ambition. Creativity is therein expressed, and the work of creativity is conducted in a peculiar interstitial environment which escapes the objectives and sanctions of most dominant, more conventional arenas. Together, genericity and the interstitial engender a novel opportunity for distributive learning, which is multi-party, multidirectional, and reciprocal. In the course of boundary crossing, research technologists acquire ideas and data from a variety of occupational specialists. As a generic method or device is re-embedded in the university, industrial, or state technical sector, practitioners learn about the workings of the generic instrument (including the elements of its lingua franca), and they further learn through tailoring the device to fit their specific requirements. Research technologists too are involved in this process, and they thereby learn about unanticipated features of their initial generic discovery.

The radical innovation associated with research technology points to the operation of an important form of non-market linkage, a category of linkage largely absent from standard sociological and economic accounts of innovation. The cognitive/technical component of generic instrumentation, expressed as instrument theory, is only loosely tied to economic market motivation. Its source lies far closer to technical curiosity and a wish for technical capability. Additionally, the boundary-crossing, multi-group communication, and the distributive learning of research technology's radical innovation are principally predicated on the drive for the validation of a generic theory, based on its embedding in multiple and highly diverse settings, on the desire for the extension and re-enforcement of technical information by all concerned, and based on the wish for greater technical control and performance by the engineering practitioners located in industry, academia, and public technical services. Yet this does not alienate research technologies from narrowly utilitarian objectives and productions, which are generally viewed as providing the ultimate epistemological proof and legitimacy of a research-technology success.

This chapter opens with a summary of certain key concepts in orthodox sociology of science and sociology of organizations/innovation related to contexts favorable for the growth of knowledge and innovation. The apparent paradox of differentiation versus integration in this corpus will receive particular attention. It will be suggested that this antipode structuring of integration/differentiation often generates environments that inhibit or block creativity and learning either through the imposition of excessive constraints or through a relaxation of boundaries to the extent that an efficient mobilization of human and material resources becomes problematic. The intent of this section is pedagogical, and the material takes the form of a literature review. Readers familiar with classical literature in the sociology of science and the sociology of organizations and innovation may wish to move quickly through this section. The second section will deal with integration-driven narrow-domain innovation. The dynamics and effects of integration are illustrated in the case of 1940s and 1950s cell biology. The third section focuses on general-purpose technologies, a form of technology that is connected to radical innovation. Research technology is one species of general-purpose technology. Its history will be explored, with reference to its internal operation. The properties of research technology will be further discussed. A now classic example of research technology is presented, followed by a discussion of how research technologies exemplify a new tendency in technical and organizational creativity and learning. Finally, it will be hypothesized that research technology constitutes a promising basis for radical and sustained innovation.

An innovation trap? The differentiation/integration paradox

To a large extent, knowledge production and innovation are framed by many sociologists of science, technology, and innovation in terms of differentiation processes. This should perhaps not come as a great surprise, since important features of modernity are embedded in the language of the division of labor, cognitive, skill, and function specialization, and community and organization building—all of which are predicated on differentiation.

The growth of knowledge: hybridization through differentiation

Robert Merton's 1938 pioneering treatise on the social roots of modern science gave central stage to societal differentiation (Merton 1938 and 1973). In seventeenth-century England there emerged a large number of new and well-defined occupations connected to increasingly critical state, economic, and technical activities, among them transport, shipbuilding, metallurgy, instrument making, and medicine. Alongside, and perhaps in some ways in response to this occupational specialization, organized science arose as a distinct occupation and community, with its own specific set of intellectual, epistemological, normative (Merton 1942), and institutional features. According to Merton, science rapidly grew into an autonomous profession, and the genesis of science was part and parcel of the more general process of societal differentiation.

Writing in the 1960s and 1970s, Joseph Ben-David extended Merton's differentiation precept to the social framework surrounding the growth of knowledge (Ben-David 1991). Here Ben-David's concern was not the distinction between science and non-science, but instead how differentiation operates inside science itself and, in particular, the functions it performs in fuelling cognitive change. He insisted that the growth of knowledge derives principally from the introduction of ever more cognitive specialties, which take the form of disciplines, subdisciplines, or new scientific fields. Within specialties, problems can effectively be formulated, protractedly studied, and institutionalized. A fresh specialty is achieved through a complex interaction between two existing specialties, states Ben-David, arguing that the relevant interactions involve two components—professional status and cognitive status. It is the aspiration of individuals to maximize both forms of status that leads to the creation of a hybrid. Hybrids listed by Ben-David include physiology, speculative and experimental psychology, and bacteriology (Ben-David 1960; Ben-David and Collins 1966).

To take the example of experimental psychology—this specialty developed in late nineteenth-century Germany at a time when there was an almost complete career blockage in physiology for young ambitious academics, as no additional university chairs were being created in the field. In order to attain an elevated academic position, and to enjoy the attendant professional status, it was thus necessary to transfer out of the intellectually stimulating and prestigious domain of physiology to some other field where there was an advantageous abundance of university chairs. In the 1870s and 1880s philosophy was then such a field, where there existed a large number of vacant chairs. Acquisition of a philosophy chair sated the thirst of professional status; it nevertheless had a negative effect, namely, philosophy had a lower intellectual status than physiology. The younger discipline was greatly revered, having become an exemplar of experimental rigor and precision measurement in an age of technical precision (Olesko 1991).

How then to combine the advantages of high professional status and high cognitive status? Ben-David claimed that the solution lay in the creation of a new hybrid specialty. In establishing experimental psychology, some of the recent beneficiaries of philosophy chairs, who had originally trained and done research in physiology, transferred the technologies and experimental protocols of their former domain to philosophy, seeking to use these tools to address psychology-oriented issues then being considered in some philosophy circles—hence experimental psychology. Scientists in this new domain enjoyed much of the intellectual respect accruing to the experimental tradition of physiology, and also enjoyed the high academic status of those holding a university chair. Ben-David equated the occupational and intellectual issues with roles. A hybrid would merge advantageous cognitive and professional roles: an individual moves from field A (intellectually desirable but occupationally unacceptable) to field B (occupationally desirable yet intellectually unacceptable). This arrangement proving inadmissible, he transfers the methods and tools of the initial field A in order better to study certain redefined problems of the domain B.

The upshot is the institutionalization of a new cognitive field C. The field C represents a differentiation with reference to specialties A and B. Ben-David stressed oppositional features of the novel domain. It exists and flourishes by distancing itself from neighboring specialties. The domain seeks to set itself off by emphasizing its uniqueness. It maximizes specialty barriers and reduces to a minimum cross-boundary movement. To do otherwise might imperil the field's separateness. Implicitly, the human, material, and cognitive resources for present and future development are to be mined from within. Reference to exogenous resources constitutes a greater danger than it does a possible source of progress. Generated through processes of differentiation, this category of cognitive innovation charts its future course in the same logic. Reaching beyond the specialty is professionally proscribed, although some circumstances might require so doing. Thus a paradox: outside inputs are useful and sometimes imperative, yet Ben-David's and Merton's differentiation bias militates against boundary crossing and even circumscribed expressions of convergence, not to speak of forms of integration.

More recently developed specialties, like radio astronomy (Edge and Mulkay 1976) and molecular biology (Abir-Am 1992) exhibit this assertive differentiationist pattern. It is perhaps Thomas Kuhn's (1962) exploration of change in science that has most fully and dramatically emphasized the operation and the extent of differentiation dynamics in cognitive innovation. Kuhn's analysis of scientific revolutions insisted on discontinuities. Each science paradigm is distinct, and even cut off, from its predecessors. The specificity may include not only cognitive and epistemological components, but also cultural, institutional, and organizational elements (Forman 1971). According to Kuhn, practitioners and groups attached to different paradigms do not address one another; in fact they cannot communicate with one another at all. This incommensurability derives from the fact, stated Kuhn, that each paradigm of perception, description, and measurement constitutes a world unto itself, possessing a specific language and set of references entirely unintelligible to those operating from within another system. In this dynamic, novelty is thus never introduced through cross-fertilization and the penetration of fresh ideas or observations from outside, but arises from noticing anomaly exclusively from within. Here is logic of implacable exclusion and assuredly not inclusion. Productively undertaking science in such a procrustean cadre has strained the credulity of many influential authors (Bechtel 1993; Galison 1997), who concede that much science and technology indeed occurs inside specialty groups, yet also point to the existence of numerous instances in which the growth of knowledge and innovation takes place through cross-boundary convergence, and sometimes even integration.

Managing innovation

Some of the fathers of the organizational sociology of innovation early recognized the special managerial and structural needs requisite to effective innovation. The organizational exigencies of industrial R&D departments figured prominently in the classic study by Lawrence and Lorsch (1967) of company structure. In *Organization and Environment: Managing Differentiation and Integration* they pointed to the particular organizational requirements of the functions composing firms—principally, production, marketing, and research and development. Organizational structure and operation, they asserted, depend importantly on elements of environment and uncertainty. Management of R&D innovation proves particularly problematic. Research work is notoriously unpredictable and hence quasi-impossible to chart. Its organization imposes a double problem. On the one hand, research entails involvement with many and diverse groups, some of which may lie outside the firm and may embody competencies which at first blush might seem to be remote from the firm's concerns. Considerable uncontrolled contact is called for here. On the other hand, the organization of R&D innovation groups must be focused on company priorities, must express a properly defined division of labor, must be stable. In effect it has to be manageable. Articulated in perhaps somewhat exaggerated terms, this sociology of innovation properly identified the essential tension between a logic of 'production' versus 'reproduction' (Bourdieu and Passeron 1970)—the former entailing occupational transversality and sometimes a form of integration, the latter embracing occupational closure. This is instantiation of the differentiation/integration innovation paradox posited in the introduction of this chapter (Shinn and Joerges 2003).

Much postmodern sociology of innovation appears less clear-cut than earlier sociology in its treatment of occupations, and particularly occupational specialties. This is due in part to a preference for analysis on a societal macro-level. One nevertheless discerns an implicit portrayal of the relationship between occupation and innovation, and I suggest that, despite appearances to the contrary, much postmodern sociology privileges a differentiated, even immensely fragmented, formula for successful innovation work. The New Production of Knowledge, postmodern perspective (Gibbons *et al.* 1994; Nowotny *et al.* 2001) is by far the most widely cited corpus (Shinn 2002). This perspective has consequently been selected for commentary.

New Production of Knowledge sociology underlines the mobile character of learning and fluidity of the professional and social relations of scientists, engineers, and technicians. Scientific and technical work, it is argued, is increasingly performed by temporary teams which assemble in order to solve a specific problem, and then disband on completion of the project. Individuals move on to participate in newly established groups that are in turn dissolved when their mission has been successfully carried out. This sociological perspective anticipates that the trend toward intellectual and human resource rapid turnover circulation will accelerate.

According to this sociological perspective, modern knowledge has been organized in the framework of two distinctively different (indeed contrasting) modes of production known as mode 1 and mode 2. Mode 1 supposedly characterized the whole of pre-1945 science. Mode 1 may continue even today to underpin some science specialties, but, particularly since the 1970s and 1980s, mode 2 has allegedly tended to move to the fore. Mode 2 science and technology include cognitive science, computers, environment studies, biotechnology, and aviation. According to some science and technology observers, mode 2 will come to supplant mode 1 learning (Nowotny *et al.* 2001).

Mode 1 is built around scientific and engineering disciplines and specialties. Entry to disciplines is carefully regulated. Practitioners receive a standardized academic education, possess a doctorate, and train under the tutelage of an acknowledged professional. Disciplines are self-referencing to the extent that they internally determine which problems are to be dealt with, how this will be done, and what constitute the criteria for evaluating the validity of research findings. This is done by disciplinary peers. New Production of Knowledge analysis stresses that mode 1 science is entirely academic. It is university based. By virtue of this alleged fact, mode 1 science is 'ivory tower' science. It is cut off from society and societal problems. Mode 1 science, it is charged, is typically deaf to society's demands. Additionally, society-based evaluation criteria for judging the propriety and worth of research findings have been historically absent.

By contrast to mode 1, mode 2 science operates outside disciplinary confines and the university. Mode 2 science and technology is interdisciplinary, allowing heterogeneous bodies and skills to coalesce freely in response to new opportunities. It is non-academic in the sense that its ties are with society and social issues. Society determines which problems are to be explored and resolved. The erstwhile evaluation criteria of disciplines (such as cold rationality and theory) are being superseded, as the crucial criterion today is social relevance. The introduction of a new epistemology, 'a socially robust epistemology' in replacement of theory has arisen for two reasons (Nowotny *et al.* 2001). First, the cognitive possibilities of theory have become exhausted. Second, as the old myths about the existence of distinctions between nature and culture and between science and society atrophy, today's citizen is at last free to impose citizen-ground reasoning practices. It is important to mention a last feature of mode 2 science. While mode 2 is intended to solve utilitarian social problems, it is interesting to note that, in much of New Production of Knowledge writing, this means in practice placing 'science' knowledge at the disposal of enterprise, in order to make business more productive (and more profitable?) (Shinn 2002).

In precisely what ways, though, is New Production of Knowledge sociology of innovation analysis grounded on tenets of intellectual and group fragmentation? In many important ways, this perspective is antidifferentiationist (Shinn and Ragouet 2005). The sociology of knowledge of R. Merton, J. Ben-David, and most other sociologists writing during the 1950s and 1960s, was differentiationist to the extent that it stressed two basic discontinuities. First, the professional practices of the scientific community distinguish it from all other social collectives. Second, the lines of demarcation between disciplines and between fields inside science itself are essential to the operation of knowledge growth. Antidifferentiation sociology of science, technology, and innovation, such as the New Production of Knowledge, sees things quite differently. Many antidifferentiationists refuse cognitive and social differentiations, and beyond. They deny the division between nature and culture, science and society, science and technology,

and between research and enterprise (Latour 1989; Callon 1986; Bijker *et al.* 1987; Bijker 1997). Boundaries between occupations are similarly minimized or denied. At first glance one might be tempted to conclude that radical antidifferentiation is equated with sweeping integration. On closer inspection, however, it becomes clear that this is assuredly not the case. Indeed the social and cognitive relations between groups are often characterized as a 'seamless web'.

New Production of Knowledge analyses, like other antidifferentiationist sociology of innovation, does not posit dynamics of integration. The concept of a radically unbounded, frontierless intellectual and social order is not necessarily an integration-rich order. Put differently, the mere absence of enforced differentiation does not automatically spell the presence of integrating connections.

The New Production of Knowledge posits atomistic learning and social interactions. This is fragmentation pushed to its extreme limit. The unceasing composition, decomposition, and recomposition of groups brought about through an unending circulation of knowledge and actors, of the sort proposed by the New Production of Knowledge mode 2, is allegedly highly adaptive to changing circumstances, but it also vehicles a high degree of instability. Integration is predicated on the intersection and eventual interpenetration (partial or thorough) of distinct units. The very existence and the effective operation of such units depends on stabilities, which are a product of a common identity, shared skills, a sense of joint mission, shared problems, and an agreed-on set of evaluation criteria. The New Production of Knowledge discounts, even abhors, such collectives, which are viewed as old-fashioned and even counter-productive in today's fast-moving world. In a word, much of postmodern sociology of innovation's penchant for atomistic social distributions and interactions is at bottom no more appreciative of the innovation potential of integration-driven social arrangements than was the pro-differentiation sociology of science and technology of the immediate post-war decades.

The recent, highly novel work by Andrew Abbott (2001) on disciplinary dynamics in the social sciences looks beyond differentiation in the search for a fresh understanding of disciplinary operations and growth. Abbott argues that the multiplication of disciplines over the last one hundred years has not come about through hybridization, which would merely constitute an association and combination of chips that have split away from previous extant disciplines. The author likens new science specialties to fractal geometries. While fractals are each distinctive, they are nevertheless all identical in important respects. According to this view, new social-science specialties are a product of a reassembling and relabelling of a limited number of components pre-existent in traditional disciplines. These, Abbott states, are the outcome of the remixing of earlier taxonomies, processes, and concepts. This new theory of disciplinary growth is interesting on numerous levels, yet it, like most of its forerunners, by and large remains mute over the issues of intellectual and social integration, and, most notably, over questions of transverse cognition, and the accompanying corollary social activity of boundary-crossing.

Integration-driven narrow domain innovation: the case of cell biology

While the description and many of the explanations of cognitive growth and of innovation dynamics have long predominantly privileged a differentiation- or fragmentation-oriented view of intellectual and occupational/organizational processes (Ben-David 1960 and 1991; Gibbons *et al.* 1994; Nowotny *et al.* 2001; Abbott 2001; Lawrence and Lorsch 1967), there nevertheless exist scattered instances of an alternative analytic path in which forces of integration versus fragmentation figure importantly. Historically, when counted one by one, the vast majority of differentiation-driven changes have mostly given rise to narrow do-

main innovations; that is, innovations whose cognitive or technological spillover has proven circumscribed, and whose diffusion and market impact have been limited. Is it reasonable to anticipate, though, that integration-driven innovation must necessarily generate broader intellectual, engineering and economic/industrial impulses? Put differently, is the innovation connected with forms of integration somehow coupled to the sweep and scope of radical innovation? The dynamics involved in the origins, evolution, and impact of the specialty field of cell biology suggests that forces of integration versus differentiation do not necessarily break the logic and limits of narrow domain innovation.

In his carefully documented article, 'Integrating Science by Creating new Disciplines: Cell Biology,' the philosopher William Bechtel (1993) insisted on the roles played by integration in the dynamics of cognitive novelty generation. For him, the emergence of a new sphere of learning is achieved precisely through a strong convergence and meshing of already existing domains. The centripetal logic of integration prevails over the centrifugal logic of differentiation for Bechtel.

Unlike the examples of sharp bipartite differentiation-ground cognitive specialty formation described by J. Ben-David, cell biology may be regarded as constituting an instance of innovation through integration. A list of the far-ranging disciplinary backgrounds of its founders is immediately revealing—Albert Claude (physician and pathologist), George Hogeboom (physician and chemist), George Palade (physician, physiologist, and microscopes), and Walter Schneider and Filip Sielevitz (biochemists).

The *Journal of Cell Biology* appeared in 1960, and the American Society of Cell Biology held its first meeting in 1961. Here follows a list of the university departments of the authors of the aforementioned journal during its first three years of existence: bacteriology, virology, cancer studies, biochemical cytology, histology, biochemistry, pediatric surgery, genetics, radiology, zoology, botany, pharmacology, biophysics, anatomy, pathology, physical chemistry, and metabolic research. The breadth exhibited here is extraordinary. It may be said from this that cell biology innovation represents a kind of loose federative integration and not simply a specialty differentiation of the sort described in the preceding pages. What are the forces that generate and propel a loose federative integration trajectory and evolution?

Cell biology explores both morphology and process at the cellular and subcellular level. Put differently, it probes the structure and the function of cellular components and connects these two aspects. Prior to the 1940s and 1950s, biochemistry had dealt with many biological processes, and the disciplines of anatomy, physiology, histology, and cytology had focused on questions of morphology. Cell biology introduces the novel capacity of combining and even fusing research on structure and process. Two new situations figured centrally in this cognitive transformation. First, the early twentieth century witnessed the birth of numerous new specialties, among them biophysics, radiology, virology, and electronics. Their appearance on the cognitive landscape opened a new space for intellectual and institutional maneuver and mobility. To some extent, this spilled over into older-established disciplines as well. Taken together, a greater sense of openness briefly characterized learning and research, and perhaps particularly so in the life sciences.

The invention of two new instruments (the ultracentrifuge and electron microscope) and their adaptation to biology constitutes the second key event. The electron microscope was developed in Germany in the 1930s, and by the early 1940s RCA had entered the market with its own version (Rasmussen 1997 and 1998). The device had initially been designed for use in physics, metallurgy, chemistry, and engineering; however RCA sought to extend the domains of application as sales were poor at best. The firm made the electron microscope available to a few US biologists in the hope that the instrument might prove effective as a research tool and that the field would then become a profitable outlet. Two early major problems hampered the microscope's utility. The initial 60 kV power level often proved too weak to explore deeper cellular components. Second, a range of accompanying technologies had to be developed (such as new microtones and stains) in order to enable the electron microscope to perform optimally.

The ultracentrifuge would prove equally crucial to cell biology. A preliminary version of this device had been used in Sweden during the 1920s (Elzen 1986), before the pioneering American research technologist, Jesse Beams (1899–1977), reworked the artifact and transformed it into a generic instrument that found its way into innumerable applications (Brown 1967; Shinn 2001b). The ultracentrifuges of the 1930s and 1940s were still technically imperfect and, most important, their spheres of utility had not yet been established. A small number of mainly European engineers and scientists glimpsed the ultracentrifuge's potential in biology, and during the 1940s and 1950s the Rockefeller Institute-based Belgian physician and biochemist, Albert Claude, steadfastly strove to adapt the instrument to a variety of biological fields, foremost of which was what became cell biology.

The ultracentrifuge and electron microscope served cell biology in several critical ways. Starting in the late nineteenth century, biochemistry had made it possible to study certain chemical/biological processes associated with cells, such as fermentation. However, it had not been possible to distinguish between processes on a fine level, and neither had it been possible to link particular processes to specific cell structure. In preparing biological materials for biochemical analysis, cell morphology was totally eradicated. By contrast, the advent of the ultracentrifuge, and its adaptation to biology, henceforth made it possible to separate cell materials from neighboring components and to concentrate them. The ultracentrifuge is a fractionation apparatus, whose fundamental instrumentation logic is specific density and gravitational units. Thanks to this technology, specific chemical processes could now be connected with identifiable cell structures.

The introduction of the electron microscope to biology, and in particular to cell biology, resulted in the observation and careful description of many hitherto undetected cellular components. The relatively long wavelength of visible light limits the optical microscope's resolving power. By contrast, the immensely shorter wave dynamics of electrons makes possible observation on a macro-molecular scale. The mitochondrion was investigated exhaustively using this new form of microscope, and the endoplasmic reticulum was similarly discovered. Research based on the ultracentrifuge and electron microscope technologies led to a detailed understanding of, for example, the cellular structures and functions of cell respiration. This required an integrated study of the rough end (versus smooth segments) of endoplasmic reticulum, specific enzymes, and particular cytoplasts.

The integration-driven socio/cognitive innovation dynamics of cell biology differ from those of differentiation-driven fields in three important ways. First, this occupational culture is inclusive versus exclusive in composition and in its dynamics. As indicated above, almost a score of specialties were involved in its inception. Additionally, an inspection of the articles of the specialty's key publication, the *Journal of Cell Biology*, shows that individuals from other fields (especially biochemistry) are still today frequently drawn into the outer atmosphere of the domain. There persist a spirit and orientation by which sustaining contact with the exterior constitutes a relevant, even an a priori, crucial source of fresh data, tools, manpower, etc.; while cell biology is, of course, itself an occupational specialty and thus predicated on insider/outsider relations, here the 'outside' nevertheless remains (as it was in the beginning), a key operating dimension of the 'inside.' Cell biology is a weak-hub culture to which diverse other specialty cultures refer, to which they may occasionally cling, and which may sometimes draw permanent outsider recruits from unexpected sites. This may be likened to a loose federative integrated framework versus a differentiated centralized exclusionary structure.

Second, the foundations of cell biology included two technologies that are 'generic technologies,' a species of technology to be discussed in detail below. The importance of generic technology to integration-based innovation processes and structures cannot be overemphasized. The electron microscope and ultracentrifuge were not transferred from one or another occupational culture, where they served as standard disciplinary instruments. To the contrary, neither technology 'belonged' to any occupational body or cognitive orientation. While this was due in part to the fact that they were both relatively historically recent observational and measurement artifacts, the more important fact is that by virtue of being generic, these two devices can serve, after appropriate redesign, the material and intellectual objectives of an almost endless variety of functions in a wealth of sectors. The technological roots of cell biology are thereby integrative, which cannot but affect the specialties membership, expectations, operating rules, and evaluation criteria.

Third, despite cell biology's specific professional/organizational particularities, it nevertheless offers an example of a heterogeneous organization as opposed to a more classic homogeneous organization (Shinn and Joerges 2003). This heterogeneity is constitutive, and forms part of what might be termed the community's 'organizational habitus.' The specialty manages to strike a fine balance between sustaining normative insider/outsider boundaries, and maintaining a gravitational field that permits associate relations of outsiders and even permits its own members to sojourn outside, and still to maintain membership.

Nevertheless, like differentiation-driven innovation, most forms of loose federative integration-driven innovation similarly tend to give rise to 'narrow domain innovation.' What is narrow domain innovation, and why does weak integration frequently generate it? What does this signify for parallels between differentiation-driven innovation and weak integrated innovation?

Much innovation can be classified as narrow domain, and it is possible that it accounts for most technical change (Shishoni 1970; Hippel 1988; Mokyr 2002). This category of novelty arises within a specific local setting. It is conceived, developed, and utilized by a single occupational set or cluster, and the spillover of technique is relatively restricted. Examples of narrow domain innovation include the cruciform screw head, coasters that fit beneath chairs, plastic nibs at the ends of shoestrings, and so forth.

In this species of novelty, the innovation is not incorporated into a system of vertical integration, nor does it spill over horizontally into endless families of end-user goods. Similarly, neither does it spread and become part and parcel of the toolbox and routine practices of engineers, technicians, or scientists in occupations distant from the one where the innovation was initially conceived and matured. The innovation is spawned and consumed in the framework of a confined circle.

The reasons for this are of two sorts. The occupational specialty that developed the technique can be so strong, in the sense of being well defined, self-identifying, and closed, that connections with other occupations are weak. Communication is minimal. Outside dealings are seen as a sort of disloyalty. In this scenario, there exist few opportunities for ideas, data, and practices to traverse the occupational frontiers. Neighboring occupational groups may glimpse the results of a technology without divining its substance or intuiting how it might be extended to their occupational pale.

In like register, the material/technical attributes of a narrow domain technology are not extensive. Certain substances, operations, finished products, and so forth simply do not physically readily fit into other substances, operations, and products. Alternatively, if there is no obstacle of fit, efficiency or cost, there may be insurmountable material barriers. In other words, often for material reasons, such technologies cannot perform the function of intermediary components that are essential to the form of complementarity needed to transform a narrow domain technique into a more general application novelty.

By comparison with our second major category of innovation, general-purpose technology (to be discussed below) (Bresnahan and Trajtenberg 1995; Helpman 1998; Joerges and Shinn 2001; Shinn 1993; Shinn and Joerges 2002), the organizational environment associated with narrow domain innovation operates as an important constraint, be it in differentiated occupational structures and even in loosely federative integrating structures. Although potential for intellectual and material extension is unarguably greater for integrative occupations than for highly differentiated ones, as exemplified in the case of cell biology, it nevertheless holds for both profiles that scope remains restricted. Occupational and material factors bar the path to diversity and distributedness. Another way of seeing this is that both weakly integrating and differentiated occupational cultures lack the measure of inherent slack more characteristic of the transverse cognitive and occupational arrangements of the sort associated with general-purpose technologies. In this important respect, loosely federative integration and differentiation-driven groups share parallel dynamics as regards narrow domain innovation.

From general-purpose technology to generic instrumentation and research technology

The concept of general-purpose technology was introduced and has been mainly developed by macroeconomists (Bresnahan and Trajtenberg 1995; Helpman 1998). To my knowledge, its general sociological, organizational, and cognitive implications have, to date, not been explored. General-purpose technology typically includes innovations like the steam engine, the electric motor (Baird 2004), chemical engineering (Rosenberg 1998), the transistor, chip and integrated circuits, the laser, computer, and so forth. General-purpose technologies are characterized by their inherent pervasive potential for technical change, and by the enhanced opportunities for rent that they offer through their technical complementarities. They spread outward both along the vertical and horizontal axes. They diffuse vertically when a general purpose technology is incorporated in a series of operations extending from primary component, through intermediate processing or production, right down to the final product. The primacy of the transistor as a fundamental electronics component in manufacturing processes and its use in many professional and consumer products is in some part due to the fact that this distribution is consistent with the simplicity, reliability, compactness, and ease of interconnection of the newly developed technology.

General-purpose technology also diffuses horizontally across a large number of sectors, as its advantages may apply to a particular function or stage in a large number of otherwise very different spheres. To take again the example of the early transistor, it figured in military devices, aviation, radios, hearing aids, etc. To repeat, the outstanding advantage of a general-purpose technology resides in its compatibility with many applications and with a variety of applications. In brief, on the vertical axis they act as intermediary components, and on the horizontal axis it is their complementarity that stands out.

Most of the discussion of general-purpose technology has focused exclusively on its role in stimulating and maintaining economic growth. It is frequently argued (Helpman and Trajtenberg 1998a and 1988b) that this involves two phases. In an initial phase, general-purpose technology induces a decline in economic development and profits, since resources are funnelled into R&D and diverted away from production. It is only in the second phase that the general-purpose technology R&D is fed productively and profitably into the economy.

Empirical studies have, however, brought to light important inconsistencies in this two-phase model, and have suggested that the general-purpose technology dynamic is considerably more complex than it is often held

to be. Aghion and Howitt (1998) stress the crucial role played by learning in general-purpose technology. The vertical and horizontal diffusion of a general-purpose technology is accompanied by learning, as specialists in different sectors and performing different functions discover which aspects of the technology are relevant, come to master the technology, and learn how to adapt it to their particular requirements. This learning process becomes a virtuous circle. The sector-based and function-based learning generates data and ideas that retro-enrich the initial general purpose technology, which, in its turn, once again reinforces local technical practice and products.

The concept of general-purpose technology raises a particularly crucial question for the study of innovation/occupation dynamics. By what mechanisms do techniques transit between occupational groups, each one of which possesses its own particular material, intellectual, and professional culture? The closed spaces of highly differentiated occupations pose an acute obstacle to transverse movement; and the dynamics of the form of weakly integrative occupations, seen above, also introduce some difficulties. In the pages that follow, I will describe a recent, essentially transverse technical movement (research technology), and will show how, through the material/epistemological characteristics of its technical artifacts (generic instrumentation), research-technology groups enable closed, differentiated occupations to enjoy a measure of horizontal communication—communication that is crucial in an age of increasing cognitive and organizational fragmentation. It will be further argued that the species of transverse arrangements represented by research technology and generic instrumentation are key elements in processes of radical innovation.

Research technology and generic instrumentation as vehicles of transversality

The generic instrumentation of research technology is conceived and matures in a particular form of organizational space. The technology is characterized by specific epistemological and theoretical features. Research technologists engage in unusual activities that foster boundary crossing of their generic techniques. The latter adopt initiatives to smooth the outward path of their generic devices and, similarly, to ensure reverse technical flow initiated by users, which renews and diversifies generic output.

The conscious and organized development of generic instrumentation arose in the late nineteenth century, first in Germany (Shinn 2001a), and then in the US (Shinn 2001b), Great Britain, France (Shinn 1993), Japan, and the USSR. The movement was spawned by Berlin-based instrument craftsmen, soon to be joined by engineers, technicians, and some scientists. Their critical idea was to move away from building devices designed to deal with a particular local technical situation. They instead sought to deal with the laws of instrumentation as opposed to the laws of nature. Their interest lay in uncovering fundamental instrument theory that could then be translated into a range of practical applications. In pursuit of this goal, the authors of such generic instrumentation set up dedicated associations, such as the Deutsche Gesellschaft für Mechanik und Optik, and sponsored generic apparatus reviews—the *Zeitschrift für Instrumentenkunde* (founded in 1883) and later *La Revue d'optique instrumentale et théorique* (1920), the *Review of Scientific Instruments* (1930), *Journal of Scientific Instruments* (1922), *Metrologia* (1965), and so forth. Stated differently, practitioners preferred to develop general instrument principles as embodied in general apparatus which could in turn be exported to multiple applications. The logic was that of transversality.

A concrete example illustrates this generic instrumentation orientation. Conventionally, instrument makers develop devices in response to highly specific problems. In instrumentation and technology trade fairs, an innovation is usually exhibited in the framework of a specific achievement (Von Hippel 1988). Hence technical novelties in the field of electronics are to be found in the electronic section, radiation innovations in the radiation section, optical improvements appear in the optical section. However, the logic of the research technologist and generic devices runs quite contrary to this pattern. The generic devices of research technologists are exhibited together, with little regard for application. By so doing, emphasis is placed on the multipurpose, multifunction, general properties of fundamental innovation. This exhibition strategy reveals the research technologist's desire to deal with particularities, but through the application of theory and principles to specific instances. This logic and form of presentation was adopted by German research technologists in the 1893 Chicago Universal Exposition, and again in Paris in 1900 and St. Louis in 1904. The new German approach to instrument research was viewed as constituting an innovative (and for some countries, a threatening) stimulant to scientific, industrial, and military development. Generic instruments developed in this spirit include automatic switching devices, the ultracentrifuge, the rumbatron (Shinn 2001b), and Fourier transform spectroscopy (Johnston 2001).

It is crucial to answer the question, how do research technologists manage to work on instrument principles, work which requires a measure of freedom and distance from interests and institutions characterized by short-term demands? Research technologists operate in an interstitial arena—acting in the spaces that occur between established organizations. Each research technologist belongs to a capital organization—university, small or large enterprise, state technical service or metrological laboratory, the military, and so forth. Nevertheless, he tends to move from institution to institution and to circulate between institutions. This does not mean that he is non-professional; rather, he is multi-professional. By being everywhere, research technologists escape bondage to any single interest. These interstitial coordinates often provide the measure of freedom and spans of time required by a commitment to long-term technical research of the sort that leads to the discovery of fundamental instrumentation principles (Nelson and Rosenberg 1994). The international research-technology venture on multi-beam interferometry, leading to the development of fast Fourier transform IR spectroscopy spearheaded by Pierre Jacquinot and his French research-technology CNRS laboratory, ran for almost thirty years (1942 to *circa* 1970). During this time, the fundamental generic new interferometry principles advanced by P. Jacquinot, P. Fellgett, J. Connes, W. Cooly, and W. Tukey became fully articulated, traversed numerous specialty boundaries, were understood, adopted, and adapted by local users. At the same time, user questions and practices were reverse-flowed back into the interstitial generic research-technology community, re-examined and modified, and again traversed specialty domain boundaries before finally being integrated into innumerable scientific, manufacturing, and metrological products and procedures. In the course of this process emerged a generic instrument lingua franca that permitted the practitioners of countless differentiated and sometimes isolated functions, domains, and uses to communicate and interact effectively (Shinn 2003).

Working out of an interstitial arena offers many important advantages. The interstitial arena is not a vested interest group. Its practitioners are hence unlikely to be perceived as competition by constituted occupations and organizations. Because they do not represent a threat, research technologists can establish comfortable interactions with occupational groups of many sorts. This is of incalculable importance, for on the social register it is this consideration that facilitates the boundary-crossing that is essential to innovation processes—boundary-crossing that does not destabilize occupational specialties. But what are the material, cognitive, and epistemological characteristics of the generic devices fashioned inside the interstitial arena?

In generating the fundamental artifacts involved in very high-speed ultracentrifuges (three million revolutions per second), semiconductor, microprocessor technology, principles of optical pumping, virtual reality software, and the rumbatron, research technologists develop artifacts, processes, and methodologies of the most general kind that one might liken to stem cells. The generic instrument embodies general theories and material or procedural relations that are not only fundamental, but that also enjoy cognitive and material linkages with many diverse and distant substances and processes. Stated differently, the research technologist labors to embed his fundamental instrument principles in a generic device. For example, building on the interferometry instrument principles of Michelson, in the 1940s, 1950s, and 1960s P. Jacquinot formulated what came to be known in optics and far beyond as the Jacquinot Advantage; and Jacquinot and colleagues then went on to embed this instrument concept in his Fourier transform infrared spectroscope. These general technical relations can be seen as being embedded in the generic instrument.

The work of research technologists is double. On the one hand, they develop their generic apparatus. On the other hand, they then dis-embed and further assist in selectively re-embedding these principles with reference to a variety of groups and diversity of techniques, functions, and usages. This is done in two ways. First, research technologists may become involved with particular occupational specialties in a search for ways in which to re-embed the generic potential in a particular application. Second, on the basis of this re-embedding experience and the lessons drawn from it during the process of technical reverse flow, where the research technologist learns from seeing how local practitioners build on their generic concepts and apparatus, they co-participate in establishing a kind of research-technology template, essential to the stabilization and standardization of the generic innovation in a trans-local setting. Here one discerns the making of a research-technology lingua franca that promotes intelligibility for and between the many distinctive and disunited groups, interests, and competencies involved in the research-technology process. This active process of generic concept, embedding, dis-embedding, re-embedding, and again embedding is best described as recursive research-technology 'cycling'. Cycling is multi-party and multicompetence and is, by its very logic, rooted in genericity, interstitiality, and metrology. Here, research technologists pass back and forth between the generic embedded, dis-embedded, and re-embedded-produced, trans-local template on the one hand, and intermittent and selective involvement in particularistic and parcelized re-embedding applications on the other. They are thus central operators in the diffusion chain and participate in occupation-centered learning processes in which the occupation temporarily and selectively opens itself to assertive exogenous inputs.

The distribution pattern of research technologists' generic productions indicates just how unlike they are to other innovation groups, and the extent and diversity of their impact. About one-quarter of research-technology written productions appears in specialized science journals and another quarter in engineering journals. This contrasts with the output of both scientists and engineers, who publish almost exclusively in their own professional forum and normally in a restricted range of publications. By contrast, research technologists circulate their generic principles in literally scores of far-ranging engineering and science specialties. In addition to production in the 'public domain,' research technologists undertake private consulting, prepare confidential reports, write in restricted circulation company news letters, and take out patents. Indeed the scope of their activities with firms and public research often goes far beyond what is habitual for most engineers—and of course for all scientists.

The trajectory of Jesse Beams illustrates this pattern. Beams's doctoral dissertation was intended to deal with the absorption time of quantum events. In order to undertake the measurement of the very short time-spans involved in such processes, he designed and engineered a series of ultra-rapidly rotating devices. This involved introducing novel semi-flexible drive mechanisms, servo devices, magnetic levitating and spinning apparatus, etc. His consummate ultracentrifuge rotated at over three million revolutions per second. It is fair to say that Beams developed the funda-

mental principles associated with high-speed rotation. This was the substance of his generic instrumentation. Through the mastery of these principles, great strides were quickly made in innumerable science, engineering, industry, metrology, and military domains—bacteriological and viral research, a high-precision gravitational constant, research on thin films, the determination of photon pressure, work on ram jet technology, purification of uranium, and so forth. Beams did not work out all of these applications of his generic instrument. The principles could, to some degree, spread independently. Nevertheless, Beams did sometimes become involved in the initial stage of diffusion work. It was here that he helped prepare a kind of training program for those who would adopt and adapt the generic device. The research technologist thereby acted as a linkage mechanism between interstitial-based fundamental innovation creation on the one hand and, on the other hand, downstream spillover into countless realms in which specialized occupational groups took on, in their particular way, the novel inputs. In the case of Beams's generic instrument, the occupational groups of physical and fluid mechanics, combustion, chemistry and chemical engineering, medicine, virology and bacteriology, metrology and quantum physics, and engineering were all involved in, and the beneficiaries of, spillover.

Reverse-flow processes are also crucial here. In the process of suggesting to a diversity of occupational specialties the potential of their generic devices, research technologists acquire ideas and technical information that feed back into the generic upgrading cycle. The generic instrument traverses an occupational boundary on its way to re-embeddings, and through the local involvement of research technologists the re-embedding inspirations and lessons are turned back into the interstitial arena. I posit that general-purpose technologies at large and, more particularly, research technology, would soon become exhausted in the absence of such reverse flow, where so much depends on inputs from occupational groups.

Yet something beyond technical data and learning opportunities is encapsulated in generic instrumentation. This category of device contains key elements that permit communication and direct linkage between all of the occupational bodies that have subscribed to a research-technology production. How does this work? As indicated above, a generic instrument incorporates fundamental expressions, in terms of materials or procedural algorithms, etc. When specialist occupations import an aspect of a generic instrument in order to generate a local innovation, the fundamental principles are also imported. These fundamental expressions may take the form of a vocabulary, standards or norms, images, or even a complete paradigm. In the case of Jesse Beams' ultracentrifuge, a language and technology of fractionation, expressed in specific density and gravitational pressures, emerged and is now used as a baseline for communication both within and between many fields and occupations. Stated differently, generic instrumentation vehicles a kind of lingua franca. This lingua franca is transverse. It allows otherwise distinct and distant occupational specialties to communicate effectively, thereby somewhat reducing the otherwise rampant consequences of ultra-postmodern specialization and fragmentation.

In what ways is research technology distinct from both the above-described differentiation and loose federative integration arenas? How is sustainable radical innovation connected to the transverse dynamics of research technology, which transcends the pale of integration and differentiation by implementing their complementarities? As indicated at the outset of this chapter, research technology supersedes the traditional antagonisms and paradox of differentiation versus integration through demonstrating that they constitute two sides of the same coin.

Research technology is transverse in four senses:

(a) In the process of technical dis-embedding and re-embedding, subtle but important and often lasting ties develop between research technologists and the engineers, technicians, and scientists of host occupations who adopt and adapt generic instruments. Outward boundary-crossing from the interstitial arena into local occupational spaces comprises the mechanism through which this component of integration operates.

(b) Processes of reverse flow further knit the integrative framework of research technology. Here the technical competence of diverse occupational specialties is conveyed to research technologists during reverse flow re-embedding, thereby being transported back into the research-technology interstitial arena.

(c) As emphasized above, the lingua franca generated during the interactions of research technologists and local occupational cultures strongly connects all of the occupations which have adopted and adapted a generic instrument to the hub research-technology community.

(d) The research-technology hub acts as a medium through which transoccupational communication flows from occupation to occupation, once again thanks to the lingua franca. The fact that research technology generates the basis for the lingua franca, participates in its articulation, implementation, and diffusion, and later sustains and nourishes the lingua franca is fundamental to its transverse status. From this we see that research technology owes its transverse action less to organizational features than to material and epistemological dynamics. In a word, research technology's transverse effects are a form of materialized reflexivity, in which segments of social action are tied to concrete technical practice.

Two factors underpin research technology's capacity for radical innovation:

A. the pronounced capacity for recursive technical combination and recombination;
B. recursive transoccupational/trans-intellectual communication.

Research technology incorporates, indeed strongly promotes, both of these features. While it is evident that, numerically speaking, most innovations consist of locally inspired, developed, and absorbed recombinations of established techniques, their potential for further recombination and for additional occupational, cognitive, and economic extension is soon exhausted, and also remains local. By contrast, the generic instrument logic and epistemology of research technology promote extensive recursive cycles, resulting in technical complementarities which are expressed in vertical and horizontal spillover on an often undreamed-of scale. For this to occur, however, occupational complementarity is similarly required. Consciousness of material complementarity results from easy communication between the founders of a template technology and the occupations that take the technology up and, equally crucial, communication between a diversity of occupational specialties. In the absence of this exchange, awareness and knowledge of complementarity would languish, or never occur. There is another key element here. As indicated above, distributive learning, achieved through boundary crossing and through reverse flow, is necessary to the maintenance of recursive recombination. Without this, transversality-fuelled sustainable radical innovation would soon degenerate into narrow domain innovation of the sort often associated with incremental innovation.

Growing innovation: a plea for plasticity

Radical innovation is neither assembled nor built. It is grown. It demands much more than either differentiation or even integration-based connections and processes. Indeed, radical innovation is grown by dint of transverse connections and combinations, which include both differentiation and integration, and also transcend these classical expressions of intellectual and organizational structuration. In view of this, it follows that the introduction and maintenance of radical innovation hereby requires a species of plasticity in artifact design and construction, community interaction, and diffusion practice of a

sort that exceeds the routines of most firms, and often escapes the analysis of most innovation experts. This is not a plea for relaxing the rules and regulations associated with the division of labor, organizational structure, or group identity. It is not a call for utter fluidity or de-differentiation. To the contrary, measured differentiation remains requisite, yet it must be compensated for by partial and selective mobility, allowing individuals and groups the freedom of temporary boundary-crossing. The growth of radical innovation, tied to transverse research-technology dynamics, revolves around distributive learning. Here, learning is multi-party, multidirectional, and reciprocal. Radical innovation engages learning both from the users of new technologies and their initiators. Such distributed learning is concentrated in the cycling processes of dis-embedding, re-embedding, reverse re-embedding, and template development associated with generic instrumentation and general-purpose technology. Indeed, the capacity to innovate is an acquired, learned skill. Radical innovation is thus 'cultivated,' in the sense that it grows out of an environment possessing particular traits.

The most cursory sociographic glance at innovation studies, and related fields, suggests that they have, by and large, been based on a differentiationist interpretation. Much of modernity has involved differentiation (the division of labor, professional and disciplinary specialties, class structure, nation-building, etc.), and it is hence understandable that much sociology has concentrated on this aspect of human action. Sociology has successfully charted the linkage between innovation and occupational differentiation, but in so doing it has often overlooked integration-driven occupations and, yet more crucially, overlooked the operation and impact of transversality—and the import of both on innovation processes.

This chapter has indicated ways in which a form of sociology sensitive to integration processes and, more specifically, to transverse material, cognitive, and occupational processes, can capture highly crucial components of innovation activities. Research technology and generic instrumentation, as mechanisms that underpin general-purpose technologies, constitute fundamental motors of sweeping contemporary technical change. Transversality-fuelled innovation is, it has been argued, often the basis for radical sustainable innovation. Radical innovation contrasts with narrow domain innovation which, I suggest, is frequently associated with differentiationist occupational culture.

The governance of research and innovation may benefit from a greater appreciation that there exists a range of innovation-related occupational formats. Each category of format possesses its specific material, cognitive, epistemological, and professional traits. It would be unwise to sacrifice some formats while privileging other formats and breeds of innovation. This would constitute a risky policy, as innovation chains are becoming increasingly complex and interdependent. Wise governance consists of striking a balance. Innovation remains fraught with uncertainty, and important contributions may have unexpected sources.

This said, the species of innovation connected to research technology and generic instrumentation possesses *ipso facto* a significant strength. It generates a lingua franca which permits transverse exchange across otherwise often closed occupational boundaries. By so doing, it mitigates occupational fragmentation; it is systematically beneficial to innovation processes by dint of its cohesive action and its capacity to provide a broad communication platform.

Note

1. *Social Science Information*, Vol 44/4 (2005). London, Thousand Oaks, CA, and New Delhi: SAGE Publications.

References

Abbott, A. (2001). *Chaos of Disciplines*. Chicago: University of Chicago Press.

Abir-Am, P. (1992). 'From Multidisciplinary Collaborations to Transnational Objectivity: International Space as Constitutive of Molecular Biology, 1930–1970.' In E. Crawford, T. Shinn, and S. Sverker (eds.) (1992), *Denationalizing Science: Sociology of the Sciences Yearbook*. Dordrecht: Kluwer Academic Publishers, 153–86.

Aghion, P., and Howitt, P. (1998). 'On the Macroscopic Effects of Major Technological Change.' In E. Helpman (ed.), *General Purpose Technologies and Economic Growth*. Cambridge, MA: MIT Press, 121–44.

Baird, D. (2004). *Thing Knowledge: A Philosophy of Scientific Instruments*. Berkeley: University of California Press.

Bechtel, W. (1993). 'Integrating Sciences by Creating New Disciplines: The Case of Cell Biology.' *Biology and Philosophy*, 8/3: 277–99.

Ben-David, J. (1960). 'Roles and Innovations in Medicine.' *American Journal of Sociology*, 65/6: 250–8.

—— (1991). *Scientific Growth: Essays on the Social Organization and Ethos of Science*. Berkeley: University of California Press.

—— and Collins, R. (1966). 'Social Factors in the Origins of a New Science: The Case of Psychology.' *American Sociological Review*, 31/4: 451–65.

Bijker, W. (1997). *Of Bicycles, Bakelites and Bulbs: Toward a Theory of Sociotechnological Change*. Cambridge, MA: MIT Press.

—— Hughes, T., and Pinch, T. (eds.) (1987). *The Social Construction of Technological Systems: New Directions in the Sociology and History of Technology*. Cambridge, MA: MIT Press.

Bourdieu, P., and Passeron, J. C. (1970). *La Reproduction: éléments pour une théorie du système d'enseignement.* Paris: Minuit.

Bresnahan, T., and Trajtenberg, M. (1995). 'General Purpose Technologies: Engines of Growth?' *Journal of Econometrics*: 65/1: 83–108.

Brown, F. L. (1967). *A Brief History of the Physics Department of the University of Virginia, 1922–1961.* Charlottesville, VA: University of Virginia Press.

Callon, M. (1986). 'Éléments pour une sociologie de la traduction: la domestication des coquilles Saint-Jacques et des marins-pêcheurs dans la Baie de Saint-Brieuc.' *L'Année Sociologique*, 36: 169–208.

Edge, D., and Mulkay, M. (1976). *Astronomy Transformed: The Emergence of Radio Astronomy in Britain.* New York: Wiley Interscience.

Elzen, B. (1986). 'Two Ultracentrifuges: A Comparative Study of the Social Construction of Artifacts.' *Social Studies of Science*, 16/4: 621–62.

Forman, P. (1971). 'Weimar Culture, Causality and Quantum Theory, 1918–1927: Adaptation by German Physicists and Mathematicians to a Hostile Intellectual Environment.' *Historical Studies in the Physical Sciences*, 3: 1–115.

Galison, P. (1997). *Image and Logic: Material Culture of Microphysics.* Chicago: University of Chicago Press.

Gibbons, M., Limoges, C., Nowotny, H., Schwartzman, S., Scott, P., and Trown, M. (1994). *The New Production of Knowledge: The Dynamics of Science and Research in Contemporary Society.* London: Sage.

Helpman, E. (ed.) (1998). *General Purpose Technologies and Economic Growth.* Cambridge, MA: MIT Press.

—— and Trajtenberg, M. (1998a). 'Diffusion of General Purpose Technologies.' In E. Helpman (ed.), *General Purpose Technologies and Economic Growth.* Cambridge, MA: MIT Press, 85–120.

—— —— (1998b). 'A Time to Sow and a Time to Reap: Growth Based on General-Purpose Technology.' In E. Helpman (ed.), *General Purpose Technologies and Economic Growth.* Cambridge, MA: MIT Press, 55–84.

Joerges, B., and Shinn, T. (eds.) (2001). *Instrumentation between Science, State and Industry.* Dordrecht: Kluwer Academic Publishers.

Johnston, S. F. (2001). 'In Search of Space: Fourier Spectroscopy, 1950–1970.' In B. Joerges and T. Shinn (eds.), *Instrumentation between Science, State and Industry.* Dordrecht: Kluwer Academic Publishers, 121–41.

Kuhn, T. S. (1962). *The Structure of Scientific Revolutions.* Chicago: University of Chicago Press.

Latour, B. (1989). *La Science en Action.* Paris: La Découverte.

—— and Woolgar, S. (1979). *Laboratory Life: The Social Construction of Scientific Facts.* London: Sage Publications [2nd edn. 1986, Princeton: Princeton University Press].

Lawrence, P., and Lorsch, J. (1967). *Organization and Environment: Managing Differentiation and Integration.* Cambridge, MA: Harvard University Press.

Merton, R. K. (1938). 'Science and the Social Order.' *Philosophy of Science*, 5: 321–37.

—— (1942). 'Science and Technology in a Democratic Order.' *Journal of Legal and Political Sociology*, 1: 115–26.

—— (1973). *The Sociology of Science: Theoretical and Empirical Investigations.* Chicago: University Press of Chicago.

Mokyr, Joel (2002). *The Gifts of Athena: Historical Origins of the Knowledge Economy.* Princeton: Princeton University Press.

Nelson, R. R., and Rosenberg, N. (1994). 'American Universities and Technical Advance in Industry.' *Research Policy*, 23/3: 323–48.

Noble, D. (1977) *America by Design: Science, Technology and the Rise of Corporate Capitalism.* New York: A. Knopf.

Nowotny, H., Scott, P., and Gibbons, M. (2001). *Re-thinking Science: Knowledge and the Public in an Age of Uncertainty.* London: Polity Press and Blackwell Publications.

Olesko, K. M. (1991). *Physics as a Calling: Discipline and Practice in the Königsberg Seminar for Physics.* Ithaca, NY: Cornell University Press.

Rasmussen, N. (1997). *Picture Control: The Electron Microscope and the Transformation of Biology in America, 1940–1960*. Stanford, CA: Stanford University Press.

—— (1998). 'Instruments, Scientists, Industrialists and the Specificity of Influence: The Case of RCA and Biological Electron Microscopy.' In J. P. Gaudillière and I. Löwy (eds.), *The Invisible Industrialist: Manufactures and the Construction of Scientific Knowledge*. London: Macmillan, 173–208.

Rosenberg, N. (1998). 'Chemical Engineering as a General Purpose Technology.' In E. Helpman (ed.), *General Purpose Technologies and Economic Growth*. Cambridge, MA: MIT Press, 167–92.

—— and Nelson, R. (1994). 'Universities and Technical Advance in Industry.' *Research Policy*, 23: 323–47.

Shinn, T. (1993). 'The Grand Bellevue Electroaimant, 1900–1940: Birth of a Research-Technology Community.' *Historical Studies in the Physical Sciences*, 24/1: 157–87.

—— (2000). 'Formes de division du travail scientifique et convergence intellectuelle.' *Revue Française de Sociologie*, 41: 447–73.

—— (2001*a*). 'The Research-Technology Matrix: German Origins, 1860–1900.' In B. Joerges and T. Shinn (eds.), *Instrumentation between Science, State and Industry*. Dordrecht: Kluwer Academic Publishers, 29–46.

—— (2001*b*). 'Strange Cooperations: The U.S. Research Technology Perspective, 1900–1955.' In B. Joerges and T. Shinn (eds.), *Instrumentation between Science, State and Industry*. Dordrecht: Kluwer Academic Publishers, 69–95.

—— (2002). 'The Triple Helix and New Production of Knowledge: Prepackaged Thinking on Science and Technology.' *Social Studies of Science*, 32/4: 599–614.

—— (2003). 'General Purpose Technologies, Generic Instrumentation, Occupations and Institutions: In Search of Linkage Mechanisms.' Paper Presented at the Berlin Technical University Workshop Entitled *Innoversity*, 18–20 September 2003, Berlin.

—— and Joerges, B. (2002). 'The Transverse Science and Technology Culture, Dynamics and Roles of Research-Technology.' *Social Science Information*, 41/2: 207–51.

—— —— (2003). 'Paradox oder Potential? Zur Dynamik Heterogener Kooperation.' In J. Glaser *et al.* (eds.), *Kooperation im Niemandsland: Neue Perspektiven auf Zusammenarbeit in Wissenschaft und Technik*. Opladen: Budrich, 77–101.

—— and Ragouet, P. (2005). *Dynamiques des sciences: sociologie des activités scientifiques*. Paris: Raisons d'Agir.

Shishoni, D. (1970). 'The Mobile Scientist in the American Instrument Industry.' *Minerva* 8/1: 58–89.

Von Hippel, E. (1988). *The Sources of Innovation*. Oxford: Oxford University Press.

15 How Markets Matter: Radical Innovation, Societal Acceptance, and the Case of Genetically Engineered Food

Eric Jolivet and Marc Maurice

Introduction

The authors of this book share a common approach to the equation: knowledge creation + collective learning = innovation. Our contribution's objective is to shed some light on the market side of innovation. Important theoretical work and ample empirical evidence have been collected to demonstrate the role of the market as a major source of learning and innovation (Rosenberg 1982; Von Hippel 1988; Bijker and Law 1992). Our interest here is more concerned with how markets learn about radical innovations. How do consumers come to form a judgment about a product or service they never heard of before and that has no equivalent so far? And how does this learning process affect their decision behaviors and consequently the diffusion pattern of the new products?

Theories of diffusion have long recognized the importance of 'information'—the knowledge consumers need to make their decisions. The influence of word of mouth (Mansfield 1961) and cross-referencing (Moore 2002) in technology adoption has been underlined in several models, stressing the interactive and collective nature of this 'information'. But the process by which this learning is done has remained to a large extent a research question to date (Geroski 2000). How does this learning process really work? How do consumers come

to form and share representations about radically new products? How do they learn about their performance and use? What factors condition the process by which radically new products are becoming a naturalized part of our daily life?

This issue is of considerable practical interest to managers and innovators, since consumers' decisions condition innovation's success or failure. Our focus will be on the cognitive dimension of consumer decisions, a fundamental key to understanding the fate of radical innovations. Our contribution will draw upon the constructivist tradition to address this question. We will analyze the knowledge creation that goes with the diffusion of a radical innovation. Drawing on Piaget's tradition, we argue that the adoption process of a radical innovation triggers the formation of a cognitive framework, thanks to which consumers make sense of the information about the new product and the society that goes with it. In turn, when this cognitive framework evolves, so does the judgment made about the innovation. Information first diffused about DDT presented it as very beneficial—'the magic insect killer.' Some twenty years after, it had changed to presenting it as a very evil chemical—'one of the most infamous' (Maguire 2004). The same product has taken on a whole different meaning according to different frames of reference.

Particularly extreme examples of this learning situation are provided by innovations based on breakthrough technologies conveying strong uncertainty about their impact on society and, as such, facing social resistance (Bauer 1995). The diffusion process is radicalized into a question of acceptance or rejection. It is increasingly labeled as 'societal acceptance.' Nuclear power, Taylorism, artificial procreation, and, more recently, genetically engineered food are obvious examples. The link between all these examples is the importance of the social changes at stake. Since they are conveying social changes that are not consensual, such innovations are becoming political issues. As such, we believe they are representing an excellent opportunity to deepen our understanding of how consumers come to evaluate radical innovations. Indeed, the sense-making and judgment-formation processes are largely elicited and discussed through confrontations often traced in media coverage, interest-group actions, and more or less institutionalized public forms of debate. In other words, the learning process usually performed in a tacit manner by consumers is more formalized and exposed. The controversies characteristic of such cases represent valuable sources of information (Rip 1986).

On the other hand, social resistance to technologies is not homogeneous. Strong variations are observable depending on the particular societies and the markets. A striking example is provided by the diffusion of nuclear power in the world (Bauer 1995). The GM food innovation represents an equally excellent case. GM food is considered as the major recent innovation in the agro-food sector. At the same time, it has raised considerable social concern and resistance. During its early stages of diffusion—although more convergence is expected in the near future as the technology becomes more entrenched and stabilized—markets around the world reacted very differently to the commercialization of GM food. The United States (US) market adopted the technology at a very fast pace. In Europe and Japan, however, GM food and, more generally, GMOs became a controversial topic: different social groups engaged in making sense of what GMOs were. They could not reach an agreement about GM food dangers, value, and use. Finally, mounting public opposition led to the delay of GMO diffusion, a situation that, in Europe, lasted until early 2004. This renders comparative analysis and comparative methodology particularly relevant to this issue.

This empirical example will be the basis for discussing how the early-stage formation of consumer judgment about innovation influences market reaction and the technology diffusion patterns. The chapter is organized in two main parts. The first one outlines the theoretical foundations of our approach, based on international comparative studies: This is Societal Analysis. After positioning this approach in the literature on innovation, we will discuss its application to the GMO case. The second part of the chapter discusses the GMO case. Tracing the controversy raised by GM food in Europe, we show how a number of coalitions of actors came to consider this innovation as risky. Key comparative elements are then discussed to account for the contrasting situations observed in the US and Europe, and we suggest a possible future research agenda to address the issue of the variety of market reactions to new technologies.

Towards a societal approach of the GMO innovation

GM food represents a rather direct application of state-of-art scientific discovery into the development of new, value-added products. This occurred in the wider context of a dramatic change of the place and role of science in societies. According to many observers, the more science and technology become central to our daily life through the shaping of innovations, the more their impact on society becomes contested.

This leads to the question—relatively understudied so far—of the social acceptance of innovation. This issue seems to have found particular relevance with the advent of science-based industries such as biotechnologies and nanotechnologies of which the GM food case is an example. (In addition, for the same innovation—nuclear power reactors, GM food, cloning—significant differences in the level and nature of diffusion have been observed among countries.)

To deal with the question, we will call upon Societal Analysis, a comparative approach developed during the 1980–90s based on international field studies comparisons, first between France and Germany (Maurice *et al.* 1986) and, more recently, between France and Japan (Lanciano *et al.* 1993; Jolivet 1999). In an attempt to shed some light on the contrasted diffusion patterns reported in the US and Europe (particularly France), this chapter aims at exploring the fruitfulness of applying this theoretical framework to the GM food case.

Before we continue, a brief reminder of the origin of Societal Analysis will be given. We will discuss the interest of the approach for the study of GM food, a stimulating case in the field of international studies and the social dimension of markets. We position Societal Analysis within the literature on theories of innovation; we then conceptually and empirically apply it to the case.

Societal Analysis of innovation

Origin of Societal Analysis and contribution to the theory of innovation

Origin

Derived from comparative research on hierarchies in French and German companies, which link modes of organization to education systems (Maurice *et al.* 1986), the approach was extended in the 1980–90s to comparative studies of innovations in France and Japan. In-depth investigations in the steel and machine tool industries showed that French and Japanese companies had very different approaches to innovation, and appropriated the technologies in fairly contrasting manners. This second generation of inquiries provided the ground for the formulation of a 'Societal Analysis of Innovation' (Lanciano *et al.* 1993).

Contribution to the theory of innovation

What are the main aspects of the societal approach contribution to our understanding of innovation? We would like to stress three main points.

First, the approach contributed to revealing the social foundations of innovation. Studying the diffusion of the same technology in different societal contexts made it clear that not just diffusion curves but the very patterns of adoption differ from country to country. It provides a qualification to the universality generally associated with technology and its use, as well as the principle of societal convergence often thought to be associated with the diffusion. Noble pointed out that technology is the product of an historically situated social process: 'technology bears the social imprint of its authors' (Noble 1986). Our research shows that innovations also bear the social imprint of their adopters. Studies performed in the chemical, software, steel, and machine tool industries all confirmed this point: the ways the same new technologies were used and the associated transformations of the workplaces were found to be societally specific.

Second, our studies clearly advocate the path-dependent character of the process of innovation. The final shape and definition of the new product/process have been conditioned by the past experience of organizations, as embodied in the actors' identities and competencies—the specific social construction of actors such as researchers, engineers, workers, and users varies from country to country—and the historically constructed organized arrangements (social 'spaces') shaping their actions and interaction such as routines and rules. In our view, path dependence is not seen as rigid and immutable, but grounded in collectively agreed principles that the actors involved consider legitimate. Observed national or area coherence might then not be treated as a pre-existing, explanatory variable, but the phenomenon that needs to be explained. International comparison is used as a framework facilitating these contextualization processes.

Third, the definition of innovation derived from these studies supports a constructivist view. In-depth observations reveal that technology was not merely transferred from one place to another, but subjected to very active work by adopters, what we call 'appropriation'. By this we are designating the learning process through which actors absorb new technologies by adapting them to their local contexts. Adopters' understanding of the technology is built on their existing beliefs, values, and experience forming their cognitive frameworks of interpretation. In the process, technology is restructured, adapted, and reinvented to match their specific contexts; adopters produce their own specific knowledge of the technology: adopters are innovators. A striking example is provided by the machine tool industry development of mechatronics. The outstanding mobility of actors in Japanese firms and their cooperative type of interactions explains the development of mechatronics—a fusion of electronics and mechanical core competencies—in 1980s Japan. The strong partitioning of French firms did not predispose them to such innovative technology fusion.

As we can see, our research has led us to conceive of innovation as a phenomenon intimately intertwined with collective learning and new knowledge creation.

Innovation as a learning process

This approach refers to a part of the literature on innovation that is particularly important in Japan (Jolivet 1996). Seminal work by Aoki (Aoki 1988), Kodama (Kodama 1995), Nonaka and Takeuchi (1995) shares a common interest in seizing learning processes underlying innovation, allowing for the diffusion of new knowledge and competencies, and for the translation of tacit knowledge embodied in actors' practices into codified knowledge.

Learning as the acquisition of new knowledge and new practices has been the subject of a wide variety of definitions, and it seems necessary to provide some clarification of its meaning in our approach.

The social dimension of learning

Our comparative societal studies have revealed the social foundations of learning. Learning dynamics are not only conditioned by past experience and values; they are also closely related to the social context in which the learning takes place. Specific actors' identities and competencies, as well as the organizations and rules shaping their actions and interactions, influence the learning process, the kind of knowledge created, and the performance reached.

The professional identity of engineers for instance, their inclusion in a wider community of practice, and the role they play in developing innovation all vary among studied countries. An example is provided by the development of pulverized coal injection in the Japanese and French blast furnaces (Jolivet 1999). In the Japanese firm studied, the codification of knowledge was mainly ensured by a socio-historical actor, the Central Engineering Department (a historical figure of the technology transfer from the west) in the form of equipment design and engineering know-how, in close relation with clients (plants). In France, another historical body, Institut de Recherche Sidérurgique (IRSID), the central research laboratory, has been the driver of knowledge codification in the form of a renewed theory of the blast furnace. The Japanese appropriated the new technology by reinforcing their design and engineering competence, and the French did so by revising their theory of the blast furnace.

One important implication regards the relation between the designers of one innovation and its users. Designers have to make critical choices not only about technology but also, more or less directly, about the future users and contexts of use, and about the 'value network' that will supply it. By so doing, designers have to develop visions/scripts about the organizational and societal context in which the product they are creating will take its place (Akrich 1992). But users are not passively enrolled in these designers' scripts. As we have seen, they create, reinvent, and adapt the proposed products or services to their specific contexts of use. Scripts for the future are therefore a natural place of negotiation and cross-learning between product designers and product users (De Laat 1996).

While the main choices and assumptions relative to the technology are made during the conception stage by designers, their becoming durable and irreversible depends on their acceptance and adoption by other actors. In that sense, we argue, new products do not really diffuse through society; they rather get through a gradual process of acceptance and embedding that we call 'appropriation' (Maurice et al. 1988).

Path dependence and diffusion

Another important implication regards the path-dependent and cumulative character of learning we noticed earlier. Leavitt and March (March 1991) have acknowledged two different dynamics of learning accounting for this cumulative character: one in which new knowledge is built within an existing cognitive framework (exploitation); one in which the cognitive framework is the very subject of development (exploration).

Evolutionist theories have stressed the cognitive foundations of technical change as being path dependent (Nelson and Winter 1982). Once a paradigm emerges, it becomes a guide for decision-making and learning. A dominant design acts as an exemplar/template, a practical standard of reference around which variations are organized (Clark 1985). Problem-definition and problem-solving activities are focused on a limited number of paths or trajectories; these lead to a number of incremental and cumulative improvements (Dosi 1982): in that sense, paradigms form shared cognitive frameworks (Metcalfe and Miles 1994).

By contrast, pre-paradigmatic periods are characterized by uncertainty and conflicts about the path to take (Garud and Rappa 1994; Lynn et al. 1996; Courtney et al. 1997). In early stages, 'there is often little agreement about technology's ultimate form or function' (Garud and Rappa 1994: 347). What is at stake is not the improvement of an existing technology along a well-defined path (exploitation), but the creation of a new technological path (exploration), of new markets, of new institutions to regulate them (Hargadon and Dougla, 2001; Garud et al. 2002). But after some time, successful products become mass-consumed, one dominant design being a shared reference for the actors of the industry. The question that then comes to mind is, how is a shared reference and framework about emerging technologies formed?

The example provided by Van den Belt and Rip (1987) in the case of the synthetic dye industry is striking. When the first synthetic dye—red aniline—was discovered, it was not recognized as a synthetic dye, despite its commercial success, but rather placed as a standard product in the color repertoire of color experts, a dominant profession in the textile industry. The qualities and possible use of new products had first been interpreted through the existing paradigm, and channeled through the very same value networks as natural colors: 'the break with the past was not immediately perceived in its full significance' (Van Den Belt and Rip 1987: 143). It was only when a shared cognitive framework—the 'azo' theory—and the synthetic dye industry emerged that its current economic significance was understood and its full qualities discovered.

Learning as an embedment process

In the societal analysis approach, learning processes are the main vectors of organizational change. Innovation, once appropriated, gets embedded into the societal contexts of which they are becoming a part. In firms, this embedding process is the subject of negotiation between actors.

In the wider context of the markets, this embedding process has also been widely observed, networks replacing organized arrangements. Social studies of innovation have enlightened the social embedding process of 'word of mouth' and 'cross-referencing' by which one dominant network is emerging. Knowledge about new technology emerges in early interactions between actors within social networks. Beliefs and interpretations about the technology-optimal shape, value, and uses are confronted as networks compete to impose alternative designs as dominant. Eventually, the diffusion process institutionalizes one 'legitimized' view and an 'established' knowledge base about the technology and its uses.

Embedment processes then regularly take the form of a competition between competing technologies supported by different networks of actors (David 1985; Cowan 1991; Garud and Rappa 1994; Garud *et al.* 2002). For Arthur, selection comes by numbers: the more an innovation is adopted, the more the incentive/returns to adopt it increase (Arthur 1989). The question is then, what conditions adoption?

Discussing the institutionalization of central utilities as the now established and agreed-upon way to deliver electricity, Granovetter's answer points to the respective power and strategies of competing networks. According to the author, the dominance of central stations 'won this battle not because his solution was the technologically correct one, but rather because he was able to construct the winning coalition' at a very early stage (Granovetter 1992: 9). By the time other arrangements, such as decentralized home-based generators, matured and their proponents networks became stronger, central stations were entrenched and the technological path irreversible.

In early times, relative qualities and performances of alternative products have been found to often be the very subject of conflicts and battles: the ultimate shape and success of one design over others is the outcome of this strategic process of negotiation and evidence building of one technology superiority (Garud and Rappa 1994; Latour 1987; Garud *et al.* 2002).

These reflections provide an insight into the considerable complexity of the learning process underlying consumers' decisions to adopt a radical innovation. Turning to the cognitive dimension of these decisions led us to emphasize the necessary knowledge creation and collective learning process accompanying the diffusion of radical innovations. As we have seen, the fate and ultimate shape of innovation are linked to their social embedding in specific societal contexts. To become mass consumed and dominant, the product has to go through an institutionalization process during which 'pre-competitive' battles strike, and alternative design supporting networks of actors—including established old technology proponents—are fighting each other. This chapter suggests that the ultimate social embedding of a technology does not occur unless a proper and legitimate cognitive framework is built to make sense of the technology.

Interest and directions for a societal analysis of GMOs

GMOs are certainly a different story from machine tools and pulverized coal, but it is our contention that the application of a similar conceptual framework would be of interest. Again, international comparison would help contextualize the learning processes occurring in each country at the early stage of GMO diffusion. Why have GMOs been quickly accepted in the US and rejected in most of Europe? Shall we not find crucial insights by studying the identity of the actors involved and the social places in which they interacted? Was the process of diffusion itself not culturally and historically situated within wider trends of the American and the European societies? Through the GMO case, is it not how food is conceived of in different societies, and the role of science and technology such as biotechnology in food production, that we need to account for? Despite its multinational character, does a firm like Monsanto not need to take account into the specificities of a market for the acceptance and use of its food products? In this chapter, we will discuss the reasons and conditions for extending the societal analysis from the comparison of organizations to the comparison of markets.

GMOs as a field for comparative studies

For sociologists, genetically engineered organisms represent a rich and complex object. Rooted in life sciences and biotechnologies, their applications and the controversies they raise nevertheless imply questions about human health and human food. This field of research, meaningful for human beings around the world, relates to cultural and ethical questions and values. Actors with different interests tend to confront one another for economic, political, and social stakes. Such situations typically convey ideological positions deeply intertwined in actor arguments and reasoning, a confusion from which even scientists and experts are not exempt.

In this situation, sociological comparative studies are a natural choice as a methodology: they are then prone both to seize the specificity of each market reaction in their cultural and historical context and to disentangle ideological aspects related to actors interests and positions. For instance, in spite of taking values and norms for granted (as in the public-perception-of-GMOs approaches based on polls naturalizing social snapshots), this method obliges researchers to look inside the sources of values and norms and the processes of their production and evolution in the US and Europe respectively: how and why did we arrive at such contrasted positions in the American and the European publics? To do so, two main biases need to be taken care of: universalism (in which the main line of explanation is that we are all the same—*mutatis mutandis*); and culturalism (we are all different—comparison is impossible). As sociologists who are experienced with international comparisons know well, the comparability of two objects need to be built methodically: in each case, what is the same and what is not needs to be clarified by investigation (Maurice 1989; Lallement and Spurk 2003). This is done by comparing phenomena situated in time and space: actors and their positions are considered as the product of their historicity and social identity (social space/network-belonging).

This raises the problem of having to compare incomparable phenomena (Maurice 1989). Among these are actors' identities, but also the respective institutions of each country.

GMOs histories in the US and Europe are neither fully separated nor fully superimposed: they are intersecting histories. These countries/areas have interacted economically, politically, and socially: for instance Joly and Marris have shown that the evolution of the GMO controversy in France and its publicization stimulated a strong evolution of the controversy in the US (Joly and Marris 2001). Multinational corporations also are essential vectors of history intersecting: Monsanto products circulated from the US context to the European context and contributed to bridging the two worlds. From this perspective, the GMO innovation processes are then built in interactions; actors in different contexts borrow, appropriate, or reject engineered goods that circulate among them.

Literature review and positioning: approaching societal acceptance

The GMO topic is certainly closely linked to the sociology of risk and risk management (Beck 1992). In this section, we will treat GMOs from a slightly different angle: the questions of the social acceptance of technology and innovation.

A recent book by Callon *et al.* (2003) took an essential step in the understanding of this problem. The authors distinguish 'confined research' (traditionally produced by academics in laboratories) and 'open research' (newly produced research with a variety of participants, among which are laypeople). The case of human genetics research in France is presented as a typical example of the emergence of open research: myopathic patients and their association, Association Française contre les Myopathies (AFM), made valuable contributions to scientific progress (Callon *et al.* 2001).

Characteristic of this new age of 'open research' are 'hybrid forums,' the open spaces where technical and scientific choices are debated and discussed before they are taken. The forums are called 'hybrid' because the participating actors and their representatives are typically heterogeneous: experts, policy-makers, business people, technicians, or laypeople. So are the questions and dimensions considered (economic, technical, or political): AIDS and nuclear waste management are good examples. Can we not assume that the networks of actors involved in the GMO controversies also participated in such research collectives? Are the field of corn and the battles around them not weaving a web of social links among actors that have hitherto ignored each other?

The reason why hybrid forums are central to open research is because they are associated with a renewed technical-decision-making process. They allow for the exploration of possible worlds associated with alternative technical paths. They submit the constructs of experts and scientists to the tests of societal robustness and acceptance. Scientific discoveries do not always raise social enthusiasm. Different actors might well be concerned by the social, economic, or political impact of science and technology applications. Adoption or acceptance of a new technology thus depends on a series of economic, political, and sociological conditions that translate into a certain state of science–society relations.

For Callon, to whose view we are very close, new technology acceptance goes with the new world it incorporates as a scenario. The adoption of a new technology is then not just a matter of product superiority: it carries an implicit collective agreement with the society that goes with it (Callon 2003). Very much as in organizations, when there are important societal changes (radical innovation), the interface between technology and society might well engender important negotiations and even disagreements and conflicts. Controversies are a major expression of these situations in which actors negotiate about the innovation and the societal context of its embedment. From the point of view of the social acceptance of technology, controversies are not just a sign of crisis, conflict, or risk, but an extremely instructive weak signal, and a social laboratory in which a new society and actors are produced.

Callon distinguishes 'cold' situations that can be managed by normal knowledge-producing institutions, and 'hot' situations characterized by the absence of stabilized knowledge that generates a profusion of approaches and 'diversified actors the list and the identity of which fluctuate in the course of the controversy' (Callon 1999). Different situations generate different learning processes with different legitimating processes. Breakthrough innovations and their strategic, disruptive societal intention typically constitute an initial condition that might evolve into a 'hot' situation, as was the case for GMOs in Europe.

In hot situations, experts or scientists alone . . . can do nothing. To trace connections, establish correlations, end up with assumptions to test, they need necessarily to go through non-specialists. The later become crucial players in the production of knowledge . . . the whole social body must agree to move on to produce accepted knowledge and measures. (Callon 1999: 417)

One critical issue is then the question of the construction of actors and knowledge within different configurations. Most of the previously reviewed literature does not clearly take into account how actors and their identity are shaped, where they come from, and how they are transformed in the innovation process. Actors seem either reified or dissolved in the continuous waves of making and unmaking of socio-economic networks. As Aggeri and Hatchuel stated, analysis of socio-technical networks from this point of view needs to be linked to 'deeper historical transformations analysis,' and their existence more closely elucidated by 'actors' history, management practices or institutional rules conditioning common actions' (Aggeri and Hatchuel 2003). Is it not desirable, depending on the kind of innovation considered, on the existing knowledge base and actors' historicity, to identify regimes of collective actions, and traditions of participation of laypeople?

Is GMO history in France not the appropriation of genetic engineering technology by historical public research actors such as

Institut National de la Recherche Agronomique (INRA) and seed and agro-chemical firms like Limagrain, Rhône Poulenc, and Novartis within the broader context of the diversification of agronomic research activities in France? Is it not embedded in the historicity of this variety of established actors and institutions and, at the same time, the social place in which a new social movement has developed and new actors have emerged? (José Bové, French farmers' union leader, was a character of a larger heterogeneous movement bringing together citizen movements, ecologist movements, 'altermondialists' (alterglobalists), assisted by established non-governmental organizations (NGOs) like ATTAC[I] and Greenpeace). Is the emergence of this hybrid forum of actors not a sign of an aspiration towards a new form of research more open to laypeople (as different experiences such as citizen conferences indicate)?

The previous reflections give an idea of what could be a societal analysis of GMOs, building on several theoretical contributions. We now need to provide some significant examples showing how it could be implemented in terms of methodology and data treatment.

Methodology and data treatment

In the following section, a societal analysis of GMOs will be sketched: our reflections at this stage are only exploratory and require further investigations and data collection. Considering the significant number of papers and work done on the subject, as well as the volume of reports and newspaper articles published, our strategy has been first to refer to existing results and data in the field in order to evaluate the relevance of our approach, as well as suggest future research programs and data-collection strategy.

The interest of such an exercise is to demonstrate, through the study of a smaller panel of actors, the extent to which an international sociological comparison such as the societal analysis might contribute to our understanding of the GMO phenomenon, and to elucidate the different market reactions to the technol-

ogy. The limits of the exercise are obvious and concern the possible bias and heterogeneity of the data and cases we will use. Our reference will include economists, sociologists, political scientists, journalists, experts, and such actors as members of associations, NGOs, or hybrid forums, as they were quoted in the media, and data that were constructed in different contexts with different instruments.

Controversy as a societal context for study

As the literature review made clear, one major issue is the application of the Societal Analysis, designed for 'confined' organizational studies, to the more 'open' world of market studies. This is a rather challenging task, as it is easier to thoroughly observe phenomena occurring in the well-defined spaces of an organization. The literature review has contributed to designating possible paths and bridges from organization studies to market studies. As social studies of science have emphasized, following the controversies is an important way to observe market learning processes and the actors involved (Latour 1987). They constitute major fields in which appropriation processes happen. They are observable places where negotiations and debates about an innovation's meaning and its impact on society are elicited: a variety of actors concerned are traceable there and their cognitive frameworks are exposed through their arguments and disagreements. In addition, the evolution of the controversy provides valuable insight regarding the process of societal embedding of the innovation: points of contention evolve over time and the closure of the controversy has been found often to correspond to the reaching of an acceptable agreement. Tracing the controversy is therefore one chosen methodological approach to observe societal acceptance phenomenon.

Scholars of social studies of science have pointed to the research interest of controversies. Callon argues that controversies represent a favourable field to 'study the mechanisms by which certain solutions, that first succeed locally, end up as solutions for the whole society' (Callon 1981). Rip underlines the socio-cognitive dimension of controversies: they are both 'informal technology assessment' processes and a source of new knowledge creation and learning: 'Society tends to learn through trial and error about the impacts of its activities: damage is experienced, small (and sometimes large) disasters occur and, gradually, measures are taken to avoid them in the future' (Rip 1986). According to Rip, tracing this socio-cognitive activity in controversies can be made by following the evolution of the main 'points of contention' and 'research agendas' linked to 'problem definition' as they are supported by different social groups. These elements evolve over time as some knowledge eventually becomes established and taken for granted: one example is the acknowledged smoking-is-bad-for-your-health warning now appearing on cigarette packages. For years, the link between smoking and health damage was highly debated. It is now commonplace. The important controversy that led to this shared knowledge has been forgotten (Rip 1986).

Actor identity and the institutional context shaping actions and interactions

Another important entry into a societal analysis of technology acceptance regards the study of the specific societal contexts of technology diffusion.

The identity of actors involved and the organized places of their interaction is a key to understanding innovation. Actors and the spaces of the interaction need to be historically situated. In our approach, the concept of actor is central and is not limited to individuals. It includes also companies or corporations like Monsanto in the US, or research organizations like INRA in France. The same is true for consumer associations, farmers' unions, or NGOs like Greenpeace or ATTAC. Beyond the identification of such actors, what is crucial to us is the way to define and treat them.

For instance, let us take the agricultural sector. For us, it is not only an economical entity designating a set of agriculture-related activities; it is also an actor (or an actor network in the Callon sense) that contributes to the social construction of other actors concerned with GMOs. In this case, the agricultural sector needs to be identified in each country, not only as a context. We need to qualify its identity, investigate what it means in the US, in France, or in the United Kingdom, and not only in economic terms but also in terms of its historicity and the symbolic values it conveys. This perspective is necessary to understand beyond statistics—certainly useful, but which need to be deconstructed to reveal the social significance of what is measured in each case—what seed companies, farmers, distributors are in the US or in France. This includes the links between the agricultural sector and others such as human food, human health, and environment which are relevant in the GMO case. The agricultural world today can hardly be considered as the world of farmers alone. Moreover, this world has probably a different shape and meaning in the US and in Europe, and even amongst European countries.

Considering our object, agricultural sectors need to be qualified (social, political, economic place), characterized in each country, with reference to its historical development, its economic and demographic weight, its industrial organization, the nature and importance of its R&D, its up- and downstream relations with other sectors (agro-food, human health, environment, agro-chemistry, seed). What is the respective importance of professional associations and unions, the orientations of the state and federal state (US and European Union (EU)) policies in each case? This approach allows for the identification of concerned actors and their social identity, a better understanding of their interests, their ability to compete and cooperate, their systems of alliances and power, and the nature of their social position. This would help to grasp their nature, the value system they follow. This is a prerequisite for an international sociological comparison that uses not statistics nor indicators but rather qualitative data that translates the nature of their embedding and social identity, and studies how the GMO innovations were shaped by actors and transformed them.

As we have seen, the Monsanto group appears as a central actor in the design and diffusion of GMOs. It is useful to our approach to situate this company in the wider American agro-food system, and compare it with European countries' agro-food systems. Without demonstrating it in these pages, is Monsanto not emblematic of an industrial society built on a highly competitive, powerful, and multinational organization strategically oriented towards international conquest and market domination? Is it not a product of the specific and particularly efficient form that capitalism, in a context of economic and political liberalism, has taken in US agro-chemistry and agriculture?

In other words, from a methodological point of view, the comparative approach excludes a piece-by-piece comparison of elements extracted from their environment. Comparability is based on observed continuities and discontinuities from one country to another. This point is essential to the good application of the societal analysis to the GMO case. As we can observe in the GMO case, there are interactions and interdependencies between actors in different countries as well as with the items and products that are associated with them (an experimental process, type of seed or plant, gene identification). It is necessary to implement wise histories integrating specificities as well as influences and imitation from country to country.

As we have just seen, provided that a certain number of precautions are taken, the application of the societal analysis framework to the GMO case seems feasible and promising. The second part of this chapter is devoted to this experimental application to the case.

The GM food case in retrospect

In 1998, under the pressure from public opinion, most European countries have banned genetically modified food, a very promising new series of products presented as the future of agribusiness. In June 1999, the European Council adopted a joint statement[2] resulting in what since has been called a de facto moratorium[3] for GM products, almost stopping the development and diffusion of agro-biotechnology in Europe for about five years.[4] This came after several years of controversial debates that started with the first commercial release of GM food in Europe in the fall of 1996.

On paper, the genetically engineered food technology is superior to existing ones: it combines the strength of a classical crop of agronomic interest (yield, adaptation to specific environments, cost) plus one additional selected quality (such as resistance to one specific pest, virus or pesticide, development of molecules of medical or nutritional interest).[5] As a result, from 1996 on, GM crops quickly replaced more traditional cotton and soy and, to a much lesser extent, corn crops in the US[6] enjoying the fastest pace of diffusion ever observed for the introduction of a new agricultural technology.[7] Economic studies about this diffusion of GM crops have put forward several explanatory variables depending on the GM crop considered: yield, profitability, better pest and weed control, and a significant simplification of farming activities (such as tillage).[8] GM products' superiority was publicized in the case of corn, cotton, and soybean.

The EU position has recently softened, as the de facto moratorium was lifted in May 2004. Although the future of GM food in Europe is still fairly unresolved to date, this might be a door towards more convergent technological trajectories.

The early-stage rejection of GM food by European markets is puzzling. To approach the issue, we will first present the strategy followed by the companies that developed the new products, namely Monsanto, which was the major player. We will review the kind of scenario built by the company to enrol adopters—mainly farmers—and control the value chain. The following section will be devoted to the analysis of European market reactions to the innovation. As we have seen, it is difficult to grasp an entity so open and large as a market. To do that, we will follow the controversy raised by GM food in Europe, with particular reference to France. The final section will discuss the main dimensions of a comparative approach between GM in the US and Europe.

Monsanto pioneers the development of GM food

As strategists and ample business examples have demonstrated, innovation is a very efficient way of beating competitors. Disruptive technology and breakthrough innovations have the power to reset the rules of competition and for innovative firms to shift a new paradigm with a considerable competitive advantage. But, as many observers have acknowledged, many companies are not aware of what is possible, or do not build the necessary competence to benefit from or initiate paradigm shifts (Bower and Christensen 1996).

Inventing the life-science company

In the 1980s, biotechnology was considered by a few pioneering firms as a good candidate for such a dramatic paradigm shift. While the whole business was still a mere projection derived from promising scientific discoveries, a few firms saw it as an opportunity to come up with radically new products and transform their competitive environment.

The most active shared in making the promise come true. They participated in the development of applied research that made it gradually possible to build genetic constructions that could be incorporated into germplasms. By far the most active company, Monsanto,[9] invested an estimated amount of $300 million R&D in ten years, both in-house and by establishing links with outside public and private research. They proved it possible to genetically embody a chosen feature—such as frost resistance, fruit ripening, specific color, insect resistance, or herbicide tolerance (HT)—in a plant; this was a fairly seductive prospect, compared with traditional chemical food engineering. By so doing, these firms not

only proved that the technology was actually feasible, they also acquired very specialized and unique competence in genetic engineering, thus taking a technological advantage and erecting efficient barriers to entry by others.

As the literature on breakthrough innovation has shown, developing the right strategic vision is one thing, but what makes the development of radical innovations a highly difficult and uncertain process (Lynn et al. 1996;[10] Courtney et al. 1997) is the difficulty of identifying markets not yet existing, in defining efficient business models to capture the value not yet created, in shaping the technology into customer-valued product attributes, given in acquiring efficient manufacturing and distributing systems to serve the markets. The commercial failure of the Calgene GM tomato provided a striking example (Harvey 1999).

Demonstrating technological superiority

But having secured a unique technological competence was only part of the picture. To harvest the fruits of long-term research, this technological advantage had to be incorporated into economically and technically superior products that would effectively transform the agro-food market and convince farmers to replace their old products and practices with GM crops. Innovation had to go from the laboratories to the market.

The first generation of GM products launched by Monsanto followed two main tracks. One first type of product incorporated the well-known pesticide properties of a bacterium, *Bacillus thuringiensis* (Bt) within the genetic arrangement of plants (cotton, potatoes, maize); thus self-defensive GM crops were constructed. A second type of product consisted of engineered plants (soybean, cotton, canola) genetically resistant to the most widespread chemical herbicides used in agriculture (Monsanto's Roundup™, Hoechst's Basta™): weed control could more accurately be managed even during plant growth, making a farmer's life easier.[11] For agro-chemical firms like Monsanto, it was a clever way to reconcile the move towards biotechnology and life science with not disrupting the cash cow business of agro-chemicals.[12]

For Monsanto, a former chemical company involved in crop protection, to be able to be first to develop and sell GM herbicide-tolerant crops compatible with glyphosate meant securing the market shares of its best-selling herbicide, Roundup. According to several observers, the crop protection industry reached a certain maturity, associated with increased R&D costs, stricter legislation, and approval procedures (Bijman 2001; Moore 1998; Joly 2003). A firm like Monsanto was typically seeking to extend the life of its cash cow product Roundup in a post-patent environment.[13] Packages like Roundup Ready crops allow for the provision of the plant together with its dedicated herbicide: plants are programmed to be more productive, thanks to better environment control. This interdependence creation is one way for agri-biotech firms to force their new GM products sales on their traditional herbicide customers and vice versa. If they have established a standard in pesticide, this is a very powerful way to introduce new products into the market (see Arthur 1989 on standards and network externalities). On the other hand, choosing a particular variety tends to bind the farmer to one unique seed and agro-chemical provider.

Gene constructions on their own were of little value since they were hardly marketable. To express themselves, genes needed to get incorporated into seeds to become proper products. To grasp an economical advantage, biotech firms soon decided to incorporate their genetic constructions into best-selling and performing (high-yield) existing varieties. They did this in a manner that some observers called 'Intel Inside,' referring to the knowledge-based, value-added method of programming plants. The additional trait provided the product with genuine qualities, such as better pest-control management and a simplification of production practices.

This combination of the technical and economic advantage of GM products proved an efficient incentive for US farmers to adopt

them in the cases of soybean, cotton, and, to a lesser extent, maize. According to some experts, these GM products enjoyed one of the fastest rates of diffusion in the history of US agriculture. US farmers transformed what was, so far, a promising technology into a demonstrated superior technology (EC 2001).

Securing profit from innovation

In his seminal work on R&D investments and innovation, Teece considers two major variables determining the ability of firms to capture the return on their innovative investments: specific complementary assets and appropriability (Teece 1986). Transforming the seed industry into a high-tech, research-intensive sector was supposed to find new ways to appropriate the return on the important research and development incorporated in the GM seeds. The search for the capture of return on R&D enlightened the important vertical and horizontal concentration between the crop protection, seed, and plant biotechnology industries and the development of collaborations to control the seed industry complementary assets (Bijman and Joly 2001). Monsanto also developed an entire strategy to enforce the property rights derived from its patents: specific yearly agreements were signed with farmers to ensure that they would use Roundup with herbicide tolerant products and pay an annual 'technology fee' corresponding to a copyright to use genetic properties conferred on plants.

One striking example is the alliance Monsanto sealed with Delta and Pine Land. Monsanto brought its property rights, technological expertise in genetic engineering in the form of genetic constructs for plants. Delta and Pine Land was number one in cotton seed production in the US. It provided Monsanto with an excellent gateway to the market for its genetic constructions. To capture the value in this particular case, Monsanto decided not to increase the price of the seed (price elasticity of demand was too high) but to distribute it through Delta and Pine Land Co., with a special agreement signed by farmers, and the billing of a special technology fee to Monsanto regarding property rights associated with the gene, because the genetic construction was subject to a brand-name strategy (Bollgard cotton). This had several advantages: the intellectual property rights were valued on a contractual basis directly with farmers, and it provided some ground for copyright control; it provided Monsanto with direct access to the cotton seed market.

When the first commercial products were launched in 1996 in the EU, the pioneers of the emerging agri-biotech sector seem to have had their campaign well prepared: they had secured a technological advantage and built the associated competence, managed to appropriate relevant intellectual property rights on living genetic constructions for capturing value, developed a number of future products for the years to come with attractive economic and technical attributes, obtained the necessary Food and Drug Administration (FDA) approval for product launch, constructed a strategic positioning to access their consumers' markets, and designed a value-creation system to capture farmers' fees on intellectual property. This sounds like the perfect strategy.

Early diffusion and the socio-cognitive foundations of rejection

As Chern, typifying many authors, observed, 'It seems odd that a good produced with a more advanced technology is less desirable to the consumer' (Chern *et al.* 2002). As we have seen, the pioneering companies that launched the first GM products on the EU markets in 1996 had a long experience of business and of European markets. The strategy employed by the major player, Monsanto, seemed to render GM food diffusion in Europe inevitable. The last episode before the new products would flow from the company laboratory to European societies was the authorization of EU regulators to commercialize GM products in

the EU markets. The first commercial product to Europe, soybean, was authorized in April 1996 and Monsanto shipped its first GM food to Europe a few weeks afterwards.

From our perspective, Monsanto had built a GM food system to channel and control the different aspects of its value network. Its strategic plan—as embodied in its product designs and their complementary assets and aimed at enrolling other actors and holding the different dimensions (technical, economic, social) of commercializing GM food—was impressive (Callon 1986). Its first experience with the US market demonstrated that this system was a good candidate to become the industry's dominant design and replace the old system of crop production worldwide. But the extension to the European markets did not unfold as expected. A controversy developed that revealed a number of hidden uncertainties and actor disagreements. Major aspects of the scaling up from the confined laboratory to the open-field society raised discussions and debates about possible externalities of the Monsanto system; they were credible enough to cast some serious doubts on the company's control over its creation.

Following the controversy

According to opinion polls taken in Europe, the level of European citizen skepticism towards genetic engineering almost doubled during the 1990s (Bauer and Gaskell 2002). In 2000, this feeling reached its acme: a specific poll carried out in 2000 on GM food showed that 70 per cent of European citizens did not want them, 60 per cent thought that they might have a negative impact on the environment, and that an enormous majority of them (95 per cent) wanted to be given the choice whether to consume them (Commission Européenne 2001). Polls are just snapshots of public opinion, but they clearly show an increased awareness of GM food as well as developing consumer concerns about their impact on health and the environment. This reflects the controversial process that accompanied the diffusion of GM food in Europe from 1996 until the 1999 moratorium. The moratorium suspended the adoption decision until the technology could be better understood and controlled.

How did so many European citizens come to consider GM food as risky for their health and the environment? What kind of knowledge creation and collective learning processes led to this collective interpretation of the new technology? To understand how it happened, we need to go back to the early-stage diffusion process of GM food: how did it disseminate from the confined world of the company laboratory to European societies (Callon *et al.* 2001)?

Although we are aware that debates and negotiations were not fully confined to this period,[14] we will consider the development of the controversy in Europe, with particular reference to the French case, between the first introduction of a commercial product in September 1996 until the June 1999 de facto moratorium. In this section, we will rely significantly on the work done by Joly *et al.*, who studied the controversy in France and in the US extensively for several years through its press coverage (Joly and Marris 2001), as well as on the investigation work done by a *Le Monde* journalist who delivered an extensive account of the unfolding GMO case (Kempf 2003). The authors provide a remarkable and detailed analysis of the way the controversy was shaped and revolved around three major points of contention they consider reality tests: the labeling, 'terminator,' and the monarch butterfly (Joly and Marris 2001). Based on this and a review of a variety of published material, we will study the process through which users participated in the shaping of the cognitive framework of GM food in France, and discuss the drivers of public interpretation of the technology.

Consumer concerns: opening the black box of GM food controversy

'*Alerte au soja fou!*' ('Look out for the mad soya.') (Kempf 2003). On 1 November 1996, *Libération*, one of the major French newspapers, played

whistleblower. A few days later, Greenpeace orchestrated a series of operations all over Europe: on 5 November 1996, the *Ideal Progress*, a ship transporting soybeans, arrived in Hamburg harbor; Greenpeace activists were there to welcome the ship and tag it with a huge X; they did the same two days later in Anvers harbor with the *Ziema Zamojska*, and in Gand Harbor on the 10th. Early in December 1996, although a large majority of the council member states were against it,[15] the European Commission (EC) stated that several tons of maize, including GM maize, had since October 1996 been imported to Europe and that it would accordingly approve GM corn products (Kempf 2003).

One major aspect of Monsanto strategy in commercializing GM soybeans, its first GM product to reach Europe, had been to confuse them with traditional soybeans by embedding them in traditional networks of distribution. It has since been argued that the rationale for it was economic: segmenting the market for GM soybeans from the field to the supermarket would have been much more costly and difficult. From the customer's viewpoint, however, this looked more like a strategic maneuver to keep them unaware and out of the decision-making process. To them, this meant that the radical novelty of the product was obscured and that GM soybeans were hidden: they could not detect GMOs or discriminate them from non-GM food with their own senses. As one journalist put it, 'None of us will know whether or when we are dining off big M's plate' (*Financial Times*, 7 December 1996).

Opponents then played the role of whistleblower, revealing what was hidden, showing and tracing the invisible, claiming that citizens were treated like cultural idiots. GM soybean was there, mixed with non-GM soybean, a very pervasive ingredient used in 60 per cent of processed food.[16] The use of the X symbol and the link with the Mad Cow crisis recalled previous experiences with invisible threats like nuclear radiation or Bovine Spongiform Encephalopathy (BSE) that European citizens had been secretly exposed to. As one analyst recalls, emotion was high as citizens felt deprived of the freedom to choose: their reaction 'had nothing to do with product safety or the impact on the environment but with consumer choice; when it appeared that GMO manufacturers had tried to introduce them secretly, the way they were perceived changed instantaneously' (Kempf 2003). Joly *et al.* consider that the right to know and freedom of choice united to bring many different interests together: consumer associations, but also several important farmers' unions and a growing number of actors in food processing and mass distribution (Joly and Marris 2001).

Labeling then became the issue. Labeling appeared to be a modus operandi to solve the question of customer information and the right to choose without rejecting the innovation: as a consensual analyst put it, 'what people really object to is being exposed to risks without their choice. This is why I support labeling, not because I think GM foods pose significant health risks, but because it is a freedom-of-information issue' (*Fortune*, 21 February 2000). From our perspective, it was a first attempt by the networks of users to change the design of the GM soybean products in order to embed them within the established contexts of shopping. A previous successful experience in the UK market with the clearly labeled Zeneca GM tomato was used as an example. Soybean is a major ingredient of the agro-food industry, and Europe the main export destination for the US (about a third). Eurocommerce (a trade union of European retailers and wholesalers) strongly supported the labeling solution and called for the segregation of US GM soybeans 'so that consumers could choose whether or not to buy products containing them.' Eurocommerce representatives claimed that consumer trust was low 'as a result of the recent mad cow "crisis" ' (*Financial Times*, 8 October 1996).

But Monsanto and its genetic constructs are a long way upstream from the consumer. And the US did not experience a comparable beef crisis. So from its perspective, labeling would just uselessly imply massive investments along the value chain to segregate GM products from traditional ones, to be able to verify and control their conformity, to transport them in separate containers, to label them on the packages, and so on. This would result in increased costs for both GM and non-GM food. Finally, this would signify that genetically engineered food was indeed different, while the company always claimed it was not. Monsanto refused to modify its established system on the basis that 'segregating the altered soybeans was impractical and unnecessary as the product has been approved as safe by EU regulatory authorities in April' (ibid.).

With labeling rejected, and the invisibility strategy being countered by opponents,[17] Monsanto's position towards European consumers turned to an educational attitude. A common idea was that first reactions were understandably emotional, but once consumers learned more about the products they would eventually end up adopting the technology because of its superiority: 'he [Monsanto CEO, Robert Shapiro] and his managers were convinced that those whose ignorance led them to reject biotechnology would eventually be swayed by Monsanto's assurances of safety and its research' (*The Economist*, 31 December 1999).[18] Other voices, including insiders and farmers, pointed to another course of action. If consumers were to be crucial decision-makers, as the controversy seems to indicate, the very conception of GM first products was ill-adapted. Functional and economic benefits were aimed at farmers, and brought no clear advantage to customers: why would they even consider eating this uncertain 'Frankenfood'? Not surprisingly, then, huge food-processing companies—like Danone, Nestlé, and Unilever—and a network of department stores—like Carrefour and Sainsbury—were the first to move and exclude GM products from their shelves before EU policymakers decided on the moratorium. Much later, the so-called second generation of GM products (*The Economist*, 26 July 2003 and 6 December 2003), requirements of food processors such as Archer Daniels Midland for segregated GM grains (*Nature*, 9 September 1999), and the EU directive 1929–1930/2003 imposing compulsory labeling (Granjou 2003) would embody the lessons of experience.

Terminator: the societal limits of Monsanto's enrollment program

On March 3, 1998, US patent 5 723 765 was issued to the US Department of Agriculture (USDA) and the Delta and Pine Land Co. for the 'control of plant gene expression.' For a few experts in the field, the patent describes an important step towards the inhibition of seed reproduction: 'the patent refers to a set of molecular "switches" that can turn genes essential for reproduction on and off ... [the] plant is forced to make a toxic protein that will sterilize its seeds after it is fully grown' (*The New Yorker*, April 10, 2000). Two actors demonstrated particular interest in this invention. On May 11, Monsanto bought the Delta and Pine Land Co. for $1.9 billion and started negotiating with the USDA for an exclusive license to use the patent (ibid.). In March 1998, the Rural Advancement Foundation International (RAFI), a Canada-based civil society organization for the defense of biodiversity and small farmers, had labeled the invention 'the terminator technology' (RAFI communiqué, March 1998). With terminator, the controversy moved to issues of industrial economics.

Robert Shapiro, CEO of Monsanto, had a visionary conception of the future of the agriculture sector: turning a mere hardware-based, low-tech, polluting industry into a software-based, high-tech, sustainable one. The cornerstone of this transformation was to design R&D-intensive, smarter products and to collect the added value that would result from their sales (Magretta 1996): as Monsanto Director of Technology, Robert Horsch, put it, 'the more you put in the seed the less you're going to rely on sophisticated agronomic practices, so plant biotech has the opportunity to be applied worldwide' (*Chemical Week*, November 6, 1996). Taking the HT and Bt seeds examples, Robert Shapiro argued, 'Up to 90 per cent of what's sprayed on crops today is wasted. If we put the right information in the plant, we waste less stuff, and increase productivity' (Magretta 1996). And Monsanto would lead this revolution: as Information Technology (IT) has its Moore's Law, biotech would have

its 'Monsanto's Law' (*Washington Post*, October 26, 1999). One major issue for Robert Shapiro was then to share this futuristic view with the rest of the world and, in the first place, with the farmers at the very center of this scenario. For Monsanto, enlisting these first users depended crucially on providing them with smarter products: GM crops featuring lasting agronomic superiority to enhance the farmer's life.[19]

In this high-technology investment business,[20] capturing return on R&D was vital. The company had to devise a transaction format that allowed it to reap the added value of its products. Traditional practices were strongly against it however; decades of open-source practices regarding seed intellectual property—under public research auspices—had made it a part of farming tradition worldwide to save a share of the harvested seeds for the following years, and even to trade them—a practice known as 'brown-bagging.'[21] It was a widely shared heritage of farmers to naturally reproduce, breed, and multiply seeds' genetic codes. In Monsanto's knowledge-based vision and appropriation strategy, it became nothing less than an illegal use of a private corporation's property rights, equivalent to throwing 'billions of dollars of investment down the drain.' It therefore had to figure out 'how to prevent farmers from obtaining its patented seeds illegally' (*Financial Times*, 13 July 1999) and make sure farmers 'buy the seeds fresh every year' (*Washington Post*, February 3, 1999).

In the company's new vision, the seed had become a mere carrier of a value-added genetic information program. Monsanto was leasing the right to use this 'software' to growers against a technology fee proportional to the planted acreage. In such a model, the company had to think of an arrangement that would render farmers' behavior compliant with the patent law. Based on its patent rights, Monsanto devised technology-use agreements as a guide for defining the legal context of use of its 'software.' Signing them, growers were committed to very restrictive use of the innovative seeds: they had to grow them within a period of one year, and agree not to keep them or trade them. In addition, GM seed buyers recognized that Monsanto had the right to perform unattended visits of their land and storage facilities and to sample and test their plants for three years after purchase. For HT varieties, they also committed to use Roundup as the only herbicide during the growth of the crop (Monsanto Technology Use Agreement 1998).

Significant misunderstanding of the definition and use of the seed by farmers paved the way to an escalating cat-and-mouse game of mutual accusation. To some farmers, these contractual arrangements were perceived as a trespassing assault against their traditional appropriation of seeds—in some cases, farmers still did their own breeding—and as a unilateral restriction of the conventional freedom and autonomy associated with their status in society.[22] As one of them explained, '[farmers] don't like to sign anything, especially anything that gives up their rights to stop trespassers' (*Washington Post*, February 3, 1999). Technology-use agreements departed greatly from the usual commodity transaction of trading seeds between independent economic agents. By defining guidelines for use, it also interfered directly with the definition of some aspects of the farming process and the authority to set the rules and control work. Some felt this was another giant step closer to being industrially salaried: as a farmer dramatized, 'we are all gonna be serfs on our own land' (ibid). Confronted with such disruptive changes, many continued seed saving and exchange as usual. But from the perspective of the St. Louis company, not ready to give away the result of its fifteen years of good work, it was necessary to change these past behaviors, and render them more consistent with the rules and laws of modern business. To enforce its property rights against 'seed pirates,' a 'gene police' was set up (ibid.).[23] The company also hired the services of a private detective agency—Pinkerton's — and advertised a controversial toll-free phone line—'tip line' 'to report on others who violate Monsanto's usage terms' (*Financial Times*, 13 July 1999). As a result, an estimated 525 cases

have been opened in Canada and the US, settlements being 'in the range of ten to hundreds of thousands of dollars each' (*Washington Post*, February 3, 1999): Monsanto felt exemplary cases had to be won to dissuade defraud.[24]

Early in 1999, in what was sketched by the press as a David and Goliath rematch, the Percy Schmeiser trial became the culminating point of this accusation process that contributed to turning many family farmers into potential opponents. Percy Schmeiser was a sympathetic, 68-year-old Canadian; a fifth-generation family farmer growing Canola. He was not a client for Monsanto's GM seeds, but, like many others, used Monsanto's Roundup herbicide. But resistant Canola plants found in his fields caused Mr Schmeiser to be sued by 'Monsanto agrobusiness giant' for seed piracy—'for doing what I have always done,' he remarked (*Washington Post*, February 3, 1999). Media coverage was very important. Percy Schmeiser always claimed not to have planted any GM seed and told a different story: 'pollen or seeds must have blown onto his farm.' Doubt remained in many people's minds: as one Monsanto representative acknowledged, 'cross-pollination occurs' (ibid.). The court ruled against Schmeiser, but the case raised considerable hostility against Monsanto all around the world, especially in areas where family farms are still powerful.

With the terminator turning point, the controversy evolved to more general points of contention about public goods and alternative industrial models to deliver them. When the Gene Protection System became an option in March 1998, Monsanto quickly understood its potential: its promise seemed just what Monsanto lacked in its farmers' enrolment plan. It was the missing piece of technology in its smarter product design. With it, there would be no more costly and unpopular juridical actions against pirates; the seeds themselves would embody an anti-piracy device and become a patent rights enforcer. The one-year lease contract would be programmed directly into the plant development process and put an end to the illegal use issue. As the USDA inventor explained: 'Our system is a way of self-policing the unauthorized use of American technology. It's similar to copyright protection' (*New Scientist*, 28 March 1998). By the same means of programming the gene's 'on and off' switch, the technology also promised to solve the embarrassing question of cross-pollination, which happened to be a flaw in the product design.[25]

Opponents pointed to the new appropriation system's relentless character to draw radically different interpretations about its impact on society and especially Third World societies (Joly and Marris 2001). Terminator technology would indeed attract private investment to new seeds and transform the industry. But what was at stake was the social impact of this transformation, and the ability of this industrial model to address social problems such as health, employment, hunger, and environmental issues, as was claimed by the technology promoters. According to an NGO like RAFI, which actively campaigned against terminator, appropriability would result in increased industry concentration and further privatization of the hitherto open-source public goods such as the biodiversity of seeds' genetic codes. Monsanto's considerable financial efforts to control the value chain from the laboratory to the market—an estimated company purchase bill of $8 billion by 1999 (*New York Times*, December 15, 1999)—supported this view. There were accusations of monopoly power and claimed risk of concentration of the future of the world; food in a few private hands led to a class-action lawsuit and a long investigation by the American Department of Justice (*The Economist*, 31 December 1999). What many opponents envisioned was a consequent reduction in seed variety around the world, the direction of research and development into the most profitable areas, the increase of commodity value, and the further industrialization of agriculture.

Although a considerable variety of opinion prevailed among farmers, some of them were afraid of excessive industry domination. As one Canadian farmer explained: 'To remain competitive internationally, farmers will be compelled to work with improved varieties covered by this terminator technology,' and the property rights enforcement system will 'ensure that most of the gains from research will accrue to companies owning the varieties and not to farmers' (RAFI communiqué, 30 March 1998).

In France, the words 'Monsanto' and 'terminator' became the perfect symbols of the new wave of global competition and industrialization transforming the agricultural sector. As one observer noticed:

One factor is worry in Europe about domination of the food chain by American companies. That partly explains why Monsanto managed to fall into the role of devil incarnate. Of course, this conveniently ignores the fact that some of the chief developers of GM technologies are based in Europe. (*Fortune*, 21 February 2000)

Recent transformation in the agricultural sector had been impacting French agriculture badly. Many French (and world) small and medium-sized traditional farmers were facing economic troubles. European agriculture was also shocked by several food and health scandals attributed to 'deregulation and over-intensive production' (*The Nation*, December 27, 1999). The recent mad cow crisis was a case in point, demonstrating the limits of the industrialized model of the food chain: 'in one terrifying package, BSE tied together the new "economic' farming practices (feeding ground-up cow carcasses to cattle), the easing of health and food industry standards, and governments' willingness to lie for the food industry at the cost of human lives' (ibid.).

Among opinion leaders that came into play as 'terminator' was announced were the small farm union leader and sheep grower José Bové (Confédération Paysanne 2d French Farmers' Union) and the alter-globalists' movement, ATTAC. Their argument was based on the denunciation of the private concentration and the industrialization of the food chain. In their perspective, GMOs and the terminator were not just mere products but a Trojan horse leading to irreversible changes. José Bové became famous in France by calling for a reflection on the industrial model the French really wanted for their food supply. Bové considered that farmers, like consumers, must have a choice: 'When were farmers and consumers asked what they think about this?', he asked during his Agen trial in January 1998 (ibid.).[26] His argument was that important industrial transformations were pushed forward by multinational corporations—the prominent role of Monsanto conveniently identified them with the US—trying to impose their model of how the food chain should be. In José Bové's view, one clear consequence was the standardization of food and tastes and the development of *malbouffe* (unhealthy/low-quality food) worldwide, irrespective of local contexts and culture. The French had a history of gastronomy and a diversified quality food model produced by traditional farming methods, he claimed, that was threatened by the global industrial model of agriculture. By accepting GMOs, French people would be agreeing to give up their tradition and their gastronomy.

Monsanto and GM food became the symbol of a long-lasting industrialization and concentration of the food sector that started decades ago. By sterilizing seeds, GM promoters went one step further in the dissociation of farmers and a certain idea of nature. For many social groups—some farmers, ecologists, believers—it was one step too far. The campaign aroused considerable emotion and on October 4, 1999, answering the friendly advice of Gordon Conway, Rockfeller Foundation President, and early supporter and founder of GMOs, Robert Shapiro publicly announced that he would not exploit the 'gene protection systems': 'I am writing to let you know that we are making a public commitment not to commercialize sterile seed technologies, such as the one dubbed "terminator" ' (Shapiro to Rockfeller Foundation, October 4, 1999).

Monarch Butterfly: substantial equivalence and the question of regulation of uncertainty

On August 10, 1998 British scientist Dr Pusztai said on the *World in Action* TV show that, based on his studies on rats, 'I would not eat GM potatoes and found it very unfair to use our fellow citizens as guinea pigs.' Preliminary results demonstrated GM potatoes can harm the immune system. Dr Pusztai was suspended on 12 August by the Rowett Institute for having released unconfirmed 'misleading information' (*The Guardian*, 12 February 1999). 'Transgenic pollen harms monarch larvae,' announced an article published in the very prestigious scientific review *Nature* (20 May 1999).

As Monsanto's strategy to commercialize GM products has consisted in confounding them with traditional products in the distribution networks, the strategic basis chosen to be granted commercial approval for them consisted in claiming GM products' similarity to their traditional counterparts: as one representative of Monsanto in Brussels said, compared with unmodified products, GM products 'are unchanged in composition, nutrition, function and safety' (*Financial Times*, 18 December 1996). For Monsanto, although the agronomic and marketing qualities of GM food are superior to existing products, they do not depart from them in terms of their chemical composition. At the end of the day, a soybean is a soybean independent of the way its genetic material was engineered and the plant grown.

This heuristic definition of GM products' status had considerable consequences in terms of the regulation adopted in the US and elsewhere in the 1990s to assess their health and environmental risks. According to a food specialist, 'What substantial equivalence did was attempt to confirm that the products of biotechnology are as safe as conventional ones' (*Nutrition Reviews*, June 2003). This concept of equivalence made it possible to interpret new foods within the framework of existing food knowledge and evaluation criteria. The practical way to deal with this has consisted in splitting the problem into two blocks to reformulate it in a way compatible with existing tests and knowledge: first, define the extent to which a GM product is similar to a traditional product—this part would not require further examination; second, subject the specific components included in the GM product—mainly protein derived from the transferred gene—to specific health assessment based on known toxic proteins: 'one must focus assessment on those unique characteristics, those unique traits, and the gene products associated with those traits.' In most cases, as far as the first generation is concerned, GM products happened to be '99.99 per cent like its conventional counterpart, but is 0.01 per cent different,' and in application of the above principle, 'these crops are substantially equivalent to their conventional counterparts except for well-defined differences' (ibid.).

For those pro-GM, 'regulatory decision on genetically-modified crops should be made on the basis of good science not hysteria' (*Farmers Weekly*, September 3, 1999). 'Substantial equivalence' was used as a reference point for the FDA approval processes: it prevented the setting of a specific approval procedure or specific labeling for GM food (apart from the recommendation to firms to test their products and the possibility of consulting FDA experts on a voluntary basis)—in accordance with the food sector tradition (*Consumer Reports*, 9/1999). In Europe, the same principle applied under directive 90/220 and was further refined in directive 258/97 on 'Novel Food and Novel Ingredients Regulation' (Granjou 2003). Companies just needed to provide scientific evidence of equivalence in order to prevent any specific review and mandatory labeling. From this angle, labeling was only to be used on GM products that were substantially different from known products; for other GM products they were considered costly and useless: 'there is no scientific basis for putting [labels] on' (*Washington Post*, June 19, 1999). For the promoters of the technology, experience and trials corroborated this safety assessment system: 'After 15 years of field study there has been no surprise, no unexpected result' (*Newsweek*, July 13, 1998). No evidence that GM products were risky for health or for the environment had been found as 'the products of ag biotech have been subjected to more scrutiny than any other products in humanity' (*Washington Post*, June 19, 1999).

The British Dr Pusztai and the US Monarch butterfly cases appeared to be puzzling anomalies in this well-orchestrated framework. By raising questions about the underlying cognitive framework used to diffuse GM, they directed the argument towards health and environment safety issues (Joly and Marris 2001). Well-publicized cases like the Monarch and the Putzai experiments became a base on which to build evidence and collective experience that the control of Monsanto and others over the GM innovation—the scaling up from confined laboratory to open field—was not exactly complete nor absolutely safe. For instance, one proponent claim regarding Bt crops was that they were targeted at specific pests—like the European corn borer—and would not harm other insects and organisms of interest. The trouble is that Bt corn expresses Bt toxin not only in the plant but in its pollen. And pollen is prone to dispersal by the wind in an uncontrolled manner. Endangered species like the Monarch butterfly are feeding in areas close to fields. Losey *et al.* demonstrated that, where exposed to GM crops' pollen, Monarch mortality was increasing dramatically (*Nature*, 20 May 1999). For opponents, such studies corresponded to scientific counter-evidence showing that their concerns were not more 'irrational' than those of the proponents of the GM technology. The Monarch butterfly case—'a sort of Bambi of the insect world'—was published in the well-reputed British review, *Nature*, by academic researchers of the prestigious Cornell University. Although several scientists either criticized (while others defended) the results of the experiment, discussed the implications of the Monarch case, or pointed to the personal interest beyond their publications (*Washington Post*, September 20, 1999), the containment practices of genetic engineering to the GM fields was called into question.

As several scientists observed, these 'anomalies' demonstrated that there are things out there to discover that Monsanto's cognitive framework can not explain, a world that requires further investigations to be unveiled and tell its truth: as John Losey put it, 'the study was not done before, and now we need to look at what it means. I take no side ...

when we did the paper, there was one other lab working on this issue. Now there are a dozen. That's the way it should be' (*New Yorker*, April 10, 2000). And indeed, experiments and publications boomed, a number of 'independent'—not financed by the industry—experiments were launched, older experiments were reviewed from a different angle, and 'evidence' started accumulating in many different directions: an allergic reaction to brazil nut soybean were discovered in 1995 by Pioneer; in 1997 gene flows between GM and wild species of oilseed rape were demonstrated by a French Institut National de Recherches Agronomiques (INRA) team; in 1998, the risk of transmission of antibiotic-resistant gene present in Novartis' GM maize to humans was exposed by a reputed French researcher of the Pasteur Institute; in 1999, bollworm resistance to Bt cotton was found by University of Arizona researchers; Roundup-resistant weeds were discovered in the US (ibid., *Financial Times*, 8 November 1997, *La Recherche*, May 1998, *Des Moines Register*, January 10, 2003). In France and worldwide, considerable dissent occurred within the scientific community (Kempf 2003). Uncertainty grew inexorably against the claimed scientific 'soundness' of the practical and theoretical models used by GM promoters to predict safety. This uncertainty then drove many to feel like 'guinea pigs' in the hands of a 'sorcerer's apprentice' using society as a laboratory: 'a huge experiment in environmental genetics is under way' (*Financial Times*, 8 November 1997). In this perspective, the fact that no safety problem had been recorded so far was no longer interpreted as a proof of GM safety but rather as a demonstration that the current knowledge, the existing instruments and theories, and the established regulatory institutions were not accurate enough to detect them.

The testing practices then became a central topic of interest: as a pro-GM expert recognized, 'carrying out full pharmaceutical-style testing on GM foods would be impossible, because low-level poisons ostensibly from GM products would not appear in ordinary toxicological testing' (*Scientific American*, May 1999). Practically, according to food experts, it was very difficult to prove food absolutely safe. Most foods we eat are not absolutely safe. Unlike the drug and health care sectors, the food sector had not built sophisticated shared knowledge and instruments to assess food innocuousness. The regulation for food commercialization was also much less constraining: 'most of what we eat has never been the object of specific regulation' (Organization of Economic and Cultural Development (OECD) Observer, March 1999). There was no established model or methodology to test human food on animals, as is the case for drugs. Rather, an internationally accepted regulatory standard is 'reasonable certainty that no harm will result from the intended uses under the anticipated conditions of consumption' (*Nutrition Reviews*, June 2003). At this point, guaranteeing the safety of a radically innovative product such as GM food consequently raised considerable difficulties: they would require a learning process to design different sets of tests and testing criteria that would depart from existing agreed-upon ones—like the WTO Agreement on Sanitary and Phytosanitary Measures.

One of the major consequences of this learning process has been the reversal of the burden of the proof: to proponents arguing that nothing proved that GM products were *not* safe, the answer was that nothing proved that they were. What was at stake was the very nature of GM food. Proponents insisted on the continuous nature of the GM innovation. Biotechnology was considered as merely 'a new name or label for a process people have used as long as we have been baking bread, fermenting wine or making cheese, or cultivating crops and breeding animals' (*Financial Times*, 22 March 2000). The process was transparent and pure, in essence the same as cross-breeding, but with more precise techniques: 'with genetic biology, you introduce one or two new genes. With crossbreeding, you introduce tens of thousands of new genes. Biotechnology is much less haphazard ... ' (Roger Beachy, *Newsweek*, July 13, 1998). For opponents, the biotech process itself was not that precise (Kempf 2003). It was the carrier of uncertainty and risks. Dr Pusztai's experiment showing damage incurred by GM potato-fed rats was highly debated. His work was, however, extended and published for his interpretation that 'damage to the rats was caused not by the protein but by the method used to put it there' (*The Economist*, 20 February 1999). In opponents' views, not only was the technology a radical innovation, and as such 'unpredictable' within the existing cognitive framework, but, as the failure to contain GM within the cultivated fields seemed to demonstrate, it threatened to be inherently irreversible. As one British respected scientist explained, 'Unlike chemicals, biological agents can multiply in the environment. There is therefore a risk that once released it will be impossible to control them' (*The Independent*, 22 May 1999). The point was made that established health and environmental assessment models were not infallible: as Robert Shapiro himself finally admitted in 2000, 'a human body is a subtle and complicated thing, it may be that only one time in a million some side effect happens. And your testing won't reveal it. It has to be out here first' (*New Yorker*, April 10, 2000). If externalities did in fact occur, and if there were a risk, then who would be accountable for it?

Scientific debates are an essential ingredient feeding the expertise on which most countries base their regulatory decisions: 'Crops should be approved when the balance of scientific opinion is that they are safe' (*The Independent*, 22 May 1999). Reviews by authorized institutions were performed in many countries. They typically aimed at such panoramic balancing: weighting the different points made by different sides, they ascertained what points were consensual, and what was uncertain and had still to be clarified. EU member states and EU regulation gradually evolved, taking more account of the uncertainty: as one influential scientist, Dr Chesson, known for being pro-GM stated, 'We will undoubtedly use novel genes that haven't faced the regulatory system before ... thought should be given now to new procedures that will have to be adapted for better scrutiny' (BBC, 7 September 1999). But these evolutions were interpreted as a lack of clarity based on political rather than technical

reasons. As James Murphy, Assistant US Trade Representative for Agriculture, argued, taking the BSE crisis as an example:

All this was compounded by the lack of an established institutional review process at the EU level that could provide sound foundation for public assurance and confidence in the safety of food products ... we support the right of countries to maintain a credible domestic regulatory structure with food safety standards that are transparent, based on scientific principles, and provide a clear and timely approval process for the products of biotechnology ... but we must ensure, without any question, that debate about the safety and benefits of biotechnology is based on scientific principles, not fear and protectionism. (*Economic Perspective*, May 1999)

At this point, two largely irreconcilable visions of regulations based on two socio-cognitive frameworks emerged out of the controversy: Europe was heading to a suspension of GM commercialization, a time for research to reduce perceived uncertainty about GM innovation and to reflect upon a regulatory framework acceptable to European public opinions; the US government was threatening to enter a trade battle against EU regulation on labeling that it considered a barrier to trade. This focus on the regulation and risk control as a major topic of opposition is meaningful. It accounts for the fact that a large majority of the heterogeneous network of actors opposed to the innovation were not strictly against GM food, but called for a stricter regulation and assurance that the products were safe.

Sketches of a societal analysis of genetically modified organisms: significant actors in the US and Europe

Media coverage of the controversy has certainly been particularly important in Europe in the second half of the 1990s. Reviews of leading US and European newspapers seems to show that EU media coverage became important as early as 1996 in many countries; in the US, it picked up later on at the very end of the 1990s, with opposition to the EU and the development of a more active movement against GMOs (Bauer and Gaskell 2002, Joly and Marris 2001). Media coverage is the product as well as a factor of mobilization though. As our analysis demonstrates, what was at stake was not primarily the access to information— or ignorance—but the frameworks of interpretation used by two different networks of actors to make sense of it. As the controversy has unfolded, two hybrid networks of actors with different positions have gradually come into place and opposed each other in their views about the innovation—continuous versus discontinuous—and how it should be regulated.

Considering the state of knowledge in 1999, both positions were perfectly rational in their own world of reference and values. One typical example is provided by the pluralistic meaning attributed to the fact that no proven case of disease had been observed: for the proponents, it was a clear demonstration of its innocuous nature; for the opponents, it was unambiguous proof that the instruments in place are inaccurate.

In tracing the controversy, we have been through a rich array of actors involved and through different socio-cognitive frameworks confronting each other in arguments. This description, although useful in understanding how the networks and their positions were shaped, does not provide a satisfying explanation of why the EU and the US markets reacted so differently, and why the network of opponents to GM food had first been so strong in Europe and so weak in the US. To go one step further in the analysis, and be able to explain such variability, a comparative approach such as Societal Analysis is necessary. More thought should be given to the respective societal contexts in which the innovation was diffused on both sides of the Atlantic.

GMO innovations in the US and Europe have developed in very different socio-political contexts. These contexts are characterized by not only their respective specific institutions, but also different conception of what public/collective good is, as well as shared, historically shaped conceptions of justice (Orléan and Aglietta 1994).

In the GMO case, it would naturally be tempting to stress differences observed between the US and Europe. We will do this to a certain extent. But we will also underline the dynamics between the two areas which, if they did not lead to convergence as such, at least exhibit a number of compromises and a negotiation process (MacNichol and Bensedrine 2003). Such compromises also indicate that histories were intersected (Joly and Marris 2001).

Farmers and the industrial model of agriculture

In our view, the previous observations point to a number of complementary questions in terms of the construction of the actors involved and their positions: why and how has a 'critical context' emerged in France? Who were the main actors involved in this criticism? Why are competitiveness and export considered as highly ranked priorities in the US, and where does that come from? How widely shared is this view and the idea that agriculture must feed the world?

The global picture of US agriculture is the result of large-scale industrialization, of very large, specialized units, with an increasing influence of agro-chemical firms. In the US, the agro-food sector seems to develop in large part following an industrial-commercial logic. As we have seen, important and costly concentration movements have occurred in the last twenty years, giving rise to powerful life-science companies and trade associations. The transformation of the seed sector into a high-technology industry also meant that R&D investment grew significantly, pressing companies to seek for return on their investment on the global markets. This results in stiff competition, dynamic exports, and even hegemonic strategies in some cases. In line with the historical development of this sector in the country and the political line of the post-Reagan era to encourage innovation and light regulation, the USDA and the US government strongly encouraged the biotechnological transformation of the food sector. After 1999 and the European crisis, increasing concerns for retailers, agri-food, and farmers of possible market loss seemed to change this univocal position. So did the official position of the US government. In the second half of 1999, the US Secretary of Agriculture, Dan Glickman, also invited companies to show a more comprehensive attitude towards consumers' concerns, the labeling issue, and creating more value-added products for consumers.

The picture in Europe is more complex. The advent of biotechnology also triggered the development of large life-science companies. A concentration movement and the transformation of the seed sector was also experienced under the favorable eyes of governments. But this emerging high-tech world is coexistent with, rather than dominant over, a wide variety of other models of production—some of which are traditional—and other actors with diverging interests along the food supply chain, including large retailers. In recent years, considerable efforts to diversify the models of production in this sector have resulted in the development of alternative kind of products based on quality, diversity, organic labels, and traceability of the process of production. This reflects the strong and diversified actors' identity involved in agriculture in most European countries. For instance, many farmers in France are peasants from a long tradition of country people, rooted in holdings often small in size; often the property is still family owned and passed down from generation to generation. Naturally, this has been evolving in the last decades. But the character of the peasant (literally 'country person' in French) remains very strong in the imagery and collective memory of the French population, and a very active group, politically speaking. This is also visible in farmers' associations and unions and influences their modes of action, oriented towards the preservation of their social and professional identity as well as the property of their production equipment. Finally, as grounded studies have observed,

belief and conformity to a number of values impregnate the way farmers produce and sell, and affect their idea of what product quality and environment conservation must be: there is a wide diversity beyond the word 'farmers' (Boisard and Letablier 1989). As MacNichol and Bensedrine (2003) summarized: 'In sharp contrast with the US, European biotech business was not strong enough to balance a pro-labeling coalition of large retailers, consumers and environmental organizations, as well as very active groups of farmers, who dominated the political scene.'

Whether one agrees or disagrees with his opinions, to understand how the iconic character of José Bové found its legitimacy, it is necessary to keep in mind the reactivation and reinterpretation of this heritage from the past, the variety of actors and models of production existing in the European food sector, and their embedment in a historically built collective memory (Halbwachs 1925). Bové's messages found an echo in the common world of those who participated in the hybrid forums on GMOs in France, and his positions affected the nature of the controversies that occurred. His book, *Des paysans contre la Malbouffe* (*Peasants against Bad Food*), rings a number of bells in the French culture and values (the peasant origins of most people, taste for gastronomically high-quality food), and more widely on the Mediterranean side of Europe, and capitalizes on a full heritage of previous debates, positions, experiences, and products (Collins 1992). The value of food quality is high, and the fight against 'malbouffe' (composed from 'malnutrition' and the difference between 'eat' and 'puff out') is symbolically strong in France. This leads us naturally to consider one very important aspect of the question, that of food quality and safety.

Path dependence and the food safety question

A look at the food quality and safety question over the last thirty years in Europe provides important insights into European reaction to

GM innovation. This history has been marked by a number of important safety and quality crises that have shaped European institutions and regulation.

According to a recent study of food safety regulation in Europe (Josling *et al.* 1999), the first food crisis confronting European institutions occurred in the 1970s and had a strong influence on the design of European regulation. It was also the beginning of a number of food-related trade conflicts between Europe and the US. In the 1970s, cases of hormonal irregularities and children born with health problems were attributed to the use of DES, a growth hormone used in veal production. DES was found in baby food, and illegal use of this growth hormone was detected in France in particular. 'The DES scare created a consumer climate in Europe that was suspicious of the use of hormones in livestock production, and fearful of the potentially harmful health effects of these practices' (Josling *et al.* 1999). After the EC council directive 88/146/EEC to ban growth hormones and hormone-grown beef, the US challenged the EU on the subject at the end of the 1980s (Josling *et al.* 1999). Later on, the European food scene was struck by several other crises in the 1990s, including salmonella, *E. coli*, and dioxin contaminations, but the most impressive crisis remains the BSE crisis. Only six months before the first GM food was released, the mad cow crisis affected European—notably British and French agribusiness—deeply. The link between food and human health became a widely shared concern as evidenced, for instance, by the increasing success of biolabeling (organic food).

The BSE crisis illustrated both the limits of extreme forms of industrial agriculture and 'short-term profit-seeking attitudes' (cows were turned into cannibals by being fed food mixed with recycled dead bodies of other cows), the limits of scientific knowledge and expertise (experts were very divided on risk assessment, estimating Kreuzfeld Jacob cases from a few to hundreds of thousands), and the failure of the safety barriers provided by policymakers and regulatory arrangements to protect consumers. It is a frequent point of reference in papers discussing GM food, and images of gigantic fires made of thousands of cows still haunt European collective memories. As Grove-White *et al.* stated: 'Thus the risk of GMO foods tended to be influenced by experiences with BSE, and were seen as the same class of risk—in terms of unnaturalness, the failure of institutions to prevent them, the long-term character of associated risks, and our ability to avoid them' (Grove-White *et al.* 1997).

Beyond consumers' mistrust, BSE and other food crises have led to important reflections and organizational changes along the food chain, especially relating to consumer information, labeling, and the traceability and quality control of products. Bovine-sector professionals and the very strong sector of hypermarkets reacted by promoting traceability and labeling as key instruments to regaining consumer trust. These innovations were rather successful and, as we have seen, the same approach was proposed when the GM controversy struck. In this context, it is understandable that French farmers' unions were favorable to transparency in the food supply chain (Fédération Nationale des Syndicats d'Exploitants Agricoles (FSNEA), Confédération Paysanne) and that traceability became a major issue in the GM controversy. In France, the connection between the mad cow crisis and GMOs has been used as a major instrument of its publicization, as in the aforementioned *Libération* article (Joly and Marris 2001). For most European consumers, GM food did not provide any additional benefit compared with more traditional food, but a higher risk.

Although consumer trust seems to be much higher in the US, several observers have noticed changes from 1999 on, especially in the question of labeling of GM food. The conjunction of the European GM rejection and food alerts such as the Starlink case rendered American consumers more sensitive to GM food and, according to an FDA study, favorable to labeling. Observers agree that, several years after their European counterparts, American consumers have tended to become more concerned about GM food (MacNichol and Bensedrine 2003). The so-called 'second generation' of GM products developed by firms takes consumers more into account by delivering products such as varieties of oil with health benefits and less environmental impact.

Food safety regulation and the institutional background as societally constructed

Anchored in a long tradition of farming, Europe as an institutional entity is at the same time very recent as compared with the US situation. The contrast and cultural variety amongst the twenty-five member states is also very striking.

One major point of confrontation during the controversy concerned the institutional arrangements put in place to protect public health and the environment. As we have seen, one of the outcomes of the diffusion of GM food in Europe and the US has been the construction of two very different regulatory systems: the US position has focused on the product/substance equivalence to known food; the EU position has tended to consider the process of production.

This important disagreement in the regulatory philosophy reflects, in part, very important administrative and political differences in the two areas. In the US, departments such as the USDA and agencies such as the FDA and Environmental Protection Agency (EPA) have a strong authority. The FDA, a long-established and legitimate, independent public organization, authorizes food products for market, and also drugs. Well known and legitimate, its advisers are respected worldwide for their scientific soundness. In time, the FDA has proven its independence, and most American citizens trust it. One of the questions raised by the controversy relates to the independence of the safety studies and data provided: companies who are asking for a market authorization are also the ones providing the data. To some extent, companies are judged innocent until proven guilty, which contrasts very much with the 'precautionary-principle' philosophy of regulation in Europe, which implies a responsibility of the state to protect citizens against vested interests and potential health dangers. But if proven guilty, sanctions against companies in the US are very severe. The Starlink crisis provided a good example. After a coalition of NGOs—Genetically Engineered Food Alert—found Cry9C corn in the human food chain (in taco shells), it was discovered that a genetic contamination had occurred with this GM product authorized only for animal feed and industrial use by the USDA and EPA. Uninformed farmers seem to have failed to maintain sufficient precaution in segregating this corn from others. The French firm Aventis was blamed by US officials, and agreed to recall large quantities of the corn. Estimates of the cost were as high as about $1 billion for the recall (300 brand products were concerned—see *Wall Street Journal*, November 3, 2000), not to mention possible settlements with consumers claiming allergic reactions to the product (Associated Press, March 8, 2002).

Food and safety policy in Europe has for a long time been the responsibility of member states, and the role of the European central administration in this matter has increased only recently; agencies such as the FDA did not exist until very recently. The European parliament also has been gaining legitimacy as a political institution in the last decades—European deputies have been elected directly since the 1970s—and endorses an active advisory role. Green political parties are well represented at the European parliament and can play the role of whistleblower on a number of environmental issues. Although it has been getting more structured over the years, the central administrative power in Europe is still relatively weak, and in cases such as food regulation, the influence of individual member states through the EC council remains very strong. Considering the great variety of political orientations, of cultures, and of regulatory systems in the different countries, it is understandable that decision-making about hot topics such as food might be relatively unstable, and often the result of compromises. This situation leaves considerable room for civilian society—non-governmental institutions—leading some authors to characterize the US as 'Laissez-Faire, Industry-dominated' and Europe as 'stakeholder corporatism' (MacNichol and Bensedrine 2003). And while the US regulation on GM food has remained almost unchanged in the last fifteen years, no fewer than six different directives were issued at the EU level alone.

One of the major food trade disputes between the US and Europe, the beef hormone dispute we mentioned earlier, contributed to a large extent to setting the international rules at the WTO level too (Josling *et al.* 1999). Started in the late 1970s, the dispute was only solved in February 1998 when the WTO Appellate Body found the EU Commission decision to ban US beef imports illegal. The EU and the US strongly disagreed on the risk assessment of this product, both providing different scientific experiences and evidence to make their case as well as different interpretations of them—including the concept of 'precautionary principle' for the European part. Interestingly, the hormone-beef case has been one of the first important food safety cases that the European institutions has had to deal with, and, as one observer noticed, non-governmental institutions (consumers and environmental groups), consumers' concerns, and compromises between the disagreeing EU Commission, EC council, and European Parliament played an important role in the position that the EU finally adopted (Josling *et al.* 1999).

No such trustable institution as the FDA existed at the European level. Records of the decision-making related to GM food in countries like France provide a very confused picture.[27] This has of course been considerably reinforced by the occurrence of a number of food, health, and environmental political crises to which many actors referred when justifying their lines of action. The contaminated blood case, the Chernobyl collapse, and the mad cow crisis have certainly contributed to propelling European societies into what Beck has called risk societies, societies in which citizen aversion to risks increased and citizen trust in senior scientists and administration decreased (Beck 1992).

As an important consequence of this, civil society tended to get highly involved in watching and discussing technological and scientific decisions. Sociologists observed the gradual development of alternative civil alert configurations and actors (Chateauraynaud *et al.* 1998). One good example of this is provided by the role played by NGOs in mobilizing European citizens: the connection of previously separated worlds with different rationales against GMOs resulted in new alliances between environmentalists (Greenpeace), consumer organizations (Bureau Européen des Unions de Consommateurs), alterglobalists (ATTAC), farmers' unions (Confédération Paysanne), and even the creation of new entity like ATTAC-GMO. This connection of actors mirrors the connection of meanings and interests they defend. Actors and the causes they fight for have converged. The weight of these stakeholders' representatives in Europe influenced the design of the 2003 EU regulation that imposed mandatory labeling, and focused on the production process of the food under consideration (Granjou 2003).

In 2004, the EU lifted the de facto moratorium on GMOs by accepting new GM products. This evolution might be attributed to a number of factors beyond US trade pressures: a new regulation in place provided a sufficient level of safety and mandatory labeling, the creation of a European Food Agency, and the reduction of the uncertainty associated with GM food. In addition, independent studies performed by different official institutions such as the Royal Society in the UK and the Royal Society of Canada stated that GM food was safe to eat for consumers, provided that adequate assessment and tests be performed (more accurate than the prevalent tests performed by companies, see *Financial Times*, 5 February 2002).

Conclusion

Dealing with GM food, as we have seen, requires a diversity of approaches and theoretical as well as interpretative debates among analysts. Our aim in this chapter has been to provide such an account of the case, viewed through the lenses of Societal Analysis, a theoretical framework developed in comparative studies of organizations. Our position is that this approach might be usefully applied to cases of international controversies such as GM food, as well as being a tool contributing to our understanding of the societal acceptance of technology.

Debates about the 'precautionary principle' and its application have been an attractive target for expressions of all kinds of judgments. Risks associated with technology and innovation are increasingly intertwined with values. Many examples have been recorded in the food and health sectors, as biotechnologies are pervasively penetrating these areas. Extensive literature is now interested in the issue, both at the national and the international level (Noiville 2003). For example, if we take a food that presents some threat to human health, causing dangerous allergies for instance, a particular state might very well restrict its trade if it considers it unacceptable in its particular territory, even if the risk has been accepted elsewhere. The French state has, for instance, forbidden one industrial dye used in cakes, arguing that the higher rate of cake consumption in France would expose French people to a specific cancer danger. As is visible from this example, the equation health—food—values—institutions embodies cultural traits that sometimes might appear as cultural exceptions. They represent the link between the risk and the values specific to one society. They also foster different kinds of adaptations and institutional innovations, like the creation of an independent agency Agence Française de Sécurité Sanitaire des Aliments (AFSSA), in charge of food benefits and risk assessment, as well as the evaluation of the cost of potential precautionary measures.

From the same perspective, the European community has voted to regulate and assess GM food (Directive 2001/18/CE). Significantly, the creation of ethical committees was added to the procedure. Equally, the Carthagena protocol on the international trade of GMOs states that a country can stop the import of one particular GM food, arguing that a state can take into account possible socio-economic impact of use, such as biodiversity.

As one American expert observed, this might sound like policy rather than science. But amidst the numerous emerging situations which cannot be proved by science with certainty, governments still have to rely on orders of justification other than industrial performance. Increasing recognition of such positions by international agencies such as the WTO might possibly give rise to a 'right to alimentary difference.' This would, in turn, imply that a certain cultural diversity might need to be recognized in international economic trade. The preceding reflections throw light on some relevant questions about the learning processes underlying the creation and diffusion of GM food in different countries. Important methodological conditions for such analysis require the identification of significant actors involved in the GM food innovation and diffusion process, as well as observable national and international institutional changes. What is at stake in sometimes hard negotiation is nothing less than tomorrow's world: is the future going to be based mainly on genetically engineered foods designed to meet the end-customer's need? Or is it going to disclose more diversity, based on the value provided to culturally and historically rooted food? Is this last possibility not linked to a different attitude towards the place of science in societies, and its links with citizens?

Notes

1. ATTAC stands for Association for the Taxation of Financial Transaction for the Aid of Citizens, in favour of the Tobin tax.

2. Commission (1999) '2194th Council meeting—ENVIRONMENT—Luxembourg, 24/25 June 1999', Press 203.
3. The term 'de facto moratorium' refers to the fact that no formal EU legislative measure has been taken to ban GM products, but member states have been blocking the approval procedure for GM products. Between 1998 and 2004, no applications for market placement of GMO products were approved.
4. According to ISAAA, in 2001, USA, Canada, and Argentina represented 96 per cent of GM crops, the EU less than 1 per cent (Spain was the main/only EU producer with about 30,000 hectares in 2001).
5. Technically, transgenesis is the transfer of one gene of interest into a germplasm.
6. They are concentrated in a few products: 81 per cent of US soy, 70 per cent of US cotton, and 40 per cent of US corn were genetically modified; see Pew Initiative (2003) US vs EU. Pew Report, August 2003.
7. James, C. (1997). ISAAA briefs.
8. GM soybean herbicide-tolerant superiority would mainly come from easier weed control and simplified production, GM Bt corn would lead to increased yield and profit (depending on the degree of corn borer infestation), GM Ht, Bt, and stake traits would increase yield and profit. For a review, see EU (2001) 'Economic Impacts of Genetically Modified Crops on the Agri-Food Sector.' General Directory Agriculture.
9. In 2002, 91 per cent of world GM crops are from Monsanto seeds. See Innovest (2003). *Monsanto and Genetic Engineering*. New York: Innovest.
10. As Lynn *et al.* stated, "The form the developing technology should take depends on how the developing markets respond to early versions of the technology, yet paradoxically, how the market respond depends on the form the technology takes." In Lynn *et al.* (1996).
11. Monsanto is a 'broad spectrum' herbicide, killing all kinds of plants. That included the crops: Roundup could only be spread before plant germination. Self-degrading on soil, it is considered relatively low polluting. The concept of Roundup ready crop + Roundup is to create a dedicated environment for the GM crop, all competing varieties being excluded.
12. Roundup was Monsanto's best-selling product, accounting for a significant share of Monsanto revenue (some estimates quote figures as high as 50 per cent: *Des Moines Register*, January 10, 2003).
13. Roundup patent in the US ended in 2000.
14. Our position is that the controversy is not yet closed, which explains the format of the temporary closure: a moratorium is indeed an option strategy, a way to suspend decision-making in time until better knowledge is acquired.
15. Under directive 90/220, Council needed to majority approve one member state decision to authorize a GM product. The European Council of June 1996 examined the approval of Novartis GM maize; only France voted for it (Kempf 2003).
16. As one journalist noticed, 'anyone who has dipped sushi in soy sauce, eaten bread, pasta, ice cream, candy, or processed meats (not to mention cornflakes) has almost certainly consumed genetically modified food. And the speed with which the products have entered our lives concerns many people' (*New Yorker*, April 10, 2000).
17. In France, Greenpeace has published a list of GM products and GM product-selling shops.
18. In June 1998, Monsanto published a number of ads in the main French newspapers claiming GM food superiority, safety, and beneficial prospects (sustainable agriculture, drug production). Part of the campaign stigmatized the French ignorance— '69 per cent of the French do not trust biotechnology, 63 per cent claim not to know what it is, hopefully, 91 per cent can read' (Kempf 2003). The impact of this campaign was rather counter-productive. An international study on consumer acceptance has shown that Austrians and Germans were amongst the most aware, yet the most reluctant to adopt GM food (Bauer, *Nature Biotechnology*, 15 March 1997). As Robert Shapiro recognized later on, 'We have probably irritated and antagonized more people than we have persuaded' (*Washington Post*, October 26, 1999).
19. Monsanto's claim of GM products' superiority seems corroborated by rapid adoption rates. Results were found to be more variable by investigators depending on the approach used, the

product considered, the location and the performance criteria selected. Many economic studies were based on experimental field data rather than actual farm-level collection. Although not as obvious as claimed by Monsanto, it is generally acknowledged that some advantages are derived by farmers either in net revenue or in ease of work (USDA 2000/2002). In Shapiro's view, this was only a first step, and the technology carried considerable potential for beneficial products: highly nutritional varieties against hunger, drug-producing varieties against cancer (Magretta 1996).

20. An estimated ten years and 300 million dollars R&D per GM product (*Washington Post*, February 3, 1999).

21. 'An estimated three quarters of the world's farmers still do so.' See *Financial Times*, 13 July 1999).

22. Accounts of small-farmer's practices in the south depict farmers' appropriation of seed as a major aspect of the food chain: selection improvement from even original commercial varieties is part of the farmer's central role in feeding society (RAFI 30 March 1998). In industrialized countries, 'for centuries, farmers have been saving seeds and breeding them over generations to make better plants' (*New Yorker*, April 10, 2000). The actual scale of seed saving is unknown: in the US, estimates based on local experiences mention figures like 20 per cent in the centre to 50 per cent in the south for soybeans. Observations in North American wheat describe generalized seed saving: commercial market buy occurs every four to five years (RAFI communiqué 30/3/1998).

23. A *cropchoice* article estimates that 75 employees and an annual budget of 10 million dollars were devoted to gene policing (**www.cropchoice.com**).

24. Probably scared by the mounting volume of the seed dispute, companies such as Agrevo opted for farmers' freedom to use the seeds.

25. Genetic drift was experienced and acknowledged for GM canola as soon as 1997, following observations by farmers. A solution to this problem was to inhibit the plant's reproduction system on demand.

26. Bové and others had destroyed a stock of Novartis Maize in January 1998.

27. In June 1996, the Juppé government supported Novartis maize demand for approval. In February 1997, the production of GM maize was forbidden in France by the same government (*Le Monde*, 14 February 1997). In November 1997, the Jospin government disapproved the culture of GM canola and approved the production of Novartis Bt maize. In September 1998, the state council suspended the government decision.

References

Abernathy W., and Clark, K. B. (1985). 'Innovation: Mapping the Winds of Creative Destruction.' *Research Policy*, 14: 3–22.

Adner, R., and Levinthal, D. A. (2002). 'The Emergence of Emerging Technologies.' *California Management Review*, 45/1: 50–66.

Aggeri, F., and Hatchuel, A. (2003). 'Ordres socio-économiques et polarisation de la recherche dans l'agriculture: pour une critique des rapports science/société.' *Sociologie du Travail*, 45: 113–33.

Akrich, M. (1992). 'The De-Scription of Technical Objects.' In W. Bijker and J. Law (eds.), *Shaping Technology/ Building Society: Studies in Sociotechnical Change*. Boston: MIT Press: 205–24.

—— Callon, M., and Latour, B. (1988). 'A quoi tient le succès des innovations?' *Annales des Mines*, 4–17 June and 14–29 September.

Anderson, P., and Tushman, M. L. (1990). 'Technological Discontinuities and Dominant Design: A Cyclical Model of Technological Change.' *Administrative Science Quarterly*, 35: 604–33.

Aoki, M. (1988). *Information, Incentives and Bargaining in the Japanese Economy*. Cambridge: Cambridge University Press.

Arthur, B. W. (1989). 'Competing Technologies, Increasing Returns, and Lock in by Historical Events.' *Economic Journal*, 99: 116–31.

Bauer, M. (1995). *Resistance to New Technology: Nuclear Power, Information Technology and Biotechnology.* Cambridge: Cambridge University Press.

—— and Gaskell, G. (eds.) (2002). *Biotechnology: The Making of a Global Controversy.* Cambridge: Cambridge University Press.

Beck, U. (1992). *Risk Society: Toward a New Modernity.* London: Sage Publications.

Bijker, W., and Law, J. (1992). *Shaping Technology/Building Society: Studies in Sociotechnical Change.* Cambridge, MA: MIT Press.

—— Hughes, T., and Pinch, T. (eds.) (1987). *The Social Construction of Technological Systems*: *New Directions in the Sociology and History of Technology.* Cambridge, MA: MIT Press.

Bijman, J., and Joly, P.-B. (2001). 'Innovation Challenges for the European Agbiotech Industry.' *AgBioForum,* 4/1: 4–13.

Boisard, P., and Letablier, M. (1987). 'Le Camembert: Normand ou Normé. Deux modèles de production dans l'industrie fromagère.' In F. Eymard Duvernay (ed.), *Entreprises et produits: Cahiers du Centre d'Études de l'Emploi,* 30: 1–30. Paris: Presses Universitaires de France.

Boltanski, L., and Thévenot, L. (1991). *De la justification: les économies de la grandeur.* Paris: Gallimard.

Bourdieu, P. (1987). *Choses dites, le sens commun.* Paris: Éditions de Minuit.

Bové, J., and Dufour, F. (2000). *Le Monde n'est pas une marchandise: des paysans contre la mal-bouffe.* Paris: La Découverte.

Bower, J. L., and Christensen, C. M. (1996). 'Disruptive Technologies: Catching the Wave.' *Harvard Business Review,* 73/1: 43–53.

Boyer, R. (2003). 'Le Paradoxe des sciences sociales.' *Current Sociology,* 47/4: 19–45.

Callon, M. (1981). 'Pour une sociologie des controverses scientifiques.' *Fundamenta Scientiae,* 2/3–4: 381–99.

—— (1986). 'Eléments pour une sociologie de la traduction: la domestication des Coquilles St Jacques et des Marins Pêcheurs dans la Baie de St Brieuc.' *L'Année Sociologique,* 36: 169–208.

—— (1999). 'La Sociologie peut-elle enrichir l'analyse économique des externalités? Essai sur la notion de cadrage débordement.' In D. Foray *et al.* (eds.), *Innovations et performances: approches interdisciplinaires.* Paris: EHESS, 399–431.

—— (2003). 'Laboratoires, réseaux et collectifs de recherche.' In P. Mustar and H. Penan (eds.), *Encyclopédie de l'innovation.* Paris: Economica, 720.

—— Lascoumes, P., and Barthes, Y. (2001). *Agir dans un monde incertain: essai sur la démocratie technique.* Paris: Éditions du Seuil.

Chern, Wen S., and Rickertsen, K. (2002). 'Consumer Acceptance of GMO: Survey Results from Japan, Norway, Taiwan and the United States.' Ohio State University Working Paper AEDE-WP-0026-02, September.

Christensen, C. M. (1997). *The Innovator's Dilemma: When New Technologies Cause Great Firms to Fail.* Cambridge, MA: Harvard Business School Press.

—— and Rosenbloom, R. S. (1995). 'Explaining the Attacker's Advantage: Technological Paradigms, Organizational Dynamics, and the Value Network.' *Research Policy,* 24/2: 233–57.

Clark, K. B. (1985). 'The Interaction of Design Hierarchies and Market Concepts in Technological Evolution.' *Research Policy,* 14/5: 235–51.

Collins, H. M. (1992). *Changing Order: Replication and Induction in Scientific Practices.* Chicago: University of Chicago Press.

Commission Européenne (2001). 'Europeans, Science and Technology.' *Eurobarometer* 55/2 (December).

Cooper, L. G. (2000). 'Strategic Marketing Planning for Radically New Products.' *Journal of Marketing,* 64: 1–16.

Courtney, H., Kirkland, J., and Viguerie, P. (1997). 'Strategy under Uncertainty.' *Harvard Business Review,* 75: 67–79.

Cowan, R. (1991). 'Tortoises and Hares: Choice among Technologies of Unknown Merit.' *Economic Journal,* 101: 801–14.

David, P. A. (1985). 'Clio and the Economics of QWERTY.' *American Economic Review,* 75/3: 332–7.

De Laat, B. (1996). 'Scripts for the Future.' PhD thesis, University of Amsterdam.

Dosi, G. (1982). 'Technological Paradigms and Technological Trajectories: A Suggested Interpretation of the Determinants and Directions of Technical Change.' *Research Policy*, 11/3: 147–62.

European Commission (2001). *Economic Impact of Genetically Modified Crops on the Agri-Food Sector*. Directory General Agriculture.

Garud, R., and Rappa, M. A. (1994). 'A Socio-Cognitive Model of Technology Evolution: The Case of Cochlear Implants.' *Organization Science*, 5/3: 344–62.

—— Jain, S., and Kumaraswamy, A. (2002). 'Institutional Entrepreneurship in the Sponsorship of Common Technological Standards: The Case of Sun Microsystems and Java.' *Academy of Management Journal*, 45/1: 196–214.

Geroski, P. A. (2000). 'Models of Technology Diffusion.' *Research Policy*, 29: 603–25.

Gold, B., Peirce, W., Rosegger, G., and Perlman, M. (1984). *Technological Progress and Industrial Leadership: The Growth of the US Steel Industry, 1900–1970*. Lexington Books.

Granjou, C. (2003). 'Traçabilité, étiquetage et émergence du citoyen consommateur: l'exemple des OGM.' Unpublished paper.

Granovetter, M. (1992). 'Economic Institutions as Social Constructions: A Framework for Analysis.' *Acta Sociologica*, 35: 3–11.

Grove-White, R. *et al.* (1997). *Uncertain World: Genetically Modified Organisms, Food and Public Attitudes in Britain*. Lancaster University: CSEC.

Hargadon, A. B., and Douglas, Y. (2001). 'When Innovation Meets Institutions: Edison and the Design of the Electric Light.' *Administrative Science Quarterly*, 46: 476–501.

Harvey, M. (1999). 'Genetic Modification as Bio-Socio-Economic Process: One Case of Tomato Purée.' CRIC Discussion Paper 31, November 1999, Manchester University.

Henderson, R., and Clark, K. B. (1990). 'Architectural Innovation: The Reconfiguration of Existing Product Technologies and the Failure of Established Firms.' *Administrative Science Quarterly*, 35: 9–30.

Institut National de la Recherche Agronomique (1998). *Les Chercheurs de l'innovation: regards sur les pratiques de l'INRA*. Paris: INRA.

Jolivet, E. (1996). 'Essai de lecture critique autour de l'innovation technologique au Japon.' *EBISU—Études Japonaises*, 14: 5–45.

—— (1999). 'L'Innovation technologique comme processus d'apprentissage industriel: analyse de la formation et de la diffusion des connaissances dans le cas des hauts fourneaux à injection en France et au Japon.' PhD thesis, LEST-CNRS, Université de la Méditérranée, Aix-en-Provence.

—— Larédo, P., and Shove, E. (2003). 'Managing Breakthrough Innovations: The Socrobust Methodology.' Presentation to *ASEAT Conference*, 2003, 'Knowledge and Economic and Social Change: New Challenges to Innovation Studies.' Manchester University, UK.

Joly, P. B., and Marris, C. (2001). 'Agenda-setting and Controversy: A Comparative Approach to the Case of GMOs in France and the US.' INSEAD workshop on *European and American Perspectives on Regulating Genetically Engineered Food*, 8–9 June 2001, INSEAD.

Josling, T., Roberts, D., and Ayesha, H. (1999). 'The Beef-hormone Dispute and its Implications for Trade Policy.' Working Paper series, Stanford Institute for International Studies.

Kempf, H. (2003). *La Guerre secrète des OGM*. Éditions du Seuil.

Klepper, S. (1997). 'Industry Life Cycles.' *Industrial and Corporate Change*, 6/1: 145–81.

Kodama, F. (1995). *Emerging Patterns of Innovation: Sources of Japan's Technological Edge*. Cambridge, MA: Harvard Business School Press.

Lallement, M., and Spurk, J. (eds.) (2003). *Stratégies de la comparaison internationale*. Paris : CNRS Éditions.

Lanciano, C., Maurice, M., Nohara, H., and Silvestre J.-J. (1993). *L'Analyse sociétale de l'innovation: genèse et développement*. Aix-en-Provence: LEST-CNRS.

Latour, B. (1987). *Science in Action: How to Follow Scientists and Engineers through Society*. Cambridge, MA: Harvard University Press.

Leonard-Barton, D. (1988). 'Implementation as Mutual Adaptation of Technology and Organization.' *Research Policy*, 17: 251–65.

Lundvall, B.-Å. (ed.) (1993). *National Systems of Innovation: Towards a Theory of Innovation and Interactive Learning*. London: Pinter.

Lynn, G. S., Morone, J. G., and Paulson A. S. (1996). 'Marketing and Discontinuous Innovation: The Probe and Learn Process.' *California Management Review*, 38/3: 8–37.

MacNichol, J., and Bensedrine, J. (2003). 'Multilateral Rulemaking: Transatlantic Struggles around Genetically Modified Food.' In M. L. Djelic and S. Quack (eds.), *Globalisation and Institutions: Redefining the Rules of the Economic Game*. Cheltenham: Edward Elgar.

Magretta, J. (1996). 'Growth through Global Sustainability: An Interview with Monsanto's CEO, Robert Shapiro.' *Harvard Business Review*, January–February: 79–88.

Maguire, S. (2004). 'The Coevolution of Technology and Discourse: A Study of Substitution Processes for the Insecticide DDT.' *Organization Studies*, 25/1: 113–34.

Mangematin, V., and Callon, M. (1995). 'Technological Competition, Strategies of the Firms and the Choice of the First Users: The Case of Road Guidance Technologies.' *Research Policy*, 24: 441–58.

Mansfield, E. (1961). 'Technical Change and the Rate of Imitation.' *Econometrica*, 29/4: 741–66.

March, J. G. (1991). 'Exploration and Exploitation in Organizational Learning.' *Organizational Science*, 21: 71–87.

Marris, C. (1999). 'OGM: Comment Analyser les Risques?' *Biofutur,* December 1999/195: 44–7.

Maurice, M. (1989). 'Méthode comparative et analyse sociétale: les implications théoriques des comparaisons internationales.' *Sociologie du Travail*, 2: 175–91.

—— Mannari, H., Takeoka, Y., and Inoki, T. (1988). 'Des entreprises françaises et japonaises face à la mécatronique: acteurs et organization de la dynamique industrielle.' LEST-CNRS, Aix-en-Provence.

—— Sellier, F., and Silvestre, J. (1986). *The Social Foundation of Industrial Power*. Cambridge, MA: MIT Press.

Metcalfe, S., and Miles, I. (1994). 'Standards, Selection and Variety: An Evolutionary Approach.' *Information Economics and Policy*, 6: 243–68.

Misa, T. J. (1992). 'Controversy and Closure in Technological Change: Constructing "Steel".' In W. Bijker and J. Law (eds.), *Shaping Technology/ Building Society: Studies in Sociotechnical Change*. Cambridge, MA: MIT Press, 109–39.

Moore, G. A. (2002). *Crossing the Chasm: Marketing and Selling High-Tech Products to Mainstream Customers*. 3rd edn. New York: Harper Collins.

Nelson, R. R., and Rosenberg, N. (eds.) (1993). *National Innovation Systems: A Comparative Analysis*. Oxford: Oxford University Press.

—— and Winter, S. G. (1982). *An Evolutionary Theory of Economic Change*. Cambridge, MA: Harvard University Press.

Noble, D. F. (1986). *Forces of Production: A Social History of Industrial Automation*. Oxford: Oxford University Press.

Nonaka, I., and Takeuchi, H. (1995). *The Knowledge-Creating Company: How Japanese Companies Create the Dynamics of Innovation*. Oxford: Oxford University Press.

Orléan, A., and Aglietta, M. (1994). *Analyse économique des conventions*. Paris: Presses Universitaires de France.

Rabeharisoa, V., and Callon, M. (2000). 'Les Associations de malade dans la recherche en France.' *Medical Science*, 16/11: 1225–31.

Rip, A. (1986). 'Controversies as Informal Technology Assessment.' *Knowledge: Creation, Diffusion, Utilization*, 8/2: 349–71.

Rosenberg, N. (1982). *Inside the Black Box: Technology and Economics*. Oxford: Oxford University Press.

Rudolf, F. (2003). 'Deux conceptions divergentes de l'expertise dans l'école de la modernité réflexive.' *Cahiers Internationaux de Sociologie*, 114: 35–54.

Sahal, D. (1985). 'Technological Guideposts and Innovation Avenues.' *Research Policy*, 14: 61–82.

Shinn, T. (2001). 'Nouvelle production du savoir et triple hélice: tendances du prêt-à-penser les sciences.' *Actes de la Recherche en Sciences Sociales*, 141–2: 21–30.

Sommier, I. (2003). *Le Renouveau des mouvements contestataires à l'heure de la mondialisation*. Paris: Flammarion.

Teece, D. J. (1986). 'Profiting from Technological Innovation: Implications for Integration, Collaboration, Licensing and Public Policy.' *Research Policy*, 15/6: 285–305.

Thévenon, O. (2001). 'La Place des femmes sur le marché du travail britannique et français: logique marchande vs logique civique.' In C. Bessy *et al.* (eds.), *Des Marches du travail équitables? Approche comparative France/Royaume-Uni*. Paris: I Peter Lang.

Tripsas, M., and Gavetti, G. (2000). 'Capabilities, Cognition, and Inertia: Evidence from Digital Imaging.' *Strategic Management Journal*, 21/10–11: 1147–61.

Utterback, J. M. (1996). *Mastering the Dynamics of Innovation*. Cambridge, MA: Harvard Business School Press.

Van den Belt, H., and Rip, A. (1987). 'The Nelson-Winter-Dosi Model and Synthetic Dye Chemistry.' In W. Bijker *et al.* (eds.), *The Social Construction of Technological Systems*. Cambridge, MA: MIT Press, 135–58.

Von Hippel, E. (1988). *The Sources of Innovation*. Oxford: Oxford University Press.

16 Prospective Structures of Science and Science Policy

Harro van Lente

Abstract

Science policy's evolution over the last decades can be understood as a strategic change of direction. Firms have a new orientation towards promises about future use of scientific findings. These promises about eventual usefulness not only legitimize the policy decisions and instruments, but also contribute to the actual development of technical-scientific fields.

A rhetorical analysis is used to highlight these dynamics. The interlocking of policies and local activities creates irreversible developments in technical-scientific fields. Using the example of the rise of genomics, this implicit and emergent coordination of knowledge production and learning is captured as what might be termed a 'prospective structure' in which science policy has to operate.

Introduction: the evolution of S&T policies

Since the early days of the British Royal Academy of Science, academic activities have had very different institutional shapes. There have been considerable changes over time, and huge variations between countries. Differences relate especially to the way new disciplines emerge and stabilize, to the relationships of private and public research activities, and to the general legitimation of science in society. Science policy and big government research projects have had different degrees of success, both in the US and in the European Community (EC) (see Kuhlmann and Shapira, this volume). Various authors have reviewed the institutional context of science and R&D, and how it has been organized since the Second World War, both in the US and the EC (Larédo and Mustar 2001; Kuhlmann 2001; Barre *et al.* 1997).

According to Mytelka and Smith (2002), science and innovation policies have co-evolved with innovation theory, and they present strong evidence of 'policy learning.' This chapter draws attention to a peculiar feature of science policy and policy learning: its existence in the realm of promises. Science and technology, with their stress on novelty, have always raised expectations, and this has become more salient in the last decade. This chapter examines this feature and reflects on its consequences; its thrust is that science policy is part and parcel of the ongoing dynamics of expectations that structures science in the first place. Thus, part of the efficacy of science policy and of policy learning must be assessed from the content of the promises that are floated in the documents and practices of science policy.

An evolution in the rationale and the instruments of science policy has taken place. Borrás (2003: 14) provides a useful summary of the evolution of these focus points in the EU in terms of a transition from 'science policy' and 'technology policy' to 'innovation policy' (Table 16.1).

After the Second World War, in both the US and in Europe, science became, in the words of the title of Vannevar Bush's famous report, 'The Endless Frontier' (*Science: The Endless Frontier*). It was seen as the source of welfare,

Table 16.1. Evolution of S&T policies in Europe

Science policy	Technology policy	Innovation policy
research	strategic industries	system applications
scientific education	RTD collaboration	building capacities
scientific infrastructures	procurement	IPRs
big science	environment technology	SMEs bioethics
	transfer standardization	social values

Source: Borrás 2003.

prosperity, and, above all, of security (Elzinga and Jamison 1995). In the EU, the first focus was on nuclear energy, which played a crucial role in the Second World War. The first Joint Research Centers (JRCs) started in the 1960s, and were dedicated to the progress of knowledge about nuclear physics and its applications.

According to Mowery (2001), the Cold War, with its urgent needs to counteract possible security threats, marked science and technology policies for a long time. The vision that integrated science with broader national missions was reified in the Bush report; the report argued for the coordination of science on a national scale. A negotiation in terms of societal promises was at stake here: in return for a granted autonomy and subsidies, science would provide society with benefits such as security and prosperity. The promise is still alive, but the institutional form that Vannevar Bush had recommended—to install a central (federal) science agency—never really succeeded. When the National Science Foundation (NSF) started in 1950, the division of the budgets along the various ministries had already taken place, especially to the benefit of the Ministry of Defense: 'In retrospect, the delay in the

establishment of the National Science Foundation was critically important in the evolution of postwar policy for research and development, not least because it cost the nation a program balanced between civilian and military patronage and purpose' (Kevles 1977: 360).

The 1984 start of the Framework Programs in the European Union marked the step from a 'science' to a 'technology' orientation. The focus was on pre-competitive research that nonetheless was supposed to bring market advantages in the future. The key was no longer the production of knowledge per se, but the development of technologies. This approach had an increasing budget (Table 16.2); important examples are the BRITE and ESPRIT programs (Cabo 1997).[1] During the mid-1990s, the technological orientation shifted to a focus on 'innovation,' that is, the successful application of the produced knowledge. Important in this third phase is the collaboration and integration of various parties.

Bozeman (2000) discusses a comparable development in the US, where three R&D policy paradigms have competed in the last decades: the market, the mission, and the cooperative paradigm.

Table 16.2. Allocation to EU Framework Programs, in million €

	FP1 1984–7	FP2 1987–91	FP3 1990–4	FP4 1994–8	FP5 1998–2002	FP6 2002–06
EU allocation	3.750	5.396	6.600	13.100	14.960	17.500

Source: Borrás 2003: 37.

Table 16.3. Three competing technology policy models

	Market failure	Mission	Cooperative technology
core assumptions on allocation	markets are the most efficient allocator of information and technology	allocation via authorized programmatic missions of agencies	markets are not always the most efficient route to innovation and economic growth
core assumption on role of government laboratories	government laboratories should be limited to market failures	government R&D is not limited to defense, to be organized in terms of missions of agencies	government laboratories and universities can play a role in developing (pre-competitive) technology
core assumption on public–private linkages	innovation flows primarily from and to private sector	government should complement but not compete with private sector	more centralized planning is required
peak influence	highly influential during all periods	1945–65; 1992–2000	1992–4
policy examples	deregulation; contraction of government role; R&D tax credits	creation of energy policy R&D, agricultural labs	expansion of federal laboratory and university roles
theoretical roots	neoclassical economics	traditional liberal governance	industrial policy theory, regional economic development theory

Source: Bozeman 2000: 631.

The market paradigm, based on the assumption that markets are the most efficient mechanisms for allocating knowledge and products, and that the competitive workings of the market see to it that the private sector responds to societal needs and fulfills them efficiently, predominates in most periods of US science policy. Hence, the government role is confined to situations where a so-called market failure may be said to exist—for example, in the case of far-reaching externalities, high transaction costs, or flaws in information flows. An important exception to the market failure paradigm is the government's role in the defense sector.

The second paradigm, mission orientation, in which encompassing societal concerns are translated into research objectives, has occasionally paralleled the market failure model.

The assumption of the mission paradigm is that the government should define missions of national interest that are not addressed by the market. In these missions, government is seen as responsible for R&D, not industry. The most important and constant such missions are defense and space R&D, occasionally stretched to areas such as energy, public health, and agriculture.

Third, and more recent, is the cooperative paradigm which aims at forming productive networks of universities and firms. Here, the role of the government is encompassing, not limited to funding R&D. Also envisaged to be important are the roles of transfer of technology, the stimulation of networks, science parks, and brokerage. This third paradigm became particularly appealing during the Clinton administration. See Table 16.3.

The institutional context of the production of knowledge has changed too. A wide range of scholars assert that we live in a knowledge-intensive society. The label accounts for various developments that have become apparent since the 1980s, such as the increasing value of knowledge and the impressive growth, both absolute and relative, of knowledge-intensive services in the economic system. In addition to the increasing weight of knowledge as a production factor, a second characteristic may even be more important: knowledge production has been distributed across various institutes and firms. A popular distinction in this respect has been suggested by Gibbons *et al.* (1994); they distinguish a new emerging mode of knowledge production, the so-called Mode 2, to be contrasted with the traditional Mode 1 (see Table 16.4). Whereas Mode 1 is discipline oriented and characterized by homogeneity in its evaluation, Mode 2 is transdisciplinary and characterized by different forms of evaluation. Organizationally, Mode 1 is hierarchical and tends to preserve its form, while Mode 2 is heterarchical and transient. The change towards a more distributed mode of knowledge production relates to, and is amplified by, the 'core business' strategy of big firms predominant in the 1980s and 1990s. This strategy has led to an intensified interdependency between industrial concerns and knowledge production. In addition, changes in the science system mean that the role of intermediaries appears to be getting more important: Van der Meulen and Rip (1998), for example, discuss the emergence of an intermediary layer between the level of individual scientists and research institutes on the one hand and the policy level on the other. The intermediary level has gained an independent role, and intermediary parties play a role in R&D developments (Bessant and Rush 1995).

Important now in the analysis of research and innovation policies is the notion of 'National Systems of Innovation,' which has two kinds of backgrounds (see the introductory chapter in this book). The first background is theoretical: innovation cannot be understood as an isolated activity of a single firm or organization, but is always the outcome of many heterogeneous and interrelated activities and resources (see Figure 16.1). The economics of technical change stresses the importance of learning processes which, in their turn, are part of ongoing social and institutional processes (Nelson and Winter 1977, 1982; Dosi *et al.* 1988). The importance of learning processes also implies that history matters and that, therefore, a dynamic analysis is needed to understand the potential and consequences of innovation. The second background is the empirical insight that countries differ in their institutional settings and that this matters a lot for science and innovation policy. The famous and seminal analysis of Freeman in *Technology Policy and Economic Performance: Lessons from Japan* (1987) showed that countries differ in their capacities for generating and implementing innovation. The analysis pointed to policy settings that are different in Japan from West-

Table 16.4 Mode 1 versus Mode 2 science

	Mode-1 Science	Mode-2 Science
context	academic context	application oriented
intellectual perspective	disciplinary	transdisciplinary
structure	homogeneous, hierarchical	heterogeneous, non-hierarchical
quality control	peer review quality control	a broader set of criteria
accountability	primarily to science	includes societal concerns
knowledge producers	academics and technicians	a wide set of actors

Source: Gibbons *et al.* 1994.

ern countries, such as the orchestrating role of the Ministry of International Trade and Industry (MITI) and the Council for Science and Technology (CST). The focus on quality at all levels of production processes, and the identification of 'technologies of the future' helped Japan, according to Freeman, to gain important advantages.

The idea of a national system of innovation has raised important questions, including ones that criticize the concept for being too broad. Larédo and Mustar (2001) differentiate the narrow from the broad thus: the narrow focuses on institutions that are directly involved in scientific and technological activities; the broad seeks to include all social institutions that are indirectly involved in the processes of science and technology. In their recent overview of research and innovation policy, they conclude that the 'landscape' of research and innovation policies has changed radically. From the detailed overview of changes in ten countries, plus an analysis of EU policy, they derive three, partly conflicting, major transformations: first, new relationships between international competitiveness and public policies have evolved; second, there is increasing attention paid to public-sector research; third, there is increasing tension between globalization of research and policy versus the importance of specificity and proximity. They conclude that the challenge for public policy is:

How ... can public interventions be developed, which are capable both of promoting the development of local links and of contributing to the organization of a global framework? (Larédo and Mustar 2001: 497)

Although the system-of-innovation perspective stresses the uniqueness of countries, the habit of comparison is still very strong. Recently, the Canadian researcher Benoit Godin analyzed the origins of the OECD's habit of comparing countries: in what he calls 'the mystique of ranking,' OECD countries are compared and given a position according to their R&D expenditures. The productivity gap has been a dominant concern since the Marshall Plan, which was launched in 1948 to reconstruct post-war Western Europe. In many increasingly sophisticated statistical studies, the disparities of productivity in industrial sectors have been made visible. In the 1960s, the dominant theme was the 'productivity gap;' in the 1970s, the attention turned to 'technological gaps' as gauged by the availability and allocation of R&D resources. Table 16.5 shows an example of such a ranking.[2]

According to Godin, the notions of gaps and the concomitant rankings 'certainly shaped political discourses, policy documents and analytical studies' (Godin 2002: 408). His analysis shows how the ranking of countries, focusing on information and communication technologies, so as to measure the

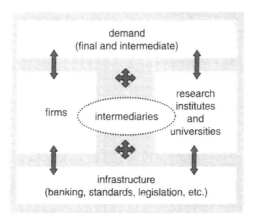

Fig. 16.1. Overview of a system of innovation
Source: Van Lente *et al.* 2003.

Table 16.5 OECD ranking of countries in 1975

Group	R&D expenditures	Countries
Group I	Large R&D and highly R&D intensive	France, Germany, Japan, United Kingdom, United States
Group II	Medium R&D and highly R&D intensive	Netherlands, Sweden, Switzerland
Group III	Medium R&D and R&D intensive	Australia, Belgium, Canada, Italy
Group IV	Small R&D and R&D intensive	Austria, Denmark, Finland, Ireland, Norway
Group V	Small R&D and other	Greece, Iceland, Portugal, Spain

Source: OECD 1975.

contribution of science and technology to economic growth, became an important task for the OECD. The European Union now assists the OECD in producing rankings of countries, and in addressing the gaps between Europe and the United States in its science and technology statistics. Godin argues: 'Emulation between countries, mimicry and convergence probably have to be accepted as an indirect effect of statistical standardization' (ibid. 409).

A trait common to the shifts discussed above is that science and technology have become increasingly 'strategic.' Since the 1970s, and especially in the 1980s, it has become fashionable in policy circles to think, talk, and plan in terms of 'strategic' science and technology (Rip 1990). In a wide range of countries, 'strategic' research programs were set up to stimulate areas of 'strategic' relevance. Science policy studies have coined the term 'strategic turn' to characterize the changes in the research system (Cozzens *et al.* 1990). A brief study of the history of the notion of strategic science illustrates that expectations and promises have gained visibility, and have become a more important part of the dynamics of science and also of technology.[3]

The notion of strategic science originally emerged to make sense of an intermediate kind of research. In the early 1970s, the distinction between 'fundamental' science, 'strategic' science, and 'tactical' science had been popular. Because of the contrast made between tactical and strategic, the term 'strategic science' took on clear connotations of longer-term development of (new) options. Both tactical and strategic research were regarded as subcategories of

applied research. In contrast, Irvine and Martin's (1984) most influential *Foresight in Science: Picking the Winners* saw strategic research as part of basic research; the other subdivision was 'pure or curiosity-orientated research.' Their definition stresses the importance of expectations:

Strategic research: Basic research carried out with the expectation that it will produce a broad base of knowledge likely to form the background to the solution of recognized current or future practical problems. (ibid. 4)

'Strategic research' differs most importantly from pure or curiosity-orientated research in the rationale behind its support, there being at least some expectation that it will contribute background knowledge required in the development of new technologies. (ibid. 3–5)

Irvine and Martin also note the changes in the research system and their impact on the way firms deal with technological options:

[Strategic research] is by no means confined solely to the university laboratory. Large science-based firms, for example, typically choose to devote a limited (but probably increasing) proportion of their R&D budgets to those areas of basic research felt most likely to provide the new knowledge required to develop the products and processes of the future.[4]

For Irvine and Martin, the notion of strategic research refers to basic research that promises to yield innovations in the long term. As they indicate themselves, it is a category that refers to the rationale behind the support of R&D: research will be supported if it is promising with regard to future innovations. They note, therefore, that 'in many respects, the actual content of such work will be little different from that of academic research funded on a "pure" or "curiosity-orientated" basis' (ibid. 5).

The notion of strategic research was also applied to the link-ups between universities and institutes to carry out fundamental research for high-level users (Johnston 1990) and was used in policy statements to indicate a new and desirable kind of scientific research. The science and technology stimulation programs of the 1970s had prepared the ground, and the increasing use of prospective assessments of the innovative potential of scientific areas is an indicator of the recognition that fundamental science can be of strategic importance. In the 1980s, 'strategic' became a regular term in science policy, and governments were willing to spend more on this kind of research—for example, by setting up strategic programs on microelectronics, biotechnology, or new materials (Roobeek 1990). The strategic turn implies a reorientation in the rationale behind the initial support, consequent shifts in budgets, and, as we will see, new forms of learning and coordination.[5]

Expectations: a rhetorical theory and method

This chapter explores the significance of the strategic turn in science policy and assesses the implications for our understanding of knowledge dynamics. Given the prominence of promises and expectations, it is timely to outline a theoretical framework of expectations that will allow us to trace the relevant dynamics.

A decade of research in knowledge dynamics shows that knowledge production is part of the surrounding social and political developments. The seminal work of Thomas Kuhn (1962) on scientific revolutions highlighted the role of socio-cognitive paradigms in the construction of science. Since then, an impressive number of studies have focused on the social foundations of knowledge dynamics. The philosophy of science, which seeks to understand the basis of the cognitive validity

of knowledge, has explored important consequences (Fuller 1988); other traditions have focused on the networks and the powers that organize them (Latour 1999), or the role of experts in the mutual alignment of science and society at large (Jasanoff 1994). Common to all these attempts are ideas that the research process is shaped by interactions, interests, and cognitive frames. My theoretical framework builds on these studies (van Lente 1993, 2000; van Lente and Rip 1998a).

The starting point is that scientific work is, and always has been, embedded in a context of expectations. First of all, science can be characterized as search processes that are guided by heuristics—that is, rules that promise, but cannot guarantee, success. As a result, scientific claims are promises about contributions to the discipline; only when they are acknowledged by peers can they be counted as 'real' contributions. They are projections into the future, to be corroborated or rejected by others. In addition, the allocation of resources within science (and within research in general—in research teams, in universities, and between national research areas) is based on potential. Thus, expectations are not science by-products that distract the attention from the core business; they inform and guide the search processes themselves.

Expectations also protect activities, even when the outcome is disappointing. Given the prevailing modus operandi of trial and error, disappointing outcomes are not rare in science and technology, so such protection is often needed.[6] Expectations help to protect the activities by making it explicit that current outcomes may be disappointing, but that final outcomes will be positive.

The third step is to appreciate the role of expectations in processes of agenda-building (van Lente 2000; van Lente and Rip 1998a). Kingdon (1984) defines 'agenda' as a list of priorities that require action. When expectations have an accepted position in the repertoire of a research group, a techno-scientific field, or at society at large, they inform the important routes and directions. In other words, shared and stable expectations will be transformed to items on agendas—at various levels. Note that agendas, by their very nature, require action or, at least, clarification why action was not taken. Earlier, I characterized this sequence as a 'promise-requirement' cycle (van Lente 2000). Processes of agenda-building are nested; thus, issues on a laboratory agenda derive their salience from their association with research-field agendas.

Because this chapter focuses on discourses that inform or require action, it uses a rhetorical analysis to unravel the methods by which audiences are seduced into beliefs or actions. An important device to interest the audience is the so-called 'funnel of interest' (Law 1986): articles typically start with a broad topic or problem that probably will catch the interest of the reader, who is then guided to the argument by subsequent smaller selections: for example, an article's opening topic might be 'global lack of water resources,' continue via 'techniques to desalinate water' and 'membrane technology as a promising technique,' to focus on a small number of experiments with membranes and solvents in a particular laboratory (van Lente and Rip 1998) about which, without the funnel, readers may not have been interested in reading. Titles are the first devices to catch the reader, so these are especially interesting to study.

The rise of genomics

One of the most strategic fields of the last decade is genomics; it is worth exploring how its rise has been accompanied by the rise of labels and terms that underpin its general promise.

The rise of the technical-scientific field of genomics is a remarkable phenomenon. Since the celebration on June 26, 2000 of the draft human genome sequence at the White House, the declaration that the Human Genome Project was fulfilled—at least in principle—and the publication of the drafts in *Nature* and *Science*, attention has shifted to all kinds of possible new subfields and applications.[7] In the previous years, a classic discovery race had been taking place. Milestones in the race were the decoding in 1995 by Craig Venter and his colleagues of the influenza microbe: this was the first time that the genome of a whole free-living cell organism had been decoded. They used the so-called 'whole genome shotgun sequencing method.' In 1998, the first complete genome of a multi-cellular organism, the roundworm C. elegans, was published. Both the International Human Genome Project and Celera Genomics, Craig Venter's private company, were able to publish working drafts of the human genome in 2001. In the last decade, the speed and accuracy of sequencing techniques has increased dramatically; the complete genomes of many organisms, such as the pufferfish, the mouse, the malaria mosquito, and two varieties of rice are now published. The latter two underpin the potential of genetic research, as they affect the health and food supply of three billion people.

Genomics starts from the acknowledgment that data from the genome sequence is not sufficient to understand its function. It is more difficult to analyze the structure of the proteins that those genes encode, while the practical application of this knowledge is, of course, much more significant. DNA has only four building blocks, but proteins are composed of amino-acids, of which there are twenty varieties. The concomitant puzzles are thus much more complex. According to most definitions (Condit 1999), what differentiates genomics from biotechnology is the use of the knowledge of how specific genes lead to specific proteins. Genomics shares with biotechnology the notion that it should be regarded as an 'enabling technology' which will affect many corners of western economies, comparable to the paradigm shift of the IT revolution (Freeman and Perez 1988). The promise of genomics is considered to be more profound than that of biotechnology, although the areas of activities are not clearly separated. Harvard professor Juan Enriquez, for instance, declared in *Science*:

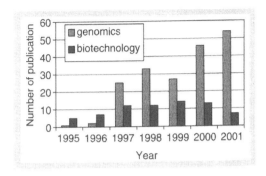

Fig. 16.2. Rise of genomics in *Science*

Genomics is not the biotech of the 1980s, which promised much and delivered little. Biotech companies tended to act alone, trying to integrate from the research bench through the drug counter. They remained relatively small, and their technology did not drive massive divestments and mergers among the world's largest corporations. The objective of a life science company is no longer to generate breakthroughs in a single area such as medicine, chemicals, or food, but to become a dominant player in all of these. (Enriquez 1998)

The rise of genomics can be illustrated in a number of ways. I will follow a rhetorical perspective here: that is, I will highlight the devices that are used to catch an audience. One way to map the increasing interest in genomics is to count the number of times it has appeared in scientific and other journals. (I have selected two specific and one set of journals: *Science*, *The Economist* and the journals of the *Nature* group (see Figure 16.4)). In Figure 16.2, we see that the term has much more frequently used since 1997. In the graph, we see an increase of

the use of the term 'genomics' at the expense of the term 'biotechnology.' Genomics became a new funnel of interest or, in another metaphor, an umbrella term preferred over others to catch the audience. *Science* is a key journal in the scientific forum; *The Economist* is a key journal that explains and explores political and business opportunities. Here we see that the term 'biotechnology' is still much more used in its articles. Nevertheless, there is also an increase in references to 'genomics' (see Figure 16.3).

In journals of the *Nature* group[8] we see an intermediate picture: 'biotechnology' was more used than 'genomics', but the situation reversed dramatically between 1999 and 2000 (see Figure 16.4). Clearly, 'genomics' is more frequently used in scientific journals than in broader journals (business and politics), where 'biotechnology' is more prevalent. Yet, the rise of the term is clear, as is shown in Figure 16.5, where the ratio between the terms is given.

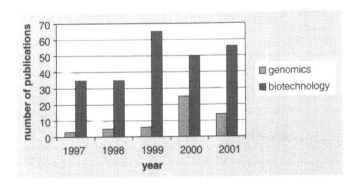

Fig. 16.3. Rise of genomics in *The Economist*

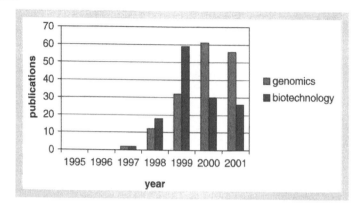

Fig. 16.4. Rise of genomics in journals of the *Nature* group

The rise of genomics as an interesting umbrella term, and the way this leads to agenda building, can be illustrated with the increasing number of science and technology policy programs that deal with this new field of research in many western countries. The Dutch government, for example, wants to lead the world in genomics through a series of concomitant programs (see Table 16.6). The country has four action plans:

- Life Science Action Plan;
- NWO program Biomolecular Informatics;
- Strategic Action Plan Genomics;
- IOP Genomics.

A central position is taken by the National Directorate Genomics (Nationaal Regieorgaan Genomics) (*Staatscourant*, 7 September 2001: 173, 11). Since 1981, 22 Innovation-Oriented Research Programs (IOPs) have been organized, starting with a program on biotechnology. The Ministry of Economic Affairs founded an Innovation-Oriented Research Program on Genomics in September 2000 (it started on October 13, 2000, in Nijmegen). It focuses on three themes:

- chronic and geriatric illnesses;
- quality and safety in food production;
- explaining biomolecular processes.

Other countries have similar programs. And on a European level, the genomics turn in biotechnology is seen as revolutionary. A clear example here is the joint statement on biotechnology by heads of government of the

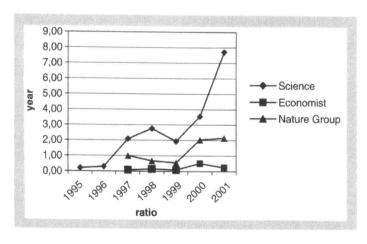

Fig. 16.5. The relative us of 'genomics' vs. 'biotechnology'

Table 16.6. The rise of genomics in the Netherlands

Date	Event
Early 2000	Start Platform Life Sciences
June 20, 2000	Strategic Action Plan Genomics (by 5 ministries). Focus on biomedical and agro-food research.
October 27, 2000	Temporary Advisory Committee Knowledge Infrastructure Genomics (= Committee Wijffels)
February 2001	Start of BioPartner by the Ministry of Economic Affairs to track and support new ventures in genomics.
April 11, 2001	Advice of the Committee Wijffels: strengthen and intensify the knowledge infrastructure; establish national coordination of efforts at NWO, the Dutch Science Organization.
July 16, 2001	Elaborated point of view and plans of the Dutch Cabinet (Kabinetsstandpunt Genomics; Kamerstukken II 2000/01 27 866 no.1). Budget 2001–6: 200 million euro.
July 2001	The Cabinet decides that a part of the huge natural gas incomes of the state will be used for genomics.
August 2001	Foundation of a National Directorate Genomics (that will coordinate and steer genomics research and investments).
May 8, 2002	Strategic Plan Genomics 2002–6 of the National Directorate Genomics
July 18, 2002	The Cabinet presents the Strategic Plan Genomics 2002–6 to Dutch Parliament.

Netherlands and the UK in advance of the Stockholm Summit in 2000:

Understanding the genome is revolutionizing bioscience research, giving new impetus to the search for cures and treatments for illness and helping to identify new ways of combating the problems of ageing. Biotechnology has the potential to help us create new drugs, with fewer side-effects, able to tackle effectively around 10,000 diseases instead of the 500 we can treat today... Our new knowledge also offers potential benefits in food safety and quality, sustainability and the environment. Agricultural biotechnology could provide the vital medical ingredients of the future. It offers potential for foods with nutritional advantages and crops which are tolerant to droughts and floods. Europe has the knowledge and skills to turn research into social and economic benefit. (Source: Stockholm Summit website)

Not only are governments excited by the new promising field of genomics, but also business parties. According to a survey of the Morpace Pharma Group (2001), pharmaceutical industry investments in genomics research were about $2 billion in 2000. Some companies are spending 20 per cent of their R&D budget on genetic research and genomics-oriented technologies.

In general, a fruitful combination is expected from the efforts of agricultural, pharmaceutical, and chemical businesses as firms position themselves in this universe of potentialities (see Figure 16.6).

The urgency of genomics

The interesting thing about the rise of genomics is not so much that we see a new field, but the sense of urgency that has built up. The attractiveness of the umbrella term 'genomics' urges governments and firms to position themselves and others in the developing scenarios.

The urgency is generated in various ways. The possibilities that genomics offers, in principle, are taken as a first step. In a famous speech in 1999, Francis Collins reaffirmed the public mission of the Human Genome Project:

Scientists wanted to map the human genetic terrain, knowing it would lead them to previously unimaginable insights, and from there to the common good. That good would include a new understanding of genetic contributions to human disease and

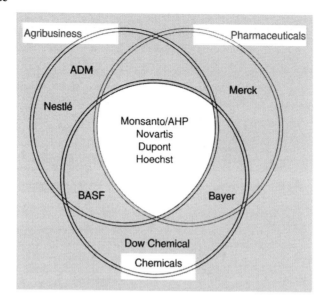

Fig. 16.6. Industrial actors in the field of genomics
Source: Morpace Pharma Group 2001.

the development of rational strategies for minimizing or preventing disease phenotypes altogether.[9]

And President Clinton related the genome endeavors to the welfare of all American families:

Chances are, every family represented in this room in our lifetime will have a child, a grandchild, a cousin, a niece, a nephew somehow benefited from the work of the Human Genome Project, which seemed nothing more than an intellectual dream just a few years ago. And one of the things that we have to do is to make sure that every American family has a chance to benefit from it.[10]

Of course, these optimistic views of the Human Genome Project were not uncontested: while a Nobel Prize winner made an explicit analogy to putting a man on the Moon, another argued in a keynote address that the scientific merits did not justify the whole project (Burris *et al.* 1998). But it is indisputable that the repertoire of genomics has all kinds of benefits for mankind, and invoking the term affords protection for those who seek support for their specific programs and projects.

Another example is from the DOE Joint Genome Institute. Senior Advisor Branscomb says:

The goal of our large-scale sequencing work is to help lay down the infrastructure that allows biological scientists to answer questions as efficiently

as possible. Genomic studies should soon reveal why some people are able to defend against the AIDS virus and others are not, for example... The genome is the basis of all life... When we get sick with an infectious disease, what's going on is a war between two sets of genes—ours and those of the virus or bacteria. Some day the medical profession will have better ways to handle these diseases—thanks to work on the genome.[11]

Urgency is also generated by creating a historical context, in which genomics is seen as a logical and not-to-be-stopped next step in the progress of knowledge. E. Pennisi (2000) in *Science*, for instance, argued that genomics is built upon many scientific victories—a useful rhetorical device (Perelman 1982):

This is a long way from the start of the 20th century, when geneticists were just rediscovering the seminal work of Gregor Mendel... It took until the 1950s for researchers to unmask DNA as the bearer of the genetic code. During the next two decades, biochemists developed the cloning and sequencing tools needed to fish out genes. By 1990, an insatiable hunger to know all the genes encoded in the DNA of humans prompted the establishment of the Human Genome Project. It was biology's first foray into big science and, by almost any measure, it has been a great success. The genome achievements this past year epitomize this century-long and decade-long quest. (Pennisi 2000)

The urgency of which this chapter speaks characterizes the end of the article:

... the allure of this knowledge has made the quest irresistible. The world eagerly awaits the published draft of the humane genome, with its genes outlined and its character explained. And, almost as eagerly, the gene searchers are chasing down the genomes of many other organisms, a quest that will tell us more about our own genome as well as about our place in the grand library of life.

How could one oppose a quest for our place in the universe? Governments feel obliged to respond, and a sense of urgency is apparent in many public discussions on technology policy. The Netherlands doesn't need to be ashamed of its position, said the Minister of Economic Affairs in July 2000, but 'I want more than we have now' (*Volkskrant*, 15 July 2000). She regrets that the Dutch's relative advantage in biotechnology in the 1980s has been lost and doesn't want to lose again with genomics. 'We lost the fine second place [after the US] and now we are at position ten. The goal of the Ministry is to belong again to the "elite" in 2005.'

Some say that the promises of genomics are more robust than those of biotechnology. Others warn that the promises need to be well balanced, otherwise the public and policy makers may be disappointed. For example, Arjan van Tunen of Plant Research International and Hans van den Berg of Akzo Nobel Pharma warn that 'now developments are so quick the Netherlands run the risk of falling behind.'[12] In general, we see two lines of reasoning, often within the same texts and often referring to the same data: the situation is favorable (so support is warranted); the situation is weak, alarming (so support is needed). The state of affairs is presented as both strong *and* weak. The first one could call a 'thesis' and the second the 'antithesis,' and indeed a synthesis—the conciliation between contradictory positions, or a 'dialectics of promises' (van Lente 1993)—is at play. The synthesis is that new forms of government funding are required. In the point of view of the Cabinet

(*Kabinetsstandpunt Genomics*) this dialectic is clearly visible:

[THESIS] Genomics is of strategic importance for the Netherlands and its citizens. It offers good chances for improvements of the quality of life ... Given its good starting point and the importance of the field, the Netherlands should have the ambition to be an important player in the field of genomics and bioinformatics.

[ANTI-THESIS] Dutch knowledge institutes have insufficiently been able to catch up with the acceleration of research. The danger, therefore, is that we will profit insufficiently from the chances that genomics offers.

[SYNTHESIS] This requires, in the light of the fundamental character of the field, coordinated and significant investments by the government.

The urgency that is assumed by the Dutch government is not a natural position, as is clear from the questions from the Parliament. Question 53 (Christenunie, the conservative Christian party) for instance is: 'Why are Dutch ambitions expressed in such strong terms?' The answer was: "The government thinks the possibilities that genomics and bioinformatics offer to Dutch society are too important not to exploit. The Dutch knowledge infrastructure has potential, provided investments are made to fully profit from these chances. The Government, therefore, urges involved parties to get involved and contribute financially.'

The urgency has its drawbacks: when genomics is adopted as a burning issue, one should be able to deliver. There are discussions about who is to blame when the Netherlands is falling behind. Some blame it on the government: in February 2001, when the Ministry of Economic Affairs provided a total amount of Dfl. 100 million (about $40 million) of financial support for new genomics firms, industry complained that the amount was ridiculously low:

When you want to be first, Dfl 100 million is not helpful. In Germany they spend billions and some government-related investment agencies invest more than that. (P. van der Meer, investment agency Gilde, *Volkskrant*, 26 February 2001)

Yet, there are also other parties to blame. Allegedly, a bigger problem is the Dutch universities, since they are not geared towards application of knowledge (a well-known complaint that appears every now and then). All kind of solutions are offered for this, such as granting patents to universities.

Scientists in general are blamed for lacking enterprise: for being too timid to start a new venture and take some financial risks, forgetting to apply for a patent. 'Scientists should be judged according to the number of patents they have applied for, just like publications.'[13] This chapter's point is not to decide who is right in this game of praise and blame, but to make clear that a game of praise and blame will start off, and that actors involved do have to be ready to be proactive.

Agenda-building

The promise of genomics requires action in the form of lists of priorities, otherwise known as agenda-building. In political science, 'agenda' is a well-defined concept. Kingdon (1984: 3, 4) defines it as

... the list of subjects or problems to which ... people ... are paying some serious attention at any given time ... Out of the set of all conceivable subjects or problems to which officials could be paying attention, they do in fact seriously attend to some rather than others. So the agenda-setting process narrows this set of conceivable subjects to the set that actually becomes the focus of attention.

When we look at expectations of technology, none is official, and there is no formally recognized forum in which the agenda is localized. But Kingdon's definition can be modified to apply to a laboratory (local agenda), or a technical-scientific domain (field agenda), or even a culture or a society at large, where some issues, topics, and ideas are held to be more important than others (cultural agenda). These require attention, generally, by self-styled spokespersons, or from those that are mandated for the particular issue or topic. Other activities and proposals can be legitimated by referring to these issues, topics, or ideas

that are held to be more important than others. Proposals that refer to the agenda are seeds that fall in fertile soil.

Various stages of agenda-building within genomics can be recognized, and these processes are even institutionalized. For instance, the launch of new organizations and new journals, such as the *Journal of Structural and Functional Genomics*, are intended to specify and coordinate the new challenges in the field. In 2002, the International Structural Genomics Organization was formed to coordinate the promotion of so-called 'structural genomics' (Stevens *et al.* 2001). Descriptions of such new fields are often in the prescriptive mode as it typically lists a set of targets and challenges:

Structural genomics requires a large number of process steps to convert sequence information into a 3D structure ... The present goals are to obtain targeted structural information reliably within a 6- to 12-month time period from DNA to structure and to reduce the cost per structure by 90 per cent. (Stevens *et al.* 2001: 90)

The sheer magnitude of this challenge in determining proteome-wide structures has necessitated the current global initiative in the academic and industrial structural genomics communities. Over the past 2 years, structural genomics consortiums have sprung up all over the world. (Stevens *et al.* 2001: 90)

Global structural genomics efforts have gotten off to a very good start, have attempted to set reasonable policies that can be adhered to, and have identified problems and challenges that need resolution in the immediate future. (Stevens *et al.* 2001: 92)

The 'problems and challenges that need resolution' according to the International Structural Genomics Organization are summarized in Figure 16.7. Here we see how the various activities are interlinked and interdependent. Note that these are activities ('challenges') to be realized in the future. So the expected outcomes of Genome Sequencing Projects are linked to advances in High-Throughput (HT) Crystallography and nuclear magnetic resonance that, in turn, depend on the advances in homology modeling and possibilities of Protein Data Banks (PDB). This figure does two things at once. First, it helps to *specify* the requirements for the various activities. Protein

Data Banks, for instance, should be able to absorb and process the data (input) and to suggest other promising targets (output). The Protein Structure Initiative, for instance, has 'the long-term goal of determining 10,000 novel protein structures over 10 years' (Stevens *et al.* 2001: 90). Second, such a map of projected activities provides opportunities for those who intend to be part of the venture, to position themselves vis-à-vis other players as they dedicate themselves to a part of the map. In other words, firms and institutes can claim a specific part of the promising territory.

Another source of agenda-building relates to the emergence of compelling and guiding metaphors. As linguistics research has shown convincingly, metaphors are not innocent. Basically, they perform two functions: they legitimate activities for outsiders (since they help to construct a compelling outlook), and they guide the insiders in their efforts (since they point to characteristics that are useful). Hellsten (2002), for instance, has shown how metaphors 'sell' the life sciences to a variety of audiences. One of the first, and still the most famous, metaphoric retellings of genetic research is the 'selfish gene,' according to which DNA segments only secure their own transmission (Dawkins 1976). The metaphor of the genome as a 'book of life' was used to sell the human genome sequencing project.

Important for the thrust of this chapter is that metaphors have research consequences too. John Avise (2001) explains in *Science* how metaphors guide genetic research. The beads-on-a-string metaphor, for instance, envisioned the notion of genes (which encode the vital proteins for an organism) neatly arranged in a row. It took some time to discover that only 2 per cent of the DNA can be seen as a bead, or coding for a protein. As a result, the metaphor shifted to protein-coding 'genomic islands' in a 'genomic desert.' But the desert metaphor suggests, misleadingly, that research on the space between the 'islands' will yield nothing of interest. So now the metaphor of a society is coming to the fore. 'Good citizen' genes are a minority in a wild society of strange characters:

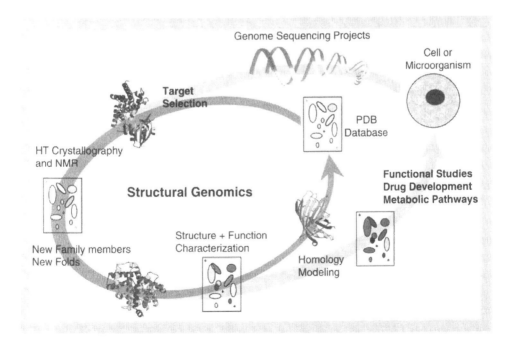

Fig. 16.7. An agenda for genomics
Source: Stevens *et al.* 2001.

The intergenic wilderness proved to be populated by a motley crew of intriguing genetic characters: active promotors and regulators of gene expression, comatose pseudogenes, descendants of immigrant DNAs (perhaps horizontally transferred from microbes), vagabond sequences, hordes of tandem short-repeats, and great armies of repetitive elements—some with hundreds of thousands of like-uniformed members. (Avise 2001: 87)

The agenda-building processes thus occur at various levels of aggregation. Within research groups, the metaphors arrange the search heuristics; at the societal level, genomics is seen as an urgent national issue. Earlier, this chapter argued that R&D occurs in a series of nested, promising scenarios that develop into agendas within firms, within technical-scientific fields, and within society (van Lente 1993; see Figure 16.8).

Expectations at higher levels of aggregation provide the fertile soil for proposals at lower levels. In genomics, for instance, we see a plethora of new promising subfields of genomics (Table 16.7).

Conclusion: the prospective structures of science and innovation policy

This chapter has studied the evolution of the institutional setting of science and technology policy. It has characterized the changes in both

the US and the EU as a 'strategic turn' in the future potential of research, so as to decide the criteria to support and select R&D activities: firms are thus oriented more towards promises about the future use of scientific findings, and this affects the nature of the policy processes. Not only do the promises about eventual usefulness legitimize the policy decisions and instruments, but the agenda-building processes also contribute to the actual development of technical-scientific fields.

The genomics case shows processes by which actors, including policy-makers, get involved because they perceive something is at stake that they do not want to miss out on. Their activities interlock because of their interest in a possible future: actors are able to refer to opportunities in proteomics, so their search processes are protected; or actors may cluster around the proclaimed future of genomics in general. Thus, we see how a stake in the future means that learning about possibilities occurs and efforts get coordinated. Because of shared interests and views informed in the first place by shared expectations, this coordination could be called 'emergent.' At different levels, activities interlock, which results in non-hierarchical role allocations and task divisions; all levels from micro to macro are involved in 'layered' coordination (van Lente 2000). This emergent coordination of activities can be seen as an outcome of collective learning (cf. the chapter by Jolivet and Maurice).

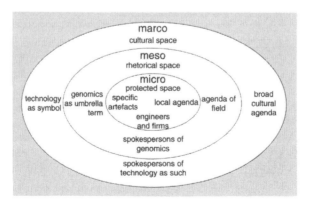

Fig. 16.8. Nested activities and agendas of genomics
Source: Adapted from van Lente 1993.

Table 16.7. New sub-genomics terms coined since 1999

Selection of new terms	
biomics	oncogenomics
CHOmics	operomics
cellomics	panomics
chronomics	pathogenomics
clinomics	peptidomics
crystallomics	pharmacomethylomics
degradomics	phylogenomics
epigenomics	phyloproteomics
fluxomics	physiogenomics
functomics	physiomics
immunomics	postgenomics
in silico transcriptomics	proteogenomics
interactomics	pseudogenomics
kinomicsligandomics:	regulomics
lipoproteomics	riboproteomics
metabolic phenomics	toxicomics
metabolomics	transcriptomics
metabonomics	vaccinomics
methylomics	variomics

Source: Cambridge Healthtech Institute (**www.chiresource.com**).

Coordination is a central phenomenon in economic theory. Three main forms are distinguished: markets, hierarchies, and networks. The 'Invisible Hand' of Adam Smith's *The Wealth of Nations*, by which man is led 'to promote an end which was no part of his intentions', summarizes the insight that actions may result in an order which is not deliberately created. The features of this order are decoupled from the intentions and characteristics of the agents. When decisions are made by atomized, rational, self-interested agents, an aggregated level emerges; at this level, the price mechanism may work as a coordinating mechanism.[14] The 'transaction cost theory' added the notion of hierarchies: since the use of the price mechanism is not free (that is, not without costs), the costs for unspecified contracts such as employer–employee may be lower than those for a long series of specified contracts (Williamson 1975). In some cases, two main forms of governance—a market and hier-

archies—minimize transaction costs. Since the 1980s, economists also talk of a third form of coordination, or at least intermediate forms of coordination, called 'clans' or, more often, 'networks.' The idea is that economic actors are bound together in many other ways than just the anonymous relationships in the market and the formal, unequal relations in a hierarchy: in networks, informal links between equal agents are emphasized. An alternative position is that networks could be considered the most general category of coordination. It is clear that the coordination that emerges through the dynamics of expectations belongs to the third ('networks') category. But a difference remains, because economics tends to deal with *static* situations. The focus tends to be on how efforts are coordinated in situations that are either static or, at least, have already become stabilized. How networks arise and stabilize is considered less important than the question whether, once they exist, they can be explained as a rational (that is, cost-saving) outcome.

Coordination is also a central phenomenon in sociology. If we follow Giddens, the discipline of sociology has a specific division of labor (Giddens 1981: 167): two streams have divided the sociological work to be done, especially after the Second World War. Roughly speaking, functionalism has dealt with structures and phenomenology and symbolic interactionism dealt with agency.[15] Functionalism tends to treat structure as some external factor working, as it were, behind the back of actors, thereby ignoring the active processes of interactions, and the social shaping of reality by agents.

Symbolic interactionism (and phenomenology), on the other hand, tends to ignore structural constraints on social processes. It focuses on face-to-face (symbolic) interaction, for example, on role-taking (Mead) or on dramaturgical presentation of self (Goffman), while there is not much effort at explaining how aggregated outcomes may result from on-going interactions. Structure enters only through the perception of constraints by actors. Several attempts have been made to redress the balance between functionalism forgetting creative actors, and interactionism forgetting constraining structure.[16] The key issues are then how actions have led to structures, and how these structures have enabled and constrained action. The analysis of expectations in science policy indicates how structure emerges and shapes action in a way in which content matters as much as traditional sociological categories of explanation. It is the content of orientations and metaphors which pulls actors together; when they act upon it, a structure emerges which shapes further action. The key point appears to be that one should recognize that structures can be prospective, and still influential.

The central phenomenon studied in this chapter is that promises and expectations help to interlock activities and to build up agendas. In other words, expectations structure activities, in the sense that shared expectations are structures to be realized. They do not yet exist, but exert force nevertheless. It is as if expectations are a script; on its basis, roles are allocated (van Lente 2000). So a new social order is possible on the basis of collective projections of the future. To emphasize this, the chapter uses the paradoxical term *prospective structure* (van Lente and Rip 1998*b*). A prospective structure is made up of links which can appear in texts. In this sense, the content of the structure matters. In the actions and reactions, the structure is filled in, modified, reshuffled, and becomes social structure in its various forms. One can now talk about the division of labor in sociology in new terms— as one of emphasis on retrospective structure versus voluntaristic actions. Prospective structure is a structure that is filled in by an agency that is itself only determined by this process: it has the power of forceful fiction, and opens up space for action. Science and innovation policy—with its stress on the strategic (and thus future) relevance—is both an expression and part of this prospective structure.

One of the most strategic fields of the last decade is 'genomics,' and this chapter has explored how its rise has been accompanied by the rise of labels and terms that underpin its general promise. The rise of this technical-scientific field is a remarkable phenomenon. The urgency requires governments to respond with programs and has led to a cascade of subfields with their own research agendas. Its study shows how efforts get coordinated due to a stake in the future. This is *emergent* coordination, emerging because of shared expectations, rather than shared interests or views. At different levels, there is interlocking of activities, resulting in non-hierarchical role allocations and task divisions; at all levels, agenda-building and collective learning is taking place. The dynamics give rise to potential new subfields that are possible under the genomics umbrella. Thus a nested dynamics of promises within promises occurs, both in policy and research activities, in which promises at a broader level allow the unfolding of more specific promises at a local level. So, in a sense, the 'promising gene' replicates itself.

Notes

1. Cabo (1997) makes a detailed analysis of the knowledge networks that respond to and are mutually shaped by Eureka and the Framework programs.
2. The groupings are based on Gross Expenditure on Research and Development (GERD) and on GERD as a percentage of Gross National Product (GNP).
3. In general, acts or decisions are strategic if they take into account that the outcomes of the choice depend on what others do. When actors are interdependent, their choice necessarily affects those of others; if the actors are aware of it and take account of it in their actions, they can be said to act strategically. Strategic action involves a kind of reflexivity, which Jon Elster (1983), in his study of forms of rationality, typifies as 'strategic rationality.' This is the case when 'the agent acts in an environment of other actors, none of whom can be assumed to be less rational or sophisticated than he is himself. Each actor, then, needs to anticipate the decisions of others before he can make his own, and knows that they do the same with respect to each other and to him' (Elster 1983: 77). Elster conceives of strategic action as accepting the other as an actor like himself, whose behavior can be interpreted or 'read,' but not predicted.
4. Irvine and Martin (1984: 3, 5). Their text continues to describe the changes in the R&D system: 'Besides hoping to produce in-house at least one or two scientific "winners," many firms use such research to develop links with the relevant academic research communities. Such links are generally essential if the firm is successfully to monitor and take advantage of the latest scientific results. They are also necessary to develop within the company the skills and techniques required to mount rapid R&D programs on new research possibilities as and when they occur.'
5. The strategic turn is not limited to science and technology policy. It is part of a larger transformation denoting an increasing awareness that 'strategic' choices are important. The origin of strategic thinking and acting should be located in warfare and in preparation for war, and it spread widely, especially after the Second World War. In industry, interest in strategic choices has grown since the fifties. The study of Ansoff (1968), *Business Strategy*, marks the beginning of a series of studies of decision problems of firms in a continually and rapidly changing environment.
6. In terms of evolutionary economics, one would say that variations need niches as protection against pressure from the selection environment (Nelson and Winter 1982).
7. See Cook-Deegan (1994) on *The Gene Wars*. Burris *et al.* (1998) gives a brief overview of the role of the reports of the National Research Council and the Office of Technology Assessment (both in 1988).
8. The *Nature* group includes: *British Dental Journal; British Journal of Cancer; Cancer Update; Materials Update; Nature Biotechnology; Nature Cell Biology; Nature Genetics; Nature Immunology; Nature Medicine; Nature Neuroscience; Nature Reviews Cancer; Nature Reviews Drug Discovery; Nature Reviews Genetics; Nature Reviews Immunology; Nature Reviews Molecular Cell Biology; Nature Reviews Neuroscience; Nature Science Update; Nature Structural Biology; Physics Portal.*
9. F. S. Collins, Shattuck Lecture—'Medical and societal consequences of the Human Genome Project,' *New England Journal of Medicine,* 341 (1999), 28–37, cited by McCain 2002 #491.
10. W. J. Clinton, *Remarks by the President*. White House Event on Genetic Discrimination in Health Insurance (14 July 1997) cited by McCain 2002 #491.
11. **www.jgi.doe.gov**
12. The Dutch newspaper, *Volkskrant*, 15 July 2001.
13. Mr P. Van der Meer of investment agency Gilde in *Volkskrant*, 26 February 2001.
14. Note that the market and economic agent, then, are not empirical entities, but concepts to understand outcomes of coordination at an aggregate level.

15. In the opposition between structure and agency, 'structure' mainly refers to the constraints which individual action faces. The idea is that these constraints stem from the coherence of an aggregated level. In this section, I will use this meaning of the term, which has, in other contexts, other connotations.

16. For example: Berger and Luckmann (social construction of reality); Giddens (structuration theory); Burns and Flam (social rule system theory); Shibutani (social processes); Strauss (social world); Boudon (transformation processes). These attempts share a belief in the necessity to overcome the dualisms and dichotomies resulting from the division of labor in sociology: structure versus agency; determinism versus voluntarism; objectivism versus subjectivism.

References

Ansoff, H. I. (1968). *Corporate Strategy: An Analytic Approach to Business Policy for Growth and Expansion.* New York: McGraw Hill.

Avise, J. C. (2001). 'Evolving Genomic Metaphors: A New Look at the Language of DNA.' *Science,* 294: 86–7.

Barre, R., Gibbons, M., Maddox, J., Martin, B., and Papon, P. (eds.) (1997). *Science in Tomorrow's Europe.* Paris: Economica International.

Bessant, J., and Rush, H. (1995). 'Building Bridges for Innovation: The Role of Consultants in Technology Transfer.' *Research Policy,* 24: 97–114.

Borrás, S. (2003). *The Innovation Policy of the European Union: From Government to Governance.* Cheltenham, UK: Edward Elgar.

Bozeman, B. (2000). 'Technology Transfer and Public Policy: A Review of Research and Theory.' *Research Policy,* 29: 627–55.

—— and Dietz, J. S. (2001). 'Research Policy Trends in the United States: Civilian Technology Programs, Defense Technology and the Deployment of the National Laboratories.' In P. Larédo and P. Mustar (eds.), *Research and Innovation Policies in the New Global Economy.* Cheltenham: Edward Elgar, 47–78.

Burris J., Cook-Deegan, R. M., Alberts, B. (1998). 'The Genome Project after a Decade: Policy Issues.' *Nature Genetics,* 20: 333–5.

Cabo, P. G. (1997). *The Knowledge Network: European Subsidized Research and Development Cooperation.* Capelle a/d IJssel: Labyrinth Publication.

Condit, C. M. (1999). *The Meaning of the Gene.* Madison: University of Wisconsin Press.

Cook-Deegan, R. M. (1994). *The Gene Wars: Science, Politics, and the Human Genome.* London: Sage.

Cozzens, S., Healey, P., Rip, A., and Ziman, J. (eds.) (1990). *The Research System in Transition.* Boston: Kluwer Academic Publishers.

Daele, W. van den (1978). 'The Ambivalent Legitimacy of the Pursuit of Knowledge.' In E. Boeker and M. Gibbons (eds.), *Conference on Science, Society and Education.* Amsterdam.

Dawkins, R. (1976). *The Selfish Gene.* New York: Oxford University Press.

Dosi, G., Freeman, C., Nelson, R., Silverberg, G., and Soete, L. (eds.) (1988). *Technical Change and Economic Theory.* London: Pinter.

Edquist, C. (1997). *Systems of Innovation: Technologies, Institutions and Organizations.* London: Pinter.

Elster, J. (1983). *Explaining Technical Change: Studies in Rationality and Social Change.* New York: Cambridge University Press.

Elzinga, A., and Jamison, A. (1995). 'Changing Policy Agendas in Science and Technology.' In S. Jasanoff, G. Markle, J. Petersen, and T. Pinch (eds.), *Handbook of Science and Technology Studies.* London: Sage.

Enriquez, J. (1998). 'Genomics and the World's Economy.' *Science,* 281/5379: 925–6.

Freeman, C. (1997). 'The Diversity of National Systems of Innovations.' In R. Barre, M. Gibbons, J. Maddox, B. Martin, and P. Papon (eds.), *Science in Tomorrow's Europe*. Paris: Economica International, 5–32.

—— and Perez, C. (1988). 'Structural Crisis of Adjustment, Business Cycles and Investment Behaviour.' In G. Dosi, C. Freeman, R. Nelson, G. Silverberg, and L. Soete (eds.), *Technical Change and Economic Theory*. London: Pinter, 38–66.

Fuller, S. (1988). *Social Epistemology*. Bloomington: Indiana University Press.

Gibbons, M., Nowotny, H., Limoges, C., Trow, M., Schwartzman, S., and Scott, P. (1994). *The New Production of Knowledge: The Dynamics of Science and Research in Contemporary Societies*. London: Sage.

Giddens, A. (1981). 'Agency, Institution, and Time-Space Analysis.' In K. Knorr-Cetina and A. V. Cicourel (eds.), *Advances in Social Theory and Methodology: Towards an Integration of Micro- and Macro-sociologies*. London: Routledge and Kegan Paul, 161–74.

Godin, B. (2002). 'Technological Gaps: An Important Episode in the Construction of S&T Statistics.' *Technology in Society*, 24: 387–413.

Harding, R. (2003). 'New Challenges for Innovation Systems: A Cross Country Comparison.' *International Journal of Technology Management*, 26/2–4: 226–46.

Hedgecoe, A. M. (2003). 'Terminology and the Construction of Scientific Disciplines: The Case of Pharmacogenomics.' *Science, Technology and Human Values*, 28: 513–37.

Hellsten, I. (2002). 'Selling the Life Sciences: Promises of a Better Future in Biotechnology Advertisements.' *Science as Culture*, 459–79.

Irvine, J., and Martin, B. R. (1984). *Foresight in Science: Picking the Winners*. Dover: Pinter.

Jasanoff, S. (1994). *The Fifth Branch: Science Advisers as Policymakers*. Cambridge, MA: Harvard University Press.

Johnston, R. (1990). 'Strategic Policy for Science.' In S. Cozzens, P. Healey, A. Rip, and J. Ziman (eds.), *The Research System in Transition*. Boston: Kluwer Academic Publishers, 213–26.

Kevles, D. J. (1977). *The Physicists: The History of a Scientific Community in Modern America*. London: Random House.

Kingdon, J. W. (1984). *Agendas, Alternatives and Public Policies*. Boston: Little, Brown and Company.

Kuhlmann, S. (2001). 'Future Governance of Innovation Policy in Europe: Three Scenarios.' *Research Policy*, 30: 953–76.

Kuhn, T. (1962). *The Structure of Scientific Revolutions*. Chicago: University of Chicago Press.

Lander, E. S. *et al.* (2001). 'Initial Sequencing and Analysis of the Human Genome.' *Nature*, 409: 860–921.

Larédo, P., and Mustar, P. (2001). *Research and Innovation Policies in the New Global Economy*. Cheltenham: Edward Elgar Publishing.

Latour, B. (1999). *Pandora's Hope: Essays on the Reality of Science Studies*. Cambridge, MA: Harvard University Press.

Law, J. (1986). 'The Heterogenity of Texts.' In M. Callon, J. Law, and A. Rip (eds.), *Mapping the Dynamics of Science and Technology*. London: MacMillan Press, 67–83.

McCain, L. (2002). 'Informing Technology Policy Decisions: The US Human Genome Project's Ethical, Legal, and Social Implications Programs as a Critical Case.' *Technology in Society*, 24: 111–32.

Morpace Pharma Group (2001). *From Data to Drugs: Strategies for Benefiting from New Drug Discovery Technologies*.

Mowery, D. C. (2001). 'The United States National Innovation System after the Cold War.' In P. Larédo and P. Mustar (eds.), *Research and Innovation Policies in the New Global Economy*. Cheltenham: Edward Elgar, 15–46.

Mytelka, L. K., and Smith, K. (2002). 'Policy Learning and Innovation Theory: An Interactive and Co-Evolving Process.' *Research Policy*, 31: 1467–79.

Nelson, R. R., and Winter, S. G. (1977). 'In Search of a Useful Theory of Innovation.' *Research Policy*, 6: 36–76.

Nelson, R. R., and Winter, S. G. (1982). *An Evolutionary Theory of Economic Change*. Cambridge, MA: Belknap Press of Harvard University Press.

OECD (1968). *Gaps in Technology:* General Report, Paris.

—— (1975). *Patterns of Resources Devoted to R&D in the OECD Area, 1963–1971*. Paris.

Pennisi, E. (2000). 'Genomics Comes of Age.' *Science*, 290: 2220–1.

Perelman, C. (1982). *The Realm of Rhetorics*. Notre Dame, IN: University of Notre Dame Press.

Peters, L., Groenewegen, P., and Fiebelkorn, N. (1998). 'A Comparison of Networks between Industry and Public Sector Research in Materials Technology and Biotechnology.' *Research Policy*, 27: 255–71.

Rip, A. (1990). 'An Exercise in Foresight: The Research System in Transition—To What?' In S. E. Cozzens, S. E. P. Healey, A. Rip, and J. Ziman, *The Research System in Transition*. Dordrecht: Kluwer Academics, 387–401.

Roobeek, A. J. M. (1990). *Beyond the Technology Race: An Analysis of Technology Policy in Seven Industrial Countries*. New York: Elsevier Science Publishers.

Stevens, R. C., Yokoyama, S., and Wilson, I. A. (2001). 'Global Efforts in Structural Genomics.' *Science*, 294: 89–92.

Van der Meulen, B., and Rip, A. (1998). 'Mediation in the Dutch Science System.' *Research Policy*, 27: 757–69.

Van Lente, H. (1993). *Promising Technology: The Dynamics of Expectations in Technological Developments* (PhD thesis, University of Twente). Delft: Eburon.

—— (2000). 'Forceful Futures: From Promise to Requirement.' In N. Brown, B. Rappert, and A. Webster (eds.), *Contested Futures: A Sociology of Prospective Techno-science*. London: Ashgate Publishing Company, 43–64.

—— and Rip, A. (1998a). 'The Rise of Membrane Technology: From Rhetorics to Social Reality.' *Social Studies of Science*, 28/2: 221–54. London: Sage.

—— —— (1998b). 'Expectations in Technological Developments: An Example of Prospective Structures to be filled in by Agency.' In C. Disco and B. van der Meulen, *Getting New Technologies Together: Studies in Making Sociotechnical Order*. New York: Walter de Gruyter, 203–29.

—— Smits, R., Hekkert, M. P., and Van Waveren, B. (2003). 'Roles of Systemic Intermediaries in Transition Processes. *International Journal of Innovation Management*, 7/3: 247–79.

Venter, J. C. *et al.* (2001). 'The Sequence of the Human Genome.' *Science*, 291: 1304–51.

Webster, A. (1999). 'Technologies in Transition, Policies in Transition: Foresight in the Risk Society.' *Technovation*, 19: 413–21.

Williamson, O. E. (1975). *Markets and Hierarchies, Analysis and Antitrust Implications*. New York: Free Press.

17 The Role of Education and Training Systems in Innovation

David Finegold

Introduction

A country's education and training (ET) system has always played a central role in shaping national patterns of innovation. In today's more knowledge-driven global economy, as innovation has become a more vital factor in economic competition, the ET system has grown even more important. Through innovation, companies are able to command premium prices and provide high-wage jobs. More routine manufacturing and administrative work, including functions such as software programming and medical services that until recently were considered highly skilled labor, are rapidly moving to low-wage nations with a surplus of well-educated workers, like India and China (Dassani and Kenney 2004). Among the other indications that education-driven innovation is growing in importance for preserving or increasing the supply of good jobs in the global economy are:

- the greater role that scientific research is playing in the technology and business strategies of companies (Lane 2002). This is true not just in the patents for high-tech sectors, such as biotechnology and aerospace, but also more traditional industries such as chemicals and food processing;
- while more and more jobs can be globally distributed with the aid of the Internet and other information technology, 90 per cent of R&D spending, and the advanced man-

power that performs this, continues to be concentrated in the country of origin (Galbraith 1998);
- rather than physical or financial capital, the largest driver of wealth creation, as measured by the shareholder value of corporations, is, increasingly, human capital and the innovation and intellectual property it generates (Lawler 1996).

This chapter is broken into two levels of analysis. The first focuses on the key attributes of national ET systems and their impact on innovation. It presents the primary types of ET systems—market-driven, corporatist, state-led, large company-based, local networks—and the forms of industrial organization and innovation associated with each in different countries. The issues involved with national ET systems are explored through a pair of cases. The first analyzes how the major, state-led reforms of the UK's ET system—from an elite to a mass higher-education (HE) system, with much greater participation in full-time schooling—have affected the nation's capacity for innovation. It also provides an illustration of the book's final key theme—the capacity for institutional change: when the right conditions converge, it is possible for the state to bring about a dramatic transformation in the ET system and its outputs in a relatively short span of time. The second national case focuses on innovations within global corporations: the spread of the Internet has provided, as new employees, the graduates of India's universities, which continue to operate a version of the traditional UK ET system.

The chapter then shifts focus to the regional level, identifying the key roles that the ET system—particularly research universities—plays in four facets of the development of clusters of innovative enterprises, or high-skill ecosystems (HSEs) that are increasingly important to innovation in the global economy. The HSE dynamics are illustrated using the case of Singapore's efforts to create a community of collective learning in the biomedical sector through coordinated changes in the ET system and other associated industrial policies.

In conclusion, the chapter lays out a framework for future research, drawing lessons from these cases and from other work on the relationship of the ET system to the capacity for innovation, and identifying key issues for policy makers.

Dominant forms of skill provision and innovation

The comparative political economy literature has long recognized that the ET system is one of a set of key institutional elements—along with financial markets, industrial relations systems, forms of corporate governance, etc.—that shape innovation strategies, relative economic performance, and labor market outcomes in different countries. For example, the UK's relatively poor economic performance throughout much of the post-war era has been attributed to a low-skill equilibrium: a mutually reinforcing combination of a poor supply of skills for the majority of the workforce emerging from the ET system linked to product market strategies, and work organization that generated a relatively low demand for skills (Finegold and Soskice 1988).

ET delivery has specific structural features that can have a significant impact on the level and type of innovation found in each country. In this analysis we will deal with four levels of innovation that are related to the ET system:

- the production of new ideas and knowledge through research and creative individuals emerging from the ET system;
- the development of new products, technologies, and services, and of improved ways of producing them;
- the creation of new firms that generate new products and processes;
- the transformation of national or regional skill-creation systems.

In particular, ET systems and surrounding institutions can affect whether a nation, or the regions within it, have a greater capacity for *radical* innovation (major, often disruptive technological changes that can bring about whole new industries), or *incremental* innovation (gradual, more continuous improvements in product/service attributes and how they are made or delivered). While countries typically feature a mix of institutional approaches to ET, each nation normally has one dominant ET approach that is closely linked to its relative comparative advantage in innovation in the global economy. In the following paragraphs, examples of each of the main forms of ET, and their associated patterns of innovation, are summarized.

State-led ET

The pattern in most countries is that the state is the main, if not sole, provider of basic schooling, which is considered a basic right and a partial public good, and thus deserving of government funding. In many nations, such as France, the state's dominance extends into further and higher education, as well as the industrial training and innovation system: all of the elite universities, and many of the dominant companies, are under either total or partial government control (Culpepper 2003). This state-led model of education and innovation has tended to work well in sectors that require very large and long-term investments, that is, in basic research or infrastructure, where the government is a key customer (for example, transportation, nuclear power, aerospace, health care).

Market-based competition

This tends to be most prevalent in adult training for those in employment. The US, however, features a wide diversity of public- and private-sector ET providers at all levels, including a growing, for-profit education sector in primary, secondary, and higher education. In the HE sector, these institutions and their faculty compete aggressively for students and funding from government, industry, and nonprofit foundations. This has produced many of the world's leading research universities, which attract top talent from across the world, and generate a healthy stream of new graduates and new intellectual property that has fueled the growth of new high-tech start-ups. The system, however, places very little emphasis on equity, and the losers in this competition are ill-prepared to operate in a knowledge economy. As an example, approximately half of all students still have little or no access to any information technology in their classrooms, and approximately 25 per cent of them drop out of high school with no form of qualification (CDE 2002).

Large company-dominated provision of skill-development

This has been the tradition across the industrialized nations. While in most countries, the number of companies offering long-term, systematic management-development programs has declined along with the decrease in employment security, in Japan this continues to be the dominant model of skill creation, following centralized, state-dominated schooling and HE. New graduates emerge from the state-school and HE system with a high level of basic skills; large Japanese firms hire the majority of them, and provide them with lengthy and carefully planned on-the-job development through a set of rotations across functions as part of the seniority-based promotion system. This system worked well from the end of the Second World War through the 1980s: the large supply of company-specific and team-based skills it produced helped foster continuous, incremental innovation that enabled Japanese

corporations to catch up with, and then pass, their Western rivals in a variety of export-oriented consumer-goods sectors (for example, autos, electronics, videogames). Since the early 1990s, when a number of related problems caused the Japanese bubble economy to burst, the dependence on large companies for skill development and the accompanying lifetime employment model have arguably been a major structural barrier to generating new, innovative companies to restore the growth of the Japanese economy (Pilling 2004). Though there has been a significant decline in lifetime-employment practices in the last five years (Japan Institute of Labor 2003: 24), over 75 per cent of firms report that the company still bears full, or almost full, responsibility for the education and training of employees, while under 4 per cent indicate that 'employees are responsible for their own education and training' (ibid. 41). Sako (2002) concludes that the continuing, almost exclusive, flow of top university graduates to large companies, which is reinforced by the still strong societal and institutional penalties for involvement in a business failure, has hindered the development of start-up enterprises, despite significant reforms to increase access to venture capital (VC). She finds that, of the twenty-nine firms listed on NASDAQ Japan for which founders' background information was available, none was a university spin-out or started by recent graduates (ibid. 28). The persistence of this single model of skill development has led some expert commentators (Schoppa 2001) to call for a 'G.I. Bill' for Japan's estimated 10 million underemployed corporate warriors within large firms, to provide them with government funding to acquire the new skills and qualifications needed to launch new careers and new ventures, just as the US government stimulated the post-Second World War economy by providing scholarships for returning soldiers to obtain college degrees.

Corporatist systems

Corporatist systems, like those found in the Germanic countries, train a large majority of young people using apprenticeships that are run through tripartite cooperation among employers' organizations, trade unions, and the state. The large supply of highly skilled craft workers and technicians that this system produces has helped Germany build and maintain a strong position in high-value-added manufacturing sectors, where incremental product and process innovation are keys to success (Culpepper and Finegold 1999). The supply of intermediate skill is closely intertwined with a set of other institutional factors—high technical qualifications of senior managers, long-term, debt-based finance, relatively high levels of employment security—that reinforce the German competitive advantage in certain sectors driven by incremental innovation (Hall and Soskice 2001). Sectoral studies (metalworking (Daly *et al.* 1985), kitchen manufacture (Steedman and Wagner 1987) and hotels and commercial banking (Keltner *et al.* 1999; Mason *et al.* 2000) that compare Germany with the UK and other countries have helped show the mechanisms through which greater levels of skill are translated into the productivity, quality, and product-design improvements that, despite high labor costs, help Germany remain competitive.

As traditional manufacturing sectors have experienced ever-increasing levels of global competition, these same corporatist ET and other institutions appear to have hindered Germany's ability to generate more radical process and product innovations. In the biotechnology and IT sectors, for example, Germany lagged behind the US and UK throughout the 1980s and 1990s. In an effort to close this gap, the government provided nearly costless capital to support new business formation at the end of the decade; while this succeeded in generating hundreds of new start-ups, many of these are now running out of funds or have already failed, along with the stock market, the Neuer Markt, that was created to help these firms go public (Casper 2000). Those firms that have survived have tended to follow lower-risk platform technology or service strategies that appear to be a closer fit with German institutions.

Local Networks or Clusters

In contrast with the formal associations and negotiations that characterize corporatist ET systems, local networks or clusters are an alternative, more informal form of cooperation in skill development that can occur among geographically clustered sets of firms and supporting actors operating in the same industry (Crouch *et al.* 1999). These local skill-creation networks have been shown to play vital roles: in ongoing, incremental innovation in ceramics, textiles, and metalworking industrial districts of Northern Italy (Bagnasco 1977; Best 1990), as well as in similar clusters in other countries—Denmark (Banke 1991), Germany (Herrigel 1996); and in stimulating more radical innovation in sectors like biotechnology and information technology in a variety of clusters in the US, UK, India, and other countries centered on leading universities (Saxenian 1994; Best 1990). These clusters promote skill development and innovation in several ways:

- stimulating HE institutions to provide specialized skills to meet the needs of a critical mass of local employers;
- encouraging individuals to take the risk of starting firms, or working for start-ups, since they know there are many other employment opportunities to use these specialized skills in the cluster;
- transferring tacit knowledge to fuel new product and process development, as individuals move among firms in the cluster, and companies within them partner and share resources.

National ET systems and innovation: the UK's supply-side skills strategy

The UK retained an elite ET system through the mid-1980s:[1] only 7 per cent of the working population possessed a university degree and participation rates in post-compulsory education were one of the lowest among Organization for Economic Cooperation and Development (OECD) countries (Hayes and Fonda 1984). In little more than a decade, however, the ET system was transformed, with dramatic increases in staying-on rates. By 1995, over 20 per cent of young people were completing a bachelor's degree, and by 2002 the percentage had grown to over 33 per cent, a greater proportion of the population than in the US's mass HE system.[2] The Blair government has set a target of 50 per cent of young people obtaining an HE qualification by the end of the decade.

A number of factors contributed to the dramatic increase in the percentage of English young people graduating from college:

- reform of the national examination system at 16 to shift from a sorting system (Ordinary (O)-levels), designed to weed out 80 per cent of young people from continuing in education, to a new, more modular exam, the General Certificate of Secondary Education (GCSE), designed to recognize what young people had learned during compulsory education;
- significant changes in the structure of the HE system, with the merging of the traditional universities and newer polytechnics into a unitary system, with funding incentives provided to those institutions that most significantly expanded enrollments at the lowest unit cost;
- the youth labor market for non-college graduates, particularly the old apprenticeship system and the skilled jobs it prepared people for, collapsed in the 1980s;
- the size of the youth cohort passing through the ET system declined by over 25 per cent from the mid-1980s through the mid-1990s (a common trend among OECD countries), enabling a signficant expansion in the percentage of young people who could attend without as large an increase in the absolute numbers of students and accompanying costs to the Treasury.

It is still very early to assess the impact that the dramatic expansion in HE, combined with other government reforms, has had on the UK's innovative capacity and economic performance, since it will take a decade or more for the sharp rise in the percentage of young people attending HE to translate into a much higher percentage of graduates in the overall workforce. Thus far, however, for the majority of UK firms and their workers, the greater supply of skills flowing from the ET system appears to have had relatively little effect on encouraging workplace redesign innovation to take advantage of these new skills (Keep and Mayhew (forthcoming)). Only 2 per cent of the establishments covered in the 1998 Workplace Employee Relations Survey (Cully et al. 1999) had adopted a set of ten or more of the sixteen practices that are associated with high-performance organizations (HPOs), and only 5 per cent had experimented with semi-autonomous work groups. Opportunities for employees to innovate in the workplace are, note Keep and Mayhew (forthcoming), 'constrained by high levels of surveillance (Collinson and Collinson 1997), tightly controlled and routinized forms of team working (Baldry et al. 1998), and strict performance monitoring' (Taylor 2002).

Perhaps the most disturbing results come from the more recent UK Skills Survey, which found that, despite noted increases in levels of worker qualifications, the percentage of employees reporting that they had 'a great deal of choice over the way they do their work' fell from 52 per cent in 1986 to 39 per cent in 2001 (Felstead et al. 2004: 13). The survey suggests that the failure to redesign work has resulted in some overqualification of the population for the jobs available. Based on the UK's National Vocational Qualification (NVQ) framework, about half of those qualified to levels 2 and 3 (semi-skilled and craft level) are in jobs that do not require these qualifications for entry, compared to 28 per cent with level 4 or above qualification and 34 per cent of graduates (ibid. 48). This is confirmed in several sector- and occupation-specific studies which have found that firms are now employing many more graduates in positions once occupied by individuals with lower-level qualifications, but that this has made little difference in the job requirements, or levels of productivity and innovation (Rodgers and Waters 2001; Mason 1998, 2001).

One of the problems with boosting the effective use of graduates appears to be the lack of emphasis on innovation in the business strategies of UK companies. UK firms' R&D spending is relatively low compared with other industrialized countries, and is heavily concentrated in just two sectors—pharmaceuticals and aerospace. This was confirmed in Michael Porter's assessment of UK competitiveness for the UK Department of Trade and Industry (Porter and Ketels 2003), which concluded that UK firms need to shift from a focus on cost control to creating unique value through innovation. Lord Sainsbury (2003), the UK Minister of Science, conceded: 'We're getting the supply side better, but struggling to know what to do on the demand side. We've done a few things, like R&D tax credits that appear to be of some value, but they're unlikely to revolutionize the system. We have not had much luck in getting up the R&D or patenting rate.'

Although the effects of rapidly expanding the supply of university graduates on private-sector innovation are still not apparent, the strain this has created within the HE sector itself is very evident. The UK Treasury has been able to finance this large increase, not only by channeling funds to those institutions that are most productive at generating graduates, but also by controlling faculty numbers and salary costs, which have risen much less than private-sector wages over the last decade. The main problem for generating new breakthrough innovations, according to Lord Sainsbury, is not the overall level of salaries, but the ability to differentiate between average and star performers: 'On average, the salary gap isn't as big as you think with the US, where there are a lot of very poorly paid academics in community colleges. The difficulty we face is that we have almost completely flat salary levels across all subjects and universities. In the US, what Harvard pays differs dramatically from your average state university, and they also differentiate by subject more clearly.' The result has been to exacerbate the 'brain drain' of the most employable academics to other countries, and to make it very difficult

to attract top UK graduates into academic careers.

Another consequence has been to create growing tension between the two primary missions of universities—teaching and research—as faculties struggle to cope with greater teaching demands alongside increased pressure to publish as a result of the government's research assessment exercise. Every five years, the research output of each department in every university is rated against its peers to determine the allocation of core research funding. This appears to have succeeded in boosting research productivity, and provides one way to reward top researchers. 'It has created a job market,' said Sainsbury. 'Vice Chancellors are now always on the lookout to hire top talent as a way to get their score up.'[3] This has enabled the HE sector to boost the output of new knowledge—as measured in publications and patents—at the same time as the growth in student numbers. For example, the number of patents from UK universities grew 26 per cent between 1999/2000 and 2000/1.

A variety of government initiatives introduced in the late 1990s to encourage greater commercialization of this university-research output also appear to be paying off. The number of new start-up companies spun out of UK universities has increased from an average of 70/year in the latter half of the 1990s to 203 in 2000–1 academic year and 248 in 2001–2. With this increase, the UK has passed the US rate of new-company creation per research dollar spent, generating one new firm for each £12 million of research spending, compared to £46 million/firm in the US (Sainsbury 2003).

In summary, the UK has been able to transition from an elite to a mass HE system in a very short period of time, with an accompanying large boost in the supply of highly qualified graduates entering the labor force. The impact of this skills-supply-led strategy on innovation and international competitiveness, however, is still too early to determine. While there are signs that it is stimulating the development of new, high-tech start-ups, and encouraging some employers to create more jobs requiring higher-level skills, the demand-side of the skill equation is slower to adjust, with many employers continuing to pursue relatively low-skill strategies that compete on cost and numerical labor flexibility, rather than innovation.

Indian ET and the Internet as drivers of new forms of industrial organization

While the UK has been busy reforming its ET system, India—the jewel of the former British empire—has retained the UK's old pyramid-like, elite ET structure: a fiercely competitive, meritocratic national examination system filters the brightest, hardest-working students to qualify for world-class technical universities. While this system has been in place since colonial days, what has changed is the global context in which this ET system is operating. With the diffusion of the Internet and the dramatic fall in communication costs, the highly educated workforce can now look beyond the civil service, or a career with a bureaucratic Indian corporation, to work for companies throughout the world without ever having to leave India. To an even greater extent than in the UK under the old system, the large majority of the Indian population lacks the education to benefit from this form of globalization; pupils leave school at an early age with limited skills, and, if they can find steady employment, are trapped in subsistence agriculture or low-wage jobs. But with a population approaching one billion, this elite education system still produces tens of millions of English-speaking graduates; particularly strong in computer programming, engineering, and other technical areas, they are having a major impact on the global economy, attracting many leading high-technology companies—for example, IBM, GE—to set up operations in Bangalore or Hyderabad to take advantage of this high-quality, but relatively low-cost, workforce.

Now virtually any work that involves the processing of information—call centers,

accounting, web design, claims processing, medical diagnoses such as reading a CAT-scan or X-ray—can be relocated to wherever the supply of labor with suitable skills can be found at the best price (Kirkpatrick 2003). Call-center operators who are paid an average of $10/hour in the US earn an average of $1.50/hour in Mumbai (Meredith 2003). Similar salary differences of 10 : 1 are found in many technical and service occupations. With this supply of high-quality, low-cost workers, Indian firms such as Infosys and Wipro are leading the way in the provision of comprehensive outsourcing services for multinational corporations. But the attraction is not only lowering labor costs. Many corporations are organizing virtual teams that take advantage of the time difference between the US, India, and Europe to offer round-the-clock technical support and expedite new product development. Making this global organization work, however, entails new innovations in organization and work-process design to facilitate virtual-team effectiveness (Gibson and Cohen 2003; Mohrman *et al.* 2003).

With offshoring made feasible by the IT infrastructure, the pressures of global competition have served to accelerate the diffusion of this organizational innovation. Already, many other countries—Malaysia, the Philippines, China, Hungary—are competing with India for multinational corporations' (MNCs) business. Estimates suggest that the US, Europe, and Japan are shifting 600,000 jobs a year to lower-wage nations. And companies which initially resisted outsourcing are now being compelled to reconsider, because they are no longer cost-competitive (Meredith 2003); electronic Data Systems (EDS), for example, plans to move 10 per cent of its workforce (close to 14,000 people) to low-wage countries in the next year. This includes not only programming and call centers, but also support functions such as helping to prepare powerpoint presentations, and maintaining the firm's computer networks in its management consulting division (Dassani and Kenney 2004). Even state and local governments, which might be expected to resist the movement of jobs away from their voter and taxpayer base, have begun to move work offshore because of budget pressure.

The global movement of work to take advantage of highly skilled and relatively low-cost labor, however, is not confined to digital information processing. In health care, a growing number of patients, who face long waiting periods and/or large bills for treatment in North America and Europe, are travelling to India for treatment that can cost one-quarter as much. For example, India's Apollo Hospitals Enterprises have grown from a single hospital in 1983 to a chain of 37 establishments; they have treated 60,000 foreign patients in the last three years (Solomon 2004). Apollo provides a range of services, from hip replacement and cardiac surgery for individuals, to clinical trials, diagnostic, and back-office services for US health care providers.

In sum, the radical innovation that the Indian education system has helped produce is not in products or services—Indian graduates have, thus far, been working mostly on relatively routine knowledge-work tasks, such as technical support, software programming, or maintenance. Instead, the radical innovation is in how multinational firms are organized, specifically the rapid diffusion of business process offshoring (Dassani and Kenney 2004).[4] But as the growth of high-tech work continues to attract many Indian technical experts and investors to return to India from the US, it appears likely that India will continue to move rapidly up the value chain, creating its own companies that will be generating new innovations in products and services.

Local networks: the role of education in high-skill ecosystems

To analyze further the multifaceted and interdependent roles that the ET system plays in innovation, we now turn our focus to one particular form of local network—high-tech industrial regions or clusters—that is widely recognized to be playing an increasingly important role in innovation and global competitiveness (Piore and Sabel 1984; Porter 1990; Saxenian 1994). The most successful of these can be described as self-sustaining, high-skill ecosystems (Finegold 1999), a form of collective-learning community that facilitates rapid transfer of the tacit knowledge essential to cutting-edge, technological innovation. Population ecologists draw an analogy between the high-skill-ecosystem framework, with its clusters of firms, and biological ecosystems (Hannan and Freeman 1977; Baum 1996): population ecology focuses on the birth, competition for survival, and death of individual firms. The HSE approach shifts the focus to a higher level: the factors that influence the chance of survival or the risk of extinction for whole HSEs (or clusters of firms) in a specific region (Young 1988). The HSE model consists of four distinct elements—catalysts, resources, supportive environment, and connectivity—that are common to the development of thriving natural ecosystems and high-technology clusters. In the discussion below we briefly describe each of these four factors and the role that ET systems, and HE in particular, play in stimulating them.

Catalysts

As with naturally occurring ecosystems, there is a strong element of historical contingency in how and where HSEs are formed (Arthur 1989). To trigger the development of successful high-technology enterprises, these regions require some catalyst, such as innovations produced by basic research, or demand from sophisticated customers. Leading universities and their faculty members often play both of these roles, producing the research breakthroughs and intellectual property that form the basis for many start-up companies and serving as the beta test sites that can experiment with new innovations before they are ready for the mass market. Zucker and Darby (1996) found, for example, that star scientists, generally based in research universities, have played a key role in the creation of the US biotechnology industry, with most start-ups locating close to the university campus that spawned them.

Nourishment

Once catalyzed, ecosystems require a sustained flow of nutrients or resources to fuel their growth. The access to highly skilled human capital supplied by the HE system is, along with financial risk-capital, the most important factor for stimulating and perpetuating HSEs. Highly skilled graduates contribute to different stages of the innovation process: advanced manpower (graduate students, PhDs, post-docs, and professors) play an active part in creating new technologies and transferring them to firms; and those with a first degree, diploma, or apprenticeship training help produce and commercialize a new product or service, and contribute to incremental process and product innovation. Top research universities not only develop the manpower in their region, but also act as magnets to attract leading talent from around the world to their locations.

Supportive environment

If the population of firms is to grow, there is a need, as in biological ecosystems, for a supportive rather than hostile environment. A region or nation's regulatory and cultural regime may either encourage or discourage individual professors and universities from working with industry to commercialize innovation. The structure of academic labor markets, intellectual property regulations, immigration laws, guidelines regarding research funding and conflicts of interest, and general societal attitudes towards entrepreneurship and risk-taking are all institutional factors that can affect the role that HE plays in industrial innovation. Universities can also take a more proactive approach to supporting new or mature enterprises through the creation of technology transfer offices and science or technology parks. An indirect

spur to innovation are the communities, attractive to top technical talent and entrepreneurs because they are rich, diverse, tolerant, and creative cultural environments (Florida 2002), that world-class universities can help to build.

Connectivity

Ecosystems that thrive over the long term have a high degree of interdependence among the many different organisms that inhabit them. Likewise, successful HSEs are populated by organizations that rely heavily on strategic alliances, partnerships, and other forms of cooperation that enable them to compete on a global scale, yet stay focused and flexible. HE institutions help foster the connectivity between these organizations by building social networks of individuals through courses (both full and part-time), alumni networks, and hosting meetings for local firms. One prominent example of this is Google, a leading Internet search engine, developed by Sergey Brin and Larry Page, two computer science graduate students who met at Stanford University and discovered a shared interest in the problem of handling large datasets. Once they came up with a novel solution, their faculty adviser introduced them to a co-founder of Sun Microsystems, a Stanford alumnus, who provided them with their initial investment.

Despite concerted government efforts around the world, HSEs are difficult to create and sustain because, like life in naturally occurring ecosystems, the absence of, or change in, just one element of the environment can have dire consequences. Many countries and regions, for example, invest heavily in infrastructure and risk capital to create companies, but lack the world-class universities to produce the research that provides clear competitive advantage or the right incentives for academics to commercialize this research. Los Angeles has the world-class research but, unlike its neighbors to the north (Silicon Valley) and south (San Diego), has failed to generate a thriving HSE outside the movie industry, perhaps because its urban sprawl has impeded the necessary connectivity among the key players (Los Angeles Regional Technology Alliance (LARTA) 2001). In the final case of Singapore, I use original, interview-based research to explore both the key elements and challenges of building an HSE.

Singapore's efforts to create a biotech HSE

Singapore's use of the ET system to drive economic development has been key to its very successful track record over the last forty years. Starting with virtually no industrial base, but the assets of a very high literacy rate, central Asian location, a stable government, good infrastructure, and an English-speaking location, the state used generous investment incentives and other policies to attract MNCs to locate manufacturing, distribution, and regional headquarters in Singapore; it then encouraged them to provide ongoing training to help the workforce move into higher-value-added jobs (Magaziner and Patinkin 1989). Companies like Apple Computer discovered that Singapore's well-educated workforce was not only very productive, but also contributed numerous suggestions for incremental innovations, helping to make their plants in Singapore among the most effective in their global operations (ibid.).

Just as quickly as it built this advantage, the government became concerned that Singapore was losing its competitiveness in IT and manufacturing to India, China, and other, much lower-cost Asian nations. To address this problem, and thereby to sustain Singapore's growth and living standards, the leaders of this city state have sought to make the leap from incremental to radical innovation through the development of the most R&D-intensive of all sectors: the biomedical industry. After intensive study of other countries' approaches, it recognized early on that building a successful biomedical industry would be a very long-term project, and that its traditional model of industrial development would need to be modified

to fit the distinctive requirements of the biomedical sector. Through the creation of a series of research institutes, it has sought to transform the higher echelons of its own ET system, while partnering with some of the leading research universities around the world to generate the innovations and manpower need to launch a biomedical industry cluster. It is still too early to evaluate the success of this strategy; but I use the Singapore case to explore the role the ET system and accompanying government policies can play in stimulating radical innovation in biotechnology, and the potential barriers that still remain to the success of this HSE.

Catalysts

As with most aspects of its economy, Singapore's move into the biomedical sciences was strongly driven by the government. The two arms of the government with responsibility for establishing Singapore as a biomedical industry hub within five years are the Agency for Science, Technology and Research (A*STAR), formerly known as the National Science and Technology Board, and the Economic Development Board (EDB). A*STAR has put in place policies, resources, and a research and education architecture intended to build Singapore's biomedical science competencies. Adopting a model similar to Germany or San Diego's successful biotech cluster, A*STAR has elected to concentrate these competencies in relatively autonomous research institutes, rather than universities. As it seeks to catch up to world leaders in biomedical research, these institutes have the advantage of a clear focus on research, rather than the multiple missions of universities, an interdisciplinary staff focused on a common problem area, and proximity to industry to create the potential for commercialization of new technologies. Singapore's first foray into the biomedical sciences sector was through the establishment of the Institute of Molecular and Cellular Biology (IMCB) in 1987 at the National University of Singapore. Other research institutes were set up between 1996 and 2002, including the Bioinformatics Center, the Genome Institute of Singapore, the Bioprocessing Technology Center, the Institute of Bioengineering, and Nanotechnology. A*STAR has already invested over S$500 million in these new research centers, with another S$1 billion in funding committed through 2006.

Nourishment

To turn basic research into radical, commercially successful innovations, the government recognized that it would need an equally radical shift in Singapore's labor supply. While strong at producing a high and uniform level of basic skills, Singapore's ET system has historically been weak at generating creative risk-takers and entrepreneurs. Creating new biotech firms requires a tolerance for failure and a free exchange of ideas among those with different viewpoints, characteristics that are not yet well incorporated into Singapore's ET system or culture, as shown in this example from an MNC manager describing a job interview with a Singapore scientist:

One standard question we use in US interviews is, 'Give me an example where you made a mistake and what you learned from it.' A common response [in Singapore] is, 'I don't make mistakes.' Once, when probing further, I was told, 'Yes, I made one mistake: I had a different opinion from my boss and told him.' When asked what was the mistake in that, [the interviewee] said, 'I learned not to have another opinion.' This is a true story and he was a PhD.

Recognizing that developing world-class scientists and having them generate new research breakthroughs take decades, the Singapore government has developed short-, medium- and long-term approaches to building the necessary skills. In the short term, A*STAR has offered generous financial incentives to attract internationally renowned scientists to set up research labs in Singapore. These include a Nobel Prize winner, Dr Sidney Brenner, a former director of the US National Cancer Institute, Dr Edison Liu, and a premier cancer

researcher from Japan, Professor Yoshiaki Ito. These foreign biomedical stars are seen as a way to provide immediate credibility to Singapore's nascent research efforts, and to serve as a magnet to attract top young scientists to work and train in Singapore. Because there is a lack of local talent with the relevant experience, the government is also helping firms recruit experienced scientific and managerial leaders from foreign bioscience firms to develop some of Singapore's new start-up companies. The medium-term strategy involves sending the top students from Singapore to the leading foreign research universities for graduate science and technology education. The government pays for their education provided that they return to Singapore when they complete their studies. In the long term, the government hopes that education reforms designed to encourage more freedom and creativity, and the expansion of its own universities and research institutes, bolstered by alliances established with top universities such as Johns Hopkins and MIT, can grow their own bioscience manpower to generate the intellectual property for future local start-ups. But by creating a very attractive environment to attract research scientists, the government may have inadvertently made it more difficult to encourage them to take part in commercializing their discoveries. Observed one local industry expert:

Researchers are pretty well off when they come to Singapore. They get good salary, housing, research subsidies, core funding for 10 years, and warm weather. A very attractive life, then someone comes along and says, 'Why don't you work hard at creating a firm?' They say, 'Why bother?'

Along with human capital, EDB has focused on filling Singapore's gap in the financial capital available for life-science firms: it has provided over S$2 billion, split between investments in venture capital funds (who provide funding for new biotech ventures) and financial incentives to attract MNCs to locate manufacturing, clinical development and, most of all, centers of biomedical research in Singapore (EDB 1999; Saywell 2001).

Supportive Environment

Three elements that support a biomedical HSE are:

- specialized infrastructure;
- regulatory policies that protect IP and support risk-taking;
- an environment attractive to knowledge workers.

In addition to excellent general infrastructure (efficient transportation, high-speed Internet network, a safe and clean city), Singapore has gone one step further by building the Tuas Biomedical park for bioscience manufacturing and the Biopolis, an ambitious 'city within a city' that specifically caters to the unique research needs of the biomedical sciences, such as a large vivarium to house the mice essential for pre-clinical studies. With its first phase opening in June 2003, the Biopolis is a S$300 million project near the National University of Singapore with the five biomedical research institutes as the anchor tenants. The Biopolis is intended to attract biomedical MNCs, start-ups, and support services such as lawyers and patent agents to locate there. The government hopes that the Biopolis, which will include plenty of restaurants, social spaces, and some living quarters, will create informal networks for knowledge-sharing, and accelerate the growth of a critical mass of biomedical expertise in Singapore.

Singapore has also put in place a regulatory environment that is very supportive of biotech in general, and stem-cell/therapeutic cloning in particular (Kong 2003). This has helped Singapore create several stem-cell companies, including ES Cell International that hired, from Scotland, Alan Colman, who took a lead role in the cloning of Dolly the sheep. Singapore also offers strong IP protection, a prerequisite for establishing research-based biotech firms, and one of the weaknesses of some of its leading potential Asian competitors (especially China and India). But while the general legal framework for IP clearly supports biotech development, the way in which IP is administered within state-funded institutions may not be as favorable. In 2002, A*STAR launched Exploit Technologies, a centralized technology-transfer office for IP management, licensing, and commercialization. Centralization of IP ownership runs directly counter to the path the US took with the landmark Bayh-Dole Act in 1982: this gave universities the freedom to commercialize federally funded research, and is widely credited with helping kick-start the biotech revolution. Retaining central control of IP risks stifling creativity, but is more in keeping with the Singapore government's traditionally direct involvement in economic development (Vig 2003). The head of a bioscience incubator explained one of the cultural reasons why research institutes are reluctant to provide significant equity positions for founding scientists:

Their concern is that if founding scientists have equity, then others may get upset, rather than seeing them as role models ... The culture is very competitive here; in the US it is more collaborative; there, if you get to be a millionaire, then I want to learn how I can do it too. It can be hard to build effective teams here if competition rather than collaboration is the culture; biotechnology is a very collaborative business.

Connectivity

As in the IP area, the government appears to exercise strong, centralized control over most aspects of Singapore's biomedical industry development. Although this top-down, coordinated approach has worked well in the past to accelerate the development of competencies in new industrial clusters (for example, the hard-disk-drive industry, see Wong 2001), and may be particularly advantageous in terms of long-term resource development for an industry like biotechnology (for example, manpower development), it runs the risk of stifling alternative approaches and marginalizing nonconforming groups. In particular, a company that is outside of this community may find it difficult to secure resources, such as funding. This may have hampered the development of independent biomedical start-ups, because very little funding exists that is not of government origin (Ginzel 2003).

Another weakness is that informal cooperation between firms and research institutes has been, so far, relatively limited. For example, one firm we interviewed mentioned approaching a research institute in the same building regarding the possibility of collaboration, beginning with a request to use an expensive piece of equipment in the laboratory that had excess capacity. It was unable to interest the laboratory in partnering, however. Several factors may explain this. First, other than IMCB, the institutes are very new, and focused on getting their research programs up and running. Next, the academic culture of research institutes focuses on basic research and publishing, and the researchers are relatively well funded by the government; thus the advantages of collaboration with firms, such as securing additional research funding or the opportunity to spin out a company, are not a high priority for the research staff. Finally, as noted above, the culture among research scientists and managers in Singapore appears to be risk averse; consequently, collaboration and sharing of information may be perceived as risky, and inter- and intra-organizational collaborations are not the norm (Stein 2003).

Progress to date

While Singapore appears to have put in place some of the elements needed for a biotech HSE, so far the bulk of the activity in the sector appears to be concentrated not in radical or incremental innovation, but rather replication of existing businesses in the Asian region. The biomedical industry in Singapore today is dominated by the manufacturing of pharmaceutical products by large, foreign-owned MNCs, with biotechnology firms still at an embryonic stage of development. While R&D expenditure and employment in the biomedical sector has grown rapidly (see Tables 17.1 and 17.2), the total level of activity is still dwarfed by the R&D operations of a single pharmaceutical firm or large research university in the US. Only a handful of biotech companies have been founded in Singapore, with the majority established in 2000 or later. Rather than domestic start-ups, most of the initial funding and attention for developing a biomedical industry focused on attracting MNCs—such as Chiron, Lilly, and GSK—to set up research facilities or new ventures to fuel innovation in Singapore. Because of the lack of well-qualified local candidates, and difficulty attracting top people from abroad, many of these organizations have struggled to fill key strategic and technical positions. These difficulties, plus the more general reluctance of MNCs to move key research operations far from their core markets, suggest that the strategy of relying on MNCs to catalyze the growth of a research-driven biomedical HSE may prove difficult to execute.

Alongside these government-backed MNCs has formed a small set of local start-up companies that represent a potential alternative path to the development of a Singapore biomedical cluster. These start-up firms bear a close resemblance to early-stage counterparts in the US or UK: they have typically been based on IP from a university or the inspiration of an entrepreneur, and are trying to fill the void between the basic research of the HE sector, and the more mature commercial technologies that venture capitalists, EDB, or foreign pharmaceutical firms are willing to fund. With limited access to these funding sources, they have struggled to grow and had to develop business models that generate revenue quickly. These firms have often located in Singapore not because of generous investment incentives from the government, but rather because that is where the founders were based, and they saw the more general business advantages of Singapore—location, strong infrastructure, and good quality of life for professionals. As Steven Fang, founder and CEO of Cygenics, a Singapore-based stem-cell company, observed:

The government is interested in stem cells and sees it as a high priority. But it has been investing mostly abroad. I understand the attitude. EDB has built a series of successful industries in the port, chemicals, and electronics. It's appealing for them to sign big deals with MNCs. It's great

Table 17.1. R&D expenditure by firms in the life-sciences industry 1993–2001

Year	No. of firms	R&D expenditure (S$m)
1993		24.8
1994		38.6
1995		34.4
1996	45	37.86
1997	50	58.34
1998	57	63.81
1999	54	89.68
2000	43	83.48
2001	48	113.58

Source: National Survey of R&D in Singapore (various years), Agency for Science, Technology and Research.

for the economy and their learning curve. We're locals, we don't have big names … It may be better for us in long term not to have (large state support) since it forces us to find ways to stand on our own.

This independent approach appears to be paying off for Cygenics, which used early revenues from cord-blood banking services to acquire a US firm and then go public on the Australian Stock Exchange. While successful in raising capital, it is skilled manpower that is now the firm's biggest constraint on growth. 'People are always the biggest challenge,' according to Fang, 'particularly in Singapore and Asia. It's hard to get good technical people and even harder to get people with bioscience business experience … We've had to go outside the industry and train them ourselves or look outside Singapore, hiring ex-pats from the US, Australia, UK, and Scandanavia.'

Table 17.2. Research scientists and engineers (RSEs) in the biomedical sector[a] (S$ million) 1993–2001

Year	No. of RSEs			(a + b)/(c) (%)
	Ph.D.	Master	Total	
	(a)	(b)	(c)	
1993	139	176	447	70.5
1994	116	157	386	70.7
1995	131	260	570	68.6
1996	158	193	507	69.2
1997	203	177	556	68.3
1998	202	203	625	64.8
1999	238	164	654	61.5
2000	300	243	1,333	40.7
2001	610	453	2,055	51.7

[a] Includes biomedical sciences and biomedical engineering.

Source: National Survey of R&D in Singapore (various years), Agency for Science, Technology and Research.

In addition to stem cells, another area where Singapore has begun to build a set of companies is bioinformatics, leveraging a strong pool of computer programming skills. As one entrepreneur, who was considering relocating his bioinformatics firm to the US, noted, however, these skills are not necessarily a product of Singapore's ET systems:

We don't need a factory of programmers, but a few good ones. Here we can get good talent, but train them for 6 months, then a year later someone poaches them away. I'd rather pay 40 per cent more in the US where people can be productive on day one, even if there will still be poaching... In terms of people in Singapore, we have had a tough time attracting any locals. The education system starts streaming early and the good people are locked in early with scholarships and bonds. The cream is stolen from the market, and then the next tranche want safe jobs. They all go to work for government or MNCs. Then the last level is a bit too thin. You get a few eccentrics who don't fit in the larger organizations, but most of the rest are foreigners. Indians are about 95 per cent of the key employees at Lilly. Same is the case with the Genomics Institute ... The Singapore government has played an active role in getting good talent into Singapore from China, India, and Vietnam.

The director of an incubator for new biomedical firms provided a good summary of the challenges Singapore has faced in moving from a model of MNC-driven, incremental innovation to generating its own firms capable of radical innovation:

The traditional Singapore industrial model doesn't work very well in the new knowledge-based economy. It comes from a top-down, logistical mindset. The policymakers in charge are used to dealing with large manufacturing-based companies ... Not a good fit are very-early-stage start-ups from universities. These types of companies are quite difficult to deal with; everyone finds them hard to create, but in Singapore, at the moment, it's even harder. Investors aren't happy funding a number of very small companies compared to making a few big investments. Singapore investors don't like the high failure rate.

Future research agenda: ET system attributes and the capacity for innovation

Each of these cases has attempted to provide new insights into how ET systems contribute to innovation in today's highly competitive global economy. This final section attempts to synthesize lessons from the cases and from other research to develop a framework for future research. The framework identifies key attributes of ET systems (whether national or regional), describes how these can be measured, and sets forth propositions on how each attribute may be related to the capacity for radical innovation. A particular emphasis is placed on the HE part of ET systems, since this is the part of the system charged with the generation of new knowledge. Table 17.3 suggests the rankings on each of these dimensions of the five largest developed economies.

Table 17.3. ET system attributes and the capacity for radical innovation

	US	Japan	Germany	France	UK
Decentralization	H	L	M	L	M
Variety in educational content/assessment	H	L	M	L	H
Diversity of HE institutions & funding	H	L	M	L	L
Diversity of students	H	L	L/M	M	M
Adult access	H	L/M	L	L	H
Work-based learning	L	H	H	L	L
Overall	H	L	L/M	L	M

H = high; M = medium; L = low.

Decentralization

ET systems can be distinguished by the extent to which local, regional, or national authorities are given control over ET provision. The degree of centralization in the ET system can be measured by the percentage of total ET expenditure coming from national government, and by the level at which the content of ET is determined (that is, is there a national curriculum, with associated testing?). Countries vary in this, from France's highly centralized system, where the Minister for Education is said to be able to tell what every schoolchild in the country is studying at a particular time and date, to the highly decentralized US system, where more than 90 per cent of the funding and most of what occurs in the classroom is determined through a mix of influences from the individual teacher, the school, the local school district, and the state. The level of decentralization is likely to be associated with higher levels of radical innovation, since it allows for more experimentation, more variation in response to local needs, and more rapid, if uneven, adjustment to changing technologies and external circumstances. The benefits of diversity and responsiveness, however, come at the expense of a lack of clear national standards that can help ensure that the majority of the population have the foundation skills needed to contribute to product and process innovation.

Educational content and measurement

Closely related to decentralization is the degree of uniformity in the curriculum across the ET system, and the way attainment of this ET content is measured. Many countries—all of the major Asian nations, France, and, to a lesser extent, Germany and the UK—have high-stake national examination systems for school leavers that measure mastery of each subject to determine admission to HE. The examinations in many countries—Singapore, Japan, South Korea, India—place a very heavy emphasis on memorization of a large volume of information. These systems tend to be very effective at pro-ducing a high and uniform level of literacy and numeracy, as measured in international comparisons of educational achievement, in the mass population, but are weaker at generating the individuals who produce radical innovations, since they focus on one uniform body of knowledge and emphasize rote learning over creativity. In contrast, the US, without a national curriculum or examination system, has the most heterogeneous results: the top-performing students not only attain a high level of foundation skills, but also have the freedom to explore creative solutions to problems; the bottom 10–15 per cent of the cohort often do not complete high school, and many that do still do not have the basic skills needed to compete effectively in a knowledge economy. Many of the Asian nations are attempting to introduce educational reforms that will better prepare their brightest students to become innovators, while not wanting to sacrifice the high level of basic skills for the whole population. So far, these have met with limited success.

Diversity of HE institutions and funding

In most countries, all or most HE institutions are public—that is, funded and governed by the state. Within these state-HE systems, some countries (for example, the UK) have pushed further to develop a unitary system of HE, removing distinctions between the polytechnics and universities. While this may produce benefits in educational participation, and reduce status differences among graduates, it may have negative consequences for radical innovation, as institutions lose a distinctive mission focus. In contrast, the US has a variety of public and private colleges and universities, including for-profit HE institutions, while Germany and France have retained types of HE institutions within their state systems that are different again. Greater institutional diversity may encourage innovation in several ways: more competition generates greater responsiveness to student needs and to economic and technological changes; greater specialization allows institutions to identify distinctive market niches, with some focusing heavily on producing world-class research, others on teaching. Many public universities, including some in the US, find commercialization more difficult, since state employees are often precluded from pursuing private-sector opportunities that may generate conflicts of interest.

The US also benefits from a diversity of funding sources for the research that drives radical innovation greater than in most other countries. While the government continues to be the largest funder of basic research in all industrialized countries, a very large and growing set of private foundations, each with a distinctive mission, provides a major, and often more flexible, source of funding for development of new ideas in the US. The Gates Foundation, for example, has been in existence less than a decade but, with an endowment of $30 billion, is already providing more money to fund innovations in global health than most national governments. And to help turn research into new products and services, US universities are able to leverage their own endowment funds to invest in new business opportunities generated by their faculty.

Diversity of HE participants

The diversity of perspectives represented on a knowledge-work team is significantly related to the creativity of the solutions they produce (Mohrman *et al.* 1997). A similar argument can be made at the level of a national ET system: the greater the blend of cultures and backgrounds participating, the higher the levels of innovation are likely to be. Although the demographics of the native population are more an input into, rather than a structural feature of, the ET system, the regulatory environment can have an impact on this variable: immigration policies, particularly the openness of the system to students from other countries, are an example. HE systems that attract a significant number of top students from other nations get the twin innovation benefits of a greater supply of individual talent and more diversity within each cohort.

Adult access

Some ET systems—such as Germany's—are heavily geared toward providing a strong foundation for entry to a particular career for as many young people as possible. Others, like the US or the UK, with its pioneering Open University created in 1971, provide less clear ET-employment links for the majority of young people, but make it easier for those who did poorly initially in ET, or for others who subsequently want or need to change careers to acquire the necessary skills. Japan falls somewhere in between, with a broader initial education and more multifunctional training than Germany, but with a heavy orientation toward gradual development along one career path in one large company. Open-access ET systems are likely both to foster more social mobility and radical innovation, since individuals are prepared to take greater personal career risk if they know they can retrain, and to respond more quickly to radical innovations in the economy, since they provide a means for retraining the existing workforce, and not waiting for a new generation of graduates.

Work-based learning

One dramatic difference among ET systems is the extent to which formal ET takes place in the workplace. As noted in the corporatist section, apprenticeships—with their heavy component of on-the-job learning—continue to be the dominant mode of ET for the majority of 16–19-year-olds in the Germanic countries. In Japan, work-based learning is also a vital part of the system, but it occurs after compulsory schooling and HE, when individuals join large firms and go through systematic programs of job rotation. Of course, work-based learning also occurs frequently in the US, UK, France, and other countries, but it is much more uneven in quality and quantity across individuals and firms because it is not part of a formal system. The multifunctional and applied skill set that these systematic, work-based learning approaches build has likely been a key factor in the sustained success of German and Japanese firms in incremental product and process innovations. The embeddedness of these ET systems in existing work processes and products, however, may mitigate more radical innovations that threaten to disrupt existing industries or require entirely new skill categories.

To summarize, ET systems that score more highly on each of these dimensions are hypothesized to have greater capacity for producing more radical product and process innovations and the new start-up companies they generate (with the exception of work-based learning, where the relationship is predicted to run in the opposite direction). Future research could expand the list of ET dimensions that may be important for innovation, and test these hypothesized relationships across a wider array of countries.

Lessons for policymakers

While the relationships in this framework still need to be tested, the case studies and other research reviewed here help identify some issues for policymakers seeking to use the ET system to increase innovation.

Tensions within the ET system

The ET system makes two vital contributions to innovation: it conducts research that can generate new IP that serves as the basis for new business opportunities; it prepares the managerial and technical talent to take advantage of these opportunities. Ideally, these two functions are mutually reinforcing, as when students' classroom experience is enhanced by exposure to the latest, cutting-edge developments in research, or when star scientists spin out companies from the university, taking with them some of their top graduates who have worked on the technology, to transfer key tacit knowledge and help turn the outputs of research into a product. In some cases, however, like UK or German HE, the two missions can come into conflict, particularly when a system is resource-constrained, and faculty

are compelled to devote most of their time to one mission at the expense of the other.

ET systems have another, potentially conflicting, set of missions to balance in the creation of capacity for innovation—the tension between providing a high level of foundation skills for as many in the population as possible, so that they can qualify for good jobs in a global economy, and creating an environment for the brightest individuals to generate the new knowledge that drives the economy. It appears that those ET systems that do the best job at generating a high level of literacy and numeracy—Singapore, South Korea, Japan—have placed a heavy emphasis on standardization and memorization that may inhibit creativity among the most academically able students. Conversely, countries like the US and UK, that have done relatively well at producing star researchers who generate new intellectual property and innovation, continue to struggle with providing adequate skills for a significant percentage of the population to participate effectively in a twenty-first-century labor market.

Insufficient raising of skill levels

Expanding the supply and quality of skills emerging from the ET system does not ensure innovation in products, processes, or ways of organizing work. As the UK experience indicates, it is possible to expand the supply of skills significantly, and yet have only a marginal immediate effect on how work is conducted in most enterprises. This lack of change may be due to institutional factors that hinder workplace innovation—for example, short-term pressure from financial markets, an adversarial industrial relations system, middle management opposition, a lack of good models of successful high-performance work organizations—as well as a time lag, as it takes a significant period for the boost in the ET-system output to change the overall composition of the labor force, and for these new graduates to reach leadership positions.

For those countries seeking to use ET as a key mechanism to develop new, innovative industries, Singapore's experience suggests that there may be an additional problem: a vicious circle may inhibit the development of a critical mass of firms needed to sustain an HSE. Scientists in universities and research institutes have little incentive to partner with industry, let alone put their careers at risk by entering industry. These scientists have employment security and generous research funding if they remain in the nonprofit sector, and are more comfortable publishing their research than commercializing it. The country's lack of experienced managers and entrepreneurs to aid scientists in building new companies further reduces the incentive to take the risk of moving to a start-up. The talent needs for a biotech HSE go far beyond the laboratory: experienced managers to develop products, manage clinical trials, and run companies, savvy life-science investors who understand the unique requirements of this sector, and life-science IP specialists who know the intricacies of structuring licensing arrangements are needed to build a successful cluster of biotech firms. In the absence of these skills, founders are forced to do it themselves, a task for which many are unprepared. The lack of proven technology, of leadership talent, and of a critical mass of new companies means that it is difficult, even for those firms that are created to attract the necessary investment.

The role of ET in the global economy

In today's global economy, the impact of the ET system on innovation extends well beyond national boundaries. In the case of the US, the HE system has been a magnet for talent, attracting highly skilled and entrepreneurial immigrants from around the world, who have played a key role in both founding and staffing the most innovative high-tech enterprises (Saxenian 2000). In India, the large supply of well-qualified, English-speaking graduates has not just spurred regional economic development, it is also a key element in the redesign and disaggregation of the organization of work in many of the world's top multinational corporations. The growth in business process out-

sourcing (BPO) that India has led is sure to be followed by other developing countries with strong educational systems, causing companies to reassess the boundaries of the firm and what activities are most effectively performed in-house.

The implications of this innovation in work organization for the advanced industrial countries' ET systems are already being felt. No longer are just unskilled and semi-skilled workers at risk of having their jobs moved to lower-wage nations. Now many highly educated professionals, managers, and service workers are competing for work with a burgeoning supply of college graduates in the developing nations. On the whole, both the advanced and developing nations should benefit from the efficiency and growth generated through freer movement of capital, information, and labor that comes with globalization; but the transitional costs are likely to be high. Those who complete higher education, and particularly those with an advanced degree, are, on average, still likely to be the winners in this global competition. But many will lose their jobs and find it difficult to find new work that offers a similar standard of living. For example, the IT sector, where this global competition appears to be fiercest, represents 8.5 per cent (2.4 million) of the 28 million US jobs that pay over $25/hour (Linden 2003). The end result may be even greater polarization in the labor market, as, wherever they reside, those who generate new innovations reap the economic benefits, while many of the new jobs created are for relatively low-skilled service work that it is still not feasible to move globally.

The regulatory environment and ET programs in the advanced industrial countries have not kept pace with the changes in work organization and BPO. For example, US IT workers laid off from high-tech companies such as IBM, Intel, and EDS because their jobs were moved to India were denied assistance by the US Department of Labor under the Trade Adjustment Assistance (TAA) program (Loftus 2003). The TAA was established in the 1960s to provide job retraining and a stipend to workers displaced as a result of imports and loss of jobs overseas; typically only those products subject to import duties are covered, offering limited eligibility to manufacturing workers. 'High-tech workers are sometimes denied certification (for TAA benefits) because they don't produce a product,' said Lorette Post, a US Department of Labor spokesperson. As governments seek to stimulate innovation and the growth of good jobs, it is vital that they not only adjust existing policies to meet the realities of the twenty-first century's global economy, but also recognize that, while ET will continue to play a central role in generating knowledge work, it is only one policy area among the many needed to foster a wider innovation system.

Notes

1. Within the UK, Scotland's education system has a somewhat different structure from that of England and Wales; it has encouraged somewhat higher levels of participation.
2. The US continues to have a much higher percentage of the population attending, close to 60%, but there is a much higher attrition rate and many students are in two-year colleges working toward associate's degrees.
3. While one university may not be able to offer much higher salaries than a rival, it can offer attractive packages, such as reduced teaching loads, to attract top researchers.
4. It is important to draw a distinction between offshoring—the movement of work from one country to another—and outsourcing, the shifting of work from inside a company to an outside supplier. In the case of India, Infosys and Wipro are supplying outsourced services to many MNCs, but others, like GE, have decided to set up their own operations in India to service their global organizations.

References

Arnold, W. (2003). 'Singapore Goes for Biotech.' *New York Times,* 26 August, W1.

Arthur, B. W. (1989). 'Competing Technologies, Increasing Returns, and Lock-in by Historical. Events.' *Economic Journal,* 99: 116–31.

Bagnasco, A. (1977). *Tre Italie,* Bologna: Il Mulino.

Baldry, C., Bain P., and Taylor, P. (1998). ' 'Bright Satanic Offices': Intensification, Control and Team Taylorism.' In C. Warhurst and P. Thompson (eds.), *The New Workplace,* London: Macmillan, 163–83.

Banke, P. (1991). *Gruppeorganisering: Fleksibel production og jobkvalitet I den syende industri,* Copenhagen: Dansk Teknologisk Institut.

Baum, J. A. C. (1996). 'Organizational Ecology.' In S. R. Clegg, C. Hardy, and W. R. Nord, (eds.), *Handbook of Organization Studies.* London, Thousand Oaks, CA: Sage Publications, 77–114.

Becker, W. E., and W. J. Baumol (eds.) (1996). *Assessing Educational Practices: The Contribution of Economics.* Cambridge, MA: MIT Press.

Best, M. (1990). *The New Competition.* Oxford: Polity Press.

Camagni, R. (1991). 'Local 'Milieu', Uncertainty and Innovation Networks: Toward a New Dynamic Theory of Economic Space.' In R. Camagni (ed.), *Innovation Networks: Spatial Perspectives.* London: Belhaven Press, 121–44.

CDE (2002). 'CTAP2 California School Technology Survey.' Sacramento: California Department of Education.

Casper, S. (2000). 'Institutional Adaptiveness, Technology Policy, and the Diffusion of New Business Models: The Case of German Biotechnology.' *Organization Studies,* 21: 887–914.

Cohen, S. S., and Fields, G. (1999). 'Social Capital and Capital Gains in Silicon Valley.' *California Management Review,* 41/2: 108–30.

Collinson, D. L., and Collinson, M. (1997). 'Delayering Managers: Time Space Surveillance and its Gendered Effects.' *Organization,* 4/3: 373–405.

Cook, P. (2003). 'The Evolution of Biotechnology in Three Continents: Schumpeterian or Penrosian?' *European Planning Studies* (Special Issue on Biotechnology Clusters and Beyond), October.

Crouch, C., Finegold, D., and Sako, M. (1999). *Are Skills the Answer?* Oxford: Oxford University Press.

Cully, M., Woodward, S., O'Reilly, A., and Dix, A. (1999). *Britain at Work: As Depicted by the 1998 Workplace Employee Relations Survey.* London: Routledge.

Culpepper, P. (2003). *Creating Cooperation.* Ithaca, NY: Cornell University Press.

—— and Finegold, D. (eds.) (1999). *The German Skills Machine in Comparative Perspective.* New York: Berghahn Books.

Daly, K., Hitchens, P., and Wagner, K. (1985). 'Productivity, Machinery and Skills in a Sample of British and German Manufacturing Plants.' *National Institute Economic Review*, 111: 48–61.

Dassani, R., and Kenney, M. (2004). 'The Next Wave of Globalization? Exploring the Relocation of Service Provision to India.' Berkeley Roundtable on the International Economy (BRIE) Working Paper.

EDB (1999). Annual Report: Life Sciences. **www.sedb.com/edbcorp/an_1999_13.jsp**

—— (2003). Joint A*STAR and EDB Biomedical Sciences Sectoral Briefing 2003, EDB website: **www.biomed-singapore.com/bms/browse_print.jsp?artid=333**

Felstead, A., Fuller, A., Unwin, L., Ashton, D., Butler, P., Lee, T., and Walters, S. (2004). 'Applying the Survey Method to Learning at Work: A Recent UK Experiment.' Paper presented to the European Conference on Educational Research, Rethymnon Campus, University of Crete, Greece, 22–5 September.

Finegold, D. (1999). 'Creating Self-Sustaining High-Skill Ecosystems.' *Oxford Review of Economic Policy*, 15/1: 1–22.

—— and Soskice, D. (1988). 'The Failure of Training in Britain: Analysis and Prescription.' *Oxford Review of Economic Policy*, Autumn: 21–51.

—— and Wagner, K. (1999). 'Transforming the German Metalworking Industry: Moves Toward Numerical Flexibility.' *Work Study*, 2, March–April.

Florida, R. (2002). *The Rise of the Creative Class: And How It's Transforming Work, Leisure, Community and Everyday Life.* Perseus Books.

Galbraith, J. (1998). 'Global Organizations.' In S. Mohrman, J. Galbraith, and E. Lawler, *Tomorrow's Organization: Creating Winning Competencies.* San Francisco: Jossey Bass.

Gibson, C., and Cohen, S. (2003). *Virtual Teams that Work.* San Francisco: Jossey Bass.

Ginzel, T. (2003). Interview with author, Singapore.

Hall, P., and Soskice, D. (2001). *Varieties of Capitalism: The Institutional Foundations of Comparative Advantage.* Oxford: Oxford University Press.

Hannan, M. T., and Freeman, J. (1977) 'The Population Ecology of Organizations.' *American Journal of Sociology*, 82: 929–64.

Hayes, C., and Fonda, N. (1984). *Competence and Competition: Training and Education in the Federal Republic of Germany, the US and Japan.* London: NEDO/MSC.

Hepworth, M., and Spencer, G. (2002). *A Regional Perspective on the Knowledge Economy in Great Britain.* Report for the Department of Trade and Industry. London: The Local Futures Group.

Herrigel, G. (1996). 'Crisis in German Decentralized Production.' *European Urban and Regional Studies*, 3/1: 33–52.

Japanese Institute of Labor (2003). *The Labor Situation in Japan 2002/2003.* Japanese Institute of Labor website, **www.jil.go.jp/eSituation/pdf**

Keep, E., and Mayhew, K. (forthcoming). 'Supply or Demand (or Both)?' In E. Keep, K. Mayhew, J. Payne, and C. Stasz (eds.), *Great Expectations: Education, Skills and the Economy.* London: Palgrave.

Keltner, B., Finegold, D., Mason, G., and Wagner, K. (1999). 'Market Segmentation Strategies and Service Sector Productivity.' *California Management Review*, 41/4: 81–102.

Kirkpatrick, D. (2003). 'Rage against Off-Shoring is Very Real.' *Fortune.com.* Monday, 23 February 2004.

Kong, H. L. (2003). Personal interview with authors, Singapore Biomedical Research Council, January.

Lane, P. (2002). 'Strategic Management under Science-based Competition.' Working Paper, Arizona State University. **http://hcd.ucdavis.edu/faculty/kenney/articles/offshoring/ bpo%20 world%20development%20ver%204.doc**

Lawler, E. (1996). *The New Logic Corporation.* San Francisco: Jossey Bass.

Linden, G. (2003). Data analysis of BLS, 2001, personal e-mail to author.

Loftus, P. (2003). 'Jobless Tech Workers Denied Benefits.' *Wall Street Journal,* 3 September, D3.

Los Angeles Regional Technology Alliance (LARTA) (2001). *Heart of Gold: The Bioscience Industry in Southern California.* Los Angeles: LARTA.

Magaziner, I., and Patinkin, M. (1989). *The Silent War.* New York: Random House.

Mason, G. (1998). *Change and Diversity: The Challenges Facing Chemistry HE.* London: The Royal Society of Chemistry.

Mason, G. (2001). 'Mixed Fortunes: Graduate Utilisation in Service Industries (Computing, Retail and Transport).' NIESR Discussion Paper 182.

—— Wagner, K., Finegold D., and Keltner, B. (2000). 'The 'IT Productivity Paradox' Revisited: International Comparisons of Information Technology, Work Organisation and Productivity in Service Industries.' *Quarterly Journal of Economics.* Also published in *Vierteljahrshefte zur Wirtschaftsforschung, 69, Jahrgang, Heft 4/2000:* 618–29.

Meredith, R. (2003). 'Giant Sucking Sound.' *Forbes,* 29 September, 58–60.

Mohrman, S., Cohen, S., and Mohrman, A. (1997). *Designing Team-Based Organizations: New Forms for Knowledge Work.* San Francisco: Jossey Bass.

—— Klein, J., and Finegold, D. (2003). 'Managing the Global New Product Development Network: A Sense-Making Perspective.' In C. Gibson and S. Cohen (eds.), *Virtual Teams that Work.* San Francisco: Jossey Bass.

Pilling, D. (2004). 'Wages Defy Recovery.' *Los Angeles Times,* 15 November, C3.

Piore, M., and Sabel, C. (1984). *The Second Industrial Divide: Possibilities for Prosperity.* New York: Basic Books.

Porter, M. (1998). 'Clusters and the New Economics of Competition.' *Harvard Business Review,* 78: 77–90.

—— (1990). *The Competitive Advantage of Nations.* New York: The Free Press.

—— and Ketels, C. H. M. (2003). 'UK Competitiveness: Moving onto the Next Stage.' DTI Economics Paper No 3.

Rodgers, R., and Waters, R. (2001). 'The Skill Dynamics of Business and Public Service Associate Professionals.' DfES Research Report 302.

Sainsbury, D. (2003). Phone interview with author.

Sako, M. (2002). 'National Institutional Change for Business Start-ups in Japan.' Oxford: Said Business School Working Paper.

Saxenian, A. (1994). *Regional Advantage: Culture and Competition in Silicon Valley and Route 128.* Cambridge, MA: Harvard University Press.

Saywell, T. (2001). 'Medicine for the Economy.' *Far Eastern Economic Review.* 15 November.

Schoppa, L. (2001). 'G.I. Bill needed for corporate Japan.' *The Yomiuri Shimbun (Tokyo),* 14 July.

Solomon, J. (2004). 'India's New Coup in Outsourcing: Inpatient Care.' *Wall Street Journal,* 26 April, A1.

Steedman, H., and Wagner, K. (1987). 'A Second Look at Productivity, Machinery and Skills in Britain and Germany.' *National Institute Economic Review,* 122: 84–95.

Stein, S. (2003). Consultant for Singapore start-ups. Personal interview with researchers.

Taylor, R. (2002). *Britain's World of Work: Myths and Realities.* Swindon: Economic and Social Research Council.

Vig, P. (2003). Consultant for Singapore start-ups. Personal interview with researchers.

Wong, P. K. (2001). 'Leveraging Multinational Corporations, Fostering Technopreneurship: The Changing Role of S&T Policy in Singapore.' *International Journal of Technology Management,* 22/5/6: 539–67.

—— (2002). 'Globalization of American, European and Japanese Production Networks and the Growth of Singapore's Electronics Industry.' *International Journal of Technology Management,* 24/7/8: 843–69.

Young, R. (1988). 'Is Population Ecology a Useful Paradigm for the Study of Organizations?' *American Journal of Sociology,* 94/1: 1–24.

Zucker, L. G., and Darby, M. R. (1996). 'Star Scientists and Institutional Transformation: Patterns of Invention and Innovation in the Formation of the Biotechnology Industry.' *Proceedings of the National Academy of Sciences of the United States of America,* 93: 12709–16.

PART IV

INSTITUTIONS AND INSTITUTIONAL CHANGE

Introduction[1]

Jerald Hage

Connecting institutional analysis to the topic of innovation is obviously appropriate because of the recent work on the national systems of innovation (Edquist 1997, and Chaminade and Edquist's contribution in this book; Edquist and Hommen 1999; Freeman and Soete 1997; Nelson 1998 and their reviews; see also the contributors to Nelson 1993). The thesis in this work is that, over time, countries develop particular styles of innovation (see, for example, Hollingsworth 1997). But this is not the only way in which the topic of innovation can be connected to the institutional literature. Perhaps the more fundamental connection, and one that has been largely neglected, is how much does innovation itself create institutional change and, conversely, can institutional change produce more innovation?

The relationship between innovation and institutional change has emerged as a central problem because the three processes of social change described in the introduction have made governments want to change their national systems of innovation. Globalization increases the number of countries that can export goods and services, including high-tech ones, and therefore the number of competitors. This is even true for scientific research, as more and more countries invest in basic and applied research (OECD 2002). Post-industrialization (Bell 1973; Hage and Powers 1992; Toffler 1981) increasingly places a premium on high-tech or knowledge-intensive products and services that are customized (Piore and Sabel 1984; Pine 1993) and state of the art; knowledge growth generates specialization in product/service markets, in scientific disciplines, and in areas of research, and it fragments both the supply chain and the idea-innovation chain in various organizations involved in facilitating collective learning (Gibbons *et al.* 1994; Hage and Hollingsworth 2000; Rammert's contribution in the second part, and the contribution of van Waarden and Oosterjick in this part). Thus the pressures on government—and firms—to increase their innovation rates are apparent. But the question is, can they?

The importance of studying institutional change goes to the heart of this question. Certainly governments have been *trying* to alter their national systems of innovation. In Europe, both the French and German governments have been quite active in attempts to create and develop a biotech industry to parallel that in the US. Some of these efforts are reviewed by Casper in this part. The European Union has also been exploring joint initiatives in what are considered to be critical high-tech areas. Some have been successful (Airbus) while others have failed (Sprite); these differences in outcome are an important topic for future research. The US altered its antitrust laws to allow competitors to form research consortia, precipitating the formation of a number of these in the US, the most famous being SEMATECH (Browning *et al.* 1995).

Most handbooks on innovation tend to ignore the problem of institutional change; we are devoting a whole part to this topic. So some consideration should be given to why this topic is so central in our thinking. Because the common theme in the many varieties of institutional analysis is how institutions constrain choices and behavior, and have the power of sanction to ensure compliance,[2] it has been neglected in the large and rapidly growing literature in economics, political science, and sociology (see references in Hollingsworth 2000: 598–9). Given this perspective, any institutional change is likely to be incremental and, in one of the major theses within the institutional literature, path dependent. Given this perspective, institutional change then should be defined as discontinuous change, so that the debate is clearly drawn.

But defining institutional change as discontinuous change still leaves unanswered what is being changed, that is, what is the nature of institutions? This is the topic for the first section: institutions are seen as rules, and coordination modes are suggested as a significant kind of institutional change, because changes in the coordination mode would be an example of a discontinuous change.

Definitions of institutions and of significant institutional change. What are institutions? Rules?

When institutionalists define the core idea in institutional analysis, it is usually the guiding rules and norms (North 1990) that constrain strategic choices and individual behavior. Included in this definition is the importance of sanctions as a method to gain compliance. However, the multitude of rules makes it difficult to develop analytical variables that could cut across the disparate possibilities. Consider all the institutional rules relevant in education: rules for student selection and evaluation, honor rules regarding exams, rules about the content of the curriculum, about who makes the decisions about the content of the curriculum, the length and number of class days, at which times of the week, vacations and their periodicity, faculty selection, faculty civil-service regulations, rules about promotion and of tenure or job security, etc. Then consider these rules at a much more abstract level—as basic norms about guiding principles such as equality vs. merit. Add the many kinds of institutional rules in other institutional sectors, such as scientific research or businesses, and the task of finding useful variables that can describe these many sets of rules, as well as determine their specific and cumulative impacts on knowledge bases, collective learning and innovation, whether in industry or in science, is daunting. But it must be done if we are to develop useful theoretical propositions and advance our understanding.

As yet, no discernible classification system of institutional rules/norms/ laws for understanding has been devised to establish which rules, specific or abstract, or collection of them, are most critical for maximizing innovation. But if the presence of some of them prevents either research organizations or businesses from developing higher rates of innovation, especially radical innovation, then policy interventions must determine which rules should be changed to be effective.

What are significant rules? Coordination modes

One critical set of rules is the laws or norms of coordination (Campbell *et al.* 1991; Hall and Soskice 2001; Hollingsworth 2000; Hollingsworth and Boyer 1997), especially the alternative non-market modes that exist: organizations, sometimes called hierarchies; associations; states; interorganizational networks; and clans or families (see Box IV.1). But even these distinctions are too simple, because each of these major modes can be subdivided. Besides competitive markets, there are oligopolistic markets, where hierarchies or organizations tend to dominate. Large organizations can be centralized or decentralized organizations, and thus the term 'hierarchy' can be misleading, with the former associated with low rates of innovation and the latter with high rates, including radical

---

BOX IV.1. A Typology of Coordination Modes (Bundles of Rules)

Distribution of Power Action Motivation	Horizontal		Vertical
Self-Interest	Markets		Hierarchy
		Interorganizational Networks	
Social Obligations	Associations Communities of practice		The State

Note: An adaptation of a figure developed by Hollingsworth & Boyer (1997: 9). This adaptation gives less emphasis to the overlap between horizontal and vertical for associations: that is, some associations can be quite vertical, or centralized.

innovations (Hage 1980). Similarly, there are different kinds of associations: states, interorganizational networks, and clans or families (Hollingsworth and Boyer 1997; Hage and Alter 1997; Kim 1998; Schneiberg and Hollingsworth 1990).

Arguing for the relevance of non-market modes of coordination does not imply that markets are irrelevant when one examines the presence of a hierarchy—or association—or state regulation. Quite the contrary. The issue is the supplementing of market mechanisms by non-market rules. For example, if an economic sector is dominated by a few large organizations, with one firm a price-leader, then prices are controlled in part by hierarchical coordination. The basic theoretical questions are: What is the division between market and non-market modes? In which sectors of the economy and of the society (of which there are many parts besides the economy, especially for the study of innovation) is this most significant? Most critically, which kind(s) of coordination modes or subcategories prevail?

The choice of this particular set of rules—coordination modes—is dictated by two criteria. First, the concept helps synthesize the three major disciplines of economics (markets), political science (the state), and sociology (organizations, associations, and interorganizational relationships). The idea of alternative modes of coordination, and their set of institutional rules (what they coordinate, how, and with what incentives or sanctions), also connects with the policy initiatives involving deregulation.

Second, non-market coordination modes appear to be more critical for the transfer of knowledge in the idea-innovation chain (Hage and Hollingsworth 2000), especially for the problem of accessing tacit knowledge. A similar set of issues is involved in the collective learning that must occur inside the organization. Again, some kinds of coordination modes might be preferable to others. The work of Nonaka and Takeuchi (1995) emphasizes the importance for collective learning of quality work circles, which in turn creates incremental innovation within the firm. And the contribution of Nonaka and Peltokorpi in Part I does the same for radical innovation.

As soon as we shift our attention to the problems of the transfer of ideas, tacit knowledge, and collective learning, we are also broadening the kinds of actors involved. It is not just economic (in the classic sense of the word) actors that are involved in the coordination modes, but also scientific, political, and educational actors, especially in the non-market modes. In this context, communities of practice, discussed in Part II, become particularly relevant.

Since the institutional rules involving coordination are so fundamental to this part of the book, we need to examine how useful Box IV.1 is for describing the differences among national systems of innovation. For example, associations in the Netherlands regulate the number of firms that can compete in a sector by controlling how many new firms can be created; associations in Germany are involved in the coordination of vocational and technical training programs (Streeck *et al.* 1987); clans have played a very important role in the economic history of Japan and, more recently, in South Korea (Whitley 1992); an interesting parallel to the role of clans is the kind of elite created by former students from the *grandes écoles* in France, who are engaged in informal coordination of many aspects of the economy; in contrast to this, and also to the special clan-like groups of interorganizational networks found in Japan (Gerlach 1992), the dominant economic coordination mode in the US has traditionally been the large corporation or hierarchy in many of the mass-production industries (Campbell *et al.* 1991).

We leave aside the historical question of why different modes of coordination develop to supplement or set limits on the market coordination of economic actors in specific countries; we ask the really critical one of what advantages (and disadvantages) they bring to innovation and, especially, radical innovation (see Whitley 1992). Further complicating this question is the variation by economic sector. For example, is state coordination the best for radical innovation in those industries, such as transportation, that require massive investment over long time periods? In contrast, is interorganizational group coordination better for industries that require continuous development of expensive, quality products such as automobiles and airplanes? Are hierarchies better for scientific research of commercial value where the science is relatively stable, such as pharmaceuticals?

Leaving aside these research agenda questions, we can now define a significant institutional change:

Significant institutional change = Discontinuous change in the coordination mode.

In other words, if interorganizational modes replace hierarchies, which appears to be the pattern in the US (see Hage's contribution), then this would be a significant change. Or if markets replace states in basic research, which appears to be the pattern in British scientific research (see the contribution by Georghiou in Part II), this would be a significant change. The experiments to replace the state with markets and democracy in the Eastern European countries are perhaps some of the most striking examples of attempts at discontinuous institutional change.

New research agendas and frameworks

The contribution by Hollingsworth provides a fundamental framework for studying how the national system of innovation impacts radical innovation in science. We have already noted the similarities and differences between this model and that of Jordan's in Part II. The key point is that this chapter starts with the basic thesis, common in institutional analysis, of path dependency, when any recognition of change is allowed. This contribution thus lays the groundwork for the first part of the relationship between innovation and institutional change, namely the delimiting of change and, hence, of innovation because of the presence of the national system of innovation.

The second part of the relationship, the impact of innovation on institutional change, including significant change defined as changes in the coordination mode, is explored in the next two contributions. In the contribution by van Waarden and Oosterwijk, the focus is on radical innovation in science and its impact. The definition of what is radical, however, is more stringent than the one used in either Hollingsworth or Jordan, since it focuses on paradigmatic shifts. The second contribution by Hage moves to the issue of how radical product and, especially, process innovation affect the shifts in coordination modes. Building upon the fundamental work of Hall and Soskice (2001), Hage demonstrates that certain kinds of coordination modes (hierarchies combined with markets) resist radical product and process changes and, because of this, the national system of innovation is altered. In contrast, other kinds of coordination modes that place an emphasis on collaboration or social obligations were more accepting, and their national systems of innovation have not changed as much.

Although both of these contributions do perceive the possibility for significant institutional change, the next two contributions that examine planned change have a more qualified perspective. Casper's contribution looks at planned government interventions to change the national system of innovation so as to have higher rates of innovation. It does not examine attempts to change the mode of coordination; however, it develops a nuanced view of how much governments are able to change, essentially suggesting that they can be more successful with increasing rates of incremental, but not of radical, innovation. Campbell moves beyond simple changes in the national system of innovation to the larger issue of changes in fundamental aspects of the society, such as movement towards the use of markets and the creation of democracy. Although he allows for the possibility of these fundamental changes occurring, he observes that the process involves more path dependency because of the use of *bricolage*. This represents another kind of qualification. On close inspection, the changes are less discontinuous than they seem.

Together, these five chapters provide a powerful framework for studying how institutions impact innovation and vice-versa; three of them (Hollingsworth, van Waarden and Oosterwijk, and Casper) focus on radical scientific innovation, while two of them examine radical industrial innovation (again, van Waarden and Oosterwijk, and Hage). A key point is that radical institutional change can be created by the absence of adoption of new industrial innovations, as indicated in the Hage contribution, as well as by the adoption of new industrial innovation, as indicated in the van Waarden and Oosterwijk chapter. Both the Campbell and the Casper contributions focus much more on the processes of planned institutional change, and suggest a much more complex perspective for studying path dependency and government interventions designed to create institutional change of one kind or another.

Notes

1. Although J. Rogers Hollingsworth does not have his name attached, it is readily apparent that his ideas have provided the foundation for this introduction. In addition, he read and commented upon it, making many helpful suggestions. I want to thank him for this effort.
2. The intellectual importance of institutional analysis is reflected by the creation in Europe of the new Association of Evolutionary Economics, and in the US by a new major section in the Professional Association of Sociology, Economic Sociology, as well as the establishment of a new interdisciplinary and international association, the Society for the Advancement of Socio-Economics.

References

Bell, D. (1973). *The Coming of Post-Industrial Society: A Venture in Social Forecasting*. New York: Basic Books.

Browning, L., Beyer, J., and Shetler, J. (1995). 'Building Cooperation in a Competitive Industry: SEMATEC and the Semiconductor Industry.' *Academy of Management Journal*, 38: 113–51.

Campbell, J., Hollingsworth, J. R., and Lindberg, L. (eds.) (1991). *Governance of the American Economy*. New York: Cambridge University Press.

Edquist, C. (ed.) (1997). *Systems of Innovation: Technologies, Institutions and Organizations*. London: Pinter Publishers/Cassell Academic.

—— and Hommen, L. (1999). 'Systems of Innovation: Theory and Policy for the Demand Side.' *Technology in Society*, 21: 63–79.

Freeman, C., and Soete, L. (1997). *The Economics of Innovation*. London: Pinter.

Gerlach, M. (1992). *Alliance Capitalism: The Social Organization of Japanese Business*. Berkeley, CA: University of California Press.

Gibbons, M., Limoges, C., Nowotny, H., Schwartzman, S., Scott, P., and Trow, P. (1994). *The New Production of Knowledge: The Dynamics of Science and Research in Contemporary Societies*. London: Sage.

Hage, J. (1980). *Theories of Organizations: Form, Process and Transformation*. New York: Wiley-Interscience.

—— and Alter, C. (1997). 'A Typology of Interorganizational Relationships and Networks.' In J. R. Hollingsworth and R. Boyer (eds.), *Contemporary Capitalism: The Embeddedness of Institutions*. Cambridge: Cambridge University Press, 94–126.

—— and Hollingsworth, R. (2000). 'Idea Innovation Networks: A Strategy for Integrating Organizational and Institutional Analysis.' *Organization Studies*, 21: 971–1004.

—— and Powers, C. H. (1992). *Post-Industrial Lives: Roles and Relationships in the 21st Century*. Newbury Park, CA: Sage.

Hall, P., and Soskice, D. (eds.) (2001). *Varieties of Capitalism*. Oxford: Oxford University Press.

Hollingsworth, J. R. (1997). 'Continuities and Changes in Social Systems of Production: The Cases of Japan, Germany, and the United States.' In J. R. Hollingsworth and R. Boyer (eds.), *Contemporary Capitalism: The Embeddedness of Institutions*. New York: Cambridge University Press, 265–310.

—— (2000). 'Doing Institutional Analysis: Implications for the Study of Innovation.' *Review of International Political Economy*, 7: 595–644.

—— and Boyer, R. (eds.) (1997). *Contemporary Capitalism: The Embeddedness of Institutions*. New York: Cambridge University Press.

Kim, L. (1998). 'Crisis Construction and Organizational Learning: Capability Building in Catching-Up at Hyundai Motor.' *Organization Science*, 9: 506–21.

Nelson, R. (ed.) (1993). *National Innovation Systems: A Comparative Analysis*. New York: Oxford University Press.

—— (1998). 'The Agenda for Growth Theory: A Different Point of View.' *Cambridge Journal of Economics*, 22/4: 497–520.

Nonaka, I., and Takeuchi, H. (1995). *The Knowledge-Creating Company*. Oxford: Oxford University Press.

North, D. C. (1990). *Institution, Institutional Change, and Economic Performance*. New York: Cambridge University Press.

OECD (2002). *Main Science and Technology Indicators, 2002/2*. Paris: OECD.

Pine, B. J. (1993). *Mass Customization: The New Frontier in Business Competition*. Boston, MA: Harvard Business School.

Piore, M., and Sabel, C. (1984). *The Second Industrial Divide: Possibilities for Prosperity*. New York: Basic Books.

Schneiberg, M., and Hollingsworth, J. R. (1990). 'Can Transaction Costs Economics Explain Trade Associations?' In M. Aoki, B. Gustafsson, and O. Williamson (eds.), *The Firm as a Nexus of Treaties*. Newbury Park, CA: Sage, 320–46.

Streeck, W., Josef, H., van Kevelaer, K.-H., Maier, F., and Weber, H. (1987). *Steuerung und Regulierung der berfulfichen Bildung: Die Rolle der Sozialpartner in der Ausbildung und beruflichen Weiterbildung in der BR Deutschland*. Berlin: Wissenschaftszentrum.

Toffler, A. (1981). *The Third Wave*. New York: Bantam.

Whitley, R. (1992). *Business Systems in East Asia: Firms, Markets and Societies*. London: Sage.

18 A Path-Dependent Perspective on Institutional and Organizational Factors Shaping Major Scientific Discoveries

J. Rogers Hollingsworth

Introduction

This chapter confronts several interrelated problems as to how the institutional environments of organizations influence their innovativeness. Using a path-dependent perspective, it addresses:

- how institutional environments influence organizational isomorphism within countries;
- how institutional environments influence both the founding of new kinds of organizations and the founding of radically new departments and divisions within existing organizations; and
- how organizational characteristics influence the making of major discoveries.

To confront these problems, I draw on some of the data from a study of 290 major discoveries (that is, radical innovations in basic biomedical science) which took place throughout the twentieth century in four countries (Britain, France, Germany, and the United States). The data relate to approximately 250 research organizations which varied in the number of major discoveries made, some having none. Because of limitations of space, most of this chapter focuses primarily on organizations in the US, but from time to time soft

comparisons will be made with the institutional environments and organizations in the other three countries. Even though the empirical research for this chapter pertains to radical innovations in the basic biomedical sciences, many of the chapter's generalizations also apply to radical innovations in other sectors, and to countries other than the four considered in this research. Two major arguments of the chapter are that:

- the path-dependent nature of the institutional make-up of societies influences variability across societies in the rate of major discoveries;
- the path-dependent culture and structure of individual research organizations influence which organizations are likely to have many, few, or no major discoveries.

Path dependency

The subject of path dependency is a tricky business. A strict determinist, who assumes that actors are totally determined by choices made in the past, or by the institutional and/or organizational environment in which they are embedded, is hard to find. If we were strict determinists, we could make confident predictions about the social world. Scholars who use the term 'path dependency' usually vary in the meaning they attach to it. At one extreme is the view that path dependency simply refers to the causal relevance of preceding events in some type of temporal sequence. For example, Sewell (1996) suggests that path dependency means 'that what happened at an earlier point in time will affect the possible outcomes of a sequence of events occurring at a later point in time.' In short, history matters: what actors do today is shaped by what they did yesterday (Pierson 2000: 252; Garud and Karnøe 2001). A different, narrower conceptualization of path dependency, but with a bit more rigor, is that offered by Margaret Levi (1997: 28): once a country, organization, or individual has 'started down a track, the costs of reversal are very high. There will be choice points, but the entrenchments of certain institutional arrangements obstruct an easy reversal of the initial choice.' As Paul Pierson (2000: 252) observes: 'The costs of exit—of switching to some previously plausible alternative—rise.' The farther along a path of developing a set of practices a society or an organization is, the more difficult it becomes to shift to alternative paths. As a result, extensive movement down a particular path, whether at the societal or organizations levels, often has a 'lock-in' effect (Arthur 1994). This is the way in which 'path dependency' is used here.

A critical issue in path dependency is in understanding how history matters. While there is considerable variation in the path-dependency literature, this chapter adopts the following perspectives:

- small events often have major consequences;
- specific courses of action, once introduced, are very difficult to reverse;
- there is a great deal of chance and contingency to the unfolding of history;
- the timing and sequence of events are very importing in shaping longer-term social processes and outcomes;
- path-dependent processes are multi-level in nature: institutional, organizational,

divisions, departments, and/or laboratories within organizations. All levels co-evolve, though components at lower levels have greater flexibility to maneuver than those at higher levels.

As Tilly (1984: 14) observes, *when* things happen in a sequence affects both how they happen and also the consequences of their occurrence (for the importance of the sequences of events, see Pierson 2000: 264; Grew 1978). In scientific organizations, where considerable emphasis is given to priority in the discovery process, being early in developing a novel technique, adopting a new type of instrumentation, or developing a new discipline may make a great deal of difference in shaping the status of an investigator or a laboratory, but adopting a process or instrument, or establishing a particular kind of discipline-based department at a much later date, may be of little consequence in the competitive discovery process.

When it comes to designing organizations, actors generally have no way of knowing a priori the consequences of their actions. Experienced and wise decision-makers are generally aware that they are gambling, that they may well be introducing components and processes which will later on prove to have undesirable consequences. Unfortunately, many social scientists who study social change tend to be excessively optimistic, rational, and functionalist in their approach to problems, and tend to exaggerate their ability to gauge the consequences of decisions by actors.

Most institutional and organizational change unfolds in processes which are somewhat blind and random (Baum and McKelvey 1999; Baum and Singh 1994). Societies that excel in being innovative in various sectors or spheres over extended periods of time do so because of their good fortune in having an institutional environment which offers them the capacity to perform well (see the discussions in Allen 2004; Hall and Soskice 2001; Hollingsworth *et al.* 1994; Hollingsworth and Boyer 1997; Hollingsworth 1997). At best, we can hope to discern retrospectively whether there are regularities in the way that history unfolds. Generally, we cannot predict what processes will definitely lead to particular outcomes but, as a result of appropriate research strategies, we hope to be able to specify those which are most likely *not* to lead to particular types of outcomes.

Institutional environments and their effects on organizations

The institutional environment of organizations provides them resources that often play a major role in shaping their behavior. How resources are allocated to organizations is inextricably bound up with the characteristics of institutional environments, as well as the relationship between organizations and their institutional environment (Aldrich 1979; Pfeffer 1981; Baum 1996; Hollingsworth *et al.* 2002). For purposes of this chapter, the analysis focuses on four aspects of institutional environments that externally constrain the behavior of research organizations. These are environmental or external control over

- the appointment of scientific personnel;
- whether or not a particular scientific discipline will exist within a research organization;
- the level of funding for the organizations; and
- the type of training needed for appointment in a research organization.

In the following analysis, societal or institutional environments are coded in terms of whether they are weak or strong.

In societies in which external controls over organizations are highly institutionalized and strong, there is less variation in the structure and behavior of research organizations. In such instances, the connectedness between research organizations and their institutional or external environments is so strong that research organizations have low autonomy to pursue independent strategies and goals. Conversely, the weaker the institutional envir-

onment in which research organizations are embedded, the greater the variation in the structure, behavior, and performance of research organizations. Where the institutional environments are more weakly developed, organizations generally have greater autonomy and flexibility to develop new knowledge and to be highly innovative. Hence, in societies where the institutional environments are most developed and rigid, there is less organizational autonomy and flexibility, there are fewer radical innovations in basic and applied science, and fewer fundamentally new products and completely new industrial sectors have emerged. In such environments, actors and organizations may not be so successful in making radical innovations; however, they are often quite successful in making incremental innovations and producing high-quality products (Hage and Hollingsworth 2000).

The data on the institutional environments of the four aforementioned countries suggest that there is a high degree of complementarity among the four concepts describing institutional environments: when one is weakly developed, the others tend to be weakly developed and vice versa. This perspective has led a number of analysts to emphasize the concept of institutional complementarity.[1]

Even though there are prototypes of strong and weak institutional environments, there can be exceptions to the way institutional environments affect types of organizations. In a weak institutional environment, which is the case with Britain, there are research establishments operated by governmental research councils or departments (the Agriculture and Food Research Council, Defence) which have had little choice about personnel, budget, or research programs. Most governmental research units in Britain have long been concentrated in a relatively small number of large organizations, have operated in a very bureaucratic manner, and have had a heavy direct-dependence on Whitehall (which determines personnel policies, research plans, and financial resources). In contrast, British universities historically have had much greater organizational autonomy and independence to shape their personnel and research policies (Ziman 1987: Chapter 2).

Institutional environments and organizational isomorphism

In weak institutional environments, there is likely to be much more heterogeneity in types of research organizations, and among organizations of the same type, than in strong institutional environments. Hence, in the United States, with a relatively weak institutional environment, there have long been many more different types of universities than has been the case in Germany, where universities have been embedded in a strong institutional environment, and are much more similar to one another. Thus, in the United States, there have been small, elite, private universities such as Rockefeller University, the California Institute of Technology, and Rice University; there have been medium-sized private universities, such as Johns Hopkins University, the University of Chicago, Vanderbilt University, Princeton; and there have been large private universities, such as Harvard, Stanford, MIT, NYU. In addition, there are the large public universities in California (Berkeley, UCLA, UCSD) and in the Midwest (Michigan, Indiana, Wisconsin, Illinois, Minnesota). Each of these kinds of universities is a distinct type of population, somewhat differentiated from the other types of research organizations, in part because their dominant competencies are not easily learned or transmitted across organizational populations (McKelvey 1982: 192; Aldrich, *et al.* 1984: 69).

Of course, in both strong and weak institutional environments every organization is unique, meaning that there is always heterogeneity within each type of organization. But organizations of the same type, and in the same institutional environment, are likely to share many of the same attributes. Even if weak institutional environments lead to more heterogeneity among types of organizations, there are forces at work that lead, over time, to organizational isomorphism both across and within organizational types. There are several bodies of literature which have provided empirical support to the idea of organizational isomorphism, even among different types of organizations in the same society. One is the varieties-of-capitalism literature (Hall and Soskice 2001; Crouch and Streeck 1997; Streeck and Yamamura 2001; Hollingsworth and Boyer 1997; Hollingsworth *et al.* 1994; Allen 2004). Another literature is that on the history of research organizations, particularly that involving universities in Britain, France, Germany, and the US. While organizational diversity persists within each of the four university systems, there nevertheless have been pressures toward organizational isomorphism. These pressures have been strongest in those countries (Germany and France) in which universities have been embedded in strong institutional environments (Clark 1993, 1995).

In addition, there is an empirical literature from the field of population ecology, though the theoretical basis for much of this literature is derived from evolutionary biology. For example, McKelvey (1982: chapter 7) argued that different populations of organizations within the same society have a set of competencies and routines which are societally specific, and as a result of these competencies, actors in both different and similar organizations engage in a great deal of common learning and socialization. Scientists, technicians, and administrators, even if from different types of organizations but in the same society, acquire a great deal of common organizational knowhow. DiMaggio and Powell (1983) some years ago picked up on these ideas when they pointed out that organizations engage in 'mimetic processes.' More recently Hodgson (2003) developed the argument that routines are organizational meta-habits which diffuse across populations of organizations within an institutional environment. As suggested above, a good bit of this insight was borrowed from evolutionary biologists (Mayr 1963, 2001) who have demonstrated that interbreeding and gene flow stabilize biological species. Picking up on ideas from biologists, Astley (1985), a population ecologist, did more than anyone else to establish clear linkages among the different literatures in biology, population ecology, and organizational isomorphism. And where there are high degrees of organizational isomorphism, organizations are not likely to diverge widely in their historical processes. In short, they are likely to share many of the same path-dependent processes.

Thus far I have raised several interrelated historical processes: how institutional environments relate to organizational isomorphism, path dependency, and innovativeness. The concept of path dependency keeps us mindful of the fact that the way things were organized yesterday—or last year, etc.—influences the way they are organized today. But institutional environments, organizations, and individual actors are always changing. The stronger the institutional environment, the greater the degree of organizational isomorphism, and the higher the degree of common path-dependent processes.

For example, throughout the twentieth century, research organizations in Germany were embedded in a relatively strong institutional environment, though the strength of that environment varied over time. Because German research organizations were embedded in a strong institutional environment, there was not as much diversity in types of research organizations as was the case in the United States, with its weak institutional environment. Within the German system are two distinctly different types of organizations—the German University and the Kaiser Wilhelm/Max Planck Institutes. But because they have been embedded in a very strong institutional environment, there have been many isomorphic pressures promoting common routines within the two types of organizations. For example, both have had somewhat authoritarian cultures, in contrast to the more egalitarian culture of American research organizations. Moreover, individual German universities and Max Planck Institutes have not had the same degree of autonomy and independence as the research organizations in the United States have had in their funding, the

criteria for appointment of senior scientists, and the development of new programs and disciplines. The individual Max Planck Institute is subjected to a complex set of bureaucratic procedures for the appointment of a new director that is quite unlike anything experienced by a research institute in the US. Moreover, German professors and the development of new disciplines and programs are subject to control by government ministers on a scale quite unlike anything in the US (Max-Planck-Gesellschaft 2003; Mayntz 2001; Ash 1997; Burchardt 1975).

But there are pressures toward organizational isomorphism even in weak institutional environments such as the United States. Moreover, within most organizations, irrespective of their institutional environments, there are pressures for differentiated internal divisions and departments to become somewhat isomorphic and to share common path-dependent processes. In short, a common organizational culture tends to become pervasive in most organizations: individuals in different departments of the same organization become socialized into common ways of addressing many problems. There are pressures both across and within organizations in the same society to emphasize homogeneous competencies. The pressures toward homogenization are especially strong when actors in highly saturated environments are competing for the same finite resources (Hawley 1950; McKelvey 1982).

Constraints on isomorphism

Isomorphism, no matter how powerful as a force, does not sweep unimpeded through history. There are counter-currents which place constraints on isomorphic tendencies. Many years ago, for example, Stinchcombe (1965) made the observation that organizations, even those of the same type, founded at different points in time, are likely to be imprinted with many of the cultural attributes of the social technologies current at the time of their creation. When Stinchcombe made his observation, social scientists had not yet explicitly developed the concept of path dependency, but his emphasis on how the history of organizations is permanently influenced by the moment of their founding is, of course, clearly suggestive of a path-dependency perspective at the level of organizations. In short, Stinchcombe was implicitly making the profound point that organizations do not necessarily track changes in their environment closely. Instead, they are somewhat inert, preserving certain non-adaptive qualities which often have deleterious effects on their capacity to be highly adaptive to their environments and to be innovative. Similarly, they offer resistance to isomorphic pressures.

There is substantial literature which suggests that continuous innovativeness in modern societies requires diversity in organizational forms, heterogeneity in organizational structures, and diversity in ideas (Garud and Karnøe 2001; Nooteboom 1999; Rizzello 1999; Rizzello and Turvani 2002). Thus, individual societies are constantly confronting contradictory pressures. They are subjected to processes which move organizational populations toward greater homogeneity and uniformity. Biologists and population ecologists alike have long realized that homeostatic forces within populations constrain evolutionary change, and thus preserve non-adaptive forms (Astley 1985: 229; Gould 1980; Mayr 2001; Baum and McKelvey 1999). But if a society is to be creative and innovative, it must have sustained variation and diversity in organizational forms and ideas. Most of the variation and diversity is shaped by path-dependent processes.

However great the force of path dependency at the institutional and organizational level, new organizational forms do emerge from time to time (Romanelli 1992). Indeed, the emergence of new organizational forms might be classified as a radical innovation. But even these evolve from processes which are path dependent in nature.

What are the conditions under which new organizational forms emerge? Unfortunately, we lack many of the theoretical tools to specify

when and where such innovations will occur. For theoretical insights into this problem, some of our best sources are the biologists who study the processes of speciation. We might think of the emergence of a new organizational form as a kind of organizational mutant. As Astley (1985: 232) reminded us, mutations occur all the time, among both biological and organizational species. However, most do not take hold since they are crowded out, are outnumbered in their population environments, and 'rapidly dissipate through the normal intermixing process' (Mayr 1963, 2001). Indeed, we know from numerous population-ecology studies that new organizations have low survival rates (Hannan and Freeman 1984, 1989). *Ipso facto*, they have little path dependency.

Thought of as a mutation, a newly emerged organizational form is more likely to survive if it occurs in environments that are sparsely populated but that have ample resources for the new type to develop; it is not crowded out by the more normal process of intermingling with other organizations. In such cases, organizational speciation has taken place. In the short term, a new form may be immune to the pressures of organizational isomorphism. In other words, environments with resources in excess of demand offer a greater opportunity for a new organizational form to survive than is the case in more competitively saturated environments (McKelvey 1982).

Using path dependency to understand the making of major discoveries[2]

Historically and geographically, those Western industrialized societies that have weak institutional environments have had more different types of organizations and lower levels of organizational isomorphism primarily because they have had environments which were not so highly saturated relative to the demand for resources. The United States was such a society during most of the twentieth century, and for that reason it was possible for new organizational forms to emerge in its research sector: private research institutes, research-oriented medical centers, small universities oriented toward research, even federally owned and operated research centers. Private research institutes such as the Rockefeller Institute for Medical Research (now, since 1964, Rockefeller University), the Salk Institute, the Carnegie Institution, and the Scripps Research Institute came into existence. The creation of the Johns Hopkins University Medical School has been much described, representing, as it did, the inauguration of a medical school which would engage in serious basic science (Hollingsworth 1986). The establishment and growth of the campus of the National Institutes of Health in Bethesda, Maryland, as a governmentally operated research institute, is another example. In short, the institutional and resource environment in the United States during the twentieth century facilitated the emergence of new and diverse forms of research organizations.

Several key factors are important for understanding why the United States had an impressive record in making major discoveries in biomedical research across much of the twentieth century. With its weak institutional environment and its abundance of resources, the United States had the conditions which made it possible for new types of organization to emerge and could then quickly adapt to the latest scientific knowledge, often to become the pace-setter in new fields of science. This pattern of the emergence of new types of organizations, able to incorporate the latest trends in science quickly, is consistent with Stinchcombe's argument (1965) about the founding and imprinting of organizations: because new organizations lack the inertia of older ones, and all other things being equal, they have greater capacity to be innovative.

Critical to our work is the definition of a major discovery. A major breakthrough or discovery is a finding or process, often preceded by numerous small advances, which leads to a new way of thinking about a problem. This new way of thinking is highly useful to numerous scientists in addressing problems in diverse fields of science. This is very different from the rare paradigm shifts analyzed by Thomas Kuhn in *The Structure of Scientific Revolutions* (1962). Major breakthroughs about problems in basic biomedical science occur within the paradigms about which Kuhn wrote. Historically, a major breakthrough in biomedical science was a radical or new idea, the development of a new methodology, or a new instrument or invention. It usually did not occur all at once, but involved a process of investigation taking place over a substantial period of time and required a great deal of tacit and/or local knowledge. My colleagues and I have chosen to depend on the scientific community to operationalize this definition, counting as major discoveries those bodies of research that have at least one of the ten criteria listed in Box 18.1.

Previous literature has not provided the theoretical tools to understand what are the particular organizational environments which facilitate major scientific discoveries, or how types of organizations, or the structures and cultures of individual organizations are associated with the making of major discoveries.[3] It is these issues that are addressed below.

As a result of an in-depth, cross-national, and cross-temporal organizational study of 290 major discoveries in Britain, France, Germany, and the United States, my colleagues (Jerald Hage and Ellen Jane Hollingsworth) and I have learned that major discoveries tend to occur in organizational contexts which have the characteristics described in Box 18.2 and Figure 18.1. The organizational contexts associated with major discoveries may exist in different types of organizations.

The few organizations where major breakthroughs occurred again and again were relatively small; they had high autonomy, flexibility, and the capacity to adapt rapidly to the fast pace of the change taking place in the global environment of science. Such organizations tended to have moderately high levels of scientific diversity and internal structures which facilitated the communication and

BOX 18.1. Indicators of major discoveries

1. Discoveries resulting in the Copley Medal, awarded since 1901 by the Royal Society of London, insofar as the award was for basic biomedical research.
2. Discoveries resulting in a Nobel Prize in Physiology or Medicine since the first award in 1901.
3. Discoveries resulting in a Nobel Prize in Chemistry since the first award in 1901, insofar as the research had high relevance to biomedical science.
4. Discoveries resulting in ten nominations in any three years prior to 1940 for a Nobel Prize in Physiology or Medicine.[a]
5. Discoveries resulting in ten nominations in any three years prior to 1940 for a Nobel Prize in Chemistry if the research had high relevance to biomedical science.[a]
6. Discoveries identified as prizeworthy for the Nobel Prize in Physiology or Medicine by the Karolinska Institute committee to study major discoveries and to propose Nobel Prize winners.[a]
7. Discoveries identified as prizeworthy for the Nobel Prize in Chemistry by the Royal Swedish Academy of Sciences committee to study major discoveries and to propose Nobel Prize winners.[a] These prizeworthy discoveries were included if the research had high relevance to biomedical science.
8. Discoveries resulting in the Arthur and Mary Lasker Prize for basic biomedical science.
9. Discoveries resulting in the Louisa Gross Horwitz Prize in basic biomedical science.
10. Discoveries in biomedical science resulting in the Crafoord Prize, awarded by the Royal Swedish Academy of Sciences, if the discovery had high relevance to the biological sciences.

[a]I have had access to the Nobel Archives for the Physiology or Medicine Prize at the Karolinska Institute and to the Archives at the Royal Swedish Academy of Sciences in Stockholm for period from 1901 to 1940. I am most grateful to Ragnar Björk, who did most of the research in the Karolinska Institute's archives to identify major discoveries according to the indicators in this Box. Because the archives have been closed for the past fifty years for reasons of confidentiality, I have used other prizes (Lasker, Horwitz, Crafoord) to identify major discoveries in the last several decades.

integration of ideas across diverse scientific fields. Moreover, these organizations tended to have scientific leaders with a keen scientific vision of the direction in which new fields in science were tending, and the capacity to develop a strategy for recruiting scientists capable of moving a research agenda in that direction. Internationally, most organizations having this kind of flexibility and autonomy in strategy have tended to be located in weak institutional environments.

To provide some sense of the path dependency of research organizations, I focus briefly on the distinctive culture of the Rockefeller Institute. Applying the criteria listed in Box 18.1, scientists in this very small organization made more major discoveries in basic biomedical science than in any other organization in

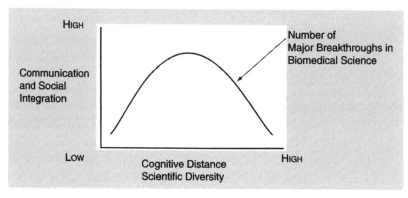

Fig. 18.1. The relationships among scientific diversity, communication/integration, and making major discoveries

---

BOX 18.2. Organizational contexts facilitating the making of major discoveries[a]

1. Moderately high scientific diversity. This existed when the organizational context had (a) a variety of biological disciplines and medical specialties and sub-specialties, and (b) numerous people in the biological sciences with research experience in different disciplines and/or paradigms. Scientific diversity exerted maximum beneficial effect when the organizational context had high depth (that is, individuals highly competent in different task areas—theoreticians, methodologists, scientists highly conversant with literature in various fields, scientists highly competent in the latest instrumentation in diverse fields).
2. Communication and social integration among the scientific community. This was the bringing together of scientists from different scientific fields through *frequent* and *intense* interaction in the following types of collective activities: (a) joint publication, (b) journal clubs and seminars, (c) team teaching, (d) meals and other informal activities.
3. Organizational leadership with the capacity to understand the direction in which scientific research was moving and the ability to integrate scientific diversity. These activities were both task oriented and socio-emotional in nature, and applied to organizational leaders who had (a) strategic vision for integrating diverse areas and for providing focused research, (b) the ability to secure funding for these activities, (c) the capacity to recruit individuals who would confront not only important scientific problems but ones which could be solved, and (d) the capability to provide rigorous criticism in a nurturing environment.
4. Recruitment. Organizational capacity to recruit individuals who internalized a moderate degree of scientific diversity.
5. Organizational autonomy and organizational flexibility. Organizational autonomy was the degree to which the organizational context where the research took place was relatively independent of its institutional environment, and organizational flexibility was the ability of the organizational context to shift rapidly from one area of science to another. To attain organizational autonomy and flexibility, it was necessary that the organizational context be loosely coupled to its institutional environment if the organizational context were an entire organization; but if the organizational context were a sub-part of a larger organization, it could attain flexibility and autonomy only if it were loosely coupled both to the larger organization and the institutional environment in which it was embedded.

[a] Terry Shinn in his chapter for this volume also addresses the concepts of differentiation/diversity and integration in facilitating scientific innovations. However, Shinn is focusing on differentiation and integration at the societal level, while this chapter is using these concepts at the level of organizations and laboratories. Despite the different levels of analysis, there is some complementarity between the two chapters.

---

the twentieth century—more than all the Kaiser Wilhelm and Max Planck Institutes combined. The variables listed in Box 18.2 and Figure 18.1 have special relevance to the Rockefeller. First is its very small size throughout its history. Second, it has not had academic departments and disciplines as we know them in the large American research universities. It has been structured around laboratories, and when the head of a laboratory retired, died, or left, the laboratory was closed; this provided the organization with the opportunity to stand back and assess what to do next. This capability provided the organization with an enormous amount of flexibility to adapt to the rapidly changing larger world of science (Hollingsworth 2002, 2003).

Most research universities are structured around departments and academic disciplines:

for that reason, they lack organizational flexibility and acquire a great deal of organizational inertia. The Rockefeller organization has always had a great deal of scientific diversity, but in contrast to universities' differentiation of diversity into departments and subspecialties, and unlike Max Planck Institutes, structured very much around a single area of research or discipline, the Rockefeller organization had a path-dependent tendency to have a great deal of scientific integration. The mechanisms for integrating diversity in organizations are present in different variations in organizations, but the emphasis here is on integration—on communication across different fields—and this can take place in a variety of ways in different organizations. During the Rockefeller's first sixty years, much of the scientific integration took place in the lunchroom. The idea was to have a fairly good lunch at tables seating generally no more than eight people, where scientists could have a single conversation about a serious problem. This took place day in and day out, with very eminent people on hand. Foreign scientists coming to America generally arrived in New York, and many of the most distinguished visited the Rockefeller organization. This added to a very exciting environment. And the lunchroom, lectures, and afternoon tea did a great deal to promote and facilitate the integration of what I call 'scientific diversity'.

For many years, Rockefeller had leaders who had a good sense of the direction in which science was moving, leaders who had an extraordinary ability to recruit people who internalized a scientific diversity and who could lead the organization in the direction in which science was moving. Finally, they had leaders who were willing to take risks. When the Institute was established, John D. Rockefeller, Sr. informed the leaders within the Institute that it would not matter if they never discovered anything of great importance. He simply wanted the Institute to do the best it could—creating an invigorating and nurturing environment, doing its best to advance the understanding of nature.

One of the things worth observing about the path-dependent culture at the Rockefeller was the development of its young scientists, a subject often overlooked. Rockefeller had more major breakthroughs in biomedical science in the twentieth century than any other organization in the world; when we focus just on the Nobel prizes that were awarded to Rockefeller scientists, we see the large number that were awarded to people who went there as very young scientists, and who made their careers there in its extraordinarily nurturing environment, where they did not have to apply for research grants, where people were encouraged to engage in high-risk research. In short, the Institute 'grew' many of their most creative scientists. Note the names of those who went there as very young scientists and eventually were awarded Nobel prizes for work they did there: Peyton Rous, Albert Claude (one of the most important people in the development of cell biology), George Palade, Wendell Stanley, John Northrop, Gerald Edelman, William Stein, Stanford Moore, Bruce Merrifield, Gunther Blobel, and Rod MacKinnon. The number of young people who went there and were ultimately awarded Nobel prizes is greater than the combined number of all Nobel prizes awarded to Harvard (or to Cambridge, UK) scientists for work accomplished there in the basic biomedical sciences. A number of other Nobel laureate scientists did their work both there and elsewhere (e.g. Karl Landsteiner, Haldan K. Hartline). But what is especially impressive is the culture in which young people were able to mature and become some of the world's most creative scientists.

On the other hand, as suggested above, there is in most societies a great deal of organizational isomorphism, even in such weak institutional environments as the United States. And most large universities in the United States, as well as in the other three countries, have tended to have the characteristics described in Box 18.3. They have been differentiated into large numbers of scientific disciplines, have had relatively little communication across scientific disciplines, and tended to have less autonomy and flexibility to adapt to the fast pace of scientific change than is the case with those organizations having the characteristics described in Box 18.2 and Figure 18.1.

Why do those organizations able to facilitate communication across diverse fields and, thus, to integrate scientific diversity, have an advantage in making major discoveries over those which have a low capacity for such communication and integration? In our study of 290 major discoveries, every single one reflected a great deal of scientific diversity. Of course, very good science can occur in those organizational environments where there is little connection across disciplines and sub-specialties, and which are highly specialized within a very narrow field. But the science which is produced in such narrow and specialized environments reflects insufficient diversity for it to be recognized as a major discovery by the scientific

---

BOX 18.3. Organizational contexts constraining the making of major discoveries

1. Differentiation. Organizations were highly differentiated internally when they had sharp boundaries among subunits such as (*a*) basic biomedical departments and other subunits, (*b*) the delegation of recruitment exclusively to departments or other subunit level, (*c*) the delegation of responsibility for extramural funding to the department or other subunit level.
2. Hierarchical authority. Organizations were coded as being very hierarchical when they experienced (*a*) centralized decision-making about research programs, (*b*) centralized decision-making about number of personnel, (*c*) centralized control over work conditions, (*d*) centralized budgetary control.
3. Bureaucratic coordination. Organizations which had high standardization of rules and procedures.
4. Hyperdiversity. This was the presence of diversity to such a deleterious degree that there could not be effective communication among actors in different fields of science or even in similar fields.

---

community, with its vast varieties of different disciplines.

Still, major breakthroughs do not only occur in those organizational environments which are small, and internally undifferentiated into departments or divisions. So how is it that major discoveries can also occur in large organizations which are internally differentiated into separate departments? First, clusters of discoveries might be explained by the rare conditions under which a 'mutant' department or division emerges and performs extraordinarily well for a relatively short period of time. Second, breakthroughs can occur in the type of organizational context described in Box 18.3, but only if the laboratory is structured quite differently from most other laboratories in Box 18.3-type organizational contexts (see Figure 18.2). In other words, the lab is headed by a scientist operating in an organizational environment which generally would not be expected to have a major discovery.

Isomorphism within organizations

In most societies, regardless of whether institutional environments are weak or strong, there are pressures for cultural homogencity and organizational isomorphism among units within organizations. However, at certain moments in time, there are exceptions to this generalization. In those organizations in which there is very little centralized control, where internal units have high levels of autonomy and good access to human, physical, and

financial resources, there is the potential that a fundamentally new discipline or scientific program could emerge in a sub-part, and which could be incorporated into a departmental structure. I equate this type of radical innovation as being a type of organizational mutation. Of course, universities are constantly establishing new departments or appointing someone with a new scientific agenda within an existing department. But when a fundamentally new—by world standards—discipline or program emerges within a particular university, this is indeed a very radical innovation. And just as we lack the theoretical tools to predict where and when a new kind of organization will emerge, neither can we predict where and when within an existing research organization a radically new program, discipline, or paradigm will emerge. However, the sociological conditions for such an emergence are somewhat similar to those under which new organizational types will emerge. The following two conditions must exist:

- the organization must be extremely decentralized (permitting the actors creating the radical innovation to have high autonomy), and
- the actors within the organization must have access to sufficient diverse types of resources so that their scientific practices and administrative routines are not crowded out by those which might already have become institutionalized within the larger environment of the host organization.

According to evolutionary logic, those in the new field must be able to escape the

homogenizing pressures in the existing organizational environment and be able to intermix,
interbreed, and reproduce their own progeny.
In short, such organizational mutations within
sub-parts of a research organization will occur
only under specific conditions.

These types of radical innovations are, of
course, very rare events. The following are a
few examples. One occurred when the University of Cambridge established its Department
of Physiology in the late nineteenth century.

Another occurred there a few years later with
the emergence of the Cambridge Department
of Biochemistry. Later, also at Cambridge, a
new research paradigm occurred in biology,
but in the Cavendish Laboratory (a physics
department) (Needham and Baldwin 1949; Hollingsworth forthcoming 2005; de Chadarevian
2002; Geison 1978). In each of these departments, a number of major discoveries emerged
in the basic biological sciences within a relatively short period of time. At Harvard, in the

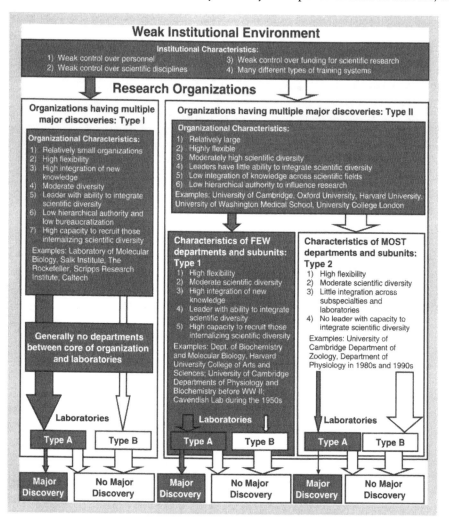

Fig. 18.2. Multi-level analysis of major discoveries

Panel one *

Note: The width of the arrows indicates the relative frequency of the specified outcome. Characteristics
in gray tend to be more associated with making major discoveries.

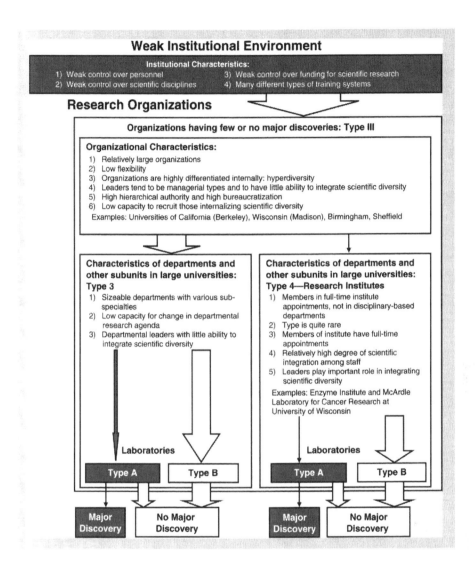

Fig. 18.2. *Continued*

Panel Two*
* **Type A Laboratories** have the following characteristics: **(1) Cognitive:** High scientific diversity; **(2) Social:** Well connected to invisible colleges (e.g. networks) in diverse fields; **(3) Material Resources:** Access to new instrumentation and funding for high-risk research; **(4) Personality of Laboratory Head:** High cognitive complexity, high confidence and motivation; **(5) Leadership:** Excellent grasp of ways scientific fields might be integrated and ability to move research in that direction.
* **Type B Laboratories** have the following characteristics: **(1) Cognitive:** Moderately low scientific diversity; **(2) Social:** Well connected to invisible colleges (e.g. networks) in a single discipline; **(3) Material Resources:** Limited funding for high-risk research; **(4) Personality of Laboratory Head:** Lack of high cognitive complexity, limited inclination to conduct high-risk research; **(5) Leadership:** Not greatly concerned with integrating scientific fields.

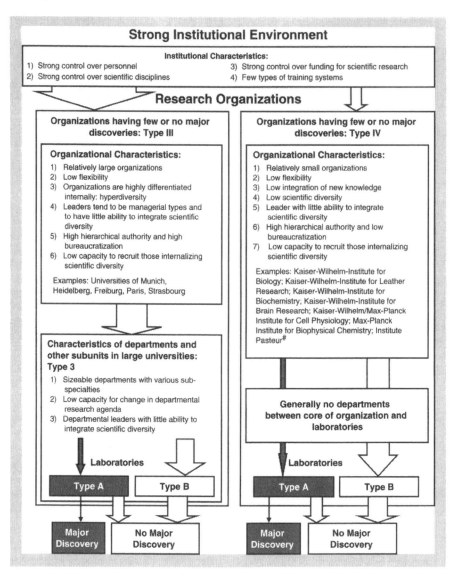

Fig. 18.2. *Continued*

Panel three

* See note and definitions on Panels One and Two.

Institut Pasteur is a bit of an anomaly within this grouping. For most of its history, it has had relatively few major discoveries. But in the first two decades of the twentieth century, and again during the late 1950s and into the 1960s, there were a number of major discoveries there. However, during the years when it had stronger connections with its strong institutional environment, it had fewer major discoveries (Hage 1998).

period between the mid-1950s and the mid-1970s, a similar innovation occurred with the establishment of two new departments: the Department of Biochemistry and Molecular Biology, and the Department of Organismic and Evolutionary Biology; again, each of these departments had a number of major discoveries (Hollingsworth et al. 2005).

Over time, however, departments have institutionalized routines, as do universities, and inertial processes set in, making it difficult for the new sub-part to continue being as innovative on the scientific world stage. The level of innovativeness of the new department eventually declines. Over time, even organizations which at one time were highly decentralized with high autonomy for each separate unit are likely to institutionalize a set of routines which slowly establish interlocking, sequential, and conditional behaviors among all of its various sub-parts and their members. Eventually, these routines establish collective capabilities and capacities which lead to the emergence of shared behavior throughout the organization (Hodgson 2003: 376).

This kind of historical process occurred in the Departments of Physiology and Biochemistry at Cambridge, and in the Departments of Biochemistry and Molecular Biology and Organismic and Evolutionary Biology at Harvard. At the Cavendish Lab, there were such strong organizational pressures for the laboratory to confine itself to the mission of physics research that the molecular biologists were strongly encouraged to leave the University of Cambridge. The biologists in the Cavendish Laboratory (for example, Francis Crick, Fred Sanger, Max Perutz, John Kendrew) were doing some of the most novel biological science of the entire twentieth century. However, they worked outside the disciplinary frameworks of existing Cambridge biological departments, and the pressures of organizational isomorphism were so great that the group, with funding from the Medical Research Council, left the university and moved to the suburbs of Cambridge, where the Laboratory of Molecular Biology (LMB) was established. The LMB eventually became one of the world's leading research centers in basic biomedical science in the latter part of the twentieth century.

Initially each of these Cambridge and Harvard departments had an outstanding leader and considerable scientific diversity, which was highly integrated—in short, the characteristics listed in Box 18.2. Even though each individual scientist tended to pursue a separate body of research, it was highly complementary to that of the research program in the entire department, which had a distinctive culture, the glue which held it together.

But eventually, for reasons which were common to all, the distinctive scientific excellence of these departments declined. Over time, the scientific agenda of the new department tended to diffuse to other organizations throughout the world; many of the members of the department either retired, died, or left the organization; scientific practices became routinized, but no new leader emerged with a radically new agenda, capable of transforming the department to being once again at the cutting edge of science; the routines of the larger organization in which the department was embedded slowly began to penetrate the department, leading to isomorphic administrative routines and practices throughout the organization. For all of these reasons, it is difficult for a research department to remain at the cutting edge of research for more than two or three decades. It may be possible for a new department with a new agenda to emerge within another part of the same organization or in a sub-part of another large organization.

These outstanding departments were very rare events—the equivalent of within-organization mutations which unpredictably were able to 'take hold'. But over the longer term the distinctiveness of the 'new species' diminishes as it interacts with the rest of the organization.

Path dependency within an organization

Institutional environments place constraints on the behavior of organizations because organizations are embedded in institutional environments, which are path dependent over long periods of time. Organizations also have path-dependent processes, but several forces operate to alter the structure and culture of research organizations: over time, because of the way the research organizations interact with each other, isomorphic pressures narrow the range of variation in their behavior and culture; and, over time, the historical record demonstrates that the institutional environments in which research organizations are embedded tend to change. Weak institutional environments—such as those in the US and Britain—have become stronger as the central governments have become more involved in funding research. Hence, the trend toward stronger institutional environments has also tended to generate greater isomorphic pressures among organizations.

Thus far, much of the discussion has focused primarily on the institutional environment of research organizations. But when we think of path-dependent processes, we also must be attentive to these processes within organizations. It is within the research organization where research and major discoveries occur. Figure 18.2 suggests the path-dependent processes which occur among institutional environments, organizations, and laboratories. The figure suggests that, in the basic biomedical sciences, there are two general types of laboratories in research organizations. Those in which major discoveries may be made are called, for simplicity, Type A, and their characteristics are:

- having a moderately high level of scientific diversity (i.e. not highly specialized);
- being well connected to invisible colleges in multiple fields of science;
- having access to instrumentation and funding for high-risk research;
- having laboratory heads who internalize high cognitive complexity, have a good grasp of the direction in which the science is moving, and a good sense of how different scientific fields might be integrated in order to move research in a chosen direction.

As Figure 18.2 suggests, laboratories can have all of these characteristics and yet have no major discoveries. In other words, a laboratory could be in an organization with characteristics associated with major discoveries, and the laboratory could have the general structural and cultural characteristics associated with major discoveries, but have no major discovery. There is a certain amount of chance and luck in the making of major discoveries (Jacob 1995; Edelman 1994: 980–6). But virtually every laboratory in our study of 290 major discoveries tended to have characteristics similar to those listed for Type A. Moreover, the organizational environments with characteristics similar to those listed in Box 18.2 were more likely to have a number of Type A laboratories.

Type B laboratories are at the opposite end of the continuum on virtually all the laboratory characteristics listed above in that they:

- have little scientific diversity;
- are well-connected to invisible colleges in a single discipline;
- have limited funding for high-risk research; and
- have laboratory heads with low levels of cognitive complexity, a tendency to avoid high-risk research, and little concern with integrating different scientific fields.

Type B laboratories hardly ever have a major discovery (as identified by the criteria in Box 18.1). As Figure 18.2 demonstrates, Type B laboratories may exist in almost any kind of research organization, but they are very common in large, highly differentiated organizations having hyper-scientific diversity.

Concluding observations

Institutions, research organizations, and their component parts co-evolve, moving along a historical trajectory which is path dependent. Even though this trajectory is important for understanding research organizations and the innovations which take place in them, this is not to suggest that there is some kind of historical determinism. Despite the fact that actors are very much constrained by their environment, those in weak institutional environments have a great deal of latitude in shaping their scientific agenda.

The importance of path dependency for understanding social processes was nicely phrased by Paul David, who is frequently credited as being one of the first to use the concept. 'It is sometimes not possible to uncover the logic (or illogic) of the world around us except by understanding how it got that way.' For David, a path-dependent sequence of events was one in which 'important influences upon the eventual outcome can be exerted by temporally remote events, including happenings dominated by chance events' (David 1985: 332; Rycroft and Kash 2002: 21–2). But as David and others (Arthur 1994; Rycroft and Kash 2002) have pointed out, small events may have modest, but lasting and important, effects, and at other times they have major consequences. Path-dependent processes tend to have both direct and indirect effects on innovativeness.

As suggested above, long-term changes in science involve path-dependent processes at multiple levels: at the macro-institutional level (the society), the meso-level (the organization), and the micro-level (the laboratory). However, these different levels are intertwined in such a way that they are part of a system with complementary parts, integrated into a social system with its own logic. Every American university and research organization is unique, with its own distinctive culture; it is also distinctly American, and one needs only the shortest of visits to a US research organization to tell from the behavior of actors (language aside) that one is not in a French or German research organization. In other words, system interdependency emerges from co-evolutionary processes which are societally specific.

Despite the path-dependent processes operating in the American science system, it is important to make several additional observations.

First, although there is a tendency for weak institutional environments to persist across time, changes in those environments are constantly occurring. In the case of the United States, the institutional environment has become somewhat stronger over time, and this alteration has increased the isomorphism among research organizations. This has also been the case in Britain, as research organizations have become increasingly dependent on funds from the central government and thus subject to governmental directives.

Second, organizational cultures and structures have a remarkable degree of stability. Hence organizational contexts with the characteristics described in Boxes 18.2 and 18.3 continue over long periods. In short, there is a high degree of organizational path dependency. Figure 18.2 suggests that the extent to which organizations make several, few, or no major discoveries over time has a distinct pattern. Even so, the structures of laboratories within organizations are somewhat indeterminate. An organization with the characteristics in Box 18.3 may have a Type A laboratory, and even an occasional scientist who makes a major discovery. But there is little likelihood that there will be multiple discoveries in such an organization. Organizations described in terms of Box 18.2 variables have more Type A laboratories, and are more likely to have multiple major discoveries. But even this type of organization may have Type B laboratories. (Type B laboratories, as previously noted, are unlikely to be places with major discoveries.)

The above analysis points out that innovations at the level of major discoveries are rare events. We cannot predict where and when they will occur. However, using path dependency, characteristics of institutional environments, organizational isomorphism, and resistance to isomorphic pressures, we can begin to address the circumstances under which major discoveries are most and least likely to occur.

Notes

I would very much like to acknowledge the help of David Gear, Jerry Hage, and Ellen Jane Hollingsworth. They have played a major role in the development of the ideas, as well as in collecting and analyzing much of the data for this chapter. David Gear, Ellen Jane Hollingsworth, and Steve Casper provided very detailed comments on an earlier draft of the chapter which were very helpful in revising.

1. On the concept of institutional complementarity, see Amable 2000; Hall and Soskice 2001; Crouch 2004; Boyer 2004; Hollingsworth and Gear 2004.

2. The research project on major discoveries summarized herein is based on a great deal of archival research, many interviews, and wide reading in many scientific fields. Archives have been used in the United States (e.g. Rockefeller Archive Center, American Philosophical Society, University of Wisconsin, Caltech, University of California Berkeley, University of California San Francisco, University of California San Diego, Harvard Medical School) and in Great Britain and Europe. I have conducted more than 400 interviews with scientists on both sides of the Atlantic as part of this research.

3. One scholar who did address some of these issues in a way quite different from that developed in this chapter was Joseph Ben-David (Ben-David 1991). Indeed, I am very much indebted intellectually to Ben-David.

References

Aldrich, H. (1979). *Organization and Environment*. New York: Prentice-Hall.

—— McKelvey, B., and Ulrich, D. (1984). 'Design Strategy from the Population Perspective.' *Journal of Management*, 10: 67–86.

Allen, M. M. C. (2004). 'The Varieties of Capitalism Paradigm: Not Enough Variety?' *Socio-Economic Review*, 2: 87–108.

Amable, B. (2000). 'Institutional Complementarity and Diversity of Social Systems of Innovation and Production.' *Review of International Political Economy*, 7/4: 645–87.

Arthur, B. (1994). *Increasing Returns and Path Dependence in the Economy*. Ann Arbor: University of Michigan Press.

Ash, M. G. (ed.) (1997). *German Universities Past and Future*. Providence, RI: Berghahn.

Astley, W. G. (1985). 'The Two Ecologies: Population and Community Perspectives on Organizational Evolution.' *Administrative Science Quarterly*, 30: 224–41.

Baum, J. A. C. (1996). 'Organization Ecology.' In C. Hardy and W. Herds (eds.), *The Handbook of Organizational Studies*. London: Sage, 77–114.

—— and McKelvey, B. (eds.) (1999). *Variations in Organization Science: In Honor of Donald T. Campbell*. Thousand Oaks, CA: Sage Publications.

—— and Singh J. V. (eds.) (1994). *Evolutionary Dynamics of Organizations*. Oxford: Oxford University Press.

Ben-David, J. (1991). *Scientific Growth: Essays on the Social Organization and Ethos of Science*, edited and with an introduction by G. Freudenthal. Berkeley, CA: University of California Press.

Boyer, R. (2004). 'Institutional Complementarity: Concepts, Origins, and Methods.' Unpublished paper presented at 16th Annual Meeting of the Society for the Advancement of Socio-Economics, George Washington University, Washington, DC, 10 July 2004.

Burchardt, L. (1975). *Wissenschaftspolitik im Wilhelminischen Deutschland*. Göttingen: Vandenhoeck and Ruprecht.

Clark, B. R. (ed.) (1993). *The Research Foundations of Graduate Education: Germany, Britain, France, United States, Japan*. Berkeley: University of California Press.

—— (1995). *Places of Inquiry: Research and Advanced Education in Modern Universities*. Berkeley: University of California Press.

Crouch, C. (2004). 'Complementarity and Innovation.' Paper prepared for Workshop on Complementarity at Max Plank Institute for Study of Societies, Cologne, Germany, 26–7 March 2004.

—— and Streeck, W. (eds.) (1997). *The Political Economy of Modern Capitalism*. London: Sage Publications.

David, P. A. (1985). 'Clio and the Economics of QWERTY.' *American Economic Review*, 76: 332–7.

de Chadarevian, S. (2002). *Designs for Life: Molecular Biology after World War II*. Cambridge: Cambridge University Press.

DiMaggio, P., and Powell, W. W. (1983). 'The Iron Cage Revisited: Institutional Isomorphism and Collective Rationality in Organizational Fields.' *American Sociological Review*, 48: 147–60.

Edelman, G. (1994). 'The Evolution of Somatic Selection: The Antibody Tale.' *Genetics*, 138: 975–81.

Garud, R., and Karnøe P. (eds.) (2001). *Path Dependence and Creation*. Mahway, NJ: Lawrence Erlbaum Associates.

Geison, G. L. (1978). *Michael Foster and the Cambridge School of Physiology: The Scientific Enterprise in Late Victorian Society*. Princeton: Princeton University Press.

Gould, S. J. (1980). 'Is a New and General Theory of Evolution Emerging?' *Paleobiology*, 6: 119–30.

Grew, R. (ed.) (1978). *Crises of Political Development in Europe and the United States*. Princeton: Princeton University Press.

Hage, J. (1998). 'Rise and Sink and Rise Again: Biomedical Science in the Institut Pasteur.' Paper presented before World Conference of the Society for the Advancement of Socio-Economics, Vienna, Austria, 15 July 1998.

—— and Hollingsworth, J. R. (2000). 'Idea Innovation Networks: A Strategy for Integrating Organizational and Institutional Analysis.' *Organization Studies*, 21: 971–1004.

Hall, P. A., and Soskice, D. (2001). *Varieties of Capitalism: The Institutional Foundations of Comparative Advantage*. Oxford: Oxford University Press.

Hannan, M., and Freeman, J. (1977). 'The Population Ecology of Organizations.' *American Journal of Sociology*, 82: 929–64.

—— —— (1984). 'Structural Inertia and Organizational Change.' *American Sociological Review*, 29: 149–64.

—— —— (eds.) (1989). *Organizational Ecology*. Cambridge: Harvard University Press.

Hawley, A. (1950). *A Theory of Community Structure*. New York: Ronald Press.

Hodgson, G. (2003). 'The Mystery of Routine: The Darwinian Destiny of an Evolutionary Theory of Economic Change.' *Revue Économique*, 34: 355–84.

Hollingsworth, J. R. (1986). *A Political Economy of Medicine: Great Britain and the United States*. Baltimore: Johns Hopkins University Press.

—— (1997). 'Continuities and Changes in Social Systems of Production: The Cases of Japan, Germany, and the United States.' In J. R. Hollingsworth and R. Boyer (eds.), *Contemporary Capitalism: The Embeddedness of Institutions*. Cambridge: Cambridge University Press, 265–310.

—— (2000). 'Doing Institutional Analysis: Implications for the Study of Innovation.' *Review of International Political Economy*, 7: 595–644.

—— (2002). 'Research Organizations and Major Discoveries in Twentieth-Century Science: A Case Study of Excellence in Biomedical Research.' Research Paper 02–003. Berlin: Wissenschaftszentrum Berlin für Sozialforschung.

—— (2003). 'Major Discoveries and Excellence in Research Organizations.' In Max-Planck-Gesellschaft, *Science between Evaluation and Innovation: A Conference on Peer Review, Max Planck Forum 6*. Munich, Germany: Max-Planck-Gesellschaft, 215–28.

—— (2005, forthcoming). 'Major Discoveries in Biomedical Research at the University of Cambridge: Twentieth Century Excellence in a Traditional Setting.'

—— and Boyer, R. (eds.) (1997). *Contemporary Capitalism: The Embeddedness of Institutions*. Cambridge: Cambridge University Press.

—— and Gear, D. (2004). 'The 'Paradoxical Nature' of Complementarity in the Quest for a Research Agenda on Institutional Analysis.' Paper prepared for Workshop on Complementarity at Max Plank Institute for Study of Societies, Cologne, Germany, 26–7 March 2004.

Hollingsworth, J. R., Schmitter P., and Streeck, W. (eds.) (1994). *Governing Capitalist Economies: Performance and Control of Economic Sectors*. New York: Oxford University Press.

—— Müller, K., and Hollingsworth, E. J. (eds.) (2002). *Advancing Socio-Economics: An Institutionalist Perspective*. Lanham, MD: Rowman and Littlefield Publishers.

—— Hollingsworth, E. J., and Hage, J. (eds.) (2005 forthcoming). *The Search for Excellence: Organizations, Institutions, and Major Discoveries in Biomedical Science*. New York: Cambridge University Press.

Jacob, F. (1995). *The Statue Within: An Autobiography*. Cold Spring Harbor, NY: Cold Spring Harbor Laboratory Press. This is a republication of the 1988 version published in English by Basic Books (translated from the French by Franklin Philip).

Kuhn, T. S. (1962). *The Structure of Scientific Revolutions*. Chicago: University of Chicago Press.

Levi, M. (1997). 'A Model, a Method, and a Map: Rational Choice in Comparative and Historical Analysis.' In M. I. Lichbach and A. S. Zuckerman, *Comparative Politics: Rationality, Culture, and Structure*. Cambridge: Cambridge University Press, 19–41.

McKelvey, B. (1982). *Organizational Systematics: Taxonomy, Evolution, Classification*. Berkeley: University of California Press.

Max-Planck-Gesellschaft (2003). *Science between Evaluation and Innovation: A Conference on Peer Review, Max Planck Forum 6*. Munich, Germany: Max-Planck-Gesellschaft.

Mayntz, R. (2001). 'Die Bestimmung von Forschungsthemen in Max-Planck-Instituten im Spannungsfeld wissenschaftlicher und ausserwissenschaftlicher Interessen: Ein Forschungsbericht.' Discussion Paper 01/8. Cologne, Germany: Max Planck Institute for the Study of Societies.

Mayr, E. (1963). *Animal Species and Evolution*. Cambridge: Harvard University Press.

—— (2001). *What Evolution Is*. New York: Basic Books.

Needham, J., & Baldwin, E. (eds.) (1949). *Hopkins and Biochemistry, 1861–1947*. Cambridge: W. Heffer and Sons Limited.

Nooteboom, B. (1999). 'Innovation, Learning and Industrial Organization.' *Cambridge Journal of Economics*, 23: 127–50.

Pfeffer, J. (1981). *Power in Organizations*. Marshfield, MA: Pitman Publishing Company.

Pierson, P. (2000). 'Increasing Returns, Path Dependence, and the Study of Politics.' *American Political Science Review*, 94: 251–67.

Romanelli, E. (1992). 'The Evolution of New Organizational Forms.' *Annual Review of Sociology*, 17: 78–103.

Rizzello, S. (1999). *The Economics of the Mind*. Cheltenham: Edward Elgar.

—— and Turvani, M. (2002). 'Subjective Diversity and Social Learning: A Cognitive Perspective for Understanding Institutional Behavior.' *Constitutional Political Economy*, 13: 210–14.

Rycroft, R. W., and Kash, D. E. (2002). 'Path Dependence in the Innovation of Complex Technologies.' *Technology Analysis and Strategic Management*, 14: 21–35.

Sewell, W. H. (1996). 'Three Temporalities: Toward an Eventful Sociology.' In T. J. McDonald (ed.), *The Historic Turn in the Human Sciences*. Ann Arbor: University of Michigan Press, 245–80.

Stinchcombe, A. (1965). 'Social Structure and Organizations.' In J. March (ed.), *Handbook of Organizations*. Chicago: Rand McNally, 153–93.

—— (1968). *Constructing Social Theories*. Chicago: University of Chicago Press.

Streeck, W., and Yamamura, K. (eds.) (2001). *The Origins of Nonliberal Capitalism: Germany and Japan in Comparison*. Ithaca, NY: Cornell University Press.

Tilly, C. (1984). *Big Structures, Large Processes, Huge Comparisons*. New York: Russell Sage Foundation.

Ziman, J. (1987). *Knowing Everything about Nothing*. Cambridge: Cambridge University Press.

19

Turning Tracks? Path Dependency, Technological Paradigm Shifts, and Organizational and Institutional Change

Frans van Waarden and Herman Oosterwijk

Introduction

Porter (1990) has shown that commercial innovative activity is not spread evenly across nations and regions. Some frequently create totally new products and new industries; others rarely do so. Some make radical innovations but fail to market them successfully; others make incremental innovations, yet have more commercial success. Furthermore, countries differ not only in the degree of innovativeness, but also in the sectors in which they are innovative. Many have developed nationally specific technological trajectories and patterns of industrial specialization.

Germany, for example, has continued to be innovative in industries in which it was already competitive before the Second World War: paper, printing, machinery, electro-technical products, motor vehicles, chemicals. Holland has done the same in horticulture and agro-food, Switzerland in pharmaceuticals and fine mechanics, Sweden in metalworking. Examples also abound of regional and municipal specialization: Delaware in corporate lawyering, Sassuolo and Fayence in ceramics, Las Vegas in gambling.

Traditional explanations by the economics of innovation have tried to account for such specializations by differences in *input*: the availability of abundant and cheap raw materials, of energy sources, markets, infrastructure, strategic location on trade routes, capital, or of the absence of restrictive regulation. These may explain the original creation of an industry in a certain place, but not its persistence over time, as product and process innovations have often made it less dependent on those resources.

'The six I's' that form a System of Innovation and make for path dependency

Once a sectoral specialization or technological trajectory has been created, other factors work towards its persistence over time and space. We identify six elements, known here as 'The Six I's':

1. Institutes. With industrial development goes the creation of organizations for a specific industry: firms, research and educational institutes, standardization bodies, quality control agencies, financial institutes, trade associations, and other organizations.
2. Interdependencies and Interlinkages. These institutes are interdependent and

interlinked through one or more forms of economic governance: markets, hierarchies, networks, clans, associations, and public-private partnerships.

3. Interests. The institutes have interests in survival and growth, for example, through the continued investment of private and public funds in these industries and their incremental development.

4. Ideas, Information, Knowledge, Competencies. Specialization implies the accumulation of competencies, both in the form of tacit knowledge embodied in personnel, and in that of codified knowledge, expressed in patents, publications, archives, or training programs. This information, knowledge, and skills, developed in specialization, provide for a competitive advantage, at least as long as these competencies are still useful for innovation.

5. Incentives. The investments in institutes and ideas motivate people (workers, entrepreneurs, and researchers) to invest further in what they already have, that is, along familiar lines, in order to exploit their competitive knowledge advantage to the utmost.

6. Institutions. The interests and incentives make the institutes create institutions—defined as social rule systems—that help perpetuate all the former items—for example, by giving preferential treatment to established institutes, ideas, or interlinkages.

These six elements form interdependent configurations, or systems, that *govern* processes of innovation. Where the sectoral specializations are specific for a country, and the institutions are nationwide, the systems can be considered national. That would be one interpretation of the concept known as 'national system of innovation' (Freeman 1987, 1991; Nelson 1987, 1988, 1993; Porter 1990; Lundvall 1992). The common assumption in this literature is that 'nation matters:' certain nationally specific characteristics favour specific industries and/or enhance specific path dependencies; and

these characteristics form some sort of system. However, such a system could also have another spatial identity: it could also be local or regional, rather than national.

The six elements translate for economists into benefits (in particular, increasing returns to scale (Pierson 2000)), but also into costs—those of choosing an alternative line of development: the cost of destroying existing competencies and institutions, of fighting established interests who resist change, etc. Together, these six elements develop an 'institutional history,' which persists as long as some minimal economic and innovative performance is realized. The discovery in nineteenth-century Germany that a wide array of pharmaceuticals and dyes could be developed from aniline sparked the development of an extensive chemical industry. In its wake, a specific industrial infrastructure developed: large firms, know-how, experience with specific product development, research laboratories, networks among different firms, a reputation for quality and reliability among customers, trust among financiers that new innovations would succeed, industrial standards, quality norms, and other regulatory protections. All these institutions gave the country a competitive advantage over other countries, as far as the development of further innovations within this sector is concerned.

Paradigm change

The systems of innovation that develop in specialization mean that a firm, region, or country that already excels in one product line can become even better at it—until the system turns from being an asset into a liability, such as when it hinders further developments within or outside the existing product lines.

Such can be the case after radical innovations, or other major technological upheavals, that make existing competencies obsolescent and, with them, their institutes, interlinkages, and institutions. However, such radical innovations may also form windows of opportunity to escape from path dependencies. As formerly useful and commercially exploitable competencies become redundant, the organizations and institutions that produced them may lose their reason for existence, also losing economic and political influence. The need for new knowledge, and new combinations thereof, may spur the creation of new institutes, interdependencies, interests, ideas, incentives, and institutions.

In this chapter we focus on two such radical innovations—we call them 'technological paradigm changes'—each in a knowledge-based industry. The industries are telecommunications and biotechnology. In telecommunications, the technological paradigm change was the change from the electromechanical to the optical-digital paradigm; in biotechnology, it was the change from classic 'trial-and-error' to modern genomic-based biotechnology. Both took place in the last quarter-century, with modern biotechnology being the more recent of the two.

We have studied the reaction to these paradigm changes in four European countries: Germany, Austria, Finland, and the Netherlands. The historical importance of the industries differs from country to country: Germany and the Netherlands have long-standing electronics/ICT (information and communications technology) sectors and agro- and pharma-industries (where biotechnology made most inroads); Finland has become strong in telecommunications more recently; Austria is interesting because it manages to combine economic prosperity with low R&D spending (the so-called 'Austrian paradox') and absence of globally competing firms in high-tech industries.

Our guiding research questions have been: What changes have taken place in the 'innovation governance systems' in telecommunications and biotechnology in these countries in the wake of the technological paradigm shifts? Have they taken advantage of the windows of opportunity opened up? Are there national differences in nature and speed of reaction to these technological changes? More specifically:

- What have been the consequences for the organizational architecture of the innovation processes?
- To what extent have the national institutions that provide basic resources—knowledge, skills, and finance—hindered or facilitated such organizational changes?

We draw upon findings of a comparative study, funded by the European Commission, in which we participated, and which focused on these questions and countries. The results can be found in a number of reports: Kaiser and Grande 2002a and 2002b; Oosterwijk 2002a and 2002b; Schienstock and Tulkki 2002; Tulkki and Schienstock 2002; Unger et al. 2002a; Unger et al. 2002b; Van Waarden 2002a and 2002b. Most of these are summaries of larger, unpublished country-sector studies. Empirical statements regarding country-sector cases are, if other references are absent, based on these documents.

Technological paradigm shifts

The two sectors we have chosen for study are both science-based industries that have experienced major technological paradigm changes in recent decades.

The telecommunications industry underwent a major shift from the analog-electromechanical to the digital-optical paradigm. The changes came in three stages. The first one was the introduction of digital technology in the mid-1980s, which generated a multitude of developments in technology, products, and services, among them the large-scale introduction of mobile telephony. Basic radio-technology had already been known for a long time, but digitalization (and miniaturization in computer technology) made it possible to increase the capacity dramatically, and reduce the size and costs of equipment. The impact of digitalization could, however, only be fully exploited because it was followed by a second revolution: the introduction of high-capacity optical transmission equipment, which allowed for long-range transmission without loss of energy, and hence maintenance of quality of signals. Thus, the combination of digital and optical technology boosted both quality and capacity of the telecommunications system. It increased the diversity of services and lowered their prices, making them accessible to a mass public. Now a third revolution is about to take place: the change from a hierarchical architecture of the phone-network to the matrix-like architecture based on Internet protocols.

The principles of biotechnology are old. For millennia, people have combined species to produce new ones, better suited to aid man in his search for food: improved strands of grains, maize, potatoes, and also domestic animals for traction, milk, meat, and even for company. But, under *classic* biotechnology, this cross-fertilization was a process of trial and error. Furthermore, cross-breeding was limited to related species, with the same number of, and similar, chromosomes. Therefore, classic biotechnology was restricted to traits already inherent in the species or related species. Modern biotechnology overcomes these limits. It uses a variety of techniques, like DNA analysis, cell fusion, bio-catalysis, bio-informatics, organ and tissue cultures, recombinant DNA, often summarized as genetic engineering, for targeted searching and combining. This toolbox has made it possible to insert genetic information from one organism into another, even into a completely unrelated species. Thus the genes of an onion have been used to improve saccharine production in sugar-beets, and the genes of flowers are used in golden rice to give an increased yield of carotene.

These technological innovations are so radical that we call them 'paradigm changes'. They are radical in many ways:

- the changes involve fundamentally different technologies and combine knowledge from a great variety of fields;
- they increase the capacity, speed, and quality of telecommunications transmissions, and the speed and quality of bio-technological innovations;

- in telecommunications, this has made possible fundamentally new applications, products, and services, such as large-scale and relatively cheap mobile telephony and data transmission (which requires more reliable, stable transmission with less noise caused by friction);
- hence, they have spurred many new innovations in their wake, both radical and incremental;
- they have had an impact on related sectors. Telecommunications innovations have fundamentally affected sectors as diverse as process industries, international finance, and education. The changes in biotechnology are affecting health care, agriculture, food production, environmental care, fine chemicals, etc.;
- the changes are also radical in that they are 'destructive:' they make existing equipment (in telecommunications especially, end-line equipment) and existing knowledge obsolete;
- the new products and services radically affect the life of people and organizations in society;
- the manner in which research and innovation is conducted in telecommunications and biotechnology has been fundamentally changed. In both sectors, the importance of software development has increased.

Finally, the radicalness has been amplified by parallel *institutional* changes in the regulatory governance regime, which themselves were made possible or prompted by the technological changes. Here we see opposite tendencies in the industries: deregulation in telecommunications versus re-regulation in biotechnology. The deregulation and privatization of telecommunications markets and monopolies, though initiated by the neoliberalist wave in public policy, have been relatively successful in this sector, because the technological changes broke down its natural monopoly character. This is underscored by the lesser success of liberalization policies in public transport, water, or energy, where natural monopolies persist. In biotechnology, by contrast, we see a wave of stricter regulation of research, of field trials, of products, and of their marketing, incited by public concern over 'Frankenstein food.'

Though these technological and regulatory changes have taken place in all countries studied, there were differences in timing and speed (or, from another perspective, in resistance to them). Finland and the Netherlands were relatively early in both digitalization and liberalization of telecommunications. Finland actually never had a telecommunications monopoly, as services were supplied by both regional companies and a national one, between which some competition existed. By contrast, Germany was late in both digitalization and liberalization, while Austria, curiously enough, was early in digitalization but late in liberalization. The changeover to modern biotechnology was also easier and less frustrated by restrictive regulation in Finland and the Netherlands, while the resistance has been greatest in Austria.

Changes in the nature of knowledge involved

The paradigm shifts have affected the nature of knowledge involved, and hence the research cycles and the architecture of the idea-innovation chain.

- The increased need for knowledge requires incentives for its production, notably intellectual property rights (IPRs). Both biotech and telecom are increasingly registering patents and copyrights: for equipment, tools, gadgets, materials, designs, software, and even genes and biodata. This is enhanced by the need for interorganizational cooperation. Firms engaged in it want to reduce the risks of knowledge sharing, the risks of opportunistic behavior of partners. Trading of knowledge requires of course that property rights be fixed.

- The increased importance of IPRs represents a trend of privatization and commodification of knowledge. A new 'enclosure' movement is happening: while once it was land, objects, and working time, now it is intellectual goods that have become exclusive and prices are charged for their use. Knowledge becomes a commodity relatively independent of the final product, and hence an object of trade, employment, careers, and prosperity. This is enhanced by market liberalization, which increases competition and forces firms to be more commercial in knowledge management.

- As knowledge gets privatized and acquires commercial value, markets for knowledge develop further. Firms earn a major share of their income from patent licensing. For Philips this amounts to half a billion euros a year. Many smaller biotech firms produce no new products but knowledge, which they sell.

- The privatization of knowledge requires that tacit knowledge get codified, in order to be registered and traded. As a consequence, knowledge systems become less local, less dependent. Tacit knowledge is tied to the holder, who can share it only in direct interaction with people. Codified knowledge can be shared over large distances. ICT contributes to making distance less relevant, as it allows for new modes of information storage, manipulation, exchange, finding and searching, combining, sharing, and diffusion.

- Though the privatization and commodification of knowledge create incentives for their production and facilitate sharing, they also have *disadvantages*:

  (a) a new product requires knowledge owned by a variety of organizations. An average cell-phone can harbor several hundred patents. This necessitates lengthy negotiations with property rights holders, though, to facilitate this, owners are pooling their patents;

(b) the 'closure' involved enhances the segmentation of knowledge communities;

(c) patent holders have an interest in exploiting their patents;

(d) all this could frustrate further innovation.

- The privatization of knowledge reduces the incentive of private actors to invest in 'public' basic research. German telecommunications firms are retreating from 'basic research' and changing towards what they call *vorlauff* research. This trend is enhanced by the shorter time horizon of commercial projects and the increased importance of short-term stock market prices. This would necessitate a greater role of the state in the production of basic knowledge in the public domain, but the opposite is happening. Governments require publicly funded research institutes to become more commercial (they are even being privatized)—to focus on more practically useful research, which can be sold to private business. Formerly public telecommunications monopolies, like the Dutch PTT (Post, Telefoon en Telegraaf), have retreated from basic research after privatization. The result could be a reduction in the production of basic research as public good, which could frustrate future radical innovation.

- This may however be offset by the blurring of the boundaries between basic and applied research. As product and, hence, knowledge life cycles are becoming shorter (in telecommunications), applied research can no longer be defined as research, with time-to-market of less than five years.

- They have increased the science base of these industries. More knowledge is needed in research and product development, but also in manufacturing, quality control, and marketing. This implies a greater need for knowledge-carrying resources from the environment: more finance to fund research, more skilled personnel.

- The knowledge required comes from an increasing diversity of sciences: in biotechnology, examples are microbiology, biometrics, infonomics, IT, food sciences, and pharmaceutics; in telecommunications, electro-mechanical engineering, material sciences, optics, and software development. This implies that competence-holders from quite different disciplines have to join forces.

- A single organization often does not preside over such a diversity of knowledge. Hence the increasing interdependence of knowledge necessitates interorganizational cooperation.

The management of knowledge and the organizational architecture of the idea-innovation network

A framework for analysis: the idea-innovation chain

Hage and Hollingsworth (2000) introduced the concept of the 'idea-innovation chain'—modeled after the concept of 'supply chain,' common in industrial organization—to typify the innovation process. It contains the logical sequential phases that an innovation passes through from initial idea to a marketable product: basic research, applied research, product development, prototype constructing, production, marketing, sales, after-sales service.

This concept built upon earlier models of the innovation process, particularly the one in which science was seen as the sole driving force for innovation, the 'science-push' model of innovation, in which an innovation passes sequentially through these phases. This model was followed by a few others: the 'market-pull' model of innovation, with the market as the major driving force; the 'interactive' model, in which the process was divided into a series of functionally distinct, but interdependent and interacting, stages (Rothwell and Zegveld 1985); the 'parallel' model, in which the innovation process was compressed and steps were no longer sequential, but simultaneous, thanks to instruments such as improved planning, simplification, or the elimination of unnecessary and overlapping steps (Eisenhardt and Tabrizi 1995).

Kamoche and Cunha (2001) provided a typology of innovation models (Table 19.1). Their 'sequential' and 'compressed' models mirror the ones already mentioned. Their 'flexible' model resembles the parallel model. It departs from the idea of a rigid sequence of phases, contains rapid and flexible iterations through system specification, detailed component design and system testing (Iansiti 1995: 2), and is geared to a turbulent environment, in which uncertainty becomes an opportunity, rather than a threat (Thomke and Reinertsen 1998). Their fourth model is the 'improvisational' one, which derives its inspiration from jazz

Table 19.1. Different models of the innovation process

Model	Sequential	Compression	Flexible	Improvization
Process flow				
Underlying assumption	Purposive rationality and predictability in stable environments.	Activities can be predetermined. Process can be adapted to the environment.	Embracing change. Absorbing uncertainty.	Action through experimentation. Improvisation is based on a template.
Process goals	Achieving efficiency. Reducing uncertainty. Providing operational guidelines.	Increasing speed while keeping low levels of uncertainty. Efficiency in time management.	Achieving flexibility. Responsiveness. Adapting to challenges.	Discovery and unrelenting innovation. Balancing between structure and flexibility in dialectical fashion.
Process characteristics	Structured, with discrete phases carried out sequentially.	Predictable series of discrete steps, compressed or removed as need be.	Variation followed by fast convergence. Overlapping procedures.	Progressive convergence within minimal structures. Emergence. Incremental evolution of product features.
Main shortcomings	Rigid. Too formal. Time consuming. Causes glitches. Difficult to achieve in reality.	Possible omission of important steps. Traps of acceleration. Quality may suffer due to shortcuts.	High uncertainty can be counterproductive. Possible delays in concept freezing. Difficult to coordinate.	Can be chaotic and ambiguous. Dialectic logic difficult to sustain. Makes a heavy demand on the appropriate culture and HR systems.
Descriptive metaphor	Relay race.	Accordion.	Rugby.	Jazz improvisation.

Source: Kamoche and Cunha 2001.

music. It allows participants to improvise creatively against a backdrop of some basic rules
regarding structure, theme, key, harmony,
rhythm, and tempo. It demands a climate of
constructive controversy and builds on dialogue and inquiry, involving colleagues and
audience.

Over time, simple, linear models of innovation have given way to more organic ones.
The latter are better able to integrate into one
process different ideas, actors, backgrounds,
departments, and organizations participating
in network-like relations. And they allow for
ideas—and their carrying actors—to move
back and forth between phases, thus increasing
the capacity for learning and problem solving.

Considering this, the idea-innovation chain
concept can be confusing, because it easily
suggests linearity. In view of this, Hage and
Hollingsworth (2000) broadened the term
'idea-innovation chain' to 'idea-innovation
networks,' a concept that catches the dynamics and interactions of innovation better.

The functions or phases in the innovation
process can be located within one firm, but
more often they are divided over different
organizations. The actors can be of a great
variety: large integrated (multinational) enterprises; specialized suppliers of raw materials,
specific services, personnel, tools, or equipment; consultancy firms; universities; research
institutes; hybrids, defined as organizations
with both company and university/research
institute characteristics (Etzkowitz 2000; Fransman 2001); business interest associations; dedicated user groups, organized by the industry;
and mediators/brokers, usually publicly
funded to promote innovation.

Change and continuity in the actor constellation of the idea-innovation networks: differentiation and specialization

Let us now use these models and distinctions
to describe the changes in the organizational
architecture of the idea-innovation networks
in our two industries.

The telecommunications sectors in our
countries were, before the paradigm change,
organized in stable hierarchies on both sides
of the telecommunications-equipment market. Except for Finland, where there was a dual-
istic structure—and even some competition—
between a national and several regional telecommunications companies, there was a monopolistic service provider (the national PTTs)
on the demand side. The concentration on
the equipment-supply side differed somewhat:
it was a near monopoly in Germany, with Siemens being dominant; the Dutch and Austrian
PTTs had several suppliers, but one main one.
The public service hierarchy ordered the market in various ways: as major equipment customer; as service provider; as national
telecommunications standard setter; as provider of other market regulations; and even as
major R&D actor.

Both PTTs and suppliers were organized as
classic hierarchies, with a linear set-up of the
innovation process which was largely technology driven. Given the mutual dependency and
cohabitation of the two monopolists, market-
pull was virtually absent; marketing by the operator was also modest. Demand largely
exceeded supply in the post-war years.

Innovation was considered the task of industry. Though some national PTTs had considerable research capacity, there was hardly any
knowledge exchange. (There were some exceptions.) Laboratory research in transmission
technology is difficult because it is costly to
imitate real-life circumstances. Any basic research done by the PTT was defensive: to
acquire knowledge about new technologies,
to be used in negotiations with suppliers. Incentives for innovation were few, and the main
stimulus was the high demand for telephone
services. However, budgets to extend the network were limited, and tariffs were set by the
state.

The paradigm change and, in its wake, market liberalization, broke up this stable market
structure. New service providers entered the
market, for the new mobile networks, but subsequently also for the fixed networks. These
were no longer tied to fixed suppliers, but

shopped around and acquired their hard- and software increasingly from foreign suppliers.

Nevertheless, the actor constellation turned out, in the long run, to be more stable than the proponents of liberalization had hoped. The paradigm changes uprooted less than expected. As so often after market liberalization—here facilitated by the paradigm changes—there was initially an increase in the number of firms entering the markets for services and for equipment. But consolidation soon followed, through mergers and acquisitions. Newly started service providers were bought up by others, mainly foreign former-monopolists. Thus in the Netherlands, Dutchtone was bought by Groupe FranceTelecom, provider Ben by T-Mobile/Deutsche Telekom, while Libertel was integrated in British Vodafone. Among equipment suppliers, newcomers were often only relatively new—they were established foreign companies who saw their chance with the opening up of formerly closed national markets, by buying up smaller, independent, national equipment suppliers. For example, Austrian Kapsch was bought by Canadian Nortel. Thus the former major international players like Siemens, Nokia, Ericcson, and Nortel still dominate the market in these countries.

The concentration movement was propelled by the huge investments needed for product and system innovation and development (given the increased science base of the industry), for repeated system building and marketing (due to the quick succession of different generations of mobile telephony) and for acquisition of GSM (Global Systems for Mobile communications) and UMTS (Universal Mobile Telecommunications System) licenses. Industry liberalization compelled (or occasioned) national governments to auction off licenses for huge sums of money.

The concentration did not imply that there was no room for smaller companies. Increasingly, the major players have outsourced activities in various phases of the idea-innovation chain. A host of smaller 'satellites' have appeared: doing specialized research, marketing, building the physical infrastructure of the networks, manufacturing end-line equipment, providing special services such as call centres or administration. The telecommunications industry has gradually withdrawn from basic research, and applied research is also hived off to specialized research institutes and universities. There is cooperation in pre-competitive research: researchers visit each other's companies, join forces in specific projects, and discuss solutions. There is the informality of network-like relations, but also the formality of contracting patent licensing, all quite different from the old regime.

Another network-like structure is found in product development, marketing, and production. Marketing, only secondary under the old paradigm, has gained in importance. The industry actively seeks cooperation with operators and customers: telecommunications manufacturers have organized mobile operators and business equipment customers in international user-groups; networks of product managers, marketers, users, and researchers discuss problems and solutions. Designers are also called upon: Nokia organized design competitions to tap into state-of-the-art design. Innovation is becoming more market-led.

This has been a common trend in the countries under analysis here. Liberalization, followed by internationalization, has led to convergence of the organization of telecommunications markets. The difference is merely that some countries host major players—Germany has Siemens and Finland has Nokia—while others have specialized in certain niches: the Netherlands in cables and Austria in speech processing and adaptation of software to nearby East European markets.

Developments in biotechnology have been more diverse, in large part because this is a less clearly demarcated sector. It includes such diverse sectors as food, pharmaceuticals, and environmental management firms. These had different structures in the various countries before the paradigm shift. Nevertheless, here too we see a continued and even increased presence of former major international players, such as Bayer, BASF, Novartis, Gist-Brocades, DSM. As in telecommunications, concentration has been fostered by the large investments needed for the development and marketing of new pharmaceuticals. But, as in telecommunications, this concentration trend went hand in hand with increased 'space' for smaller satellites, especially in research, product development, field trials, etc. Often the major players stimulated and nurtured start-ups for high-risk endeavours in the first phases of the idea-innovation chain. They provided venture capital, and acted as an incubator.

In both sectors we found new entrants, but a different pattern of specialization. In telecommunications, entrepreneurial firms are located at the midstream and downstream phases of the idea-innovation chain. Satellites build new applications and specialized software on generic platforms of hardware technology. In biotechnology, new entrants, specialized suppliers, or hybrid firms are concentrated on the upstream phases in research.

This is related to the different lengths of product-development cycles. The telecommunications market is volatile, with short time-to-market and increasingly shorter product life-cycles. This forces researchers to concentrate on the demand side, rather than the underlying science. In biotechnology, the time-to-market is much longer, especially in pharmaceuticals. The time from idea to a marketable product can be fifteen years, and involves enormous investments. Furthermore, the biotechnology revolution has just begun: there is much work to do in basic and applied research. The tendency in telecommunications is towards integration of platform technologies and variation in end-user products and services. In biotechnology, it is towards specialization and variants in each field.

With the increased science base of these industries, the production of knowledge itself has become more differentiated. In the past, idea-innovation chains were mainly seen for products. With knowledge itself becoming a tradeable commodity, complete idea-innovation chains have developed for the input of specific knowledge, for example, in biotech-

nology for the development of research tools and platform technologies, or for the deciphering of specific DNA codes.

Complexity and interdependence

The differentiation has led to an increased complexity and interdependence of the idea-innovation chains. This had already been caused by the system character of these industries. The telecommunications sector has become one large globe-spanning network of voice and data communication, including end-of-line, transmission, and switching equipment and a host of related technical and content services. The parallel in biotechnology has been the tracing of phenotypes of diverse biological species to a common genotype root, DNA structures, and the possibilities of combining traits of rather different species. This has led to interlinkages between formerly distinct industries such as food and pharmaceuticals.

However, the complexity and interdependence has also been increased by the differentiation in the production of knowledge and its related privatization. Main idea-innovation chains have become differentiated in sub-chains, which have become interdependent as well, resulting in complex webs of chains. The complexity has been further augmented by the fact that one finds different types of innovation (strategic, developmental, adaptive, fashion (Whitley 2000)) in these different chains.

Coordination and integration

The differentiation and increased interdependence of different idea-innovation chains and their various stages raise the need for coordination and integration. In the 'varieties of capitalism' literature (Hollingsworth and Boyer 1997; Hall and Soskice 2001), national, sectoral, and regional economies are characterized by their dominant principle of coordination. In our sectors, this dominant principle changed with the paradigm shift, to the greatest degree in telecommunications, which was formerly coordinated by both public hierarchies ('the state') on the service-provider side, and private hierarchies on the equipment-supplier side. With liberalization, 'state' and 'hierarchy' were partly replaced by 'markets' and 'networks.'

The interdependence of stages of different idea-innovation chains, located in different independent organizations rather than, as previously in a few large hierarchies, implies that organizations need to coordinate activities more. Inter-firm relations are becoming more important, intra-firm ones less so. Informants from the industry frequently mentioned that departments communicated more with those in other firms than with other departments in their own. (This calls into question the assumption of transaction economics that hierarchies are instruments for coordination and for reduction of transaction costs: if there are no transactions between parts of the same hierarchy, then there can be no transactions costs; and hierarchies cannot justify their existence by reduction of transaction costs if there are none.)

Coordination within the webs of idea-innovation chains is increasingly done through the 'market.' The organizational units in the chains trade knowledge and services as 'intermediary products.' This is reflected in the aforementioned tendency to patent knowledge. However, the interdependencies are often too complex for market coordination. The distribution of chains and stages over different organizations poses problems for the 'tight coupling' and feedback necessary for innovation. Product development, design, and marketing have to work in parallel and in concert, adjusting flexibly to each other like jazz improvisers. Therefore, they have to invest in competencies to be able to absorb knowledge from partners. Market interactions, with the risks of exit, asset specificity, and hold-up, do not allow that. Therefore we see ever larger,

more or less stable supply and cooperation chains developing. Major players surround themselves with suppliers, to which they externalize risks, but with whom they also cooperate in innovation. Down the supply chain, Nokia has more than 300 direct suppliers, and many more indirect ones; upwards, it is integrating among service providers, organizing turnkey projects, and providing financial, technical-implementation, and maintenance services.

Various new embryonic and hybrid forms of organization have developed between formally independent organizations: joint ventures, user groups, product teams, patent pools, collective trademarks, technology clusters, partnerships, alliances, or virtual firms (networks of contracts that do not produce anything themselves). These hybrid forms differ regarding:

- the existence of mixed or joint investments, with bi- or multilateral dependence;
- the structure for coordination and control, through, for example, an authority, technical or regulatory standardization, or a system of mutual quality control;
- the rules on incentives—that is, rules on rent sharing, or on 'fairness,' to limit the chance of opportunism. They may be informal or formal/contractual.

The degree of inter-firm cooperation differs among our four countries. According to the Community Innovation Survey (CIS), it is more common in Finland (78 per cent of all firms report cooperation) and the Netherlands (62 per cent), than in Austria (17 per cent) and Germany (5 per cent). Furthermore, the country-sector studies report differences in the mobility of research personnel. This is again high in Finland and the Netherlands, and low in Austria and Germany. Finnish and Dutch interviewees reported that people in the sector meet often in committees, project groups, seminars. Both countries have the advantage of smallness. In the Germanic countries, hierarchy and traditional social distances may form hindrances.

The importance of space

Within networks of stable customer-supplier relationships, spatial proximity—that is, regional clustering—is an asset. Where a critical mass of suppliers and knowledge providers has formed, other organizations are attracted as well, or cannot afford to be absent because, in such cluster areas, the most up-to-date knowledge and information circulate, in part through informal communication and the inter-firm mobility of personnel. Annalee Saxenian (1994) found this for Silicon Valley. We found similar effects of biotechnology clusters in the Munich area and in Finland. Both Ericcson and Siemens reported that they remained in Finland because of the attraction of the large number of suppliers and knowledge providers that have formed around competitor Nokia.

The division of labor within the webs of idea-innovation chains is increasingly international. This holds for the relations between firms—for example, between Nokia and its worldwide suppliers, but also for the relations within firms. Thus the R&D department of Siemens in Vienna works for the whole Siemens company, including its subsidiaries around the world. Our countries increasingly specialize on both the beginning and end of the chains: research, development, design, brand management, adaptation to local requirements, marketing, after-sales service.

In telecommunications, as in biotechnology, the major players—Nokia, Siemens, Philips, Ericcson, Alcatel—are concentrating more and more on their so-called core competencies, namely, product development, engineering, design, marketing and brand management, and, to a lesser extent, manufacturing. In-house manufacturing gets outsourced; first the accessories and parts, increasingly also whole products, and final assembly work. Outsourcing is also extended to the increasingly important software development. As Schienstock and Tulkki (2002) write, Nokia is now mainly specializing in research, software production, final product design, and brand management. It has largely outsourced actual production; 30 per cent of its employees worldwide work in research.

Supporting or hampering institutions

The idea-innovation networks are embedded in 'institutional environments' that facilitate or hamper innovative performance and change (Archibugi and Michie 1997). Formal institutions provide resources and regulate the access of actors to them. They also regulate interaction and communication. Three environments, from which resources are drawn, deserve attention: the sources of regulation, of capital, and of qualified labor.

The 'legal environment' outlines the field of play for innovation. It expresses the meta-institutional environment (culture, tradition, history, ethics, norms, values, development paths). Laws and regulations reflect what is considered appropriate in a given space and time frame (Van Waarden 2001). Thus they influence sectoral organizational structures. Through regulations, telecommunications services used to be organized as a state monopoly. Its natural monopoly character was only one argument for this; another was its being conceived of as a 'public service,' providing basic societal infrastructure, general availability, and uniform access. This provided both constraints and opportunities for the idea-innovation networks.

Regulatory changes cause organizational ones. In both sectors there is a relation between regulatory and paradigm change, but it is not so clear what came first: technological or institutional change. In telecommunications, regulatory change seems to have preceded technological change. The antitrust suits that eventually liberalized US telecommunications markets—and from which European market liberalization drew its inspiration—started in 1974, years before digitalization. This seems to suggest that deregulation sparked innovation. But continued liberalization was unthinkable without digitalization and miniaturization, which helped in breaking down the natural monopoly character of telecommunications. In biotechnology, it was re-regulation rather than deregulation that happened, and it followed rather than preceded paradigm change. It placed considerable constraints on innovation, but facilitated spatial division of labor and, hence, organizational changes of the networks.

Capital is the second resource that idea-innovation networks require from the environment. The manner in which it is provided differs among countries. There are credit-based systems (Germany), credit-based systems with strong government influence (Japan, France), and market-based systems (US, UK) (Christensen 1992). Important also is the availability of risk-seeking capital, such as seed and venture capital. Casper (1999) has argued that the risk-averse German credit-based system, weak on venture capital, has favored incremental rather than high-risk innovation, as exemplified by German biotechnology's concentration on platform technologies.

As long as public budget issues influenced the availability of resources for state monopolies, finance was a constraint in telecommunications; liberalization and privatization alleviated this. In biotechnology, finance can be more of a constraint. Much is needed, given the long and research-intensive product-development cycles. This is a particular problem for spin-offs from universities—they are too small for the stock market, and too unknown to attract international venture capital. Finance is sometimes provided by major international players, like Baxter, Bayer, or Novartis, but more often this is to start-ups splitting off from these companies. In risk-averse credit-based systems, like Germany and Austria, banks have been hesitant to invest in high-risk early phases of biotechnology development. Public support of biotechnology differs by countries. It is highest in Finland (8.1 per cent of government R&D spending) and Germany (6.7 per cent). The Dutch (2.5 per cent) and Austrian (1.5 per cent) governments promote this industry less. In Austria, the two

science-based industries were never in the centre of technology policy, perhaps because they were the 'territory' of other government ministries.

Thirdly, idea-innovation networks require qualified labor: researchers, engineers, marketers, brokers. Countries have historically developed different institutions to satisfy this need: universities, and vocational training institutes in various forms. German and Austrian universities are relatively traditional, hierarchic and rigid, low in international orientation, and score below average on a number of OECD indicators for performance of university education. They have been relatively slow in adjusting to the changed labor demands of biotechnology. Thus German firms report a shortage in students of bio-informatics. And mobility between universities and business is hindered by rigid career patterns and requirements (dissertation, habilitation (second dissertation), publications) and very long recruitment procedures at universities. As these discourage the return of scientists from business back to universities, academics already employed at universities do not easily leave for a temporary career in business. Typically, founders of German and Austrian start-ups in biotechnology have attended American universities.

The German/Austrian dual vocational training and secondary engineering schools (*Hoehere Technische Lehranstalten* and *Fachhochschulen*), combining theory with practical training in the workplace, are better suited to bridge business and educational institutes. Thus in telecommunications, for which these schools are important institutes, there is more mobility, and less shortage of qualified labor.

By contrast, Finland and the Netherlands have more flexible higher education institutions; they have been more responsive to changing demands following the paradigm changes, and have thus facilitated them. Finland has newly established fourteen university graduate schools for biotechnology, integrated in regional centres of biotechnology expertise. Telecommunications engineering institutes have answered rapidly the new needs of telecommunications technology, and their increasing numbers and high qualifications of graduates have attracted international players like Siemens and Ericcson to locate R&D departments in Finland (TIEKE 1999: 11–14).

Our case studies indicate how national institutions influence the flexibility and performance of a sector. The concerted Finnish efforts in public policy and education for biotechnology and telecommunications have paid off. By contrast, restrictions of finance, regulation, and absence of technology policy contribute to a relatively poor Austrian performance. However, in both sectors we perceive a decline in importance of national institutions. Personnel can and do get recruited from elsewhere—some pharmaceutical companies have employees from thirty-five countries; and regulatory restrictions can be evaded by outsourcing sensitive research abroad.

The time dimension: path dependency

We started with identifying long-term sectoral specializations or technological trajectories in the economic structure of countries. Did paradigm changes disrupt such path dependencies? Did they provide opportunities for change, for escape from path dependency? After all, they do tend to disrupt the power of established institutes, create new interests and incentives, give those with new ideas an advantage, and allow for new institutions (for example, liberalization).

In telecommunications the changes that interrupted path dependencies were more radical than in biotechnology. There the technological paradigm shift was followed by major breaks in organizational architecture and regulatory regimes due to the privatization and liberalization policies made possible, at least in part, by the technological changes. Such organizational changes did not follow the paradigm shift in biotechnology. The sector was already a market economy dominated by private firms. Biotechnology did experience changes in regulatory regimes, but these went in the opposite direction of those in telecommunications: less rather than more liberalization.

The starting positions in telecommunications before the technological paradigm shift were more or less comparable in the different countries. Except for Finland (where there was a duopoly) we found everywhere a public monopolist PTT as service provider on protected domestic markets, which was also a monopsonist on the domestic equipment market. As to the supplier side of this market, there was a slight variation between the countries. Though most markets were protected, with long-time, stable suppliers, the number of suppliers varied from basically one in Germany to several in the others. Austria's four was the largest number, but there was hardly any competition among them as each one's market share was relatively fixed and the subject of collective negotiation. These large hierarchies on both sides of the market organized the idea-innovation chains in telecommunications, which were to some extent nationally segmented, as each national PTT set its own standards. Research was done both by the PTT (in the Netherlands) and the equipment supplier (for example, in Germany).

The timing of the radical transformations differed. Austria and the Netherlands introduced new digital technology relatively early on, though realization was relatively slow in Austria. However, in all countries it took some time until the real promise of digitalization was understood. Dekker, then CEO of Philips, mentioned that, originally, managers and researchers were sceptical about digitalization. In Germany, 'uniform technology' (*Einheitstechnik*) was an important value, and equipment producers were uncertain how digitalization would affect that (Werle 1990). The speed of the organizational and regulatory changes also differed. Countries which already had some competition—Finland among service providers, the Netherlands among equipment suppliers—were also the first to privatize and liberalize. Apparently they already had a more liberal regulatory tradition. In Germany and Austria, the monopolies persisted longer, and organizational and regulatory changes were delayed.

With the liberalization and opening up of national markets, the large country, Germany, had an advantage. Its hitherto larger, protected domestic market allowed for the development of a large telecommunications equipment industry, led by Siemens, which had already become a world player before the paradigm change. The smaller countries were at a disadvantage, and most of them could not pull it off.

In Austria, the organization of the industry did not change much. The typical Austrian corporatist *Proporz* (proportional representation), already present in the negotiated market share division between the four equipment suppliers, was maintained in the early years of digitalization, when the country chose two different technical systems in parallel (at greater inefficiency and higher costs): that of Kapsch/Nortel, and of Siemens. Foreign domination—originally by Siemens—increased after the paradigm change when the two domestic equipment suppliers, Kapsch and Schrack, were bought by Nortel and Ericcson. The country could not use the opportunity provided by the paradigm change to develop its own domestic industry, as the Finns did (see below). While the Finnish story is one of 'from foreign domination to global strength of a domestic supplier' (to paraphrase Schienstock and Tulkki 2002), the Austrian story is the opposite: to stronger foreign domination. One reason may have been that the Austrian government considered Siemens Austria as something of an Austrian company. This is a curious case of historical path dependency, set off by historical incident. Siemens Austria was nationalized in the first decade after the war, to prevent the Russians from carting off the stocks of tools as war spoils. Eventually, Siemens Austria was returned to Siemens Germany, but the Austrian government continued to think of the company as a national industry.

Neither did the Netherlands profit from the opportunity. Perhaps it tried it too early. Its domestic provider of telecommunications equipment, Philips, otherwise a major electronics multinational, had already begun to internationalize its telecommunications equipment activities when most European markets were still protected. Its joint venture with AT&T, created for this purpose, was not a success, and the failure to get major contracts from the Austrian and, especially, the French PTT strengthened Philips in the belief that it could not yet penetrate nationally protected markets, even with technological, organizational, and marketing support from a leading American equipment supplier. The company backed out, and concentrated on supplying parts for telecommunications equipment (for example, chips and LCD screens), a choice it had also made in the computer industry. There is something to be said for this strategic decision. Parts customers did not cloud commercial considerations with national sentiments. Furthermore, there have been some minor, path-induced successes for the Dutch industry, notably in the niche of telecommunications cables. This may be related to the country's strong position in civil engineering, hydraulics, and pipe laying.

Among the smaller countries, Finland is a curious exception. It is a case, at least from a broader perspective, of successful escape from path dependency. The country seemed locked into some relatively narrow industrial specializations: the cluster of forestry, wood-working, pulp and paper, based on the abundance of the northern woods; and exports to the former USSR. It managed to reduce its dependency on these products and markets, by profiting from the telecommunications paradigm change. In a relatively short time it managed to create a major telecommunications industry. It was no coincidence that it was in the new subsector of mobile telecommunications: Nokia currently has a 40 per cent share in the world market for mobile telecommunications equipment.

This development was facilitated because, in that sector, there was much less of a path laid out, and hence less need for escape. Relations between operator and industry were looser than elsewhere, giving the industry more room for maneuver. Furthermore, the domestic industry's specialization in networking technology made it easier to perceive the advantages of digital technology than if there had been a heavy economic and intellectual investment in analog switching technology.

Schienstock and Tullki (2002) mention a number of factors that, together, are responsible for this success. Finland managed to turn a disadvantage—great distances, thinly populated spaces, the threat of migration from the northern territories to the south-west—into an advantage. These favored an early development of radio communications, and when digitalization made greater density of such communications possible, Finland took the lead in developing a mobile telecommunications standard, subsequently the basis of the Nordic standard and the GSM world standard. In addition to these geographic factors there were economic ones, such as early exposure to competition. There was already some technological competition between the public national PTT and the private local TelCos, and hence between their suppliers. Subsequently, Finland liberalized its telecommunications market early. Schienstock and Tullki (2002) also list cultural factors: the Finnish openness to new technology; the techno-nationalism among Finnish engineers; the presence of entrepreneurship; and a global orientation.

Important also is the traditional close relation between universities, businesses, and public research institutes, in a number of regional clusters; so is a concerted and activist technology policy, which enhanced the supply of skilled personnel, focused research, and investment funds, and amplified the entrepreneurial spirit. Typical of this state support is that Finland granted UMTS-licences for free which, considering the enormous fees telecommunications firms had to pay elsewhere in Europe, was a huge indirect subsidy.

Behind these concerted public-private efforts to escape from path dependency was an acute sense of vulnerability and crisis, tipped off by the loss of the Soviet market. It is reminiscent of the sense of crisis that the Netherlands experienced after the Second World War

and the loss of its colonies in East Asia. The Dutch then developed an active industrialization and wage-moderation policy, which reduced its dependence on agriculture, food, and trade, and, among others, helped the expansion of Philips.

Biotechnology is a different story. It experienced less radical change, if only because technological change was not followed by liberalization. Rather, regulatory changes went in the opposite direction, for example, the stricter rules on genetically modified organism (GMO) testing. Thus the paradigm change may have been less upsetting for existing path dependencies. However, the differentiation in paths, and in industrial-technological trajectories, is greater between our countries. This is because biotechnology can be applied to different industries (agriculture and food, pharmaceuticals, environmental care, also known as green, red, and grey biotechnology), and our countries had a different strengths of tradition in these industries. Germany developed a strong pharmaceutical industry, which now has a 40 per cent share in world trade in pharmaceuticals. The Netherlands has for centuries specialized in agriculture and food production; and Finland had a strong orientation in forestry and fisheries. Austria had none.

Biotechnology developed from these different bases. In Germany it made its major inroad in pharmaceuticals, though it was late in coming, and precisely for path-dependent reasons. The organizations, institutions, and interests of the German pharma-industry were tightly bound to the familiar, traditional chemical paradigm of drug development. This made for a certain conservatism. Only when it became clear that biotechnology offered revolutionary new possibilities for developing drugs did they get involved, and with all the resources they had at their disposal. This happened, however, not without extensive public support for small start-up offshoots from research and academia, less burdened by the chemical paradigm tradition. Catching up took place first in the less risky platform technologies, to gain experience, and for start-ups to earn income;

it then spread to the development of consumer drugs (Kaiser and Grande 2002b).

While German industry is tilted towards red biotechnology, the Dutch have a stronger profile in green biotechnology, befitting the country's orientation towards agriculture and food industries. Dutch turnover in these sectors is more than twenty times that in pharmaceuticals (Oosterwijk 2002b). The country is the world's largest exporter of cut flowers, cheese, milk powder, margarine, beer, pork, chicken, tomatoes, eggs, seeds, and potato starch (Jacobs et al. 1990) and has an extensive infrastructure for public and private research on the processing of these foods and their derivatives. The Dutch were quicker to realize the possibilities of biotechnology than the Germans. For them, the existing path-dependent sectoral specialization was less of a hindrance, as modern biotechnology differs less from classic biotechnology than from the chemical method of drug development, while the advantages of genetic engineering over trial-and-error technology are more readily visible.

However, the path-dependent-induced specialization of the Dutch on green biotechnology became eventually a disadvantage, and explains why the country has started to lag behind in this field. Applying biotechnology to food production turned out to be more risky. The public is more willing to accept biotechnology in life necessities, like drugs, than in the, relatively speaking, luxury sector of foods, especially if the innovations do not directly profit the consumer, but the farmer and the environment (higher crop yields, greater resistance to diseases, less need to spray pesticides). The best perspectives are perhaps still in food-related products that are not directly consumed, such as seeds and starch.

The smaller biotechnology sectors in Finland and Austria are less strongly specialized. Green biotechnology is more important in Finland than in Austria. Both countries have made efforts to develop biotechnology, but the sectors are still relatively small, in part because there were no big sectors on which the industry could build. In Finland, the large forestry industry would have been a candidate, but slow-growing wood is a less likely candidate for biotechnology application than fast-growing garden crops. Nevertheless, the country tried to repeat the success story of Nokia in this other new, major science-based industry. The success has been limited, but is still greater than in Austria, where the institutional preconditions were less favorable: absence of venture capital, and very tight regulation on biotechnology in food production. In many ways Austria is the opposite of Finland:

- Finland has embarked on a concerted and activist technology policy; Austria has not;
- Austria scores low in inter-firm cooperation and interaction between universities, business, and public research institutes, while Finland scores high on these indicators;
- Austria missed the sense of crisis which Finland experienced, and which could boost a concerted public-private effort to capitalize on the opportunity provided by the paradigm changes in these science-based industries. Its long-standing and relatively good economic performance provided fewer incentives to innovate its industrial base.

Space: are there still *national* systems of innovation?

The concept of the idea-innovation network has the advantage over the concept of national systems of innovation of being neutral towards the spatial dimension of innovative activities. That allows us to treat as an open empirical question whether idea-innovation chains are (still) located within national boundaries, and in which parts.

The paradigm changes have globalized our sectors further. The industries where biotechnology is being applied—pharmaceuticals, chemicals, and food processing—were already globalized, but they are now even more so. Telecommunications markets used to be nationally segmented, but technological paradigm changes, privatization, and deregulation have broken up these national markets and integrated them into world markets, facilitated also by the global network character of telecommunications.

The paradigm changes may have provided opportunities for new firms to develop; yet both sectors are dominated by a limited number of world players: in telecommunications, Siemens, Alcatel, Ericsson, AT&T, Philips (as parts supplier), Motorola, and Nokia; in biotechnology, Bayer, Hoechst, Novartis, Baxter, and, lesser known but dominant in its subsector of starch, the Dutch AVEBE.

These leading firms organize their idea-innovation networks worldwide, both internally and externally, that is, with lots of suppliers and research laboratories in many different parts of the world. Nokia has forty-four research centers in twelve countries. World players locate research facilities in those regions where certain activities are concentrated. Thus Siemens does most of its Internet research in Silicon Valley, but has its mobile Internet facility in Finland. These establishments work for the whole enterprise. Thus the Austrian research center of Siemens does 95 per cent of its work for the head office of Siemens in Munich. Similarly, product development, design, or manufacturing are spread around the world, but embedded in world-spanning idea-innovation networks of the company. Thus the international division of labor is no longer concerned with only final products, with one country producing bananas, and another mobile phones; it is concerned also with the stages in the idea-innovation networks—of parts of individual products, and of their parts too.

Of course, that is not really new. Design, manufacture, and adaptation to local markets have been spread over different countries for a longer time. What is new is that, as idea-innovation networks have been extended, complicated, and further differentiated, so too has their spatial distribution become more complex and more differentiated, involving more and more countries.

As a consequence, there is often no direct causal link between the various activities of

one multinational enterprise (MNE) in a single country. Siemens Austria has a number of researchers in a research centre; and it has a turnover in Austria. But this turnover is not produced by these researchers. The researchers work for the head office, which pays their salary, either directly or indirectly. The Austrian turnover of Siemens concerns the sale of—among other things—telecommunications equipment, imported from Siemens establishments elsewhere in the world, and adapted to local needs and standards. It would, therefore, be ridiculous to compare Siemens' input (workers, salaries) and output, or to calculate a 'productivity' of the Siemens workers. What holds for the individual firm may to some extent hold also for the aggregated data of a sector in a certain country.

This is a problem in comparing indicators, traditionally aggregated at the national level. Previous research (Nelson 1993; Lundvall 1992) has focused on three measures of national innovativeness: money invested in research; number of patents; export trade balances for designated high-tech areas. But what has not been well appreciated is that these measures are related to different components of idea-innovation networks: money invested in research is an input indicator in the idea-generation phase; patents registered indicate ideas actually generated, even if not (yet) commercialized; and trade-balance scores are evidence of the successful commercialization of innovative ideas.

Thus there may be no causal relation between input and output indicators of various stages of idea-innovation networks. Does it make sense to compare national R&D expenditures with national patent registrations? Often these patents do not reflect the inventiveness of the researchers in that country. MNEs tend to patent in countries where patenting is easier, cheaper, faster, more important, or just where the head office of the MNE is located. Thus Philips patents many inventions made by its overseas research centers in the Netherlands.

This brings us to the so-called European paradox (EU 1993, 1995): high R&D input, but relatively low output in terms of the production of commercially successful innovations (Andreasen *et al.* 1995; Coriat 1995). Is this not the result of this international division of labor in the idea-innovation networks? Most likely, R&D is done in Europe, actual production outside of it. Whether the R&D is successful or not is not apparent from production/export statistics concerning final goods. To measure this, R&D-input indicators should be related to indicators of knowledge export. But other than what can be gleaned through details of licensing, internal transfers of knowledge—information and tacit knowledge—within one MNE do not show up in international statistics. And public policy measures that are guided by such indicators may misfire—the high patent-score of the Netherlands could induce complacency among policy-makers, just as the apparently low commercialization could induce nervousness and policy measures.

Is this international division of labor in idea-innovation networks really a problem for the countries involved? Should the European paradox deserve the nervousness it apparently creates? If the international division of labor in the production of goods is efficient, according to the theory of comparative advantages, why would this not hold for a division of labor along the idea-innovation network? Can countries not increase their wealth by concentrating on research—and employing and paying researchers—even if the final goods made with this research are produced by other countries? Knowledge has become a tradeable commodity in itself, almost a final product. Is it not more efficient to do actual production where manufacturing wages are lower? Or to locate after-sales service, like call centers, in countries where people speak foreign languages? Conversely, is low R&D investment something to be concerned about? Could not a country import technology developed elsewhere—either as licences (as the Netherlands seems to do), or as embodied knowledge in equipment (as Austria does)—and use it to produce value in other sectors, or stages of idea-innovation networks, such as logistics or marketing?

Finally, should the conclusion be that it no longer makes sense to speak of 'national systems of innovation'? On the contrary: there must be a reason why international firms locate certain activities of idea-innovation networks in some countries and others elsewhere; why companies in Western Europe focus more and more on the beginning and end of the idea-innovation chain; why Siemens maintains a research center in Austria and Nokia concentrates only on research, software production, final-product design, and brand management in Finland. Siemens and Nokia do so because both countries have a good supply of software engineers, and in Austria they are cheaper than in neighboring Germany. There still is a 'national system of innovation,' namely those institutions that provide attractive resources—finance, skilled personnel, a favorable legal environment, a good communication infrastructure—for the stages in idea-innovation networks that produce the knowledge for innovations. The complementary 'system of innovation' in, for example, Taiwan consists of those organizations and institutions that provide attractive resources for the manufacturing phase, and that induce international firms to locate manufacturing in that country. In an increasingly knowledge-based economy, it may make sense to concentrate on the knowledge-producing phases of idea-innovation networks; to nurture and develop national institutions that attract research, engineering, and design employment, at the front of the chain, and marketing, adjustment to local needs and standards, and after-sales service at the end of the chain, leaving actual manufacturing of the final goods to other 'national systems of production.'

One disadvantage of such spatial distribution of different phases in idea-innovation networks around the globe could be that direct and informal feedback between phases in the chain, for example, from manufacturing to design, becomes more difficult. It could complicate the 'jazz improvisation' style of innovation. But here perhaps information, and communications technology, the output of our idea-innovation networks, could provide the technical solution.

References

Andreasen, L. E., Coriat, B., den Hertog, F., and Kaplinsky, R. (eds.) (1995). *Europe's Next Step: Organizational Innovation, Competition and Employment.* London: Frank Cass.

Aoki, M. (1988). *Information, Incentives, and Bargaining in the Japanese Economy.* Cambridge: Cambridge University Press.

Archibugi, D., and Michie, J. (eds.) (1997). *Technology, Globalization and Economic Performance.* Cambridge: Cambridge University Press.

Bierly, P., and Hämäläinen, T. (1995). 'Organisational Learning and Strategy.' *Scandinavian Management Journal,* 11/3: 209–24.

Casper, S. (1999). 'National Institutional Frameworks and High-Tech Innovation in Germany: The Case of Biotechnology.' Working Paper for the Netherlands Institute for Advanced Studies, Wassenaar.

Christensen, J. L. (1992). 'The Role of Finance in National Systems of Innovation.' In B.-Å. Lundvall (ed.), *National Systems of Innovation: Towards a Theory of Innovation and Interactive Learning.* London: Pinter.

Coriat, B. (1995). 'Variety, Routines and Networks: The Metamorphosis of Fordist Firms.' *Industrial and Corporate Change,* 4/1.

Daft, R. L., and Lengel, R. H. (1986). 'Organisational Information Requirements, Media Richness, and Structural Design.' *Management Science,* 32: 554–71.

Ebers, M. (ed.) (1997). *The Formation of Inter-Organizational Networks.* Oxford: Oxford University Press.

Eisenhardt, K. M., and Tabrizi, B. N. (1995). 'Accelerating Adaptive Processes: Product Innovation in the Global Computer Industry.' *Administrative Science Quarterly,* 40: 84–110.

Etzkowitz, H., Webster, A., Gebhardt, C., and Terra, B. R. C. (2000). 'The Future of the University and the University of the Future: Evolution of the Ivory Tower to Entrepreneurial Paradigm.' *Research Policy,* 29: 313–30.

EU (1993). *White Paper on Growth, Competitiveness, Employment. The Challenges and Ways Forward into the 21st Century.* Brussels: EU.

—— (1995). *Green Paper on Innovation.* Brussels: EU.

Fransman, M. (2001). 'Designing Dolly: Interactions between Economic Technology and Science, and the Evolution of Hybrid Institutions.' *Research Policy,* 30: 263–73.

Freeman, C. (1987). *Technology and Economic Performance: Lessons from Japan.* London: Pinter.

—— (1991). 'Networks of Innovators: A Synthesis of Research Issues.' *Research Policy,* 20: 499–514.

Grandori, A., and Soda, G. (1995). 'Inter-firm Networks: Antecedents, Mechanisms and Forms.' *Organisation Studies,* 16/2: 183–214.

Granovetter, M. (1982). 'The Strength of Weak Ties.' In P. Marsden and N. Lin (eds.), *Social Structure and Network Analysis.* Beverly Hills, CA: Sage, 105–30.

Grant, W. (1996). 'Different EU and US Perspectives on Biotechnology and their Implications for Trade Relations.' In O. T. Solbrig, R. Paarlberg, and F. di Castri (eds.), *Globalization and the Rural Environment.* Cambridge, MA: Harvard University, David Rockefeller Center for Latin American Studies.

Hage, J., and Alter C. (1997). 'A Typology of Inter-organisational Relationships and Networks.' In J. R. Hollingsworth and R. Boyer (eds.) (1997), *Contemporary Capitalism: The Embeddedness of Institution.* Cambridge: Cambridge University Press.

—— and Hollingsworth, J. R. (1999). 'Idea Innovation Networks: A Strategy for Research on Innovations.' Unpublished Discussion Paper.

—— —— (2000). 'A Strategy for the Analysis of Ideas: Innovation Networks and Institutions.' *Organization Studies,* 21/5: 971–1004.

Hall, P., and Soskice, D. (2001). *Varieties of Capitalism: The Institutional Foundations of Comparative Advantage.* Oxford: Oxford University Press.

Hämäläinen, T. (1993). 'The Organizational Consequences of Increasing Specialisation and Division of Labor.' *GSM Working Paper Series, 93/25.* Newark, NJ: Rutgers University.

—— and Schienstock, G. (2001). 'The Competitive Advantage of Networks in Economic Organisation: Efficiency and Innovation in Highly Specialised and Uncertain Environments.' In OECD, *Innovative Networks: Co-operation in National Innovation Systems,* Paris: OECD.

Hollingsworth, J. R. (2000). 'The Institutional Context of Innovation.' Presentation given at the Research Seminar of the Finnish National Fund for Research and Development, Helsinki.

—— and Boyer, R. (eds.) (1997). *Contemporary Capitalism: The Embeddedness of Institutions.* Cambridge: Cambridge University Press.

Iansiti, M. (1995). 'Shooting the Rapids: Managing Product Development in Turbulent Environments.' *California Management Review,* 38/1: 1–22.

Jacobs, D., Boekholt, P., Zegveld, W. (1990). *De Economische Kracht van Nederland.* The Hague: SMO.

Johannison, B. (1987). 'Beyond Process and Structure: Social Exchange Networks.' *International Studies of Management and Organisation,* 17/1.

Jorde, T. M., and Teece, D. J. (1990). 'Innovation and Cooperation: Implications for Competition and Antitrust.' *Journal of Economic Perspectives,* 4/3: 75–96.

Kaiser, R., and Grande, E. (2002a). 'From the Electro-mechanical to the Opto-digital Paradigm: Organizational Change and the Management of Research in the German Telecommunication Sector.' In van Waarden 2002b: 1–23.

—— —— (2002b). 'The Emergence of the German Pharmaceutical Biotech Industry and the Role of the National Innovation System.' In F. van Waarden (ed.), *Building Bridges between Ideas and Markets. Part II: Country-Sector Summary Reports.* Final report Part II of the EU-TSER-funded project, National Systems of Innovation and Networks in the Idea-innovation Chain in Science-based Industries (Contract number: Soe1-ct98-1102; Project number: PI. 98.0230), Utrecht: Utrecht University; Brussels: European Commission 2002b: 184–202.

Kamoche, K., and Cunha, M. P. (2001). 'Minimal Structures: From Jazz Improvisation to Product Innovation.' *Organisation Studies,* 22/5.

Lundvall, B.-Å. (ed.) (1992). *National System of Innovation: Towards a Theory of Innovation and Interactive Learning.* London: Pinter Publishers.

Metcalfe, S. (1995). 'The Economic Foundations of Technology Policy: Equilibrium and Evolutionary Perspectives.' In P. Stoneman (ed.), *Handbook of the Economics of Innovation and Technological Change.* Oxford: Blackwell Publishers.

Metze, M. (1997). *Let's Make Things Better.* Nijmegen: Sun.

Nahapiet, J., & Ghoshal, S. (1998). 'Social Capital, Intellectual Capital, and the Organizational Advantage.' *Academy of Management Review,* 23/2: 242–66.

Nelson, R. R. (1987). *Understanding Technical Change as an Evolutionary Process.* Amsterdam: North-Holland.

—— (1988). 'Preface' and 'Institutions Supporting Technical Change in the United States.' In G. Dosi, C. Freeman, R. Nelson, G. Silverberg, and L. Soete (eds.), *Technical Change and Economic Theory.* London: Pinter.

—— (ed.) (1993). *National Innovation Systems: A Comparative Analysis.* New York: Oxford University Press.

Nonaka, I., and Takeuchi, H. (1995). *The Knowledge Creation Company.* New York: Oxford University Press.

Nooteboom, B. (1999). *Inter-firm Alliances: Analysis and Design.* London: Routledge.

OECD (1997). *National Innovation Systems.* Paris: OECD.

Oosterwijk, H. (2002a). 'Switching Technology through Five Decades: Dutch Telecommunications under Change.' In Van Waarden 2002b: 75–137.

—— (2002b). 'Hesitant Entrepreneurship in the Dutch Biotech Industry: Sectoral Patterns of Development and the Role of Path-Dependency.' In Van Waarden 2002b: 231–72.

Pavitt, K. (1994). 'Key Characteristics of Large Innovating Firms.' In M. Dodgson and R. Rothwell (eds.), *The Handbook of Industrial Innovation.* Aldershot: Edward Elgar.

Pierson, P. (2000). 'Increasing Returns, Path Dependency, and the Study of Politics.' *American Political Science Review*, 94/2: 251–68.

Polt, W. (2001). *Forschungs- und Technologiebericht 2001. Studie der Arbeitsgemeinschaft TIP im Auftrag des Bundesministeriums für Verkehr, Innovation und Technologie und des Bundesministeriums für Bildung, Wissenschaft und Kultur.* Vienna: Austrian Research Centers (ARCS), WIFO, and Joanneum.

Porter, M. (1990). *The Competitive Advantage of Nations.* New York: Macmillan, Free Press.

—— and Fuller, M. B. (1986). 'Coalitions and Global Strategies.' In M. Porter (ed.), *Competition in Global Industries.* Boston, MA: Harvard Business School Press, 315–44.

Prahalad, C., and Hamel, G. (1990). 'The Core Competencies of the Corporation.' *Harvard Business Review*, May–June, 79–83.

Rothwell, R. (1991) 'External Networking and Innovation in Small-and Medium-Sized Manufacturing Firms in Europe.' *Technovation*, 11/2: 93–112.

—— (1994). 'Industrial Innovation: Success, Strategy, Trends.' In M. Dodgson and R. Rothwell (eds.), *The Handbook of Industrial Innovation.* Aldershot: Edward Elgar.

—— and Zegveld, W. (1985). *Re-industrialization and Technology.* Harlow: Longman.

Sarmento-Coelho, M. (2000). 'Innovation and Quality in the Service Sector: Application to SMEs.' In P. Conceição, D. Gibson, M. Heitor, and S. Shariq (eds.), *Science, Technology, and Innovation Policy: Opportunities and Challenges for the Knowledge Economy.* London: Quorum Books.

Saxenian, A. (1994). *Regional Advantage: Culture and Competition in Silicon Valley and Route 128.* Cambridge: Harvard University Press.

—— (1996). 'Inside-Out: Regional Networks and Industrial Adaptation in Silicon Valley and Route 128.' *Cityscape: A Journal of Policy Development and Research*, 2/2.

Schibany, A., and Polt, W. (2001). 'Innovation and Networks: An Introduction to the Theme.' In OECD, *Innovative Networks: Co-operation in National Innovation Systems.* Paris: OECD.

Schienstock, G., and Tulkki, P. (2002). 'From Foreign Domination to Global Strength: Transformation of the Finnish Telecommunications Industry.' In van Waarden 2002b: 138–83.

Steinmüller, E. W. (1994). 'Basic Research and Industrial Innovation.' In M. Dodgson and R. Rothwell (eds.), *The Handbook of Industrial Innovation.* Aldershot: Edward Elgar.

Sydow, J. (1992). *Strategische Netzwerke, Evolution und Organisation.* Wiesbaden: Gabler Steinmüller.

Thomke, S., and Reinertsen, D. (1998). 'Agile Product Development: Managing Development Flexibility in Uncertain Environments.' *California Management Review*, 41/1: 8–30.

TIEKE Information Technology Development Center (1999). *IT Cluster in Finland Review.* Espoo.

Tulkki, P., and Schienstock, G. (2002). 'On the Threshold of a New Technology: Finnish Life Science Industries at the Beginning of the 21st Century.' In van Waarden 2002b: 273–99.

Unger, B., with Oosterwijk, H., and Rossak, S. (2002a). 'Austrian Biotechnology: Where to Find it on the Map?' In Van Waarden 2002b: 203–30.

—— with Giesecke, S., Rossak, S, and Oosterwijk, H. (2002b) 'Telecommunications and the Austrian Paradox.' In Van Waarden 2002b: 24–74.

Van Waarden, F. (2001). 'Institutions and Innovation: The Legal Environment of Innovating Firms.' *Organization Studies*, 22/5: 765–95.

—— (ed.) (2002a). *Building Bridges between Ideas and Markets. Part I. Summary Findings.* Final Report Part I of the EU-TSER-funded project, *National Systems of Innovation and Networks in the Idea-Innovation Chain in Science-Based Industries.* (Contract number: Soe1-ct98-1102; Project number: pl 98.0203), Utrecht: Utrecht University; Brussels: EV.

—— (ed.) (2002b). *Building Bridges between Ideas and Markets. Part II. Country-Sector Summary Reports.* Final report Part II of the EU-TSER-funded project *National Systems of Innovation and Networks in the Idea-Innovation Chain in Science-Based Industries* (Contract number: Soe1-ct98–1102; Project number: pl 98.0230). Utrecht: Utrecht University; Brussels: EU.

Werle, R. (1990). *Telecommunication in der Bundesrepublik: Expansion, Differenzierung, Transformation.* Frankfurt am Main: Campus Verlag.

Whitley, R. (2000). *Divergent Capitalisms: The Social Structuring of Business Systems.* Oxford: Oxford University Press.

20

Institutional Change and Societal Change: The Impact of Knowledge Transformations

Jerald Hage

Introduction

In the introduction to this book, there is an equation that connects the knowledge base, collective learning, with the creation of new knowledge, or innovation. This equation can be applied at different levels: at the organizational level, at the organizational-population level, and at the societal level. Brief mention has been made of the feedback effect of new knowledge or innovation on the knowledge base, on entire populations of organizations, and on the society. But as yet, these feedback effects have not been considered in detail.

This chapter argues that the accumulation of knowledge transformations, as measured by the number of radical innovations in both process technologies (such as flexible manufacturing) and in products (such as whole new industries or new standard designs in existing industries), is creating a considerable amount of institutional and societal change. Globalization is aiding these knowledge transformations by lowering the tariff walls. However, the institutional arrangements and, specifically, the dominant coordination mode of the advanced industrial societies, are filtering the impact of these knowledge transformations in surprising ways, so that path dependency occurs in some countries but not in others.

A major stumbling block in any discussion of institutional and societal change is defining the concepts of institutional change and of path dependency. Hollingsworth's chapter broaches this problem. In this chapter, institutional change is defined as changes in the dominant mode of coordination that supplements market coordination in a society—for example, moving from a hierarchical arrangement to an interorganizational network. In contrast, path dependency is defined as the continuation of the dominant mode that supplements market coordination across a number of industrial sectors in society: associations, clans, interorganizational relationships, etc. Although a number of other ways of defining institutional arrangements exist, the focus in this chapter is on the *dominant* coordination mode that *supplements* the market mechanisms. Societal change, in contrast to institutional change, is simply the addition of new elements to various institutional sectors of the society, such as new diploma programs in education, new research arenas in science, new governmental responsibilities, new industrial sectors, or new health and welfare programs. Obviously, societal change occurs much more frequently than institutional change.

This chapter argues that path dependency has *not* occurred in the liberal market economies (Hall and Soskice 2001) because many of the major hierarchical companies that dominated their industrial sectors have disappeared, along with most, if not all, of the organizational populations in those sectors. Instead, there has been a rapid spread of inter-organizational networks. This is a new mode of coordination for these countries. In the coordinated market economies, in contrast, the various forms of coordination have filtered the negative impacts of the knowledge transformations and of globalization. As a consequence, they have followed a path-dependent model. Although interorganizational networks are increasing there as well, the pace of institutional change is not as rapid (Harbison and Pekar 1998). Moreover, the use of interorganizational networks in the coordinated market economies is not an institutional change, because they have been a common mode for a century or more.

There is a qualification to this thesis: it does not apply to all sectors. One of the major problems in the broad comparative institutional literature is the tendency to focus on certain exemplars and then argue that they typify the entire society. The arguments about path dependency and about institutional change in this chapter are qualified by the nature of the sectors or technological regimes (Archibugi and Pianta 1992; Campbell *et al.* 1991; Guerrieri and Tylecote 1998; Hagedorn 1993; Kitschelt 1991: 460; Malerba and Orsenigo 1993, 1997; Pavitt 1984).

Theoretical framework

Explaining the differential responses of liberal market economies and of the coordinated market economies to the twin social forces of knowledge transformations and of globalization requires a somewhat complicated framework:

- sources or pressures for institutional and societal change;
- a typology of institutional arrangements defined as coordination modes, and specifically the differences between the liberal market economies and coordinated market economies (Hall and Soskice 2001);

- a typology of sectors or sets of organizational populations to explicate the specific qualifications.

Sources or pressures for institutional and societal change

One way of thinking about potential candidates for sources of, or pressures for, change is to ask what would produce change powerful enough to affect many sectors of the society, not just a few economic ones. As Schumpeter (1934) has observed, the first industrial revolution is a good example of a major change that precipitated many other changes. During this period of time, large-scale companies and the assembly-line method of production spread from one sector to another, starting with railroads and moving to steel, standard food products, cigarettes, elevators, etc. (Chandler 1977). Radical process or product innovation affect a specific sector, but this knowledge transformation is not important enough to bring about, as did the first industrial revolution, such pervasive change. Changes of sufficient magnitude to qualify for this category are those that shift the rules of competition across a number of sectors in the economy, with consequences for the non-economic parts of society.

My argument is that the world has experienced two simultaneous, major shifts in the competitive rules:

- post-industrialization or the development of the new economy;
- economic globalization or the increase in the number of countries providing goods and services.

These terms have been defined in the introduction to this book. Post-industrialization is driving the movement towards interorganizational networks of various kinds, ranging from joint ventures, to research consortia, to global alliances (Alter and Hage 1993; Dussauge and Garrette 1999; Doz and Hamel 1998; Harbison and Pekar 1998; Häkansson 1990; Jarillo 1993; Kogut *et al.* 1993; Lundvall 1992; O'Doherty 1995). This is seen most along the idea-innovation chain/ network (see the contribution by van Waarden and Oosterwijk). Globalization is, paradoxically, also driving firms into interorganizational networks, more typically along the supply chain (see contribution by Meeus and Faber) and, in particular, in what are called 'commodity chains' (Gereffi and Korzeniewicz 1994), because of the radical reduction in prices. Occurring together, post-industrialization and economic globalization have shocked many of the traditional businesses out of their comfort under the old rules of competition: mass markets protected by tariffs, and specialized products protected by patents.[1] There are many who have argued that we should be talking about 'the third Industrial Revolution' or 'the New Economy' or 'Postmodern Society,' all names reflecting some fundamental societal, if not necessarily institutional, changes.

In our analysis of institutional change and path dependency, both post-industrialization and globalization created a wave of radical product and process innovations in one sector or organizational population after another. The term 'knowledge transformations', as in the title of this chapter, is defined as radical product or radical process innovations *within* an organizational population or sector, whether economic or non-economic. Thus, knowledge transformations do not necessarily produce any institutional change, and only in some cases do they reflect societal changes. When the radical product innovation represents a new industry, a new education program, a new scientific research arena, or a new government responsibility, it is, in our definition, a societal but not necessarily an institutional change.

Defining institutional change as changes in coordination modes

A new set of research findings has emerged in what might be called macro-institutional theory, represented in such literatures as the varieties of capitalism, business systems, and social systems of production (Edquist and Hommen 1999; Hall and Soskice 2001; Hollingsworth 1997; Lundvall 1992; Nelson 1993; Whitley 1992a, 1992b, 1999). These literatures have emphasized the *institutional differences* between countries, providing a comparative framework for the analysis of economic institutions.[2] Given this solid foundation, one can address the problem of which institutional arrangements changed, whether they did so as a consequence of knowledge transformations and economic globalization, and which did not change.

This macro-institutional literature is quite rich and, as has been indicated in the introduction Part IV, complex models have been developed to describe the differences between the institutional arrangements found in various countries of Western Europe, the US, Japan, and other East Asian societies. Typically, these arrangements include discussions of financial markets, the state, educational institutions, and labor–management relationships (see the contributions by Casper and Finegold). Since these descriptions are vast and complex, it becomes more difficult to perceive institutional change, and the thesis of incremental change along path-dependent ways becomes more appealing.

Again, I propose to limit the analysis of institutional change to one important characteristic of institutional arrangements: how market coordination is supplemented in a nation (Campbell *et al.* 1991; Hollingsworth and Boyer 1997). Building upon Hall and Soskice (2001), I am especially interested in contrasting the institutional changes in coordination modes in the liberal market economies with the coordinated market economies. With this focus, it becomes easier to observe how a specific kind of institutional arrangement has either facilitated or hindered the process of organizational responses to economic globalization, knowledge transformations, or state interventions.

What justification is there for focusing on coordination, especially given the richness of the existing models of capitalism or of business systems? Let me suggest at least two reasons.

Many aspects of economic life, besides prices, wages, supply and demand, and other variables that form the core of economic analysis, are coordinated. Market coordination of these attributes is supplemented by non-market forms of coordination. The extension of this variety allows us to explain several aspects of the institutional arrangements that are important in the explanation of the differences between societies. In particular, these coordination mechanisms deal with labor conflict and with the connection between education and employment (see the contribution by Finegold). Both of these are important for understanding the adoption of radical process technologies, such as flexible manufacturing.

In Table 20.1 is a simple typology of coordination mechanisms that captures the essence of much of the institutional literature on this topic. (It obviously does not represent all kinds of institutional coordination mechanisms that supplement the market, but it does cover the major ones). It moves beyond Hall and Soskice (2001) in a number of ways. First, it attempts to place the coordination modes of the developed countries within a historical perspective that allows for the inclusion of fascism and communism. Second, it deconstructs coordinated market economies into several types, including interorganizational networks and associations. Third, its two dimensions—of state involvement and of emphasis on collective orientation, or what might now be called social capital—have not been given enough attention.

Several observations about this table are to the point. It is a snapshot of the several decades from the 1950s through the 1960s *before* the impacts of economic globalization and of post-industrialization began to be experienced. The table specifies very clearly that the typology only includes medium levels of state involvement and of collective arrangements. The high levels would cover authoritarian or socialist states where the hierarchies were completely owned or managed by the state. Finally, some subtypes have not been specified, in particular small family networks that are to be found in some Asian societies and in middle Italy. Nor does this include the very special topic of industrial districts, another form of coordination, but typically found in parts of societies coordinated by other means.[3]

The countries that typified these examples prior to the 1970s or 1980s are:

- associational supported hierarchies, as found in Germany, Austria, the Netherlands, and Sweden;
- interorganizational supported hierarchies, as found mainly in Japan, but increasingly in other societies where the state is encouraging the development of idea-innovation chains;
- corporate hierarchies, as found in liberal market economies such as the US, UK, Italy, Canada, and Australia;
- state-supported hierarchies, as found in France and South Korea.

Regardless of various limitations, the typology provides a clear focus on how to define institutional change. Did the dominant coordination mode that supplements the market alter? The thesis that is advanced here is that this has

Table 20.1. A typology of coordination mechanisms that supplement markets

Degree of state involvement	Emphasis on collective orientation		
	Low	Medium	High
Low	Corporate hierarchies	Association hierarchies	
Medium	State-supported hierarchies	Interorganizational networks	
High			

happened in the liberal market economies and, to a lesser extent, in the coordinated market economies. The reasons for this are provided below in the next section. But as has already been suggested, this has not occurred uniformly across all sectors within the economy. Thus, we need to add a typology of sectors.

Defining exceptions and qualifications with a typology of organizational populations or sets of sectors

Our simple, but extreme, examples in Table 20.2 of the exports of shoes vs. commercial aircraft highlight the obvious fact that not all products/services are the same. Footwear is a highly differentiated market with a small knowledge base, frequently described as a craft art, while commercial aircraft is the opposite extreme, essentially a global mass market, but one with a large knowledge base. When we use the term 'knowledge base', we are including the variety of occupations/disciplines/paradigms, the amount of tacit knowledge, and the embodiment of knowledge in machines, software, people, and theories or ideas/information.

Less obvious is how to create a relatively simple typology that delineates the major differences between product/service types in a way that is relevant to our analysis of economic globalization and of knowledge

Table 20.2. Classification of product/service, organizational populations

Attribute	Type of organizational population			
	Craft	Mass	Large science	Small science
1. Knowledge base size	small	small	large	large
2. Market size	small	large	large	small
3. Size of organizations	small	large	large medium	small medium
4. Number of organizations	many	few	some	many
5. Per cent sales R&D	little	1–5%	6–15%	20% +
6. Type of innovation	process	process	product/process	product
7. Product life	fashion	long	medium	short
8. Impact on market	replace	replace	divide	divide
9. Impact on competencies	add	destroy	add	add

Notes:
- The craft/artisan set of organizational populations include the following: house remodelling and construction, footwear, restaurants, boutiques, police, fire, primary and secondary education, social welfare, agriculture, furniture, etc.
- The mass-market sectors include the following: bulk materials such as steel, coal, aluminium; consumer durables including office machines, household appliances, consumer electronics; glass products; containers; elevators; cigarettes; processed food products; and various services including railroads, insurance, banks, social-security system, prisons, etc.
- The large-science set of organizational populations include the following: generic software, semiconductors, automobiles, airplane construction, airline services, computers, chemicals, drugs, electrical products, hospital health services, large governmental laboratories, etc.
- Examples of small science include the following: medical instruments, biotechnology, material sciences, alternative energies, alternative transportation, specialized software, specialized machine-tools, robots, university departments, clinic health services, etc.

transformations. This poses an intriguing question. By what criteria would one choose a particular typology of organizational populations or sets of sectors? Obviously, the criteria that are selected in part determine the answer that one obtains in the analysis.

The typology suggested in Table 20.2 is a combination of those proposed by Hage (1980) and Pavitt (1984). It cross-classifies the size of the knowledge base with market size, and together these two dimensions predict the distinctive innovation patterns. A large number of examples—services as well as products—are provided, which is important, as the advanced industrialized countries are essentially becoming service economies and the problem of globalization increasingly affects services.

Several comments about this typology are worth making. First, the impact made by innovation in products or processes is different in each of the four quadrants. In the craft and mass markets, it results in replacement. (In mass markets, it destroys competencies but in craft markets it simply means an adjustment via new learning on the job.) In the two science markets, the impact of innovation is usually to differentiate the market place even more. As a consequence it does not destroy competencies but simply adds to the skills and routines that already exist. As Pavitt (1984) observed, in big science, the radical process and product innovation are usually combined. In small science, the emphasis is on product innovation.

Second, knowing the number of firms allows one to predict a large number of organizational characteristics of interest to industrial economists, management, and organizational sociologists, and found in these organizational populations (Hage 2006). In other words, this typology goes far beyond a simple categorization of patterns of innovation.

The disappearance of most of the organizations within a particular sector represents a major kind of societal change. When dealing at the population level, it is always possible that one or two companies can survive even as the population has largely disappeared, for example, a single US rubber-tire company, or

Harley-Davidson in the motorcycle industry. Here our concern is with the decline in the number of national firms producing a specific product/service in nationally owned plants in the liberal market economies such as the US and the other Anglo-Saxon countries. Nor does the movement of American production offshore into commodity chains represent a success, but rather a change in the mode of coordination from a corporate hierarchy to an internal interorganizational network along the supply chain. In fact, we find a continuation of the same policies that have led to the failure of the corporate hierarchy as an institutional coordination arrangement. A good example is the criticism of exploitation made against Nike, which is a prime example of a corporate hierarchy in a commodity chain (Gereffi and Korzenewicz 1994).

While the disappearance of most of the organizations within a sector reflects a societal change, it does not necessarily mean an institutional change. An institutional change would mean either the disappearance of the dominant coordination mode (whether this be a corporate hierarchy, an associational support hierarchy, or state-supported hierarchy), or its transformation into another kind of institutional coordination mode (such as an interorganizational network along the supply chain or the idea-innovation chain, an interdependent network, or a commodity chain in which a large firm dominates). Changes in the coordination mode must occur over a number of sectors before we could argue that there has indeed been an institutional change in a specific society.

Institutional change in the liberal market economies and path dependency in the coordinated market economies

As indicated above, our concern is with examining different types of sectors or organizational populations, because the impact of knowledge transformations and of economic globalization has been different in these sectors. Furthermore, it is not the case that the dominant institutional coordination mode is the same across all sectors. Typically, the dominant mode is most characteristic of the mass-production and service sectors, less so of the large-science sectors, and still less of the relatively new, small-science sectors. In addition, we want to be aware of exceptions—examples that deny the basic trend—because these can inform the general thesis.

Mass-production and service-provision sectors in the liberal market economies

As Chandler (1977) has demonstrated, the corporate hierarchy became the dominant mode in most of the liberal market economies for coordinating prices, wages, and supply of products during the first industrial revolution. The reasons vary across societies, but in general it was because market mechanisms were allowed to regulate, and the most productive firms gradually became monopolies or oligopolies. Each of them developed a dominant design and a characteristic production method—the assembly-line.

During the 1960s and the 1970s, because of growing demand for different kinds of models (Hage and Powers 1992), many of these mass markets became more differentiated. Flexible manufacturing developed as a radical solution for meeting a variety of market demands (Piore and Sabel 1984). A fact less frequently emphasized has been that, in some of these sectors, radical new designs emerged that also represented a major challenge to the hitherto most successful firms. Failing to adopt new process technologies in the US were most of these firms, most notably those in the steel industry. Jakimur (1986) in his study of flexible manufacturing makes clear that, in those organizations where it was introduced, it was poorly implemented; gains in productivity were not

achieved; most critically, none of the potential for flexibility was utilized. In the Foster *et al.* chapter, we have another example, that of American paper-producing mills, which failed to adopt the new technologies and are disappearing as a consequence, frequently being purchased by foreign firms that have adopted them. Other examples include the railroad industry (high-speed trains), rubber tires (radial tires), the marriage of electronics in sewing machines, and a number of mass-production industries. In most instances, this has meant that most, if not all, of the firms in these organizational populations have disappeared: today there remains only one American company that produces rubber tires; no American companies produce televisions. America now buys these products from overseas.

What is true for the corporate hierarchy in the US is also largely true for this coordination arrangement in the other liberal market economies. Zammuto and O'Connor (1992) report that many industries in these countries have failed to adopt flexible manufacturing.

Why did this institutional arrangement fail? Why did it not adopt the new process technologies and the new radical designs? Utterback (1994) has, of course, provided a number of reasons why the most successful firms fail to adopt new radical innovations: all investments had been made into the old, and the impact of the new was to destroy their competency (Anderson and Tushman 1990). But these arguments do not fully explain the failure to adopt flexible manufacturing. Why? The characteristics of corporate hierarchies that the comparative institutionalists isolate as explaining why America should have more radical innovation are precisely the ones that indicate why corporate hierarchies fail to adopt radical process technologies such as flexible manufacturing. Corporate hierarchies do not take risks: because they have flexible labor markets, it is easier for them to downsize and shift work overseas. Furthermore, given the history of labor conflict in the liberal market economies, the spirit of cooperation that is necessary to adopt radical process technologies or radical new standard designs in these industries is absent. Shaiken (1985) discovered that when corporate hierarchies did adopt flexible manufacturing, they used it as a control device with the hope of deskilling workers previously deemed skilled, rather than upgrading their skills and relying upon their tacit knowledge to exploit the advantages of this technology.

But this is only part of the answer; the explanation is more complex than this. Zammuto and O'Connor (1992) show that these corporate hierarchies are centralized and bureaucratic. This makes them slow to adopt new innovations, as a large literature on innovation has shown (Hage 1980). The previous institutionalization of success makes it difficult for management to recognize fundamental changes in the nature of the market or the technology. The prevalence of a kind of groupthink eliminates any questioning of the situation until it is too late.

As always there are some interesting exceptions. The major ones are those mass-production firms that followed a strategy of product innovation and maintained a basic and/or applied research center. For example, General Electric adopted flexible manufacturing in its consumer-products division, specifically washing machines, allowing it to become a much larger player in this mass-production sector. Light bulbs are another mass-production product that has adapted new radical designs and new process technologies. GE has played an important role in this industry, too. But then GE is not a corporate hierarchy. It is decentralized and has invested in basic research. When the first energy crisis occurred, it set as its goal the reduction of energy consumption in all of its products by 30 per cent, and achieved it by developing a number of new designs in many of its product areas.

Consistent with this argument are the mass-production firms of Corning Glass and Proctor and Gamble, which have for many decades pursued policies of aggressive, radical product innovation. Even if the population at large dies by failing to adopt radical innovations, single organizations in this category can survive. But these exceptions prove the rule.

Another important qualification is that some of the American corporate hierarchies in the first part of the twentieth century followed a strategy of aggressive product and process innovation, which then largely weakened and even disappeared in the post-Second World War period. RCA and General Motors are two notable examples. This shift away from a strategic policy of innovation needs further exploration by economic historians working in the Chandlerian tradition. I would suggest that one of the reasons was the rise in the power hierarchy of the accountant that led to the emphasis on profits and short-time horizons that are used to categorize the limitations of the corporate hierarchy. A probable reason for this strategic shift is the kind of training received in the most prestigious business and engineering schools in the US, which has tended to emphasize general models of thinking and the importance of profitability.

Mass-production and service-provision sectors in the coordinated market economies

It would, however, be wrong to conclude that the failure to adopt radical process technologies, such as flexible manufacturing, or radical product innovations, such as new product designs, has been uniform across the developed economies. The differences are explainable by the nature of the institutional context and, to a lesser extent, by the nature of the state interventions. As a summary statement, corporate hierarchies did not respond well, whereas associational supported hierarchies and interorganizational networks along the supply chain did respond well *and for the same reason*.

A good example of the associational supported hierarchy is Germany. Flexible manufacturing has been adopted not only in the mass-production sectors, but in many of those industries considered craft or artisan industries (Steedman and Wagner 1987, 1989). And radical new designs, such as high-speed trains, have been taken up quickly. But it also must be said that, in many of the mass-production markets, the associational supported hierarchies have positioned themselves in high-quality niches such as specialized machine tools, instruments, electrical products, specialty chemicals, etc.

Germany tends to be the country most frequently studied in the comparative capitalism literature (Hall and Soskice 2001; Hollingsworth 1997). But the advantages of associational hierarchies are not limited to Germany. Walton (1987), in his pioneering study of the shipping industry, has demonstrated that this institutional arrangement led to the adoption of radical shipping technologies, even when it slightly reduced employment in the Netherlands. In the contribution by Foster *et al.*, we observe that another associational hierarchy country, Sweden, has more quickly adopted the new production technologies in paper manufacturing.

The dramatic issue involved in flexible manufacturing is its impact on employment. Adoption of radical knowledge transformation thus requires some kind of institutional arrangement that allows for negotiation between labor and management. This is where associational supported hierarchies have an advantage. They provide a forum where the adoption's impact on employment can be reduced, making the radical process technology more acceptable.

Another important difference is that associational supported hierarchies have coordinated technical and vocational education. This has increased the size of the knowledge base, easing the adoption of radical new process technologies and new designs (Streeck *et al.* 1987), because their skilled labor forces reduce the costs of retraining, or at least render this task easier. One could argue that, in some technical sense, because of the greater reliance on technical education and the production of specialized market niches, these are not mass-production industries. In terms of the framework in the introductory chapter, some firms in the same sector may have a large knowledge base and, because of this, adopt radical new

process technologies and new standard designs more quickly and easily. Streeck *et al.* (1987) would stress that lifetime guarantees of work ensure that the impact of the new technology on employment is kept to the minimum, and provide a framework for how the adjustments are made. But I would suggest that these guarantees of lifetime work are a consequence of the associational supported hierarchical method of coordination.

The same arguments can be applied to the Japanese interorganizational networks along the supply chain, where technical education appears to be less of a factor. Contrary to the common assumption that the Japanese do not have radical innovation, the same study that demonstrated the US's failure to adopt flexible manufacturing found that Japan had adopted it in many more industries and with much greater gains in productivity and flexibility (Jakimur 1986). Enormously successful have been the new standard design in the railroad industry, and both new standard designs and radical process technologies in the consumer-electronics industry. And in the contribution of Nonaka and Peltokorpi to this book, we have the account of how the staid Japanese telephone company pioneered a successful radical product innovation.

Perhaps the most telling example of radical product innovation in Japan is the case of robots in manufacturing. Although the US developed many of the patents, it is the Japanese who have steadily improved on the basic design, and now American firms import their robots primarily from Japan and secondarily from Germany. The really interesting question is why American companies did not pursue this radical innovation more determinedly. It appears, yet again, that the corporate hierarchies were unwilling to develop this product further until it was commercially profitable. This illustrates the unwillingness to accept risks that is characteristic of this institutional mode of coordination.

In other words, even though the institutional coordination arrangements are somewhat different and have a different historical basis, Japan's case is quite parallel to

Germany's. As in Germany, there has been a policy of lifetime employment that increases the security of the worker. The interorganizational supply-chain networks provide a cooperative arrangement in which the various problems of the adoption of radical product innovations can be discussed, and their negative consequences mitigated.

The case of state-supported hierarchies lies somewhere in between the associational and interorganizational, network-coordinated market economies and the liberal market economies because, while the state does coordinate, it has some of the same myopia associated with the corporate hierarchies in the liberal market economies. In the new high-tech areas, the state-supported hierarchies have been slow to develop. In those sectors where the state controls a major interest, then new process technologies tend to be adopted: the French state, for example, has pioneered the development of radically new designs in telecommunications, in urban transport, in high-speed trains, in nuclear energy and long-distance utility lines. France may be exceptional in its support of some of the country's major hierarchies in their expansion overseas when they have purchased other companies to gain more global-market share: it supported Thomson when it purchased the television-making facilities of RCA and GE in the US. In nickel steel, the French company L'Oréal has achieved a world monopoly. There are probably other cases of which I am not aware.

Another important qualification is that not all large firms in a specific organizational population in a particular country are necessarily the same, or have the same institutional coordination mode, whether it be corporate hierarchies, associational supported hierarchies, or state-supported hierarchies, or interorganizational networks along the supply chain. Examples are firms that followed a policy of investing in R&D by having a corporate research unit, even though this is not typical within these organizational populations, as we have already observed in the case of the US. A case in point is Michelin, the French rubber-tire company, which has avoided contacts with the French state and therefore is not a state-supported hierarchy; another French example is L'Oréal.

Large-science product and service sectors in the liberal market economies and in the coordinated market economies

The importance of making distinctions by sector is amply illustrated when we move from the mass-production to the large-science sectors or sets of organizational populations. Firms with large knowledge bases and mass markets emerged in the second industrial revolution (Landes 1969). The German dye companies, for instance, established R&D research departments between 1877 and 1883 and by 1900 had diversified into special dyes, pharmaceuticals, photographic products, plastics, and artificial fibres. In America, GE's first national laboratory was created in 1900, while AT&T and Kodak founded their industrial laboratories between then and 1914.[4] DuPont founded its first laboratory in 1902 and by 1927 had begun performing basic scientific research in polymer chemistry that led to its many radical innovations such as nylon, rayon, corfam, gortex, etc. This means that these sectors evolved quite differently from the previous ones: rather than consolidating into a few firms, they expanded as more and more firms moved into the new markets created by the R&D, as was suggested in Table 20.2.

Two general comments about this sector. First, in the liberal market economies, not all of these sectors are dominated by the institutional coordination mode of corporate hierarchies. Nor do we necessarily have either associations or interorganizational networks in some of these sectors in the coordinated market economies. Second, in some sectors and in both types of economies, this mode has changed, or is in the process of changing, towards interorganizational networks. Let us compare four industrial sectors where there has been an evolution from corporate hierarchies towards interorganizational modes in the liberal market economies, and compare their performance with the equivalent industries in the coordinated market economies to highlight the consequences of different kinds of institutional coordination modes for societal change.

First, let us consider the automobile industry. This industry has probably seen some of the most striking increases in R&D expenditures across time from a few per cent in the 1950s and 1980s to now more than 5–7 per cent. The radical innovations have tended to be in new components (seat belts, electronic controls on the engine, catalytic converters), although there have also been some radical new product designs, such as SUVs, and most recently the hybrid car, as illustrated in the contribution by Nonaka and Peltokorpi. Finally, it should be noted that most of the innovations in this industry have come from either the Japanese or the German automobile industry, examples of coordinated market economies, while the liberal market economies have played catch-up and, in one important case, Britain, completely disappeared.

Even though the US corporate hierarchies in this industry did spend money on R&D, they were more like their counterparts described above: centralized and slow to move in the direction of more interorganizational networks along the supply chain. Since the mid-1980s, this US industry has been steadily losing market share, and now the two remaining American-owned companies account for less than 50 per cent of the cars produced in the US. In the 1950s, General Motors alone had this percentage of the market. Not surprisingly the two countries now accounting for the majority of cars sold in the US are Germany and Japan, with their different kind of institutional coordination arrangement. Furthermore, the Japanese automobile industry has moved the furthest towards the development of mass customization.

The same failure of adaptation to changing market and technological conditions also occurred in the corporate hierarchies in the British automobile industry. This provides additional evidence about my theory that corporation hierarchies do not respond well to innovation. This is potentially a very powerful argument for the theory, namely if there is no evolution in the amount of money spent on R&D and in structural change, then the industry in that country disappears.

Second, the same basic pattern observed in the automobile industry, especially in the US institutional pattern, has been duplicated in the commercial aircraft industry. Boeing, created as one large company (though with some interorganizational linkages) used to dominate; Airbus, created as an interorganizational network, has now become the dominant player, with more than 50 per cent of the market. But it is a very special case, because it is a multi-state-supported interorganizational network along the supply chain and the idea-innovation chain.

Third, a different pattern has emerged in another very important industry: semiconductors. In the beginning, American firms excelled in this industry as they had previously in automobiles and aircraft construction, then lost their leadership to Japan with its interorganizational network institutional coordination mechanism. But in this industry, the pace of radical product and process change in semiconductors is much faster than in the previous two, occurring at a regular interval of less than two years. In this industry, American firms evolved structurally and were able to recapture their dominance. They were saved, thanks to the research consortia SEMATECH (Browning et al. 1995), an example of an interorganizational network for research among competitors, which makes it different from the interorganizational supply-chain hierarchies found in Japan, the major competitor that dominated in this sector until the founding of SEMATECH.

Why did the American firms evolve towards research consortia among the competitors, and why have the Japanese firms found this difficult to do? And why have the American firms in semiconductors not had the same problems of evolving structurally as did those in the automobile industry or the aircraft-construction industry? The reasons appear to be that, in this much newer sector, there was not enough time for a rigid corporate hierarchy to develop, and the rapid pace of innovation meant that these firms were more flexible. And they had been, from the beginning, much more decentralized than their counterparts in the automobile and aircraft-construction industries. Meanwhile, though Japanese semiconductor companies entered into various kinds of global alliances, they were unable to create research consortia such as SEMATECH within Japan. In other words, the interorganizational network along the supply chain has difficulty creating research consortia between these supply chains or interorganizational networks because of competitive pressures (Aldrich and Sasaki 1995). As a consequence the Japanese semiconductor industry has been declining. It is not just because of the resurgence of the US; South Korea and Taiwan, other countries that have relied upon interorganizational networks along the idea-innovation chain, and in quite different ways, have shown strength too.

Fourth, the telephone industry, which has become the telecommunications industry, has had essentially the same pattern of institutional control in both liberal market economies and in coordinated market economies. In the former case, it was a corporate hierarchy carefully regulated by the state, AT&T being the example, while in most other countries it has been a state-supported hierarchy that tightly regulated the industry as well. However, the innovation pattern in the telecommunications industry has been quite different from that in semiconductors.[5] Rather than keeping a more or less constant and rapid rate of innovation over several decades, this industry has exploded with a number of radical product and process innovations in the last decade. A proliferation of end-of-the-line uses, including fixed and mobile phones, faxes, computers, as well as interlinked systems including voice, data, and mobile voice, date and text capabilities, has occurred. At the same time, a number of the radical process technologies have been developed, including digitalization, optical fibers, the internet, and GSM. Sector boundaries between computer science, telecommunications, and multi-media have become blurred; the products themselves are much more complex and depend upon a variety of scientific areas. All of this required major structural changes, among them the creation of both supply-chain and idea-innovation-chain networks.

In the US, AT&T was an associational hierarchy and had the same difficulty in responding. Perhaps the most interesting example of failure in this institutional pattern is that Bell Labs developed many of the radical innovations mentioned above, but then never implemented them effectively into the telephone service. What helped the US industry was the deregulation of the company that created a number of 'Baby Bells,' and these new companies adopted a number of new innovations.

In other countries, the telephone service was a state-controlled hierarchy with a lack of response to the radical product and process innovations. What changed the state-owned companies in Europe was the impact of innovations in the other industrial sectors, and the particular influence of the mobile phone and Internet services. The response was to separate the telephone service from the postal service, and allow it to function as a separate company. It rapidly became a technological leader. Thus, we observe that, before the separation of the telecommunication services from the postal service, the state-owned companies had all the same characteristics as corporate hierarchies, indicating the impact of centralization on innovation.

Let us now consider some industries that, in the liberal market economies, have not had corporate hierarchies, but instead relied more upon their own basic research and frequently interorganizational networks. The computer and software companies are probably the best examples, and they have done quite well. A major reason for this different institutional pattern is because of Silicon Valley (Saxenian 1994) and, to a lesser extent, Texas Instruments centered in Austin, Texas.

The one interesting exception is the former corporate hierarchy, IBM. This company was rapidly declining in the late 1980s, but quickly moved towards interorganizational networks in the early 1990s and arrested its decline, although in the process it shed about 25 per cent of its employment worldwide. Again, we have an exception that proves the rule. Corporate hierarchies can survive if they change their corporate hierarchy to another form of institutional coordination, more particularly, an interorganizational network along the idea-innovation chain if they are located in the large-science sector. Today, most of the major players in the sector of computers have moved towards idea-innovation networks, usually of the global kind, especially when it involves the development of new product that requires a global standard (Gomes-Casseres 1996).

A second sector that has been successful in both the liberal market economies and in the coordinated market economies, but is also an exception in both circumstances because it is neither controlled by corporate hierarchies nor by associational hierarchies, is the sector that spends the most money on R&D, the pharmaceutical industry. It has made the greatest structural evolution towards an idea-innovation chain, and thus is probably a good predictor of what will happen in other industries as their R&D expenditures increase. The American firms have had much higher knowledge bases, are decentralized, and have been organic (Burns and Stalker 1961; Zammuto and O'Connor 1992; Hage 1999) in form.

The paradigmatic shift in biotechnology, which has led to the possibility of manipulating genes in both what is called the 'red biotechnology area' (therapies for humans) and the 'green biotechnology area' (that is for food consumption), has meant that pharmaceutical companies have had to rely on small biotech companies to do the trial-and-error research associated with new therapies. In turn, the pharmaceutical companies engage in very expensive and multiple clinical trials (see the contribution by van Waarden and Oosterwijk). Here, the US companies have been quite successful in moving towards the idea-innovation chain network. In contrast, the associational supported hierarchy has not acted as quickly.

Why is structural evolution necessary? The growth in R&D expenditures has entailed a growth in the complexity of the product/service and the need for more specialized knowledge (Hage and Hollingsworth 2002). Both processes increase the need for interorganizational networks that integrate the component parts in a complex assembled product or the sources of basic and applied science, and usually both. As evidence of this differential evolution, the locus of R&D research remains corporate in the chemical and automotive industry, but has moved to interorganizational networks in biopharmaceutical, semiconductor/electronics, information technology, and telecom hardware and services. Furthermore, the number of research partnerships continues to grow at a steady pace, averaging about 600 per year.

Small science product and service sectors in the liberal market economies and in the coordinated market economies

Since these sectors are new ones, we tend not to find corporate hierarchies in the liberal market economies; in the coordinated market economies, the major message is the attempt of the state to jump-start new industries in this area, not always with success. The classic example is biotech which, as we have seen, is connected into interorganizational networks, usually, but not always, with universities and with pharmaceutical companies. Casper reviews the various attempts by the coordinated market economies to start or increase their biotech sector.

In many instances in these new sectors, we find that a number of the firms are concentrated in what might be called 'research industrial districts' that have interorganizational networks. The best example for the liberal market economies is Silicon Valley, but is not the only one.

Depending upon whether one considers the technologies involved in film-making to be high-tech or not, corporate hierarchies in the movie industry have given way to interorganizational networks, as Storper (1997) has observed. Also Scott (1998) documents the emergence of the mixed-media high-tech area, which is also located in California.

These various examples suggest that small high-tech sectors do well when they are connected into interorganizational networks in the liberal market economies, and this appears to be a development that can occur without state intervention. But in the coordinated market economies, it is more typically the state that intervenes to foster the development of these industries. Besides the examples provided in Casper's contribution, there are:

- state-sponsored networks of small high-tech companies, as found in Taiwan (Matthews 1997) and in the Netherlands (Meeus *et al.* 1999);
- state-sponsored technical parks that have a number of joint ventures, but have not yet emerged as interorganizational networks as such (Monck 1988);
- state-sponsored research centres located at universities that involve the participation of private-firm researchers, with academic researchers focused on some fundamental science/technology area, as found in the US.

The important point about this list is that it is not only the coordinated market economies that are facilitating small high-tech companies, but the liberal market economies as well.

Small-craft product and service sectors in the liberal market economies and in the coordinated market economies

Not much can be said about the small-craft product and service sector, because it is much less studied. But in the few cases that are available, the comments made in our discussion of the mass markets about the differences between the liberal and the coordinated market economies apply too. In general, many small services are local and therefore not controlled by corporate hierarchies in the liberal economies, but may be controlled by associations in the coordinated market economies.

On the product-sector side, a number of the associated organizational populations have been decimated in both the liberal and coordinated market economies because of globalization. The imports of shoes, textiles, toys, etc. have largely eliminated these American companies. The major exception is, of course, middle Italy with its industrial districts and interorganizational networks that Piore and Sabel (1984) made so famous in their second industrial divide. Other examples are lesser-known industrial districts that fit the same pattern, such as haute couture in Paris. The minor exception is those corporate hierarchies that evolved into interorganizational networks along the supply chain, that is, into commodity chains. These exceptions would appear, again, to support the general rule about the disappearance of corporate hierarchies in the liberal market economies, as well as in many of the coordinated market economies, unless there are industrial districts with associations and interorganizational networks. This tends to be less typical in this set of organizational populations.

In summary, although centralized corporate hierarchies have evolved towards interorganizational networks in some sectors and in some countries, they have been slow to do so and have lost market share to associational supported hierarchies and interorganizational supply-chain networks. Decentralized and organic hierarchies have tended to evolve structurally much more quickly, and are doing well. In contrast, interorganizational networks along the supply chain have not been able to move as easily towards research consortia. Nor have associational networks been able to move as easily towards the idea-innovation chain.

Discussion and managerial implications

The chapter began with a distinction between societal and institutional change. The former is more common, and is represented by the addition of academic disciplines, government responsibilities, industrial sectors, and health and welfare programs. In contrast, institutional change that is defined as changes in the nature of the coordination mode is much rarer.

The basic thesis is that the liberal market economies have experienced a considerable amount of institutional change, as corporate hierarchies have largely disappeared and been replaced by interorganizational networks, not only along the supply chain but increasingly across the idea-innovation chain as well. In contrast, the coordinated market economies have followed more of a path-dependent model because of the competitive advantages of associational hierarchies and interorganizational networks along the supply chain.

In providing evidence for this somewhat bold assertion, four distinctive kinds of industrial sectors have been surveyed. This was necessary because much of the literature comparing the liberal market economies and the coordinated market economies focuses primarily on either the mass-production sectors or the large-science sectors, that is, the large rather than the small organizations, with the notable exception of biotech. In the liberal market economies, most of the firms associated with the mass-production sectors have either disappeared or have been purchased by foreign capital. In contrast, their counterparts in the coordinated market economies have survived much better. The reason for this is that the associational hierarchy and the interorganizational network along the supply chain have adopted the radical process technologies, such as flexible manufacturing and new radical product designs, much more quickly and completely, giving them a competitive edge in productivity and therefore in price.

When large-science-sector firms in the liberal market economies have had interorganizational networks, they have competed effectively. In some sectors, they have been evolving towards this, but perhaps not fast enough to save the firms from being either closed or purchased by foreign capital. In the coordinated market economies, the large-science firms have had these institutional arrangements, and have accordingly performed reasonably well.

Does this mean that there is no institutional change in the coordinated market economies? The answer is that there is some, as the state in these countries attempts to encourage the development of new industrial sectors in the small-science sectors and, perhaps more critically, interorganizational relationships along the idea-innovation chain.

Both liberal market economies and coordinated market economies have been losing some of their craft industries as a consequence of globalization. Again, the exception proves the general rule. Industrial districts such as those in middle Italy, with their associational hierarchies and interorganizational networks along the supply chain, have survived.

What are the managerial implications of this? First, that rapidly adopting radical process technologies and new product designs is fundamental for the survival of the firm. Second, to do this requires a decentralized organization that is developing cooperative relationships at least along its supply chain, if not also along the idea-innovation chain in those sectors where basic scientific research is an important element in product innovation. Third, the exact configuration of interorganizational networks does vary by sector.

Appendix: Definitions of major concepts

Institutional change	Any substantial alteration in the dominant mode of coordination that supplements markets
Institutional path dependency	Continuity in the dominant mode of coordination that supplements markets
Societal change	An addition (or subtraction) of programs the major institutional sectors of science and education, government and the military, the economy, and health
Knowledge transformations	Radical product and/process innovation

Notes

1. The role of patents is decreasing for three reasons: (1) the pace of change makes the expense of patents less worthwhile; (2) small changes in many product sectors allow competitors to largely duplicate the same product; and (3) some countries such as India are refusing to pay for patents, especially in the drug industry where they have been most profitable.
2. And this is an important limitation. Ideally one would like to have a similar effort relative to political institutions and the varieties of states that exist.
3. However, there are some who argue that Denmark, for example, is nothing but one large industrial district and, certainly, a similar case can be made for Singapore.
4. Here an important distinction is being made between the research work of the entrepreneur innovator such as Bell, Edison, or Siemens and the creation of a research laboratory with the strategy of developing new products.
5. Much of the material in these paragraphs comes from Oosterwijk and van Waarden 2003.

References

Aldrich, H., and Sasaki, T. (1995). 'R&D consortia in the United States and Japan.' *Research Policy*, 24: 301–16.

Alter, C., and Hage, J. (1993). *Organizations Working Together: Coordination in Interorganizational Networks*. Beverly Hills, CA: Sage.

Anderson, P., and Tushman, M. (1990). 'Technological Discontinuities and Dominant Designs: A Cyclical Model of Technological Change.' *Administrative Science Quarterly* 35: 604–33.

Archibugi, D., and Pianta, M. (1992). 'The Technological Specialization of Advanced Countries.' A report to the EEC on International Science and Technology Activities. Boston: Kluwer.

Browning, L., Beyer, J., and Shetler, J. (1995). 'Building Cooperation in a Competitive Industry: SEMATEC and the Semiconductor Industry'. *Academy of Management Journal*, 38: 113–51.

Burns, T., and Stalker, G. M. (1961). *The Management of Innovation*. London: Tavistock.

Campbell, J., Hollingsworth, J. R., and Lindberg, L. (eds.) (1991). *Governance of the American Economy*. New York: Cambridge University Press.

Chandler, A. (1977). *The Visible Hand: The Managerial Revolution in American Business*. Cambridge, MA: Harvard University Press.

Daly, A., Hitchens, D. M., and Wagner, K. (1985). 'Productivity, Machinery and Skills in a Sample of British and German Manufacturing Plants.' *National Institute Economic Review*. February: 48–61.

Doz, Y. L., and Hamel, G. (1998). *Alliance Advantage: The Art of Creating Value through Partnering*. Boston: Harvard Business School Press.

Dussauge, P., and Garrette, B. (1999). *Cooperative Strategy: Competing Successfully through Strategic Alliances*. Chichester, NY: Wiley.

Edquist, C., and Hommen, L. (1999). 'Systems of Innovation: Theory and Policy for the Demand Side.' *Technology in Society*, 21: 63–79.

Gereffi, G., and Korzeniewicz, M. (eds.) (1994). *Commodity Chains and Global Capitalism*. Westport, CN: Preager.

Gomes-Casseres, B. (1996). *The Alliance Revolution: The New Shape of Business Rivalry*. Cambridge, MA: Harvard University Press.

Guerrieri, P., and Tylecote, A. (1998). 'Interindustry Differences in Technical Change and National Patterns of Technological Accumulation.' In C. Edquist (ed.), *Systems of Innovation: Technologies, Institutions and Organizations*. London: Pinter, 108–29.

Hage, J. (1980). *Theories of Organizations: Form, Process, and Transformation*. New York: Wiley.

—— (1999). 'Organizational Innovation and Organizational Change.' *Annual Review of Sociology*, 25: 597–622.

—— (2006) 'Knowledge and Societal Change: Institutional Coordination and the Evolution of Organizational Populations.' Forthcoming in Arnaud Sales and Marcel Fournier (eds.), *Knowledge, Communication and Creativity*. Sage Studies in International Sociology. London: Sage.

—— and Hollingsworth, J. R. (2002). 'A Strategy for Analysis of Idea Innovation Networks and Institutions.' *Organization Studies* (Special issue: *The Institutional Dynamics of Innovation Systems*), 5: 971–1004.

—— and Powers, C. (1992) *Post Industrial Lives*. Newbury Park, CA: Sage.

—— Collins, P., Hull F., and Teachman, J. (1993). 'The Impact of Knowledge on the Survival of American Manufacturing Plants.' *Social Forces*, 72: 223–46.

Hagedoorn, J. (1993). 'Strategic Technology Alliances and Modes of Cooperation in High-Technology Industries.' In G. Grabher (ed.), *The Embedded Firm: On the Socioeconomics of Industrial Networks*. London: Routledge, 116–38.

Häkansson, H. (1990). 'Technological Collaboration in Industrial Networks.' *European Management Journal*, 8: 371–379.

Hall, P., and Soskice, D. (eds.) (2001). *Varieties of Capitalism*. Oxford: Oxford University Press.

Harbison, J. R., and Pekar, P. P. (1998). *Smart Alliances: A Practical Guide to Repeatable Success*. San Francisco: Jossey Bass.

Hollingsworth, J. R. (1991). 'The Logic of Coordinating American Manufacturing Sectors.' In J. L. Campbell, J. R. Hollingsworth, and L. Lindberg (eds.), *The Governance of the American Economy*. Cambridge: Cambridge University Press, 35–73.

—— (1997). 'Continuities and Changes in Social Systems of Production: The Cases of Japan, Germany, and the United States.' In J. Rogers Hollingsworth and Robert Boyer (eds.), *Contemporary Capitalism: The Embeddedness of Institutions*. New York: Cambridge University Press, 265–310.

—— and Boyer, B. (eds.) (1997). *Contemporary Capitalism: The Embeddedness of Institutions*. New York: Cambridge University Press.

Inkpen, A. C., and Dinur, A. (1998). 'Knowledge Management Processes and International Joint Ventures.' *Organization Science*, 9: 454–68.

Jakimur, R. (1986). 'Postindustrial Manufacturing.' *Harvard Business Review*, 64: 69–76.

Jarillo, J. C. (1993). *Strategic Networks: Creating the Borderless Organization*. Oxford: Butterworth-Heinemann.

Kitschelt, H. (1991). 'Industrial Governance, Innovation Strategies, and the Case of Japan: Sectoral Governance or Cross-national Comparative Analysis?' *International Organization*, 45: 453–93.

Kogut, B., Shan, W., and Walker, G. (1993). 'Knowledge in the Network and the Network as Knowledge: The Structuring of New Industries.' In G. Grabher (ed.), *The Embedded Firm: On the Socioeconomics of Industrial Networks*. London: Routledge, 67–94.

Landes, D. (1969). *The Unbound Prometheus: Technological Change and Industrial Development in Western Europe from 1750 To Present*. Cambridge: Cambridge University Press.

Lundvall, B.-Å. (1992). *National Systems of Innovation: Towards a Theory of Innovation and Interactive Learning*. London: Pinter.

Malerba, F., and Orsenigo, L. (1993). 'Technological Regimes and Firm Behavior.' *Industrial and Corporate Change* 2: 45–71.

—— —— (1997). 'Technological Regimes and Sectoral Patterns of Innovative Activities.' *Industrial and Corporate Change*, 6: 83–117.

Matthews, J. A. (1997). 'A Silicon Valley of the East: Creating Taiwan's Semiconductor Industry.' *California Management Review*, 39: 26–55.

Meeus, M., Oerlemans, L., and Hage, J. (1999). 'Interactive Learning: Varieties of Learning Relationships.' Unpublished paper, Eindhoven Centre for Innovation Studies, Faculty of Technology Management, Eindhoven University of Technology, Eindhoven, the Netherlands.

Mockler, R. J. (1999). *Multinational Strategic Alliances*. Chichester, NY: Wiley.

Monck, C. S. P. *et al.* (1988). *Science Parks and the Growth of High-Technology Firms*. London: Croom Helm.

Nelson, R. R. (ed.) (1993). *National Innovation Systems: A Comparative Study*. Oxford: Oxford University Press.

O'Doherty, D. (ed.) (1995). *Globalisation, Networking, and Small Firm Innovation*. London: Graham and Trotman.

Oosterwijk, H., and van Waarden, F. (2003). 'The Architecture of the Idea-Innovation Chain.' Unpublished paper, School of Public Policy, Utrecht University.

Pavitt., K. (1984). 'Sectoral Patterns of Technical Change: Towards a Taxonomy and a Theory.' *Research Policy*, 13: 343–73.

Piore, M., and Sabel, C. (1984). *The Second Industrial Divide: Possibilities for Prosperity*. New York: Basic Books.

Powell, W. W. (1998). 'Learning from Collaboration: Knowledge and Networks in the Biotechnology and Pharmaceutical Industries.' *California Management Review*, 40: 228–41.

Schumpeter, J. (1934). *The Theory of Economic Development*. Cambridge, MA: Harvard University Press.

Scott, A. J. (1998). 'From Silicon Valley to Hollywood: Growth and Development of the Multimedia Industry in California.' In H.-J. Braczyk, P. Cooke, and M. Heidenreich (eds.), *Regional Innovation Systems*. London: UCL Press, 136–62.

Shaiken, H. (1985). *Work Transformed: Automation and Labor in the Computer Age*. New York: Holt, Rinehart and Winston.

Steedman, H., and Wagner, K. (1987). 'A Second Look at Productivity, Machinery and Skills in Britain and Germany.' *National Institute Economic Review*, November: 84–95.

—— —— (1989). 'Productivity, Machinery and Skills: Clothing Manufacture in Britain and Germany.' *National Institute Economic Review*, 128: 40–57.

Storper, M. (1997). *The Regional World: Territorial Development in a Global Economy*. New York: Guilford Publications.

Streeck, W., Josef, H., van Kevelaer, K-H., Maier, F., and Weber, H. (1987). 'Steurung und Regulierung der Beruflichen Bildung: Die Rolle der Sozialpartner in der Ausbildung und Beruflichen Weiterbildung in der BR Deutschland.' Berlin: Wissenschaftszentrum.

Utterback, J. M. (1994). *Mastering the Dynamics of Innovation: How Companies Can Seize Opportunities in the Face of Technological Change*. Boston: Harvard Business School Press.

Van de Ven, A. H., and Polley, D. (1992). 'Learning While Innovating.' *Organization Science*, 3: 92–116.

Walton, R. (1987). *Innovating to Compete: Lessons for Diffusing and Managing Change in the Workplace*. San Francisco: Jossey Bass.

Whitley, R. (1992a.) *Business Systems in East Asia: Firms, Markets and Societies*. London: Sage.

—— (1992b). *European Business Systems: Firms and Markets in their National Context*. London: Sage.

—— (1999). *Divergent Capitalisms: The Social Structuring and Change of Business Systems*. Oxford: Oxford University Press.

Zammuto, R., and O'Connor, E. (1992). 'Gaining Advanced Manufacturing Technology Benefits: The Role of Organizational Design and Culture.' *Academy of Management Review*, 17: 701–28.

21

Exporting the Silicon Valley to Europe: How Useful is Comparative Institutional Theory?

Steven Casper

Introduction

This chapter explores the link between national systems of innovation and country performance in the new economy, focusing on small entrepreneurial technology firms and the Silicon Valley model.

The Silicon Valley model is a template of company organization in industries characterized by rapid technological change, high returns for new product innovators, and, typically, strong intellectual property regimes. These characteristics prompt races among large numbers of entrants to invent, patent, and commercialize new technologies successfully. Firms competing in new technology markets must engage in a common set of activities, including accessing novel technologies from universities, recruiting and motivating highly skilled scientists and engineers, and obtaining high-risk finance. The key claim examined in this chapter is that the viability of each of these activities is dependent on the architecture of institutional factors that are primarily national, and which surround the organization of university research and technology transfer systems, labor markets, and financial systems.

A leading body of institutional theory, the framework known as 'varieties of capitalism,' makes strong predictions about country performance in developing patterns of innovation associated with the Silicon Valley model. Countries organized into what Hall and Soskice (2001) call 'liberal market' economies— such as the US and UK— should excel in developing entrepreneurial technology firms, while those with 'coordinated market economies,' such as Germany or Sweden, should fail in new economy industries such as biotechnology and software.

This chapter will examine this argument and then confront it with evidence suggesting that entrepreneurial activity in new technologies has been on the rise in Europe. Policymakers and entrepreneurs have proved far more optimistic than most academic commentators about importing the Silicon Valley model to Europe: they have assumed, despite the institutional obstacles to supporting new technology firms (such as the lack of appropriate financial institutions to back venture capital) noted in policy debates, that the necessary institutions can be created, and have enacted a wide-ranging series of institutional reforms and technology policies to spur entrepreneurialism.

This leads to a general puzzle: is it possible to reconcile the apparent success of European economies in importing the Silicon Valley model with institutional theory? Can comparative research on national innovation systems say anything useful on this topic? In particular, can this perspective push forward the debate on issues such as the drivers of entrepreneurial activity, the impact of institutions on sustaining innovative activity, and, more broadly, the types of public policies that can be effective in importing new models of economic organization into heterogeneous institutional environments? The chapter will address these themes, drawing on recent studies from an ongoing project on the development of the software and biotechnology industries in Germany, the UK, and Sweden, and from additional research on other European countries when available.

The chapter begins by exploring the varieties-of-capitalism argument: it summarizes how the framework links institutions to different types of commercial innovation, paying special attention to new technology companies associated with clusters such as Silicon Valley; it also briefly surveys evidence of the recent upsurge of entrepreneurial activity in Europe that appears to invalidate the theory. The chapter then examines two waves of more recent research that, in some respects, reinvigorate institutional perspectives on innovation.

The first approach draws on 'sectoral systems of innovation' research (Malerba 2004). One way to validate varieties-of-capitalism theory is to suggest that the characteristics of European companies in the new economy actually resemble those suggested by the theory. A key contribution from the theory is the development of micro-foundations linking institutions to innovation. This research links the technological characteristics of various market places to a number of organizational or competency dilemmas that are faced by firms, and then suggests that institutional characteristics influence the credibility of commitments made between managers, employees, and investors of firms. This research provides a more subtle understanding of the mechanisms by which institutions impact the activities of firms and other actors within the economy. One of its important results is to create a more sophisticated conceptualization of radical versus incremental types of innovation, showing that most sectors of the new economy, such as biotechnology and software,

have subsectors with different innovative characteristics. Though some important exceptions exist, patterns of subsectoral specialization across publicly listed new technology firms in the UK, Germany, and Sweden conform to expectations of varieties-of-capitalism theory.

While the sectoral research into systems of innovation helps, there are examples of significant clusters of firms appearing to adopt the wrong type of subsector specialization. Recent research on early stage (pre-IPO) firms has exposed, for example, a large number of radically innovative biotechnology firms in Germany, while studies of the Internet software industry routinely point to Sweden as home to many of the industry's most innovative companies. Many of these firms are failing (supporting institutional theory in a perverse way). But why do so many of these firms even exist?

The second wave of more recent research explores these problematic cases. These studies help point to an important problem with institutional theory: actors are not institutionally reflexive enough—they do not develop strategies or organizational characteristics according to the logics prescribed by institutional analysis. The drivers of activity often differ from those specified by the theory. For example, large companies in the software industry appear to have had a dominant role in creating conditions in coordinated economies supportive of innovations by small firms. Moreover, when institutional factors appear to be important, the drivers are often different institutions from those specified in the theory. In the case of German biotechnology, academic research-system incentives appear to be dominant in determining the technology strategies of start-up firms. The chapter will explore both these cases, focusing on a variety of drivers of sustainable labor-market organization within high-technology clusters (Almeida and Kogut 1999).

The chapter concludes with a summary of key issues for institutional research raised by the discussion, and then ends with a discussion of public-policy implications of the analysis.

The Silicon Valley model as an arena to study institutions and innovation

One way to analyze the types of entrepreneurialism found in technological hotbeds such as Silicon Valley is through examining the types of competency development found consistently within these firms (for overviews of Silicon Valley, see Saxenian 1994; Kenney 2000). The entrepreneurial business models organized within small innovative firms are associated with the development of four key competencies: the management of high-risk finance, the development of human resources within a competency-destroying environment, the creation of sufficiently high-powered motivational incentives for personnel, and access to primarily university-based technology. Briefly examining these competencies helps clarify possible roles of institutional frameworks in their governance.

Managing high-risk finance

Successful technology start-ups often create enormous financial returns. However, high technological volatility, reliance on often unproven business models, and uncertainty surrounding the ability to capture returns from R&D can produce substantial financial risks. The large costs of R&D and marketing, coupled with low profitability in the phases of start-up and expansion, generate high burn-rates for new technology firms. To obtain funding, most entrepreneurial technology firms use equity-based financing schemes—trading equity within the firm for finance at different periods in the firm's development. At early stages, equity deals are made with venture capitalists; later they are made through the investment banking community and third-party investors through stock offerings.

To enable funding of high-risk ventures, managers of entrepreneurial technology firms must manage complex relationships with venture capitalists, investment bankers, and other financiers. This usually necessitates the creation of business strategies that can accommo-date milestones negotiated with venture capitalists to justify further funding. However, the viability of equity-leveraged financial plans is also strongly dependent on likely exit options for financiers (both to close out unsuccessful investments quickly but, more importantly, to get out of successful ones through IPOs, mergers, or acquisitions). Moreover, knowing that the investors can (and will) pull out if projects underperform puts managers of firms under constant pressure to demonstrate at key milestones that their projects have met growth or earnings targets that justify ongoing capital investments.

Developing human resources within a competency-destroying environment

Attracting and retaining staff and managers to work in the risky and dynamic environments of technology start-ups is a second challenge facing most new technology ventures. Hiring and firing is frequent. A large number of projects fail on technological grounds, or are cut for commercial reasons (because of the failure of surrounding business models), or change focus over time. When competency destruction is high, managing human resources becomes an important organizational problem (Bahrami and Evans 1995). To achieve flexibility, managers of technology firms must have the ability to develop new research and development competencies quickly, while cutting others. To do this, they must have access to a pool of scientists, technicians, and other reputed specialists who can quickly be recruited to work on projects. If the flexibility of labor markets is limited, or if there is a cultural stigma attached to failing or changing jobs regularly, engineers and managers may choose not to commit to firms with high-risk research projects: the project's failure could damage the value of their engineering and/or management experiences.

Organizing high-powered motivational incentives for personnel

Managers of technology start-ups must motivate staff to commit to what are often demanding, competitive, and time-intensive work environments. For this, they often employ performance-based incentive schemes which they must be able to maintain as credible and high powered: the prospect of large financial rewards helps align the private incentives of engineers and scientists with those of commercial managers (see generally Miller 1992). In addition to salary increases and performance-related pay, companies have, over the last decade, primarily used share-options packages, made attractive by the expectation that share value will multiply many times if the company goes public, or is sold at a high valuation to another firm.

Technology acquisition

A large percentage of technology start-ups, particularly in science-intensive fields such as biotechnology, are originally spun out of academic research laboratories within universities. Zucker *et al.* (1998) have argued that, in addition to accessing university-based intellectual property through licensing agreements, most new technologies developed by 'star scientists' are tacit in nature, necessitating collaboration with university laboratories to transfer and begin commercializing new technologies. Such collaborations include close relationships between founding scientists and firms, scientists taking equity ownership in the firm, and the movement of more junior scientists involved with the project from academic laboratories to the firm. Such relationships are often difficult to organize because of the different motivations of academic scientists and commercial managers.

Institutions: the varieties-of-capitalism perspective

Institutional arrangements influence the types of organizational risks that firms can easily govern. To develop this argument, we draw upon extensively elaborated typologies of national business systems developed by scholars working within the varieties of capitalism field (Hall and Soskice 2001; Whitley 1999). Based on a relatively simple dichotomy between 'liberal market economies' or LMEs (the US, UK, or Canada) and 'coordinated market economies' or CMEs (Germany, Sweden, or Japan), these scholars explain how differences in the historical development of key business institutions' governance, industrial relations, finance, labor markets, and inter-firm relations influence patterns of industrial organization within an economy. Institutional frameworks influence the activities of firms through providing templates or tool-kits that firms may use to structure activity. The tools in these kits do not fix all dilemmas.

Table 21.1 highlights some of the primary institutional differences across CMEs and LMEs. It focuses on Germany and the United States, and explains how contrasting the patterns of employment and ownership relations that evolve within these institutions provides both incentives and constraints in governing risks associated with different types of entrepreneurial technology firms. This leads to hypotheses pertaining to patterns of comparative institutional advantage across CMEs and LMEs.

The successful orchestration of each of these competencies within technology start-ups is strongly influenced by the national institutional environments within which firms are embedded. Firms situated in economies with abundant high-risk venture capital, with follow-on capital market institutions, with robust labor markets for highly trained personnel, and with a company law and industrial-relations environment conducive to both the orchestration of high-powered performance incentives, and to the facilitation of university technology transfer systems, should prosper; firms in impoverished institutional environments should not.

A brief survey of the US case supports this argument. In the financial area, US-based technology start-ups have been able to organize financial resources through turning to a huge market for high-risk venture capital embedded within supportive, facilitative financial institutions. Most importantly, through the NASDAQ

Table 21.1. Institutional framework architectures in CMEs and LMEs

	CMEs	LMEs
Labor law	Regulative (coordinated system of wage bargaining; high redundancy costs of laying off employees); bias towards long-term employee careers in companies	Liberal (decentralized wage bargaining; few redundancy costs of laying off employees); few barriers to employee turnover
Company law	Stakeholder system (two-tier board system, plus co-determination rights for employees)	Shareholder system (minimal legal constraints on company organization)
Skill formation	Organized apprenticeship system with substantial involvement from industry. Close links between industry and technical universities in designing curriculum and research	No systematized apprenticeship system for vocational skills. Links between most universities and firms almost exclusively limited to R&D activities and R&D personnel
Financial system	Primarily bank-based with close links to stakeholder system of corporate governance; no hostile market for corporate control	Primarily capital-market system, closely linked to market for corporate control and financial ownership and control of firms
University-industry relations	Focused primarily on long-term relationships between applied research departments and existing, large companies. Legal frameworks grant IP from government funded research to professors; few tech transfer resources to start-ups	Primarily focused on shorter-term transactions—licensing university-owned IP to established companies or professor-led spin-offs. Laws and tech-transfer offices facilitate university spin-offs

exchange, large capital markets exist in which thousands of technology firms have successfully taken listings. A viable exit option allows early-stage investors to adopt a portfolio strategy by diversifying risks across several investments. It also creates a viable refinancing mechanism for venture capitalists (Lerner and Gompers 1999). In the matter of competency destruction in the US, deregulated labor markets within clusters of high-technology firms that have adopted complementary human resource policies generally create extremely active markets for engineers and managers (see Saxenian 1994; Hyde 1998). Clusters of radically innovative firms can develop more easily when the career risk faced by talented new personnel has been lowered (Almeida and Kogut 2000; see discussion below). More-

over, the prospect of large financial rewards through realistic IPO scenarios for successful firms, coupled with a series of stock-option, friendly finance, and industrial-relations laws, help US technology start-ups craft high-powered performance instruments easily. This is a prime reason why US high-tech firms have become associated with extremely long work-weeks and general dedication to projects. Finally, the US has developed a variety of technology transfer regulations (for example, the Bay–Dole legislation) and, through usually large university endowments, gathered the resources needed to foster the widespread commercialization of science (see Mowery *et al.* 1999).

Until recently, none of these institutional characteristics existed in most continental

European economies. Because of its importance within varieties-of-capitalism research, Germany is this chapter's focus. Its economy has long been categorized as 'organized' or 'coordinated.' German institutions facilitate the organizational competencies needed by firms that are active in sectors of established industries (such as many segments within the metalworking, engineering, and chemicals sectors) that are characterized by incremental innovation processes (Streeck 1992). Deep patterns of vocational training within firms, consensual decision-making, long-term employment, and patient finance are all linked to the systematic exploitation of established technologies in a wide variety of niche markets, a strategy Streeck (1992) labels 'diversified quality production.' On the other hand, the regulative nature of German economic institutions, combined with pervasive non-market patterns of coordination within the economy, create constraints on the organization of industries that best perform within shorter-term, market-based patterns of coordination (Soskice 1997).

A brief survey demonstrates this weak institutional support for competency development within German entrepreneurial technology firms.

High-risk finance

Germany's traditionally credit-based financial system excels at providing patient finance to firms in traditional sectors with relatively small long-term risk, but provides obstacles towards the financing of more risky entrepreneurial projects (Edwards and Fischer 1994). Data from 1996, for example, reveals that market capitalization as a percentage of GDP in Germany was only 26 per cent, compared to 121 per cent in the US and 151 per cent in the UK (Deutsche Bundesbank 1998). Venture capital is hard to sustain in countries without large capital markets willing to support IPOs. In addition to often-discussed financing gaps in high-risk capital within bank-centered financial systems, the lack of experienced venture capitalists with in-depth industry knowledge and contacts creates additional difficulties (see Tylecote and Conesa 1999).

Human resource development

In Germany, stake-holder-based company laws combine with high costs of employee dismissals to promote long-term employment within firms. Labor law cedes a formal right for staff at firms with more than five employees to form a works council, which holds important bargaining rights over personnel policy, training, and overtime. Within German manufacturing firms, works councils usually demand long-term employment guarantees in return for flexibility in work organization and overtime negotiations (see Streeck 1984). This helps the management of German firms to convince their workers to invest in skills or knowledge that are often tacit or firm-specific, and thus difficult to sell on the open labor market. Competency enhancement within organizations is strong within Germany, and it systematically inhibits the creation of the active labor markets needed to create incentives for firms and their employees to embark on high-risk projects. Similarly, limits on hiring and firing make it difficult for firms to compete in rapidly developing fields in which vital research competencies change quickly.

Employee financial motivation

Germany's bank-centered financial system tends to dampen ownership-related incentives through muting the effectiveness of share-dispersal schemes. Without a realistic possibility of an IPO, the performance incentive provided by stock options or outright share dispersals is weakened (though merger activity or management buy-outs provide weaker exit options). Prior to 1999, legal restrictions on firms buying and selling their own shares complicated matters further by creating technical difficulties on the organization of stock-option plans. Moreover, German works councils strongly resist efforts by management to institute performance-related pay, especially on an individual basis. This has constrained its introduction within large German firms. Most small German technology firms established during the early 1990s, and during the recent boom, have not created work councils (*Wirtschaftswoche* 2000); if they do begin to form, similar restraints on performance-related pay and related schemes might emerge.

University-industry relations

While Germany has a complex system of relationships between universities and industry, this system is oriented mainly towards the diffusion of applied technologies to existing firms. Well-funded organizations, such as the Fraunhofer and Steinholz Institutes, support this system. Until recently, professors owned the intellectual property of most publicly funded research, aiding the formation of relationships between faculty and established companies (companies would receive IP in exchange for research funding and consulting fees). Lacking ownership of IP, universities had no incentive to develop technology-transfer offices to promote more upstream commercialization activities within basic research fields. Moreover, as publicly funded institutions, few universities in Germany have developed endowments for use in funding technology-transfer offices (see Abramson *et al.* 1997 for a detailed overview).

In sum, core German market institutions are primarily geared toward the creation of the firm-level competencies needed to create sustained, incremental innovation patterns in industries with lower scientific intensity. The result during the 1980s and early 1990s was poor performance in most sectors that had technological profiles best advantaged through the creation of entrepreneurial business models (see Casper *et al.* 1999). Germany lacks institutions to nurture the development of entrepreneurial competencies systematically. While the German institutional system may have the orientation most clearly hostile to promoting new technology firms, similar institutional arrangements exist in many other European economies (see for example, Kogut 2003).

Evidence: a remarkable upswing in new technology firms across coordinated market economies

Beginning in the mid-1990s, the success of Silicon Valley and other US technology clusters promoted calls that Europe faced an innovation crisis and needed to shift resources (people, finance, infrastructure) away from declining industries and towards those represented by the new economy, particularly biotechnology and software. What is remarkable about these debates is a widespread recognition that at least some institutional factors—particularly in the spheres of financial and technology transfer—strongly impact the viability of start-up companies. These debates often informed the creation and nurturing of technology clusters, usually built close to well-known universities. Public and private actors differ from most institutional scholars in their optimistic belief that, through strong technology policies, European economies could relatively easily create an environment strongly supportive of technology firms in new industries (see Lehrer 2000 for a good summary of the debate in Germany).

Through the decade, widespread venture capital and technology-transfer policies were introduced to foster technology start-ups. Venture-capital programs usually entailed the development of state-run venture-capital funds which, with matching funds from the private sector, were used to promote initial investments in start-up companies. These programs were most widespread in Germany, but exist in Sweden, the Netherlands, France, and elsewhere (Lehrer 2000). Venture-capital subsidy programs were usually coupled with initiatives to develop technology-oriented stock markets (mimicking NASDAQ), to allow firms to raise further funds and, importantly, create the viable exit options for the venture capitalists and other investors widely perceived as necessary to sustain long-term investing in new technology sectors. In Germany the government introduced financial and tax regulations aimed at facilitating the use of stock options within companies; other countries, such as France, introduced similar reforms, but with social provisions mandating relatively lengthy ownership of options (usually three years) before tax advantages could be realized (see Trumbull 2004 for a good overview of French reforms).

The second area of sustained reform surrounds the commercialization of university science. In this area, a variety of sector-specific reforms were developed, typically surrounding the funding of technology-transfer offices, incubator programs, and technology parks in or nearby major universities. Again, the most sustained policies exist in Germany, though similar programs were developed elsewhere in Europe. These programs aimed to orchestrate the commercialization of basic research technologies, focusing in particular on biology, but also including other new sectors such as software. Resources were provided to consult with professors over the commercialization of research, pay for patenting of technologies, and to help nascent firms develop business plans and other materials needed to apply for funds through 'public venture capital' programs (often coordinated through the same network of offices). In Germany these initiatives began in 1995 with a well-funded 'BioRegio' initiative that led to the establishment of twenty-two local technology transfer initiatives. Later programs were introduced in other new industries. Finally, beginning in the late 1990s, many European economies introduced legislation essentially copying the US Bay–Dole Act, transferring ownership from professors to universities.

These initiatives were substantial in terms of government expenditure. One recent estimate of Germany's spending on the promotion of biotechnology industry (taking into account all public venture capital, spending on technology parks and incubator laboratories, and research grants to firms) is about 3 billion euro (Casper et al. 2004). Not surprisingly, the German Science and Education Minister Ruettgers announced early on in the program that the goal of the program was to develop, virtually from scratch, Europe's leading biotechnology

industry. Interestingly, no major reforms to labor-market regulations, industrial relations systems, and corporate governance (or company law) institutions were introduced in Germany. Policymakers implicitly believe that the new economy could be effectively isolated from the old, governed with unique institutional rules.

By all accounts, these programs were successful in stimulating the widespread entry of new technology firms. These technology policies also coincided with the Internet boom, leading to substantial interest in funding new companies in this sector, and the mimicking of virtually all successful US firms by new entrants in most European economies (see Kogut 2003 for a good overview of the global internet economy). Hundreds of new biotechnology firms were created in Germany (Schitag Ernst and Young 2002), and most European economies had established a presence in the sector by 2001 (Ernst and Young 2001). During the same period, a boom in venture capital occurred, sustained by the successful introduction of NASDAQ-inspired stock markets in most major European economies. The German *Neuer Markt* was the most successful, sponsoring nearly 300 IPOs by its peak in 2001. This market was particularly strong in software, with over sixty firms listed. Because of longer development cycles, fewer biotechnology firms (16) successfully took *Neuer Markt* listings; a variety of Internet and new media stocks and firms from various other technology niches filled out the list.

The rapid downturn in the US technology market, particularly in the Internet sector, has adversely affected Europe's new technology market place. In Germany, a majority of *Neuer Markt* firms lost most market capitalization and were delisted, and numerous companies, particularly in the Internet and new media sectors, failed. In early 2003, the *Neuer Markt* was consolidated into a technology-oriented segment of the mainline Frankfurt Stock Exchange, and there have been virtually no new IPOs in Germany or most other European economies during the 2001–4 period. Despite this, governmental enthusiasm for the new economy still exists in Europe, and there are no signs that the venture capital or technology programs will end. New firms continue to be funded, with the expectation that a viable infrastructure to sustain new economy firms has been achieved.

The new economy sector in the UK has performed much better than its nascent continental European cousin, supporting the varieties-of-capitalism perspective. But the widespread entry of biotech and software companies (imitating key aspects of the Silicon Valley model) in Germany and other continental European economies starkly contradicts core predictions of the theory. The following sections use case studies from Germany, the UK, and Sweden to explore more closely the connection between institutional frameworks and innovation in new technology segments. While the usefulness of varieties-of-capitalism theory (and comparative institutional research more generally) is validated, the results of these studies suggest a more nuanced, firm-level approach is needed to gain analytic leverage.

Sectoral systems of innovation: the subsector specialization argument

One way to salvage varieties-of-capitalism theory is to suggest that the characteristics of European new-economy companies actually resemble those suggested by the theory. A key contribution from the theory is the development of micro-foundations linking institutions to innovation. This framework links the technological characteristics of various market places to a number of organizational or competency dilemmas faced by firms; it then suggests that institutional characteristics influence the viability of these organizations' being governed.

Recent research on 'sectoral systems of innovation' (Malerba 2004) can help in this regard, in particularly by expanding upon the meaning of 'radical' versus 'incremental'

innovation to demonstrate that most sectors of the new economy, such as biotechnology and software, comprise *subsectors* with different innovative characteristics. In biotechnology, for example, most drug-discovery research has innovative characteristics considered radical, while other important subsectors, such as much of the platform technology, have technological characteristics resembling those associated with diversified quality production (or incremental innovation). The subsector specialization argument draws on the idea that most competitive market places are underpinned by relatively consistent 'technology regimes' (Malerba and Orsenigo 1993). These regimes represent different constellations of the technological and market risks that face firms within particular industry segments. Three characteristics have been shown to be particularly important for new technology industries: the appropriability regime; levels of technological cumulativeness; and the extent to which knowledge or skill sets are generic to most firms in an industry, or are firm-specific. The following examines how these characteristics differ across ideal-typical examples of radical and incremental innovation within the biotechnology and software industries.

Radically innovative industries, such as the therapeutic discovery segment of biotechnology, the packaged software sector, and Internet communications or 'middleware' software (see Casper and Glimstedt 2002), are associated with relatively tight appropriability regimes, meaning that the intellectual property protection is strong (see Teece 1986). Furthermore, most skill-sets within such industries are generic, or industry-wide (for example, relatively standard laboratory methods within biotechnology, or use of generic programming languages within standard software). However, unless firms can develop the capabilities to pursue their chosen paths of research and development successfully, they face high levels of technology risk (Woodward 1965; Perrow 1985).

Viewed in terms of company capabilities, the level of technological cumulativeness relates to the rate at which specific technological assets change during the evolution of an industry. Low cumulativeness implies that Schumpeterian patterns of competency destruction are high within an industry. Technological competencies within a firm have a high probability of failing (as they are tried and found to be inappropriate for resolving particular research and development problems). In industries where cumulativeness is low, firms must have sufficient financial resources, or must develop a capacity to adjust their technological assets quickly; failure is common. High technological risk denotes a significant probability of either outright failure or rapidly changing R&D trajectories that necessitate hire-and-fire personnel policies. Skilled employees may refuse to work within such firms if doing so poses a high risk of long-term unemployment, or a risk that the large percentage of skills acquired while working within the firm are not saleable on open labor markets (see Saxenian 1994).

Firms in new technology industries that are more incrementally innovative, such as most areas of enterprise software, and many platform biotechnologies, face higher levels of cumulativeness. However, these sectors often contain market risks created by difficulties in capturing value from innovations. For example, work-arounds for patents may exist, leading to widespread entry of new competitors, and to the acquiring of a generic quality for the assets developed from innovation. Following Teece (1986), when appropriability regimes are relatively weak, technological assets developed by the firm are generic and may be easily mimicked by competitors. In this case, Teece has suggested that, to capture value from innovations, firms must develop complementary assets that are both specific to the firm and tied to the generic assets. This often involves creating co-specialized assets used to customize products for particular clients, as often occurs in the machine tool industry or, more broadly, strategies focused on marketing and distribution.

Firms developing co-specialized assets tend to create more complex organizational structures than firms innovating within tight appropriability regimes. A key attribute of a firm's competitive success can be its ability to develop an organizational culture, or set of routines, enabling different types of professional employees to work well in cross-functional teams. For employees, this primarily represents firm-specific, often tacit, knowledge that is difficult to sell on the open labor market. Employees' concerns that their careers will be held up once firm-specific knowledge investments are made may create managerial dilemmas; and employees also may worry about managers pursuing opportunistic employment policies, such as holding wages below industry norms, once extensive firm-specific knowledge investments have been made (see Miller 1992). Without assurance from managers that they will not exploit them, employees could refuse to make such long-term knowledge investments within cross-functional teams, creating patterns of sub-optimal work organization that could hurt the performance of the firm.

Table 21.2 summarizes the different technology regimes surrounding these two broad systems of innovation.

Returning to the discussion of national institutional frameworks (see also Casper and Whitley 2004; Casper and Soskice 2004), the subsector explanation predicts that in radically innovative subsectors (such as therapeutic segments of biotechnology, standard software, and middleware software), countries with LME frameworks will develop comparative institutional advantages; in incrementally innovative subsectors (such as platform biotechnologies and enterprise software) they will perform less well (see Casper and Whitley 2004 for a full description of these subsectors). CME countries, on the other hand, should develop comparative institutional advantages in incrementally innovative subsectors, but perform poorly in the more radically innovative ones. Drawing on data developed by Casper and Whitley (2004) on companies listed on European new technology stock markets in Sweden, the UK, and Germany in early 2003, Tables 21.3 and 21.4 examine patterns of subsector specialization across these countries.

In general, these results support varieties-of-capitalism theory. Eighty-eight per cent of the

Table 21.2. Technology regimes for radical and incremental innovation

	Radically innovative sectoral systems (e.g. discovery-based biotechnology, standard software)	Incrementally innovative sectoral systems (e.g. platform biotechnology, enterprise software)
Appropriability regime	Tight	Loose
Level of cumulativeness	Low	High
Degree of generic versus firm-specific knowledge	Generic knowledge	Firm-specific knowledge

German companies are in subsectors characterized by incremental innovation: German companies appear to migrate to industry segments with company organizational patterns that, in many respects, are best governed by German institutions overseeing work and company organization. One interesting facet of this result is that German technology policies have done little to change normal patterns by which institutions structure company organization. German new-economy companies may draw on venture capital and technology-transfer opportunities (and these companies may be growing more quickly as a result), but they have gravitated to more incrementally innovative segments in which hire-and-fire and individualized performance incentives are not

prerequisites for success. German firms do use stock options (Casper and Vitols 2006). However, if distributed uniformly across the firm, these options may act as a collective, rather than individual, incentive mechanism.

The results for the UK are also supportive of comparative institutional theory: 88 per cent of UK firms are in radically innovative sub-segments. However, it is interesting to observe that few UK firms appear to be succeeding in the middleware segment of software.

Finally, the Swedish results are puzzling. While the biotechnology results are supportive, the Sweden software industry is strong in both standard and middleware software. This result is consistent with other recent studies (see Glimstedt and Zander 2003), arguing

Table 21.3. Subsector distribution of software companies

	Germany		United Kingdom		Sweden	
	No.	%	No.	%	No.	%
Incremental innovation technologies						
Enterprise software	54	71	23	18	20	35
Platform biotechnologies	13	17	6	4	8	14
Radical innovation technologies						
Therapeutics-based biotech	3	4	34	27	3	5
Standard software	3	4	58	45	16	28
Internet (middleware) software	3	4	7	5	10	18
Total	76	100	128	100	57	100

Source: Casper and Whitley 2004. Classification based on company web-pages, annual reports, and IPO prospectuses.

Table 21.4. Clinical trial data of German and UK biotech firms

	Preclinical	Stage 1	Stage 2	Stage 3	Total
Germany[a]	19	10	2	1	32
UK (Public Companies)[b]	32	37	46	13	128

Source: [a] author data, [b] Ernst and Young 2001.

that an extremely vibrant and radically innovative software cluster exists in Stockholm. This is explored in more detail below.

Institutional reflexivity: what are the drivers of change?

Aided by sectoral systems of innovation research, comparative institutional theory appears useful in explaining why types of innovative activity cluster across different types of economies. The comparison between the US and UK on the one hand, and Germany on the other, appears particularly robust, but problems remain. The public-company results for Sweden were contradictory to the theory. And they focus only on public companies. In some instances this is appropriate, for going public is a measure of success. However, the majority of Silicon Valley-type start-ups never go public, either because they fail or are bought out by other companies. As reported below, studies of early-phase German biotechnology companies reveal very different patterns of specialization from the public-only companies. Most early-phase German companies are in radically innovative segments, and most of them also appear to be on the road to failure, which supports the varieties-of-capitalism research in a perverse way. However, the theory implicitly makes strong claims of 'reflexivity'—actors are presumed to develop strategies and competencies in the shadow of the institutional incentives and constraints they face. Institutional reflexivity may be low, leading to a waste of resources as many firms die, and a few firms adopting institutionally appropriate specializations survive.

This section focuses on this problem. Drawing on case studies of the Swedish software and German early-stage biotechnology industry, it helps identify the drivers of new technology entrepreneurialism. The results are strongly supportive of the general idea that environmental factors strongly affect the characteristics of firms. However, the drivers of innovation may include the activities of large companies (in the Swedish case) or institutional domains not anticipated in varieties of capitalism theory (in the German case).

Both cases draw on recent research connecting the success of regional technology clusters to the composition and, in particular, mobility of scientists and engineers in regional labor markets. A recent group of studies has provided one of the clearest, and best empirically documented theories linking the innovative performance of locally clustered companies to the external labor-market environment. The core idea behind these studies, first presented in Saxenian's (1994) account of the success of Silicon Valley's semiconductor industry, is that firms embedded in flexible labor markets can more easily sustain innovation strategies with high risks of failure than firms embedded in less flexible labor markets. This theory is persuasive as it reconciles the interests of talented employees with those of firms. Skilled employees will, rationally, only join a high-risk firm if the career risk of failure is low, and the reward for success is high. Flexible labor markets, with dense inter-firm networks of ties across skilled personnel, serve both needs: employees of failed firms can tap local labor networks to find new jobs; successful ones use the availability of job offers elsewhere to develop upward wage pressure on existing employers. Flexible labor markets also help local technology companies: they can more easily hire and, if necessary, fire, lowering the transaction costs of developing assets needed to innovate.

In addition to developing a clear theoretical mechanism linking labor-market environments to innovation, this approach helps explain why regional varieties of innovative performance exist *within* countries. Through extensive qualitative research, Saxenian argued that inter-firm mobility helps account for the higher innovative intensity of Silicon Valley's semiconductor firms than that of a rival region, the Route 128 area of Boston. Almeida and Kogut (1999) added credence to the explanation with their cross-sectional analysis of inter-firm mobility with patent data: of twelve regions with high concentrations of semiconductor firms, only Silicon Valley had high inter-firm mobility and correspondingly higher rates of innovation. Varieties-of-capitalism research predicts that radically innovative firms should be able to develop anywhere within a liberal market economy such as the US or UK. The literature on regional labor market dynamics helps explain why this is not the case.

Institutional explanations can help explain the origins of labor-market flexibility-both regionally and, in some cases, nationally. For example, California has been shown to have dramatically stricter laws governing non-compete clauses within labor-market contracts than most US states; this helps to drive the creation of flexible labor markets in Silicon Valley, but not elsewhere (Gilson 1999). More broadly, a key tenet of varieties-of-capitalism research is that labor-market mobility is much lower within organized economies such as Germany, Japan, and France, because of long-term employment norms within large firms, buttressed by employment laws, social policies, and industrial relations systems that create incentives against hire and fire.

The research on inter-firm mobility within technology clusters helps identify an important problem: what are the *mechanisms* that lead to high inter-firm mobility in some areas, but not others? It is relatively clear that institutions can be a brake on the development of mobility. But we will present extensive evidence that firms with radically innovative intentions exist in large numbers within several German technology clusters. What are the mechanisms—institutional or otherwise—which apparently have led these firms to access highly skilled scientific personnel?

Swedish telecommunications software: large firms as drivers

Within some industry segments, human resource policy and, at times, the corporate venturing strategies of local large technology firms can promote flexible labor markets within particular regions, facilitating higher-risk innovation strategies. This argument appears most straightforward within liberal market economies. In such countries, normal patterns of labor-market regulation promote hire-and-fire policies and significant career mobility across firms. It is plausible that large firms might act as catalysts for the development of local agglomerations of scientists and engineers. A recent large-scale study of networks of innovators within Silicon Valley (Flemming and Juda 2004) found a striking result of this kind. Based on a study of job changes across patenting scientists, Flemming discovered that, during the mid-1980s, over 30 per cent of job mobility in Silicon Valley could be traced to a large post-doctoral training system maintained by IBM. IBM would attract leading engineers to the program; after two years, most would move to other firms in the region.

The Swedish software case suggests that a dominant firm, Ericsson, could catalyze high labor-market mobility within a coordinated market economy. This result suggests that local labor markets can strongly tilt away from normal patterns, and that the activities of companies can trump institutional incentives. This case has been developed in detail by Henrik Glimsted and colleagues (see Glimstedt 2001; Glimstedt and Zander 2003; Casper and Glimsted 2001); it is briefly summarized here to draw out key implications. The key issue exemplified by this case is this: in order to induce engineers, managers, and financiers to make commitments to projects that are normally extremely risky within their societal contexts, what constellation of policies must the large firm adopt? Can dominant actors take actions to tip labor-market institutions in a direction contrary to the normal institutional incentives with an economy?

Ericsson's success in helping to create a cluster of middleware software companies focused on wireless technologies is partly driven by important network externalities within this subsector. Governments have at times played important roles within telecommunication standards (see Glimstedt 2001); but within much of the middleware software sector, most firms are dependent upon large corporations—typically telecommunication equipment manufacturers and established companies active in network-intensive standard software products—for the provision of standards to help products become inter-operable (see Casper and Glimstedt 2001). Examples of the former include large network equipment manufacturers such as Cisco Systems, Lucent, or Ericsson; Microsoft, Sun, or Oracle exemplify the latter. Each of these firms has been involved in the creation of technology platforms for emerging network communication markets. These firms hope to provide technology platforms that function as club goods to middleware software companies, enticing them to develop a variety of follow-on technologies aimed at eventually creating new software platforms. Large firms are self-interested when providing these standards: by controlling emerging network communication protocols, they hope to secure large markets for equipment and software using the standards.

Large firms can help stabilize technologies through attracting middleware firms to create applications for their standards. As a result, middleware software firms are most likely to exist within technology clusters dominated by large companies; these can entice them to commit to a technical standard, either through a reputation of past success or through other

means such as financial incentives or technical support. By locating within regional economies dominated by such firms, middleware firms can plausibly hope to insert their software engineers into emerging technical communities surrounding new platforms. Privileged access to such communities can provide a competitive advantage for middleware firms: they can, for example, supplement codified technical knowledge (protocols, languages) with tacit knowledge surrounding their efficiency.

Within Sweden, Ericsson has become the dominant provider of end-to-end wireless communication systems and, as of 2003, had about 40 per cent of all orders for third-generation wireless equipment (Glimstedt and Zander 2003). Other major telecommunications equipment players, such as Nokia, have set up development centers in Stockholm, and Microsoft recently opened an R&D center for wireless software. Hundreds of software firms focusing primarily on wireless Internet technologies have developed in the Stockholm area of Sweden, most in technically intensive middleware technologies (see Glimstedt and Zander 2003: 128–34).

Through the 1980s and early 1990s, Ericsson in many ways resembled Siemens, Alcatel, and other European telecommunications equipment manufacturers. Operating as a quasi-monopoly equipment provider in a highly regulated domestic telecommunications market, it developed the capabilities needed by large systems to design early digital switching technologies designed primarily for voice traffic. As the only significant telecommunications equipment manufacturer in Sweden, it could attract the country's best engineering graduates, who were then offered stable, long-term careers in Ericsson. The company developed proprietary protocols and systems integration languages. The core of Ericsson's programming staff, for example, were experts in Ericsson's in-house systems integration language, Plex, a computer language used nowhere else. While the convergence of data communication and voice-based digital communication technology has forced Ericsson to adopt new languages for its next-generation telecommunications gear, several thousand employees have been retained for their expertise in Plex, which is still used to update legacy equipment.

During the late 1990s, data-communications networking devices began to converge with traditional switching equipment. The increased use of Internet protocol-based switching has forced firms like Ericcson to increasingly adopt connectivity standards developed for data-communications networks. An issue for such firms is how this influences internal product development. In designing switching equipment, base-tower systems, and related capabilities for Ericsson's Internet-compatible wireless equipment, a small group of system engineers within the company developed a new systems-integration language, called Erlang. As with Plex, Ericsson's initial strategy was to make this technology proprietary. However, unlike Plex, Erlang is a systems-development language based on standardized, object-oriented programming tools; they have the potential to help firms in a number of industries develop software to manage complex technological systems.

To help promote technology spillovers into the Stockholm economy, Ericsson made two strategic moves. First, it decided to make Erlang an 'open source' development language. Using its protocols ensures that enhancements to Erlang by third parties would flow back into Ericsson. More importantly, however, it helped to create industry-specific, rather than firm-specific, skills among engineers involved in large-scale systems integration. Sponsorship of emerging wireless connectivity standards, such as Bluetooth and WAP, or widely used mobile scripting languages like UML, produces a similar effect. Standardization of development tools, protocols, and connectivity standards dramatically increases the portability of skills across local firms working in wireless technology areas.

Second, Ericsson has changed its personnel policy towards engineers who leave long-term careers at Ericsson to work in start-up firms. It had formerly shunned them, signaling that they would not be re-employed by Ericsson in the future. Now, through a corporate venture-capital program, it allows engineers to leave to try their hand at technology entrepreneurialism. Given that most wireless start-ups within the Stockholm area are involved in the development of Ericsson-sponsored standards, and in many cases are using its systems-development language, local start-up ventures are working primarily to develop technologies compatible with Ericsson's next-generation wireless technologies. If individual firms fail, their managers can now easily return to work within Ericsson, perhaps having developed new managerial skills or career perspectives through working in a start-up. If start-up firms are successful, Ericsson benefits through its sponsorship of key technologies, and has close links with the management of the new companies.

In sum, the existence of industry-specific, rather than firm-specific, standards reduces the career risk for engineers leaving established large firms for start-ups. Industry-specific standards ensure that investments made by programmers and engineers in skill and knowledge are portable. It allows managers of high-tech firms to recruit highly skilled technical talent knowing that competence destruction and accompanying hire-and-fire risks are high. Within normally conservative Swedish labor markets, this employment insurance is key to creating the extremely active labor markets necessary to sustain competence-destroying technology strategies.

German early-stage biotechnology: the dominance of the research system as an institutional driver

As noted earlier, the German government has spent billions of euros promoting biotechnology. A number of studies (Giesecke 2000; Casper 2000) have echoed the subsector specialization results discussed earlier: *successful*

German biotechnology companies are primarily specialized in platform technology segments of the industry that have competency requirements similar to traditional German industry—machine tools for biotechnology, in many respects. However, recent comparative research on early-stage companies suggests that German university spin-outs in biotechnology have technology trajectories remarkably similar to US or UK firms (the following draws from Casper and Murray 2004 and Casper *et al.* 2004). This research analyzed forty-five German biotechnology firms located in the four large regional clusters. The research used bibliometric methods to capture information on these firms' scientific linkages with universities, a good indicator of relative scientific intensity. The German firms had similar numbers of academic publications, collaborations with university scientists. An assessment of the scientific output of scientists working within firms revealed similar findings: German firms had recruited scientists with broadly similar levels of prestige (measured by publications and experience) as scientists working in biotech firms in Europe's leading cluster in Cambridge, UK. Moreover, 80 per cent of the German firms were actively engaged in drug-discovery research—a primary indicator of radical innovation used in other studies.

So early-stage German biotechnology companies are decidedly not following the dictates of the theory of comparative institutional advantage. Their reflexivity to the types of institutions identified as important by varieties-of-capitalism research is low. If the performance of these companies were strong, this would strongly invalidate varieties-of-capitalism research. Table 21.4 presents data on the therapeutic pipelines of the forty-five German companies examined in the Casper and Murray study, and approximately thirty UK companies assessed in an Ernst and Young study. These results should be viewed as rough, as the UK companies are generally older than the German firms (giving their drug candidates more time to progress through trials). However, despite this drawback, the German firms are clearly being outperformed by those in the UK; only

three candidates have made it to relatively advanced stage 2 or 3 trials, compared to fifty-nine UK drugs.

In sum, large numbers of German biotech firms, backed in part through public venture capital and resources provided by technology transfer policies, have launched high-risk biotech firms, but then appear to be failing to perform. Can institutional theory help resolve this puzzle? Using bibliometric research to track the careers of publishing scientists within each firm, it is possible to examine the career histories of these scientists and, in doing so, reveal a partial picture of the human resource composition of the firms. Table 21.5 displays the most recent job for scientists working within German and Cambridge/UK firms. These results help explain the impact of normal varieties-of-capitalism institutions on these firms, and point to a likely driver of their relatively radical technology strategies.

Clear differences exist across the German and Cambridge biotechnology firms. The German biotechnology spin-offs tend to employ significantly fewer scientists with prior experience in commercial therapeutic research than do the Boston biotechnology spin-offs. Of the 299 German scientists in our sample, only 11 per cent were directly recruited from either a biotechnology firm (4 per cent) or a pharmaceutical firm (7 per cent). In contrast, of the 79 Cambridge scientists in our sample, 43 per cent were directly recruited from either a biotechnology or pharmaceutical firm.

This data helps explain why the German firms are performing poorly. Markets for *downstream* assets—experienced industry scientists who can work on pharmaceutical development processes—have remained untapped by biotechnology entrepreneurs. As a result, German firms, once founded, have had a hard time recruiting commercial development capabilities, and could very well begin to fail as a result.[1] The normal, relatively tight, German labor markets for mid-career scientists, which is the result of the well-documented German system of long-term employment in large companies, seem to be tied to this. Of particular note is the small number of scientists moving

Table 21.5. Previous jobs of biotechnology scientists

	Germany	Cambridge/UK
Founding lab	101 (34%)	15 (19%)
Other academic lab	166 (55%)	30 (38%)
Biotechnology firm lab	12 (4%)	11 (14%)
Pharmaceutical firm	20 (7%)	23 (29%)
Total	299	79

from large pharmaceutical companies to biotech start-ups. Most companies do not have access to experienced industry personnel. Research on the biotechnology industry in the United Kingdom has shown that, during the initial formation of clusters in Cambridge and elsewhere, British firms recruited extensively from British large pharmaceutical firms.

Another striking result is that German biotechnology companies tend to recruit a large number of scientists from the laboratory from which they were initially spun off: 34 per cent of scientists working for German biotechnology firms come directly from the founding laboratory, as opposed to 19 per cent of scientists working for Cambridge biotechnology firms. Most biotechnology firms contain a cadre of scientists from the founding academic lab; these scientists are usually involved in transferring tacit knowledge from the lab to firm. However, when second and third prior jobs are taken into account, fully half of the German scientists in this dataset were previously employed in the firm's academic founder laboratory. Strong ties to senior German scientists are the core mechanism by which German firms recruit scientists. Cambridge companies appear to make broader use of referral networks: most employees do not have a prior employment relationship to founders.

Founding academic laboratories are serving as a substitute for normal labor markets in the German biotechnology industry. The academic science background of most employees in these firms, coupled with the lack of experienced industry scientists, helps explain why these companies have moved into research areas centered on high-risk therapeutics. Scientists are probably following the trajectories of firms founded by colleagues in other countries, pursuing strategies then in vogue in the global industry. Only later, when more experienced personnel are needed (primarily in pharmaceutical development activities), do the constraints of being in German industrial labor markets become apparent.

One of the key questions revealed through this research is why German professors have invested so heavily in these firms; why, given the problems involved with attracting scientists with downstream product-development expertise, they moved scientists to their firms from their publicly funded laboratories in the first place. The study of early-phase German biotechnology firms reveals that incentives within the academic research system—an institutional environment largely ignored by research on innovation (though see Whitley 2003)—has strongly influenced the trajectory of the German industry. Through moving far more laboratory workers into firms than is normal in the UK or US, German professors appear willing to invest substantially in spin-outs from their laboratories. How does this complement broader goals of these laboratories, such as performing well in international science, or finding suitable employment for graduate students and post-docs trained within German laboratories?

Concluding discussion: informing public policy

Varieties-of-capitalism research is a useful tool for investigating the development of new models of innovation, in this case the Silicon Valley model, into heterogeneous business systems. Above all else, the theory helps specify a number of general competencies associated with most entrepreneurial technology firms, and clearly posits a framework linking national institutional frameworks to their governance. Armed with the idea of subsector specialization, varieties-of-capitalism theory does a good job of explaining patterns of industry specialization in core new economy sectors in the UK and Germany (two of the cases around which the theory was built). The portability of the general findings to other countries is an important area for future research, as seen through the less persuasive general findings on Sweden.

Moreover, recent in-depth case studies suggest that the drivers of innovative activity might differ from those specified in the theory. Research on labor markets suggests that, within regional economies, large firms and the activities of laboratories within academic research systems can, in a sense, circumvent normal institutional incentives within coordinated market economies. In the Swedish Internet software case, this is probably a good outcome, as Stockholm has emerged as a world leader in software development for the global mobile telephone industry. In Germany the activities of German research laboratories in pushing large numbers of scientists into local biotech firms, and orienting these firms towards radically innovative activities, appears to have led to less desirable outcomes. Normal German labor markets, dominated by employment within large pharmaceutical firms, appear to exist for mid-career scientists, leading to a lack of experience in development activities downstream, and poor longer-term performance. In this case, the circumvention of German labor-market institutions appears incomplete, with predictable consequences. Moreover, this case provides a warning that, given ample resources during the start-up phase, there is little reason to suppose that managers of firms will be institutionally reflexive to downstream concerns.

Varieties-of-capitalism theory may give us more useful tools to explore the governance of firm-level capability building; at its present stage of development, it is not a strongly predictive theory, except at the most macro-levels. With this in mind, how can comparative institutional theories on innovation usefully inform public policy?

One of the clearest dictates from the theory is that public policy must be incentive-compatible with overarching institutions in an economy. A major problem with German technology policies is that they are strongest in areas in which a sectoral focus is possible-venture capital for new firms and technology transfer. In other institutional domains, such as labor-market regulation and industrial relations, little change has occurred. As a result, when the needs of firms in the new economy (a minority of companies in Germany) collide with those of established firms (the majority), the institutions and practices of the latter will prevail. This is why there is an ample labor market for junior academic scientists in the German biotech sector, but practically none at all for experienced scientists, most of whom have long-term careers in established firms.

On the other hand, institutional environments may be much more malleable than supposed in the theory. In other words, institutional complementarities may be relatively low; institutions may be loosely coupled in terms of their effects on firms. This may help explain why policies towards finance and technology transfer were so successful in promoting start-ups across most Western European economies. The German early-stage biotech case may be interpreted as one of failing to recognize institutional complementarities in the labor market. Moreover, a key area for future research is to continue to examine patterns of subsector specialization and the long-term performance of firms. The subsector evidence suggests that strong policies may have hastened the entry of large numbers of companies into new industries, but that these companies may have developed anyway due to pre-existing comparative institutional advantages.

A related policy issue concerns the hybridization of policies-mixing attributes of institutions needed to support the Silicon Valley model with local rules, for example, often driven by a political need for compromise between the old and new. Though not discussed in this chapter, research on the French new-economy experience is evocative of this issue (see Trumbull 2004 for a thorough overview). Examples include reforms to legitimize and widely promote stock options, but social restraints on cashing them in to rein in opportunism and foster long-termism.

Finally, a key issue brought forward in this review suggests that a variety of drivers of adoption and change within economies may exist. In hindsight, such drivers of labor-market flexibility as large firms, or as the research system in pushing forward science-based firms, may be obvious. But policy must be flexible enough to account for unanticipated drivers of entrepreneurial activity and, particularly when dealing with new industries and actors in an economy, an institutional myopia when it comes to thinking about longer-term competency development. Policies aimed at reducing such myopia and increasing the reflexivity of firms to their institutional environment could improve the longer-term performance of firms entering an economy on the waves of new technology.

Note

1. Some German firms seem to have attempted to overcome their lack of commercial development capabilities by purchasing foreign firms in the United Kingdom or the United States which have acquired these capabilities (e.g. Cellzome's acquisition of GlaxoSmithKline's Cell Map Unit in the United Kingdom in 2001).

References

Abramson, H. N., Encarnacao, J., Reid, P. P., and Schmoch, U. (eds.) (1997). *Technology Transfer Systems in the United States and Germany*. Washington, DC: National Academy Press.

Adelberger, K. E. (1999). 'A Developmental German State? Explaining Growth in German Bio-technology and Venture Capital.' Berkeley Roundtable on the International Economy Working Paper 134.

Almeida, P., and Kogut, B. (1999). 'Localization of Knowledge and the Mobility of Engineers in Regional Networks.' *Management Science*, 45/7.

Aoki, M. (2001). *Toward a Comparative Institutional Analysis*. Cambridge, MA: MIT Press.

Asakawa, K., and Lehrer, M. (forthcoming). 'Pushing Scientists into the Marketplace: German and Japanese Efforts to Promote Science Entrepreneurship.' *California Management Journal*, 46/1: 55–76.

Bahrami, H., and Evans, S. (1995). 'Flexible Re-cycling and High-Technology Entrepreneurship.' *California Management Review*, 37: 62–88.

Bates, R. H., Grief, A., Levi, M., Rosenthal, J.-L., and Weingast, B. R. (1998). *Analytic Narratives*. Princeton: Princeton University Press.

Boyer, R., and Hollingsworth, J. R. (1997). *Contemporary Capitalism: The Embeddedness of Institutions*. Cambridge: Cambridge University Press.

Casper, S. (2000). 'Institutional Adaptiveness, Technology Policy, and the Diffusion of New Business Models: The Case of German Biotechnology.' *Organization Studies*, 21/5.

—— and Glimstedt, H. (2001). 'Economic Organization, Innovation Systems, and the Internet.' *Oxford Review of Economic Policy*, 17: 265–81.

—— and Matraves, C. (2003). 'Institutional Frameworks and Innovation in the German and UK Pharmaceutical Industry.' *Research Policy*, 32/10.

Casper, S., and Murray, F. (2004). 'Careers and Clusters: Analyzing Career Network Dynamics of Biotechnology Clusters.' *Journal of Engineering and Technology Management*, 22: 51–70.

—— and Soskice, D. (2004). 'Sectoral Systems of Innovation and Varieties of Capitalism: Explaining the Development of High-Technology Entrepreneurship in Europe.' In F. Malerba (ed.), *Sectoral Systems of Innovation: Concepts, Issues, and Analysis from Six Major Sectors in Europe.* Cambridge: Cambridge University Press.

—— and Vitols, S. (2006). 'Managing Competencies within Entrepreneurial Technologies: A Comparative Institutional Analysis of Software Firms in Germany and the United Kingdom.' In M. Miozzo and D. Grimshaw (eds.), forthcoming, *Knowledge Intensive Business Services: Organizational Forms and National Institutions.* Oxford: Oxford University Press.

—— and Whitley, R. (2004). 'Managing Competences in Entrepreneurial Technology Firms: A Comparative Institutional Analysis of Germany, Sweden and the UK.' *Research Policy*, 33/1: 89–106.

—— Jong, S., and Murray, F. (2004). 'Commercializing German Science: How the Organization of Professional Labor Markets in Germany Has Impacted the Development of the German Biotechnology Industry.' Manuscript, Keck Graduate Institute.

—— Lehrer, M., and Soskice, D. (1999). 'Can High-Technology Industries Prosper in Germany? Institutional Frameworks and the Evolution of the German Software and Biotechnology Industries.' *Industry and Innovation*, 6: 6–23.

Deutsche Bundesbank (1998). *Monthly Report*, April.

Edwards, J., and Fischer, K. (1994). *Banks, Finance and Investment in Germany.* Cambridge: Cambridge University Press.

Ernst and Young (2001). *European Life Sciences 2001.* London: Ernst and Young International.

—— (2003). *Beyond Borders: The Global Biotechnology Report.* London: Ernst and Young International.

Flemming, L., King C., and Juda, L. (2004). 'Small Worlds and Innovation.' Manuscript, Harvard Business School.

Giesecke, S. (2000). 'The Contrasting Roles of Government in the Development of Biotechnology Industry in the US and Germany.' *Research Policy*, 29: 205–23.

Gilson, R. J. (1999). 'The Legal Infrastructure of High Technology Industrial Districts: Silicon Valley, Route 128, and Covenants not to Compete.' *New York University Law Review*, 74/3.

Gittelman, M. (2000). 'Mapping National Knowledge Networks: Scientists, Firms and Institutions in Biotechnology in the United States and France.' PhD Dissertation, University of Pennsylvania, Ann Arbor: UMI Dissertation Services.

—— and Kogut, B. (2003). 'Does Good Science Lead to Valuable Knowledge? Biotechnology Firms and the Evolutionary Logic of Citation Patterns.' *Management Science*, 49/4.

Glimstedt, H. (2001). 'Competitive Dynamics of Technological Standardization: The Case of Third Generation Cellular Communications.' *Industry and Innovation*, 8: 49–78.

—— and Zander, U. (2003). 'Sweden's Wireless Wonders: The Diverse Roots and Selective Adaptations of the Swedish Internet Economy.' In B. Kogut (ed.), *The Global Internet Economy.* Cambridge: MIT Press, 109–52.

Goldthorpe, J. H. (ed.) (1984). *Order and Conflict in Contemporary Capitalism.* Oxford: Clarendon Press.

Granovetter, M. (1973). 'The Strength of Weak Ties.' *American Journal of Sociology*, 78/6: 1360–80.

Hall, P. A., and Soskice, D. (eds.) (2001). *Varieties of Capitalism: The Institutional Foundations of Comparative Advantage.* Oxford: Oxford University Press.

Herrigel, G. (1993). 'Power and the Redefinition of Industrial Districts: The Case of Baden-Württemberg.' In G. Grabher (ed.), *The Embedded Firm.* London: Routledge.

Holmstrom, B., and Milgrom, P. (1994). 'The Firm as an Incentive System.' *American Economic Review*, 84/4: 972–91.

Hyde, A. (1998). 'Employment Law after the Death of Employment.' *University of Pennsylvania Journal of Labor Law*, 1: 105–20.

Kenney, M. (ed.) (2000). *Understanding Silicon Valley: Anatomy of an Entrepreneurial Region.* Palo Alto, CA: Stanford University Press.

Kneller, R. W. (1999). 'University-Industry Cooperation in Biomedical Research in Japan and the U.S.' In L. M. Branscomb, F. Kodama, and R. Florida (eds.), *Industrializing Knowledge: University-Industry Linkages in Japan and the United States.* Cambridge: MIT Press.

Kogut, B. (ed.) (1993). *Country Competitiveness: Technology and the Organizing of Work.* Oxford: Oxford University Press.

—— (2003). *The Global Internet Economy.* Cambridge: MIT Press.

Lehrer, M. (2000). 'Has Germany Finally Fixed its High-Tech Problem? The Recent Boom in German Technology-Based Entrepreneurship.' *California Management Review,* 42/4: 89–107.

Lerner, J., and Gompers, P. (1999). *The Venture Capital Cycle.* Cambridge: MIT Press.

Locke, R. M. (1995). *Remaking the Italian Economy.* Ithaca, NY: Cornell University Press.

Malerba, F. (2004). *Sectoral Systems of Innovation: Concepts, Issues, and Analysis from Six Major Sectors in Europe.* Cambridge: Cambridge University Press.

—— and Orsenigo, L. (1993). 'Technological Regimes and Firm Behaviour.' *Industrial and Corporate Change,* 2: 45–71.

Meyer, J. W., and Rowan, B. (1977). 'Institutional Organizations: Formal Structure as Myth and Ceremony.' *American Journal of Sociology,* 83/2.

Miller, G. (1992). *Managerial Dilemmas.* Cambridge: Cambridge University Press.

Mowery, D., and Nelson, R. (eds.) (1999). *Sources of Industrial Leadership: Studies of Seven Industries.* Cambridge: Cambridge University Press.

—— —— Sampat, B., and Ziedonis, A. (1999). 'The Effects of the Bay–Dole Act on U.S. University Research and Technology Transfer: An Analysis of Data from Columbia University, the University of California, and Stanford University.' In L. M. Branscomb, F. Kodama, and R. Florida (eds.), *Industrializing Knowledge: University-Industry Linkages in Japan and the United States.* Cambridge: MIT Press.

Nelson, R. (ed.) (1991). *National Innovation Systems: A Comparative Analysis.* Oxford: Oxford University Press.

Perrow, C. (1985). *Normal Accidents.* Princeton: Princeton University Press.

Powell, W. W. (1998). 'Learning from Collaboration: Knowledge and Networks in the Biotechnology and Pharmaceutical Industries.' *California Management Review* (Special Issue on Knowledge and the Firm), 40/3: 228–40.

—— and DiMaggio, P. J. (1983). 'The Iron Cage Revisited: Institutional Isomorphism and Collective Rationality in Organizational Fields.' *American Sociological Review,* 48/2: 147–60.

Saxenian, A. (1994). *Regional Advantage: Culture and Competition in Silicon Valley and Route 128.* Cambridge, MA: Harvard University Press.

Schitag Ernst and Young (2002). *German Biotechnology Report 2002.* Stuttgart: Schitag Ernst and Young.

Schropp, C., and Conrad, J. (1999). 'Can Your Company Benefit from German Financing?' *Nature Biotechnology,* 17.

Sheridan, C. (2003). 'Germany Biotech Gets Second Chance.' *Nature Biotechnology,* 21/12.

Sorge, A. (1988). 'Industrial Relations and Technical Change: The Case for an Extended Perspective.' In R. Hyman and W. Streeck, *New Technology and Industrial Relations.* Oxford: Basil Blackwell.

Soskice, D. (1997). 'German Technology Policy, Innovation, and National Institutional Frameworks.' *Industry and Innovation,* 4: 75–96.

Streeck, W. (1984). *Industrial Relations in West Germany: A Case Study of the Car Industry.* New York: St Martin's Press.

—— (1992). *Social Institutions and Economic Performance: Studies of Industrial Relations in Advanced Capitalist Economies.* London: Sage Publications.

Teece, D. (1986). 'Profiting from Technological Innovation: Implications for Integration, Collaboration, Licensing, and Public Policy.' *Research Policy,* 15: 285–305.

Trumbull, G. (2004). *Silicon and the State: French Innovation Policy in the Internet Age.* New York: Brookings Institute Press.

Tylecote, A., and Conesa, E. (1999). 'Corporate Governance, Innovation Systems, and Industrial Policy.' *Industry and Innovation,* 6: 25–50.

Whitley, R. (1999). *Divergent Capitalisms: The Social Structuring and Change of Business Systems.* Oxford: Oxford University Press.

—— (2003). 'Competition and Pluralism in the Public Sciences: The Impact of Institutional Frameworks on the Organization of Academic Science.' *Research Policy,* 32/6.

Wirtschaftswoche (2000). 'Aktien statt Mitbestimmung.' 9 March 2000.

Woodward, J. (1965). *Industrial Organization: Theory and Practice.* Oxford: Oxford University Press.

Zucker, L. G., Darby, M. R., and Brewer, M. B. (1998). 'Intellectual Human Capital and the Birth of U.S. Biotechnology Enterprises.' *American Economic Review,* 88/1.

22 What's New? General Patterns of Planned Macro-institutional Change

John L. Campbell

Introduction

Institutional analysis has been revitalized during the last twenty-five years or so.[1] Its long history stretches back at least to the work of Max Weber, Karl Polanyi, Émile Durkheim, and other social scientists of the late nineteenth and early twentieth century. But today there are vast new literatures in sociology, political science, and economics that build on these older traditions; Jerald Hage reviews some of them in the introduction to this section of the volume. Many of these literatures seek to explain how institutions constrain behavior, create incentives, influence decision-making and, in turn, affect a wide variety of social phenomena, including economic performance, technological innovation, politics, and more.[2]

Given the increasing importance placed on institutions by scholars, it is ironic how little attention has been paid to how institutions themselves are created and how they change. Indeed, institutional analysis has been repeatedly criticized for having an inadequate understanding of institutional change (for example, Lieberman 2002; Scott 2001: 181; Thelen and Steinmo 1992: 15). This is not to say that scholars have completely ignored the issue of institutional change. For instance, some economists have argued that when the costs of monitoring market transactions become prohibitive, actors will modify market institutions or build entirely new institutions, such as complex, long-term subcontracting relation-

ships or corporate hierarchies, that better control these transaction costs (Williamson 1985; see also North 1990). But even in such carefully specified theories as transaction cost economics, the *process* whereby institutions are modified or created is largely ignored. We are told that shifts in transaction costs trigger institutional innovation, but the innovation process itself remains a mystery. So institutional innovation seems like an automatic and inevitable reaction to certain transaction-cost conditions, rather than a process involving deliberation and planning by the actors involved.

Similarly, population ecology theorists argue that institutional innovation occurs as new principles and practices diffuse through fields of organizations and survive (or not) as a result of competitive selection (Carroll and Hannan 1989). Business historians have also emphasized the importance of the invisible hand of competitive selection in their analyses of institutional and organizational change (for example, Chandler 1977). In neither case, however, is it necessarily clear exactly how this selection process occurs, or what the underlying mechanisms actually are. As I have discussed elsewhere at length, lack of attention to process and mechanisms is a problem for many theories of institutional innovation (Campbell 2004: ch. 3). In fact, contributors to this book have also identified the failure to specify change processes as a problem in many studies of technological innovation (Edquist this volume; Hatchuel *et al.* this volume).

This chapter sheds light on the process by which *planned* institutional innovation and change occur. Not all instances of institutional innovation are planned and deliberative (for example, Hage this volume). For example, judge-made law is often said to evolve in ways that result eventually in important shifts in legal institutions through the cumulative effects of many small judicial decisions, but without much planning or deliberate design. The same is true for change that stems from competitive selection (Rutherford 1994: chap. 5). In other words, while actors may be purposive, the institutional changes that follow from their actions may be unintended (for example, Nelson and Winter 1982). But, important though unplanned institutional change may be, it is not my concern here.

Why, in a volume that focuses largely on *technological* innovation and learning, should we be concerned about *institutional* change? As several chapters in this volume assert, institutions affect technological innovation and learning, and, in turn, the competitiveness of industrial sectors and national economies (for example, Casper; Chaminade and Edquist; Finegold; Hage). This is also an argument found in the broad literature on the varieties of capitalism, which shows that variations in national political and economic institutions have had significant impacts on innovation, learning, and economic performance (Hall and Soskice 2001; Hollingsworth and Boyer 1997; Ziegler 1997). From a policy standpoint, it is important to recognize that we can manipulate the institutions that underlie the innovation process, in addition to manipulating the flow of money, expertise, and other resources, to facilitate technological innovation and learning. Changes in economic and environmental regulations, tax law, patent law, and property rights in general are just a few examples of institutions that can be adjusted in order to spark technological innovation (for example, Campbell and Lindberg 1990; Foster *et al.* this volume). Moreover, when budgets are tight and resources scarce, changing institutions provides an alternative policy lever for stimulating technology innovation. If we want to use this policy lever, then we need to understand how planned institutional change occurs in the first place.

The chapter presents a series of concepts and propositions designed to lay the foundation for a theory of planned institutional change. It is certainly not the last word on the subject. Instead, it is intended to provoke others to think more carefully about how institutional change happens, and how it should be conceptualized and studied in the future. First, I introduce the concept of *bricolage* and argue that this is the basic process whereby planned institutional change typically occurs. It is a process through which actors combine, in new ways, the already existing institutional elements they find at their disposal. Second, I introduce the concept of 'translation', which is closely related to *bricolage*. Translation is a process whereby actors take new institutional principles and practices that diffuse to them from external sources, and blend and fit them into their local institutional contexts in ways that facilitate institutional change. Third, I specify the processes of *bricolage* and translation more carefully. I pay close attention to the kinds of problems that trigger episodes of planned institutional change, the central role of institutional entrepreneurs in the change process, and the social, institutional, resource, political, and other constraints within which these creative entrepreneurs innovate. Indeed, this process can best be described as one of 'constrained innovation'— a process, as we shall see, that involves both structure and agency.

One clarification is in order. I do not claim that *bricolage* and translation are the only processes by which planned institutional change necessarily occurs. For instance, it is possible that new institutions are occasionally imposed more or less in their entirety from the outside, with little evidence of *bricolage* or translation at all. West Germany's imposition on former East Germany of a set of political institutions may be a case in point. However, this is a rather unique case, even among transforming East European countries. The point is that, while there may be other processes by which institutional change can occur (for example,

Sztompka 1993), I suspect that *bricolage* and translation are the most common. As a result, they take center stage in this chapter.

Bricolage

Planned institutional innovation is typically a deliberate effort to reorganize or otherwise modify already existing institutional arrangements. By 'institutions' I mean formal and informal rules, monitoring and enforcement mechanisms, and systems of meaning that define the context within which individuals, corporations, universities, research laboratories, labor unions, governments, and other organizations operate and interact. Institutions are settlements that are forged through bargaining and struggle. As a result, institutions reflect the resources and power of the people who made them. In turn, once they are created, they affect the distribution of resources and power among people associated with them. Institutions are powerful external forces that help determine how people make sense of their world and how they behave; they channel and regulate conflict, and thus tend to ensure stability in society. In this sense, I depart from Hage who argues in the introduction to this part of the volume that knowledge transformation is the key determinant of institutional change. For me, institutional change is determined largely by struggles over resources and power.

Note that this definition of institutions is at odds with the more popular understanding of the term, that of 'well-established organizations', as in the phrase, 'Maxim's brasserie is a local Parisian institution.' From my point of view, Maxim's is really an organization, a group of people that produces goods and services. It exists within a set of institutions that make up its surrounding environment, such as the rules established and enforced by government regarding the restaurant's health, accounting, labor, and other practices, as well as the taken-for-granted local customs regarding the appropriate way to treat customers,

employees, and suppliers. That said, organizations themselves contain institutions, such as charters and other formal and informal rules specifying how people are supposed to act within the organization. Maxim's, for example, certainly has a set of rules and norms regarding the organization of work in the kitchen, and the customers, and the quality of food it serves. Thus, when we speak of institutions, we may do so at several levels of analysis ranging from macro- to meso- to micro-analysis.[3]

Institutions consist of several dimensions, each with their own principles and practices. There is some debate about what the most important ones are. For example, according to W. Richard Scott (2001), institutions include three dimensions: the regulative dimension consists of legal, constitutional, and other rules that constrain and regularize behavior; the normative dimension involves principles that prescribe the goals of behavior and the appropriate ways to pursue them; the cultural-cognitive dimension entails the culturally shaped, taken-for-granted assumptions about reality and the frames through which it is perceived, understood, and given meaning. Douglass North (1990: 45) describes institutions as embodying formal rules and procedures, as well as informal codes of conduct. Even formal property rights, the regulative institution with which he is often most concerned, consist of a bundle of dimensions that can be conceptually disaggregated into rights of ownership, usage, and appropriation (Barzel 1989; Campbell 1993). However, the point here is not to quibble over how to specify these dimensions (although this is not a trivial matter), but to recognize two things: first, institutions provide a repertoire of already existing institutional principles and practices that actors can use to innovate; second, planned institutional innovation entails the reorganization or modification of any or all of these principles and practices in a particular institutional setting.

The process of combining and recombining already existing institutional elements to bring about institutional reorganization and modification became known as *bricolage* (Levi-Strauss

1966; Douglas 1986). The notion is very similar to the idea that Joseph Schumpeter (1983) had when he described industrial innovation as being caused by entrepreneurs who recombined technology, capital, and other factors to create profitable new products and market opportunities.

It is worth mentioning that *bricolage* is not a process that is restricted to institutional change. A very similar process has been noted for scientific and technological innovations. For instance, important innovations in hearing-aid technologies stemmed from creative individuals combining several relatively minor technical discoveries in ways that transformed cochlear implant technology (Van de Ven and Garud 1991). In this volume, even though they do not use the term, the chapters by Armand Hatchuel *et al.*, Terry Shinn, and Chaminade and Edquist describe innovation processes that involve *bricolage*. And major intellectual breakthroughs in philosophy often have resulted from philosophers being located in places where several different intellectual currents converged, thereby affording them an opportunity to blend these seemingly disparate ideas in new and profound ways that changed the course of intellectual history (Collins 2000).

Bricolage tends to result in path-dependent, evolutionary change. Why? On the one hand, previously created institutions provide a repertoire of principles and practices with which actors can innovate creatively through recombination. On the other hand, the finite nature of this repertoire constrains the range of possible innovative combinations available to them. Hence, institutions simultaneously enable and constrain innovation. Furthermore, given the finite nature of this repertoire, innovations differ from, but still resemble, those that preceded them. An architect may build many different houses by combining the wood, nails, glass, and other materials available to him in different ways. Each house may be different in one way or another as he learns from his mistakes and improves on previous designs, but there will still be a strong resemblance among them all, because they are built with the same materials, and because the architect tends to subscribe to a loose set of aesthetic principles and practices regarding the form and function that he views as appropriate. The same is true for institutional entrepreneurs who recombine the available basic institutional elements in their repertoires in ways that are innovative, but still bear a strong resemblance to those that came before.[4]

Two important implications follow from this. First, even apparently dramatic and radical institutional innovations are less revolutionary than they might seem initially. This was certainly true in post-communist Europe during the first half of the 1990s. For instance, new quasi-corporatist labor-market institutions were devised; they were modifications of old tripartite bargaining arrangements that involved managers of state enterprises or industrial sectors, directors of communist unions, and state planning officials (Pedersen *et al.* 1996). Similarly, some new enterprise structures and ownership arrangements were derived from the recombination of old communist-era exchange relationships among enterprises (Stark 1996). In both cases, an apparently revolutionary change—the shift from a command to market economy—actually involved an important evolutionary and path-dependent dynamic, as new institutions were crafted from the bits and pieces of old ones that were inherited by the institutional entrepreneurs involved.

Second, the distinction between revolutionary and evolutionary change is a tricky one. How can we tell how revolutionary or evolutionary an episode of change really is? It depends on how many dimensions of an institution change over a given period of time. Revolutionary change consists of simultaneous change across most, if not all, dimensions of an institution; evolutionary change consists of change in only a few of these dimensions; and stability consists of the absence of change in most, if not all, of these dimensions. Thus, change can be located on a continuum ranging from stability on one end, through different degrees of evolutionary change in the middle, to more revolutionary change on the other end.[5]

Do not infer from the examples of post-communist institutional change that *bricolage* occurs only during times of great political or economic upheaval. It is much more common than that. Insofar as institutions consist of substantive and symbolic principles and practices (March and Olsen 1989), we can conceive of at least two types of *bricolage*.

Much institutional change is undertaken in order to achieve various substantive goals. For economic institutions, these goals include such things as reducing transaction costs, increasing market share, managing labor-relations problems, improving product quality, and so on. 'Substantive *bricolage*' involves the recombination of already existing institutional principles and practices in order to address these sorts of problems, and thus follows a logic of instrumentality (March and Olsen 1989). For instance, Taiwanese entrepreneurs built hierarchically organized conglomerates after the Second World War by combining the institutional principles of large multidivisional business firms that had already started to develop in Taiwan with the institutional principles of family honor that had persisted in Taiwan for centuries. During the 1950s, owners of private firms began to recognize that survival and growth depended on building larger and more far-flung corporations; they also realized that managing these conglomerates would become increasingly difficult, in particular due to principal-agent monitoring problems. They branched out into new and unrelated lines of business by extending the multidivisional form, but they placed close family members (siblings, sons, daughters, and in-laws) in top divisional posts to ensure that the operations were run by people whom they could trust (Lin 1995). Thus two well-established institutional principles (bureaucratic organization and family honor) were combined in order to solve a substantive managerial dilemma. The new arrangements that resulted helped transform the Taiwanese economy into a vibrant, innovative, and internationally competitive one.

Institutional change also involves the recombination of symbolic principles and practices through a process of 'symbolic *bricolage*.' In this sense, *bricolage* involves a logic of appropriateness (March and Olsen 1989). This is particularly important insofar as the solutions that actors devise must be acceptable and legitimate within their broad social environment. Social scientists have recognized that for new institutions to take hold they must be framed with combinations of existing cultural symbols that are consistent with dominant normative and cognitive institutions. The utilization of symbolic language, rhetorical devices, lofty and culturally accepted principles, and analogies to what is believed to be the natural world are central to this framing exercise (for example, Douglas 1986; Snow *et al.* 1986; Swidler 1986). A case in point was the rapid industrialization of the South Korean economy during the post-war period, which was marked by the creation of massive conglomerate firms that were run with a strong, hierarchical, and often authoritarian hand. In order to legitimize these practices to employees, directors deployed various taken-for-granted symbolic elements of traditional South Korean culture: they argued that the firm was like the hierarchically organized family; employees owed the firm their allegiance because it provided for their livelihood, just as sons owed allegiance to their fathers who supported them during their childhoods. Directors also drew on the country's strong nationalist ideology, rooted in a long history of political and economic struggles against more powerful countries like Japan and the United States, to convince workers that their acquiescence to the firm's policies was tantamount to supporting the national interest (Janelli 1993). In sum, directors were *bricoleurs* combining bits and pieces of Korean culture in innovative ways that created symbolic support for their managerial and organizational approach.[6] All of this contributed to South Korea's capacity to compete successfully in international markets.

Of course, both substantive and symbolic elements may be involved simultaneously in a *bricolage* (for example, Haveman and Rao 1997). The emergence of total quality management (TQM) systems in Japan during the 1980s and 1990s provides an illustration. In an effort to improve the capacity for product innovation (including product quality), increase productivity, reduce employee absenteeism and turnover, alleviate bureaucratic paralysis, and resolve a variety of other technical problems, Japanese managers combined elements of three quite different organizational models with which they were familiar. First, employees and managers were taught production management, statistical control, and other substantive techniques from scientific management to improve efficiency and quality in product design, manufacturing, distribution, and sales. Second, drawing on the human relations model, managers used attitude surveys, teamwork, quality-control circles, and the like to improve worker motivation, cooperation, and responsiveness. Finally, firms learned from structural analysis models to reduce the levels of bureaucracy, to form federations of companies, and to increase divisional autonomy to eliminate bureaucratic rigidity. All of this was framed within the ideology, symbolism, and rhetoric of the human relations approach, which emphasized personal improvement and job satisfaction. Thus, TQM evolved from the recombination of both the substantive and symbolic principles and practices of older organizational models. The result was a new form of corporate organization, but one that shared a strong, albeit eclectic, resemblance to its predecessors (Guillén 1994a). It was also a form that proved to be important for improving the capacity of Japanese firms to become more innovative and, thus, more competitive internationally.

Translation

Bricolage involves a recombination of already given institutional elements. Sometimes new elements may be added to the institutional repertoire through diffusion from outside the local setting. This increases the chances that institutional change will be relatively more revolutionary than evolutionary, and that it

will not bear as strong a resemblance to the institutions of the past as would be the case if institutional entrepreneurs simply worked within their already existing repertoires. For example, much has been written about how new institutional models for organizing government ministries, citizenship entitlements, and human rights law have diffused through the community of nation states with profound transformative effects on national-level institutions (Boli and Thomas 1999; Risse *et al.* 1999). Similarly, models of corporate management have diffused internationally (Guillén 1994b).

Rarely, however, is a newly introduced element or idea adopted in toto and unchanged. Once a diffusing element arrives at a local institutional location, it is modified in order to fit with already existing local institutional arrangements. I call this process 'translation'. It is similar to *bricolage* except for the fact that some of the elements available to the institutional entrepreneur have been introduced from the outside. The more that new principles and practices diffuse to a given locale and are translated fully into practice, the more likely it is that change will be more revolutionary than evolutionary, and that it will diverge sharply from past precedents and legacies. To be sure, there is a vast literature on the diffusion of institutional practices, but discussion of the process of translation is virtually absent from it (for example, Strang and Soule 1998; Wejnert 2002).

The rise of neoliberalism during the late twentieth century is a good illustration of translation. Neoliberalism is a set of cognitive and normative principles that promised solutions for the twin macroeconomic problems of economic stagnation and inflation that befell the advanced capitalist economies during the 1970s and 1980s. Integral to the neoliberal project was a shift away from Keynesian economic ideas, which emphasized the political management of aggregate demand, to a more conservative approach based on supply-side, monetarist, and rational expectations theories (Heilbroner and Milberg 1995). Central here was a call for reductions in economic regulation, taxes, and government spending, especially for welfare programs. All of this was based on the assumption that relatively unbridled markets were more effective than government policies for stimulating the kinds of innovation and investment that would enhance national economic competitiveness (Campbell and Pedersen 2001: ch. 1). To a large extent, neoliberalism emanated from the United States to other parts of the world, often with the strong support of international organizations like the OECD, International Monetary Fund (IMF), and World Bank (Wade and Veneroso 1998*a*, 1998*b*).

Many countries adopted bits and pieces of the neoliberal model in order to change their institutions. However, these adoptions and changes were not uniform across countries. Instead, they were tailored to local institutional contexts. In Denmark, for instance, decision-makers searched for ways to incorporate neoliberal principles into traditional Danish institutions where industrial policy had long been the norm, a product of elaborate negotiations between the state and private actors at the national level. These formal bargaining institutions were reinforced by the pervasive Danish belief that the only appropriate way to conduct policy-making was through inclusive, corporatist negotiations. Indeed, it was difficult for Danes to imagine doing things differently after so many years of negotiation (Nielsen and Pedersen 1991; Pedersen 1993). Nonetheless, they were attracted to neoliberalism's emphasis of the decentralization of political control over the economy and of a more efficient way to manage economic activity. So, rather than abandoning industrial policy and absolving the state of its responsibilities for industrial development, as the hard-core neoliberal view would suggest, Danish leaders reorganized their institutions by establishing a more decentralized set of institutional links between the government, local authorities, business, labor, and other private organizations. The result was a new, decentralized, but still negotiated and corporatist form of decision-making that encouraged firms to adopt new technologies and production practices (Kjaer and Pedersen 2001; Kristensen forthcom-

ing). Neoliberalism was translated into traditional Danish practices rather than replacing them (see also Martin 2002).

This sort of decentralized but negotiated approach has proven to be a source of successful technology innovation in Denmark. For instance, Denmark has become the world's leading manufacturer of electricity-generating wind turbines. Success stemmed in large part from a decentralized, bottom-up, and often negotiated approach to technology innovation, where a group of small machine-shop-like firms began to experiment with turbine technologies, and often cooperated in solving various shared problems, such as technology testing, which was required by the Danish government. The government also encouraged the fledgling industry's development and technological innovation through a variety of incentives and subsidies. Trial-and-error experimentation and incremental innovation resulted, and eventually produced a cutting-edge technology that set the industry standard worldwide. Other countries, notably the United States, engaged in a much more centralized, hierarchically organized, top-down approach that ended up developing a technology that was scientifically complex but that experienced a variety of technical difficulties that undermined its viability in the market (Karnøe 1995).

There is much evidence to support the notion that the international diffusion of ideas is mediated by translation processes. It has been documented for the development of new rules and norms regarding citizenship (Soysal 1994), human rights (Keck and Sikkink 1998), and corporate regulation (Vogel 1996) as well as new fiscal policies, property rights, and other public institutional practices (Campbell 2001). Nationally specific translation is also evident in the diffusion of common-market directives from transnational authorities to member states in both the European Union and Mercosur (Duina 2003, 1999). In all of these cases, new ideas have been translated into local contexts in ways that result in considerable institutional change, but with much variation across countries, and without complete breaks from each country's institutional legacy. That is, although change occurred through diffusion, the process of translation ensured a degree of continuity between past and present—a more evolutionary, path-dependent process of institutional change than would be expected typically by the diffusion literature.

Of course, diffusion and translation are not restricted to the public sector. They also occur in the private sector, especially as firms engage in economic activity that is more global (Whitley 1997: 255). When firms open operations in other countries, they often adapt their practices to local ones. For example, when US and Japanese automobile manufacturing firms set up subsidiaries in Mexico, they imported many aspects of their human resource management systems, such as employee training practices, which continued to differ from each other and mirror practices back home in the headquarters country. However, labor relations in both subsidiaries were stamped with a distinct Mexican imprint. Notably, labor relations are typically more flexible in Mexico than in the United States, but less so than in Japan. Both the US and Japanese subsidiaries had to adjust for this, and conform to some local labor practices (Hibino 1997). And when foreign firms listed their stock on the New York Stock Exchange and NASDAQ, they retained most of their old ways, but occasionally blended in a few elements of the US model. Perhaps not surprisingly, the firms that were most likely to adopt the full range of US institutional practices were new start-ups that had not yet had a chance to develop well-institutionalized practices and belief systems of their own (Davis and Marquis forthcoming).

Specifying the process of *bricolage* and translation

An important question remains. Given the range of possible combinations available to someone engaging in institutional *bricolage*, why do they make one *bricolage* rather than another? This is a complicated question that

has received virtually no attention from scholars. It raises a number of issues that require more research in the future. What follows is a preliminary effort to answer this very basic question by specifying more closely the processes of *bricolage* and translation. The argument is summarized in Box 22.1 as a series of propositions about the problems that actors face, the actors themselves, and the constraints under which they operate. By focusing on actors as well as the constraints they face, the argument explicitly addresses the call issued in this volume's introductory chapter to pay close attention to both structure and agency in our accounts of innovation.

To begin with, we need to think carefully about the conditions under which actors seek institutional change in the first place. It can be triggered by either exogenous or endogenous factors. Exogenous factors include things like war, economic catastrophe, and other calamities, as well as abrupt shifts in prices, transaction costs, and state policies, dramatic technological innovations, pressure from outside organizations, and the like. However, if we accept the notion that institutions are multidimensional entities that are composed of different institutional principles and logics guiding action, then we should expect that there may be much inconsistency among these dimensions and logics. Institutions may create potentially contradictory incentives and opportunities for action. Such inconsistencies may generate enough tension, friction, and other problems to cause actors to seek new institutional arrangements. In sum, constellations of institutions may themselves generate endogenous pressures for change (Friedland and Alford 1991; Lieberman 2002; Orren and Skowronek 1994; Schneiberg 1999).

For instance, when the US Atomic Energy Commission (AEC) was established, Congress gave it a dual mandate. On the one hand, it was supposed to facilitate the development of nuclear technology that could be used by private utility companies to generate and sell electricity. On the other hand, it was also supposed to ensure that the nuclear power industry operated safely. Eventually, the logic of technol-

---

BOX 22.1 Seven Propositions about Institutional Change[a]

Problems
Proposition 1: Institutional change occurs in response to problems (either endogenous or exogenous to the institution) that threaten the fundamental distribution of resources or power of people operating within the institution.
Actors
Proposition 2: Institutional entrepreneurs are the key actors to the process of institutional change. They are the ones who suggest how to recombine institutional elements in innovative ways; the ones who frame situations as problems; and the ones who frame innovations as promising solutions.
Constraints
Proposition 3: Entrepreneurs that are located at the borders and interstices of several social networks, organizations, and institutions are more likely to be exposed to the diffusion of new ideas, which then become part of their repertoire, and thus lead to relatively more revolutionary than evolutionary changes, than entrepreneurs who are not located at these borders and interstices. Social location determines the repertoire of ideas, principles, and practices with which entrepreneurs work.
Proposition 4: Entrepreneurs have to *fit* their proposed innovations to the prevailing institutional situation, if they are to convince decision-makers to use their innovative ideas. They have to make it seem that their innovations are practical (regulative), appropriate (normative), and sensible (cognitive) given the surrounding institutional context. Again, this is very much about framing innovations in ways that appeal to decision-makers. Fitting innovations to existing institutional arrangements tends to constrain the degree to which innovations will be more revolutionary than evolutionary.
Proposition 5: Entrepreneurs need tangible resources (money, clout). Without them, their ideas often fail to take hold and become institutionalized.
Proposition 6: Innovation is a process whereby entrepreneurs mobilize political support for their innovative ideas. Without political mobilization, they will fail.
Proposition 7: Entrepreneurs are more likely to convince decision makers to adopt their ideas if decision-makers have the organizational capacities to implement and sustain the innovation.

[a] For a more comprehensive list of propositions, see Campbell 2004: ch. 6.

---

ogy development and commercialization collided with the logic of safety regulation. Developmental considerations took primacy over safety considerations, political scandal ensued, the AEC was embroiled in a major legitimation crisis; as a result, its authorizing legislation was rewritten and the agency was replaced by two new ones, each assigned one of the AEC's original mandates (Campbell 1988: ch. 4).

Regardless of whether the origins of problems are exogenous or endogenous, I suspect that problems lead to institutional change when they threaten the fundamental distribution of resources or the power of people operating within these institutions. Remember that institutions are settlements that are born from bargaining and struggle, and that reflect the resources and power of the people who made them. It follows that anything that threatens to disrupt these settlements and upset the distribution of resources and power is likely to trigger a struggle over institutions themselves. For example, during the 1970s and 1980s, the

advent of new production technologies among European steel manufacturers threatened the domestic market share for US steel producers, who then successfully urged Congress to pass protectionist trade legislation (Scherrer 1991). Another example is from the time of engagement in the First World War, which caused the US War Department to experience supply shortages, and led to the creation of temporary, corporatist-style planning boards that were designed to improve industrial governance and increase production for the war effort (Cuff 1973). And the AEC's contradictory institutional mandate to promote and regulate commercial nuclear power triggered conflict between government bureaucrats and scientists over each other's jurisdiction regarding nuclear reactor safety policy—a conflict that sparked a legitimation crisis for the agency and an overhaul of regulatory institutions. In each case, situations developed that threatened actors' access to resources or power (markets, customers, profits; military supplies;

regulatory authority) and then precipitated institutional change (new rules regarding international trade; industrial governance; regulatory policy).[7]

Two clarifications are in order. First, just because a problem develops does not necessarily mean that institutional change will result. Actors may disagree how to resolve problems, stalemates may result, inertia may set in, and problems may fester for a long time without steps being taken to resolve them. Actors may also try to handle them without resorting to changing institutions. Second, actors must perceive that there is a problem, that it requires an institutional solution, and that there are possible solutions available. We have much to learn about how actors develop these perceptions, but central to this process are institutional entrepreneurs (DiMaggio 1988; Fligstein 1997, 2001; Lounsbury and Glynn 2001). As suggested earlier, these are the ones who suggest how to recombine institutional elements in innovative ways. They are also the ones who frame situations as problems and who frame innovations as promising solutions. Suffice it to say that unless problems and solutions can be framed in terms that convince decision-makers to seek institutional change, not much is likely to happen.

This very much parallels arguments about technology innovation made elsewhere in this volume. For example, Shinn emphasizes the leading role that research technologists (that is, entrepreneurs) play in the innovation process through their recombinatory efforts. And Hatchuel and his colleagues theorize the important role that designers have in the innovation process when they create fictions, conceptualizations, and cognitive fabrications (that is, frames) about new technological possibilities and then move to make these fictions a reality. In other words, technological entrepreneurs engage in very similar practices to those of institutional entrepreneurs, including *bricolage* and the framing of problems and solutions.

To return to the question raised earlier, how do entrepreneurs decide on one *bricolage* rather than another? I submit that understanding how this happens does not begin with an assessment of their individual qualities, like genius or charisma, but with an appreciation of their position within a set of social relationships and institutions. Several things are involved. The following discussion sets forth a series of propositions worth further consideration and research in the future.

First, I propose that entrepreneurs with more diverse social, organizational, and institutional connections tend to have more expansive repertoires with which to work, and tend to be exposed to more ideas about how to creatively recombine elements in their repertoires. Similarly, entrepreneurs located at the borders and interstices of several social networks, organizations, and institutions will be more likely to be exposed to the diffusion of new ideas, which then become part of their repertoire and, thus, lead to institutional changes that are relatively more revolutionary than evolutionary.[8]

Consider the computer industry. Fledgling entrepreneurs in the Silicon Valley region of northern California often operated simultaneously in several organizational environments. For instance, they frequently had links with Stanford University's electrical engineering departments and other research organizations, as well as various small electronics companies in the valley. As a result, when the industry began to take off in the 1970s, these entrepreneurs cognitively envisioned all sorts of innovative formal and informal institutional arrangements that facilitated information sharing and collective problem-solving among firms and other organizations. These were quite open and unique institutional innovations in contrast to typical US business practice in which such information-sharing was viewed as a threat to proprietary knowledge and, thus, to competitive advantage. Yet these new institutions were made possible by the fact that entrepreneurs had broad repertoires of ideas about how technology, firms, and the industry overall might be organized and function. In contrast, entrepreneurs on the East Coast around Boston were much more insulated. They operated generally within a single, large, hierarchically organized firm. In turn, they had a relatively limited repertoire of institutional principles and practices at their disposal. So when the industry began to develop on the East Coast, entrepreneurs stuck to the traditional, centralized, closed, and relatively rigid institutional arrangements with which they were familiar. In the end, the advantage went to West Coast entrepreneurs, who enjoyed considerable success (Saxenian 1994). Of course, the idea that the structure of interpersonal and interorganizational networks affects the possibilities for successful technological innovation and learning is a theme that resonates with several chapters in this volume.[9]

Second, all of this occurs, of course, within a broader institutional milieu that can have additional effects on how creative entrepreneurs are likely to be. Following Scott's threefold distinction, if an entrepreneur's institutional location limits the range of innovations that can be imagined cognitively, as I just suggested, then it may also limit the range of innovations that will be normatively appropriate or legitimate. For instance, in the United States during the 1980s, proposals for state-sponsored industrial policy, which were offered as a way to spark technological and industrial innovation, were never adopted because opponents argued convincingly that industrial policy represented a form of state intervention that was tantamount to socialism and, therefore, antithetical to basic American norms and values (Graham 1992). Similarly, the entrepreneur's location within a set of regulative institutions will limit the range of innovations he or she is likely to pursue. In particular, the ability of entrepreneurs to reorganize forms of economic governance in business has long been contingent on their ability to convince the courts and regulatory agencies that these innovations fit existing law (Campbell and Lindberg 1990). The more that entrepreneurs can demonstrate that their innovations fit the prevailing institutional situation, the greater will be their capacity for innovation and the likelihood that their innovations will stick. This implies that having to fit innovations into existing institutional arrangements tends to constrain the degree to which innovations will be more

revolutionary than evolutionary. Hence, a conservative dynamic results, whereby institutional change tends to be evolutionary rather than revolutionary.

Third, access to tangible resources, like money and political clout, is also important. If entrepreneurs don't have access to these resources, then their innovative ideas, no matter how brilliant, will often fail to take hold and become institutionalized (Aldrich 1999: 76).[10] For instance, during the late nineteenth century, Thomas Edison wanted to institutionalize his concept of centralized generation of electrical power. His ability to pursue this technological innovation depended on his ability to shift from a network of American financiers, led by J. P. Morgan, who favored a decentralized approach, to a network of European financiers who supported his idea of central-station power generation (McGuire *et al.* 1993). Similarly, nineteenth-century merchants and industrialists in the United States were able to create huge corporate organizations because they had the resources needed to win legal and legislative battles that reduced local business regulation, limited personal liability in the event of bankruptcy, and swept away other political and institutional obstacles to the growth of these enormous organizations (Perrow 2002). As it turned out, these huge organizations lacked the capacities for quick technological and product innovation and flexibility that many scholars have argued are of paramount importance in today's new economy (Best 1990; Piore and Sabel 1984). Had Edison not been able to free himself from the Morgan group, and had nineteenth-century entrepreneurs not been able to change the law, a different set of innovations would likely have prevailed. Others in this volume have made similar arguments about the importance of resource constraints on the technology innovation process (for example, Mohrman *et al.*).

Thus, while entrepreneurs' social, organizational, and institutional location affects their capacity for creative innovation, they face institutional and resource constraints that affect their capacity to make their innovations stick. Recognizing all of this is important because, without an understanding of these constraints and how they limit the range of creative opportunities available to entrepreneurs, we could incorrectly assume that entrepreneurs can create whatever innovation and *bricolage* they please, and that institutional innovation is simply a matter of their individual cleverness.

Fourth, the balance of power and political mobilization among entrepreneurs, competing for the attention of decision-makers and other actors, also affects institutional innovation. Indeed, political struggles frequently occur over matters of institutional change. For instance, the rules and regulations governing technology development and innovation have often been contested. The AEC's efforts to develop and regulate commercial nuclear power, mentioned above, are a case in point. So too are the recent controversies over policies intended to regulate human cloning and the genetic engineering of agricultural products. And the political and legal battle that occurred during the 1970s over whether microwave telecommunications companies, such as MCI, should be granted access to AT&T's local telephone exchanges was a protracted struggle over regulatory institutions. Once these institutions were changed, and local access was granted, innovations in telecommunications technologies flourished (Bickers 1991). Not only are the processes of *bricolage*, translation, and planned institutional change creative, they are also very much political processes, infused with conflict and struggle.

Fifth, and closely related to the preceding point, adoption of an institutional innovation may also depend on whether powerful external actors in the surrounding environment can coerce or otherwise convince local actors to adopt an institutional change. To return to the case of neoliberalism, the degree to which countries were in debt, and thus dependent on the IMF, World Bank, and other financial institutions in both post-communist Europe and East Asia during the 1980s and early 1990s, considerably affected the degree to which these countries embraced neoliberalism or not. Those with great debt burdens, or who had come to rely heavily on financial aid from the international financial community, were more likely to accept the tough neoliberal approach (for example, Guillén 2001; Wade and Veneroso 1998a, 1998b). Poland and Hungary, for instance, were more susceptible to pressure from the IMF immediately after the communist regimes fell because they had inherited large external debts from the old regimes. Czechoslovakia did not. This was one reason why the Czechoslovakian version of neoliberalism was much gentler than the Hungarian and, especially, the Polish versions in the beginning. The Czechoslovakian version was marked by relatively aggressive labor-market policies, including job training and worker-relocation programs rather than just unemployment benefits, and relatively lenient bankruptcy policies, both of which were designed to minimize unemployment (Campbell 2001).

Sixth, organizational characteristics may affect institutional change. Scholars have drawn on the policy-implementation literature to argue that organizations that are exposed to new institutional principles and practices through diffusion are more likely to translate them into practice substantively, rather than just symbolically, if leaders inside the organization are sympathetic and ideologically committed to the new practice. Substantive translation is also more likely if the organization itself has the financial, administrative, and other implementation capacities necessary to support the new practice (Hironaka and Schofer 2002; Westney 1987; Zald *et al.* 2002). A similar argument has been made about nation states. In particular, scholars have argued that small states that are especially open to international pressures—both economic and geopolitical—tend to develop capacities that enable them to respond quickly and flexibly to these pressures, in part by pursuing the sorts of industrial policies, noted above, for example, in Denmark, that the United States has not followed (Campbell *et al.* forthcoming; Garrett 1998; Katzenstein 1985).

In sum, the degree to which innovations are, or are not, implemented and, therefore, precipitate planned institutional change that is more or less evolutionary or revolutionary, depends on local institutional contexts, power struggles, leadership support, and implementation capacities. This is all demonstrated by Marie-Laure Djelic's (1998) analysis of the diffusion of the American model of industrial production to France, West Germany, and Italy after the Second World War. The American model involved several dimensions, including large, hierarchically organized, multidivisional firms, oligopolistic markets, and dispersed patterns of ownership. Shifting to the American model required corresponding changes in legal institutions, such as, respectively, laws of incorporation, antitrust law, and limited liability law. While no European country completely abandoned all of their old, nationally unique production systems and legal institutions, there were important differences in outcomes. Briefly, France and West Germany experienced relatively radical shifts along these dimensions toward the American model, but change in Italy was much more modest. Particularly in the northern region, Italian production continued to be based largely on small- and medium-size, family-owned firms, linked through cooperative networks. Why? First, Italy had weaker cross-national ties to the United States through which the diffusion of ideas could flow, and so Italian political leaders at the national level were less supportive and accommodating of the American model than leaders in other European countries. Second, governments at the local level, particularly Christian Democratic and communist ones, opposed the American model because it threatened their political interests. The American model threatened to eradicate the Christian Democrats' petit bourgeois electoral base and to create large-scale capitalist firms, which were anathema to the communists. Third, local governments had the administrative capacities to block implementation of the American model, such as by offering tax breaks and infrastructural supports only to small and medium-size firms, not to large ones. This created incentives for owners not to expand, vertically integrate, or otherwise transform their operations (see also Weiss 1988). As a result, compared to France and West Germany, in Italy the

translation of various organizational and legal aspects of the American model into practice was much less complete, the original institutional context was preserved the most, and a more modest, evolutionary transformation occurred.

Conclusion

Earlier I suggested that the degree to which institutional change is relatively more evolutionary than revolutionary depends on how many dimensions of the institution in question change in a given period of time through the processes of *bricolage* and translation. The degree to which such change occurs is heavily mediated by all of the factors I have described above. Certainly the creativity, brilliance, drive, and resourcefulness of institutional entrepreneurs is important, but their capabilities in this regard are constrained in important ways by the factors I have discussed. As such, the process I have described can best be characterized as one of constrained innovation—a process that involves both structure and agency. But where does all of this point in terms of directions for future research? A full discussion of this issue is beyond the scope of this chapter, but a few brief remarks are in order.

To begin with, perhaps the most important challenge for us is to think more carefully about the conditions under which planned institutional change is likely to be more or less evolutionary or revolutionary. In other words, the question is not whether *bricolage* and translation actually change the fundamental institutional parameters of society, but under what conditions they are likely to do so, and to what degree. This constitutes an important direction for future research because it will help us develop better theories of different types of institutional change, rather than monolithic theories that seek to explain all types at once.[11]

A second direction for future research involves the notion of 'fit.' I have said a lot about the necessity for translating and fitting innovative programs into local institutional contexts. We need to learn much more about this process too. What does it take for an innovation to fit, or not? Is fit really a function of the amount of political or organizational support an innovation has, such that greater levels of support mean that people will simply try harder to make it work and be more forgiving when the innovation creates problems? Or is fit a function of something else? Researchers have begun to generate careful studies of the translation process, particularly in the area of comparative political economy (Djelic 1998; Duina 1999; Marjoribanks 2000), but we need more of this sort of work. Again, it would be extremely useful to know more about how people who devise and implement innovations take the issue of fit into account. Do they try to anticipate problems, take steps to avoid them pre-emptively, garner support in advance from constituents for translation, and perhaps make adjustments in already existing local institutions to prepare for translation of an innovation? If so, how do they do these things? Until we know more about translation and fit, our arguments about institutional change will remain poorly specified.

Finally, many of the examples I have used here focus on macro-level political and economic institutions. There is, of course, a large and impressive literature that shows how these institutions affect macroeconomic performance, and how variation in these institutions across countries accounts for much of the difference in their performance as well. This 'varieties-of-capitalism' literature, as well as the closely related literature on national systems of innovation, which is discussed by Meeus and Hage in the introduction to this volume, tends to conceive of macro-level institutions as rather rigid and impervious to much change (Hall and Soskice 2001; Hollingsworth and Boyer 1997). In this view, for example, institutions act as constraints on change and innovation, and little credence is given to the possibilities of convergence as a result of common problems and pressures, such as globalization (for example, Berger and Dore 1996). While I am sympathetic to this view, it is excessively deterministic insofar as it neglects how national institutions are made more dy-

namic and malleable by the capacities and efforts of institutional entrepreneurs. Varieties of capitalism are real, and are likely to survive for the foreseeable future, but they are also likely to evolve in response to problems associated with globalization and other problems in path-dependent ways through the processes of *bricolage* and translation. Why?

Analytic typologies notwithstanding, real countries have long consisted of combinations of different elements of different varieties of capitalism (for example, Campbell *et al.* 1991). Britain and the United States, two economies that are often described as being rather liberal in the sense that they have rather limited levels of state economic intervention, illustrate the point. Britain, unlike other liberal market economies, has a national health service. The United States, unlike some liberal market economies, has a vast state sector, notably in defense and health care, that helps shape the organization and functioning of private market activity. In fact, there is much more of this hybridization within OECD countries than is generally recognized (Casper, this volume; Zeitlin 2003). Such hybridization should not be surprising. After all, it is entirely consistent with my earlier observations that, first, institutions consist of various elements, dimensions, principles, practices, and logics, and, second, that these things are often not internally consistent.

How varieties of capitalism and national systems of innovation evolve is a question that provides a third direction for future research. Will they converge on a single hybrid type that looks more or less the same everywhere? Or will nationally specific characteristics persist despite such hybridization? All of this is especially important if we take seriously the notion that other forms of innovation (technological and organizational), discussed elsewhere in this volume, are embedded in institutions, and that this embeddedness has significant effects on the nature of these other innovation processes. If this is true, then understanding the evolution of varieties of capitalism and national systems of innovation better will also help us better understand these other types of innovation.

Notes

1. The arguments in this chapter are developed at greater length in my *Institutional Change and Globalization* (Princeton University Press, 2004).
2. For reviews of this literature see, for example, Campbell (2004: ch. 1; 1997), Peters (1999), and Rutherford (1994).
3. For further discussion of the distinction between organizations and institutions, see Perrow (1986: 167–77; 2002, ch. 1).
4. For a similar argument about the path-dependent nature of technological innovation, see Hatchel *et al.* (this volume).
5. For further discussion of how to determine how much change occurs in a given empirical episode, see Campbell (2004: ch. 2). Recognizing that change can be conceptualized as a continuous rather than as a dichotomous variable is also an important insight for studies of technological innovation that tend to accept the dichotomous distinction between radical/ revolutionary and incremental/evolutionary change (for example, Jordan this volume). Hatch- uel *et al.* (this volume) illustrate the importance of considering carefully the temporal dimen- sion when seeking to determine how much change has occurred in studies of technological innovation.
6. Economists tend to forget that institutional innovations require symbolic framing (Douglas 1986: 46; Hodgson 1988: 156). Even the process of market creation requires framing (Flig- stein and Mara-Drita 1996).
7. Organizational sociologists have argued that an important impetus to institutional change is the desire that organizations have to reduce uncertainty in their environment (for example, Fligstein 1990). But it would be a mistake to infer from this that uncertainty is a motivation for institutional change that is distinct from concerns over resources and power. In fact, what organizations are often uncertain about when they seek institutional change is precisely whether their resources or power, and the activities associated with procuring them, are currently in jeopardy. They engage in institutional change to reduce the uncertainties associ- ated with obtaining and retaining the resources and power that they desire. Hence, for example, US corporations pressed for clarification and revision of antitrust law during the late nineteenth and early twentieth centuries in order to reduce the uncertainty then sur- rounding corporate mergers and other activities that they wanted to pursue in order to increase future profits (Kolko 1963).
8. Inspiration for this idea comes in part from social movements research where scholars report that institutional and structural location affects the repertoire and innovative capacities of social movement entrepreneurs (Ganz 2000; Mansbridge 1986; Morris 2000: 450). Similar arguments have been made by organizational studies scholars (Aldrich 1999: 81–5; Morrill forthcoming; Uzzi 1996).
9. Shinn (this volume) makes a very similar argument about the importance for technology innovation of research technologists being located at the interstices of different occupational groups precisely because, in his view, this affords them an opportunity to gather new and potentially innovative ideas to which they would otherwise not have been exposed.
10. An important part of obtaining resources involves the ability of entrepreneurs to develop trust with those from whom they require resources. They must cultivate an image of their innov- ation as something that naturally should be taken for granted, and an image of themselves as risk oriented but responsible. The more they do this and secure legitimacy and support from those around them, the more likely they will be to make innovations that are sharp departures from the past and, thus, represent truly revolutionary changes. Their ability to generate trust also increases the chances that they can expand their network and institutional locations, increase the breadth of their repertoires, and improve their access to resources. For further discussion, see Aldrich (2000; 1999: 87) and Lounsbury and Glynn (2001).

11. I suspect that specifying the conditions under which innovation is relatively more evolutionary or revolutionary is also something that students of technology and organizational innovation ought to take seriously.

References

Aldrich, H. E. (1999). *Organizations Evolving*. Thousand Oaks, CA: Sage.

—— (2000). 'Entrepreneurial Strategies in New Organizational Populations.' In R. Swedberg (ed.), *Entrepreneurship: The Social Science View*. New York: Oxford University Press, 211–28.

Barzel, Y. (1989). *Economic Analysis of Property Rights*. New York: Cambridge University Press.

Berger, S. and Dore, R. D. (eds.) (1996). *National Diversity and Global Capitalism*. Ithaca, NY: Cornell University Press.

Best, M. (1990). *The New Competition*. Cambridge, MA: Harvard University Press.

Bickers, K. (1991). 'Transformations in the Governance of the American Telecommunications Industry.' In J. L. Campbell, J. R. Hollingsworth, and L. N. Lindberg (eds.), *Governance of the American Economy*. New York: Cambridge University Press, 77–107.

Boli, J., and Thomas, G. M. (eds.) (1999). *Constructing World Culture: International Nongovernmental Organizations Since 1875*. Stanford, CA: Stanford University Press.

Cameron, D. (1978). 'The Expansion of the Public Economy: A Comparative Analysis.' *American Political Science Review*, 72: 1243–61.

Campbell, J. L. (1988). *Collapse of an Industry: Nuclear Power and the Contradictions of US Policy*. Ithaca, NY: Cornell University Press.

—— (1993). 'Property Rights and Governance Transformations in Eastern Europe and the United States.' In S.-E. Sjostrand (ed.), *Institutional Change: Theory and Empirical Findings*. Armonk, NY: M. E. Sharpe, 151–70.

—— (1997). 'Recent Trends in Institutional Political Economy.' *International Journal of Sociology and Social Policy*, 17/7/8: 15–56.

—— (2001). 'Convergence or Divergence? Globalization, Neoliberalism and Fiscal Policy in Postcommunist Europe.' In S. Weber (ed.), *Globalization and the European Political Economy*. New York: Columbia University Press, 107–39.

—— (2004). *Institutional Change and Globalization*. Princeton: Princeton University Press.

—— and Lindberg, L. N. (1990). 'Property Rights and the Organization of Economic Activity by the State.' *American Sociological Review*, 55: 634–47.

—— and Pedersen, O. K. (eds.) (2001). *The Rise of Neoliberalism and Institutional Analysis*. Princeton: Princeton University Press.

—— Hollingsworth, J. R., and Lindberg, L. N. (eds.) (1991). *Governance of the American Economy*. New York: Cambridge University Press.

—— Hall, J. A., and Pedersen, O. K. (eds.) (forthcoming). *The State of Denmark: Small States, Corporatism, and the Varieties of Capitalism*. Montreal: McGill University Press.

Carroll, G. R., and Hannan, M. T. (1989). 'Density Dependence in the Evolution of Populations of Newspaper Organizations.' *American Sociological Review*, 54: 524–48.

Casper, S. (2003). 'Institutions and the New Economy in Europe: Case Study–German Biotech.' Conference on Innovation, Learning and Macro-Institutional Change, Utrecht University, Netherlands.

Chandler, A. D., Jr. (1977). *The Visible Hand: The Managerial Revolution in American Business*. Cambridge: Harvard University Press.

Collins, R. (2000). *The Sociology of Philosophies: A Global Theory of Intellectual Change*. Cambridge, MA: Harvard University Press.

Cuff, R. D. (1973). *The War Industries Board: Business-Government Relations during World War I*. Baltimore: Johns Hopkins University Press.

Davis, G. F., and Marquis, C. (forthcoming). 'The Globalization of Stock Markets and Convergence in Corporate Governance.' In V. Nee and R. Swedberg (eds.), *The Economic Sociology of Capitalism*. Princeton: Princeton University Press.

DiMaggio, P. J. (1988). 'Interest and Agency in Institutional Theory.' In L. G. Zucker (ed.) *Institutional Patterns and Organizations: Culture and Environment*. Cambridge: Ballinger, 3–21.

Djelic, M-L. (1998). *Exporting the American Model: The Postwar Transformation of European Business*. New York: Oxford University Press.

Douglas, M. (1986). *How Institutions Think*. Syracuse: Syracuse University Press.

Duina, F. (1999). *Harmonizing Europe: Nation States within the Common Market*. Albany: State University of New York Press.

—— (2003). 'National Legislatures in Common Markets: Autonomy in the European Union and Mercosur.' In T. V. Paul, G. J. Ikenberry, and J. A. Hall (eds.), *The Nation State in Question*. Princeton: Princeton University Press, 183–212.

Edquist, C. (2003). 'The Role of Policy in Stimulating Product and Process Innovation.' Conference on Innovation, Learning and Macro-Institutional Change, Utrecht University, Netherlands.

Finegold, D. (2003). 'The Role of Education and Training Systems in Innovation.' Conference on Innovation, Learning and Macro-Institutional Change, Utrecht University, Netherlands.

Fligstein, N. (1990). *The Transformation of Corporate Control*. Cambridge, MA: Harvard University Press.

—— (1997). 'Social Skill and Institutional Theory.' *American Behavioral Scientist*, 40/4: 397–405.

—— (2001). 'Social Skills and the Theory of Fields.' *Sociological Theory*, 19: 105–25.

—— and Maria-Drita, I. (1996). 'How to Make a Market: Reflections on the Attempt to Create a Single Market in the European Union.' *American Journal of Sociology*, 102: 1–32.

Foster, J., Hildén, M., and Adler, N. (2003). 'The Complicated Story of Induced Innovation: Experiences from the Role of Public Interventions in the Pulp and Paper Industry.' Conference on Innovation, Learning and Macro-Institutional Change, Utrecht University, Netherlands.

Friedland, R., and Alford, R. A. (1991). 'Bringing Society back in: Symbols, Practices and Institutional Contradictions.' In W. Powell and P. DiMaggio (eds.), *The New Institutionalism in Organizational Analysis*. Chicago: University of Chicago Press, 232–64.

Ganz, M. (2000). 'Resources and Resourcefulness: Strategic Capacity in the Unionization of California Agriculture, 1959–1966.' *American Journal of Sociology*, 105: 1003–62.

Garrett, G. (1998). *Partisan Politics in the Global Economy*. New York: Cambridge University Press.

Graham, O. L., Jr. (1992). *Losing Time: The Industrial Policy Debate*. Cambridge, MA: Harvard University Press.

Guillén, M. (1994a). 'The Age of Eclecticism: Current Organizational Trends and the Evolution of Managerial Models.' *Sloan Management Review*, 36/1: 75–86.

—— (1994b). *Models of Management*. Chicago: University of Chicago Press.

—— (2001). *The Limits of Convergence: Globalization and Organizational Change in Argentina, South Korea, and Spain*. Princeton: Princeton University Press.

Hage, J. (2003). 'Institutional Change and Societal Change: The Impact of Globalization and of Knowledge Transformations.' Conference on Innovation, Learning and Macro-Institutional Change, Utrecht University, Netherlands.

Hall, P. A., and Soskice, D. (eds.) (2001). *Varieties of Capitalism: The Institutional Foundations of Comparative Advantage*. New York: Oxford University Press.

Hatchuel, A., Weil, B., and Lemasson, P. (2003). 'Building Innovation Capabilities: The Development of Design-Oriented Organizations.' Conference on Innovation, Learning and Macro-Institutional Change, Utrecht University, Netherlands.

Haveman, H. A., and Rao, H. (1997). 'Structuring a Theory of Moral Sentiments: Institutional and Organizational Coevolution in the Early Thrift Industry.' *American Journal of Sociology*, 102: 1606–51.

Heilbroner, R., and Milberg, W. (1995). *The Crisis of Vision in Modern Economic Thought*. New York: Cambridge University Press.

Hibino, B. (1997). 'The Transmission of Work Systems: A Comparison of US and Japan Auto's Human Resource Management Practices in Mexico.' In R. Whitley and P. H. Kristensen (eds.), *Governance at Work: The Social Regulation of Economic Relations*. New York: Oxford University Press, 158–70.

Hironaka, A., and Schofer, E. (2002). 'Decoupling in the Environmental Arena: The Case of Environmental Impact Statements.' In A. J. Hoffman and M. J. Ventresca (eds.), *Organizations, Policy, and the Natural Environment: Institutional and Strategic Perspectives.* Stanford, CA: Stanford University Press, 214–34.

Hodgson, G. M. (1988). *Economics and Institutions: A Manifesto for a Modern Institutional Economics.* Philadelphia: University of Pennsylvania Press.

Hollingsworth, J. R., and Boyer, R. (eds.) (1997). *Contemporary Capitalism: The Embeddedness of Institutions.* New York: Cambridge University Press.

Janelli, R. L. (1993). *Making Capitalism: The Social and Cultural Construction of a South Korean Conglomerate.* Stanford, CA: Stanford University Press.

Jordan, G. B. (2003). 'Alternative Kinds of Research Organizations and Outputs: A Theory of Diversity of Research Organizations.' Conference on Innovation, Learning and Macro-Institutional Change, Utrecht University, Netherlands.

Karnøe, P. (1995). 'Institutional Interpretations and Explanations of Differences in American and Danish Approaches to Innovation.' In W. R. Scott and S. Christensen (eds.), *The Institutional Construction of Organizations.* Thousand Oaks, CA: Sage, 243–76.

Katzenstein, P. J. (1985). *Small States in World Markets.* Ithaca, NY: Cornell University Press.

—— (2002). 'Small States and Small States Revisited.' Unpublished manuscript, Department of Government, Cornell University.

Keck, M., and Sikkink, K. (1998). *Activists beyond Borders: Advocacy Networks in International Politics.* Ithaca, NY: Cornell University Press.

Kjaer, P., and. Pedersen, O. K. (2001). 'Translating Liberalization: Neoliberalism in the Danish Negotiated Economy.' In J. L. Campbell and O. K. Pedersen (eds.), *The Rise of Neoliberalism and Institutional Analysis.* Princeton: Princeton University Press, 219–48.

Kolko, G. (1963). *The Triumph of Conservatism.* Chicago: Quadrangle.

Kristensen, P. H. (forthcoming). 'The Danish Production System: Transforming toward a New Economy.' In J. L. Campbell, J. A. Hall, and O. K. Pedersen (eds.), *The State of Denmark: Small States, Corporatism, and Varieties of Capitalism.* Montreal: McGill University Press.

Levi-Strauss, C. (1966). *The Savage Mind.* Chicago: University of Chicago Press.

Lieberman, R. C. (2002). 'Ideas, Institutions, and Political Order: Explaining Political Change.' *American Political Science Review,* 96: 697–712.

Lin, A. (1995). 'The Social and Cultural Bases of Private Corporate Expansion in Taiwan.' Unpublished PhD dissertation, Department of Sociology, Harvard University.

Lounsbury, M., and. Glynn, M. A. (2001). 'Cultural Entrepreneurship: Stories, Legitimacy, and the Acquisition of Resources.' *Strategic Management Journal,* 22: 545–64.

McGuire, P., Granovetter, M., and Schwartz, M. (1993). 'Thomas Edison and the Social Construction of the Early Electricity Industry in America.' In R. Swedberg (ed.), *Explorations in Economic Sociology.* New York: Russell Sage Foundation, 213–46.

Mansbridge, J. J. (1986). *Why We Lost the ERA.* Chicago: University of Chicago Press.

March, J. G., and. Olsen, J. P. (1989). *Rediscovering Institutions: The Organizational Basis of Politics.* New York: Free Press.

Marjoribanks, T. (2000). *News Corporation, Technology and the Workplace: Global Strategies, Local Change.* New York: Cambridge University Press.

Martin, C. J. (2002). 'Activating Employers.' Unpublished manuscript, Department of Political Science, Boston University.

Mohrman, S., Galbraith, J. R., and Monge, P. (2003). 'Network Attributes Impacting the Generation and Flow of Knowledge within and from the Basic Research Community.' Conference on Innovation, Learning and Macro-Institutional Change, Utrecht University, Netherlands.

Morrill, C. (forthcoming). 'Institutional Change through Interstitial Emergence: The Growth of Alternative Dispute Resolution in American Law, 1965–1995.' In W. W. Powell and D. L. Jones, *How Institutions Change.* Chicago: University of Chicago Press.

Morris, A. (2000). 'Reflections on Social Movement Theory: Criticisms and Proposals.' *Contemporary Sociology,* 29: 445–54.

Nelson, R. R., and Winter, S. G. (1982). *An Evolutionary Theory of Economic Change*. Cambridge, MA: Harvard University Press.

Nielsen, K., and Pedersen, O. K. (1991). 'From the Mixed Economy to the Negotiated Economy: The Scandinavian Countries.' In R. M. Coughlin (ed.), *Morality, Rationality, and Efficiency: New Perspectives on Socio-Economics*. New York: M. E. Sharpe, 145–67.

North, D. C. (1990). *Institutions, Institutional Change and Economic Performance*. New York: Cambridge University Press.

Orren, K., and Skowronek, S. (1994). 'Beyond the Iconography of Order: Notes for a "New Institutionalism".' In L. D. Dodd and C. Jillson (eds.), *The Dynamics of American Politics*. Boulder, CO: Westview Press, 311–30.

Pedersen, O. K. (1993). 'The Institutional History of the Danish Polity: From a Market and Mixed Economy to a Negotiated Economy.' In S.-E. Sjöstrand (ed.), *Institutional Change: Theory and Empirical Findings*. New York: M. E. Sharpe, 277–300.

—— Ronit, K., and Suhij, I. (1996). 'The State and Organized Interests in the Labor Market: Experiences from Postcommunist Europe.' In J. L. Campbell and O. K. Pedersen (eds.), *Legacies of Change: Transformations of Postcommunist European Economies*. New York: Aldine de Gruyter, 109–36.

Perrow, C. (1986). *Complex Organizations: A Critical Essay*. 3rd edn. New York: Random House.

—— (2002). *Organizing America: Wealth, Power, and the Origins of Corporate Capitalism*. Princeton: Princeton University Press.

Peters, B. G. (1999). *Institutional Theory in Political Science: The 'New Institutionalism.'* London: Pinter.

Piore, M. J., and. Sabel, C. F. (1984). *The Second Industrial Divide*. New York: Basic Books.

Risse, T., Ropp, S. C., and Sikkink, K. (eds.) (1999). *The Power of Human Rights: International Norms and Domestic Change*. New York: Cambridge University Press.

Rutherford, M. (1994). *Institutions in Economics: The Old and the New Institutionalism*. New York: Cambridge University Press.

Saxenian, A. (1994). *Regional Advantage: Culture and Competition In Silicon Valley and Route 128*. Cambridge, MA: Harvard University Press.

Scherrer, C. (1991). 'Governance of the Steel Industry: What Caused the Disintegration of the Oligopoly?' In J. L. Campbell, J. R. Hollingsworth, and L. N. Lindberg (eds.), *Governance of the American Economy*. New York: Cambridge University Press, 182–208.

Schneiberg, M. (1999). 'Political and Institutional Conditions for Governance by Association: Private Order and Price Controls in American Fire Insurance.' *Politics and Society*, 27/1: 67–103.

Schumpeter, J. A. (1983[1934]). *The Theory of Economic Development*. New Brunswick: Transaction Books.

Scott, W. R. (2001). *Institutions and Organizations*. 2nd edn. Thousand Oaks, CA: Sage.

Shinn, T. (2003). 'Occupational Innovation: Sustainable Innovation through Integration.' Conference on Innovation, Learning and Macro-Institutional Change, Utrecht University, Netherlands.

Snow, D. E., Rochford, B., Worden, S., and Benford, R. (1986). 'Frame Alignment Processes, Micromobilization, and Movement Participation.' *American Sociological Review*, 51: 464–81.

Soysal, Y. (1994). *Limits of Citizenship*. Chicago: University of Chicago Press.

Stark, D. (1996). 'Recombinant Property in East European Capitalism.' *American Journal of Sociology*, 101: 993–1027.

Strang, D., and Soule, S. A. (1998). 'Diffusion in Organizations and Social Movements: From Hybrid Corn to Poison Pills.' *Annual Review of Sociology*, 24: 265–90.

Swidler, A. (1986). 'Culture in Action: Symbols and Strategies.' *American Sociological Review*, 51: 273–86.

Sztompka, P. (1993). *The Sociology of Change*. Cambridge: Blackwell.

Thelen, K., and Steinmo, S. (1992). 'Historical Institutionalism in Comparative Politics.' In S. Steimo, K. Thelen, and F. Longstreth (eds.), *Structuring Politics: Historical Institutionalism in Comparative Analysis*. New York: Cambridge University Press, 1–32.

Uzzi, B. (1996). 'The Sources and Consequences of Embeddedness for the Economic Performance of Organizations: The Network Effect.' *American Sociological Review*, 61: 674–98.

Van de Ven, A. H., and Garud, R. (1991). 'Innovation and Industry Development: The Case of Cochlear Implants.' *Research on Technological Innovation, Management and Policy*, 5: 1–46.

Vogel, S. K. (1996). *Freer Markets, More Rules: Regulatory Reform in Advanced Countries*. Ithaca, NY: Cornell University Press.

Wade, R., and Veneroso, F. (1998a). 'The Asian Crisis: The High Debt Model Versus the Wall Street-Treasury-IMF Complex.' *New Left Review*, 228: 3–24.

—— —— (1998b). 'The Gathering World Slump and the Battle over Capital Controls.' *New Left Review*, 231: 13–42.

Weiss, L. (1988). *Creating Capitalism: The State and Small Business Since 1945*. New York: Basil Blackwell.

Wejnert, B. (2002). 'Integrating Models of Diffusion of Innovations: A Conceptual Framework.' *Annual Review of Sociology*, 28: 297–326.

Westney, E. D. (1987). *Imitation and Innovation: The Transfer of Western Organization Patterns to Meiji Japan*. Cambridge, MA: Harvard University Press.

Whitley, R. (1997). 'The Social Regulation of Work Systems: Institutions, Interest Groups, and Varieties of Work Organization in Capitalist Societies.' In R. Whitley and P. H. Kristensen (eds.), *Governance at Work: The Social Regulation of Economic Relations*. New York: Oxford University Press, 227–60.

Williamson, O. E. (1985). *The Economic Institutions of Capitalism*. New York: Free Press.

Zald, M. N., Morrill, C., and Rao, H. (2002). 'How Do Social Movements Penetrate Organizations? Environmental Impact and Organizational Response.' Paper presented at the conference on Organization and Social Movement Theory, University of Michigan.

Zeitlin, J. (2003). 'Introduction: Governing Work and Welfare in a New Economy: European and American Experiments.' In J. Zeitlin and D. M. Trubek (eds.), *Governing Work and Welfare in a New Economy; European and American Experiments*. New York: Oxford University Press, 1–31.

Ziegler, J. N. (1997). *Governing Ideas: Strategies for Innovation in France and Germany*. Ithaca, NY: Cornell University Press.

23 Insights for R&D Managers

Parry M. Norling

There is nothing more difficult ... more perilous to conduct, or more uncertain in its success, than to take the lead in the introduction of a new order of things. Because the innovator has for enemies all those who have done well under the old conditions, and lukewarm defenders in those who may do well under the new. This coolness arises partly from fear of opponents who have the laws on their side, and partly from the incredulity of men, who do not readily believe in new things until they have had a long experience of them. (Machiavelli: *The Prince*, Chapter VI: 1515)

Introduction

To create competitive advantage, growth, and value for their firms, R&D managers must do more than develop strategies, manage budgets, deal with difficult scientists, prepare reviews and reports, and act as all-knowing scientists. But what should they do? How should they do it? They might find some answers in the management literature, learn on the job, attend conferences or courses on technology management (Burgelman *et al.* 2001; Prather and Gundry 1995), or, through membership in various associations, get advice from others in similar situations. They might also learn from the growing literature in Innovation Studies (Fagerburg 2004) and related work represented in this volume. But this literature may not speak directly in practical terms to innovation practitioners, and may not be relevant to their day-to-day concerns (Meeus 2005).[1] Here we seek to overcome some barriers and show how these chapters are indeed relevant and applicable for R&D managers and, further, are complementary to studies in the more familiar management literature.

As the focus of R&D managers moves from R&D to the entire innovation process, they readily recognize the links between the industrial innovation process, scientific research, knowledge dynamics, and institutional change. They can therefore come to appreciate all four parts of this volume.

We shall examine in turn the crosscutting themes or connections that tie these sections together, as discussed by Meeus and Hage (Introduction, Table I.1), and do this in a logical progression: first, perspectives on or models of the innovation process itself with links to the other processes; second, studies of linkages between and among innovators; third, examination of forces in the environment (including government policies) that govern, control, shape, coordinate, or facilitate innovation; finally, explorations of societal and institutional change and its impact upon the innovation process.

Models: different perspectives on the innovation process

Authors in this volume can each give R&D managers a piece of the industrial innovation puzzle. R&D managers, however, may be no different from practitioners in any field who initially see little relevance to their interests and work in academic theories or models. They are initially wary when Kuhlmann and Shapira speak of understanding practice first through theory and then testing that theory with comparative evidence. R&D managers would prefer to go directly to the comparative case studies.

Christensen *et al.* (2004), however, point out the relevance of models. They maintain that such theories, models, or understanding of the innovation process can be used to predict industry change and guide the decision-making processes of industry leaders and R&D managers. They look at performers (competitors who may be incumbents or attackers), at their strategic choices, at the interplay of non-market forces, especially governmental involvement, and at innovation in different industrial sectors 'to show how theory helps to explain why things in the past happened as they did and what is likely to happen in the future.' They maintain that using theory 'in a meticulous, rigorous fashion can shine a light where darkness once prevailed ... and bring an end to an era when hucksters made their livings selling splendid tales to desperate disciples ... Using theory allows us to see the future more clearly and act more confidently to shape our destiny.' Maybe here we can help R&D managers understand in their terms how models or process descriptions can indeed be seen as relevant to their major concerns.

What is a meaningful model for an R&D manager?

From the perspective of a chemist turned R&D manager, a meaningful process model (as differentiated from an architectural model) would be a detailed description (analogous to a chemical process) that included descriptions of six model elements:

(1) inputs or resources (raw materials or reactants);
(2) steps, activities, and practices employed in the process, transforming inputs into outputs and the relationships (design, structure, or architecture) among the steps, sub-processes and supporting processes (the unit operations, process flows, and kinetics with recycles all controlled by plant operators);
(3) the key players and how they are related one to another (the reaction intermediates);

(4) the external forces that can affect the process (time, temperature, pressure, catalysts);
(5) the motivators, incentives, and objectives for the process (process and product design);
(6) the nature of the outputs and their acceptance into other societal processes (product quality).

We would also want to know both the controllable and uncontrollable elements in the system or process, and the context in which the process is operated.

Such descriptions of the innovation process would help R&D managers understand how something valuable is created in the innovation process, by and for whom, and for what purpose, and that it can help R&D managers understand what forces, resources, practices, and decisions can be deployed to create value for the firm and society in their particular situation.

Models have changed over time

Process models of the idea-innovation chain or networks have changed from the linear science-push, to the market-pull, to the interactive model, and then to the parallel model. For van Waarden and Oosterwijk, the sequential, compression, flexible, and improvisational models, each with assumptions, goals, characteristics and shortcomings, also reflect changes in an increased understanding of innovation in practice, actual changes in product life-cycles, and hence the innovation process itself. Amidon (1997; Table 23.1) has described five generations of R&D management as the R&D process has evolved into the innovation process, including third-generation R&D (Roussel *et al.* 1991) and fourth-generation R&D (Miller and Morris 1999). Miller's Fourth-Generation R&D is essentially a fusion of Amidon's fourth and fifth generations of R&D.[2] (See also Geisler 2000, 2001; Pavitt 2004; Brown and Svenson 1988.)

These models may simply reflect recommended sets of good management practices rather than a picture of what is really happening, but reports from firms practicing Open Innovation (Chesbrough 2000), Value Inno-

Table 23.1. Amidon's five generations of R&D management

| | R&D Generation | | | | |
	1st	2nd	3rd	4th	5th
Core asset→ *Other features*	Technology as the Asset	Project as the Asset	Enterprise as the asset	Customer as the asset	Knowledge as the asset
Core strategy	R&D in isolation	Link to business	Technology/ Business integration	Integration with customer R&D	Collaborative innovation system
Change Factors	Unpredictable serendipity	Interdependence	Systematic R&D management	Accelerated discontinuous global change	Kaleidoscopic dynamics
Performance	R&D as overhead	Cost sharing	Balancing Risk/ reward	'Productivity paradox'	Intellectual capacity/ impact
Structure	Hierarchical, functionally driven	Matrix	Distributed coordination	'Multidimensional communities of practice'	Symbiotic networks
People	We/They competition	Proactive cooperation	Structured collaboration	Focus on values and capacity	Self-managing knowledge workers
Process	Minimal communication	Project-to-project basis	Purposeful R&D/ portfolio	Feedback loops and information 'persistence'	Cross-boundary learning and knowledge flow
Technology	Embryonic	Data based	Information based	IT as competitive weapon	Intelligent knowledge processors

vation (Dillon *et al.* 2005), Integrated Science (Connelly 2005), and Competitive Technology and Business Intelligence (Norling *et al.* 2000) indicate that the innovation process as currently practiced reflects the trends as described, is more complex, has required some different strategies and tactics in playing different innovation games, increasingly involves knowledge creation and transformation, and is involving many more players within and outside the firm.

In the earlier chapters, R&D managers can find helpful descriptions and analyses of the six process elements at different performing levels (individual research, organizational R&D, inter-firm innovation, scientific disciplines and technology domains, sectoral innovation, or national systems of innovation).

Inputs

The raw materials for innovation processes are explored tangentially in this volume, showing how different inputs can affect the final output: Chaminade and Edquist, van Lente, and Georghiou on R&D funding, Finegold on educational and training systems, Nonaka and Peltokorpi on tacit knowledge and ideas, Hatchuel *et al.* on research targets and market opportunities, Jordan on research objectives, Kuhlmann and Shapira on venture capital and research talent, and Casper on experienced research talent in biotechnology.

Process steps, activities, and practices

Nonaka and Peltokorpi link the steps of the knowledge-creation process to radical innovation at Toyota (Figure 23.1), defining a spiral knowledge-creation process—the SECI model (socialization, externalization, combination, and internalization),[3] saying that 'leaders have to facilitate the differentiation and interweaving among seemingly distant and disconnected *ba* and synthesize the knowledge that emerges from the larger *ba* (a shared context or reality that is changing, evolving, or is "in motion")' such as the various technical or design divisions of the larger *ba,* the project team; all are involved in the knowledge conversion process. (Rammert also sees innovation depending upon many processes of knowledge production distributed over various institutional settings.)

Shinn reinforces this contention that managing R&D requires both integration and differentiation processes. Studies of innovation have examined processes of change with reference to integration versus differentiation or specialization, but industrial R&D and the work of the R&D manager itself involve the integration of disparate ideas, talents, and the work of collaborating partners. The research-technology-driven radical innovation at Toyota involves both integration and differentiation; the implications for R&D managers are clear.

Depending on the project size and its focus on incremental/evolutionary or radical/revolutionary innovation (Jordan), innovative processes in scientific research, their outputs, and their measurement will differ from one R&D profile to another. Different goals or strategies call for R&D managers to adopt different organizational structures, management approaches, and measures of success, especially as projects or project phases can spread over the four profiles (Norling 1997; Ranftl 1980). Jordan provides some guidance for R&D managers on what structures and management approaches are appropriate for each profile. A recent five-year study of radical innovation projects (Leifer *et al.*

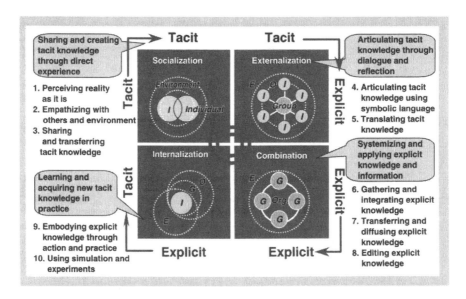

Fig. 23.1. Nonaka: Spiral knowledge-creation process

2000; Rice *et al.* 1996) reinforces Jordan. Many of the standard management techniques for incremental innovation were not effective with such breakthrough efforts; management had to find new ways to capture the initial radical ideas, keep the project on track, gain understanding of the target markets, resolve uncertainty in the business model to market the product, process, or service, respond to competency and resource gaps as the direction for the project became clearer, accelerate the transition from project to operating status, and continue engaging individual motivation when the future looked bleak. Miller and Floricel (2004) go even further in a study 'Managing R&D for Growth' that involved senior R&D managers. Twenty-seven generic management practices (drawn from an initial set of 105) were first identified. It was then found that success was related to the firms' adapting their capabilities and specific practices of the twenty-seven to the requirements for creating value in one of the eight innovation 'games' in which the firm was involved, such as battling for architectures in software and telecommunications, racing to the patent and regulatory offices, delivering safe science-based products, or developing complete solutions for problems of large customers. You will note that the playing of the first two games in Sweden, the UK, and Germany was shaped in quite different ways by national institutional frameworks, rather than by the specific management practices employed as described by Casper.

Chaminade and Edquist define, provisionally, the ten important activities in systems of innovation:

1. provision of R&D creating new knowledge as the technology base for the firm;
2. competence-building in the research, manufacturing, and marketing communities;
3. formation of new product markets and industries;
4. articulation of quality requirements and end-user needs;
5. creating and changing organizations needed for the development of new fields of innovation—enhancing entre-

preneurship to create new firms and intrapreneurship to diversify existing firms, creating new research organizations and policy agencies;
6. networking through markets and other mechanisms, integrating new knowledge developed outside with elements already available inside the innovating firm;
7. provision of institutions that influence innovating organizations and innovating processes by providing incentives or obstacles to innovation;
8. incubating activities providing resources for new innovating efforts;
9. funding of innovation processes and the commercialization of their outcomes;
10. provision of consultancy services for technology transfer, commercial information, and legal advice.

In systems of innovation, organizations or individuals perform these activities; institutions provide the incentives and obstacles influencing these activities. Within firms, however, R&D managers need to be involved in all. Do R&D managers agree? Would they add others such as structuring alliances and collaborations; optimizing the R&D portfolio to support business plans; or energizing and motivating researchers?

Key players or actors

In innovation processes, performers can act as individuals, in groups, in inter-group collaborative networks (Mohrman *et al.*), in a number of interorganizational relationships (Meeus and Faber), or in more extended relationships called 'systems' (Chaminade and Edquist). Organizations are innovators in that they are not only conducive to innovative behavior by individuals, but they are responsible for generating technological innovations. Firms are unique in their role in the innovation process. They are often no longer the only source of innovation-relevant knowledge, but they remain the only entity to combine the many disparate kinds of knowledge to produce a new product, process, service, or business; governments are innovators when their policies impact the performers (Metcalfe) and encourage innovation with funding, standard setting, regulations (Foster *et al.*), and formation of networks and cooperative arrangements.

External forces

Throughout the volume, we see the discussion of forces such as globalization, trends in knowledge creation, development of markets calling for high-tech, complex, knowledge-intensive products, and institutions of many types that can impact aspects of the innovation process (Jordan and Hage).

History is another force that matters; the past has shaped the institutions that may constrain or support innovation (Hollingsworth); R&D managers should therefore understand their country-, corporate-, business-, or organizational history, and use that understanding to build missions and strategic plans.[4] Hollingsworth, like Chaminade and Edquist, points out that innovative activities vary from society to society, organization to organization, and sector to sector (Malerba 2004), in part because of differences in their institutional environments that have been shaped over time. Experience has shown that it is insufficient for an organization to have creative and talented researchers in producing valuable innovations. For more effective performance from the organization, R&D managers need to deal with:

(*a*) the dominant norms, rules, habits, and conventions in the innovator's environment;
(*b*) the governance arrangements which coordinate relationships among innovators;
(*c*) the structure and processes of institutions associated with innovative activity;
(*d*) the culture of the organizations in which innovation occurs.

The challenge for R&D managers is to find what changes need to be made and the techniques to make those changes or mitigate such external forces.[5]

Motivators, incentives, objectives; design in the process

Innovative capability in a firm is not simply nurturing creativity, building a capacity for research and technology development, forming effective teams, or being good at networking; innovativeness is an organizational capability in 'design strategies and design work.' Hatchuel *et al.* challenge R&D managers to understand how innovators create design strategies to promote the simultaneous generation of innovations and knowledge or competencies within firms, as design activities become the core regulators of the innovation process. The innovation process is always described with metaphors that belong to the Design tradition: architecture, mapping, framing, or patterns. But the limited influence on R&D management practice of the literature on innovation studies may be due to the failure to consider a link between organizational theory and design theory. Left with no clear organizing principles, no clear meaning of what is effective mapping, framing, or networking, the development of innovation capability has been identified in management practice as the development of project, platform, portfolio, or knowledge management processes. Hatchuel *et al.* maintain that R&D managers can better understand how an organization develops the capability to innovate by understanding the dynamics of design activities in providing direction for the organization. A number of companies with a clear capacity to innovate repeatedly over long periods of time struggled simultaneously to design lineages of products and lineages of competencies. They point out that a dominant design is usually the standard that comes out ahead in the market place (Utterback 1994); but, from another perspective, a dominant design is a 'design strategy': a selected combination of design choices and related competencies that allow for long-term and large-scale product development and improvements. In the Tefal case-history, identifying design strategies was an intentional management process, learned through the years when management attention was given to:

(1) generating and maintaining scientific domains and research concepts or targets that can yield generations of innovative

products, processes, and services. The more the research target calls for the generation of knowledge far removed from current experience, the greater the chance for creative insight and significant and sustainable innovation (more on this later from Rammert);

(2) structuring and creating work groups that simultaneously develop innovations and increased research or scientific skills;

(3) detecting market opportunities for applying knowledge and technologies that have been or will be generated;

(4) using the design processes for the growth and development of those in the innovation process.

Lester and Piore (2004) use the design theme, pointing out a missing dimension in the innovation process, namely, interpretation. Unlike analysis or problem-solving, interpretation embraces and exploits ambiguity, a source of creativity, and discovers new meanings. By emphasizing interpretation, and showing how these two radically different processes can be combined, they give R&D managers and designers the concepts and tools for developing new products and services. The focus is a process which is open-ended—a 'conversation' between product designers and future customers. Metcalfe agrees; different interpretations of information are not in the presentation of the information but in the different minds of those presenting and those receiving the information, those involved in the conversations. The growth of knowledge depends on such divergent interpretations. All innovations, including scientific breakthroughs, are based on disagreement, on different readings or interpretations of information. Rammert may be dealing with aspects of analysis and interpretation when he contrasts 'explicitation' (the explanation and exploitation of codified knowledge) with exploration (the tacit circulation and informal integration of implicit and explicit knowledge).

The tension between analysis and interpretation, explication and exploration, integration and differentiation, short-term and long-term

focus, incremental and radical innovation, or certainty and ambiguity is inevitable, unavoidable, and an important management problem that R&D managers must learn to confront. The R&D Profiles Theory (Jordan) captures a number of these major tensions that R&D managers must face, and links these tensions to structural and other management practices in a way that facilitates looking at trade-offs all at once and balancing them as needed.

Outputs

Damanpour and Aravind review the determinants of product and process innovation and the extent to which product or process innovation prevail in firms with differing characteristics. While some correlations—such as process innovation being advantageous for large firms—did seem to exist, results were mixed: studies have not distinguished among industry types, generational or adaptive R&D, and radical versus incremental innovation. They also point to studies that show the interrelatedness of process and product innovation and the work of Abernathy and Utterback (1975) on the product life-cycle model: here the rates of product and process innovation change over the three phases of the development of a family of products, with product innovation leading process innovation, followed by increased process innovation, and ending with much of a balance between the two. This has now been extended by Moore (2004). The type of innovation practiced in the firm depends on the point in the life cycle of the product or technology: from disruptive innovation to application, product, process, and experiential innovation (doing such things as streamlining the supply chain and delighting customers with small modifications of products); marketing innovation; business model innovation; and structural innovation. R&D managers should consider the phases of a market's life-cycle. Different types of innovation produce greater value at different points in the life-cycle. Disruptive innovation, for example, is rewarded most during the earliest phase. Once the life-cycle advances to the next phases, other types of innovation yield better returns. Attempts to change a firm's direction are often thwarted by the inertia that success creates. To overcome the inertia, R&D managers, while aggressively extracting resources from legacy (from their past), must introduce new types of innovation. R&D managers will then run the two efforts in parallel, often in opposition to business managers.

Using the models

Porter and Stern (2002) described why some nations are much more innovative than others. They used three main determinants:

(a) the common innovation infrastructure that supports innovation in the economy as a whole (including investment in basic science);

(b) the cluster-specific conditions that support innovation in particular groups of interconnected industries;

(c) the strength of the linkages among them (such as the ability to connect basic research to companies and the contribution of corporate efforts to the overall pool of technology and skilled personnel).

They developed a mathematical model of national innovation systems; it quantified these determinants, giving an overall Innovation Index. This was not a measure of near-term competitiveness, but a benchmark of a nation's potential to sustain productivity growth and competitiveness in the long run. The measures in the index include total R&D personnel, total R&D investment, the percentage funded by private industry, the percentage performed by the universities, spending on higher education, the strength of IP protection, openness to international competition, and the nation's per capita GDP. The index uses statistical modeling to examine how, as measured by international patenting and subsequently correlated with economic growth, these measures have affected innovative output across countries and over time. The statistical analysis yields a weighting of the relative importance of the measures, as applied to each of the countries' actual resources, and policy

choices to determine its index value. The Index measures innovative capacity on a per capita basis, not per dollar of GDP, and has been used in developing the National Innovation Initiative for the United States—a set of recommendations for governmental policies and actions throughout society to promote innovation and competitiveness (Council on Competitiveness 2004).

Linkages between and among innovators: networks, alliances, and collaborative arrangements

Many authors in this volume explore the links and interactions among the various actors in the innovation process, between the actors and society, and links among scientific domains.

The organization of basic scientific research affects the innovative outputs; basic research communities (or eco-systems) are self-organizing and self-renewing networks of researchers and knowledge (Mohrman, Galbraith, and Monge). To draw upon this knowledge, R&D managers will need to make sure that their researchers can link to and participate in specific external scientific networks, as well as a firm's networks or communities of practice that may be either self-emergent or organized by R&D management (Norling 1996; Miller *et al.* 1997; Chester 1994; Miller 1995; Sakkab 2002; Amidon 1997). Such networks link technology and technologists, and promote information flow along the technology supply chain. Mohrman *et al.* point to studies showing that such knowledge communities grow and are transformed on the basis of three evolutionary principles: variation, selection, and retention. R&D managers may need to understand these principles in action when they act as entrepreneurs seeking opportunities to develop new knowledge communities around intellectual or scientific innovations. This requires the development of networks of knowledge and communication or, at least, linking into such networks. Mohrman *et al.* further discuss the flow or transfer of complex knowledge across communities. This may take place in a process of two or more steps, and may require the establishment of intermediary organizations or gatekeepers, as described by the former head of Bell Laboratories as part of his 'Barrier and Bonds' theory of innovation. For successful innovation, there must be one bond and one barrier. If two activities take place in one location, people meet face-to-face and develop special bonding relationships; transfer of knowledge is easy; little new innovation, however, transfers. If activities take place in different places and different organizations, innovations happen; it is, however, difficult to transfer the knowledge across two barriers. The R&D manager must then eliminate one of the barriers, by organizational or location changes.

Reinforcing Mohrman *et al.* was the finding that the real difference between star and average researchers at Bell Labs was not in their IQ but in their capabilities in nine work strategies. One of the most important of these was networking: getting direct and immediate access to co-workers with technical expertise and sharing one's own knowledge with those who need it (Kelly and Caplan 1993).

Mohrman *et al.* point to the practice of building centers and communities of practice into the design of R&D project portfolios and structure to maintain knowledge flows and creation of value: examples of this are Bell Labs, a health research institute, and pharmaceutical companies. They raise questions about such networks that will need to be answered by future research and by R&D managers. Given the organization into networks, what policy and managerial approaches should be taken to generate a greater flow of value through the investment in basic research? If basic research occurs largely within self-organizing communities, what kinds of measures will cause the ongoing self-renewal activities in this overall eco-system to be heedful of the ways in which increased value can be created and focused? The R&D manager might echo Rammert: what techniques allow for the sensible integration, coordination, and utilization (such as patent filings) of such dispersed sources of knowledge?

Such networks are created both to exploit common but distributed resources and to explore areas of knowledge that have a common interest but come from diverse perspectives. Exploitation-focused networks tend to develop corporate structures; exploration-focused networks are always in the forming, interactively learning in new areas of knowledge (Rammert). R&D managers will need to learn to utilize both.

Meeus and Faber ask what effects interorganizational relations (such as networks, alliances, or collaborative arrangements) have on innovative behavior of firms or organizations, and what induces such relationships. Such relationships can be valuable: they provide needed resources and skills to be shared, important links between basic science and innovation, and also between the innovative firm and customers with needs. They focus on the exchange and learning processes between innovator firms and their partners, looking at partnerships such as those between buyer and supplier, industry with universities, and reporting on strategies to overcome the lack of information on the competencies and reliability of potential partners. R&D managers will seek partners, form alliances, and nurture networks considering five driving forces:

(*a*) business strategies and environmental pressures;
(*b*) limits on resources such as funding, patents, and technology and knowledge base;
(*c*) costs in forming technical links;
(*d*) the extent of technological change and complexity in the firm's industrial sector;
(*e*) the present or potential position that the firm occupies in the collaborative network.

Roberts and Berry (1985), using the second driving force, provided collaboration guidance for R&D managers. Based on the extent of firms' knowledge of both the target markets and the technology being developed, what type of collaborative interorganizational relationships are warranted for different R&D projects: internal developments, acquisitions, licensing, internal ventures, joint ventures or alliances, use of venture capital, and educational acquisitions? Meeus and Faber point out that interorganizational collaborative relationships offer serious benefits for both the adoption of innovation and innovative performance. The position of the firm in the network, the size of the network, and the profile of participants all foster innovative output and knowledge exchange.[6] Furthermore, network concentration in high-tech industries is much higher.[7] In the early stages of an emerging or complex technology, the R&D manager often does not have a choice between going it alone or collaborating. Here competitive technology intelligence must be called upon to find the right partners and manage the risks in collaboration (Norling 2004).

One aspect of networks is probably overlooked by R&D managers and their marketing associates: the role of networks in the market place in adopting an innovation (Meeus and Faber). Networks of consumers (or advisers to consumers such as physicians) can aid in the adoption or rejection of a new product, technology, or service. R&D managers face the difficult task of detecting such networks and using them as a positive force rather than the negative force in the chapter by Jolivet and Maurice as outlined.

Intellectual links among areas of expertise

A type of radical innovation termed 'research-technology-driven innovation' for which the outputs are general-purpose technologies (also called 'enabling technologies') impact and provide links with many other intellectual, technical, and economic domains; general-purpose technologies such as instrumentation have even enabled the development of entire scientific domains such as cell biology (Shinn). Such innovation is contrasted to 'differentiation-driven' or 'narrow domain innovation.'[8]

Data-mining or citation-mining techniques are especially effective in analyzing such links between scientific communities that could be created by 'general-purpose technologies' as well as other technologies common to the communities. Klavans (1997, 2005) creates a map of science—a visual picture of relationships between areas of science using co-citation analysis. A recent map for 2003 covers 96,000 research communities or clusters of papers representing groups of researchers working on the same problem. These are the networks discussed by Mohrman *et al.* Additional data can allow one to identify the science that is of special interest: possibly the presence of a general-purpose technology or the creation of new disciplines by general-purpose technologies that can yield commercially valuable intellectual property.

Forces in the environment that influence innovation processes

Six elements in society form interdependent configurations or systems that govern or influence innovation processes that make for path dependency (or resistance to change based on history). They are made memorable in van Waarden and Oosterwijk's chapter by being known as the Six 'I's:

(1) institutes: industrial development creates organizations for particular industries such as research and educational institutes, standardization bodies, financial institutions, and trade associations;

(2) interlinkages: these institutes are interlinked through one or more forms of economic governance such as markets, networks, or associations of various kinds;

(3) interests: the institutes have special interests for survival and growth through continual funding of the industries for development;

(4) ideas, information, knowledge, competencies; specialization brings the accumulation of a set of competencies, knowledge and skills that, for a while at least, can provide competitive advantage;

(5) incentives: the investments in institutes and ideas motivate individuals to invest further;

(6) institutions that help perpetuate all the other 'I's.

A seventh 'I' may be implied: infrastructure—the physical buildings, communication pathways, research laboratories (Kelly *et al.* 2003) and architectural designs that may be more appropriate to one industrial specialty than an other and pose challenges for communities seeking economic development through a new industrial specialization. Understanding institutional history may help managers go with the flow of history or break the path dependency—whichever is needed in managing for success.

Metcalfe and other economists see modern capitalism as a particular kind of knowledge-based economic system in which innovation, enterprise, and competition are governed by systems of market and non-market institutional forces, and which are involved in the development and growth of an economy—a restless, continuous process of change and transformation, of creative destruction. Such modern economic systems are not chaotic, but highly structured, ordered by the workings of market and non-market institutional forces. Surprisingly, economic and social order is maintained and coordinated by these institutions; but these same ordering processes give rise to the opportunities and the growth of knowledge that come to transform the existing order and further redefine economic possibilities. The economic institutions permit the prevailing pattern of activity to be invaded by disruptive innovations where the attacker or outsider may have the advantage (Foster 1986; Christensen 1997). Innovations, a matter of experimentation, are still seen as surprises, novelties, and unexpected consequences of a particular kind of knowledge-based capitalism. Market forces shape the return on the innovation for the business and, in turn, influence the eventual outcomes of innovation and the ability of the business to continue to innovate. That ability to innovate is influenced by per-ceived opportunities in the market, available resources, economic and other incentives, and the capabilities to manage the process. These often require multiple trade-offs between efficiency, investment, and innovation itself. Metcalfe calls on R&D managers to articulate each of these factors. While each is well recognized by most R&D managers, executing the balancing act distinguishes the outstanding from the mediocre manager.

Public Policy

Kuhlmann and Shapira model policymaking within an innovation system as a process of competition, networking, and consensus-building among various communities or stakeholders (one of which is industry), and then examine the development of innovation policy and its impact in four case studies. They show how six governance variables affected innovation performance in Germany and the US: for example, the ways in which growth of biotechnology in Germany and the US was hampered in one case by discipline-aligned universities (with an inability to produce the needed talent) and supported in the other by decentralized governance with multiple actors that encouraged flexibility, responsiveness to change, tolerance of research risk, and the ability to embrace emerging technologies. R&D managers can use this model to see where and why industry might be involved in the policy-making process.

Similarly, Chaminade and Edquist discuss the role that public policy plays in an innovation system. Systems of innovation evolve over time in a largely unplanned manner, but innovation policy, a conscious activity, can aid to a limited extent in the development of such systems. Large-scale and radical technology advances rarely take place without public intervention, because markets and firms perform least efficiently in new activities where uncertainty and the risks are large. They go on to list the various actions that can be taken by governments to strengthen innovations systems.

R&D managers' role in policymaking

R&D managers are becoming ever more aware of the need to work with or lobby policy makers and regulatory agencies to ensure that such actions, policies, and regulations foster their innovations in a fair manner. They must recognize that certain societal concerns over perceived risks can make specific technologies, products, or processes unacceptable or irrelevant in the market place. R&D managers, especially those with scientific or engineering training, can overlook the social factors affecting their innovations. They need to be aware that institutional strategies may at times dominate technology strategies.

Here the study of Jolivet and Maurice is relevant. They examine the influence that markets can have in shaping innovations, in this case the acceptance, or lack thereof, of genetically engineered food in Europe, Japan, and the US, where collective learning processes proceeded quite differently. They re-emphasize the importance of R&D managers, considering, at both the design and commercialization stages, cultural and social factors as well as economic and technological ones when targeting research towards potential markets. R&D managers might ask how Monsanto should have proceeded differently in Europe.

Society or government may influence innovation by motivating research organizations. Foster *et al.* examined the role that public policy plays in providing incentives and motivators for firms to innovate. Large and resource-rich companies, over time, reduce their capacity for generating innovations, developing core rigidities and inertia (Leonard-Barton 1995; and Meeus and Hage Introduction); they will, over time, be more successful in process than in product innovations; the success that built them is what also limits their capability to innovate.

Foster *et al.* help R&D managers recognize cases where governmental action has overcome this inertia, inducing innovation in environmental technologies. The environmental economics literature has argued that technological innovation, rather than resource reallocation, is the key to the effective solution of environmental problems. Therefore, the creation of regulatory incentives for innovation is essential. This has been recognized in the US and only more recently in the European Union (EU). But there is considerable debate about whether certain regulations can act as barriers to innovation or can, in fact, induce innovation. The study of the pulp and paper industry in the US, Sweden, and Finland showed that regulations in general did induce technological innovation aimed at solving environmental problems. However, in some cases, market disincentives for major investments to implement technological advances overrode such inducements. Similar case studies have also studied the possible role of governmental action upon innovations in 'green chemistry' (Lempert *et al.* 2003) and upon the environmental R&D efforts at DuPont, Intel, Monsanto, and Xerox (Resetar 1999).

Hage examined non-market coordination mechanisms: institutional arrangements, structures, or contexts of businesses, and the responses of the associated organizational populations to two sources of institutional and societal change—economic globalization and knowledge transformations such as radical process or product innovation. R&D managers should study his findings on what kinds of organizational structures and types of businesses seem to be surviving in the face of economic globalization and rapid technological change, and which structures and characteristics of businesses make them likely to disappear. These findings might contribute to both offensive and defensive strategies in gaining competitive advantage, or in just surviving.

The impact of government funding of research

Governmental funding of scientific research has come to be greatly influenced by promises and expectations related to the future use of the scientific findings (van Lente). To obtain funding, researchers need to demonstrate that the research is strategic—that is, research carried out with the expectation that it will produce a broad base of knowledge that can be used to solve practical problems.[9] Such basic research can be use-inspired as well as curiosity driven (Stokes 1997). Van Lente explores the power of promises and expectations through his examination of the development of the field of genomics. Promises at a broader level allow the unfolding of more specific promises at a local level. These all promote funding, development of programs, agenda-building, creation of subfields, and an element of coordination or structure for the scientific field. R&D managers well recognize their role in motivating and leading by their expectations and aspirations, and the role played by promises in gaining project funding, approvals, and in developing R&D plans in support of business plans—or even shaping a new business plan.

Georghiou discusses the positive and negative consequences as industry, universities, and government laboratories move from differentiated funding sources to one of competing as sellers of contract research. Positive consequences include efficiency through competition, elimination of poor performers, and scientific advice that can be contested, and is not confined to one government ministry or agency acting as a promoter or regulator. Negatives include loss of coverage and variety, movement from a laboratory's original mission, possibly compromised scientific advice, and difficulty in getting investment in facilities and resources in the face of uncertain future funding. There are also consequences for collaboration which, in research and innovation, has been empirically shown to depend upon complementarity. Partners seek competencies and characteristics that they do not possess. Convergence means that similar organizations find it more difficult to cooperate. It is concluded that policy for a research and innovation systems should create conditions in which all parts of the system are fully networked, but preserve the specialized functions of the individual components and encourage the formation of new differentiated activities. These trends in funding could create barriers

for R&D managers as they build alliances and collaborative relationships.

Casper asks how 'comparative institutional theories on innovation' can help in framing public policy. He explores the link between innovation systems and country performance in growing small entrepreneurial technology firms. Such firms must engage in a similar set of activities; these include accessing novel technologies from universities, recruiting and motivating talented scientists and engineers, and obtaining high-risk finance. The viability of each of these is dependent on the institutional factors related to university research, technology transfer systems, labor markets, and financial systems. He points out that institutional theory, especially the varieties-of-capitalism framework, makes predictions that liberal market economies should excel at developing such entrepreneurial firms, while 'coordinated' economies should fail in developing new economy industries such as biotechnology and software. He then gives a set of examples that show that predicting the influence that institutions can have on the success or lack of success of biotech and telecommunication start-ups is not necessarily certain. In some of his cases, strategies were developed to deal with the negative effects of the institutions affecting the labor input and technical capabilities available to the start-ups. He concludes for R&D managers and entrepreneurs that, while a number of drivers for the adoption of new technologies and change may exist within economies, governmental policies may turn out to be flexible enough to account for unanticipated drivers of entrepreneurial start-ups and the overcoming of 'institutional myopia.'

Societal and institutional change

Peter Drucker, one of the earlier students of innovation after Machiavelli and Schumpeter, observed:

Innovation is the specific tool of entrepreneurs, the means by which they exploit change as an opportunity for a different business or a different service ... Entrepreneurs need to search purposefully for the sources of innovation, the changes and their symptoms that indicate opportunities for successful innovation. (Drucker 1986)

Change sparks creativity and then creative destruction. R&D managers need to look for the changes that will create opportunities and changes that need to be made to make innovation possible and effective.

Campbell lays the foundation for a theory of planned institutional change and routes for implementation to which R&D managers should give special attention. He details the process by which planned institutional innovation and change occur, and reviews the outputs of the process. Planned institutional change involves the combining of existing institutional elements in new ways (*bricolage* or architectural innovation). New principles and practices may be added as appropriate. Campbell gives a detailed model of how institutional change can take place, stressing the role of social entrepreneurs (Bornstein 2004) as the change-makers; he sees planned institutional innovation as a deliberate effort on their part to change institutional arrangements such as formal and informal rules, monitoring and enforcement mechanisms, and systems of meaning that define the context within which individuals, firms, labor unions, nations, and other organizations operate and interact with each other. Campbell puts forward seven propositions on how institutional change takes place that can also be studied and applied by R&D managers.

Creation and Change of Organizations and Creation and Change of Institutions are two of the ten activities within an innovation system as discussed by Chaminade and Edquist. They suggest a possible division of labor between the private and public sectors for certain actions, helping R&D managers understand their possible role and ways to gain benefits from actions in the public sector. The state, for example, can facilitate private activities by simplifying the rules of the game, by creating tax benefits and new R&D organizations, developing alternative patterns of learning and innovation, and nurturing emerging sectoral systems of innovation.

Rammert's chapter is rich with advice and lessons: he points out that, to choose particular policies and procedures, it is important for R&D managers to know about the different patterns of knowledge production, distinct styles of knowing and of knowledge regimes, and how they are changing. R&D managers must deal with increasing knowledge specialization and fragmentation, with the acceleration of the tempo of knowledge production, and with the limitations of the past's linear and sequential modes of integration and coordination. Explication by specialists requires disciplinary communities with specialized organizations and networks. Distributed exploration requires heterogeneous expert communities in interdisciplinary research organizations with industry–university collaborations. To optimize knowledge flows and integration in such R&D organizations, networks and society, R&D managers need a different approach: the parallel-interactive coordination approach described by Rammert.

The ways in which education and training systems grow and change as they develop research talent within a society may affect the extent to which that society is innovative. Finegold maintains that certain structures of education and training systems can develop more creative individuals who will, in turn, generate radical products, process innovations, and start up companies in that society. He looks at the elements of decentralization, variety in educational content and process, diversity of higher education institutions and students, and the extent to which adults have access to further education. For example, the educational system in India, with its meritocratic national examination system, filters out the brightest, hardest-working students from a population approaching one billion, and sends them to world-class technical universities. With the Internet and globalization, this educated workforce is now available to multinational corporations to provide services, or as part of virtual teams for technical support or new product development. He further shows how educational and training systems play an important role in the development of high-skill ecosystems (HSEs) or clusters of firms in a specific region, often start-up companies built from research results of university scientists; highly skilled graduates then provide the continuing nourishment for continued innovation. Universities further support these new or mature enterprises through the creation of technology transfer offices and science and technology parks. Because they are rich, diverse, tolerant, and creative cultural environments, they also help spur innovation-building communities where top technical talent and entrepreneurs want to live. Higher-education institutions help foster the connectivity between these organizations by building social networks of individuals through courses, alumni networks, and meetings for local firms. Finegold demonstrates, through several examples, a lesson in economic development: it takes many different elements in the environment to create and maintain vibrant HSEs from Singapore to the EU or the US.

Summary

R&D managers completing this volume can take away some very practical lessons.

1. An understanding of the innovation process, and how processes differ from industry to industry, profile to profile, nation to nation, institutional setting to institutional setting, and sector to sector will enable R&D managers to analyze their particular situation to shape R&D strategies, project plans, marketing approaches, to assemble a set of best practices, and to obtain resources and move ahead even in the face of objections by business managers.

2. Attention to design strategies can facilitate the simultaneous development of product offerings and competencies in the organization, leading to lineages of product families as well as lineages of skills and talent within the firm.

3. As knowledge creation is more fragmented and specialized, R&D managers need a new sensitivity to the roles of networks: to the ways in which innovators are linked; the ways in which some new techniques and practices that have been described can integrate the work among knowledge generators and with knowledge users; the ways in which creativity can be encouraged by seeking knowledge in areas beyond their experience, and by combining or seeing knowledge in new ways. R&D managers are given advice on building interorganizational collaborative relationships—when, with whom, how, and why; they need to learn how to analyze potential network effects, integrating this analysis into business and technology strategies. They may need to take specific steps to facilitate technology transfer through reorganizations or relocating scientists or development teams.

4. To an extent not seen before, today's R&D managers will need to deal with external forces. New talents, far removed from those developed as a researcher, will be needed to bring about institutional changes, to participate in political processes, or overcome the forces of history and tradition. Opportunities for innovation are found in change; implementing innovation may require a degree of societal change.

5. To manage knowledge generation, transformation, and integration, R&D managers must reconcile a number of opposing perspectives or thinking processes: analysis and interpretation; broad and narrow focus; explication and exploration; divergence and convergence; short-term and long-term focus; creation and destruction; and incremental and radical innovation. Although there are no simple prescriptions, the astute R&D manager will have found several important hints about what steps to take.

Notes

1. We note little concern for example over (1) how to pursue a promising radical innovation requiring major investment that poses a political threat to business managers who won't meet near-term earnings targets and get that next promotion; (2) how to get the needed analytical equipment or laboratory building on the capital forecast versus investments in new manufacturing equipment; (3) how to find time to run plant tests in manufacturing the new product when the plant superintendent has a sold-out plant; or (4) how to select researchers for 'downsizing' when the CEO calls for cost savings.

2. '...the management of innovation in 4th generation R&D is the synthesis of many threads including the management of knowledge from many diverse sources, expeditionary marketing (with mutually dependent learning), integration of both explicit and tacit knowledge development of robust models for competitive architecture and organizational capability, new organizational models, new approaches to finance, decision-making and accounting, the management of technology represented in the form of intellectual property, a new innovation process, the process and tools through which these elements are integrated—and an understanding of the dynamics of the changing dominant design for businesses, markets, and technologies...' (Miller and Morris 1999).

3. This in fact has become the core process of Fourth Generation R&D.

4. The history of R&D at DuPont has helped R&D managers as well as CEOs to appreciate their heritage and structure strategies consistent with organizational culture (Hounshell and Smith 1988; Miller 1997*a*).

5. The importance of dealing with the realities of the external environment was seen in a study of government and nonprofit organizations in Minnesota—the 'Surviving Innovation Project' where they were found to have the capability to transform the single, occasional act of innovating into an everyday occurrence able to forge a culture of natural innovation. They faced the outside world and its institutions, embraced the volatility they saw, and used crises as wake-up calls, riding the turbulence to new ideas and public support with a commitment to controlling their external environments (rather than the other way around) (Light 1998).

6. In developing a network to develop the technology for Freon® chlorofluorocarbon alternatives, the R&D manager at DuPont was at the center of the 'spokes and wheel' network in which the links were only between the other participants and manager; few if any links were between the other participants, allowing DuPont to control intellectual property and knowledge integration (Miller 1997*b*; Norling 2004).

7. Professor George Whitesides of Harvard has observed that three models have emerged for governmental/industrial/academic partnerships[0]. The microelectronics industry, for example, has successfully pursued a triumvirate model, relying on all three partners. Similarly, the biotechnology industry has expanded on this model, incorporating medical schools and venture capitalists. Historically, the chemical industry has had a more limited, linear approach, working with academia or with government, but not melding the two into a single, focused effort (Connelly 2005).

8. R&D managers would at this point consider the technique for idea generation and problem-solving—TRIZ—a Russian acronym: *Theoria Resheneyva Isobretatelskehuh Zadach* (Theory of Solving Problems Inventively)—that emerged from Russia, based on the assumption (from studies of the patent literature) that there are only a limited number of solutions to the world of problems. Problems can be classified in various ways and a computer search can identify possible ideas or innovative solutions that have solved similar problems—but in far removed applications and domains (Braham 1995; Altschuller 1999). Might TRIZ be applied to a narrow domain innovation, taking a 'narrow niche novelty' and by adaptation, or application of some of the principles of the technology, turn it into an integration-driven innovation—the integration being performed well after the initial development, broadening applications and providing even more intellectual links among scientific disciplines and scientific communities and enabling the development of new scientific domains?

9. Today, to be included in the US Federal Budget, research programs must meet three criteria, the first being relevance—why the program is important, relevant, and appropriate, and expectations for societal benefits. The other criteria are quality (how the funds will be appropriately allocated to ensure quality R&D) and performance (demonstration of how well the investment is performing) (Marburger and Daniels 2002).

References

Abernathy, W. J., and Utterback, J. M. (1975). 'A Dynamic Model of Product and Process Innovation.' *Omega*, 3/6.

Altschuller, G. (1999). *The Innovation Algorithm: TRIZ, Systematic Innovation and Technical Creativity* (translated from Russian by Lev Shulyak and Steven Rodman). Worcester, MA: Technical Innovation Center. See: **www.TRIZ.org**

Amidon, D. M. (1997). *Innovation Strategy for the Knowledge Economy: The Ken Awakening*. Boston: Butterworth-Heinemann.

Bornstein, D. (2004). *How to Change the World: Social Entrepreneurs and the Power of New Ideas*. New York: Oxford University Press.

Braham, J. (1995). 'Inventive Ideas Grow on "TRIZ".' *Machine Design,* 12 October; **www.ideationtriz.com**

Brown, M. G., and Svenson, R. A. (1988). 'Measuring R&D Productivity.' *Research-Technology Management,* July–August: 67–71.

Burgelman, R. A., Modesto, A. M., and Wheelwright, S. C. (2001). *Strategic Management of Technology and Innovation*. 3rd edn. New York: McGraw-Hill.

Chesbrough, H. (2003). *Open Innovation: The New Imperative for Creating and Profiting from Technology*. Boston: Harvard Business School Press.

Chester, A. N. (1994). 'Aligning Technology with Business Strategy.' *Research-Technology Management,* 37/1: 25–32.

Christensen, C. M. (1997). *The Innovator's Dilemma: When New Technologies Cause Great Firms to Fail*. Boston: Harvard Business School Press.

—— Anthony, S. D., and Roth, E. (2004). *Seeing What's Next: Using Theories of Innovation to Predict Industry Change*. Boston: Harvard Business School Press.

Connelly, T. M. (2005). 'Innovation for Growth.' In F. Budde, U.-H. Felcht, and H. Frankenmolle (eds.), *Value Creation: Strategies for the Chemical Industry*. Frankfurt: Wiley-VCH.

Council on Competitiveness (2004). *Innovate America: National Innovation Initiative Report*. Washington, DC.

Dillon, T. A., Lee, R. K., and Matheson D. (2005). 'Value Innovation: Passport to Wealth Creation.' *Research-Technology Management,* 48/2: 22–42.

Drucker, P. F. (1986). *Innovation and Entrepreneurship: Practice and Principle*. New York: HarperBusiness Edition.

Fagerberg, J. (2004). 'Innovation: A Guide to the Literature.' In J. Fagerberg, D. C. Mowery, and R. R. Nelson (eds.), *The Oxford Handbook of Innovation*. London: Oxford University Press.

Foster, R. N. (1986). *Innovation: The Attacker's Advantage*. New York: Summit Books.

Geisler, E.(2000). *The Metrics of Science and Technology*. Westport, CT: Quorum Books.

—— (2001). *Creating Value with Science and Technology*. Westport, CT: Quorum Books.

Hounshell, D. A., and Smith, Jr, J. K. (1988). *Science and Corporate Strategy: DuPont R&D, 1902–1980*. Cambridge: Cambridge University Press.

Kelley, R., and Caplan, J. (1993). 'How Bell Labs Creates Star Performers.' *Harvard Business Review,* July–August: 128–39.

Kelly T. K., Kofner, A., Norling, P. M., Bloom, G., Adamson D. M., Abbott M., and Wang, M. (2003). *A Review of Reports on Selected Large Federal Science Facilities: Management and Life-Cycle Issues*. MR-1728-OSTP. Santa Monica, CA: RAND. Available at **www.rand.org**

Klavans, R. A.(1997). 'Identifying Research Underlying Technical Intelligence.' In R.A. Ashton and W. B. Klavans (eds.), *Keeping Abreast of Science and Technology: Technical Intelligence for Business* (1997): 23–47. See **www.mapofscience.com** (2005).

Leifer, R., McDermott, C. M., O'Connor, G. C., Peters, L., Rice, M. P., and Veryzer, R. (2000). *Radical Innovation: How Mature Companies Can Outsmart Upstarts.* Boston: Harvard Business School Press.

Lempert, R. J., Norling P. M., Pernin C., Resetar, S., and Mahnovski, S. (2003). *Next Generation Environmental Technologies: Benefits and Barriers.* MR-1682-OSTP. Santa Monica, CA: RAND.

Leonard-Barton, D. (1995). *Wellsprings of Knowledge: Building and Sustaining Sources of Innovation.* Boston: Harvard Business School Press.

Lester, R. K., and Piore, M. J. (2004). *Innovation: The Missing Dimension.* Cambridge, MA: Harvard University Press.

Light, P. C. (1998). *Sustaining Innovation: Creating Nonprofit and Government Organizations that Innovate Naturally.* San Francisco: Jossey Bass.

Malerba, F. (2004). 'Sectoral Systems: How and Why Innovation Differs across Sectors.' In Jan Fagerberg, David C. Mowery, and Richard R. Nelson (eds.), *The Oxford Handbook of Innovation.* London: Oxford University Press.

Marburger III, J. H., and Daniels, M. (2002). 'FY 2004 Interagency Research and Development Priorities.' *Memorandum for the Heads of Executive Departments and Agencies,* 30 May, The White House, Washington, DC.

Meeus, M. (2005). 'From R&D Management to Management of Innovation.' In J. Sundbo *et al.* (eds.), *Contemporary Management of Innovation: Are We Looking at the Right Things?* Basingstoke: Palgrave Macmillan, 21–37.

Miller, J. A. (1997*a*). 'Basic Research at DuPont.' *CHEMTECH,* April: 12–16.

—— (1997*b*). 'Upset the Natural Equilibrium.' In R. M. Kanter, J. Kao, and F. Wiersema (eds.), *Innovation: Breakthrough Thinking at 3M, DuPont, GE, Pfizer, and Rubbermaid.* New York: Harper Collins.

—— Norling, P. M., and Collette, J. W. (1997). 'Research/Technology Management.' *Kirk-Othmer Encyclopedia of Chemical Technology* (4th edn.), xxi. 263–91. New York: John Wiley and Sons.

Miller, R., and Floricel, S. (2004). 'Value Creation and Games of Innovation.' Research-Technology Management, November–December: 25–37.

Miller, W. L. (1995). 'A Broader Mission for R&D.' *Research-Technology Management,* 38/6: 24–36.

—— and Morris, L. (1999). *Fourth Generation R&D: Managing Knowledge, Technology, and Innovation.* New York: John Wiley and Sons.

Moore, G. A (1991, 1995). *Crossing the Chasm; Inside the Tornado.* New York: Harper Business. See **www.chasmgroup.com**

—— (2004). 'Darwin and the Demon: Innovating within Established Enterprises.' *Harvard Business Review,* 82/7/8: 86–92.

National Science Board (NSB), National Science Foundation, Division of Science Resources Statistics(2004), *Science and Engineering Indicators 2004,* Arlington, VA: NSB 04–01. Available at: **www.nsf.gov/sbe/srs/seind04/pdfstart.htm**

Nicholson, G. (2003). Email to Emeriti Representatives in the Industrial Research Institute, 2 December.

Norling, P. M. (1996). 'Network or Not Work.' *Research-Technology Management,* Jan.–Feb: 42–8.

—— (1997). 'Structuring and Managing R&D Work Processes: Why Bother?' *CHEMTECH,* October: 12–16.

—— (2004). 'Competitive Technology Intelligence: Improving Your Odds of Success.' Keynote presentation at Competitive Technical Intelligence 2004 meeting of the Society for Competitive Intelligence Professionals, 28 October 2004, Boston, MA.

—— Herring, J. P., Rosenkrans, W. A., Stellpflug, M., and Kaufman, S. B. (2000). 'Putting Competitive Technology Intelligence to Work.' *Research-Technology Management,* September–October: 23–8.

Pavitt, K. (2004). 'Innovation Processes.' In J. Fagerberg, D. C. Mowery, and R. R. Nelson (eds.), *The Oxford Handbook of Innovation.* London: Oxford University Press.

Porter, M. E., and Stern,S. (2002). 'National Innovative Capacity.' Report from the Institute for Strategy and Competitiveness, Harvard Business School. Available at **www.isc.hbs.edu/ innov_9211.pdf**

Prather, C. W., and Gundry, L. K. (1995). *Blueprints for Innovation: How Creative Processes Can Make You and Your Company More Competitive.* New York: American Management Association.

Ranftl, R. M. (1980). 'R&D Productivity.' *CHEMTECH,* November 1980: 661–9.

Resetar, S. (1999). *Technology Forces at Work: Profiles of Environmental Research and Development at DuPont, Intel, Monsanto, and Xerox.* Santa Monica, CA: RAND, MR-1068/1-OSTP.

Rice, M. P., O'Connor, G. C., Peters, L. S., and Morone, J. G. (1996). 'Managing Discontinuous Innovation.' *Research-Technology Management,* Jan.–Feb.: 52–8.

Roberts, E. B., and Berry, C. A. (1985). 'Entering New Businesses: Selecting Strategies for Success.' *Sloan Management Review,* 26/3.

Roussel, P. A., Saad, K. N. and Erickson, T. J. (1991). *Third-Generation R&D: Managing the Link to Corporate Strategy.* Boston: Harvard Business School Press.

Sakkab, N. Y. (2002). 'Connect and Develop Complements Research and Develop at P&G.' *Research-Technology Management,* 45/2: 38–45.

Schumpeter, J. A. (1934). *The Theory of Economic Development.* Cambridge, MA: Harvard University Press,

Stokes, D. E. (1997). *Pasteur's Quadrant: Basic Science and Technological Innovation.* Washington, DC: Brookings Institute Press.

Utterback, J. M. (1994). *Mastering the Dynamics of Innovation.* Boston: Harvard Business School Press.

Glossary

architecture a characteristic of systems including organizations of people, markets, and businesses; includes rules that guide practice such as development, operations, and use or maintenance (Miller 1995).

capability a characteristic of an organization of people with specific knowledge, tools, technology, and processes ready to perform work or to learn (Miller 1995).

community of practice a sustained, cohesive group of people with a common purpose, identity for members, and a common environment using shared knowledge, language, interactions, protocols, beliefs, and other factors not found in job descriptions, project documentation, or business processes (Miller 1995).

development the process of converting some knowledge into something tangible and useful.

discovery to find something that already existed.

innovation turning an idea into something new and tangible that has value: "turning knowledge into money" (Nicholson 2003).

innovation systems the broad array of institutions, organizations, and relationships involved in scientific research, the accumulation and diffusion of knowledge, education, and training, technology development, and the development and distribution of new products, processes, services, and organizational arrangements. Different types of innovation are:

- *Incremental*: adaptation, refinement, and enhancement of existing products and services or production/delivery systems.
- *Radical*: development of entirely new product and service categories or production and delivery systems, or significant changes in product or service functionality or significant reduction in production costs.
- *Architectural*: reconfiguration of the system of components that constitute a product, product, service, or institution.
- *Open*: use of both external and internal ideas, and internal and external paths to the market to advance technology.

institutions sets of common habits, routines, established practices, rules, or laws that regulate the relations and interactions between individuals, groups, and organizations.

intellectual property protected know-how such as patents, trademarks, trade secrets.

invention to produce something useful for the first time using imagination or experimentation.

knowledge the capacity to reproduce or replicate findings, products, or programs and the processes that produce them, something that is incorporated in organizational routines, seen more as a competence to do something than as a tangible good, and is not independent from its incorporation in skilled bodies, trained brains, or technical media of representation. Knowledge is a relational and practical term, relating the knowing person or collective with what is known.

learning process in which all kinds of knowledge are (re)combined to form something new.

learning organization an organization skilled at creating, acquiring, interpreting, transferring, and retaining knowledge, and at purposefully modifying its behavior to reflect new knowledge and insights.

management the direction or carrying out of business affairs.

model a human construct to help in the understanding of real-world processes, systems, and structures. Models usually have information inputs or resources, a description of the processing of the inputs and then an output of expected results. Five types of models: conceptual, physical analogues, mathematical, statistical, and visualization models.

organizations formal structures that are consciously created and have an explicit purpose.

R&D intensity R&D expense divided by revenues (sales).

research to search or investigate, to increase understanding or knowledge, often about WHY things work; 'turning money into knowledge' (Nicholson 2003).

science systemized knowledge attained through study—in the natural sciences various disciplines: WHY things are as they are.

strategy the art of devising and employing plans and resources to achieve certain objectives.

tactics methods of employing resources.

technological innovation combined activities leading to new marketable products and services or new production and delivery systems.

technology theoretical and practical knowledge, skills, and artifacts that can be used to develop products, services, and their production and delivery systems, applied science; methods of achieving a practical purpose; converting the WHY to HOW; know-how.

Conclusion

Jerald Hage and Marius T. H. Meeus

Rethinking theories of innovation and theories of knowledge production and exchange

This book began with the claim that there is a need for a new theory of innovation. Its chapters are pieces of the puzzle that is a multi-level, multi-sector theory of knowledge production and innovation. This final chapter begins the task of putting together the pieces that make up that new theory.

This book looked first at current theory's critique of innovation in the organizational sociology and management literatures; this final chapter summarizes how those contributions dealt with these criticisms, corrects the basic ideas in the theory, changes the definitions of the key concepts of complexity, integration and high-risk strategies, and adds others.

Then this book looked at the need to construct a theory of knowledge production, and presented an equation that connected knowledge, learning, and innovation as a basic foundation for the beginnings of such a theory; this chapter assesses how the book contributes to the construction of such a theory. The aim is, of course, to integrate what our chapters have discussed, but also to point out exceptions and qualifications. And this chapter moves on to the feedback consequences of the aforementioned equation. An unresolved issue is how much product or scientific innovation itself produces discontinuous institutional or societal change. While there are no definite answers to this question, we set the stage for making the theory of innovation and the

theory of knowledge production more dynamic by considering the feedback consequences on these two themes. The chapter ends with a discussion of the relative utility of a new knowledge paradigm as a supplement to existing paradigms.

Rethinking innovation theory in organizational sociology and management

This book suggests that this theory can be applied to each of the analytical levels observed in its Introduction: internal organization level, interorganization and interpersonal, sectoral, and societal. At each level, we ask what is the degree of complexity, the extent of the integration and whether or not high-risk strategies are being pursued. Before we rethink each of these concepts, and how their content changes as we shift levels, it is useful to revisit briefly the essential critique of the organizational theory of innovation, and how this book has addressed this critique.

Lacunae in innovation theory

We observe that the focus of early research on the internal workings and performance of innovator organizations has created lacunae: the neglect of economic variables; almost no attention to research laboratories or units; ignorance of the context and, in particular, differences between sectors and societies. Many of the contributions deal with these lacunae. Damanpour and Aravind study whether or not various economic variables could explain differences in the rates of product or process innovation. They do not find consistent differentiation and, in many instances, they find inconsistent results. They do not simultaneously explore the impact of the organizational and management variables such as complexity, organic structure, or integration and high-risk strategies that might have made the economic variables more potent in explaining innovation, and, in their conclusion, also observe the necessity of studying the impact of economic variables by industrial sector.

A second of the literature's major criticisms is the absence of theory and research on research laboratories (except for some work of Latour, Woolgar, Knorr-Cetina, and others in the sociology of science) where many of the innovative ideas are developed, and which is now so central to the problem of organizational learning. The contributions of Jordan and Hollingsworth speak to the importance of developing a theory about how research laboratories and projects should be organized and managed so as to achieve particular goals. In a considerable improvement over the previous organizational and management work on innovation, which stressed only one model, the organic model, Jordan suggests that there are four different profiles for describing and analyzing research laboratories, or four structural variations on the organic theme: small and specialized, small and complex, large and specialized, large and complex. Each of these is associated with different kinds of strategy; this refines another aspect of the extant theory on innovation, which has focused only on high-risk strategies, and has not made a distinction on the basis of the scale or systemic nature of the innovation. Jordan suggests the need to make a distinction between small- and large-scope strategies as well as low- versus high-risk ones part of the conclusion.

Hollingsworth reports on studies of research laboratories, especially those concentrating on science. Like Jordan, he observes the need for a multi-level model involving the major concepts in the organizational theory of innovation. Unlike Jordan, he builds up the context of the research project with descriptions of the research organizations and the national system of scientific research. Essentially, he suggests that there is more radical innovation when the research laboratory is located in a relatively small research organization, and in a national system that does not have strong institutions. What is particularly interesting is how he uses the major concepts in organizational and management theory at multiple levels, specifically the research project and the research organization. However, in comparing his research work with Jordan's theory, it is important to recognize that his focus is on small-scale radical innovations, not large-scale ones where the organizational size of either the research organization or a consortium of organizations becomes significant.

The importance of studying the differences between sectors and between societies is reflected in all the contributions that report comparative research findings: those of Foster, Hildén, and Adler; Jolivet and Maurice; Finegold; van Waarden and Oosterwijk; Hage; Casper; and Meeus and Faber. Several of the contributions in this book indicate that the exploration of industrial sectors would be a helpful way of refining current economic theories about both innovation and management theory. For example, Hage, echoing the earlier work of Pavitt (1984), observes that there are considerable differences in responses to innovation by industrial sector, within countries and across them. This has been a major thesis in the work of the institutionalists interested in the problem of innovation (see Hollingsworth 1997; Hall and Soskice 2001). But this leaves unanswered how these sectors should be described theoretically, a point to which we return below.

Expanding the concepts of complexity, organic structure or integration, and high-risk strategy

The theory of innovation in sociology and management is a relatively simple one. Given a diversity of perspectives (complexity) that is integrated (organic structure), and provided the leadership takes a high-risk strategy, radical innovation is more likely. Not well studied is the implication that, absent one of these, incremental innovation is more probable. The critique above begins to lay the groundwork for expanding these concepts and enriching the theory with other ideas so that it can be generalized.

Rethinking the concept of complexity

Let us start with the concept of complexity. It has been measured primarily in the literature by diversity of occupations, level of training, and providing papers at conferences (Hage 1980, 1999). All of these measures are largely internal or reflect organizational policies about the management of knowledge. The Hollingsworth contribution, except for its consideration of the national system of innovation or of scientific research, is primarily internal; however, it provides a number of refinements to the measures of complexity. One of these is the idea of needing a certain depth of experts in a particular area. This kind of complexity is best represented in the research consortia of competitors, which have grown in such large numbers in recent years in the US. The Jordan contribution echoes this in noting the importance of studying the complexity of the machines used in scientific research, as best illustrated in radio astronomy, oceanographic

ships, linear accelerators, the system of weather satellites, and, most expensive of all, the space station.

Clearly, the thrust of many of the contributions in this book is that innovation now involves external relationships of many kinds. As organizations search for the expertise that they need, they discover that it resides outside the organization. But the concept of complexity can be applied to various levels of analysis and, in particular, to horizontal relationships. Hence measures of complexity should include the following ideas:

- diversity of occupations outside the organization involved in interpersonal flows of knowledge or communities of practice;
- diversity of occupations outside the organization involved in interorganizational relationships;
- characterizing sectors by the inherent levels of complexity;
- characterizing modes of coordination by their implicit levels of complexity;
- variety of organizational pressures.

Each of these ideas needs to be discussed along with the particular contribution that illustrates the point.

The Mohrman, Galbraith, and Monge chapter clearly highlights the importance of the interpersonal flows of knowledge in the development of radical innovation. To obtain an accurate count of complexity, one would need to know what diversity and depth are added by the interpersonal relationships attached to the specific research project. Hargadon (2003) argues that both Edison and Ford had quite elaborate interpersonal relationships that monitored scientific developments relative to their research interests very successfully. A quite different perspective on how complexity is increased via interpersonal flows is the Shinn contribution on how generic technologies link together diverse pools of research, enriching each of them. An updated version of innovation theory must take into consideration these communities of knowledge that mesh together via relationships or technologies. The current attempt to study citations of both papers and patents is an obvious movement in this direction, even if there are a number of problems with these kinds of measures.

Why is there an important distinction made between interpersonal flows of knowledge and knowledge communities, and interorganizational relationships and research consortia as being really two distinct kinds of analysis? The former are more ephemeral, while the latter, because of their relatively greater structural permanence, create a different set of issues, the most critical being the problem of organizational autonomy.

Interorganizational relationships and their consequences for innovation are studied in two contributions to this book: Meeus and Faber's, and van Waarden and Oosterwijk's. In the former, the authors note the paradox that what seems to make the focal firm look for more external sources of knowledge in interorganizational relationships is the complexity of a project, and a medium strength of knowledge resources. Weak and strong resource stocks are both associated with lower levels of external collaboration. In the latter contribution, radical transformations of knowledge caused by new paradigms hasten the movement towards interorganizational relationships, as well as other modes of coordination. There are two different ways in which complexity can be measured in these contributions. One is by examining the diversity of occupations, their specialties, their technologies, etc.; the other is by asking to what extent different organizations are involved. The van Waarden and Oosterwijk contribution makes clear that, with the separation of organizations as a consequence of their specializing in basic research, applied research, product development, manufacturing research, quality-control research, and marketing research, the problem of innovation has become much more complex. Handling a variety of diverse organiza-

tional entities introduces another kind of complexity into the mix, including the idea of complexity's being knowledge based, because different organizations, even within the same sector, have quite different areas of expertise, as well as distinctive organizational cultures. This aspect of complexity is to be found in those high-tech sectors with a number of research organizations, which leads naturally to our next level of analysis.

In the four-sector model of industrial sectors suggested in Hage's contribution are several variables that can be used to characterize the degree of complexity of specific sectors. As the knowledge base and per cent sales allocated to R&D increase, then the sector is classified as more complex. But the sector can also be considered to be more complex because, as the lives of products become shorter and shorter, it requires a continual search for new expertise to create new products. It might be noted that this typology of four kinds of industrial sectors also includes economic variables, a criticism of the literature that we observed above, and suggests that different kinds of innovation are likely to be found in these sectors.

Measures of complexity are also implicit in the various contributions that focus on modes of coordination and the varieties of environmental pressures. The various modes range from a non-complex system, such as market coordination, to somewhat more complex systems, such as vertical hierarchies that control a market, or state hierarchies that attempt to coordinate either a sector or even all of the economy, to still more complex ones such as interorganizational relationships and associations. Above, we have described the ways in which the complexity of interorganizational relationships can be discerned; the same logic applies to associations. But how much complexity is involved in these various modes of coordination has to be carefully determined and varies a great deal across situations within national systems of innovation, as well as between them. The empirical cases of van Waarden and Oosterwijk provide a way of studying how one can do this.

Rethinking the concept of integration

For a number of years, except for arguing that centralized organizations with a lot of bureaucratic rules tended to reduce the amount of innovation, the organizational and management research on innovation largely ignored the problem of integration. This was perceived to be the antithesis of the organic structure (Burns and Stalker 1961) that informed much of this stream of research. In the past ten years, the problem of integration has moved to the forefront; this is partly because of the recognition of the difficulties of making explicit tacit knowledge (see Nonaka and Peltokorpi's chapter) and partly because of the call for more research projects involving Pasteur's Quadrant (see Jordan's), that integrates basic and applied research.

The distinctions made above about levels of analysis—knowledge or practice communities, interorganizational relationships, sectoral and societal levels—also apply to our discussion of integration. Indeed, the same problem exists. If we perceive that there is more complexity in a particular sector or interorganizational relationship, then the question is, is this complexity integrated or not? The contributions to this book provide a number of insights about these different levels of analysis, and whether or not they are integrated. But let us begin with the problem of internal integration, and then move to external integration.

Both Jordan's and Hollingsworth's contributions provide a bridge between these two perspectives. Each is concerned, in similar ways, with the integration of research projects and, in somewhat dissimilar ways, with their integration with the external world; each begins with the idea of the organic structure: that is, decentralization, with project autonomy and the elimination of formal bureaucratic rules. Jordan, however, observes that large-scale research projects necessitate some form of coordination, and suggests that clearly defined research goals and strategies, along with well-executed strategic planning, are essential for the large-scale research project.

The Nonaka and Peltokorpi contribution illustrates the importance of tight integration of individuals with the correct skills, and how this can lead to rapid development (in this case of a radical new car, the hybrid Toyota Prius). Another important effect of a high level of integration is observed here: a shorter time to market. What is particularly striking about this development is that, rather than involving the usual coordination along the vertical *keiretsu* chain typical in Japanese companies, and especially automobile firms, it was accomplished entirely in-house at Toyota.

Both Jordan and Hollingsworth move into the discussion of integration with external sources of information, but in different ways. The former discusses the importance of external relationships; the latter focuses on the rapid integration of external knowledge. Hollingsworth stresses the importance of leadership in integrating new knowledge.

A more explicit discussion of external integration is involved in each of the contributions tabulated in the row that is labeled 'networks' in the introductory chapter (Table 1.2). As we have observed in the Introduction and elsewhere, the growth in knowledge and the movement towards specialization across both supply chain and the idea-innovation chain/network have created a problem of integration across organizational boundaries. Both interpersonal and interorganizational networks provide a partial solution. A common measure of how much integration there is in each of the different kinds of networks—interpersonal (Mohrman, Galbraith, and Monge; Shinn) and interorganizational (Meeus and Faber), and modes of coordination (van Waarden and Oosterwijk)—is indicated in the Meeus and Faber contribution, where they discuss the frequency of exchange and the amount of information and knowledge transfer. In other words, the greater this amount, then the more integrated is the particular kind of external relationship.

At another level of analysis, one can discuss to what extent their integration is sectoral or societal. The Shinn paper develops the theme that a large section of the world of science can be integrated with the use of common or generic technologies. This theme of integration is central to Rammert's contribution, which illustrates a variety of ways in which this is achieved. In a different way, the contributions of van Lente and Jolivet and Maurice refer to integration via the development of consensus, the former about a particular research program, the latter about the pros and cons of a specific radical innovation.

The discussion of modes of coordination in the context of complexity and integration for the purposes of innovation is interesting for a number of reasons. As has been suggested above, markets are neither complex, nor, except in the very minimal sense of supply and demand, integrated. Sometimes the non-market modes of coordination, particularly state interventions, represent attempts to create integration; they reflect conscious attempts to bring more coherence. This can be of special importance for innovation, as has been argued in the contributions of Meeus and Faber on inter-organizational relationships, and with different kinds of modes in the contributions of Casper, Finegold, Hage, and van Waarden and Oosterwijk. The latter contribution makes quite explicit the dominant ideas of complexity and integration within the context of idea-innovation networks. One of the more interesting extensions of the theory of innovation is the positive role of associations for adopting radical product and process innovations, as indicated in the contributions of van Waarden and Oosterwijk, and Hage. Furthermore, we have observed at various points how business firms have been investing less and less in basic research. If it were not for the state, this critical function in the idea-innovation network would not be represented. This does not mean that state policies are always effective, a topic that we return to later in this chapter. And at this point we turn to the rethinking of high-risk strategies.

Rethinking the concept of high-risk strategies

Technology strategy in companies has received considerable attention in the case of companies, but has been understudied in research organizations. The role of strategy in research organizations has been rethought in several of the contributions. A recurrent theme is that, rather than thinking only about one kind of high-risk strategy, we need to recognize that there are different kinds with varying consequences for the research organizations and firms that are involved. Jordan correctly calls attention to the fundamental distinction between high-risk strategies that involve large-scale radical innovations, and those that involve small-scale. This same distinction is found in Hage's typology of sectors.

Perhaps the most dramatic example of the rethinking of strategy is involved in the Hatchuel, Lemasson and Weil paper that focuses on the importance of the reciprocal relationship between design strategies and the development of knowledge. The authors argue from several case studies that, given what has already been accomplished, there is an inherent competence enabling the development of new products, and that innovations occur in the process of searching for the relevant competencies, which is an increasing of complexity. Conversely, the creation of new knowledge creates new opportunities for existing product lines, as long as this is part of the explicit strategy. This latter idea is akin to that of Hollingsworth; although his terms are different, his contribution stresses that absorption of new knowledge is a critical aspect of the strategy of the research organization.

The concept of high-risk strategies has not been used in the contributions on networks. Yet it could easily be. Typically, the reasoning has been that researchers in organizations search for competencies, interorganizational relationships are formed, and within them decisions might be made about high-risk strategies. But the opposite line of reasoning is also in the literature, albeit not in the specific contributions in this book. Global alliances (Gomes-Casseres 1996) and research consortia (Browning et al. 1995) are formed as a part of a high-risk radical innovation strategy.

The relevance of high-risk radical innovation strategies becomes most prominent when one shifts from market to non-market modes of coordination. Hierarchies, whether private or state and, to a lesser extent and sometimes, associations, may explicitly adopt high-risk innovation strategies. The notion and impact of high-risk strategies are most relevant for state modes of coordination, because frequently their interventions are designed to stimulate radical innovations such as the creation of new industry. Foster, Hildén, and Adler, Finegold, and van Waarden and Oosterwijk describe here various aspects of state policy that attempted to increase the rates of innovation, including radical innovation, in several societies. Some examples are also provided in Hage.

In our Introduction, we suggested that we wanted to move towards a multi-sector, multi-level theory of innovation. One step in this direction has been to generalize the theory of organizational innovation by extending the ways in which the concepts of complexity, integration, and high-risk strategies are measured, and then applying this theory to multiple levels and different sectors of society. At each level, and in each sector, one can determine the degree of complexity, the level of integration, and whether or not high-risk strategies are being employed. In addition, this provides the beginning of a diagnosis for societies that, in particular sectors, may have lower rates of innovation than they desire.

Towards a theory of knowledge production and exchange

As we indicated in the first chapter, one of the objectives of this book is not just to develop a more elaborate theory of innovation than has been advanced in the organization and management literature. One of our especial aims in a theory of knowledge production and exchange is the inclusion of science, knowledge trajectories, and institutional change. Our starting point is this simplified equation:

$$\text{Knowledge} + \text{Collective Learning} = \text{Innovation or New Knowledge}$$

This equation not only provides a way of connecting the two literatures, but also allows us to suggest how most of the contributions in this book add to a general theory of knowledge production. It is thus an integrating device. Let us start with the generalization of the organizational theory of innovation discussed in the previous part, which is a theoretical way of stating the same equation. The hypothesis is that complexity, when integrated and combined with high-risk strategies, leads to more radical innovation; it makes connections among a particular kind (diversity) of knowledge with collective (integrated) learning and a specific kind (high-risk or radical) of innovation.

The reader immediately recognizes that the organizational theory that was generalized in the previous part is too simple a statement, as the various contributions in this book have made clear. But the equation allows us to move into a discussion of each of the contributions, perceiving them as defining, facilitating, or hindering each of these three concepts in the equation, thus adding the necessary qualifications and subtleties to the theory and, at the same time, moving us in the direction of a theory of knowledge production and of exchange.

Definitions and discussions of kinds of knowledge

As has already been suggested, the degree of complexity is a measure of the diversity of knowledge involved at some level of analysis. But we can also distinguish three different usages of the concept of knowledge in this book:

(1) kinds of research or knowledge creation;
(2) collective perspectives or paradigms;
(3) competencies or qualifications in the knowledge base of the organization or labor force.

Part II calls attention to the simple distinction between basic scientific research and industrial research, which has not been a major focus of study except as an input in an equation, which we observed in Damanpour and Aravind's chapter and in the introduction to Part II. Jordan's chapter makes this distinction more complicated by observing that both basic and applied research can be organized in different ways, leading to different kinds of collective learning and different outcomes in terms of the kinds of innovation.

But Part II on science is not the only way in which kinds of research as a form of knowledge creation can be distinguished. The van Waarden and Oosterwijk chapter differentiates the kinds of knowledge into the following six arenas: basic research, applied research, product development, manufacturing research, quality-control research, and marketing research. Again, our equation above is made much more complicated because now we understand that, to have industrial innovation as the output, we need to be concerned about the different kinds of collective learning occurring between these different kinds of research or knowledge-producing arenas. Furthermore, as the authors make clear in their contribution, this is not a simple linear relationship but instead is one that moves back and forth between organizations involved in one or another of these arenas.

As we observed in the Introduction, the equation above is a static one. But it becomes dynamic when one begins to discuss how changes in knowledge and/or changes in collective learning lead to changes in the nature of innovation. This allows us to explore yet another way in which the concept of knowledge has been discussed in this book, namely in the sense of a paradigm or unified perspective. Van Waarden and Oosterwijk report on an exploration of this dynamic version of the equation: in their study of two sectors (telecommunications and biotechnology in a very broad sense) and four countries (Austria, Finland, Germany, and the Netherlands),

they observe how radical changes in paradigms led to radical changes in the ways in which collective learning occurred (not only in the ways in which different arenas of research are connected, but also in the nature of the modes of coordination), to a higher level of collective learning (see the next section of this chapter for more discussion of this), and, in turn, to changes in the rates of product and process innovations and in the diversity of market sectors in which this occurs. The most striking consequence of these paradigmatic changes in knowledge is that they destroy knowledge: that is, as well as making technologies and their equipment obsolete, they destroy competency. So changes in knowledge can be subtractions as well as additions. This is an important qualification to the above equation.

The discussion of competency destruction leads naturally to the third way in which the concept of knowledge is explored in this book. The Finegold contribution focuses on the training of competencies, or education, raising questions of how a society or institutions of education influence the nature of the knowledge base, both of the society and of particular organizations within it. Finegold examines how changes in the level of knowledge in the workforce can impact the nature of innovation. Of course, one could do a finer-grained assessment, and ask in what sense are individuals trained in each of the different kinds of research, and how quickly the education and training system responds to the need created by a paradigmatic shift for new kinds of qualifications. The cases of India and Singapore in Finegold's contribution are interesting from this viewpoint.

The contributions of Metcalfe on the evolution of the economy and of Rammert on the evolution of knowledge move beyond observing the consequences of paradigmatic changes in knowledge and the destruction of competencies to the dynamic aspects of the knowledge-production equation. Both present a dynamic and self-organizing view of the world that is directly linked to the changes in the nature of knowledge.

Kinds of collective learning as in interorganizational networks and knowledge communities

The duality of ideas continues in our discussion of collective learning. Part II of the book paid much attention to its structural analog—integration. In Part III, the emphasis is on the process and amount of learning that occur with integration. It is apparent that this varies, depending, in part, upon the degree of integration: there has been a tendency to use indicators of one for the other. Just as there are different kinds of knowledge, there are disparate kinds of collective learning. Indeed, one of the objectives of this book has been to expand the idea of organizational learning by recognizing not only that there are other kinds of collectives but that, depending upon the kind of linkage that is being integrated, the nature of the learning itself can vary.

The traditional thought on organizational learning was that it was largely internal and incremental. A good example of this perspective is the Nonaka and Takeuchi (1995) work on the impact of quality work circles and the continuous improvements in the productivity of Japanese companies. Nonaka and Peltokorpi's contribution in this volume shifts toward another focus of non-incremental or radical product innovation, one effect of which is a radical reduction in product development time. This is achieved by reducing the size of the normal circle of engineers, and the companies that they represent, involved in the development of a new car to a small inner circle of diverse experiences.

Though these terms are not used, this same perspective on organizational learning is also represented in Hollingsworth's chapter in Part

IV. Again, a restricted but moderately diverse circle of individuals leads to radical innovations in small organizations. More than Nonaka and Peltokorpi, who present a single case, Hollingsworth is arguing the limits to collective learning created by a too diverse or complex a set of knowledge in a much broader number of organizations (although these are restricted to the area of biomedicine and, thus, small-scale research). Again, this is an important qualification to the previous section and to the basic equation. While Hollingsworth argues that it is important for leaders to incorporate new knowledge, the external aspects of collective learning are not emphasized in his work.

In sharp contrast to these perspectives, most of the other contributions in this book stress the necessity of collective learning, more specifically, knowledge exchange occurring because of external sources, which also changes the idea of organizational learning into extra-organizational learning combined with organizational learning. This is especially true for radical innovation, as suggested in the chapters of Jordan, of Mohrmann, Galbraith, and Monge, and of Meeus and Faber. In particular, Jordan suggests that collective learning must be in part external if true radical innovation is to occur, a point to which we return below. The different kinds of extra-organizational learning are easily contrasted by the kinds of linkages involved; among others there are:

- interorganizational relationships (Meeus and Faber; Hage; Rammert; van Waarden and Oosterwijk, etc.);
- knowledge communities (Mohrman, Galbraith, and Monge; van Lente);
- generic technologies (Shinn);
- modes of coordination (Georghiou; van Waarden and Oosterwijk; Casper; Hage);
- broad publics (Jolivet and Maurice).

The linkages are not always the same even within the same general category. For example, Mohrman, Galbraith, and Monge emphasize the role of individuals, whereas van Lente discusses the role of rhetoric, as in a slogan that captures the imagination of the scientific community. This is more akin to the work of Jolivet and Maurice on the processes by which public judgments are made about a particular technology such as genetically modified food.

Which linkages are best for which kinds of knowledge? There is some implication that knowledge communities and generic technologies are more likely to be associated with basic research, less with applied research. In contrast, the examples used in these contributions show that interorganizational relationships and modes of coordination are likely to be associated with areas of research more typical of industrial innovation. One can carry this thought further and suggest that perhaps the breakdown that occurs in national systems of innovation is a change from the style of knowledge communities to the style of interorganizational relationships. Are there organizations that can act as midwives between these two kinds of linkages; is this the appropriate role for technological research centers?

From this follows the insight that there has been almost no research on the overlap between these kinds of linkages and the consequences this might have for the industrial innovation of a society. This is one of the major reasons why we have brought these ideas together in the same book: to highlight the importance of studying whether or not these linkages reinforce each other, and in which sense they are all required for effective industrial innovation.

Finally, we suggested above that the concept of organizational learning had to be expanded to include extra-organizational learning. But more than this coupling, one also needs to begin to analyze how much learning has occurred at other levels of analysis:

- within knowledge communities (are some more fertile?);
- within interorganizational relationships (joint organizational learning);

- within and between sectors or organizational populations (population learning);
- within society (societal learning).

These suggest quite interesting new areas of research about forms of collective learning.

Kinds of radical innovation

As a reader might expect from a book about innovation, we have provided quite a rich array of examples. One of the major emphases has been on radical product or process innovation. Many contributions in this book (of Damanpour and Aravind; Nonaka and Peltokorpi; Foster, Hilden, and Adler; Hatchuel, Lemasson, and Weil; Jolivet and Maurice; van Waarden and Oosterwijk; Hage; and Casper—or approximately 40 per cent of the contributions in this book) treat this topic, and in a rich variety of countries and industrial sectors.

Our major contribution has been to add a relatively ignored kind of innovation, namely scientific research, and argue that the same distinctions of product and process innovation can be applied to the study of basic and applied research. Reconceptualizing the role of innovation within science has clearly been the agenda of the contributions of Jordan and Hollingsworth. In addition, Jordan also distinguishes between small- and large-scale radical innovation. Thus her contribution can be seen as a series of corollaries under the main equation indicated above.

In research, process innovations are reflected most notably in the Shinn contribution on generic technologies. But more than that, we have linked research and industrial innovation with the discussion of the idea-innovation chain/network; as we have seen, this represents a way of thinking not only about different kinds of knowledge but also about different kinds of innovation. The terms can be applied to each of the stages in the chain/network.

Independent of the impact of the institutional environment or other context on knowledge and collective learning, innovation is also influenced by the nature of its context. For example, in science, it is the areas in which research is allowed or permitted. The van Lente contribution calls attention to the importance of rhetoric and expectations in creating availabilities of research funds for certain kinds of research. The cause célèbre in the US is stem cell research. The issue of rhetoric and its impact should be part of any theory of innovation in science or industry. Why do some slogans resonate at particular moments in specific cultures, and others not? The same reasoning applies to the public acceptance or rejection of major technologies such as nuclear reactors, genetically modified crops, supersonic transport.

The modes of coordination are not only mechanisms that facilitate or hinder collective learning; they can also influence the relative success of particular innovations beyond the issue of public acceptance. The most interesting example of this is Georghiou's contribution about markets dampening scientific innovation. Hage demonstrates how vertical hierarchies dampen rates of implementation of radical process innovations, whereas associations encourage their acceptance. The whole role of the state in influencing rates of innovation is a special topic that is treated in the next section.

The simplified equation for the production of knowledge has been made systematically more complicated and, we hope, useful by distinguishing kinds of knowledges, kinds of collective learning, and kinds of innovation. In the process, we have built in various qualifications to the generalized theory elaborated above. An important subsidiary theme has been how the institutional level affects the kinds of knowledge, the kinds of collective learning, and the kinds of innovation that one has, and thus the production of knowledge.

We have also made the equation dynamic by observing how radical changes in knowledge lead to changes in the nature of collective learning and in changes in the kinds of innovation. But this is not the only dynamic generated by knowledge changes. The other dynamic is the consequences for society of

innovation or new knowledge, especially radical innovation. Without this, any theory of innovation and of knowledge production would be incomplete. How is the society altered as a consequence? This chapter now turns to the issue of institutional change.

New knowledge (innovation) and radical societal change

Why is the topic of institutional change so critical for a book on industrial and scientific innovation? The reasons are quite straightforward. We have seen the argument in several contributions that the institutional level constrains or prevents radical change and even product innovation or scientific discovery. Indeed, frequently, the state attempts to change the institutional order so as to increase the rate of industrial innovation and/or, even more so now, the rate of scientific breakthroughs. So it is not just for the sake of theory that one has to grapple with these issues, but for the efficacy of state policy. The reverse set of issues—how does radical industrial or scientific innovation affect the nature of society?—are equally important, but more ignored. We should study the feedback of industrial or scientific innovation, especially radical innovation, on the society. Under what circumstances does this also produce radical institutional change? Developing the theories about these disparate feedback effects that allow for a theory of innovation and of knowledge production and exchange not only completes the cycle but also begins to confront the problem of evolution.

These issues of radical societal change have become more urgent, especially in the context of a theory of innovation, because of what might be called a certain paradox. On the one hand (as presented in the book's Introduction, the Introduction to Part I, and in several of its chapters), there is clearly a steadily accelerating rate of radical innovation, especially at the organizational level. On the other hand, there is considerable evidence for path dependency in the way in which institutions change, as we have seen in Part IV. How does one reconcile this paradox?

One way of beginning to resolve the problem, and to provide more clarity to the discussion, is to perceive that radical societal change can exist at multiple levels. Its origin can start at either with a new form of organization or a new set of institutional rules. Indeed, a main theme of this book has been how these two literatures have not considered a multi-level approach to the issues that concern them. This is another example of the same problem. In this last part of the concluding chapter, we first define what is a radical societal change and do this for each level. Then we consider a typology of radical societal changes that distinguishes between the origin of the change and its consequences. A particularly important subcategory of radical societal change are the attempts, some successful and some unsuccessful, of the state to produce radical change. We conclude with some observations.

as a consequence of paradigmatic changes in knowledge. Finally, Hage illustrates the disappearance of organizational populations, along with their dominant institutional mode of co-ordination, because of the failure to adopt radical process technologies.

At the institutional level, radical societal change means discontinuity in the nature of the dominant coordination mode that supplements the market. Again, several of the contributions in Part IV illustrate this (Casper; van Waarden and Oosterwijk; and Hage). More fundamentally, the issue remains whether the growing importance of interorganizational relationships as a coordination mode reflects a fundamental institutional change for most post-industrial societies. One must also recognize that, at the institutional level, one can have mixed cases, as reflected in Campbell's analysis of *bricolage*.

These are not the only kinds of radical societal changes, but they are important ones and easily recognized.

What is radical societal change?

Let us begin with a clear definition at each level of what is radical societal—that is, discontinuous—change. At the organizational level, this means discontinuity in the basic organizational form. A number of the contributions in this book illustrate new organizational forms, even though this was not their explicit focus. Both Jordan and Hollingsworth define new research organization forms designed to produce radical innovation. And although Campbell argues that the new organizational forms in Japan, South Korea, and Taiwan are examples of *bricolage* because they combine new elements with existing ones, these still represent new organizational forms within these societies and, therefore, are examples of radical societal change at the organizational level. Casper's contribution illustrates the emergence of new high-tech forms in industry as a consequence of state policies, while van Waarden and Oosterwijk's contribution does the same

Kinds of radical societal change

Considering our definition of radical societal change as existing at either the organizational or the institutional level suggests that we need to combine both levels to consider the kinds of radical societal change that there are. The change can originate at either level, and can have consequences for either level. The differences between the origin of change and what is change provide us with a simple fourfold typology that describes four kinds of radical societal change, each of which is represented in this book. Societal change can originate at the level of the institution and involve either organizational change or institutional change. Likewise, societal change originating at the level of the organization can also have both consequences. In the process of describing and illustrating these four kinds of societal change, we can expand the debate of Part IV of the book, as well as summarize some of the contributions in this new light.

The contribution of Georghiou focuses on the radical change in the market coordination mechanism of basic research in England and thus is an example of where both the origin and effect are at the same level. Examples of how organizational change can produce radical institutional change, that is, changes in the institutional modes of coordination, are found in varying degrees in the contributions by van Waarden and Oosterwijk and Hage. Radical change in the paradigms of knowledge have produced considerable institutional change in the form of new modes of coordination, such as interorganizational relationships, as well as changes in the relative frequency of other modes. One might question whether paradigmatic changes are organizational changes, but they have been coded this way because they are the products of basic research organizations rather than some institutional effort, except insofar as they reflects a major effort upon the part of the state, which does sometimes occur.

Institutional changes that produce organizational changes are found in Finegold, who indicates how state policies resulted in a number of changes in educational organizations. Furthermore, many of the examples of the contributions involving state policy imply or involve examples of this, including Foster, Hildén, and Adler, and Kuhlmann and Shapira.

Finally, examples of how organizational changes result in radical organizational changes are not really represented in any of these contributions, perhaps because it would be the most obvious case and therefore of less interest. The best example of this is the move by organizations to create interorganizational relationships, as in the contribution by Meeus and Faber.

The role of the state: successes and failures

Any theory of innovation must directly confront the issue of when the state can successfully transform the society and, for that matter,

itself, so that there can be more industrial innovation and scientific breakthroughs. The state is a major actor and is the most obvious source of discontinuous change. Although increasing rates of industrial and scientific innovation may be the objective of the state, it is not the only kind of transformation that we have witnessed, especially in the last fifteen years. Movements towards democracy and towards competitive markets have been common efforts, some of which have been the basis for Campbell's review of the evidence about the possibilities for radical societal change. He argues against this.

The review of the evidence provided in this book, especially relative to the state's efforts, indicates that the more radical the desired institutional change, the more likely there is to be some component of path dependency mixed with a certain amount of institutional change—a deflection in the path. This argument is most notable in the Campbell contribution. The theme of *bricolage* indicates how radical changes are transformed by being combined with prevailing institutional models, adapted to the prevailing norms. But the theme is also contained in Casper's contribution, which shows how state interventions designed to create a biotech industry capable of radical innovation have been partially successful. A number of small firms were created, but as yet they have not produced radical innovations.

The Finegold contribution has a more optimistic note, in that state intervention to change the British educational system and, in particular, the relative emphasis on vocational and technical education, have been successful. But again, the objective of a more innovative economy has still to be realized. Of course, the Singapore case of microelectronics is certainly a success story.

The juxtaposition of the Finegold contribution with the Georghiou contribution is interesting because they both involve the United Kingdom, and suggest that the British government may have a contradictory innovation policy in developing more trained personnel, but setting in motion policies that diminish basic research. Again, we are led to the use of the idea-innovation network concept for evaluating the success of state interventions. Furthermore, there is an interpretation that flows from this. Are state policies more likely to be unsuccessful if they move in the direction of trying to create more simple modes of co-ordination, such as markets, rather than more complex modes that would speak to problems of integration and collective learning that we have discussed in the previous two sections?

If this is the case, then we are led to a counter-intuitive proposal regarding the success of states in encouraging innovation. Despite the wide-spread discussions of the role of market com-petition and the need to deregulate and to downsize government, there appears to be a strong movement towards the need for state sup-port of basic research and the construction of various integrative mechanisms between basic and applied research, to encourage the forma-tion of knowledge communities and the con-stuction of interorganizational relationships.

Concluding observations

We have not resolved the paradox of whether the higher and higher rates of radical product and process innovation are producing more and more discontinuous change at either the organizational level (in the form of new organ-izational forms, or populations of organiza-tions) or at the institutional level (with shifting patterns of coordination modes or other kinds of changes in the patterns of insti-tutional rules). What we can say is that radical societal change exists at both levels, and the typology is useful for focusing on this problem.

One theoretical possibility for resolving the paradox is that decoupling between the two levels of analysis may be increasing; that there is more and more discontinuous change at the organizational level, without necessarily discontinuous change at the institutional level. A second theoretical possibility is that the boundaries between organizational entities become less prominent in thinking and strate-gizing, more specifically because innovation is such a multidisciplinary and multilevel activ-ity. Firms cannot ignore the institutional environment: they need institutional informa-tion to set goals and design strategies with acceptable risk; they must also, for their own survival, create organizational designs that are able to adapt to both changing technological and economic conditions. This makes innov-ation management a dual activity that inte-grates institutional and technical constraints and options in order to assess the feasibility of innovation strategies. We conclude our book, therefore, with an important area for future research.

References

Browning, L., Beyer, J., and Shetler, J. (1995). 'Building Cooperation in a Competitive Industry: SEMATEC and the Semiconductor Industry.' *Academy of Management Journal*, 38: 113–51.

Burns, T., and Stalker, G. M. (1961). *The Management of Innovation*. London: Tavistock.

Gomes-Casseres, B. (1996). *The Alliance Revolution: The New Shape of Business Rivalry*. Cambridge, MA: Harvard University Press.

Hage, J. (1980). *Theories of Organizations: Form, Process, and Transformation*. New York: Wiley.

—— (1999). *Organizational Innovation (History of Management Thought)*. Dartmouth: Dartmouth Publishing Company.

Hall, P., and Soskice, D. (eds.) (2001). *Varieties of Capitalism* . Oxford: Oxford University Press.

Hargadon, A. (2003). *How Breakthroughs Happen: The Surprising Truth about how Companies Innvate*. Boston, MA: Harvard Business School Press.

Hollingsworth, J. R. (1997). 'Continuities and Changes in Social Systems of Production: The Cases of Japan, Germany and the United States.' in J. R. Hollingsworth and R. Boyer (eds.), *Contem-*

porary Capitalism: The Embeddedness of Institutions. Cambridge: Cambridge University Press, 265–310.

Nonaka, I., and Takeuchi, H. (1995). *The Knowledge-Creating Company*. New York: Oxford University Press.

Pavitt, K. (1984). 'Sectoral Patterns of Technical Change: Towards a Taxonomy and a Theory.' *Research Policy*, 13: 343–73.

INDEX

Printed in Great Britain
by Amazon